P9-AEY-712

THE BIOLOGY OF THE COCCIDIA

THE BIOLOGY OF THE COCCIDIA

Edited by

Peter L. Long, D.Sc., Ph.D.
Professor
Department of Poultry Science
University of Georgia College of Agriculture
Athens, Georgia

Edward Arnold

First published in the United States of America in 1982
by University Park Press.

First published in the United Kingdom in 1982 by Edward Arnold (Publishers)
Limited, 41 Bedford Square, London WC1B 3DQ.

Printed in the United States of America.

Library of Congress Cataloging in Publication Data
Main entry under title:

The Biology of the Coccidia.

Includes index.
1. Coccidia. I. Long, Peter L. [DNLM: 1. Coccidia. 2. Coccidiosis. 3. Sporozoa.
QX 123 B615]
QL368.C59B57 636.089'6016
ISBN 0 7131 2845 3 AACR2

Contents

Contributors

H. D. Chapman, D.Phil.
Senior Scientific Officer
Houghton Poultry Research Station
Houghton, Huntingdon, Cambs.
England PE17 2DA

Bill Chobotar, Ph.D
Professor
Department of Biology
Andrews University
Berrien Springs, Michigan 49103

David J. Doran, Ph.D
United States Department of Agriculture
Science and Education Administration
Animal Parasitology Institute
Poultry Protozoan Diseases Laboratory
Beltsville, Maryland 20705

Ronald Fayer, Ph.D.
Laboratory Chief
Ruminant Parasitic Diseases Laboratory
United States Department of Agriculture
Animal Parasitology Institute
Beltsville, Maryland 20705

M. A. Fernando, M.B., B.S.
Professor
Department of Pathology
Ontario Veterinary College
University of Guelph
Guelph, Ontario
Canada

Thomas K. Jeffers, Ph.D.
Research Scientist
Lilly Research Laboratories
Greenfield Laboratories
P.O. Box 708
Greenfield, Indiana 46140

Leonard P. Joyner, B.Sc., Ph.D., D.Sc, F.I. Biol.
Head
Department of Parasitology
Central Veterinary Laboratory
M. A. F. F. New Haw, Weybridge
Surrey, England

Norman D. Levine, Ph.D.
Professor
College of Veterinary Medicine
Agricultural Experiment Station
University of Illinois at Urbana-Champaign
Urbana, Illinois 61801

Larry R. McDougald, Ph.D.
Associate Professor
Department of Poultry Science
University of Georgia
Athens, Georgia 30602

W. Malcolm Reid, Ph.D.
Professor Emeritus
Department of Poultry Science
University of Georgia
Athens, Georgia 30602

M. Elaine Rose, Ph.D., M.R.C.V.S.
Head
Department of Parasitology
Houghton Poultry Research Station
Houghton, Huntingdon, Cambs.
England PE17 2DA

Erich Scholtyseck, Ph.D.
Professor
Zoologisches Institut der
Universität Bonn
Poppelsdorfer Schloss
5300 Bonn, West Germany

Martin W. Shirley, Ph.D.
Senior Scientific Officer
Houghton Poultry Research Station
Houghton, Huntingdon Cambs.
England PE17 2DA

Ching Chung Wang, Ph.D.
Senior Investigator
Merck Institute for
Therapeutic Research
P.O. Box 2000
Rahway, New Jersey 07065

Preface

There have been a number of books on the coccidia, but few of them have been concerned with reviewing the full range of biological characteristics and host-parasite relationships. The late Dr. Datus M. Hammond, Dr. Erich Scholtyseck, and I thought that it was impossible for a single author to review the information adequately. Accordingly, in planning **The Coccidia: *Eimeria, Isospora, Toxoplasma, and Related Genera*** (Hammond, D. M., and Long, P. L. 1973), we enlisted the services of 10 scientists with special knowledge to write separate chapters on the taxonomy, biology, host-parasite relationships, pathology, pathogenicity and immunity, ultrastructure, in vitro culture, biochemistry, toxoplasmosis, and techniques. The present book includes similar subjects, except that there is neither a specialized chapter on *Toxoplasma,* so ably written by Dr. J. K. Frenkel in the previous book, nor a special chapter on techniques.

The coccidia are protozoa of the phylum Apicomplexa. Those belonging to the suborder Eimeriorina include the "true" coccidia and are the subject of the considerations of this volume. The coccidia cause coccidiosis, which is a disease of major economic importance in domestic animals. Parasites belonging to the genera *Eimeria, Isospora, Toxoplasma,* and *Sarcocystis* are discussed in depth, because these are the cause of disease and economic loss in animals. *Toxoplasma* is, in addition, an important parasite of man; *Sarcocystis* is potentially important. Information on *Toxoplasma, Sarcocystis,* and coccidia other than *Eimeria* are covered in each of the chapters, so that information on all of these interesting parasites may be discussed and compared.

Dr. Hammond and I soon realized that the exclusion of chapters on chemotherapy and control of coccidiosis from the previous book was a major error and, accordingly, chapters on these subjects are included in this volume. Also included is a chapter on "Genetics, Specific and Infraspecific Variation," because a great deal of information has been accumulated on this subject in the last few years. I am delighted that some of the authors who contributed to the previous book have written chapters in this volume. Also, I am pleased that Drs. L. P. Joyner, C. C. Wang, B. Chobotar, M. A. Fernando, T. K. Jeffers, L. R. McDougald, M. W. Shirley, H. D. Chapman, W. M. Reid, and R. Fayer have contributed to the present book.

In December, 1979, Dr. Erich Scholtyseck suffered a severe heart attack and was unable to complete the chapter alone. Dr. Bill Chobotar kindly agreed to help and by spending much time in Bonn in March, June, and July of 1980, completed Chapter 4.

Coccidiosis is an active area of research; several thousand scientific papers have been published on the subject. The aim of this book is to review the important advances that have been made in recent years. I am grateful for such a favorable response from the contributors who have set aside their research to write these reviews. My task has been made easier by the quality of their writing.

Each author was asked to discuss work published up to the end of 1979, although there are many references to work published in 1980. As in the last book, they were asked to discuss their own work and to interpret the work of others in their field of interest. We have tried to avoid repeating much of the information given in the 1973 book, but some discussion of that information is essential if useful reviews are to be achieved.

The classification and terminology used for the coccidia has been under constant review these past few years as new information has accumulated. There has been a great deal of controversy regarding this matter. However, the authors have used the terminology to which they are accustomed and have referred to published work exactly as it was cited. I have encouraged the contributors to avoid getting "bogged down" by these matters and they have responded by producing stimulating reviews. Perhaps my views on some of the taxonomy and terminology should be clarified. In dealing with generic and species names of parasites, I am sure that it is quite possible to use old names for parasites found today, and that, on the whole, fewer problems will arise by doing this. However, problems did arise in the terminology of some species of *Sarcocystis* when some authors used different species names in referring to the same parasite. With the objective of clarifying the situation, the names of the hosts were used as species names. It may seem useful to link the names of two hosts in the species name, but such a method is only valuable if a single species of *Sarcocystis* shares these hosts. The use of the name *S. bovicanis* has already lost much of its merit because two or possibly three bovine-canine species have been discovered.

Accordingly, I believe that the current international code of zoological nomenclature should be used and that the names *S. fusiformis* and *S. tenella* for the species *S. bovifelis* and *S. ovifelis,* respectively, are correct.

The generic name *Cystoisospora* has recently been used for parasites thought to differ from *Isospora* because sporozoites may have a dormant phase in a paratenic host. This might eventually be useful; however, at this stage more knowledge is required on the extent to which dormant stages occur in *Isospora* and other coccidia.

The book is intended for advanced undergraduates, graduate students, research workers, teachers with interests in both basic and applied aspects of the subject, and veterinary and medical scientists interested in the treatment and control of coccidiosis.

I am grateful to the authors for their cooperation throughout and to Dr. Levine for compiling a very useful glossary. I wish to acknowledge Geoffrey Mann, University Park Press, for his hard work, especially in the early stages, in helping me organize this new book and Loretta Cormier and Michael Treadway for their efforts in the production of this volume. Thanks are also due to Mrs. Jane Blount for the secretarial work. Finally, I wish to thank my wife, Verna, for helping me with the seemingly endless checking which was necessary.

1

Taxonomy and Life Cycles of Coccidia

Norman D. Levine

The history of our knowledge of the coccidia has been given by Levine (1973a, 1974), and will not be repeated here. The first coccidia were seen in 1674, when Leeuwenhoek found the oocysts of *Eimeria stiedai* in rabbit bile. However, they were not described until 1839 when Hake thought that they were a new form of pus globule. Lindemann (1865), who thought they were gregarines, named them *Monocystis stiedae.*

When it was recognized that the coccidia were protozoa, it was not realized that they had a complicated life cycle involving both sexual and asexual reproduction. As a consequence, the oocysts of the same species were placed in one genus (*Coccidium*) and the meronts (schizonts) in another (*Eimeria, Isospora,* etc.). The two stages were separated even more widely: Labbé (1899) put them in different suborders, and Minchin (1903) in different families. Stiles (1902) and Lühe (1902), whose view was eventually accepted, pointed out that they belonged to the *same genus and that the name Eimeria,* given by Schneider (1875), had precedence over the name *Coccidium,* given by Leuckart (1879). *Coccidium* is commemorated now only in the name of the groups to which *Eimeria* and its relatives belong and in the name of the disease that they cause. Books on the coccidia include Hammond and Long (1973) and Pellérdy (1974).

TAXONOMY

Coccidia are protozoa belonging to the phylum Apicomplexa and within it to the suborders Adeleorina and Eimeriorina. Some authors include all members of the subclass Coccidiasina; however, this definition is too broad. The suborder Eimeriorina, which comprises 10 families, 37 genera, and about 1,500 named species, includes the "true" coccidia, which are considered in this book. The diagnoses of the Apicomplexa, Eimeriorina, and related groups are given below, so that one can see how the true coccidia relate to the other groups and to each other. The diagnoses to suborder are those of the Committee on Systematics and Evolution of the Society of Protozoologists (Levine et al., 1980). The diagnoses of lower groups are those of this author. This classification is somewhat different from that given only a few years ago (Levine, 1973a). However, knowledge gained since then has necessitated the changes.

Phylum Apicomplexa Levine, 1970

Apical complex, generally consisting of polar ring(s), rhoptries, micronemes, conoid, and subpellicular microtubules, present at some stage; micropore(s) generally present at some stage; chilia absent; sexuality by syngamy; all species parasitic; about 4,000 named species.

Class Perkinsasida Levine, 1978

With flagellated zoospores (sporozoites?); zoospores with anterior vacuole; conoid forms

1

incomplete truncate cone; sexuality absent; homoxenous; about 1 named species.

Class Sporozoasida Leuckart, 1879

If present, conoid forms a complete truncate cone; reproduction generally both sexual and asexual; oocysts contain infective sporozoites which result from sporogony; locomotion by body flexion, gliding, undulation of longitudinal ridges or flagellar lashing; flagella present only in microgametes of some groups; pseudopods ordinarily absent, if present, used for feeding, not locomotion; homoxenous or heteroxenous; about 4,000 named species.

Subclass Gregarinasina Dufour, 1828

Mature gamonts extracellular, large; mucron or epimerite ordinarily present in mature organism, the mucron being formed from the conoid; syzygy of gamonts generally occurs; gametes generally similar (isogamous) or nearly so; similar numbers of male and female gametes produced by gamonts; zygotes form oocysts within gametocysts; life cycle characteristically consists of gametogony and sporogony; parasites of digestive tract or body cavity of invertebrates or lower chordates; generally homoxenous; about 1,430 named species.

Subclass Coccidiasina Leuckart, 1879

Gamonts ordinarily present; mature gamonts small, typically intracellular; conoid not modified into mucron or epimerite; syzygy generally absent but, if present, involves gametes; anisogamy marked; life cycle characteristically consists of merogony, gametogony, and sporogony; most species in vertebrates; about 2,420 named species.

Order Agamococcidiorida Levine, 1979

Merogony and gamonts both absent; in marine annelids; about 3 named species.

Order Protococcidiorida Kheisin, 1956

Merogony absent; in marine invertebrates; about 14 named species.

Order Eucoccidiorida Léger and Duboscq, 1910

Merogony present; in vertebrates and/or invertebrates; about 2,400 named species.

Suborder Adeleorina Léger, 1911

Macrogamete and microgamont usually associated in syzygy during development; microgamont produces 1 to 4 microgametes; sporozoites enclosed in envelope; endodyogeny absent; homoxenous or heteroxenous; about 440 named species.

Family Adeleidae Mesnil, 1903

Zygote inactive; sporocysts formed in oocysts; in epithelium of intestine and its appended organs, chiefly in invertebrates; about 40 named species.

Genus *Adelea* Schneider, 1875

Oocysts ellipsoidal or ovoid, with thin wall; oocysts with 6 to 48 flattened sporocysts, each with 2 sporozoites; in chilopods and mollusks; about 4 named species.

Genus *Adelina* Hesse, 1911

Oocysts spherical or subspherical, with thick wall; oocysts with 3 to about 20 spherical or ellipsoidal sporocysts, each with 2 sporozoites; in annelids, chilopods, and insects; about 17 named species.

Genus *Klossia* Schneider, 1875

Oocysts with quite numerous spherical sporocysts, each with 4 sporozoites; in mollusks and perhaps birds and mammals; about 9 named species.

Genus *Orcheobius* Schuberg and Kunze, 1906

Oocysts with 25 or more sporocysts, each with 4 sporozoites; in annelids; about 3 named species.

Genus *Chagasella* Machado, 1911

Oocysts with 3 sporocysts, each with 4 to 6 (or more) sporozoites; in intestine of insects; about 4 named species.

Genus *Ithania* Ludwig, 1947

Oocysts with 1 to 4 sporocysts, each with 9 to 33 sporozoites; in insects; about 1 named species.

Family Legerellidae Minchin, 1903

Zygote inactive; no sporocysts formed in oocysts; in diplopods, nematodes, or insects; about 6 named species.

Genus *Legerella* Mesnil, 1900

With the characters of the family; about 6 named species.

Family Haemogregarinidae Léger, 1911

Zygote active (ookinete), secreting a flexible membrane which is stretched during development; heteroxenous, life cycle involving 2 hosts, one vertebrate and the other invertebrate; merogony in various cells of vertebrates; gamonts in vertebrate blood cells; sporogony in invertebrates; gamonts with about 70 to 80 subpellicular microtubules; about 380 named species.

Family Klossiellidae Smith and Johnson, 1902

Zygote inactive; typical oocyst not formed; a number of sporocysts, each with many sporozoites, develops within a membrane that is perhaps laid down by the host cell; 2 to 4 nonflagellated microgametes formed by a microgamont; homoxenous, with gametogony and merogony in different locations in the same host; in kidney and other organs of host; about 12 named species.

Genus *Klossiella* Smith and Johnson, 1902

With the characters of the family; in mammals; about 12 named species.

Suborder Eimeriorina Léger, 1911

Macrogamete and microgamont develop independently; syzygy normally absent; microgamont typically produces many microgametes; zygote not motile;

1

sporozoites typically enclosed in a sporocyst; endodyogeny absent or present; homoxenous or heteroxenous; about 1,500 named species.

Family Spirocystidae Léger and Duboscq, 1915

Meronts vermicular, curved, with 1 end markedly narrowed; mature meronts coiled like a snail shell, with numerous nuclei; gametes dissimilar, nonflagellate; 1 oocyst per gametocyst; gametocysts and oocysts only in chlorogogen cells; oocysts very thick-walled, ovoid or piriform, with micropyle; each oocyst contains 1 coiled, vermicular naked sporozoite; about 1 named species.

Genus *Spirocystis* Léger and Duboscq, 1911

With the characters of the family; in oligochaetes; about 1 named species.

Family Selenococcidiidae Poche, 1913

Meronts develop as vermicules in host intestinal lumen; meronts with myonemes and a row of nuclei; about 1 named species.

Genus *Selenococcidium* Léger and Duboscq, 1910

With the characters of the family; in lobster; about 1 named species.

Family Dobellidae Ikeda, 1914

Male and female gamonts produced by micro- and macroschizogony, respectively; syzygy present; about 1 named species.

Genus *Dobellia* Ikeda, 1914

With the characters of the family; in sipunculids; about 1 named species.

Family Aggregatidae Labbé, 1899

Development in host cell proper; oocysts typically with many sporocysts; most genera heteroxenous, with merogony in one host and gametogony in another; syzygy absent; about 29 named species.

Genus *Aggregata* Frenzel, 1885

Oocysts large, with many sporocysts; sporocysts with 3 to 28 sporozoites; heteroxenous, with merogony in a decapod crustacean and gametogony in a cephalopod mollusk; about 17 named species.

Genus *Angeiocystis* Brasil, 1904

Oocysts with 4 sporocysts, each with about 16 sporozoites; gamonts at first in form of large sausage; possibly heteroxenous; known stages in polychaete; about 1 named species.

Genus *Merocystis* Dakin, 1911

Oocysts with many sporocysts, each with 1 sporozoite; merogony unknown; presumably heteroxenous; about 1 named species.

Genus *Pseudoklossia* Léger and Duboscq, 1915

Oocysts with no or many sporocysts, each with 2 sporozoites (if sporocysts occur); merogony unknown; presumably heteroxenous; known stages in mollusks; about 5 named species.

Genus *Grasseella* Tuzet and Ormières, 1960

Oocysts with many sporocysts, each with 2 sporozoites; in ascidians; about 1 named species.

Genus *Ovivora* Mackinnon and Ray, 1937

Oocysts with many sporocysts, each with up to 12 (?) sporozoites; homoxenous; in eggs of echiuroids; about 1 named species.

Genus *Selysina* Duboscq, 1917

Oocysts with no sporocysts but with a variable number of heliospores consisting of many sporozoites arranged in a circle around a residuum like the petals of a flower; in ascidians; about 3 named species.

Family Caryotrophidae Lühe, 1906

Oocysts without definite wall; sporocysts with 8 or 12 sporozoites; homoxenous; in polychaetes; about 2 named species.

Genus *Caryotropha* Siedlecki, 1902

Sporocysts with 12 sporozoites; about 1 named species.

Genus *Dorisiella* Ray, 1930

Sporocysts with 8 sporozoites; about 1 named species.

Family Cryptosporidiidae Léger, 1911

Development just under surface membrane of host cell or within its brush border and not in the cell proper; syzygy absent; oocysts and meronts with a knob-like attachment organelle at some point on their surface; oocysts without sporocysts, with 4 naked sporozoites; microgametes without flagella; homoxenous; about 10 named species.

Genus *Cryptosporidium* Tyzzer, 1907

With the characters of the family; in vertebrates; about 10 named species.

Family Pfeifferinellidae Grassé, 1953

Oocysts without sporocysts, with 8 naked sporozoites; fertilization of macrogamete through a vaginal tube; homoxenous; about 2 named species.

Genus *Pfeifferinella* von Wasielewski, 1904

With the characters of the family; in mollusks; about 2 named species.

Family Lankesterellidae Nöller, 1902

Development in host cell proper; syzygy absent; oocysts with or without sporocysts, but with 8 or more sporozoites; heteroxenous, with merogony, gametogony, and sporogony in the same vertebrate host; sporozoites in blood cells, transferred without developing by an invertebrate (mite, mosquito, or leech); infection by ingestion of invertebrate host; microgametes with 2 (?) flagella; about 30 named species.

1

Genus *Lankesterella* Labbé, 1899

Oocysts produce 32 or more sporozoites; sporozoites with about 30 subpellicular microtubules; in amphibia; known invertebrate hosts leeches; about 4 named species.

Genus *Atoxoplasma* Garnham, 1950

Similar to *Lankesterella,* but in birds; sporozoites in leukocytes; known vectors are mites; about 17 named species.

Genus *Schellackia* Reichenow, 1919

Oocysts produce 8 sporozoites; in lizards; merogony in small intestine, connective tissue, and/or reticuloendothelial system; vectors are mites or Diptera; about 8 named species.

Family Eimeriidae Minchin, 1903

Development in host cell proper; without attachment organelle or vaginal tube; syzygy absent; oocysts with 0, 1, 2, 4, or more sporocysts, each with 1 or more sporozoites; homoxenous or at least without asexual multiplication in nondefinitive host; merogony within host, sporogony typically outside; microgametes with 2 or 3 flagella; without metrocytes, in vertebrates or invertebrates; about 1,340 named species.

Genus *Tyzzeria* Allen, 1936

Oocysts without sporocysts, with 8 naked sporozoites; in vertebrates; about 9 named species.

Genus *Eimeria* Schneider, 1875

Oocysts with 4 sporocysts, each with 2 sporozoites; in vertebrates and a few invertebrates; about 1,040 named species.

Genus *Mantonella* Vincent, 1936

Oocysts with 1 sporocyst containing 4 sporozoites; about 3 named species.

Genus *Cyclospora* Schneider, 1881

Oocysts with 2 sporocysts, each with 2 sporozoites; about 9 named species.

Genus *Caryospora* Léger, 1904

Oocysts with 1 sporocyst containing 8 sporozoites; about 22 named species.

Genus *Isospora* Schneider, 1881

Oocysts with 2 sporocysts, each with 4 sporozoites; usually in vertebrates; about 200 named species.

Genus *Diaspora* Léger, 1898

Oocysts unknown, sporocysts each with 1 sporozoite; sporocysts without bivalved wall, without longitudinal dehiscence suture; in invertebrates; about 1 named species.

Genus *Dorisa* Levine, 1979

Oocysts with variable number of sporocysts, each with 8 sporozoites; about 9 named species.

Genus *Wenyonella* Hoare, 1933

Oocysts with 4 sporocysts, each with 4 sporozoites; about 15 named species.

Genus *Octosporella* Ray and Ragavachari, 1942

Oocysts with 8 sporocysts, each with 2 sporozoites; about 2 named species.

Genus *Hoarella* Arcay de Peraza, 1963

Oocysts with 16 sporocysts, each with 2 sporozoites; about 1 named species.

Genus *Sivatoshella* Ray and Sarker, 1968

Oocysts with 2 sporocysts, each with 16 sporozoites; about 1 named species.

Genus *Pythonella* Ray and Das Gupta, 1937

Oocysts with 16 sporocysts, each with 4 sporozoites; about 2 named species.

Genus *Barrouxia* Schneider, 1885

Oocysts with many sporocysts, each with 1 sporozoite; sporocysts with bivalved wall, with a longitudinal dehiscence suture; about 10 named species.

Genus *Gousseffia* Levine and Ivens, 1980

Oocysts with 8 sporocysts, each with many sporozoites; about 1 named species.

Genus *Skrjabinella* Machul'skii, 1949

Oocysts with 16 sporocysts, each with 1 sporozoite; about 1 named species.

Family Sarcocystidae Poche, 1913

Heteroxenous, producing oocysts following syngamy; syzygy absent; oocysts with 2 sporocysts, each with 4 sporozoites, in intestine of a definitive host; with asexual stages in an intermediate host; about 105 named species.

Subfamily Sarcocystinae Poche, 1913

Obligatorily heteroxenous; asexual multiplication in intermediate (prey) host; last generation meronts (sarcocysts) in intermediate host form metrocytes, which give rise to bradyzoites, which are infectious for definitive (predator) host; oocysts sporulate in predator host tissues; sporulated sporocysts in its feces.

Genus *Sarcocystis* Lankester, 1882

Last generation meronts typically in striated muscles; merozoites elongate; about 90 named species.

Genus *Frenkelia* Biocca, 1978

Last generation meronts typically in central nervous system; merozoites elongate; about 2 named species.

Genus *Arthrocystis* Levine, Beamer, and Simon, 1970

Last generation meronts typically in striated muscle, jointed like bamboo; merozoites spherical; about 1 named species.

1

Subfamily Toxoplasmatinae Biocca, 1956

Complete life cycle obligatorily heteroxenous, but asexual stages usually transmissible from 1 intermediate host to another; metrocytes not formed; oocysts do not sporulate in host tissues; about 14 named species.

Genus *Toxoplasma* Nicolle and Manceaux, 1908

Meronts in many types of cell; host cell nucleus outside meront wall; about 7 named species.

Genus *Besnoitia* Henry, 1913

Meronts in fibroblasts and probably other cells; host cell nuclei within meront wall; about 7 named species.

Suborder Haemospororina Danilewsky, 1885

Macrogamete and microgamont develop independently; conoid ordinarily absent; syzygy absent; microgamont produces about 8 flagellated microgametes; zygote motile (ookinete); sporozoites naked, with 3-membraned wall; endodyogeny absent; heteroxenous, with merogony in vertebrate host and sporogony in invertebrate; pigment (hemozoin) visible with the light microscope may or may not be formed from host cell hemoglobin; transmitted by blood-sucking insects; about 460 named species.

Subclass Piroplasmasina Levine, 1961

Piriform, round, rod-shaped or amoeboid, without conoid; without oocysts, spores, or pseudocysts; flagella absent; most genera without subpellicular microtubules; with polar ring and rhoptries; locomotion by body flexion, gliding or in sexual stages (in Babesiidae and Theileriidae, at least) by large axopodium-like "Strahlen"; asexual and probably sexual reproduction present; parasitic in erythrocytes and sometimes also in other circulating and fixed cells; heteroxenous, with merogony in vertebrate and sporozoitogony in invertebrate; sporozoites with 1-membraned wall; the vectors are thought to be ticks; about 150 named species.

The Apicomplexa include five principal groups of protozoa (gregarines, haemogregarines, coccidia, malaria parasites and their relatives, and piroplasms) plus a rather large number of transitional or dead-end groups or species. Only the coccidia, of which there are quite a few genera, are reviewed in this book. However, so little is known about many of the species that only the more important or common ones are discussed.

LIFE CYCLES OF SELECTED FAMILIES AND GENERA

The life cycles of the eucoccidia are quite similar. The basic life cycle is shown in Figure 1. The oocysts are ingested by a host animal, and the sporozoites are released from them. The sporozoites enter host cells, turn into meronts, and multiply by merogony, forming a variable number of merozoites by multiple fission. These enter new host cells, turn into meronts, and again multiply by merogony. There is ordinarily a small, fixed number of asexual generations. The last generation merozoites enter new host cells and become gamonts. Some (the macrogamonts) develop into macrogametes without further multiplication, while others (the microgamonts) divide asexually by multiple fission to form a large number of flagellated microgametes. Syngamy takes place to form a zygote; a wall is laid down around it, and it becomes an oocyst. The oocyst ordinarily passes out of the host and does not develop further (i.e., sporulate) until it reaches the outside. However, sporulation takes place within the host's body in *Sarcocystis* and *Frenkelia* (and also in some species of *Eimeria* and *Isospora* in fish and reptiles). Once the oocysts have left the body, the infection is over. Some immunity develops as a result of infection, but it is not usually so great as to prevent reinfection.

In some species of *Isospora,* a sporozoite may enter a lymph node or other cell of a host animal species in which it cannot mature, and grow into a giant sporozoite, which Markus (1978) has denominated a hypnozoite. Here it remains for weeks, months, or years and infects its natural host when the latter eats the foreign (transport) host.

There are three multiplications in the coccidian life cycle—merogony, gametogony, and sporogony. Coccidia are haploid throughout their life cycle, except in the zygote stage. The first sporogonous division is meiotic, reestablishing the haploid condition.

The life cycles of selected genera of coccidia are given below, some of which were discussed by Hammond (1973). Therefore, duplication of much of his material is avoided.

Adeleidae

Adelina cryptocerci Yarwood, 1937 lives in the intestine of the woodroach *Cryptocercus punctulatus.* Sporulated oocysts are ingested by the woodroach. Sporozoites are released in its intestine. They pass through the intestinal wall and enter the fatty tissues around the intestine, where they form meronts. These each produce 8 to 40 merozoites by the 25th day. Each merozoite enters a new fat-containing cell and forms an average of 16 to 32 new merozoites. Merogony may be repeated several times. The last generation merozoites enter new fat cells and become macrogametes or microgamonts. The microgamonts enter into syzygy with the macrogametes, and a thin gametocyst wall is secreted around them. The macrogamete and microgamont grow, and the latter produces four motile microgametes. One of these unites with the macrogamete, forming a zygote. A wall forms around the zygote, and its nucleus divides, first by meiosis and then by mitosis, to form 5 to 21 uninucleate sporoblasts. Each sporoblast lays down a wall around itself and becomes a sporocyst. Each sporocyst forms two sporozoites and a residuum within itself. The sporozoites invade new hosts (Yarwood, 1937).

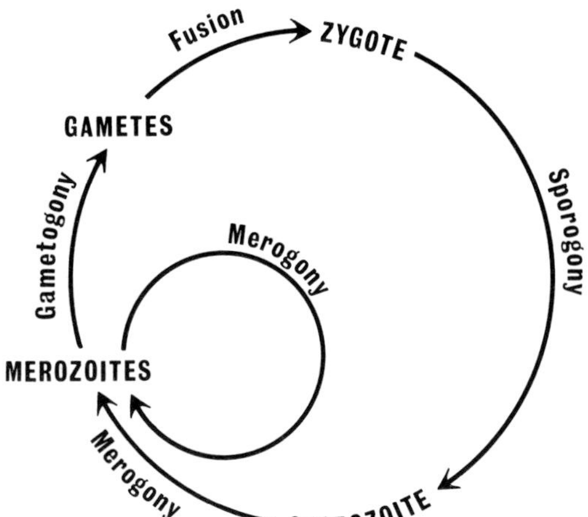

Figure 1. Basic life cycle of the order Eucoccidiorida. Oocysts (which are essentially zygotes with a secreted wall around them) are passed from the body, undergo development (sporogony) in the environment and come to contain infective sporozoites. The sporulated oocysts are ingested by a new host, the sporozoites emerge, enter host cells, and form several generations of similar organisms (merozoites) by multiple fission (merogony, schizogony). The number of these asexual generations is ordinarily limited. The last generation merozoites become macrogametes or microgamonts. The latter form microgametes, usually by multiple fission (gametogony). The gametes unite in syngamy to form a zygote, which lays down a wall around itself and passes out of the body. Reprinted by permission from: Corliss, J. O. Classification and phylogeny of the Protista. *Actualités Protozoologigues* 1:251–264 (1974).

Klossiellidae

Klossiella is the only genus in this family. Its members are essentially nonpathogenic and have usually been found by workers who were looking for something else. Most species occur in the kidneys. *Klossiella equi* Baumann, 1946 is found in the kidneys of the horse, ass, and zebra. The oocysts are in the kidney tubules, where they contain up to more than 40 sporocysts, 6–10 × 4–6 μm, each with 8 to 15 sporozoites. The sporocysts pass out in the urine and infect new hosts when they are ingested. The sporozoites enter the endothelial cells of Bowman's capsules, where they form meronts 8–12 μm in diameter containing 20 to 30 merozoites. These merozoites enter the epithelial cells of the proximal convoluted tubules and produce second generation meronts 15–23 μm in diameter with 15 to 20 merozoites 8 μm long. The second generation merozoites enter kidney tubule cells in the thick limb of Henle's loop and become macrogametes 10–15 μm in diameter, or microgamonts. The latter produce up to 10 rounded, nonflagellate microgametes 2–3 μm in diameter. Syzygy apparently does not occur. A microgamete fertilizes a macrogamete, and the resultant zygote becomes an oocyst which proceeds to form sporocysts and sporozoites (Baumann, 1946; Schiefer, 1967; Vetterling and Thompson, 1972).

Spirocystidae

This is an example, but perhaps an extreme one, of how little we know about coccidian taxa with only a few members. The family contains a single species *Spirocystis nidula* Léger and Duboscq, 1911. Léger and Duboscq (1911) found it quite common about 1909 in a little stream at Domène near Grenoble, France; however, it then disappeared, so that their 1915 paper was based on their earlier findings. To my knowledge, it has not been seen since. I have placed it in the suborder Eimeriorina; however, Grassé (1953) said that he did not know whether it was a coccidium or a gregarine. Gametocysts are formed in the chlorogogen cells of the earthworm *Lumbriculus variegatus*. Each gametocyst contains a single, very thick-walled, ovoid or piriform oocyst 35 μm long, which has a micropyle. Within each oocyst is a coiled, vermicular, naked sporozoite up to 40 μm long. Infection presumably occurs by ingestion of oocysts, perhaps in the dead and decaying body of another earthworm. The sporozoites enter host intestinal cells and form curved, vermicular meronts with one end markedly narrowed. The meronts are about 35 μm long and are found in the body cavity. They are coiled like a snail shell and contain numerous nuclei. Syzygy is apparently absent. Dissimilar, nonflagellate gametes are formed in the chlorogogen cells and in cells of the somatic and visceral peritoneum. The macrogametes are ovoid or spherical, 11 μm in diameter, and the microgametes are fusiform or ovoid, 7 × 3 μm (Léger and Duboscq, 1911, 1915).

Aggregatidae

This family contains seven genera and about 29 named species, but the life cycle has been worked out for only a few. Most species are presumably heteroxenous, with merogony in one host and gametogony in another. All occur in marine invertebrates. An example is *Aggregata eberthi* (Labbé, 1899) Léger and Duboscq, 1906. It is the only member of this genus of 17 named species whose complete life cycle is known. It was worked out by Léger and Duboscq (1906) and Dobell (1925). Merogony occurs in the intestine of the crab, *Portunus depurator,* and gametogony and sporogony in the intestine of the cuttlefish, *Sepia officinalis.* Merozoites infect the cuttlefish when it eats the crab. They enter the submucosa of the cecal and intestinal epithelium and develop into macrogametes and microgamonts. The latter produce scores to thousands of flagellated microgametes. Fertilization takes place to form a zygote, which later develops into an oocyst by forming an outer wall. There is no oocyst residuum. The zygote has 12 chromosomes. Reduction division takes place, and then the sporont divides (by repeated binary fissions, according to Dobell) to form a large number of spherical sporoblasts each of which becomes a sporocyst. Three sporozoites, 15×1.5 μm, plus a small residuum form in each sporocyst. The sporulated oocysts pass out in the cuttlefish's feces and are ingested by a crab. The sporocysts, which resemble 2 watchglasses joined together by a longitudinal suture, open at the suture, and release the sporozoites. These enter the subepithelial connective tissue, where they grow into meronts and produce a large number of merozoites. The mature meronts go to the crab's celom, where they remain until a cuttlefish eats the crab (Léger and Duboscq, 1906; Dobell, 1925).

Cryptosporidiidae

This family contains a single genus, *Cryptosporidium,* with about 10 named species, which occur in mammals and birds. These are very small coccidia and occur in the brush border or just under the surface membrane of the host gastrointestinal epithelial cell. Oocysts have been seen in the feces only, by Brownstein et al. (1977) and Pohlenz et al. (1978), in Giemsa-stained smears of a snake *Cryptosporidium* and *Cryptosporidium bovis,* respectively. Since the oocysts contain four naked sporozoites, sporocysts of *Sarcocystis* have been thought by some workers to be *Cryptosporidium* oocysts, and have been named accordingly. Vetterling et al. (1971) reviewed the genus; however, this was before it was known that *Sarcocystis* had *Isospora*-like oocysts. Consequently, they thought that some sporocysts that were actually those of *Sarcocystis* were those of *Isospora.* An example of this genus is *Cryptosporidium muris* Tyzzer, 1912, which lives in the gastric glands of the laboratory mouse. Infection is presumably by ingestion of sporulated oocysts. These are ellipsoidal, 7×5 μm, and contain four slender, fusiform sporozoites 12–14 μm long with one end pointed. The meronts are, at the most, 7 × 6 μm and produce eight banana-shaped merozoites and a small residuum. The microgamonts are never more than 5 × 4 μm and produce 16 rod-shaped, nonflagellate microgametes 1.5–2.0 μm long. The oocysts are presumably formed by zygotes after fertilization (Tyzzer, 1907; Vetterling et al., 1971; Pohlenz et al., 1978).

Lankesterellidae

This is perhaps the most puzzling family among the coccidia. Its members are supposed to be heteroxenous, with merogony, gametogony, and sporogony in the same vertebrate host. Sporozoites are said to occur in the blood cells, and infection is said to be transferred from one vertebrate to another by an invertebrate (a mite, mosquito, or leech) which sucks the blood. Infection occurs when the definitive host eats the vector. The principal controversy in this family revolves around the genus *Atoxoplasma* Garnham, 1950, which is common in the leukocytes of various birds. Lainson (1959) said that it is a synonym of the genus *Lankesterella,* which occurs in amphibia. Working in England, Lainson transmitted *Lankesterella adiei* of the house sparrow by means of the common red mite *Dermanyssus gallinae.* However, Box (1967, 1970), working in Galveston, Texas, where sparrows do not have *D. gallinae* and where *Lankesterella* is extremely common in sparrow blood, was unable to transmit the organism with mites, and gave evidence that *L. adiei* might really be a parenteral stage of the common house sparrow coccidium *Isospora lacazei.* Baker et al. (1972) were unable to resolve this controversy. Černá (1973) thought that *Atoxoplasma* was a stage in the life cycle of *Isospora* of birds. Box (1973, 1975, 1977) said that canaries *Serinus canarius* had two species of *Isospora* oocysts in their feces. One, *Isospora canaria* Box, 1975 was found only in the intestinal epithelium, while the other, *Isospora serini* (Aragão, 1933) Box, 1975 occurred both there and in mononuclear phagocytes in the blood.

I am, for the present accepting three genera in this family: *Lankesterella* in amphibia, with about four named species, presumably transmitted by leeches; *Atoxoplasma* in birds, with about 17 named species, presumably transmitted by mites; and *Schellackia* (syns., *Lainsonia, Gordonella*) in reptiles, with about eight named species, presumably transmitted by mites and Diptera. It is hoped that, in the future, the matter will be resolved.

An example of *Lankesterella is L. minima* (Chaussat, 1850) Nöller, 1912 of the frog *Rana esculenta,* whose life cycle was worked out by Nöller (1912, 1913, 1920). The sporozoites occur in the red blood cells. They are 10–15 μm long when free. The vector is the leech *Placobdella* (syn., *Hemiclepis) marginata.* It ingests the sporozoites when it sucks blood. The sporozoites do not develop in the leech, but are transferred to the frog when the leech sucks blood. Merogony takes place in the endothelial cells and leukocytes of the blood and lymph capillaries in the spleen, liver, pancreas, lung, and bone marrow. The meronts are bean-shaped and about 10–12 × 3 μm; they produce perhaps 30 to 40 merozoites. There may or may not be a residuum. The gamonts also occur in these cells, and the microgamonts occur especially in the capillaries of the pancreas, mesentery, and kidneys. Gametes are found 25–29 days after inoculation. Zygotes are found primarily in the endothelial cells of the liver and kidney capillaries. They produce oocysts that mature to contain 32 or more sporozoites without a sporocyst. The first stages of sporogony occur 40 days after inoculation. The sporozoites enter the red blood cells, and the life cycle starts again.

The life cycle of *Schellackia bolivari* Reichenow, 1919 is typical of the genus. It was worked out by Reichenow (1919). Merogony, gametogony, and sporogony occur in the lizards *Acanthodactylus vulgaris* and *Psammodromus hispanicus.* When a lizard eats a vector mite *(Liponyssus saurarum),* the sporozoites enter epithelial cells in the midgut and turn into meronts, which grow and produce about 10 to 16 merozoites. The host cells die and enter the intestinal lumen, where the merozoites emerge, enter new epithelial cells in the midgut, and undergo merogony again. After 3–4 weeks, depending on the temperature, the asexual cycle ends and all the merozoites become macrogametes or microgamonts. The latter produce a large number of slender, biflagellate microgametes. These fertilize the macrogametes, which have developed in the subepithelial connective tissue. The resultant zygote develops a thin wall, becoming an oocyst. Reduction division then takes place, and eight naked sporozoites and a residuum are formed in the oocyst. The sporozoites soon leave the oocysts and either enter the blood capillaries and then the erythrocytes or enter lymphocytes in various tissues and then pass into the blood stream. They occur in the erythrocytes in *A. vulgaris* and the lymphocytes in *P. hispanicus.* When the mite sucks blood, the blood cells in its gut are taken up by the gut phagocytic cells and the *Schellackia* sporozoites become more rounded and possibly a little larger, but they do not change otherwise. When a lizard eats the mite, the sporozoites are freed in its midgut, enter the lizard's intestinal epithelial cells, and begin to develop afresh.

Eimeriidae

This family contains the classical coccidia. There are 16 genera and some 1,340 named species in the family, of which the most important are *Eimeria, Isospora,* and, to a lesser degree, *Tyzzeria. Eimeria* contains well over 1,000 named species, *Isospora* about 200, and the other 14 genera about 90. The family (and the genus *Eimeria* itself) occurs in both vertebrates and invertebrates, but mostly in vertebrates. The genera are differentiated on the basis of the number of sporocysts in each oocyst and of sporozoites in each sporocyst. Only the oocysts (which are passed in the feces) are known for the great majority of species. Members of this family are intracellular, and most undergo merogony in the intestinal cells of their hosts.

1

Eimeria

This is the largest genus in the family. An example is the common chicken coccidium *Eimeria tenella* (Railliet and Lucet, 1891; Fantham, 1909; see Figure 1, McDougald, Chapter 9, this volume) which is found in the ceca (see Levine, 1973a). The oocysts are passed in the feces; at this time they contain a single cell, the sporont, which is diploid. In the presence of oxygen, this cell undergoes reduction division, throwing off a polar body, and then it sporulates by sporogony. Now haploid, the sporont divides into four sporoblasts, and each turns into a sporocyst containing two sporozoites. This process takes about 2 days at ordinary temperatures. If the mature oocysts are eaten by a chicken, the sporozoites emerge from them when exposed to bile and trypsin, and enter the wall of the intestine. They pass through the basement membrane into the lamina propria, are engulfed by macrophages, and are carried by them to the glands of Lieberkühn in the cecum. Here they leave the macrophages and enter the epithelial cells of the glands, where they come to lie beneath the host cell nucleus (i.e., on the side away from the lumen). They round up and turn into first generation meronts. Each meront forms about 900 first generation merozoites, about 2–4 μm long, by multiple fission (schizogony, merogony). The first generation merozoites enter the lumen of the cecum about 2.5–3 days after inoculation. Each one enters a new host cell, rounds up and becomes a second generation meront, which lies above the host cell nucleus. It forms about 200 to 350 second generation merozoites about 16 μm long about 5 days after inoculation. Some of them enter new intestinal cells, round up to form third generation meronts, which lie beneath the host cell nuclei and produce 4 to 30 third generation merozoites about 7 μm long. Others, however, enter new host cells and begin the sexual phase of the life cycle, known as gamogony or gametogony. Most of these round up and develop into female gametes (macrogametes). Others become male gamonts (microgamonts). The macrogametes and microgamonts may occur above or below the host cell nuclei. Each microgamont produces a large number of biflagellate microgametes plus a residuum of cytoplasmic material. The microgametes fertilize the macrogametes to form zygotes. These have eosinophilic granules (wall-forming bodies) around their periphery, which flatten out, fuse, and form a three-layered wall around the zygote. The resultant structure is an oocyst. It passes out of the body in the feces, and the life cycle begins again.

The prepatent period from the time of inoculation to the appearance of the first oocysts is slightly over 5 days. Oocysts are discharged for a number of days, but in the absence of reinfection the infection is self-limiting.

Each oocyst is theoretically able to produce 2,520,000 ($8 \times 900 \times 350$) second generation merozoites, each of which can develop into a macrogamete or microgamont. However, the actual number of oocysts produced per oocyst fed is considerably lower than the theoretical number (see Levine, 1973b).

Isospora

The life cycle of *Isospora* is similar to that of *Eimeria.* The number of asexual generations is limited. An example is the life cycle of *Isospora felis* of the cat, which was worked out by Shah (1971). Infection occurs after the ingestion of sporulated oocysts, which contain two sporocysts, each with four sporozoites. All endogenous stages lie above the host cell nucleus in the epithelial cells of the villi in the ileum and occasionally in the jejunum and duodenum. There are three asexual generations. The first generation meronts are 11–30 × 10–23 μm when mature and contain 16 to 17 banana-shaped merozoites 11–15 × 3–5 μm. They become mature 4 or, sometimes, 5 days after ingestion. The first generation merozoites enter new host cells, round up, and form second generation meronts. These form 2 to 10 spindle-shaped second generation merozoites, which are uninucleate at first, but by 6 days after inoculation are larger, multinucleate, and some have lost their elongate shape and have become ovoid. These are third generation meronts, which are about 12–16 × 4–5 μm with a mean of 14 × 5 μm. Each forms up to six banana-shaped merozoites 6–8 × 1.2 μm with a mean of 7 × 2 μm. The third generation meront is still within the second generation meront, which is often called a merozoite-shaped schizont. The third generation merozoites are not all formed at the same time, so that second generation meronts may be found containing both fully formed third generation merozoites and intermediate stages. Mature "cysts" containing only third generation merozoites are found 6–9 days after inoculation, and are most abundant at 7 days. By this time, they contain 36 to 70 or more merozoites that are not arranged in any particular fashion. Thus, the third generation meronts and merozoites develop within the same host cell and parasitophorous vacuole as the second generation meronts and merozoites.

The third generation merozoites break out of their host cell, enter other epithelial cells, and round up to form macrogametes or microgamonts. Mature microgamonts are 24–72 × 18–32 μm and contain a central residuum and a large number of biflagellate microgametes 5–7 × 0.8 μm. Mature macrogametes are 16–22 × 8–13 μm and have a large central nucleus and nucleolus. Gametes are most numerous 8–9 days after inoculation. Fertilization takes place; the resultant zygote lays down a wall around itself and becomes an oocyst, which is passed in the feces. The prepatent period from the time of inoculation until the time oocysts occur in the feces in 7–8 days, and the patent period (period of oocyst discharge) is 10–11 days. Then the life cycle is over, although reinfection may take place.

At present, the genus *Cystoisospora* is considered a synonym of *Isospora.* This genus is said to differ from *Isospora* because it has hypnozoites in its life cycle, whereas it is thought that *Isospora* does not. However, the few species of *Isospora* that have been checked for this character have been found to have it. One cannot assume that a species that has not been examined lacks hypnozoites in its life cycle, although in the future this may be decided for one or more species of *Isospora*. Consequently, *Cytoisospora* must be considered a synonym of *Isospora*, until it is proven otherwise.

Tyzzeria

The life cycle of *Tyzzeria* is similar to those of *Eimeria* and *Isospora*. The only species for which it is known is *Tyzzeria perniciosa* Allen, 1936, a pathogen of the domestic duck. It was worked out by Allen (1936). The oocysts are ellipsoidal, colorless, and 10–13 × 9–11 μm. They sporulate on the ground in 1 day, forming eight naked, banana-shaped sporozoites about 10 × 3.5 μm plus a large residuum. Ducks become infected by ingesting the sporulated oocysts. There are at least three asexual generations, all in the mucosal and submuscosal cells of the small intestine. The first generation meronts are about 12 × 8 μm and contain relatively few, small merozoites. The later meronts are about 15–16 × 14–15 μm and contain more and larger merozoites than those of the first generation. Merogony continues long after the formation of gametes. The macrogametes are somewhat irregular in shape. The first microgamonts appear 2 days after inoculation. They are about 7.5 × 6 μm and produce a large number of tiny microgametes. Oocysts appear in the feces 6 days after inoculation.

Cyclospora

This genus occurs in reptiles, millipedes, and moles. The only one whose life cycle is known is *Cyclospora caryolytica* of moles. Schaudinn (1902) worked it out in the Old World mole *Talpa europaea*, while Tanabe (1938) did so in the Asian mole *Talpa micrura coreana*. Their accounts differ considerably; therefore, I am including this genus because it illustrates the lack of agreement frequently encountered by those working with coccidia. Immature oocysts are passed in the feces, and sporulate in 3–5 days. According to Schaudinn, the merozoites that will form male and female gamonts are different from the start. There are several generations of merogony, in all of which these merozoites differ. Those that will form macrogamonts are long and thin, 12–15 × 1.5–2 μm, have a nucleus in the posterior third of the cell, and lack a clear globule in the cytoplasm. Those that will form microgamonts are shorter and broader, 10–12 × 2–3 μm, have a nucleus in the anterior third of the cell, and have a clear globule in the posterior half of the body. However, Tanabe (1938) said that he saw no such differentiation, and thought that Schaudinn had confused the life cycles of *C. caryolytica* and *Eimeria scapani*, both of which were present in Tanabe's moles. Tanabe said that each meront produces 16 merozoites measuring 3–5 × 1–1.2 μm in smears. He did not know the number of microgametes produced by each microgamont, but said that they are biflagellate. According to Schaudinn (1902), they are 9–10 × 1–1.5 μm; according to Tanabe (1938), they are 3–5 × 0.8 μm in smears.

Sarcocystidae

The basic life cycle of members of this family is the same as that of the Eimeriidae, but it differs in that the species are heteroxenous and the sexual and asexual stages occur in different hosts. All members of this family have a predator-prey host life cycle pattern, with the oocysts in a predator and the asexual stages in a prey animal.

Sarcocystis

The complete life cycle of *Sarcocystis* was discovered recently (Fayer, 1972; Rommel et al., 1972; Heydorn and Rommel, 1972a, b; Rommel and Heydorn, 1972). From a list of 91 named species, both definitive and intermediate hosts of only 26 are known. The life cycle of *Sarcocystis cruzi* (Hasselmann, 1926) Wenyon, 1926, an ox-dog species, is given as an example. However, the first 2 generations of meronts occurring in *S. cruzi* have not been seen in most other species; the only asexual stages known in most species are the sarcocysts.

Sporulated sporocysts and a few sporulated oocysts are passed in the feces of the dog; other known definitive hosts are the coyote *Canis latrans,* wolf *Canis lupus,* red fox *Vulpes vulpes,* and raccoon *Procyon lotor.* When eaten by cattle, the sporozoites emerge and invade endothelial cells of small and middle-sized arteries in the cecum, colon, kidney, pancreas, and cerebrum (Fayer, 1977). They are up to 52×28 μm and produce 100 or more first generation merozoites (tachyzoites). Fayer (1977) found only nuclei and no merozoites in calves killed 15–16 days after having been fed sporocysts from dogs. The first generation merozoites enter the endothelial cells of the capillaries in many organs, especially the kidney glomeruli (Fayer and Johnson, 1973, 1974; Mehlhorn, et al., 1975), round up and become second generation meronts about 15×9 μm. Immature ones can be found 26 to 33 days after inoculation. They produce up to 50, or perhaps more, second generation merozoites (tachyzoites) $7\text{–}8 \times 2\text{–}3$ μm. These enter skeletal and heart muscle cells, where they can be found beginning 34 days after inoculation. At this time, they are very small and contain only metrocytes. The metrocytes divide by endodyogeny to form merozoites (bradyzoites). Both metrocytes and merozoites reproduce by endodyogeny. At first, only metrocytes are present; later, there are both metrocytes and merozoites, and by 76 days only merozoites are present. It takes about 3 months for the sarcocysts to become fully mature and ready for transmission to dogs. The mature sarcocysts are up to 1 cm long. They are compartmented when mature, and contain merozoites about 10 μm long.

Dogs can be infected only by mature bradyzoites, not by tachyzoites of the first two merozoite generations. Gamonts, gametes, zygotes, oocysts, and sporocysts develop in the lamina propria of the villi of the small intestine, especially the distal jejunum and proximal ileum (Fayer, et al., 1973; Fayer, 1974). Sporulation occurs in the host. Oocyst production

1

begins in the dog 8 days after it has ingested sarcocysts (Fayer, 1974). The prepatent period is 8–33 days, and the patent period is 3–70 or more days.

The genus *Frenkelia* differs from *Sarcocystis* in that its last generation meronts occur in the brain rather than in the muscles. However, this is a small genus and no species is considered important enough to be mentioned in this chapter.

Toxoplasma

The only important species of *Toxoplasma* is *T. gondii.* The literature on this coccidium is vast. Jira and Kozojed (1970) published a two-volume bibliography with 7,763 entries for the period 1908–1967, and many more papers have been published since. *T. gondii* was first found by Nicolle and Manceaux (1908) in the gondi *Ctenodactylus gundi,* an African rodent, and since has been found in perhaps 200 species of mammals and a few species of birds. These are all intermediate hosts. The type definitive host is the domestic cat, but other felids can be infected and also produce oocysts (Miller et al., 1972; Jewell et al., 1972; Marchiondo et al., 1976). Man is an intermediate host; about one-third of the population, of the U. S. have serum antibodies against *T. gondii.*

Unsporulated oocysts are passed in the feces of cats and other felids. These sporulate in 2–3 days at 24 °C, or longer at lower temperatures (Frenkel et al., 1970), and are then infective for the intermediate hosts. When the sporulated oocyst is ingested by a susceptible animal (which may be almost any mammal or one of a considerable number of birds), the sporozoites emerge and pass to the parenteral tissues via the blood and lymph; any type of cell may be invaded. Here they multiply by endodyogeny. The stage in which this occurs is known as a group stage, and the merozoites within it are tachyzoites ("fast," i.e., rapidly developing zoites) (Frenkel, 1973) and multiply by endodyogeny. The group stage with its tachyzoites is the stage found in the leukocytes in peritoneal exudates, but it also occurs in other locations such as the liver, lungs, and submucosa; this is the stage that occurs in acute toxoplasmosis.

There is an indefinite number of tachyzoite generations. Eventually, they enter other cells and form the structure generally called a cyst; it is actually a pseudocyst or meront. Within it a large number of bradyzoites ("slow" zoites, i.e., slowly developing zoites) (Frenkel, 1973) are formed by endodyogeny. The meronts and bradyzoites are much more resistant to trypsin and pepsin than the tachyzoites, and they may remain viable in the tissues for years. This is the stage found commonly in the brain, but it also occurs in other tissues such as muscle; it is the stage found in chronic infections.

To my knowledge, the above meront is the end of the life cycle in all animals except felids. In the cat and other felids, the bradyzoites enter the intestinal epithelium and multiply. Dubey and Frenkel (1972) recognized five types of multiplication in these cells. Type A is present 12–18 hr after feeding meronts and multiplies by endodyogeny. Type B is present 12–54 hr after feeding meronts and multiplies by endogenesis (endodyogeny and endopolygeny) (Vivier, 1970; Piekarski et al., 1971). Type C is present 28–54 hr after feeding meronts and multiplies by schizogony. Type D is present 32 hr to 15 days after feeding meronts and multiplies by schizogony, endopolygeny, or splitting. Type E is present 3–15 days after feeding meronts and multiplies by schizogony. The merozoites are 5–8 × 1–2 μm.

The intestinal zoites form male and female gamonts in the intestinal epithelial cells. The macrogamonts are haploid and simply grow. The microgamonts produce 12 to 32 slender, crescentic microgametes about 3 μm long, which have two flagella plus the rudiments of a third (Pelster and Piekarski, 1971). Fertilization takes place, and the resultant zygotes form walls around themselves, becoming oocysts, and are released into the intestinal lumen.

Infection may take place either by ingesting sporulated oocysts or infected tissues of animals containing meronts, bradyzoites, or tachyzoites; by injection of meronts, bradyzoites, or tachyzoites; or congenitally via the placenta. Congenital toxoplasmosis of the newborn resulting from infection of the mother while she is pregnant is probably the most common form in man and perhaps sheep. Mice can be infected congenitally for generation after generation; Beverley (1975), for instance, reported that *T. gondii* was transmitted congenitally through at least nine successive generations in Swiss mice. Thus, *T. gondii* may be transmitted directly from one intermediate host to another via a partial life cycle, or it may be transmitted through its complete life cycle via a definitive host. The generic name *Hammondia* is a synonym of *Toxoplasma.*

Besnoitia

The life cycle of this genus is quite similar to that of *Toxoplasma,* except that the meront (pseudocyst) wall is relatively thick and contains several flattened giant host cell nuclei rather than a single one. An example is *Besnoitia besnoiti* of cattle. The first meronts are found in the endothelial cells of the blood vessels. They produce merozoites that enter the cells of the skin, the scleral conjunctiva, the nasal mucosa, and other sites, and become thick walled meronts containing thousands of merozoites but no metrocytes. The sexual stages and probably the last generation of meronts are apparently in the intestine of felids. The oocysts are ovoid, 14–16 × 12–14 μm, and unsporulated when passed. They appear in the feces of cats 4–25 days after the cats have been fed meronts from cattle (Rommel, 1975). The patent period in cats is 3–15 days (Peteshev et al., 1974). Sporulation occurs outside the host's body. Transmission may occur from one intermediate host to another or via the complete life cycle (i.e., from one intermediate host to a definitive host to another intermediate host, etc.)

The life cycle of *Besnoitia wallacei* is rather similar except that transmission from one intermediate host to another does not occur. It has been studied by Frenkel (1977) and Ito et al. (1978), but their findings were somewhat different and it is possible that two species are involved. The known intermediate hosts are rats, Mongolian gerbils, mice, rabbits, voles *(Microtus montebelli)*, and (to a slight extent) golden hamsters. Little more is known about the life cycles of *Besnoitia*. Further research is indicated.

IMPORTANT SPECIES OF COCCIDIA

There are about 1,560 named species of coccidia. It would be impossible in a book this size to discuss them all. Indeed, so little is known about most of them that it would not be worthwhile even to name them. Suffice it to say that about 710 species of *Eimeria* and *Isospora* are known in mammals, 220 in birds, 150 in reptiles, 40 in amphibia, 90 in fishes, and 11 in invertebrates, mostly centipedes and millipedes. There are no species of *Isospora* in fishes. The following are among the important species.

Enteric Coccidia

Ox The ox *Bos taurus* has about 15 named species of intestinal coccidia. Most of these species also occur in the zebu *Bos indicus*. The following species are known: *Eimeria alabamensis* Christensen, 1941; *E. auburnensis* Christensen and Porter, 1939; *E. bovis* (Züblin, 1908) Fiebiger, 1912; *E. brasiliensis* Torres and Ramos, 1939: *E. bukidnonensis* Tubangui, 1931;*E. canadensis* Bruce, 1921; *E. cylindrica* Wilson, 1931; *E. ellipsoidalis* Becker and Frye, 1929; *E. illinoisensis* Levine and Ivens, 1967; *E. kosti* Elibihari and Hussein, 1974 (questionable); *E. pellita* Supperer, 1952; *E. subspherica* Christensen, 1941, *E. thianethi* Gwéléssiany, 1935; *E. wyomingensis* Huizinga and Winger, 1942; and *E. zuernii* (Rivolta, 1878) Martin, 1909.

The most important species are *E. auburnensis, E. bovis,* and *E. zuernii. E. auburnensis* is moderately pathogenic. It occurs in the middle and lower thirds of the small intestine of the ox, zebu, and water buffalo. Its oocysts are elongate ovoid, usually smooth, and about 36–41 × 22–26 μm. It has two asexual generations, the first consisting of giant meronts 78–250 × 48–150 μm in the epithelial cells lining the crypts of Lieberkühn or in the cells of the reticular connective tissue of the jejunum and ileum, and the second consisting of small meronts 6–12 × 5–9 μm in the cells of the lamina propria. Its gamonts are in the subepithelium of mesenchymal mesodermal cells.

E. bovis is probably the most common cause of coccidiosis in cattle. It occurs in the ox, zebu, and water buffalo. The oocysts are ovoid, smooth, and are about 27–29 × 20–21 μm. It also has two asexual generations. The first consists of large meronts averaging 303 × 281 μm in the endothelial cells of the lacteals within the villi of the small intestine. The second generation meronts are in the epithelial cells of the villi of the cecum and colon and average 10 × 9 μm. Its gamonts are in the epithelial cells of the intestinal villi of the cecum and colon, extending into the terminal ileum in heavy infections.

E. zuernii is highly pathogenic, causing a bloody diarrhea. It occurs throughout the large and small intestines of the ox, zebu, and water buffalo. The oocysts are subspherical or ovoid, smooth, and average about 18 × 15 μm. There is probably more than one asexual generation, but only one is known. The meronts are in the epithelial cells of the villi and average about 13 × 10 μm. The gamonts also occur there.

Sheep The sheep *Ovis aries* has about 16 named species of intestinal coccidia. The following species are known: *Eimeria ahsata* Honess, 1942; *E. crandallis* Honess, 1942; *E. danielle* Dida, 1970 (dubious); *E. faurei* (Moussu and Marotel, 1902) Martin, 1909; *E. gilruthi* (Chatton, 1910) Reichenow and Carini, 1937 (oocysts unknown); *E. gonzalezi* Bazalar and Guerrero, 1970; *E. granulosa* Christensen, 1938; *E. hawkinsi* Ray, 1952 (questionable); *E. intricata* Spiegl, 1925; *E. marsica* Restani, 1971; *E. ovinoidalis* McDougald, 1979; *E. ovina* Levine and Ivens, 1970; *E. pallida* Christensen, 1938; *E. parva* Kotlan, Mocsy, and Vajda, 1929; *E. punctata* Landers, 1955; and *E. weybridgensis* Norton, Joyner, and Catchpole, 1974.

The most important species are *E. ahsata, E. ovinoidalis,* and *E. ovina. E. ahsata* occurs in both domestic sheep, the Rocky Mountain bighorn sheep *Ovis canadensis* and the mouflon *Ovis musimon.* It occurs in the small intestine and is markedly pathogenic, perhaps the most so of any sheep coccidia. The oocysts are elipsoidal to somewhat ovoid, smooth, and have a micropyle and micropylar cap; they are 29–44 × 17–2 μm. Only a single asexual generation has been recognized. Its meronts are mostly in the mucosa of the central part of the small intestine and average 184 × 165 μm. The gamonts are mostly in the epithelial cells lining the intestinal glands.

E. ovinoidalis was formerly called *E. ninakohlyakimovae,* but the latter species occurs in goats and cannot be transmitted to sheep (McDougald, 1979). It parasitizes the small intestine, especially the posterior part, and also the cecum and colon. It is markedly pathogenic. The oocysts are ellipsoidal or subspherical to somewhat ovoid, smooth, 16–28 × 14–23 μm, and have neither a micropyle nor micropylar cap. The life cycle is uncertain, because different authors have reported different sizes and numbers of endogenous stages.

E. ovina was once called *E. arloingi,* until it was realized that the species in sheep and goats are different. *E. arloingi* occurs in goats, and *E. ovina* in domestic sheep and also Rocky Mountain bighorn sheep, the mouflon, and the argali *Ovis ammon.* It occurs in the small intestine, and is not very pathogenic. It is extremely common, however, and has in the past been mistaken for *E. ahsata* and credited with the latter's pathogenicity. Its oocysts are quite similar to those of *E. ahsata,* being ellipsoidal to ovoid, smooth, and having a micropyle and micropylar cap, but are smaller, 23–36 × 16–24 μm. Only a single generation of meronts is

1

micropylar cap, but are smaller, 23–36 × 16–24 μm. Only a single generation of meronts is known; these are in the endothelial cells lining the central lacteals of the villi and are 122–146 μm in diameter. The gamonts occur in the epithelial cells of the villi.

Goat The goat *Capra hircus* has about 15 named species of intestinal coccidia: *Eimeria arloingi* (Marotel, 1905) Martin, 1909; *E. christenseni* Levine, Ivens, and Fritz, 1962; *E. crandallis* Honess, 1942 (perhaps actually *E. hirci); E. faurei* (Moussu and Marotel, 1902) Martin, 1909 (?); *E. gilruthi* (Chatton, 1910) Reichenow and Carini, 1937 (oocysts unknown); *E. granulosa* Christensen, 1938 (?); *E. hawkinsi* Ray, 1952 (questionable); *E. hirci* Chevalier, 1966; *E. intricata,* 1925; *E. ninakohlyakimovae* Yakimoff and Rastegaieff, 1930; *E. pallida* Christensen, 1938 (?); *E. parva* Kotlan, Mocsy, and Vajda, 1929 (?); *E. punctata* Landers, 1955 (?); *E. skrjabini* Dashnyam, 1961 (questionable); and *E. tirupatiensis* Sivanarayana and Venkataratnam, 1969.

So little is known about the goat coccidia that one cannot say which are the most important species. Until recently it had been thought that the same species occur in both sheep and goats, and it is not known for most species whether they can be transmitted between these hosts. *E. arloingi* occurs commonly in the small intestine of the domestic and various wild goats. Nothing is known of its pathogenicity. Its oocysts are ellipsoidal or slightly ovoid, smooth, with a micropyle and micropylar cap, and are 22–36 × 16–26 μm. It presumably forms giant meronts 280 × 150 μm in the endothelial cells of the lacteals of the jejunum, and other meronts about 10–14 × 9–10 μm in the epithelial cells of some glands of the jejunum. Its gamonts are in the mucosa of the small intestine and upper colon.

Swine The pig *Sus scrofa* has about 14 named species of intestinal coccidia: *Eimeria betica* Martinez and Hernandez, 1973; *E. debliecki* Douwes, 1921; *E. guevarai* Romero and Lizcano, 1971; *E. neodebliecki* Vetterling, 1965; *E. perminuta* Henry, 1931; *E. polita* Pellérdy, 1949 (syn., *E. cerdonis* Vetterling, 1965); *E. porci* Vetterling, 1965; *E. residualis* Martinez and Hernandez, 1973; *E. scabra* Henry, 1931; *E. spinosa* Henry, 1931; *E. suis* Nöller, 1921; *Isospora almaatensis* Paichuk, 1950; *I. neyrai* Romero and Lizcano, 1971 (dubious validity); and *I. suis* Biester, 1934.

We know almost nothing about the pathogenicity of any of these species. Perhaps the commonest is *E. debliecki.* It occurs in the small intestine. Its oocysts are generally ellipsoidal, smooth, and 20–30 × 14–20 μm, without a micropyle. There are two meront generations, both in the epithelial cells of the villi. The first generation meronts are only in the jejunum, and the second generation meronts are in both the jejunum and ileum. They are 8–12 and 13–16 μm, respectively, in diameter.

Rabbit The domestic rabbit *Oryctolagus cuniculus* has about 12 named species of intestinal coccidia plus one *(Eimeria stiedai)* in the bile ducts. The intestinal species are: *Eimeria coecicola* Kheisin, 1947; *E. elongata* Marotel and Guilhon, 1941; *E. exigua* Yakimoff, 1934; *E. intestinalis* Kheisin, 1948; *E. irresidua* Kessel and Jankiewicz, 1931; *E. magna* Pérard, 1925; *E. matsubayashii* Tsunoda, 1952; *E. media* Kessel, 1929; *E. nagpurensis* Gill and Ray, 1961; *E. neoleporis* Carvalho, 1942 (possibly a synonym of *E. elongata); E. perforans* (Leuckart, 1879) Sluiter and Swellengrebel, 1912; and *E. piriformis* Kotlan and Pospesch, 1934.

The most important species are *E. stiedai* and *E. perforans*. The most common and most pathogenic species is *E. stiedai*. It is the scourge of rabbitries and is difficult to eliminate from laboratory animal colonies. It lives in the epithelial cells of the bile ducts, which multiply tremendously and form large arborescent folds so that the ducts are grossly visible as yellowish nodules in the liver. Its oocysts are ovoid, sometimes ellipsoidal, smooth, with a micropyle, and 26–40 × 16–25 μm. At least five or six asexual generations of meronts are known, and the number is probably indefinite. There are two types of meronts, none of which is giant. One type produces a small number of plump merozoites, and the other many slender ones. The gamonts are also in the bile duct epithelial cells.

E. perforans is perhaps the most common intestinal coccidium of domestic rabbits, and is mildly to moderately pathogenic. It is found throughout the small intestine and also in the cecum. Both meronts and gamonts are in the epithelial cells of the villi. The oocysts are broadly ovoid, smooth, without a micropyle, and 14–31 × 9–25 μm. There are two meront generations.

Dog The dog *Canis familiaris* has about five species of *Isospora* of its own (and can be infected experimentally with two more), and seven of *Sarcocystis* (and can be infected experimentally with at least one more). Oocysts of several species of *Eimeria* have also been found in dog feces; however, they are probably pseudoparasites and live normally in prey animals that the dog has eaten. The *Isospora* species are: *I. bahiensis* de Moura Costa, 1956; *I. burrowsi* Trayser and Todd, 1977; *I. neorivolta* Dubey and Mahrt, 1979; *I. canis* Nemeséri, 1959; and *I. ohioensis* Dubey, 1975; plus (experimentally) *I. arctopitheci* Rodhain, 1933 of several New World monkeys and *I. vulpina* Nieschulz and Bos, 1933 of foxes. The *Sarcocystis* species are *S. bertrami* Doflein, 1901; *S. cruzi* (Hasselmann, 1926) Wenyon, 1926; *S. equicanis* Rommel and Geisel, 1975; *S. fayeri* Dubey, Streitel, Stromberg and Toussant, 1977; *S. levinei* Dissanaike and Kan, 1978; *S. miescheriana* (Kühn, 1865) Labbé, 1899; and *S. tenella* (Railliet, 1886) Moulé, 1886. In addition, *S. hemionilatrantis* Hudkins and Kistner, 1977 of the coyote can be transmitted to the dog experimentally.

Although coccidiosis due to *Isospora* may be important in dogs, little is known about the species involved; however, *I. canis* is quite common. It occurs in the small and large intestines, and is mildly to moderately pathogenic. Its oocysts are broadly ellipsoidal to slightly ovoid,

32–42 × 27–33 μm, without a micropyle, and smooth. There are three asexual generations, usually just beneath the epithelium in the lamina propria. The first generation meronts are about 25 × 21 μm, the second generation about 15 × 11 μm, and the third generation about 24 × 17 μm. The gamonts are in the epithelial cells or subepithelial connective tissue.

I. ohioensis is also quite common; it occurs in the epithelial cells of the small intestine and is usually nonpathogenic. Its oocysts are ellipsoidal to ovoid, 20–27 × 15–24 μm, smooth, and without a micropyle. There are at least two asexual generations of meronts, both 17–24 × 12–15 μm.

The sporocysts of all dog species of *Sarcocystis* apparently look alike, and it is not known which species is most common. This presumably depends on the geographic locality. Apparently, none of the species is pathogenic for the dog, although *S. cruzi* is known to be pathogenic for its ox intermediate host, and *S. tenella* for its sheep intermediate host. The intermediate host of *S. cruzi* is the ox, of *S. bertrami, S. equicanis,* and *S. fayeri* the horse, of *S. miescheriana* the pig, of *S. tenella* the sheep, of *S. levinei* the water buffalo, and of *S. hemionilatrantis* the mule deer. Oocysts of all species sporulate in the intestinal wall, and sporulated sporocysts are passed in the feces. They are ellipsoidal, 11–18 × 7–15 μm, and have a residuum but no Stieda body. Merogony does not occur in the dog.

Cat The cat *Felis catus* has about two species of *Isospora,* eight of *Sarcocystis,* two of *Toxoplasma* and three of *Besnoitia* in its intestine: *Isospora felis* Wenyon, 1923; *I. rivolta* (Grassi, 1879) Wenyon, 1923; *Sarcocystis cuniculi* Brumpt, 1913; *S. cymruensis* Ashford, 1978; *S. fusiformis* (Railliet, 1897) Bernard and Bauche, 1912; *S. gigantea* (Railliet, 1886) Ashford 1977; *S. hirsuta* Moulé, 1888; *S. leporum* Crawley, 1914; *S. muris* (Railliet, 1886) Labbé, 1899; *S. porcifelis* Dubey, 1976; *Toxoplasma gondii* (Nicolle and Manceaux, 1908) Nicolle and Manceaux, 1909; *T. hammondi* (Frenkel and Dubey, 1975) Levine, 1977; *Besnoitia besnoiti* (Marotel, 1913) Henry, 1913; *B. darlingi* (Brumpt, 1913) Mandour, 1965; and *B. wallacei* (Tadros and Laarman, 1976) Dubey, 1977. Oocysts of several species of *Eimeria* have also been reported in cat feces, but they are probably pseudoparasites that live normally in prey animals that the cat has eaten.

While coccidiosis may be important in the cat, little is known about the species that cause it. *I. felis* occurs commonly in the epithelial cells of the villi, primarily of the ileum; it is slightly, if at all, pathogenic. Its oocysts are ovoid, 38–51 × 27–39 μm, with a smooth wall and without a micropyle. There are three asexual generations. The first generation meronts are 11–30 × 10–23 μm; the second generation meronts are ovoid and contain a number of third generation meronts 12–16 × 4–5 μm.

I. rivolta occurs commonly in the epithelial cells of the villi and glands of Lieberkühn of the small intestine, cecum, and colon. It is presumably only slightly, if at all, pathogenic. Its oocysts are ellipsoidal to somewhat ovoid, 21–28 × 18–23 μm, smooth, and without a micropyle. It has three asexual generations of meronts. The first generation meronts are 6–13 × 3–6 μm, the second generation 9–18 × 9–13 μm, and the third generation 7–24 × 4–21 μm.

As in the dog, sporocysts of all cat species of *Sarcocystis* are similar in size and shape. All their oocysts sporulate in the intestinal wall, and sporulated sporocysts are passed in the feces. The sporocysts are ellipsoidal 9–17 × 6–11 μm, with a residuum but without a Stieda body. Merogony does not occur in the cat. The intermediate host of *S. cuniculi* is the domestic (European) rabbit *Oryctolagus cuniculus, of S. cymruensis* the Norway rat *Rattus norvegicus,* of *S. fusiformis* the water buffalo, of *S. gigantea* the sheep, of *S. hirsuta* the ox, of *S. leporum* the cottontail rabbit *Sylvilagus,* of *S. muris* the house mouse, and of *S. porcifelis* the pig. None is pathogenic for the cat or, apparently, for their intermediate hosts.

T. gondii oocysts are not particularly common in the cat, nor are the intestinal stages pathogenic. However, many mammals and some birds may be intermediate hosts of this coccidium; it is common in them, and it may also be pathogenic in them. It oocysts are unsporulated (like those of *Isospora*) when passed, and are subspherical, 11–14 × 9–11 μm, smooth, and without a micropyle. Its life cycle has already been described.

B. besnoiti is enzootic in parts of southern Africa, Kazakhstan, and perhaps the Middle East; however, it does not occur in North America. It is apparently not pathogenic in the cat, but may be highly pathogenic in its ox intermediate host; it is apparently not pathogenic in its antelope intermediate hosts. Its oocysts are unsporulated like those of *Isospora* when passed. They are ovoid, 14–16 × 12–14 μm, smooth, and apparently without a micropyle. The life cycle has already been described. *B. wallacei* has so far been reported only from Hawaii and possibly Japan.

Man *Homo sapiens* has about two species of *Isospora* and two of *Sarcocystis* in his intestine: *Isospora belli* Wenyon, 1923; *I. natalensis* Elsdon-Dew, 1953; *Sarcocystis hominis* (Railliet and Lucet, 1891) Dubey, 1976; and *S. suihominis* (Tadros and Laarman, 1976) Heydorn, 1977.

I. belli occurs in the small intestine and may cause a mucous diarrhea, but it is usually only slightly, if at all, pathogenic. It is not common. Its oocysts are elongate ellipsoidal, 20–33 × 10–19 μm, smooth, and sometimes with a very small micropyle. Its endogenous stages are not known.

S. hominis occurs in the small intestine and may cause a mucous diarrhea, but it also is only mildly, if at all, pathogenic. It is apparently more common than *I. belli,* but is seldom seen. Its intermediate host is the ox. Its oocysts are sporulated when passed. They are about 20 × 15 μm, and the sporocysts within them are ellipsoidal, with one side flattened, about 15 × 10 μm, without a Stieda body but with a residuum. Its endogenous stages in man are unknown.

S. suihominis is similar to *S. hominis*, except that its intermediate host is the pig. It is pathogenic for both little pigs and man. Rommel (1978) found it so pathogenic for man that studies in this host were discontinued. Its oocysts are 18–20 × 12–15 μm and are sporulated when passed in the feces. Its sporocysts are 12–17 × 10–12 μm, ellipsoidal, with a residuum but without a Stieda body. The endogenous cycle in man is unknown.

Norway Rat About six species of coccidia have been found in the feces of the Norway or laboratory rat: *Eimeria alischerica* Musaev and Veisov, 1965; *E. bychowskyi* Musaev and Veisov, 1965; *E. contorta* Haberkorn, 1971 (probably a mixture of *E. nieschulzi* and *E. falciformis;* see Stockdale et al. (1979)); *E. miyairii* Ohira, 1912; *E. nieschulzi* Dieben, 1924; *E. separata* Becker and Hall, 1931; and *Isospora ratti* Levine and Ivens, 1965. *E. nieschulzi* and *E. separata* are frequently used for laboratory studies.

E. nieschulzi may be common in some localities. It usually occurs in the epithelial cells of the villi of the small intestine, particularly the middle part, and is more or less pathogenic. Its oocysts are ellipsoidal to ovoid, 16–26 × 13–21 μm, smooth, and without a micropyle. There are four asexual generations of meronts.

E. separata is fairly common. It occurs in the epithelial cells of the cecum and colon, and is slightly, if at all, pathogenic. Its oocysts are predominantly ellipsoidal, 10–19 × 10–17 μm, smooth, and without a micropyle. There are three asexual generations of meronts.

House Mouse About 13 species of coccidia have been found in the feces of the house mouse *Mus musculus: Eimeria arasinaensis* Musaev and Veisov, 1965; *E. baghdadiensis* Mirza, 1970 (perhaps *E. falciformis); E. contorta* Haberkorn, 1971 (probably a mixture of *E. nieschulzi* and *E. falciformis;* see Stockdale et al. (1979)); *E. falciformis* (Eimer, 1870) Schneider, 1875; *E. ferrisi* Levine and Ivens, 1965; *E. hansonorum* Levine and Ivens, 1965; *E. hindlei* Yakimoff and Gousseff, 1938; *E. keilini* Yakimoff and Gousseff, 1938; *E. krijgsmanni* Yakimoff and Gousseff, 1938; *E. musculi* Yakimoff and Gousseff, 1938; *E. papillata* Ernst, Chobotar, and Hammond, 1971; *E. paragachaica* Musaev and Veisov, 1965; *E. schueffneri* Yakimoff and Gousseff, 1938; and *E. vermiformis* Ernst, Chobotar and Hammond, 1971. Of this impressive number, only *E. falciformis* and *E. ferrisi* have been studied considerably, and the original culture of *E. falciformis* was probably a mixture; this is unfortunate, because it is the type species of the genus.

E. falciformis is apparently quite common in Europe, but rare in the United States. It occurs in the epithelial cells of the small and large intestines, and may cause diarrhea and even death if enough are present. Its oocysts are broadly ovoid to ellipsoidal, subspherical or spherical, 16–21 × 11–17 μm, smooth, and without a micropyle. The number of meront generations has not been determined.

Chicken About 11 species of coccidia have been named from the intestine of the domestic chicken *Gallus gallus: Eimeria acervulina* Tyzzer, 1929; *E. brunetti* P. P. Levine, 1942; *E. hagani* P. P. Levine, 1938; *E. maxima* Tyzzer, 1929; *E. mitis* Tyzzer, 1929; *E. mivati* Edgar and Seibold, 1964; *E. necatrix* Johnson, 1930; *E. praecox* Johnson, 1930; *E. sporadica* Plaan, 1951 (dubious validity); *E. tenella* (Railliet and Lucet, 1891) Fantham, 1909; and *Wenyonella gallinae* Ray, 1945. Because the chicken is a favorite research animal, hatching free of all microbes, and because some of these coccidian species are pathogenic, some have been studied intensively.

Thousands of papers have been written on *E. tenella.* It is a common species and most pathogenic in chickens. It occurs in the villar epithelial cells and submucosa of the ceca, and is highly pathogenic, causing hemorrhagic enteritis and even death in young birds. Its oocysts are broadly ovoid, 14–31 × 9–25 μm, smooth, and without a micropyle. There are three asexual generations of meronts.

E. necatrix is also a common species. Its first and second generation meronts occur in the small intestine, and its third generation meronts and its gamonts are in the cecum; it is also highly pathogenic. It causes the small intestine mucosa to become thick; this thickness remains after the coccidia are gone. The oocysts are oblong ovoid, 12–29 × 11–24 μm, smooth, and without a micropyle.

E. acervulina is perhaps the most common species seen. It occurs in the epithelial cells of the villi and, to a lesser extent, in the gland cells of the anterior small intestine. Some strains are only slightly pathogenic, while others are mildly to moderately pathogenic, if a large number of oocysts are given. Its oocysts are ovoid, smooth, 12–23 × 9–17 μm, and without a micropyle. There are four asexual generations of meronts, which are 5–11, 5–5.5, 4.5–5, and 7–9 μm, respectively, in diameter.

E. maxima is also common. Its meronts occur in the epithelial cells of the villi of the small intestine, and its gamonts are displaced toward the center of the villi and come to lie in their interior. Its oocysts are ovoid, smooth, or somewhat roughened, 21–42 × 16–30 μm, and without a micropyle. There are two asexual generations of meronts, which are both about 10 × 8 μm.

Turkey About seven species of coccidia have been named from the intestine of the turkey *Meleagris gallopavo: Eimeria adenoeides* Moore and Brown, 1951; *E. dispersa* Tyzzer, 1929; *E. gallopavonis* Hawkins, 1952; *E. innocua* Moore and Brown, 1952; *E. meleagridis* Tyzzer, 1927; *E. meleagrimitis* Tyzzer, 1929; and *E. subrotunda* Moore, Brown, and Carter, 1954. One of these species, *E. dispersa,* is unusual in that it is mesoxenous, occurring not only in the turkey but also in the bobwhite quail and ringneck pheasant and being transmissible to the chukar partridge and chicken.

E. adenoeides is quite pathogenic. It occurs in the epithelial cells of the villi in the posterior ileum, cecum, and colon. Its oocysts are ellipsoidal, sometimes ovoid, smooth,

19–31 × 13–21 μm, and sometimes have a micropyle. There are two asexual generations of meronts, the first being about 30 × 18 μm and the second about 10 × 10 μm.

E. meleagrimitis occurs in epithelial cells of the villi of the anterior half of the small intestine, and is probably the most pathogenic species in the turkey. Its oocysts are subspherical, smooth, 16–27 × 13–22 μm, and have no micropyle. There are three asexual generations of meronts, the first about 17 × 13 μm and the second and third about 8 × 7 μm.

Extraenteric Coccidia

Ox The ox has three species of sarcocysts in its muscles. The definitive host of *S. cruzi* (Hasselmann, 1926) Wenyon, 1926 is the dog. This species has microscopic sarcocysts and is pathogenic for cattle; its life cycle has been given earlier; its first two generations of meronts are pathogenic for cattle. The definitive host of *S. hirsuta* Moulé, 1888 is the cat. It has macroscopic sarcocysts and is not pathogenic for cattle. The definitive host of *S. hominis* (Railliet and Lucet, 1891) Dubey, 1976 is man. It is probably not pathogenic for cattle. Especially in Africa, *Besnoitia besnoiti* (Marotel, 1913) Henry, 1913 meronts occur in the skin and other sites of cattle, and may cause serious disease in them. The life cycle has been given earlier. Its definitive host is the cat.

Sheep Sheep have two or perhaps more species of sarcocysts in their muscles. The definitive host of *Sarcocystis tenella* (Railliet, 1886) Moulé, 1886 is the dog. It has small sarcocysts and is pathogenic for sheep, especially lambs. It is the equivalent in sheep of *S. cruzi* in cattle. The definitive host of *S. gigantea* (Railliet, 1886) Ashford, 1977 is the cat. It has macroscopic sarcocysts and is not pathogenic for sheep.

Swine Pigs have three species of sarcocysts in their muscles. The definitive host of *Sarcocystis miescheriana* (Kühn, 1865) Labbé, 1899 is the dog; nothing is known of its pathogenicity for the pig. The definitive host of *S. porcifelis* Dubey, 1976 is the cat; nothing is known of its pathogenicity for the pig. The definitive host of *S. suihominis* (Tadros and Laarman, 1976) Heydorn, 1977 is man; it is pathogenic for both the pig and man.

Man *Sarcocystis lindemanni* Rivolta, 1878 sarcocysts have been found in the muscles, in a few cases. I suspect that the organisms that have been reported under this name actually belong to several other species that are parasitic in lower animals and that they are only accidental parasites of man. Its definitive host(s) and pathogenesis are unknown.

T. gondii (Nicolle and Manceaux, 1908) Nicolle and Manceaux, 1909 meronts occur in a wide variety of mammals and in a few birds. It definitive hosts are the cat and other felids. It is pathogenic in a small proportion (usually newborn) of the animals that it infects. The life cycle has been given earlier.

LITERATURE CITED

Allen, E. A. 1936. *Tyzzeria perniciosa* gen. et sp. nov., a coccidium from the small intestine of the Pekin duck, *Anas domesticus* L. Arch Protistenk. 87:262–267.

Baker, J. R., Bennett, G. F., Clark, G. W., and Laird, M. 1972. Avian blood coccidians. Adv. Parasitol. 10:1–30.

Baumann, R. 1946. Beobachtungen beim parasitären Sommerbluten der Pferde. Wien. Tieraerztl. Mschr. 32:52–55.

Beverley, J. K. A. 1975. Vertical transmission of *Toxoplasma gondii.* Prog. Protozool. 4:41.

Box, E. D. 1967. Influence of *Isospora* infections on patency of avian *Lankesterella (Atoxoplasma,* Garnham, 1950). J. Parasitol. 53:1140–1147.

Box, E. D. 1970. *Atoxoplasma* associated with an isosporan oocyst in canaries. J. Protozool. 17:391–396.

Box, E. D. 1973. Comparative development of *Atoxoplasma* and *Isospora* in the canary. Prog. Protozool. 4:59.

Box, E. D. 1975. Exogenous stages of *Isospora serini* (Aragao) and *Isospora canaria* sp. n. in the canary (*Serinus canarius* Linnaeus). J. Protozool. 22:165–169.

Box, E. D. 1977. Life cycles of two *Isospora* species in the canary, *Serinus canarius* Linnaeus. J. Protozool. 24:57–67.

Brownstein, D. G., Strandberg, J. D., Montali, R. J., Bush, M., and Fortner, J. 1977. *Cryptosporidium* in snakes with hypertrophic gastritis. Vet. Pathol. 14:606–617.

Cerná, Z. 1973. Le cycle de développement et la spécificité des isospores des oiseaux. Prog. Protozool. 4:82.

Corliss, J. O. 1974. Classification and phylogeny of the Protista. Actual. Protozool. 1:251–264.

Dobell, C. 1925. The life history and chromosome cycle of *Aggregata eberthi* (Protozoa, Sporozoa, Coccidia). Parasitology 17:1–136.

Dubey, J. P., and Frenkel, J. K. 1972. Cyst-induced toxoplasmosis in cats. J. Protozool. 19:155–177.

Fayer, R. 1972. Gametogony of Sarcocystis sp. in cell culture. Science 175:65–67.

Fayer, R. 1974. Development of *Sarcocystis fusiformis* in the small intestine of the dog. J. Parasitol. 60:660–665.

Fayer, R. 1977. The first asexual generation in the life cycle of *Sarcocystis bovicanis.* Proc. Helminth. Soc. Wash. 44:206–209.

Fayer, R., and Johnson, A. J. 1973. Development of *Sarcocystis fusiformis* in calves infected with sporocysts from dogs. J. Parasitol. 59:1135–1137.

Fayer, R., and Johnson, A. J. 1974. *Sarcocystis fusiformis:* Development of cysts in calves infected with sporocysts from dogs. Proc. Helminth. Soc. Wash. 41:105–108.

Fayer, R., Mahrt, J. L., and Johnson, A. J. 1973. The life cycle and pathogenic effects of *Sarcocystis fusiformis* in experimentally infected calves. J. Protozool. 20:509 (film).

Frenkel, J. K. 1973. Toxoplasmosis: Parasite life cycle, pathology, and immunology. In: D. M. Hammond, and P. L. Long (eds.), The Coccidia: *Eimeria, Issospora, Toxoplasma,* and Related Genera, pp. 343–410. University Park Press, Baltimore.

Frenkel, J. K. 1977. *Besnoitia wallacei* of cats and rodents: With a reclassification of other cyst-forming isosporoid coccidia. J. Parasitol. 63:611–628.

Frenkel, J. K., Dubey, J. P., and Miller, N. L. 1970. *Toxoplasma gondii* in cats: Fecal stages identified as coccidian oocysts. Science 167:893–896.

Grassé, P.-P. 1953. Classe des coccidiomorphes (Coccidiomorpha Doflein, 1901—Coccidies des auteurs). In: P.-P. Grassé (ed.), Traité de Zoologie 1(2):691–797.

Hammond, D. M. 1973. Life cycles and development of coccidia. In: D. M. Hammond and

P. L. Long (eds.), The Coccidia: *Eimeria, Isospora, Toxoplasma,* and Related Genera, pp. 45–79. University Park Press, Baltimore.

Hammond, D. M., and Long, P. L. (eds.). 1973. The Coccidia: *Eimeria, Isospora, Toxoplasma,* and Related Genera. University Park Press, Baltimore.

Heydorn, A. O., and Rommel, M. 1972a. Beiträge zum Lebenszyklus der Sarkosporidien: II. Hund and Katze als Überträger der Sarkosporidien des Rindes. Berl. Muench. Tieraerztl. Wochenschr. 85:121–123.

Heydorn, A. O., and Rommel, M. 1972b. Beiträge zum Lebenszyklus der Sarkosporidien: IV. Entwicklungsstadien von S. fusiformis in der Dünndarmschleimhaut der Katze. Berl. Muench. Tieraerztl. Wochenschr. 85:333–336.

Ito, S., Tsunoda, K., and Shimura, K. 1978. Life cycle of the large type of *Isospora bigemina* of the cat. Natl. Inst. Anim. Health Q. (Tokyo) 18:69–82.

Jewell, M. I., Frenkel, J. K., Johnson, K. M., Reed, V., and Ruiz, A. 1972. Development of *Toxoplasma* oocysts in neotropical Felidae. Am. J. Trop. Med. Hyg. 21:512–517.

Jira, J., and Kozojed, V. 1970. *Toxoplasmose—1908-1967.* G. Fischer, Stuttgart, Germany. (2 vols.)

Labbé, A. 1899. Sporozoen. *Das Tierreich.* Vol. 5.

Lainson, R. 1959. *Atoxoplasma* Garnham, 1950, as a synonym for *Lankesterella* Labbé, 1899: Its life cycle in the English sparrow *(Passer domesticus domesticus,* Linn.). J. Protozool. 6:360–371.

Léger, L., and Duboscq, O. 1906. L'évolution d'une *Aggregata* de la seiche chez le *Portunus depurator* Leach. C. R. Soc. Biol. (Paris) 58: 1001–1002.

Léger, L., and Duboscq, O. 1911. *(Spirocystis nidula).* Bull. Soc. Zool. Fr. 1911:62–63.

Léger, L., and Duboscq, O. 1915. Etude sur Spirocystis nidula Lég. et Dub. schizogrégarine du Lumbriculus variegatus Mull. Arch. Protistenk. 35:199–211.

Leuckart, K. 1879. *Algemeine Naturgeschichte der Parasiten.* Winter, Heidelberg.

Levine, N. D. 1973a. Introduction, history and taxonomy. In: D. M. Hammond, and P. L. Long (eds.), The Coccidia: *Eimeria, Isospora, Toxoplasma,* and Related Genera, pp. 1–22. University Park Press, Baltimore.

Levine, N. D. 1973b. Protozoan Parasites of Domestic Animals and of Man. 2nd Ed. Burgess Publishing Co., Minneapolis, Minn.

Levine, N. D. 1974. Historical aspects of research on coccidiosis. Proceedings of the Symposium on Coccidia and Related Organisms, Guelph, Ontario, pp. 1–10. University of Guelph, Guelph, Ontario, Canada.

Levine, N. D., Corliss, J. O., Cox, F. E. G., Deroux, G., Grain, J., Honigberg, B. M., Leedale, G. F., Loeblich, A. R., III, Lom, J., Lynn, D., Merinfeld, E. G., Page, F. C., Poljansky, G., Sprague, V., Vavra, J., and Wallace, F. G. 1980. A new revised classification of the Protozoa. J. Protozool. 27:37–58.

Lindemann, K. 1865. Weiteres über Gregarinen. Bull. Soc. Imp. Nat. Moscou. 38(2):381–387.

Lühe, M. F. L. 1902. Ueber Geltung und Bedeutung der Gattungsnamen *Eimeria* und *Coccidium.* Zentralbl. Bakt. I. Abt. Orig. 31:771–773.

McDougald, L. R. 1979. Attempted cross-transmission of coccidia between sheep and goats and description of *Eimeria ovinoidalis* sp. n. J. Protozool. 26:109–113.

Marchiondo, A. A., Duszynski, D. W., and Maupin, G. O. 1976. Prevalence of antibodies to *Toxoplasma gondii* in wild and domestic animals of New Mexico, Arizona and Colorado. J. Wildl. Dis. 12:226–232.

Markus, M. B. 1978. Terminology for invasive stages of protozoa of the subphylum Apicomplexa (Sporozoa). S. Afr. J. Sci. 74:105–106.

Mehlhorn, H., Heydorn, O., and Gestrich, R. 1975. Licht- und elektronenmikroskopische

Untersuchungen an Cysten von *Sarcocystis fusiformis* in der Muskulatur von Kälbern nach experimenteller Infektion mit Oocysten und Sporocysten von *Isospora hominis* Railliet et Lucet, 1891: I. Zur Entstehung der Cyste und Cystenwand. Zentralbl. Bakt. Hyg. I. Orig. A. 231:301–322.

Miller, N. L., Frenkel, J. K., and Dubey, J. P. 1972. Oral infections with *Toxoplasma* cysts and oocysts in felines, other mammals, and in birds. J. Parasitol. 58:928–937.

Minchin, E. A. 1903. Protozoa: The Sporozoa. In: E. R. Lankester (ed.), A Treatise on Zoology, Part 1, Fasc. 2, pp. 150–360. Black, London.

Nicolle, C., and Manceaux, L. 1908. Sur une infection a corps de Leishman (ou organismes voisins) du gondi. C. R. Acad. Sci. (D) (Paris) 147:763–766.

Nöller, W. 1912. Über eine neue Schizogony von *Lankesterella minima* Chaussat (= *Lankesterella ranarum* Lank.). Arch. Protistenk. 24:201–208.

Nöller, W. 1913. Die Blutprotozoen des Wasserfrosches und ihre Übertragung. Arch. Protistenk. 31:169–240.

Nöller, W. 1920. Kleine Beobachtungen an parasitischen Protozoen. (Zugleich vorläfige Mitteilung über die Befruchtung und Sporogonie von *Lankesterella minima* Chaussat). Arch. Protistenk. 41:169–189.

Pellérdy, L. P. 1974. Coccidia and Coccidiosis. 2nd Ed. Akademiai Kiado, Budapest and Paul Parey, West Berlin.

Pelster, B., and Pierkarski, G. 1971. Elektronenmikroskopische Analyse der Mikrogametenentwicklung bei *Toxoplasma gondii.* Z. Parasitenkd. 37:267–277.

Peteshev, V. M., Galuzo, I. G., and Polomoshnov, A. P. 1974. Koshki—definitivnye khozyaeva besnoitii *(Besnoitia besnoiti).* Izv. Akad. Nauk. SSSR Biol. 1:33–38. (University of Illinois College of Veterinary Medicine, English Transl. No. 50).

Piekarski, G., Pelster, B., and Witte, H. M. 1971. Endopolygenie bei *Toxoplasma gondii.* Z. Parasitenkd. 36:122–130.

Pohlenz, J., Moon, H. W., Cheville, N. F., and Bemrick, W. J. 1978. Cryptosporidiosis as a probable factor in neonatal diarrhea of calves. J. Am. Vet. Med. Assoc. 172:452–457.

Reichenow, E. 1919. Der Entwickelungsgang der Hämococcidien *Karyolysus* und *Schellackia* nov. gen. Berlin Sitzbericht Ges. Naturf. Freunde. 1919:440–337.

Rommel, M. 1975. Neue Erkenntnisse zur Biologie der Kokzidien, Toxoplasmen, Sarkosporidien und Besnoitien. Berl. Muench. Tieraerztl. Wochenschr. 88:122–117.

Rommel, M. 1978. Summary of recent research on sarcosporidiosis. Invited Paper, IV International Congress of Parasitology, Warsaw, Poland.

Rommel, M., and Heydorn, A. O. 1972. Beiträge zum Lebenszyklus der Sarkosporidien: III. *Isospora hominis* (Railliet and Lucet, 1891) Wenyon, 1923, eine Dauerform der Sarkosporidien des Rindes und des Schweins. Berl. Muench. Tieraerztl. Wochenschr. 85:143–145.

Rommel, M., Heydorn, A. O., and Gruber, F. 1972. Beiträge zum Lebenszyklus der Sarkosporidien: I. Die sporozyste von *S. tenella* in den Fäzes der Katze. Berl. Muench. Tieraerztl. Wochenschr. 85:101–105.

Schaudinn, F. 1902. Studien über krankheitserregende Protozoen: I. *Cyclospora caryolytica* Schaud., der Erreger der perniciosen Enteritis den Maulwurfs. Arb. Gesundh. 18:368–416.

Schiefer, B. 1967. *Klossiella equi* in der Niere eines Zebra. Berl. Muench. Tieraerztl. Wochenschr. 80:63–65.

Schneider, A. 1875. Contributions a l'histoire des grégarines d'invertebrés de Paris et de Roscoff. Arch. Zool. Exp. Gen. 4:493–604.

Shah, H. L. 1971. The life cycle of *Isospora felis* Wenyon, 1923, a coccidium of the cat. J. Protozool. 18:3–17.

Stiles, C. W. 1902. *Eimeria stiedae* (Lindemann,

1865), correct name for the hepatic coccidia of rabbits. U. S. Dept. Agric. Bur. Anim. Ind. Bull. 25:18–19.

Stockdale, P. H. G., Tiffin, G. B., Kozub, G., and Chobotar, B. 1979. *Eimeria contorta* Haberkorn, 1971: A valid species of rodent coccidium? Can. J. Zool. 57:264–270.

Tanabe, M. 1938. On three species of coccidia of the mole, *Mogera wogura coreana* Thomas, with special reference to the life history of *Cyclospora caryolitica*. Keijo J. Med. 9:21–52.

Tyzzer, E. E. 1907. A sporozoan found in the peptic glands of the common mouse. Proc. Soc. Exp. Biol. Med. 5:12–13.

Vetterling, J. M., Jervis, H. R., Merrill, T. G., and Sprinz, H. 1971. *Cryptosporidium wrairi* sp. n. from the guinea pig *Cavia porcellus*, with an emendation of the genus. J. Protozool. 18:243–247.

Vetterling, J. M., and Thompson, P. E. 1972. *Klossiella equi* Baumann, 1946 (Sporozoa, Eucoccidia, Adeleina) from equids. J. Parasitol. 58:589–594.

Vivier, E. 1970. Observations nouvelles sur la reproduction asexuée de *Toxoplasma gondii* et considérations sur la notion d'endogenèse. C. R. Acad. Sci. (Paris) 271(D):2123–2126.

Yarwood, E. A. 1937. The life cycle of *Adelina cryptocerci* nov. sp., a coccidian parasite of the roach *Cryptocercus punctulatus*. Parasitology 29:370–390.

2

Host and Site Specificity

Leonard P. Joyner

Modern discussions of the concept of species as biological taxa have drawn attention to the physiological basis of the definition of species as groups that are reproductively isolated from one another (Mayr, 1969). From the practical point of view, the biologist must decide how such groups are to be recognized when the lack of interbreeding implied in this definition cannot be tested. For diagnostic purposes, certain morphological, geographical, and physiological characters can be studied. The problem then arises as to how closely these correspond to the systematic taxa. For the coccidia, affinities between parasites and particular hosts, and even individual tissues within them, are strong and contribute to the biological distinctions between the parasites. These host and tissue specificities are of value to the diagnostician and form the basis of much of the work on species identification.

This chapter is concerned with current aspects of these characters of certain genera of the Eimeriidae and Sarcocystidae. Because the *Eimeria* have been studied in greatest detail, most of the available data concern these species. However, a new dimension to concepts of host specificity has been introduced by the discovery of heteroxenous life cycles in the Sarcocystidae involving prey-predator systems in which asexual and sexual development are separated in different hosts. Extensive accounts of the different life cycles are given in recent reviews by Dubey (1977a) and Fayer (1980).

To be sure that the parasites in a given host are one phase of the same parasite in another host requires that the complete life cycle be known. This has been achieved for only a minority of the species of the Sarcocystidae, and in some cases the nomenclature used has been questioned. In this review, an attempt is made to appraise the current situation.

IDENTIFICATION OF PARASITES—SYSTEMATIC CHARACTERS AND LIFE CYCLES

Early work on the systematics of coccidia assumed that species could be identified by the nature of the oocysts and the host in which they occurred. To some extent, this still is true of many species of the *Eimeria* with their tetrasporic oocysts and dizoic sporocysts. This group generally completes the life cycle in one host (monoxenous) and is limited to one species of host (stenoxenous). However, host specificity and tissue specificity within the host can be studied in detail and may be used in the identification of individual species. Oocyst characters are of special systematic significance for the *Eimeria.* This is discussed in a later section of this chapter.

Previously, it was believed that parasites of the genus *Isospora* developed dicystic oocysts with tetrazoic sporocysts and displayed strong host specificity, so that development was also both monoxenous and stenoxenous. The demonstration that tissues of animals containing stages of *Sarcocystis* or *Toxoplasma* when fed to certain carnivorous mammals, birds, or reptiles resulted in a gametogony cycle typical of the coccidia, culminating in the formation of oocysts

2

with two sporocysts similar to those of *Isospora,* raised great problems of classification and identification. The new type of life cycle involves two hosts—the final or definitive host in which the sexual cycle is completed and the intermediate host in which only asexual development occurs. Such a cycle which involves development of different stages in different hosts is termed heteroxenous.

In *Toxoplasma gondii,* asexual stages are transmissible from one intermediate host to another and a number of intermediate host species can be infected with oocysts from the final host (cat). There is thus a facultative alternation of hosts in this cycle, but the completion of the life cycle takes place only in the feline host. This species may therefore be either monoxenous or heteroxenous. In all *Sarcocystis* species studied, the alternation of hosts seems to be obligatory, and the carnivorous habit of the final definitive host becomes highly significant. Unlike *Toxoplasma* or the related *Besnoitia,* oocysts or sporocysts of *Sarcocystis* are fully sporulated when shed.

In *Eimeria* species, the complete life cycle, schizogony and gametogony, is completed in one host. Host specificity is thus a readily available character for the differentiation of species. The heteroxenous life cycle of the other genera makes the identification of individual species much more complicated because in each case oocysts from one host need to be related to tissue stages in another. In only a relatively few cases has this been achieved. Many species are known only by the tissue cysts which occur in a given host.

THE OOCYST

Specific Morphological Characters

The oocyst is the most easily accessible phase of any coccidian, and many species are known only by the morphology of the oocysts and by the identity of the host in which they were found. There are many descriptions in protozoological texts of the anatomy of the oocyst, which present a number of features which may be defined in quantitative or qualitative terms; the possibilities of variation and combination of specific features are considerable. Color, size, shape, surface texture, the presence or absence of a polar cap, the presence or absence of a clearly visible micropyle and its structure, the shape of the sporocysts, and the presence and nature of various defined structures such as residual bodies, polar granules, and Stieda bodies are recognizable and subject to variation between species. It is upon the description of such characters and the identification of the host species that the majority of species of *Eimeria* have been defined.

In some hosts, however, the oocysts of species of *Eimeria* are so similar that they cannot be reliably differentiated by morphology and size. In the fowl, for example, only oocysts of

Eimeria maxima can be recognized reliably solely by appearance (see Long et al., 1976). In ruminants and especially cattle and sheep, however, oocyst morphology can be used to differentiate species (Joyner et al., 1966; Levine and Ivens, 1970; Norton et al., 1974; Catchpole et al., 1975) and to provide a basis for epidemiological studies. The range of morphological types from sheep, for example, is given in Figure 1.

In making identifications from oocyst morphology, one should note the possibility of variation within individual species. Catchpole et al. (1975) observed the tendency in heavy infections, with some of the species of *Eimeria* in sheep, to produce "bizarre forms" including "double oocysts" containing eight sporoblasts and two polar caps. Toward the end of patency, when oocyst numbers are diminishing as the hosts' immune mechanisms begin to operate, deformed oocysts become increasingly common. Cheissin (1947) made similar observations in a detailed analysis of the factors affecting oocyst variability in *Eimeria magna* in the rabbit. More recently, Coudert et al. (1979) noted the variation in oocysts of *Eimeria perforans* in infections derived from single oocysts. Cheissin (1957) further noted that in the case of *Eimeria intestinalis,* shape could vary from broad pear shape to subspherical.

It has been known for many years that oocyst size is not necessarily constant. Joyner and Long (1974) listed several studies which showed that oocyst size and conformation varied with the stage of patency. Despite this variability, oocyst dimensions are frequently quoted as specific characters. When this is done, some statistical statement of variability should be included.

Shape, on the other hand, tends to be more constant and, although deformities may occur, the characteristic shape of most oocysts is retained throughout the range of dimensions. Customarily, a "shape index" is calculated as a ratio of the mean length to mean breadth. A more satisfactory method of recording this ratio was used by Norton and Joyner (unpublished data) in comparing oocysts of *Eimeria acervulina* and *Eimeria mivati.* By this method, length was plotted against breadth for each oocyst and the slope of the line so obtained reflected the shape of the oocysts throughout the range of dimensions and could be used for the comparison of the species (Figure 2).

Most descriptions of oocysts give some indication of surface appearance, and in a few cases obvious features such as thickness, color, or surface texture are sufficiently distinctive to be of systematic value.

Speer et al. (1979) recently reviewed studies of the ultrastructure of the oocyst wall, which has been shown to consist of one to three prominent layers. The innermost layer is generally composed of fine granular electron-lucent material. The outer layers seem to be more variable. The oocyst wall of *Eimeria tenella* consists of a single layer similar to the inner layer of other species. Oocysts of a number of other species of *Eimeria* and *T. gondii* have an inner electron-lucent layer and an outer electron-dense layer. Three distinct layers have been described in the oocyst wall of *Isospora canis, Isospora canaria,* and *Isospora serini,* which are, respectively,

2

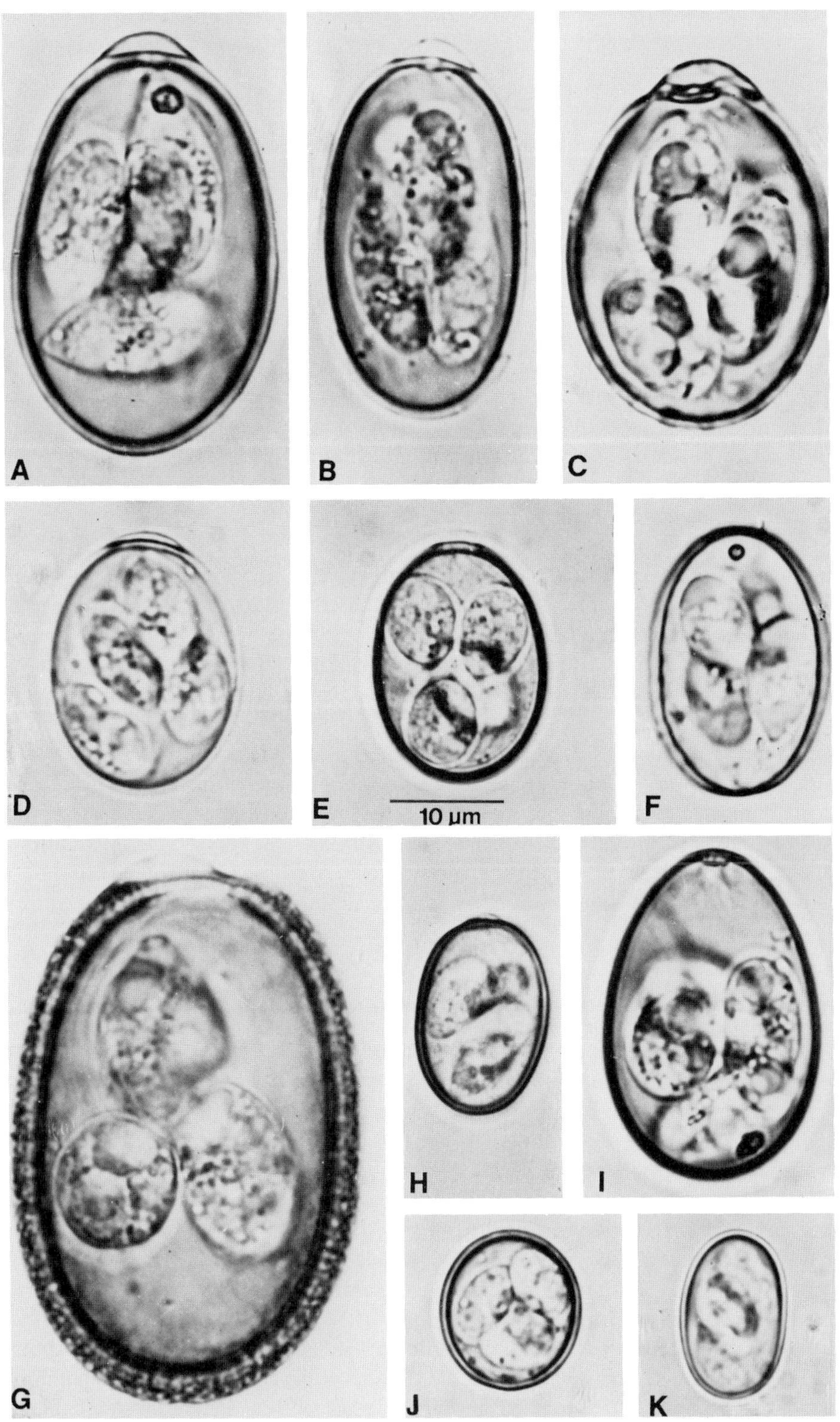

Figure 1. Sporulated oocysts from sheep. *A, Eimeria ahsata; B, E. ovina; C, E. granulosa; D, E. weybridgensis; E, E. crandallis; F, E. ovinoidalis; G, E. intricata; H, E. marsica; I, E. faurei; J, E. parva; K, E. pallida.*

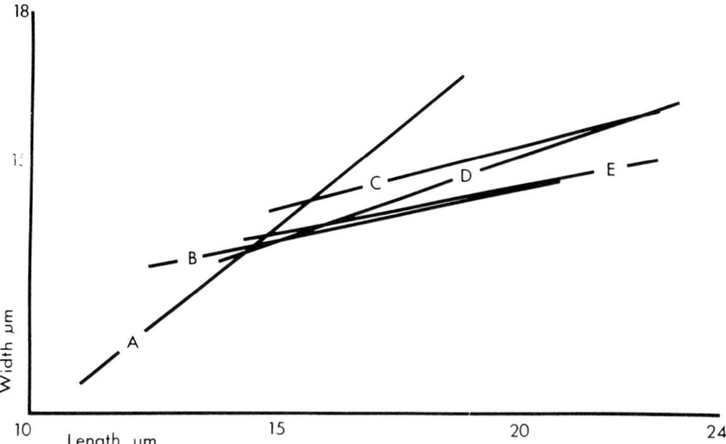

Figure 2. The range of oocyst dimensions and regression lines for width against length for strains of two species of *Eimeria. A, E. mivati; B, C, D,* and *E, E. acervulina.*

electron-lucent, -dense, and -lucent. In their paper, Speer et al. (1979) found that *Isospora lacazei* from sparrows has four prominent layers. Layers one, two, and three were electron-lucent, -dense, and -lucent, respectively, and the outermost fourth layer was discontinuous and was comprised of electron-dense spheroid bodies. These authors considered that ultrastructural differences in the oocysts walls may be of value to distinguish between species that are almost identical by light microscopy. Further studies under strictly comparable conditions would be necessary to establish this principle.

Cheissin (1947) considered that the range of variability is genetically controlled and that the individual characteristics of oocysts cannot be used as criteria for the classification of species without adequate analysis of variability. Certainly, it should not be assumed that because similar oocysts are present, different hosts are necessarily parasitized by the same species of coccidium. Some of the fallacies that have arisen in this way are mentioned elsewhere in this review.

A typical example of the practical difficulties of isolating and identifying oocysts of a pure line is provided by the work of Coudert et al. (1979) with *E. perforans* from the rabbit. These workers list the published measurements of oocysts of this species and further demonstrate the pleomorphism in figures of subspherical, elliptical, and subcylindrical oocysts all derived from a single oocyst. The strain eventually obtained was characterized by a prepatent period of 5 days, which is relatively short for species of *Eimeria* in the rabbit, and a sporulation time of 21 hr at 26 °C.

Sporulation

The process of sporulation—division of the sporoblast and the formation of sporozoites within sporocysts—requires oxygen and is temperature-dependent. The speed of sporulation is often quoted as a species characteristic; Reid (1973) has summarized some of the published data for species of *Eimeria* from the chicken. This character may be satisfactory for comparative purposes when oocysts of one species can be compared with those of another under identical conditions, but generally the conditions of measurement have not been standardized sufficiently for its status as a species characteristic to be properly assessed.

In the first place, there is little agreement on a standard temperature; 28 °C is commonly quoted but is not universally applied. Furthermore, it is commonly assumed that sporulation is arrested at low temperatures and fresh oocysts for sporulation studies are frequently collected in ice water. Storage temperatures below 0 °C are deleterious, but very little is known about changes which may take place between 0° and 5 °C, which is the normal range of laboratory refrigerators. Coudert et al. (1973) drew attention to earlier work of Pérard, which demonstrated that *Eimeria stiedai* completed sporulation, although slowly, at 5–6 °C; they themselves found sporulation to be completed after 13 days at 10 °C.

2

Although difficult to judge, the precise time at which sporulation is completed is often assumed to be when the Stieda bodies or the sporocystic refractile globules are clearly visible; however, this may not necessarily coincide with the completion of development to the infective stage which, in turn, is unlikely to be synchronized for all oocysts under observation. Edgar (1955) attempted to resolve this problem by inoculating samples of oocysts at regular intervals into susceptible chicks. This method determines the time of appearance of the first infective oocyst. Edgar and Seibold (1964) used this minimum sporulation time in the comparison of a number of species from the fowl.

Metabolism is not constant throughout the process of sporulation. Initially, polysaccharides are utilized, but later, lipid metabolism plays a greater part (Wilson and Fairbairn, 1961). Wagenbach and Burns (1969) recorded variations in oxygen consumption and correlated them with different stages in sporulation. Some of the most detailed studies of the sporulation process have been made on oocysts of *E. tenella* and *E. stiedai* by Yvoré and Coudert (1972) and Coudert et al. (1973), respectively. They separated the sporulation process into four different morphological stages and related them to cumulative oxygen consumption. It was found that the quantity of oxygen necessary to complete a given stage or for completion of sporulation was constant and independent of temperature. These authors considered that the correlation between morphological change in the oocyst and oxygen consumption is sufficiently precise to be of systematic value. This approach seems to be worthy of further study.

SPECIFICITY IN THE *EIMERIA*

Host Specificity

The host specificity of species of *Eimeria* is strong; it is rare for one of these parasites to occur naturally in more than one host genus. Frequently, studies on host specificity have relied upon oocyst recognition. Similar oocysts from different animals have sometimes been interpreted to indicate the presence of one coccidial species in more than one host. However, as previously stated, oocyst morphology is not always a reliable guide to speciation, and after more critical investigation the host specificity of some species has had to be revised.

For example, *Eimeria arloingi* previously was reported in both sheep and goats. Levine and Ivens (1970) drew attention to the weakness of oocyst size as a basis for differentiating species. They discussed the identity of *E. arloingi* in sheep, which was first described from goats by Marotel (1905). Christensen (1938) identified it on the basis of oocyst characteristics and considered it to be the most common species in sheep. Other authors have since studied it in this host. However, Levine and Ivens (1970) pointed out that *E. arloingi* from the goat has never been seen to complete its development in the sheep. Because there seem to be detectable differences in the oocysts from the two hosts, they proposed that *E. arloingi* should be retained as the name for the parasites in goats only. *Eimeria ovina* was the name proposed for the comparable parasites in sheep.

McDougald (1979) has considered further the *Eimeria* of sheep and goats and drawn attention to the status of *Eimeria ninakohlyakimovae*, which was first described from the goat by Yakimoff and Rastegaieff (1930). He found that oocysts corresponding to this species isolated from sheep failed to infect goat kids, thus confirming the experience of some previous workers. This is a major pathogen in sheep, but because of the priority of the corresponding parasite in goats, McDougald (1979) has renamed it *Eimeria ovinoidalis*. Reports of *E. ninakohlyakimovae* from the mouflon, the argali, and the gazelle, based upon the morphology of the oocysts, should be regarded as incorrect (McDougald, 1979).

A further note of caution is sounded by the work of Stockdale et al. (1979). These workers studied an isolate of *Eimeria contorta*, which was described by Haberkorn (1971) as having the unusual ability to infect both rats and mice. They passaged the parasites in two lines through mice and rats. Those recovered after passages in rats gave cross-protection against *Eimeria nieschulzi* and those from mice protected against *Eimeria falciformis*. Stockdale et al. (1979) concluded that the original isolate was impure and comprised a mixture of *E. falciformis* from mice and *E. nieschulzi* from rats. They drew attention also to the experience of Černá (1975), who reported that *E. contorta* of rat origin, when passaged through mice, gave identical antigenic reactions in the IF (immunofluorescence) test to *E. falciformis*. It was suggested that this isolate also may have been impure.

Clearly, purity of parasite lines is a fundamental requirement for studies on host specificity, although it is unusual for parasites from one host to be contaminated with parasites of another.

Sometimes the classification of hosts has been inaccurate and assumed host specificity has led to incorrect nomenclature of the coccidia. For example, Levine and Ivens (1972) in a review of the coccidia of Leporidae have pointed out that much confusion has arisen in their nomenclature because all rabbits and hares were equated and a parasite species that occurred in one was thought to occur in all others. Even when it became recognized that the coccidia of hares were different from those of rabbits, it was still thought that the species in the wild rabbit of Europe *(Oryctolagus cuniculus)* and the wild rabbit of the New World *(Sylvilagus* sp.) were the same. Levine and Iven's review makes it clear that only *E. stiedai* and *Eimeria neoleporis* are shared naturally by the two rabbit genera. A few other species of *Eimeria* have been transmitted experimentally from one host to another but generally each lagomorph genus has its own species of *Eimeria*.

In experiments on host specificity, it is usual to inoculate a given species with a large number of oocysts, and the appearance of further oocysts in the feces is judged to indicate successful parasitization. An obvious fallacy arises when sporulated oocysts derived from the

2

inoculum are detected in the feces. Most diagnostic laboratories have detected oocysts from rabbits in the feces of grazing animals, such as horses or cattle. Cross-infection experiments, therefore, should be monitored quantitatively.

There is, however, a quantitative aspect with *Eimeria mohavensis,* the natural host for which is *Dipodomys panamintinus mohavensis,* which Marquardt (1973) discussed in citing the work of Doran (1953). Doran (1953) compared the oocyst production in a number of species and subspecies of *Dipodomys* and found that in at least two other hosts in which *E. mohavensis* is not normally found, more oocysts were produced than in the natural host.

Eimeria dispersa is primarily a parasite of the turkey, in which it was frequently seen in a recent survey by Long and Millard (1977). It has been recorded from several avian species, but Pellérdy (1974) regarded some of the reports with uncertainty. Doran (1978) studied a strain of *E. dispersa* isolated from turkeys, in chickens, chukar partridge, ring-necked pheasant, and bobwhite quail. He showed that the parasite was poorly adapted to chickens and pheasants. In the partridge and quail, the prepatent period was shorter, and the early schizonts were larger and contained more merozoites. Norton (1979) passaged a turkey strain of *E. dispersa* in turkeys and pheasants. Oocyst production in the latter remained consistently low and showed no sign of adaptation. Very light infections were produced in young chickens with this strain, but not in Japanese quail.

Eimeria colchici from the pheasant is not completely host-specific. Infections may be produced in turkeys given large numbers of oocysts; Norton (1967) failed to maintain *E. colchici* in this host.

E. tenella evidently is highly host-specific, and when Vetterling (1976) inoculated it into eight species of gallinaceous birds, development was completed only in the chicken. A few macrogametocytes were found in the ceca of *Alectoris graeca* (chukar), but oocysts were not formed. In vitro, however, this strong host specificity is not sustained. Doran and Augustine (1973) found that cell cultures from the kidneys of the turkey, ring-necked pheasant, chukar, Japanese quail, and guinea fowl all supported the complete life cycle of *E. tenella,* although development was best in cells from chickens.

In summary, strong host specificity is a characteristic of the *Eimeria,* and the identity of the host is generally a good guide to speciation. With precise identification of the parasite, the high degree of host specificity is usually confirmed. It is rare for a species of *Eimeria* to infect more than one host genus and, with few exceptions, they are restricted to closely related species or subspecies.

There may, however, be variation within the host species so that genetical differences may be expressed in different susceptibilities to coccidial infections. Long (1968, 1973a) discussed the evidence that different breeds of chickens show differing susceptibilities to *Eimeria* species, which are partially controlled by hereditary factors. A further aspect of this was discussed by Marquardt (1976), who quoted attempts to infect mice with *Eimeria separata* from the rat. The genetical constitution of the mice was a significant factor in determining whether the life cycle could be completed in the foreign host.

Tissue Specificity

Most species of *Eimeria* have a characteristic site of invasion generally in epithelial cells of endodermal origin. In the domestic fowl, where site specificity has been most completely studied, parenteral injection of oocysts or sporozoites by intravenous, intramuscular, or intraperitoneal routes resulted in the development of parasites in normal sites in the intestine (Davies and Joyner, 1962; Sharma, 1964). These locations are so characteristic that they are often used as diagnostic features of the individual species (Reid, 1973).

Under certain extreme conditions, site predilections may be distorted. Normally, *E. tenella* is found only in the cecum of the chicken, although occasionally with heavy infections it may invade the adjoining intestine. Similarly, when the cecal tubes were removed, Leathem (1968) found that following heavy infections with this species, the proximal large intestine and the small intestine adjoining the cecal diverticula were parasitized. Under normal circumstances, the schizogony of *Eimeria necatrix* takes place in the intestine, and subsequent gametogony occurs in the ceca. Horton-Smith and Long (1965) showed that if sporozoites of this species are introduced directly into the ceca, the parasite will complete the whole life cycle in that organ. Similarly, *Eimeria brunetti* and *E. mivati,* which normally parasitize posterior intestinal tissues, readily developed in the ceca, whereas *E. acervulina* and *E. maxima,* which primarily invade the duodenum and anterior intestine, respectively, did not. *E. acervulina* is characteristically a parasite of the duodenum. Joyner and Norton (1972), however, showed that if large numbers of sporozoites were directly injected in the ceca, limited development would take place in the latter site, an observation subsequently confirmed by Long (1973b). Despite several determined attempts, Joyner and Norton (1972) were unable to maintain their strain of *E. acervulina* in the cecum.

In mammals also, species of *Eimeria* invade certain tissues. Haberkorn (1970) parenterally injected sporulated oocysts of *E. falciformis* into mice and observed developing parasites in the colon and cecum, which are the usual sites for this species. Similarly, Pellérdy (1969) found that intravenous inoculation of rabbits with sporozoites, oocysts, and merozoites of *E. stiedai* resulted in the development of the parasite in the liver, the normal site. This type of experiment, however, is not always successful; for example, Fitzgerald (1965) was unable to infect calves with *Eimeria bovis* by the intraperitoneal route.

There are a number of species of *Eimeria* that form gametocytes in sites different from those occupied by schizonts. One of the best known examples is *E. necatrix,* in which schizogony takes place in the intestine and gametogony occurs in the ceca. The recent redescription of *E. flavescens* from the rabbit provides a further example of site specificity within the life cycle. Here the first generation schizonts are formed deep in the glands of the lower small intestine, whence the merozoites migrate to the cecum and colon where second, third, and fourth generation schizonts develop in the superficial epithelium. A further generation of schizonts and gametocytes are formed in the glandular epithelium (Norton et al., 1979).

The results of infection by parenteral injection of oocysts or sporocysts suggest that sporozoites or merozoites have appreciable powers of migration. Macrophages may play a role in this migration, although these cells may be capable of destroying the parasites, especially in foreign hosts. However, the parasites can be engulfed by macrophages and remain viable at least for limited periods (Huff and Clark, 1970; El-Kasaby and Sykes, 1973). After intraperitoneal or intravenous inoculation, trophozoites or early schizonts are recovered from the liver (Long, 1970; Long and Millard, 1976). It is possible that they reach this site via the macrophages and thence to the intestinal tract via the portal system.

It has been suggested that the speed of excystation is a factor determining the region of the intestine to be parasitized. In a comparison of excystation times of species of *Eimeria* from the chick, Farr and Doran (1962) showed that oocysts of *E. tenella*, which parasitize the ceca, released the sporozoites appreciably slower than those of *E. acervulina,* which parasitize the duodenum. Similarly, in the turkey, the excystation time of *Eimeria gallopavonis* from the cecum was longer than that of *Eimeria meleagrimitis* from the duodenum. This phenomenon is likely to be of minor importance, because the parenteral injection studies suggest that the determining factors in site specificity may possibly be related to local physiology of tissues, which may or may not meet the requirements of a given species.

Cell Specificity

Parasite host-cell relationships have been studied in most detail in the genus *Eimeria,* where with few exceptions parasites develop in epithelial cells, particularly of the intestinal villi. Notable exceptions are the few species that develop in the epithelial cells of other organs, such as the bile duct (*E. stiedai* in the rabbit), the kidney tubules (*Eimeria truncata* in the goose), or the uterus (*Eimeria neitzi* in the impala; McCully et al., 1970). Schizonts and gametocytes of *Eimeria* species have a clear disposition to parasitize epithelial cells. However, there has been some debate on the origin of the cells in the lamina propria, which are parasitized by the second generation schizonts of *E. tenella* or *E. necatrix.* These resemble macrophages morphologically; however, Bergmann (1970) concluded that aggregates of these cells were joined by desmosomes and presumably, therefore, epithelial cells. Stockdale and Fernando (1975) and Fernando et al. (1974) discussed the greatly changed appearance of the parasitized cells, which have an increased DNA content, with an enlarged nucleus resembling fibroblasts. These authors noted that the cells were joined by desmosomes and they, too, concluded that they were most likely epithelial cells.

In chicks treated with dexamethasone, or in chick embryos, *E. tenella* will develop to the second generation schizont stage in the liver (Long 1970, 1971). It was thought that the schizonts were in epithelial cells of the bile ducts. In the chick embryo liver, gametocytes were found within cells thought to be hepatocytes. From an electron microscopical study of the infections in chick embryos, Lee and Long (1972) concluded that the second generation merozoites invaded endothelial cells, hepatocytes, fibrocytes, and macrophages. Schizonts developed in all these different cells except macrophages in which the merozoites seemed to be destroyed. While most invaded cells are of endodermal origin, the fibrocytes lining the sinusoids provide a possible example of parasitized mesodermal cells.

Sporozoites of *E. tenella* are capable of invading a variety of cells in vitro, but schizonts develop mainly in epithelial cells, although occasionally in fibrocytes. Gametocytes seem to develop in cells of epithelial type (Strout and Ouellette, 1969; Doran, 1970), but these may become modified to resemble fibroblasts.

Evidently, sporozoites and merozoites are capable of invading a number of different cell types. Their subsequent development which seems to be limited mainly to epithelial cells, depends upon interactions between parasite and host cell and results in changes in the morphology and physiology of the latter. This is an area in which research could be very rewarding.

Merozoites of *E. necatrix* invade epithelial cells in the glandular part of the tissue where cells are proliferating. Fernando et al. (1974) have shown that the host cells undergo extensive hypertrophy with at least a 4-fold increase in the DNA and RNA content of the nucleus. They suggested that the additional DNA synthesis is induced by the coccidial infection and may be due to a blockage of mitosis. Gametocytes of *E. maxima,* however, infect the villus epithelial cells which do not replicate, and in this situation there was extensive cellular enlargement without nuclear hypertrophy or enhanced DNA synthesis.

Pasternak et al. (1977) attempted to explore this phenomenon further with *Eimeria zuernii* from calves. This parasite at the first generation schizont stage generally infected non-proliferative cells, but not exclusively. Nuclear hyperplasia occurred in a majority of infected cells. However, this could not be accounted for by a proportional increase in DNA synthesis. In this case, the two processes were not necessarily related.

Pathogenicity

The pathology of coccidial infections and the pathogenicity of species of *Eimeria* have been exhaustively reviewed by Long (1973a) and Ryley (1975). Some of the biochemical aspects of pathogenicity were further discussed by Ryley (1973). Many of the features of disease in the host are expressions of the characteristics of individual species of *Eimeria* and as such are often used as a guide to identification.

The species of *Eimeria* in the fowl have been studied in great detail. Most of them are pathogenic to varying degrees, and by using the characters of the lesions, their location, and association with developmental stages of the parasites, it is possible to construct diagnostic keys for the important pathogenic species. An example of such a key, subject to modification in the light of new knowledge about *E. mivati,* is given by Long et al. (1976). *E. mivati* was included as a variant of *E. acervulina,* but it is now recognized that these two species can be distinguished by electrophoretic enzyme separation and the ability of *E. mivati* to develop in the chick embryo. Both have similar pathogenicity patterns and typically produce reductions in growth without severe mortality. *E. acervulina* produces characteristic lesions in the duodenum associated with the gametocytes, whereas *E. mivati* invades epithelium throughout the intestine, including the ceca. Lesions due to the latter are not visible to the naked eye, but severe disorganization of villous structure can be detected microscopically (Norton and Joyner, 1980). Without more specialized laboratory tests, it would not be possible to detect this species by routine postmortem techniques. Similarly, *Eimeria praecox,* which is only mildly pathogenic, produces no easily recognizable lesions.

In a few cases, pathogenicity has been defined in terms of physiological changes. Yvoré et al. (1972) have reviewed the effect of infections due to *E. acervulina* ascribing pathogenesis to two basic effects, namely, a modification of intestinal structure and activity leading to disturbances of absorption and permeability, and an indirect effect resulting in reduced feed and water consumption. Joyner et al. (1975) confirmed from radio tracer studies that anorexia and protein leakage are major factors in the pathogenesis of infections due to this species. Yvoré and Mainguy (1972) further showed that even very low levels of infection with *E. acervulina,* which may not produce other clinical symptoms, can result in disturbed storage and transport of carotenoid pigments. With *E. tenella* and *E. necatrix,* intestinal or cecal hemorrhage are the major symptoms of acute infections, and blood haematocrit measurement are valid guides to the severity of disease (Joyner and Davies, 1960).

In sheep, cattle, and rabbits, a large number of coccidial parasites have been identified. In each host, acute disease is associated with particular species, and the pattern of pathogenicity is an important guide for the diagnostician. Thus, in cattle, acute hemorrhagic diarrhea is associated with two species, *E. zuernii* and *E. bovis.* Both species cause lesions in the colon and rectum, but the size and shape of the oocysts enable them to be differentiated. In sheep, the most severe disease is caused by *E. ovinoidalis (E. ninakohlyakimovae),* characterized by severe hemorrhagic diarrhea and high mortality. In heavy infections, acute symptoms may coincide with the development of the large schizonts in the ileum, and animals may die without producing many oocysts. In sheep, disease is also associated with *Eimeria crandallis* and possibly with *Eimeria faurei* and *Eimeria ahsata.* These can be identified by the diagnostician from the characters of the oocysts, but generally the disease is not so acute as that caused by *E. ovinoidalis*.

In the rabbit, *E. stiedai* is generally confined to the liver. It is the only species found in this organ under normal conditions and gives rise to hypertrophy with focal necrosis around the bile ducts leading to a chronic wasting condition of varying severity. The pathogenesis of the disease was described by Martine and Yvoré (1974). Many wild rabbits carry this infection often with only mild liver lesions and few clinical signs. In domestic rabbits, susceptibility seems to vary with breed, and our experience indicates that New Zealand white strains are more susceptible than Dutch.

Acute intestinal coccidiosis is caused by some species of *Eimeria* in rabbits, but from published reports it is difficult to associate disease with particular species because of uncertainty in the identification of the parasites. Our experience with defined infections indicates that *E. flavescens* and *E. intestinalis* are particularly pathogenic for rabbits. Licois et al. (1978a, b) used these infections in their study of diarrhea in this host. Major changes due to *E. intestinalis* occur in the posterior small intestine and lesions due to *E. flavescens* coincide with the development of gametocytes in the cecum.

Pathogenicity is a specific property of species of *Eimeria,* which is often a valuable guide for the diagnostician to the etiology of a particular enteropathy. Mechanisms of pathogenesis are beyond the scope of this chapter, but it should be noted that the pattern of disease may be altered by the state of the host. In addition to breed susceptibilities, the dietary state of the host may influence the effects of parasitism. Thus, in chickens, a deficiency of *p*-aminobenzoic acid in the diet of the host will reduce the severity of infection due to *E. tenella* (Joyner, 1963), or a deficiency of vitamin K may exacerbate it (Ryley and Hardman, 1978).

While it is possible to ascribe typical lesions and effects to particular parasites, which may be of value in identification, under field conditions, animals generally are infected with more than one species of *Eimeria.* Interactions between species can result in an exacerbation of total effects or modifications in activity of individual species. Hein (1976) investigated the pathogenicity of *E. acervulina, E. brunetti,* and *E. maxima* in various combinations. Generally, weight loss increased with multiple infections, but competition tended to reduce oocyst production attributable to a given species in the presence of another. Such competition was not apparent with *E. acervulina* and *E. brunetti,* which have different predilection sites. In rats, Duszynski (1972) found enhanced effects due to *E. separata,* when it was superimposed on an existing *E. nieschulzi* infection.

Working with *Eimeria weybridgensis, E. ahsata, E. ovina,* and *E. ovinoidalis (E. ninakohlyakimovae)* in sheep, Catchpole, et al. (1976) found that in mixed infections patency and oocyst production of the first three species only were increased. *E. ovinoidalis* was not affected in this way. Only lambs receiving this infection demonstrated pathogenic effects.

2

Patterns of pathogenicity are valid guides to the identification of species of *Eimeria.* They are of particular value to the pathologist, but it must be remembered that as pathogens, they rarely occur in isolation and that the host is being used as an indicator. Variations in the hosts may be reflected in differences in the apparent activity of the parasite.

Infections in Abnormal Hosts

The criterion of a successful host is the completion of the parasitic life cycle with the production of normal oocysts. In fact, oocysts will often excyst in the intestine of a host in which subsequent development may be limited. Haberkorn (1970) observed this with *E. tenella* from the chicken and *E. falciformis* from the mouse. Both these parasites excysted in the respective foreign host—*E. tenella* in the mouse, and *E. falciformis* in the chick, and the sporozoites invaded the intestinal epithelium and developed into trophozoites. *E. falciformis* did not develop further in the foreign host but *E. tenella* in the mouse proceeded into nuclear division. In the rat, which is more closely related to the normal host, *E. falciformis* developed further into second generation schizonts, but gametogony did not occur.

Treatment of rats *(Rattus norvegicus)* with the corticosteroid dexamethasone rendered them susceptible to infection with *E. vermiformis* from the mouse. Todd and Lepp (1972) demonstrated the presence of sexual stages of the parasite in the lower small intestine of treated rats, which is one of the sites they occupy in mice. However, they were not seen in the cecum or large intestine where they also occur normally. Oocyst production started in the rats on the 11th day and continued until the 28th day after inoculation. The prepatent period for this species in mice is 7 days. Rose and Millard (personal communication) were unable to establish infections due to this species in athymic (nude) rats, nor could patent infections due to *E. nieschulzi* be established in athymic mice.

Péllerdy and Dürr (1969) reported numerous attempts to break down the host specificity of *E. stiedai* from the rabbit and *E. tenella* from the chicken. Treatment with x-rays, hydrocortisone, ethionine, or deficient diets failed to make rats or guinea pigs susceptible to *E. tenella.* McLoughlin (1969), however, administered daily injections of dexamethasone to chickens and succeeded in infecting them with *E. meleagrimitis* from the turkey. Similar treatment of turkey poults failed to make them susceptible to infection with *E. tenella.*

In the chick, it also seems that tissue specificity may be broken down following immunodepressant therapy. Long (1970) showed that *E. tenella* could develop to the second generation schizont stage in the liver of chickens treated with dexamethasone.

Fitzgerald's (1970) observation that *E. stiedai* from the rabbit will develop in the chorioallantoic membrane of the chick embryo suggests that specificity may not be so precise in embryonic tissues as in the adult. Long and Millard (1978) reported that *Eimeria grenieri* from the guinea fowl formed schizonts and a few oocysts in chick embryos and that *E. tenella* developed in guinea fowl embryos. However, not all species of *Eimeria* from the fowl will develop in the chick embryo in which, to date, only *E. tenella* and *E. mivati* can be satisfactorily propagated. It is interesting to note that embryos from different strains of White Leghorn chickens did not show the same patterns of relative susceptibilities to infection with *E. tenella* as the adult birds. Long (1970) suggested that host resistance factors develop during the first few weeks after hatching.

Long and Millard (1979) studied in some detail the fate of sporozoites in foreign hosts. Oocysts of *E. maxima* from the chicken given via the crop to guinea fowl *(Numida meleagris)* excysted and were recoverable from the intestinal mucosa 6–12 hr later. Some of the sporozoites reached the liver and persisted there for 48 hr. By contrast, *E. maxima* sporozoites survived in the intestinal mucosa of chickens for appreciably longer periods, up to 96 hr, and even in immune birds viable parasites were recoverable from the intestine 48–72 hr after inoculation (Rose and Hesketh, 1976). It seems from this that the mechanism by which parasites are rejected from the foreign host is different from that in the immune one.

It is possible that macrophages may play a role in the rejection process, for although *E. tenella* is not destroyed by macrophages of its normal host, the fowl (Long and Rose, 1976), Long and Millard (1979) showed that they were destroyed by macrophages from guinea fowl and that survival in turkey macrophages was much reduced. Sporozoites of *E. grenieri* from the guinea fowl did not survive in macrophages from chickens.

Apart from acquired immunity, anticoccidial therapy, or treatment with immunodepressants, the host's reaction to the parasite may be modified by changes in the gut flora. The response seems to depend upon the species of *Eimeria.* In the fowl, *E. tenella* seems to require the presence of intestinal microorganisms for the expression of its typical pathogenicity (Visco and Burns, 1972a), whereas with *E. brunetti* and *E. acervulina* no such relationship obviously exists (Hegde et al., 1969; Ruff et al., 1974). The susceptibility of a normal host may be further modified by nutrition (Joyner, 1963; Murillo et al., 1976) and by strain variation (Long, 1968; Visco and Burns, 1972b).

There are thus a few treatments that will modify the parasites' performance, but it is almost impossible to disturb their host specificity to any significant degree.

SPECIFICITY IN THE *ISOSPORA*

Although it is considered that host specificity generally is highly developed in the coccidia, it has often been thought that this characteristic is not so well expressed in *Isospora* species. In recent years, this has been judged to accord with the relationships of this genus with the tissue-bearing coccidia. However, a true assessment has been difficult because species have not always

2

been correctly differentiated and systematic relationships have been the subject of extensive revision.

A typical example is provided by *Isospora rivolta* (Grassi, 1879) Wenyon, 1923 which was first named from the cat. Wenyon (1923) described it from the dog and considered that the two parasites were identical, although cross-infection experiments between dogs and cats had not been performed. Later, Mahrt (1967) studied the life cycle of *I. rivolta* in the dog and failed to transmit it to cats. Dubey (1975a) separated the canine and feline parasites on the basis of host specificity. It was proposed that the original name *I. rivolta* be retained for the species in the cat, and the parasite from the dog was named *Isospora ohioensis*.

The parasites described from the dog by Dubey were assumed to be the same as those recorded by Mahrt (1967), but although the oocysts were similar, there were differences between the endogenous stages of the coccidia studied by the two workers. Dubey and Mahrt (1978) concluded that the two species in the dog are distinct and proposed the new name *Isospora neorivolta* for the coccidium studied by Mahrt (1967). The main difference between the two species is in their sites of development. *I. neorivolta* develops predominantly in the lamina propria of the posterior half of the small intestine, whereas *I. ohioensis* develops only in the epithelium and the infection occurs throughout the small intestine. A third species, *Isospora burrowsi* Trayser and Todd, 1978 is differentiated from the other two by the larger size of the late merozoites and their blunt anterior end. The size of the oocysts (20 × 17 mμ) overlaps that of *I. ohioensis* (they are no longer available for comparison). *I. canis, Hammondia heydorni,* and *Sarcocystis* species can be differentiated from other canine coccidia by the size or structure of the oocysts (Dubey, 1977b).

Originally, *Isospora* species were thought to follow a monoxenous type of life cycle, confined to intestinal tissues of the host animals. Frenkel and Dubey (1972), however, showed that mice, rats, and hamsters could act as intermediate hosts following ingestion of oocysts from infected cats. In these hosts, multiplication is limited and although other tissues may be involved, single parasites are found mainly in the lymph nodes. For these stages which electron microscopy has shown to be enclosed in a parasitophorous vacuole (Mehlhorn and Markus, 1976), Markus (1978a) has introduced the term "hypnozoite." In this state, they can persist for considerable periods—in the case of *I. rivolta,* up to 23 months. A detailed description of the life cycle of *I. rivolta* in cats and mice has been published by Dubey (1979a).

The *Isospora* of cats and dogs, *Isospora felis, I. rivolta, I. canis,* and *I. ohioensis* may follow a monoxenous oral-fecal life cycle. The intermediate hosts are generally rodents but additionally cats have been shown to act as intermediate hosts of *I. canis* of dogs, and dogs and chickens as intermediate hosts of *I. rivolta* of cats (Dubey, 1975b).

Life cycle and host specificity studies are important in the systematics of this group of coccidia. It might be thought that host-specificity in the *Isospora* would be most precise for the cycle in the definitive host; this is not necessarily so. *Isospora arctopithece* is a parasite of New

World primates which has been redescribed by Hendricks (1974, 1977). Oocysts recovered from marmosets *(Saguinus geoffroyi)* are transmissible to a remarkably wide range of hosts, which includes different genera of several families included in both primates, marsupials and carnivores. In the experiment quoted by Hendricks (1977), oocysts were recovered from species of six genera of primates native to Panama included in the families Cebidae and Callithricidae (marmoset *S. geoffroyi)*. The carnivores included domestic dogs and cats and Panamanian wild mammals of the Procyonidae and Mustelidae families. The didelphid marsupial, the common opossum *(Didelphys marsupialis)*, was also infected. Through this wide range of animal types, prepatent periods were remarkably consistent and in most cases the recovered oocysts produced patent oocyst infections in the marmoset, the original host species.

This species of *Isospora* produces extraintestinal infections in both rodents and domestic chickens (Hendricks and Walton, 1974), which thus serve as intermediate hosts.

Dubey (1977b) proposed a new generic name—*Levineia*—for those isosporan parasites such as *I. felis, I. canis, I. rivolta*, etc., of cats and dogs in which development is not limited to the intestinal tract and may pass through an intermediate host. This would distinguish them from organisms like *I. canaria*, which follows a direct life cycle limited to the intestine (Box, 1977). However, it leaves uncertain the position of *I. serini*, also from the canary, which undergoes division in mononuclear phagocytes but for which no intermediate host is known. The identification of parasites in this group at present depends upon the demonstration of the life cycle; this has been done for relatively few species. This essential step must be completed before any significance can be accorded to morphological features. Where life cycles are established, different parasites in the same definitive host can sometimes be distinguished morphologically. Dubey (1976) has published pictures and diagrams of the different oocysts occurring in the feces of cats and dogs.

Oocysts of the genera *Toxoplasma, Sarcocystis, Frenkelia*, and *Besnoitia* all conform to the disporocystic *Isospora* type. Černá et al. (1978b) have pointed out that they differ from the true *Isospora* in the absence of a Stieda body in the sporocyst. Members of the genus *Sarcocystis* (and *Frenkelia*) produce sporulated oocysts in the host.

SPECIFICITY IN THE SARCOCYSTINAE—LIFE CYCLES AND HOST SPECIFICITY AS SYSTEMATIC CHARACTERS

Toxoplasma

In this volume, Levine (Chapter 1) defines the family Sarcocystidae and divides it into two subfamilies, the Sarcocystinae and the Toxoplasmatinae. They are obligatorily or facultatively heteroxenous. Problems of species identification arise because in only a small proportion of them is the complete life cycle known.

2

Levine (1977a) named seven recognized species of *Toxoplasma* from mammals, birds, reptiles, and amphibia. For only two of these are the complete life cycles and the definitive hosts known—*T. gondii* and *T. hammondi* (syn. *Hammondia hammondi*) the gametogony phases of which occur only in felids. To this date, the latter species has been described only in cats and mice.

T. gondii, however, occurs widely in at least 200 species of mammals and birds. Usually the definitive and intermediate hosts show few clinical signs of infection. Intermediate hosts may acquire infection by ingestion of tissue cysts in infected meat or by contamination of food with oocysts from the feces of cats. Strains may vary in their pathogenicity and those avirulent for one species may be virulent for another. In man, most infections are benign, but in the developing fetus, the effects of congenitally acquired infections can be serious and reflect pathological changes in neural tissue. In adults, symptoms commonly arise from transient lymphadenopathy. In animals, severe symptoms are caused in pigs, sheep (when abortion occurs), and young dogs and cats (Dubey, 1977a).

The presence of parasites is detectable by serological tests and the available methods are sufficiently specific to distinguish between *Toxoplasma* and other tissue cyst-forming coccidia. By comparison of results of different techniques, currently active cases may be distinguished from carriers. The parasites may also be identified histologically in preparations of lesions and by inoculation into laboratory animals, of which mice seem to be the most susceptible. In this host, *T. gondii* can be maintained by serial passage of the infected tissue or fluids, a character which distinguishes it from *T. hammondi,* which is obligatorily heteroxenous.

In addition, *T. gondii* can survive in earthworms, cockroaches, and muscid flies, which are exposed to the oocysts. These invertebrates are thus able to serve as paratenic hosts (Dubey et al., 1970; Wallace, 1971, 1972).

Following the ingestion by the cat of parasites from the intermediate hosts, they may disseminate to both intestinal and extraintestinal tissues, where further divisions occur and where they may persist for long periods. Gametogony in the intestinal epithelium results in the production of unsporulated oocysts 3–10 days after ingesting tissue cysts or 20 days or longer after consuming oocysts. The oocysts are spherical to ellipsoidal measuring $11–14 \times 9–11$ μm.

An extensive account of the parasite and its effects is given by Frenkel (1973), from which it is seen that *T. gondii* has been studied in great detail. Of all the members of the Sarcocystinae, *T. gondii* is probably the most widely distributed but precisely defined.

Sarcocystis

For many years, protozoa of the genus *Sarcocystis* were known only by the organisms that typically occur in cysts in the muscles of a wide range of vertebrates. In the list published by Kalyakin and Zasukhin (1975), nine named species of *Sarcocystis* are recorded in 11 species of reptiles, 15 species in 72 species of birds, and 45 species in 105 to 109 species of mammals. A majority of species are unnamed and identified only from the species of the host in which they were found. Synonymy seemed to be very confused, and the data really only served to indicate the wide occurrence of muscle cysts in many species of vertebrates. Cysts were not recorded in fish or amphibia. A similar survey of Australian wildlife also showed a widespread prevalence of cysts in 73 species of mammals, including marsupials, cetaceans, and rodents. Among the birds examined in Australia, muscle sarcocysts were most prevalent in carnivorous, omnivorous, or insectivorous species; there was a low prevalence in water fowl. Of 14 species of reptiles examined, sarcocysts were found in the muscles of Varanidae (goannas) and Scincidae (skinks) (Munday et al., 1978, 1979).

With the demonstration of the two-host life cycle, the systematics of the group had to be revised. In only a relatively few cases has the complete life cycle been demonstrated and the species fully defined. Markus (1978b) and Dubey (1977a) have given competent reviews, and Levine (1977b) has critically examined the systematic position. Mehlhorn and Heydorn (1978) have reviewed data on fine structure.

The literature is complicated by the fact that different names have often been given to the fecal and tissue stages of the same parasite. Furthermore, it has been assumed that parasites in the same host are the same species. This is now known not to be so. An example of the importance of information on the life cycle for the correct identification of species of *Sarcocystis* is given by the species found in the muscles of the pig from which three species have been reported, each with a different definitive host.

The dog is the final host for *Sarcocystis miescheriana.* Its life cycle was described in West Germany by Rommel and Heydorn (1972). It is not transmissible to cats (Rommel et al., 1974). It was shown in the USSR that the cat is the final host (Golubkovan and Kisliakova, 1974) for *Sarcocystis porcifelis* Dubey, 1976 and that man is the definitive host for *Sarcocystis* suihominis. S. porcifelis is reported to be pathogenic to pigs. In addition to transmission studies, these species in the intermediate may only be identified from the cyst walls which may be distinguished by the type of surface projections seen under the electron microscope.

Mehlhorn et al. (1976) compared the ultrastructure of the cyst wall of 13 *Sarcocystis* species. They concluded that the formation of the primary wall, that is, the thickened border of the transformed parasitophorous vacuole, was a typical feature of every *Sarcocystis* species examined and its morphology was identical in all mature cysts of the same species wherever they were located. For example, the mature cysts of *Sarcocystis hominis (S. bovihominis)* in cattle always had large palisade-like protrusions with fibrilar structures in their interior, whereas no such protrusions were seen in mature cysts from the mouse. Mehlhorn et al. (1976) concluded that cyst-wall morphology alone is not a valid criterion for species differentiation because they found several intermediate host species with morphologically similar sarcocysts, for example,

2

whale, mouse, and monkey. Nevertheless, the character could be used to differentiate between the different cysts in sheep and cattle after infection from definitive hosts. Dubey (1979b) attempted to identify the porcine species of *Sarcocystis* found in pigs in Ohio, by feeding infected tissues to cats and dogs. Neither of these hosts could be infected. Thus, it was presumed that the species was *S. suihominis,* although its infectivity for man was not tested.

Some of the difficulties of identifying *Sarcocystis* infections in the absence of knowledge of the complete life cycles is discussed by Beaver et al. (1979), who reviewed recorded cases of *Sarcocystis* in the skeletal muscles or heart of man. No regular predators of man occur. Thus, it follows that *Sarcocystis* infection in man is zoonotic and that the cysts should be referable to species found in the muscles of wild or domestic animals. These authors reviewed the descriptions of 40 cases. From the structure of the cysts, the size, shape, and orientation of zoites, four types of sarcocysts were distinguished. Three of these closely resembled species found commonly in monkeys.

Similar morphological differences led Mehlhorn et al. (1976) to conclude that there is more than one species of *Sarcocystis* in rhesus monkeys. Kan (1979), too, felt that the differences which he observed were distinct and consistent. He examined four species of rats and a bandicoot from Malaysia and found three types of cyst in a single rat. Most host species contained more than one type of sarcocyst. These observations suggested that individual species of rodents can act as intermediate hosts for more than one species of *Sarcocystis.* Bergmann and Kinder (1975) observed four ultrastructurally different types of cyst wall in cattle, whereas Levine (1977b) gives the names of only three known species in this host.

Further studies to develop means of identifying species of *Sarcocystis* in the intermediate host on the above lines, or perhaps by isoenzyme analysis, as suggested by the work of Kepka and Rezaeian (1976), would be worth pursuing. Indeed, without an additional means of identification, the resolution of much systematic confusion will be difficult.

Markus (1978b) has indicated that certain species of *Sarcocystis* display a characteristic distribution of cysts in different tissues. The genus *Frankelia* is distinguished by the location of the cysts in brain tissue of the intermediate host. *Frenkelia microti* develops in the brain tissue of the short-tailed vole *(Microtus agrestis).* The final host is the European buzzard (*Buteo buteo).*

Krampitz and Rommel (1977) found that the parasite from voles could not be transmitted by feeding to the long-eared owl (*Asio otus*), a barn owl (*Tyto alba*), a tawny owl (*Strix aluco*), or a kestrel (*Falco tinnunculus*). Two out of three buzzards (*Buteo buteo*) produced sporocysts in the feces following ingestion of *Frenkelia,* that is, infected brains of *M. agrestis.* Direct transmission between buzzards was not possible.

The gametogony phase of the life cycle of this species therefore seems to be strongly host-specific. On the other hand, a range of rodent genera seems to be susceptible to infection with oocysts from the buzzard (Krampitz and Rommel, 1977). These include the field mouse, golden hamster, laboratory rat, chinchilla, and domestic rabbit. Merogony, therefore, does not seem to be so strongly host-specific. Among the experimental hosts, Rommel and Krampitz (1978) found the multimammate rat to be the most susceptible.

In the vole, the final stages in asexual development are the lobulate cysts in the brain. In this host, however, the first asexual multiplication occurs in the liver, where schizonts are present 5–8 days after infection (Geisel et al., 1979).

The buzzard (*B. buteo*) is also the definitive host for *Frenkelia clethrionomyobuteonis* from the bank vole (*Clethrionomys glareolus).* Sporocysts of this species were not infective for *Microtus arvalis* (European vole), M. agrestis (short-tailed vole), *Apodemus sylvaticus* (long-tailed field mouse), all of which are intermediate hosts for *F. microtus* (Krampitz et al., 1976).

Černá et al. (1978a) considered that the location of cysts in the brains of rodents infected with *Frenkelia* species constitutes no more than a subgeneric character. The location of extra-muscular schizonts in the liver and their common antigenic structure suggest that *Frenkelia* and *Sarcocystis* belong to the same genus, *Sarcocystis,* a view which accords with that of Levine (Chapter 1, this volume). These workers studied two species of *Sarcocystis* from rodents, with birds of prey as definitive hosts and with evident strong host specificity.

Sarcocystis cernae will infect the kestrel (*F. tinnunculus*) but not the tawny owl (*S. aluco*). The common vole (*M. arvalis*) is the intermediate host; development does not take place in the mouse (*M. musculus*) (Černá et al., 1978a).

The barn owl (*T. alba*) is the definitive host for *Sarcocystis dispersa,* which will also develop to a lesser degree in the long-eared owl (*A. otus*), but not in *S. aluco* or *B. buteo.* In this case, the intermediate host is the mouse *M. musculus;* the vole *M. arvalis* cannot be infected (Černá et al., 1978b).

A further species of *Sarcocystis* from the common vole *M. arvalis,* with the weasel *Mustela nivalis* as the definitive host, has also been described. This species will also infect the stoat *Mustela erminae* and the ferret *Mustela putorius furo.* Bank voles, hamsters, mice, and rats were refractory as intermediate hosts to this infection (Tadros and Laarman, 1976).

It seems that the gametogony phase of species of *Sarcocystis* is restricted to closely related definitive hosts. Further examples are provided by two species from man. *S. suihominis,* for which the pig is the intermediate host, has the chimpanzee (*Pan troglodytes*) and rhesus and cynomolgus monkeys (*Macaca mulatta* and *Macaca irus*), as well as man as definitive hosts (Fayer et al., 1979). Cattle (*Bos taurus*) are intermediate hosts and man, baboons (*Papio cynocephalus*), and rhesus monkeys (*M. mulatta*) are definitive hosts (Heydorn et al., 1976) for *S. hominis.*

Markus (1978b) has summarized available data on these and other species of *Sarcocystis* in domestic animals and man. Definitive and intermediate host specificity seems to vary, and knowledge of the life cycles is incomplete; therefore, no definite conclusions can yet be drawn.

2

LITERATURE CITED

Beaver, P. C., Gadgil, R. K. and Morera, P. 1979. *Sarcocystis* in man: A review and report of five cases. Am. J. Trop. Med. Hyg. 28:819–844.

Bergmann, V. 1970. Electronenmicroskopische Untersuchungen zur Pathogenese der Blinddarmkokzidiose der Hühnerküken. Arch. Exp. Vet. Med. 24:1169–1184.

Bergmann, V., and Kinder, E. 1975. Ultrastruktur der Zystenwand von Sarkozysten in Muskelfasern spontan infizierter Rinder. Monatshefte für Veterinärmedizin 24:945–947.

Box, E. D. 1977. Life cycles of two *Isospora* species in the canary *Serinus canarius* Linnaeus. J. Protozool. 24:57–67.

Catchpole, J., Norton, C. C., and Joyner, L. P. 1975. The occurrence of *Eimeria weybridgensis* and other species of coccidia in lambs in England and Wales. Br. Vet. J. 131:392–401.

Catchpole, J., Norton, C. C., and Joyner, L. P. 1976. Experiments with defined multispecific coccidial infections in lambs. Parasitology 72:137–147.

Cheissin, E. M. 1947. Variability of oocysts of *Eimeria magna* Pérard. Zoologicheskii Zhurnal 26:17–30.

Cheissin, E. M. 1957. Variability of the oocysts of *Eimeria intestinalis* Cheissin, 1948 parasite of the domestic rabbit. Vestnik Leningradskogo gosudarstvennogo Universiteta Leningrad. Biological series 9(2):43–52.

Christensen, J. F. 1938. Species differentiation in the coccidia from the domestic sheep. J. Parasitol. 24:453–465.

Černá, Ž. 1975. On the problem of antigenic identity between the coccidian *Eimeria contorta* Haberkorn, 1971 and *E. falciformis* Eimer, 1870. J. Protozool. 22(suppl.):176.

Černá, Ž., Kolarova, I., and Sulc, P. 1978a. *Sarcocystis cernae* Levine, 1977 excystation, life cycle and comparison with other heteroxenous coccidians from rodents and birds. Folia Parasitol. 25:201–207.

Černá Ž., Kolarova, I., and Sulc, P. 1978b. Contribution to the problem of cyst-producing coccidians. Folia Parasitol. 25:9–16.

Coudert, P., Licois, D., and Streun, A. 1979. Characterization of *Eimeria* species: I. Isolation and study of pathogenicity of a pure strain of *Eimeria perforans* (Leuckart, 1879; Sluiter and Swellegrebel, 1912). Z. Parasitenkd. 59:227–234.

Coudert, P., Yvoré, P., and Provot, P. 1973. Sporogonie d'*Eimeria stiedai* (Lindemann, 1865) Kesskalt et Hartmann, 1907. Ann. Rech. Vét. 4:371–388.

Davies, S. F. M., and Joyner, L. P. 1962. Infection of the fowl by the parenteral inoculation of oocysts of Eimeria. Nature (Lond.) 194:996–997.

Doran, D. J. 1953. Coccidiosis in the kangaroo rats of California. Univ. Calif. Publ. Zool. 59:31–60.

Doran, D. J. 1970. Eimeria tenella: From sporozoites to oocysts in cell culture. Proc. Helminth. Soc. Wash. 37:84–92.

Doran, D. J., 1978. The life cycle of *Eimeria dispersa* Tyzzer, 1929 from the turkey in gallinaceous birds. J. Parasitol. 64:882–885.

Doran, D. J., and Augustine, P.C. 1973. Comparative development of *Eimeria tenella* from sporozoites to oocysts in primary kidney cell cultures from gallinaceous birds. J. Protozool. 20: 658–661.

Dubey, J. P. 1975a. *Isospora ohioensis* sp.n. proposed for *I. rivolta* of the dog. J. Parasitol. 61:462–465.

Dubey, J. P. 1975b. Experimental *Isospora canis* and *Isospora felis* in mice, cats and dogs. J. Protozool. 22:416–417.

Dubey, J. P. 1976. A review of *Sarcocystis* of domestic animals and of other coccidia of cats and dogs. J. Am. Vet. Med. Assoc. 169:1061–1078.

Dubey, J. P. 1977a. *Toxoplasma, Hammondia, Besnoitia, Sarcocystis* and other tissue cyst-forming coccidia of man and animals. In: Parasitic Protozoa, Vol. III. pp. 101–237. J. P. Kreier (ed.), Academic Press, New York.

Dubey, J. P. 1977b. Taxonomy of *Sarcocystis* and other coccidia of cats and dogs. J. Am. Vet. Med. Assoc. 170:778, 782.

Dubey, J. P. 1979a. Life cycle of *Isospora rivolta* (Grassi, 1879) in cats and mice. J. Protozool. 26:433–443.

Dubey, J. P. 1979b. Frequency of *Sarcocystis* in pigs in Ohio and attempted transmission to cats and dogs. Am. J. Vet. Res. 40:867–868.

Dubey, J. P., and Mahrt, J. L. 1978. *Isospora neorivolta* sp.n. from the domestic dog. J. Parasitol. 64:1067–1073.

Dubey, J. P., Miller, N. L., and Frenkel, J. K. 1970. Characterization of the new fecal form of *Toxoplasma gondii.* J. Parasitol. 56:447–456.

Duszynski, D. W. 1972. Host and parasite interactions during single and concurrent infections with *Eimeria nieschulzi* and *E. separata* in the rat. J. Protozool. 19:82–88.

Edgar, S. A. 1955. Sporulation of oocysts at specific temperatures and notes on the prepatent period of several species of avian coccidia. J. Parasitol. 41:214–216.

Edgar, S. A., and Seibold, C. T. 1964. A new coccidium of chickens. *Eimeria mivati* sp.n. (Protozoa: Eimeriidae) with details of its life history. J. Parasitol. 50:193–204.

El-Kasaby, A., and Sykes, A. H. 1973. The role of chicken macrophages in the parenteral excystation of *Eimeria acervulina.* Parasitology 66:231–239.

Farr, M. M., and Doran, D. J. 1962. Comparative excystation of four species of poultry coccidia. J. Protozool. 9:403–407.

Fayer, R. 1980. Epidemiology of protozoan infections: The coccidia. Vet. Parasitol. 6:75–103.

Fayer, R., Heydorn, A. O., Johnson, A. J., and Leek, R. G. 1979. Transmission of *Sarcocystis suihominis* from humans to swine to nonhuman primates *(Pan troglodytes, Macaca mulatta, Macaca irus)* Z. Parasitenkd. 59:15–20.

Fernando, M. A., Pasternak, J., Barrell, R., and Stockdale, P. H. G. 1974. Induction of host nuclear DNA synthesis in coccidia-infected chicken intestinal cells. Int. J. Parasitol. 4:267–276.

Fitzgerald, P. R. 1965. The results of parenteral injections of sporulated or unsporulated oocysts of *Eimeria bovis* in calves. J. Protozool. 12:215–221.

Fitzgerald, P. R. 1970. Development of *Eimeria stiedae* in avian embryo. J. Parasitol. 56:1252–1253.

Frenkel, J. K. 1973. Toxoplasmosis: Parasite life cycle, pathology, and immunology. In: D. M. Hammond and P. L. Long (eds.), The Coccidia: *Eimeria, Isospora, Toxoplasma,* and Related Genera, pp. 343–410. University Park Press, Baltimore.

Frenkel, J. K., and Dubey, J. P. 1972. Rodents as vectors for feline coccidia *Isospora felis* and *Isospora rivolta.* J. Infect. Dis. 125:69–72.

Geisel, O., Kaiser, E., Vogel, O., Krampitz, H. E., and Rommel, M. 1979. Pathomorphologic findings in short-tailed voles *(Microtus agrestis)* experimentally infected with *Frenkelia microti.* J. Wildl. Dis. 15:267–270.

Golubkovan, D. I., and Kisliakova, Z. I. 1974. The sources of infection for swine *Sarcocystis.* (In Russian). Veterinariia 11:85–86.

Haberkorn, A. 1970. Zur Empfanglichkeit nicht spezifischer Wirte für Schizogonie-Stadien verschiedener *Eimeria*-Arten. Z. Parasitenkd. 35:156–161.

Haberkorn, A. 1971. Zur Wirtsspezifität von *Eimeria contorta* n.sp. (Sporozoa: Eimeriidae). Z. Parasitenkd. 37:303–314.

Hegde, K. S., Reid, W. M., Johnson, J., and Womack, H. E. 1969. Pathogenicity of *Eimeria brunetti* in bacteria-free and conventional chickens. J. Parasitol. 55:402–405.

Hein, H. 1976. *Eimeria acervulina, E. brunetti* and *E. maxima:* Pathogenic effects of single or mixed infections with low doses of oocysts in chickens. Exp. Parasitol. 39:415–421.

Hendricks, L. D. 1974. A redescription of *Isospora arctopitheci* Rodhain, 1933 (Protozoa:

Eimeriidae) from primates of Panama. Proc. Helminth. Soc. Wash. 41:229–233.

Hendricks, L. D. 1977. Host range characteristics of the primate coccidian *Isospora arctopitheci* Rodhain, 1933 (Protozoa: Eimeriidae). J. Parasitol. 63:32–35.

Hendricks, L. D., and Walton, B. C. 1974. Vertebrate intermediate hosts in the life cycle of an isosporan from a non-human primate. Proceedings of the Third International Congress on Parasitology. Munich, 1974. Section A6. pp. 96–97. Facta Publication, Vienna.

Heydorn, A. O., Gestrich, R., and Janitschke, K. 1976. Beiträge zum Lebenszyklus der Sarkosporidien: VIII. Sporozysten von *Sarcocystis bovihominis* in den Fäzes von Rhesusaffen (Macaca rhesus) und Pavianen (*Papio cynocephalus*. Berl. Muench. Tieraerztl. Wochenschr. 89:116–120.

Horton-Smith, C., and Long, P. L. 1965. The development of *Eimeria necatrix* Johnson, 1930 and *Eimeria brunetti* Levine, 1942 in the caeca of the domestic fowl *(Gallus domesticus)*. Parasitology 55:401–405.

Huff, D., and Clark, D. T. 1970. Cellular aspects of the resistance of chickens to *Eimeria tenella* infections. J. Protozool. 7:35–39.

Joyner, L. P. 1963. Some metabolic relationships between host and parasite with particular reference to the *Eimeriae* of domestic poultry. Proc. Nutr. Soc. 22:26–32.

Joyner, L. P., and Davies, S. F. M. 1960. Detection and assessment of sublethal infections of *Eimeria tenella* and *Eimeria necatrix*. Exp. Parasitol. 9:243–249.

Joyner, L. P., and Long, P. L. 1974. The specific characters of the *Eimeria*, with special reference to the coccidia of the fowl. Avian Pathol. 3:147–157.

Joyner, L. P., and Norton, C. C. 1972. The development of *Eimeria acervulina* in the caeca of young fowls. Parasitology 64:479–483.

Joyner, L. P., Norton, C. C., Davies, S. F. M., and Watkins, C. V. 1966. The species of coccidia occurring in cattle and sheep in the Southwest of England. Parasitology 56: 531–541.

Joyner, L. P., Patterson, D. S. P., Berrett, S., Boarer, C. D. H., Cheong, F. H., and Norton, C. C. 1975. Amino-acid malabsorption and intestinal leakage of plasma-proteins in young chicks infected with *Eimeria acervulina*. Avian Pathol. 4:17–33.

Kalyakin, V. N., and Zasukhin, D. N. 1975. Distribution of Sarcocystis (Protozoa: Sporozoa) in vertebrates. Folia Parasitol. 22:289–307.

Kan, S. P. 1979. Ultrastructure of the cyst wall of *Sarcocystis* spp. from some rodents in Malaysia. Int. J. Parasitol. 9:475–481.

Kepka, O., and Rezaeian, M. 1976. Chemotaxonomische Untersuchungen an Sarcocystidae (Sporozoa: Apicomplexa) Z. Parasitenkd. 50:210.

Krampitz, H. E., and Rommel, M. 1977. Experimentelle Untersuchungen über das Wirtsspektrum der Frenkelien der Erdmaus. Berl. Muench. Tieraerztl. Wochenschr. 90:17–19.

Krampitz, H. E., Rommel, M., Geisel, O., and Kaiser, E. 1976. Beiträge zum Lebenszyklus der Frenkelien: II. Die Ungeschlechtliche Entwicklung von Frenkelia clethrionomyobuteonis in der Rotelmaus. Z. Parasitenkd. 51:7–14.

Leathem, W. D. 1968. Organ specificity of *Eimeria tenella* in cecectomized chickens. J. Protozool. 15(suppl):18.

Lee, D. L., and Long, P. L. 1972. An electron microscopal study of *Eimeria tenella* grown in the liver of the chick embryo. Int. J. Parasitol. 2:55–58.

Levine, N. D. 1977a. Taxonomy of *Toxoplasma*. J. Protozool. 24:36–41.

Levine, N. D. 1977b. Nomenclature of *Sarcocystis* in the ox and sheep of fecal coccidia of the dog and cat. J. Parasitol. 63:36–51.

Levine, N. D., and Ivens, V. 1970. The coccidian parasites (Protozoa: Sporozoa) of ruminants. Illinois Biological Monographs. Vol. 44. University of Illinois Press, Urbana.

Levine, N. D., and Ivens, V. 1972. Coccidia of the Leporidae. J. Protozool. 19:572–581.

Licois, D., Coudert, P., and Mongin, P. 1978a. Changes in hydromineral metabolism in diarrhoeic rabbits: 1. A study in the changes in water metabolism. Ann. Rech. Vét. 9:1–10.

Licois, D., Coudert, P., and Mongin, P. 1978b. Changes in hydromineral metabolism in diarrhoeic rabbits: 2. Study of the modifications of electrolyte metabolism. Ann. Rech. Vét. 9:453–464.

Long, P. L. 1968. The effect of breed of chickens on resistance to *Eimeria* infections. Br. Poult. Sci. 9:71–78.

Long, P. L. 1970. Development (schizogony) of *Eimeria tenella* in the livers of chickens treated with corticosteroid. Nature (Lond.) 225: 290–291.

Long, P. L. 1971. Schizogony and gametogony of *Eimeria tenella* in the livers of chick embryos. J. Protozool. 18:17–20.

Long, P. L. 1973a. Pathology and pathogenicity of coccidial infections. In: D. M. Hammond and P. L. Long (eds.), The Coccidia: *Eimeria, Isospora, Toxoplasma,* and Related Genera, pp. 253–294. University Park Press, Baltimore.

Long, P. L. 1973b. Studies on the relationship between *Eimeria acervulina* and *Eimeria mivati.* Parasitology 67:143–155.

Long, P. L., Joyner, L. P., Millard, B. J., and Norton, C. C. 1976. A guide to laboratory techniques used in the study and diagnosis of avian coccidiosis. Folia Vet. Lat. 6:201–217.

Long, P. L., and Millard, B. J. 1976. Studies on site finding and site specificity of *Eimeria praecox, Eimeria maxima,* and *Eimeria acervulina* in chickens. Parasitology 73:327–336.

Long, P. L., and Millard, B. J. 1977. Coccidiosis in turkeys: Evaluation of infection by the examination of turkey broiler litter for oocysts. Avian Pathol. 6:227–233.

Long, P. L., and Millard, B. J. 1978. Studies on *Eimeria grenieri* in the guinea fowl. *(Numida meleagris).* Parasitology 76:1–9.

Long, P. L., and Millard, B. J. 1979. Rejection of *Eimeria* by foreign hosts. Parasitology 78:239–247.

Long, P. L., and Rose, M. E. 1976. Growth of *Eimeria tenella in vitro* in macrophages from chicken peritoneal exudates. Z. Parasitenkd. 48:291–294.

McCully, R. M., Basson, P. A., de Vos, V., and de Vos, A. J. 1970. Uterine coccidiosis of the impala caused by *Eimeria neitzi* spec. nov. Onderstepoort J. Vet. Res. 37:45–58.

McDougald, L. R. 1979. Attempted cross-transmission of coccidia between sheep and goats and description of *Eimeria ovinoidalis* sp.n. J. Protozool. 26:109–113.

McLoughlin, D. K. 1969. The influence of dexamethasone on attempts to transmit *Eimeria meleagrimitis* to chickens and *E. tenella* to turkeys. J. Protozool. 16:145–148.

Mahrt, J. L. 1967. Endogenous stages of the life cycle of *Isospora rivolta* in the dog. J. Protozool. 14:754–759.

Markus, M. B. 1978a. Terminology for invasive stages of Protozoa of the Subphylum Apicomplexa (Sporozoa) S. Afr. J. Sci. 74:105–106.

Markus, M. B. 1978b. *Sarcocystis* and Sarcocystosis in domestic animals and man. In: C. A. Brandly and C. E. Cornelius (eds.), Advances in Veterinary Science and Comparative Medicine, pp. 159–193. Academic Press, New York.

Marotel, G. 1905. La Coccidiose de la chevre et son parasite. Bull. Soc. Sci. Vet. Med. Comp. Lyon 8:52–56.

Marquardt, W. C. 1973. Host and site specificity in the coccidia. In: D. M. Hammond and P. L. Long (eds.), The Coccidia: *Eimeria, Isospora, Toxoplasma,* and Related Genera, pp. 23–42. University Park Press, Baltimore.

Marquardt, W. C. 1976. Some problems of host and parasite interactions in the coccidia. J. Protozool. 23:287–290.

Martine, G., and Yvoré, P. 1974. Aspects pathogéniques de la coccidiose hépatique du lapin domestique. Cah. Med. Vet. 43: 147–158.

Mayr, E. 1969. The biological meaning of species. Biol. J. Linn. Soc. 1:311–320.

Mehlhorn, H, Hartley, W. J., and Heydorn, A. O. 1976. A comparative ultrastructural study of the cyst wall of 13 *Sarcocystis* species.

Protistologica 12:451–467.
Mehlhorn, H., and Heydorn, A. O. 1978. The Sarcosporidia (Protozoa: Sporozoa): Life cycle and fine structure. In: W. H. R. Lumsden, R. Muller, and J. R. Baker (eds.), Advances in Parasitology, Vol. 16. pp. 43–91. Academic Press, New York.
Mehlhorn, H., and Markus, M. B. 1976. Electron microscopy of stages of *Isospora felis* of the cat in the mesenteric lymph nodes of the mouse. Z. Parasitenkd. 51:15–24.
Munday, B. L., Hartley, W. J., Harrigan, K. E., Presidente, P. J. A., and Obendorf, D. L. 1979. *Sarcocystis* and related organisms in Australian wildlife: II. Survey findings in birds, reptiles, amphibians and fish. J. Wildl. Dis. 15:57–73.
Munday, B. L., Mason, R. W., Hartley, W. J., Presidente, P. J. A., and Obendorf, D. 1978. *Sarcocystis* and related organisms, in Australian wildlife: I. Survey finding in mammals. J. Wildl. Dis. 14:417–433.
Murillo, M. G., Jensen, L. S., Ruff, M. D., and Rahn, A. P. 1976. Effect of dietary methionine status on response of chicks to coccidial infection. Poult. Sci. 55:642–649.
Norton, C. C. 1967. *Eimeria colchici* sp.nov. (Protozoa: Eimeriidae), the cause of caecal coccidiosis in English covert pheasants. J. Protozool. 14:772–781.
Norton, C. C. 1979. Coccidiosis of pheasants and turkeys: Recent experiments with pheasant coccidia at Weybridge. International Symposium on Coccidia and Further Prospects of Their Control. Prague, November, 1979. Proceedings. pp. 174–177. Czechoslovak Academy of Sciences, Prague.
Norton, C. C., Catchpole, J., and Joyner, L. P. 1979. Redescriptions of *Eimeria irresidua* Kessel and Jankiewicz, 1931 and *E. flavescens* Marotel and Guilhon, 1941 from the domestic rabbit. Parasitology 79:231–248.
Norton, C. C., and Joyner, L. P. 1980. Studies with *Eimeria acervulina* and *E. mivati:* Pathogenicity and cross-immunity. Parasitology 81:315–323.
Norton, C. C., Joyner, L. P., and Catchpole, J. 1974. *Eimeria weybridgensis* sp.nov. and *Eimeria ovina* from the domestic sheep. Parasitology 69:87–95.
Pasternak, J., Fernando, M. A., Stockdale, P. H. G., and Weber, D. 1977. Nuclear size and DNA content in host cells during first generation schizogony of *Eimeria zuernii.* Parasitology 74:199–203.
Pellérdy, L. 1969. Parenteral infection experiments with *Eimeria stiedai.* Acta Vet. Acad. Sci. Hung. 19:171–182.
Pellérdy, L. 1974. Coccidia and coccidiosis. 2nd Ed. Parey, Berlin.
Pellérdy, L., and Dürr, U. 1969. Orale and parenterale Ubertragungsversuche von Kokzidion auf nicht spezifische Wirte. Acta Vet. Acad. Sci. Hung. 19:253–268.
Reid, W. M. 1973. A diagnostic chart for nine species of fowl coccidia. University College of Georgia, College of Agriculture Experimental Station Research Report 163.
Rommel, M., and Heydorn, A. O., 1972. Beitrage zum Lebenszyklus der Sarkosporidien: III. *Isospora hominis* (Railliet und Lucet, 1891) Wenyon, 1923 eine Dauerform der Sarkosporidien des Rindes und des Schwiens. Berl. Muench. Tieraerztl. Wochenschr. 85: 143–145.
Rommel, M., Heydorn, A. O., Fischle, B., and Gestrich, R. 1974. Beiträge zum Lebenszyklus der Sarkosporidien: V. Weitere Endwirte der Sarkosporidien von Rind, Schaff und Schwein und die Bedeutung des Zwischenwirtes für die Verbreitung dieser Parasitose. Berl. Muench. Tieraerztl. Wochenschr. 87:392–396.
Rommel, M., and Krampitz, H. E. 1978. Weitere Untersuchungen uber das Zeischenwirtsspektrum und den Entwicklungszyklus von *Frenkelia microti* aus der Erdmaus. Zentralbl. Vet. Med. B. 25:273–281.
Rose, M. E., and Hesketh, P. 1976. Immunity to coccidiosis: Stages of the life cycle of *Eimeria maxima* which induce and are affected by the response of the host. Parasitology 73:25–37.
Ruff, M. D., Johnson, J. K., Dykstra, D. D., and

Reid, W. M. 1974. Effects of *Eimeria acervulina* on pH in conventional and gnotobiotic chickens. Avian Dis. 18:96–104.

Ryley, J. F. 1973. Cytochemistry, physiology, and biochemistry. In: D. M. Hammond and P. L. Long (eds.), The Coccidia: *Eimeria, Isospora, Toxoplasma,* and Related Genera, pp. 145–181. University Park Press, Baltimore.

Ryley, J. F. 1975. Why and how are coccidia harmful to their host? In: A. E. R. Taylor and R. Muller (eds.), Pathogenic Processes in Parasitic Infections. Symposia of the British Society of Parasitology Vol. 13. Blackwell Scientific Publications, Ltd., Oxford.

Ryley, J. F., and Hardman, L. 1978. The use of vitamin K deficient diets in the screening and evaluation of anticoccidial drugs. Parasitology 76:11–20.

Sharma, N. N. 1964. Response of the fowl (*Gallus domesticus*) to parenteral administration of seven coccidial species. J. Parasitol. 50:509–517.

Speer, C. A., Marchiondo, A. A., Mueller, B., and Duszynski, D. W. 1979. Scanning and transmission electron microscopy of the oocyst wall of *Isospora lacazei.* Z. Parasitenkd. 59:219–225.

Stockdale, P. H. G., and Fernando, M. A. 1975. The development of the lesions caused by second generation schizonts of *Eimeria necatrix* Res. Vet. Sci. 19:204–208.

Stockdale, P. H. G., Tiffin, G. B., Kozub, G., and Chobotar, B. 1979. *Eimeria contorta* Haberkorn, 1971: A valid species of rodent coccidium? Can. J. Zool. 57:264–270.

Strout, R. G., and Ouellette, C. A. 1969. Gametogony of *Eimeria tenella* (coccidia) in cell cultures. Science 163:695–696.

Tadros, W., and Laarman, J. J. 1976. Sarcocystis and related coccidian parasites: A brief general review, together with a discussion on some biological aspects of their life cycles and a new proposal for their classification. Acta. Leiden. 44:1–107.

Todd, K. S., and Lepp, D. L. 1972. Completion of the life cycle of *Eimeria vermiformis* Ernst, Chobotar and Hammond, 1971, from the mouse *Mus musculus* in dexamethasome-treated rats *Rattus norvegicus.* J. Parasitol. 58:400–401.

Trayser, C. V., and Todd, K. S., Jr. 1978. Life cycle of *Isospora burrowsi* n. sp. (Protozoa: eimeriidae) from the dog *Canis familiaris.* Am. J. Vet. Res. 39:95–98.

Vetterling, J. M. 1976. *Eimeria tenella:* Host specificity in gallinaceous birds. J. Protozool. 23:155–158.

Visco, R. J., and Burns, W. C. 1972a. *Eimeria tenella* in bacteria-free and conventionalized chicks. J. Parasitol. 58:323–331.

Visco, R. J., and Burns, W. C. 1972b. *Eimeria tenella* in bacteria-free chicks of relatively susceptible strains. J. Parasitol. 58:586–588.

Wagenbach, G. E., and Burns, W. C. 1969. Structure and respiration of sporulating *Eimeria stiedae* and *Eimeria tenella* oocysts. J. Protozool. 16:257–263.

Wallace, G. D. 1971. Experimental transmission of *Toxoplasma gondii* by filth flies. Am. J. Trop. Med. Hyg. 20:411–413.

Wallace, G. D. 1972. Experimental transmission of *Toxoplasma gondii* by cockroaches. J. Infect. Dis. 126:545–547.

Wenyon, C. M. 1923. Coccidiosis of cats and dogs and the status of *Isospora* of man. Ann. Trop. Med. Parasitol. 17:231–288.

Wilson, P. A. G., and Fairbairn, D. 1961. Biochemistry and sporulation in oocysts of *Eimeria acervulina.* J. Protozool. 8:410–416.

Yakimoff, W. L., and Rastegaieff, E. F. 1930. Zur Frage uber Coccidien der Ziegen Arch. Protistenk. 70:185–191.

Yvoré, P., and Coudert, P. 1972. Étude de la respiration endogène et de la segmentation de l'oocyste d'*Eimeria tenella* durant la sporogonie. Ann. Rech. Vét. 3:131–143.

Yvoré, P., Dubois, M., Sauveur, B., and Aycardi, J. 1972. Pathogénie de la coccidiose duodenale a *Eimeria acervulina.* Ann. Rech. Vét. 3:61–82.

Yvoré, P., and Mainguy, P. 1972. Influence de la coccidiose duodenale sur la teneur en carotenoides du serum chez le poulet. Ann. Rech. Vét. 3:381–387.

3

Genetics, Specific and Infraspecific Variation

Thomas K. Jeffers and Martin W. Shirley

A review of the biology of any group of organisms must consider the genetic contributions to the enormous and complex diversity which is invariably found within them. Biological studies on parasitic organisms must further consider that host-parasite relationships are conditioned by genetic influences of both the parasite and the host. Thus, the existing information on genetic factors in the host affecting susceptibility to coccidia is reviewed in this chapter.

Our interest in the genetics of the parasite was predicted long ago by Tyzzer (1929) when he stated that, "It is quite possible that the coccidia of the chicken may furnish material favorable for work in genetics, since we have in a single host a number of related species of protozoa, showing sex-differentiation." Nevertheless, despite Tyzzer's prophetic assessment, active research on the genetics of the coccidia is a relatively recent development. The results of these studies are reviewed and particular attention is given to the applicability of various methods of genetic investigation for identifying variation within and between different species of coccidia.

GENETIC ASPECTS OF THE HOST RESPONSE TO COCCIDIAL INFECTION

Response in Vivo

Eimeria The fact that animals differ in susceptibility to infectious diseases has been known since the early stages of domestication. There is no doubt that Gowen's (1937) belief that "No investigator who has adequately sought inherited host differences in disease response has failed to find them" has been borne out by studies conducted in the ensuing years. Genetic variation was found in resistance to many infectious diseases in animals, thus providing a basis for selective breeding intended to increase resistance to disease and to improve the economics of animal production (Hutt, 1958, 1964; Fredeen, 1965). The importance of the domestic fowl as a source of animal protein stimulated extensive studies on the genetic bases for more efficient poultry production. Genetic variation in resistance to a number of poultry diseases was demonstrated during the course of these studies (Hutt, 1949).

Johnson (1927) first suggested that resistance to coccidiosis in the domestic fowl might be increased through selective breeding. Although early reports of genetic differences in resistance to avian coccidiosis were also provided by Krallinger and Chodziesner (1931, reported by Rosenberg et al., 1948) and by Mayhew (1934), Herrick (1933–1934) first demonstrated that selection was effective in increasing resistance to coccidiosis in chickens. Several investigators reported differences in sensitivity to coccidiosis among different breeds of fowl (Rosenberg, 1941; Edgar et al., 1951; Long, 1968). However, a general ranking of breeds according to susceptibility is not possible because each of these studies included only a single strain of each breed. Variation in resistance to coccidiosis among strains within breeds may in fact exceed that between breeds (Buvanendran and Kulasegaram, 1972).

3

Strain variation in response to coccidial infection is often found despite an absence of intentional selection for differences in host susceptibility. This variation is attributed to random shifts in the frequency of genes controlling differential susceptibility, which may occur during the development and differentiation of different strains. Thus, Carson (1951), Jeffers et al. (1969), and Long (1970a, 1973c) independently described wide differences in susceptibility to *Eimeria tenella* among lines of fowl intentionally selected for differences in susceptibility to Marek's disease. There was, however, no apparent association between the selected trait and resistance to cecal coccidiosis. Jeffers et al. (1970) found that the percentage of survival after inoculation with 10^5 *E. tenella* oocysts ranged from 20.4% to 78.0% among 13 unselected inbred lines of fowl. These differences in resistance to cecal coccidiosis were not linked to genetic associations between susceptibility to *E. tenella* and other traits being selected and maintained in these lines, and were thus attributed to chance differences in gene frequency established during the course of inbreeding. Variation in susceptibility to *E. tenella* infection was also found among different egg production stocks obtained from commercial sources, despite the probable absence of intentional direct selection for this trait by the breeding organizations producing the parent stocks of the chicks tested (Challey et al., 1968).

Rosenberg (1948), Rosenberg et al. (1948), and Champion (1951a, b) developed through intentional selection the Wisconsin R (resistant) and S (susceptible) lines from poultry stocks maintained at the University of Wisconsin, and clearly demonstrated that the susceptibility of chickens to cecal coccidiosis could be rapidly altered by genetic selection. Direct selection of resistant breeders from birds surviving a standardized survival test was used to establish the resistant line, while indirect selection of susceptible breeders from sibs of the most susceptible families was used to establish the susceptible line. After three generations of selection, the respective survival rates of the Wisconsin R and S lines after infection with 10^5 *E. tenella* oocysts were 55.6% and 12.2%. The differential susceptibility of these lines to *E. tenella* infection was apparently established as a strain characteristic after only four generations of selection and remained relatively stable in the absence of continued selection (Jeffers, 1968). Three subsequent independent selection experiments as reported by Palafox et al. (1949), Edgar et al. (1951), Moultrie et al. (1953–1954), Rosenberg et al. (1954), and Klimes and Orel (1969, 1972) confirmed the effectiveness of genetic selection in altering resistance to *E. tenella*.

Studies on the susceptibility of progeny resulting from reciprocal F_1 and F_2 crosses of resistant and susceptible lines provided a basis for establishing the mode of inheritance of resistance to *E. tenella* (Champion, 1951a, b, 1954; Rosenberg et al., 1954). Differences in susceptibility were attributed to multiple genetic factors, lacking dominance and acting in an additive manner. Sex linkage, maternal effects, or cytoplasmic inheritance were not considered important factors in determining resistance or susceptibility to infection with *E. tenella*. However, the results of other studies in which the susceptibility of progeny from reciprocal crosses of unselected inbred lines was evaluated suggest that maternal effects and sex-limited factors

(W-chromosomal inheritance) may also contribute to differences in susceptibility to cecal coccidiosis (Jeffers et al., 1970; Jeffers, 1978).

The physiological basis for differences in genetic resistance to coccidiosis has not been elucidated. Challey (1966) found that the Wisconsin R and S lines differed in corticosterone levels. The adrenal corticosterone concentrations during the acute phase of *E. tenella* infection were highest in the S line of chickens, while adrenal effluent levels were much greater in the R line. He hypothesized that these differences resulted from differential rates of adrenal function and secretion, which may be involved in the host's resistance to infection. Differential responses to *E. tenella* infection as measured by different survival rates cannot be attributed to variation in the host's ability to limit the growth of the parasite, because genetic groups, although differing widely in their ability to survive, exhibit heavy infections with extensive cecal lesions (Jeffers et al., 1970).

Genetic crosses in which each parent is derived from a line of distinctly different origin may produce progeny exhibiting improved performance, that is, heterosis or hybrid vigor. Heterosis in the progeny of such crosses is reflected in differences between their performance and the average mid-parent performance. Although a universally accepted explanation of the genetic basis for this phenomenon has not been forthcoming as yet, the general principle of producing heterosis by combining gametes of diverse origin is widely used to produce improvements in animal and plant production. Millen et al. (1959) first demonstrated that heterosis may be a significant factor in determining the resistance of chickens to coccidiosis. Challey et al. (1968) likewise found that chicks derived from crosses of different breeds or inbred lines maintained by commercial poultry breeding firms were generally more resistant to *E. tenella* infection than were chicks derived from a random-bred control population. The responses of the commercial stocks in this study were probably influenced by heterotic effects, but this cannot be definitively ascertained because the response of the parent lines used in the breeding combinations was not measured. Buvanendran and Kulasegaram (1972) determined the resistance to *Eimeria necatrix* infection of progeny produced by two strains each of Light Sussex and Rhode Island Reds and all reciprocal crosses between them. A heterotic response was found whereby resistance of strain and breed-cross progeny was higher than mid-parent averages for each cross-combination.

In an extensive attempt to detect the possible effects of nonadditive gene interaction on host resistance to *E. tenella* infection, Jeffers et al. (1970) derived progeny from reciprocal crosses of several unselected inbred lines of fowl and tested them for response to challenge with 10^5 *E. tenella* oocysts.

Nine inbred lines produced a total of 2,314 pure-line and 2,890 reciprocal line-cross progeny for resistance testing. The survival rates of pure-line progeny ranged from 13.3–86.5% (mean = 49.5 ± 3.8%), while those of the line-cross progeny ranged from 28.8–83.0% (mean = 55.3 ± 1.8%). The response of line-cross progeny significantly exceeded the mid-parent average in five of the nine breeding combinations studied, confirming that heterosis may significantly influence resistance to infection with *E. tenella.* On the contrary, two of the inbred lines produced line-cross progeny exhibiting negative responses when compared to the mid-parent average, indicating that under certain circumstances genes for susceptibility having nonadditive effects on the resistance of hybrid progeny may also be important.

There is a paucity of research on genetic variation in mammalian host response to *Eimeria* species. This is surprising because genetically characterized lines of certain mammals (e.g., mice and rats) are readily available and easily maintained. Furthermore, these mammalian hosts have relatively short generation intervals, making them appropriate models for studies on genetic variation in response to coccidial infection. Liburd (1973) studied some of the aspects of host response to *Eimeria nieschulzi* infection in inbred and outbred strains of rats. Although he reported no difference in the pathogenic effects of this coccidian on the two different strains, inbred rats required a higher immunizing dose of oocysts than did outbred rats in order to develop resistance to reinfection. The severity of infection or development of resistance to reinfection was not influenced by the sex of the host.

Klesius and Hinds (1979) reported differences in susceptibility to infection with *Eimeria ferrisi* among seven inbred and two random-bred strains of mice. They found that a range of infective dose levels was important in evaluating susceptibility, because strain differences which may be easily discerned at intermediate dose levels may not be apparent at high or low infective dose levels. The F_1 progeny resulting from crosses of the highly susceptible C57BL/6 strain with each of three other moderately susceptible inbred strains were more resistant than either of the respective parent strains. Although Klesius and Hinds (1979) attributed this finding to a dominant expression of resistance, the possibility that heterosis plays an important role in determining the resistance of strain-cross progeny to infection with *E. ferrisi* should not be ruled out. No apparent association was found between genes determining susceptibility to *E. ferrisi* infection and gene loci in the major murine histocompatibility region (H-2), a chromosomal region also determining mixed leukocyte reactivity, skin graft rejection, and specific immune responses (Benacerraf and McDevitt, 1972).

Response to Other Genera of Coccidia Genetic variation in host responses to coccidia developing in a predator-prey life cycle (e.g., *Sarcocystis* and *Toxoplasma*) can best be measured in the "prey" host because it harbors the asexual stages of the parasite, which generally produce more distinctive pathological changes than do the sexual stages. Araujo et al. (1976) reported significant differences in susceptibility to infection with *Toxoplasma gondii* trophozoites among seven inbred lines of mice. They also emphasized the importance of the size of the inoculum used for evaluation of comparative resistance. For example, the least susceptible strain given the lowest dose of trophozoites (5×10^3) used in their study was the most susceptible when a larger inoculum (1×10^5) was used.

These findings of genetic variation in susceptibility to *T. gondii* were confirmed in a subsequent study by Williams et al. (1978), who used four congenic strains of mice and their F_1 and F_2 progeny. Their data suggested that differences in murine susceptibility to *T. gondii* are under multigenic control and are most easily revealed within an intermediate range of challenge, or when a relatively avirulent strain of *T. gondii* is used to produce the challenge. Furthermore, at least one of the genes determining susceptibility was found to be linked to the major murine histocompatibility locus (H-2). Olisa et al. (1977) also reported significant differences in the survival time of four inbred strains of mice after inoculation with *T. gondii*, strongly suggesting genetic variation in resistance to the lethal effects of this coccidium. Kamei et al. (1976) identified an inbred strain of mouse having particularly high susceptibility to *T. gondii* strains of low virulence. Inocula thought to contain one cyst of the Beverley strain of *T. gondii* produced an acute fatal infection in this strain of mouse, while causing only latent chronic infection in a less susceptible strain. The high susceptibility to toxoplasmosis could not be attributed to poor antibody-producing ability, because strains differing greatly in susceptibility exhibited no significant differences in antibody response after infection with a strain of *T. gondii* having low virulence.

Response in Ovo

Long (1965) was the first to demonstrate that *E. tenella* completes its endogenous development in the chorioallantoic membrane (CAM) of the chick embryo after inoculation of sporozoites into the allantoic cavity (see Doran, Chapter 6, this volume). Inoculation of embryonated eggs with sporozoites of *E. tenella* produces dose-related mortality commencing about 90 hr. Thus, embryonic response to this coccidian can be quantitatively determined. Jeffers and Wagenbach (1970) evaluated the response of embryos from mating groups selected for differential susceptibility to cecal coccidiosis and found that genetic differences in embryonic susceptibility to *E. tenella* exist among different strains of fowl. However, the resistance of embryos to infection with *E. tenella* was not significantly related to the resistance of chicks from the same mating groups, suggesting that inoculation in ovo with sporozoites of *E. tenella* is a relatively ineffective method for detecting genetic differences in resistance to cecal coccidiosis. A general lack of association between the responses in vivo and in ovo of the Cornell S, C, and K strains of White Leghorns to *E. tenella* infection was also reported by Long (1970a, 1974b). Female embryos suffer significantly higher mortality during the acute hemorrhagic phase of the infection than do male embryos (Jeffers and Wagenbach, 1969).

Long (1970b) described a method for evaluating anticoccidial drugs in ovo by counting focal lesions on the CAM associated with schizogony of *E. tenella*. This parameter of measurement is probably more sensitive to small differences in the severity of infection than is mortality, and thus may also be useful in detecting subtle genetic differences in embryonic response to this parasite.

3

Prospects for Control through Selective Breeding

Three of the four envisaged approaches to the control of coccidiosis (genetic, chemotherapeutic, immunological, and eradication) have been effective when evaluated under controlled conditions. Genetic variation in host response to coccidia, which might be exploited through selection to increase host resistance, has been demonstrated in a number of experiments. Likewise, considerable progress was realized in the development of highly effective chemotherapeutic agents, making this the most widely accepted approach to the control of coccidiosis (see McDougald, Chapter 9, this volume). Although improved host resistance may also be obtained through immunity developed following exposure to controlled dosages of coccidia, immunological control measures may involve a risk of uncontrolled infections. Eradication of the coccidia seems unlikely in the foreseeable future, but the threat of coccidiosis can nevertheless be reduced through improved management. Thus, the most effective means of minimizing losses due to coccidiosis in domestic animals may require the coordinated use of more than a single approach. The various ways of controlling coccidiosis are fully discussed in Chapter 11 (Fayer and Reid, this volume).

The prospect for control of coccidiosis through selective breeding is best applied to the domestic fowl, in which genetic variation in host response to coccidia is abundant and the effectiveness of selection in altering host resistance has been clearly demonstrated. Nevertheless, it is unlikely that poultry breeders will soon exploit this genetic variation. Genetic improvement in the overall performance of poultry requires the breeder to practice selection on several traits in tandem, or simultaneously, by using a selection index in which each trait is weighted according to its relative economic importance. A serious objection to the incorporation of genetic resistance to coccidiosis as a trait in a selection index is the fact that increases in the number of traits selected result in reductions in the intensity of selection practiced on each individual trait. Thus, breeders must concentrate on a limited number of the traits most directly related to the profitability of poultry production.

Gavora and Spencer (1978) suggested breeding for general disease resistance, which they define as "An ability to resist any alteration of the state of the body by external causes (microorganisms and/or stress) which interrupts or disturbs proper performance," as a possible solution to this dilemma. They offer three possible approaches to selection for general disease resistance: 1) selection based on challenge of the breeding stock with many disease agents; 2) the use of progeny tests based on challenge of separate groups of sibs or progeny with many disease agents while keeping the breeding stock in a relatively disease-free environment; and 3) indirect selection based on characters that are measurable without the use of deliberate exposure to disease. The first approach is unacceptable because of possible losses of valuable breeding stock, while the second alternative, although more acceptable, would be very costly. Although indirect selection would thus seem the most appropriate method, measurable traits indicative of general disease resistance have not been identified. Thus, poultry breeders must continue to concentrate on those traits that are of great economic importance and that can be effectively improved principally through genetic selection, while depending upon the current immunological, chemotherapeutic, and management control measures to reduce losses in production due to coccidiosis.

GENETIC CONSTITUTION OF COCCIDIA

Life Cycle

All developmental stages of the coccidia are presumed to be haploid with the exception of diploid nuclei (zygotes) formed after fusion of the male and female gametes (see, e.g., Dobell and Jameson, 1915; Canning and Anwar, 1968). Canning and Morgan (1975) measured, by microdensitometry, the quantity of DNA present in the nuclei of different stages. For *E. tenella,* mean values of 13.95 and 29.75 standard units were obtained for microgametes and pre-first division zygotes, respectively. These values were taken to represent haploid and diploid amounts of DNA.

In a light microscope study of nuclear division in freshly discharged oocysts of *Eimeria maxima,* Canning and Anwar (1968) observed completion of meiosis (reduction division) by 9 hr. Two subsequent mitotic divisions occupied the next 2 hr, and the final maturation of the sporocyst and sporozoite involved no further nuclear division and took another 14 hr.

The coccidia possess relatively few chromosomes, although studies of nuclear division are few and most have been concerned with the largest representatives which are generally more amenable to cytological investigation. Dobell and Jameson (1915) found a haploid number of six chromosomes in *Aggregata eberthi,* and other authors (cited by Kheysin, 1972) found eight in *Adelina cryptocerci* (Yarwood, 1937) and five in *Barrouxia schneideri* (Wedekind, 1972). Vetterling (1966) observed a haploid number of four chromosomes in *Eimeria debliecki* from the pig, and Canning and Anwar (1968) confirmed the finding of Scholtyseck (1953) that the haploid number in *E. tenella* and *E. maxima* was five. Canning and Anwar (1968), furthermore, did not observe chromatids during reduction division and, with the concomitant lack of chiasmata, postulated that because crossing over of genes did not occur, genetic variation was achieved solely through the random assortment of chromosomes.

Sex Determination

The development of viable progeny from single sporulated oocysts (as first described by Tyzzer, 1929) indicated that in the coccidial life cycle the sexes diverge somewhere between the diploid nucleus of the zygote and the gametes.

3

Until recently, direct evidence as to the sex of sporocysts and sporozoites was not available, because only a few attempted infections with these stages had been reported. Long (1959), for example, failed to infect a chicken with a single sporocyst of *E. maxima* (he was successful when 16 and 50 sporocysts were given), and Reyer (1937) (cited by Canning, 1962) was unable to observe a patent infection in the centipede *(Lithobius forficatus)* given a single sporocyst of *B. schneideri*. He, too, was successful when an oocyst was administered. (Unlike eimerian parasites whose sporocysts contain two sporozoites, those of *B. schneideri* contain only one; Reyer, therefore, attempted a single sporozoite infection.)

The sex of sporozoites and merozoites has been investigated by indirect methods, and some evidence suggests that sexually differentiated merozoites are produced during infection. In a description of the life cycle of *Eimeria irresidua, Eimeria magna, Eimeria media,* and *Eimeria perforans* from the rabbit, Rutherford (1943) observed two types of merozoites in each species and considered that those he referred to as type A produced macrogametocytes, while those designated as type B produced microgametocytes. Similar findings were presented for *Eimeria stiedae* by Pellérdy and Durr (1970): type A schizonts contained two relatively thick merozoites which, through more generations of large merozoites, subsequently produced macrogametes, and type B schizonts, which contained five to eight slender merozoites, eventually produced microgametes. In a study of the endogenous development of *B. schneideri,* Canning (1962, 1963) observed two types of merozoites after periodic acid-Schiff (PAS) tests for carbohydrate. She concluded that the type which stained poorly and was characterized by few and small (if any) granules of carbohydrate developed into microgametocytes, whereas the other deeply stained merozoites which contained considerable carbohydrate reserves developed into macrogametocytes. Additionally, a conspicuous polar granule present in the "female" merozoite was traced through to the macrogametocyte and, similarly, the absence of a nucleolus in the "male" merozoite was also observed in the microgametocyte.

The PAS staining reaction was also used by Klimes et al. (1972) to study the development of sporozoites of *E. tenella* in cultured kidney cells. Mature second and third generation schizonts and merozoites liberated from them were either very strongly or very weakly PAS-positive. They postulated that the former developed into macrogametocytes and the latter into microgametocytes.

Klimes et al. (1972) also considered whether sexual differentiation was of genotypic (endogenous) or environmental (exogenous) origin. The latter possibility was excluded on the premise that if sexual differentiation was influenced only by environmental conditions (ultimately the host cell), all the gametocytes occurring in a single cell ought to be of the same sex. In fact, cells containing both macro- and microgametocytes were observed; it was concluded that this was evidence for genetic determination of sex. The same conclusion was reached by Hammond (1973) when reviewing the earlier results of Hammond et al. (1961), who found both macro- and microgametocytes of *Eimeria auburnensis* in the same cell. Canning (1963) similarly considered that the sexuality of Adeleid coccidia was not influenced by any external factors, because gametocytes of both sex were observed in close association. Canning (1962) had earlier taken the argument a step further and postulated that if sex determination was genotypic, the male and female characters must be separated at nuclear division either during schizogony or sporogony. Thus, depending on the separation of the sexes at one of three possible stages, the sex of the sporozoite would vary. If separation occurred: (*a*) at zygotic reduction division, the sporocysts would be unisexual and the sporozoites within would be of one and the same sex; (*b*) within the sporocyst, each sporocyst would contain one female and one male sporozoite; and (*c*) at schizogony, the sporozoite would be bisexual. Canning's (1962) results, with those of Reyer (1937), were taken to indicate that zygotic reduction division was the point of sexual differentiation. Kheysin (1972) also believed that the sexes differentiated during reduction division in sporozoans. He made the assumption that half of the subsequent haploid products are female and the other half male, that is, an eimerian oocyst contains two female and two male sporocysts. However, the results of Haberkorn (1970) were contrary to this assumption, because he recovered oocysts from 16 of 32 mice each given a single merozoite of *Eimeria falciformis* via the ceca. The first oocysts were recovered between 3 and 11 days after infection and some sporulated. Haberkorn (1970) concluded that not only was the further development of merozoites into gametocytes or schizonts not determined (because of the delay in oocyst production with some infections) but also sex determination must be environmental. The latter conclusion has now been substantiated by further direct evidence. Independent studies by Shirley and Millard (1976) and Lee et al. (1977), who obtained viable oocysts of *E. tenella* from infections of eggs with single sporozoites, established the bisexual nature of the haploid eimerian sporozoite. McDougald and Jeffers (1976a) also suggested that sporozoites were bisexual from their observations of macrogametes and microgametes within isolated clusters of parasites presumed to be the progeny of a single sporozoite/merozoite. Further evidence for environmental sex determination was provided by Pfefferkorn et al. (1977) working with *T. gondii*. They demonstrated the production of oocysts in cats infected with cysts (bradyzoites) derived from individual tachyzoites cultured in human fibroblast cells.

Cloning

It is now possible to "clone" some coccidia, and the term has been approved by Joyner et al. (1978) for populations of oocysts derived from single sporozoites or single merozoites. The technique is presently limited to studies in vitro; to date, clones of a field strain of *E. tenella* (Lee et al., 1977) and embryo-adapted strains of *E. tenella* and *Eimeria mivati* (Shirley and Millard, 1976; Shirley, 1977, respectively) have been established from single sporozoites inoculated into embryonated eggs. Pfefferkorn et al. (1977) similarly established their clones of *T. gondii* after a single trophozoite had initially developed in cell culture.

3

If future research indicates that the two sporozoites within an eimerian sporocyst are genetically identical, it will then be a relatively easy matter to develop sporocyst clones of at least all the species from the chicken. Shirley (in press) serially passaged 13 times the Houghton (H) strain of *E. maxima* as a single sporocyst, and in an examination of the progeny of oocysts from the first and 13th passage found no differences in their pathogenicity or antigenicity.

GENETIC MARKERS—THE BASIS FOR VARIATION WITHIN AND BETWEEN SPECIES

Mutation is the original source of the genetic variation from which evolutionary changes, including the establishment of species differences, are accumulated. All genetic changes, except those produced by gene recombination, are mutations in the broad sense. Thus, mutations may be: 1) gene mutations produced by substitution, addition, or deletion of nucleotides in the DNA of the gene; 2) changes in chromosomal structure (deficiency, duplication, translocation, inversion, or transposition) affecting the arrangement of genes in the chromosomes; or 3) numerical changes (aneuploidy, haploidy, or polyploidy) affecting the number of chromosomes. Mutations are generally considered to be random events in that they represent an imperfect transmission of hereditary information and are thus undirected in terms of their usefulness to the organism in which they occur. The fact that most mutations appear deleterious stems from their occurrence in a genetic background already contributing to optimum adaptiveness in the environment in which the organism normally exists. Nevertheless, these mutations may contribute favorably to adaptiveness should the organism be placed in a novel environment, as has been demonstrated by microorganisms exposed to different chemotherapeutic agents.

Wide differences of opinion exist on both the definition and genesis of species. We adhere to the biological species concept which considers the species to be a collection of genetically similar populations which are capable of interbreeding and which, through genetically or environmentally determined isolating mechanisms, are evolving in a pattern distinct from other such collections of populations (Mayr, 1963, 1970; Grant, 1971). Thus, specific variation is that which exists between such collections of populations, while infraspecific variation is that which exists within them. Because there are many proposed patterns of species formation, it is not possible to detail the nature of each of them here. Nevertheless, each of these proposed concepts of the genesis of species has a common endpoint in the eventual differentiation of reproductively isolated groups of populations which may exist in the same environment (sympatric species) or separate environments (allopatric species). Although species usually differ genetically as a consequence of selection for optimum fitness in differing environments, allopatric species could theoretically be genetically identical, because interspecific hybridization among them may be prevented only by geographic isolation. The host specificity of many allopatric species of coccidia (see below) may contribute to reproductive isolation between them inasmuch as development in a "foreign" host does not proceed to gametogony, thus preventing contact of the gametes of different species. Sympatric species must be reproductively isolated to prevent gene exchange between them, for they would otherwise fuse into a single species.

The sympatric existence of different species is proof of their reproductive isolation. Prevention of gene exchange through interspecific hybridization of sympatric species requires genetic differences between them expressed either as prezygotic mechanisms preventing the formation of hybrid zygotes or postzygotic mechanisms reducing the viability of fertility of hybrid zygotes. Thus, reproductively isolated sympatric populations are necessarily different genetically.

Infraspecific variation is a consequence of genetic differences among strains and differences in the environments to which they are exposed. Although the genetic components of infraspecific variation are our primary concern, identification of environmental contributions is also important. Genetic variation among strains can be more easily evaluated by stabilizing those environmental variables which can be experimentally controlled. The genetic basis of infraspecific variation is provided by changes in gene frequency and, consequently, genotypic frequencies. Gene frequencies may be changed through systematic processes (migration, mutation, and selection) and the dispersive process arising from sampling in small populations. Among these processes, selection is usually the most important contributor to changes in gene frequency.

In order to conduct genetic analyses of sexually reproducing organisms, it is necessary to make controlled matings between strains within species and to follow the transmission of genetic traits from the parent strains to the progeny produced by crossing them. The parent strains used in these studies must exhibit stable heritable traits or genetic markers, which can be used to differentiate them from one another and to conclusively identify the recombinant progeny resulting from interstrain crosses. Four such characters, namely drug resistance, developmental rate, enzyme variation, and temperature sensitivity have thus far been described in the coccidia.

Drug Resistance

The economic importance of the development of resistance to drugs being used for the prevention of coccidiosis accounts for this trait being the most widely studied genetic marker in the coccidia. Resistance to almost every anticoccidial drug has been developed through intentional selection and has been found after the use of these drugs under commercial conditions of animal production. Because the results of these studies have been reviewed elsewhere (Joyner, 1970; McLoughlin, 1970b; Chapman, 1978; Chapman, Chapter 10, this volume), they will not be detailed here.

3

Weppelman et al. (1977) provided the first estimates of mutation rates in coccidia, thus confirming the basis for genetic variation in drug resistance in unselected populations. Using a modified bacterial fluctuation test to isolate drug-resistant mutants of *E. tenella*, they found that the frequency of mutants resistant to amquinate and glycarbylamide was 5.8×10^{-8} and 2.4×10^{-7} per wild type oocyst, respectively. They were, however, unable to isolate mutants resistant to amprolium, monensin, nicarbazin, or robenidine, suggesting that mutants resistant to these drugs are less frequent than 7.5×10^{-9} per wild type oocyst. Nevertheless, the response of coccidia to intentional selection for resistance to three of these drugs (amprolium, nicarbazin, and robenidine) confirms the existence of genetic variation in the drug sensitivity of previously unselected populations (McLoughlin and Gardiner, 1967, 1968; McLoughlin and Chute, 1978).

In order to be useful in genetic crosses, a strain must exhibit a homogeneous expression of drug resistance. McLoughlin and Chute (1974a) characterized the response to amprolium of isolants derived from single oocysts of a strain of *E. tenella* resistant to amprolium and showed that all individuals in the population may not be equally resistant. Thus, derivation of the strain from a single oocyst, or if possible as a clone from a single sporozoite (see above) will help to assure a homogeneous response. Likewise, by optimizing the relationship between the intensity of selection and the expression of resistance, one can develop a strain exhibiting a very high degree of resistance (Jeffers, 1978). Jeffers (1974a) used both of these procedures to develop homogeneous strains of *E. tenella* exhibiting a very high degree of resistance to amprolium and decoquinate. These strains were then useful in genetic recombination experiments (see below).

Developmental Rate

The prepatent period is generally regarded as a species-specific characteristic of *Eimeria*. However, the use of this parameter for the definitive identification of species is limited, because there may be an overlap in the prepatent periods of different species of *Eimeria* infecting the same host. Although the length of the prepatent period may be slightly modified by various environmental factors such as the age of the parasite, the immune status of the host, or the exposure of the parasite to certain anticoccidial drugs, the control of the developmental rate is largely intrinsic in the parasite. This fact was clearly demonstrated by Roudabush (1935) and Levine (1940) in studies on infections resulting from the transfer of merozoites of the rat coccidium *Eimeria miyairii* and the chicken coccidium *E. necatrix*, respectively. When the infection was interrupted in one host and transferred to another coccidia-free host, the time required for development of oocysts was equal to that required for infections maintained in a single host, indicating that the length of the life cycle was determined not by increasing resistance of the host, but that it was inherent in each species of coccidia. The recent findings of Rose et al. (1979) support this conclusion. Nude (athymic) rats failed to become resistant to infection with *E. nieschulzi*, and the course of development of this parasite in three successive infections, as measured by the duration of patency, was the same as the primary infection in immunologically competent rats.

Jeffers (1975) altered the developmental rate of *E. tenella* by selecting for precociousness (early oocyst production). Selection through 46 generations reduced the prepatent period of the selected strain (Wis-F) to about 72 hr, while that of the parent strain from which it was derived is approximately 120 hr. McDougald and Jeffers (1976a,b) found that this abbreviated prepatent period resulted from the reduction of second generation schizogony in the selected strain. The ability of first generation merozoites of this strain to proceed directly to gametogony is a stable genetic trait which is useful in separating strains of *E. tenella*, and is thus a suitable genetic marker for use in genetic recombination experiments (see below).

Enzyme Variation

Enzymes are primary products of structural genes and therefore make extremely useful genetic markers. They are particularly amenable to characterization by electrophoresis, and the application of selective staining (the zymogram technique) facilitates a direct visual comparison of variant forms of homologous enzymes which, although catalyzing the same reaction, may differ in their size, shape, and molecular configuration. The expression of enzyme variation in haploid organisms is particularly stable because alternative alleles in homologous chromosomes are not available to mask any deleterious mutations that may arise. Thus, mutations that lead to the production of nonfunctional or less effective protein molecules will be lost if, as a result, the organism cannot fully develop and reproduce.

Variation in the electrophoretic mobility of several enzymes in all species of *Eimeria* from the chicken, and in some species from the rabbit and sheep, is now recognized (Rollinson, 1975; Shirley, 1975; Shirley and Rollinson, 1979). With the avian coccidia, at least, identification of different species and the recognition of different strains is now possible by consideration of these genetic markers (see below and Figures 2 to 4).

Temperature Sensitivity

Conditional mutants that determine a loss of ability of organisms to grow and multiply at restrictive temperatures have been particularly useful in genetic studies on microorganisms. Wild type organisms, from which temperature-sensitive mutants are obtained, develop at both restrictive and permissive temperatures, thus facilitating the detection of the temperature-sensitive mutants by incubation at the restrictive temperature. Temperature-sensitive mutants presumably produce a defect in an essential protein, rendering it either thermolabile, or nonfunctional at the restrictive temperature. Although temperature-sensitive mutants were previously described in the free-living protozoan, *Paramecium aurelia* (Hipke and Hanson,

3

1974), Pfefferkorn and Pfefferkorn (1976) offer the only evidence of this type of mutant in intracellular parasitic protozoa. They used *N*-methyl-*N'*-nitro-*N*-nitrosoguanidine to induce temperature-sensitive mutants in *T. gondii*, which were unable to form plaques in human fibroblast tissue cultures when incubated at the restrictive temperature (40°C), but grew well in the cultures incubated at the permissive temperature (33°C). The RH strain from which they were derived produced plaques in the fibroblast cultures at incubation temperatures ranging from 30°C to 41°C. The virulence of two of the temperature-sensitive mutants, as measured by the severity of infections produced in mice, was compared with that of the highly virulent RH (parent) strain. The mutants were at least 10^4-fold less virulent than the parent strain.

Pfefferkorn and Pfefferkorn (1979) subsequently evaluated the efficacy of various mutagenic agents on extracellular and intracellular *T. gondii* in inducing mutants resistant to 5-fluorodeoxyuridine (FUDR). Among the mutagens studied, the alkylating agents *N*-methyl-*N'*-nitro-*N*-nitrosoguanidine and ethylmethane sulfonate were the most efficient, increasing the frequency of FUDR-resistant mutants to more than 200 mutants per 10^6 parasites. This is the first reported quantitative assessment of intentional mutagenesis in the coccidia. Pfefferkorn and Pfefferkorn (1976) pointed out that treatment of the asexual haploid forms of coccidia with mutagenic agents simplifies the detection and isolation of recessive mutations, which might otherwise be masked in the diploid heterozygous state. Although the utility of temperature-sensitive mutants as markers in genetic recombination experiments with coccidia has not as yet been demonstrated, they should be amenable to this use since temperature sensitivity is a stable genetic trait which can be used to differentiate potentially interbreeding strains.

GENETIC RECOMBINATION

The genetic basis for observed variation in the phenotypic expression of measurable traits within species is confirmed by transmission of the traits from one strain to another and the consequent recombination of two or more traits in a single strain. Jones (1932) was the first to attempt experimental hybridization in coccidia by trying to cross *Eimeria acervulina* and *E. maxima.* She found no suggestion of hybridization, as evidenced by oocysts with intermediate size among progeny produced by infections with a mixture of species, a finding that she correctly interpreted as being due to reproductive isolation between the two species.

Jeffers (1974a) provided the first evidence of genetic recombination in coccidia by crossing strains of *E. tenella* differing in resistance to amprolium and decoquinate. Cultures of the parent strains and a culture derived from a mixture of the parent strain oocysts were propagated through drug barriers in an experimental design appropriate for definitively separating the respective parental and recombinant phenotypes. Some of the progeny of the strain-cross culture reproduced in the presence of a simultaneous or tandem double drug barrier, showing that they were recombinant phenotypes produced through cross-fertilization of gametes of the parent strains. Comparative evaluations of the drug sensitivity of the recombinant progeny confirmed that they were simultaneously resistant to amprolium and decoquinate. These findings showed that certain traits which may be used to phenotypically differentiate strains of coccidia are appropriate genetic markers in conventional genetic recombination experiments. Jeffers (1974a) also noted the utility of this experimental design for rapidly developing strains with multiple drug resistance for use in studies on anticoccidial drug resistance.

Joyner and Norton (1975) subsequently demonstrated the genetic transfer of sulfaquinoxaline (SQ) resistance between two strains of *E. maxima*. However, they were unable to obtain genetic recombinants of *E. maxima* resistant to clopidol and methylbenzoquate (MB) (Joyner and Norton, 1975, 1976). Although they suggested that the lack of genetic recombination of these traits was due to their determination by genetic factors on the same chromosome, they were nevertheless able to develop a strain simultaneously resistant to a synergistic mixture of methylbenzoquate and clopidol through genetic selection for resistance first to one drug and then to the other (Joyner and Norton, 1978). It was much more difficult to develop resistance simultaneously to both drugs through propagation in the presence of the drug mixture (Norton and Joyner, 1978). Joyner and Norton (1977) also reported genetic recombination of resistance to methylbenzoquate and sulfaquinoxaline in *E. maxima* after the inoculation of merozoites of the parent strains, suggesting that genetic recombination is mediated through cross-fertilization at zygote formation. In this study, they demonstrated genetic recombination of drug resistance between *E. maxima* (Weybridge (W) strain) and *E. maxima* (*indentata*), the latter described as a new species by Fernando and Remmler (1973b). This confirmed the view of Long (1974a) that *E. indentata* was not a new species. Ryley and Hardman (1978) obtained doubly drug-resistant recombinants of *E. acervulina* by using drug mixture barriers to differentiate parental and recombinant phenotypes. However, under the same conditions, they failed to demonstrate genetic recombination between methylbenzoquate-resistant *E. mivati* and sulfaquinoxaline- or robenidine-resistant *E. acervulina*. They concluded that these two organisms are reproductively isolated, thus supporting the status of *E. mivati* as a distinct species.

Jeffers (1976b) also demonstrated that precociousness (see above) is a useful marker to differentiate strains of coccidia in genetic studies. Cultures of the decoquinate-sensitive and precocious Wis-F strain and a normally developing decoquinate-resistant strain (368) of *E. tenella*, as well as a culture derived from a mixture of these strains, were propagated through a simultaneous drug and developmental time barrier to differentiate parental and recombinant phenotypes. Those organisms present in the strain-cross culture which were able to traverse this simultaneously imposed drug and development time barrier were shown to be concurrently precocious and decoquinate-resistant. This study also showed that the precocious Wis-F strain of *E. tenella*, although markedly different from other strains in its developmental cycle, is nevertheless reproductively compatible with them.

A convenient third genetic marker for genetic recombination studies utilizing drug resistance has proved to be enzyme variation (see above). The two reports of genetic recombination to consider these strain markers have both involved crosses between established laboratory strains and parasites originally isolated from the Malaysian jungle fowl, namely, *E. mivati* (*diminuta*) (Shirley, 1979a), originally described as *E. diminuta* (Fernando and Remmler, 1973a), and *E. maxima* (*indentata*) (Long, 1974a), originally described as *E. indentata* (Fernando and Remmler, 1973b).

Shirley (1978a) described recombination between *E. mivati* (H) and *E. mivati* (*diminuta*), having earlier found that the strains differed in their variant forms of lactate dehydrogenase (LDH) and in their response to SQ (Shirley et al., 1977). *E. mivati* (H) was characterized by LDH-1 and resistance to 0.05 mg of SQ per egg, whereas *E. mivati* (*diminuta*) had LDH-6 and full sensitivity to SQ at 0.05 mg of SQ per egg. (Embryo-adapted lines of these strains were used and the experiments were conducted using chicken embryos.) The two lines were allowed to cross-fertilize in the absence of SQ, and a selection was then made for drug resistance. Oocysts produced despite the drug barrier were subjected to enzyme electrophoresis, and the presence of both LDH-1 and LDH-6 was clear evidence for a mixture of parasites with the parental phenotype (LDH-1 and drug resistance) and the recombinant phenotype (LDH-6 and drug resistance) (see Figure 1). Similar results from experiments in chickens were obtained by Rollinson et al. (1979) who demonstrated the transfer of MB resistance between resistant and sensitive lines of *E. maxima* (*indentata*) and *E. maxima* (W). The enzyme marker was phosphoglucomutase (PGM), and *E. maxima* (*indentata*) and *E. maxima* (W) were characterized by variants designated as PGM-1 and PGM-2, respectively. Judged by the presence of both PGM-1 and PGM-2 in the progeny of crosses made between appropriate lines passaged first in the absence of drug and then in the presence of drug, recombination was demonstrated in two of three attempts. The first successful cross was between MB-sensitive *E. maxima* (W) and MB-resistant *E. maxima* (*indentata*), but more clearly defined was the cross between MB-sensitive *E. maxima* (*indentata*) and MB-resistant *E. maxima* (W), which had been allowed to cross-fertilize in the ratio of 4:1.

Results of the genetic recombination experiment by Rollinson (1976) are difficult to interpret in view of subsequent data confirming the delineation of *E. acervulina* and *E. mivati* (Ryley and Hardman, 1978; Shirley, 1979a). Rollinson "crossed" *E. mivati* (H) and *E. acervulina* (W), which differed in their electrophoretic form of glucose phosphate isomerase (GPI) and in their ability to develop in embryonated eggs. *E. mivati* (H) was characterized by GPI-1 and growth in eggs, whereas *E. acervulina* (W) had GPI-2 and would not grow in eggs. Recombination was demonstrated when progeny of the cross passaged in eggs were found to be

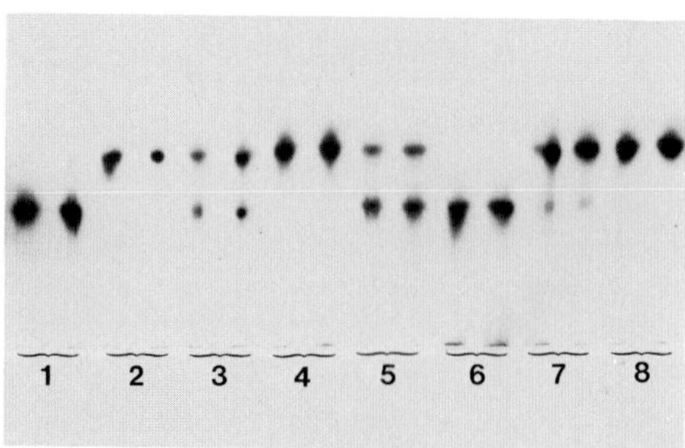

Figure 1. Cross fertilization between *E. mivati* (MA) and *E. mivati (diminuta)* (DA) as demonstrated by drug resistance and enzyme electrophoresis. 1 = *E. mivati* (MA) + *E. mivati (diminuta)* (DA) allowed to cross-fertilize in a ratio of 1:10. Untreated; (LDH-1, LDH-6). 2 = *E. mivati* (MA) + lysed sporozoites of *E. mivati (diminuta)* (DA). A control group treated with SQ (LDH-1). 3 = *E. mivati* (MA) + *E. mivati (diminuta)* (DA) allowed to cross-fertilize in a ratio of 1:1. Untreated; (LDH-1, LDH-6). 4 = *E. mivati* (MA). Untreated; (LDH-1). 5 = *E. mivati* (MA) + *E. mivati (diminuta)* (DA) allowed to cross-fertilize in a ratio of 1:10. Treated with SQ; (LDH-1, LDH-6). 6 = *E. mivati (diminuta)* (DA). Untreated; (LDH-6). 7 = *E. mivati* (MA) + *E. mivati (diminuta)* (DA) allowed to cross-fertilize in a ratio of 1:1. Treated with SQ; (LDH-1, LDH-6). 8 = *E. mivati* (MA). Treated with SQ; (LDH-1). Reprinted by permission from: Shirley, M. W. Electrophoretic variation of enzymes: A further marker for genetic studies of the *Eimeria. Zeitschrift für Parasitenkunde* 57:83–87 (1978).

characterized by GPI-2 (and GPI-1) (i.e., the recombinant phenotype of GPI-2 and growth in eggs). However, problems in the interpretation of these results arise because first, the culture of *E. mivati* (H) used by Rollinson was predominantly *E. acervulina* (see Shirley, 1979a) and second, no change in enzyme pattern accompanied the removal of *E. acervulina* and the recovery of pure *E. mivati* when "*E. mivati*" (H) was passaged in chicken embryos.

Pfefferkorn and Pfefferkorn (1980) have demonstrated genetic recombination between drug-resistant mutants of *T. gondii*. Mutants resistant to adenine arabinoside (ara-A) or to FUDR were used to infect mice in order to produce brain cysts containing bradyzoites. Kittens fed a mixture of cysts that contained both mutants produced *T. gondii* oocysts containing approximately 12% recombinant doubly resistant (ara-A–FUDR) organisms. Recombinants sensitive to both drugs were also isolated. Two of the FUDR mutants used in these studies were defective in oocyst production. These mutants did not recombine with a fertile ara-A mutant, indicating that the inability to produce oocysts results from a defect in the production of both macrogametes and microgametes.

IDENTIFICATION OF COCCIDIA

Specific Variation

Eimeria

Morphology of Oocysts The number of sporocysts and the distribution of sporozoites within the oocyst are considered useful characters for differentiating certain genera of coccidia (Levine, 1973). Furthermore, certain characteristics of the oocyst are useful taxonomic tools in describing different species of coccidia. Among these are: 1) the dimensions, shape, and color of the oocyst; 2) the characteristics of the surface of the oocyst wall and the appearance of a polar cap or micropyle; 3) the characteristics of the sporocyst; and 4) the presence or absence of granules or residual bodies in the oocyst or sporocyst. Although certain groups of coccidia occurring in the same host are readily differentiated by oocyst characters (e.g., the *Eimeria* species of sheep and cattle), others (e.g., the *Eimeria* species of the domestic fowl) are much more difficult to discern because of the apparent similarity of the oocysts of different species. Additional limitations on the use of oocyst characters for the definitive identification of species stem from various environmental effects on oocyst dimensions. An increase in the size of oocysts passed during patency has been reported by Fish (1931) for *E. tenella*; by Jones (1932) for *E. acervulina*, *E. maxima*, and *E. tenella*; by Becker et al. (1955) for *E. brunetti*; by Becker et al. (1956) for *E. necatrix*; by Rommel (1970) for *E. scabra* and *E. polita*; by Duszynski (1971) for *E. separata*; and by Long and Millard (1979a) for *E. dispersa*. The morphology of oocysts may also be related to the degree of infection (Rommel, 1970; Kheysin, 1972; Catchpole et al., 1975). This variation is apparently a function of the development of more than a single macrogamete within individual host cells in heavy infections. Although infraspecific variation in oocyst dimensions has also been reported (Jeffers, 1978), these differences do not exceed differences between species having distinctively different oocyst dimensions (e.g., *E. acervulina* and *E. maxima*). Long et al. (1977) found marked differences between the dimensions of oocysts of the Yorkshire and Weybridge B strains of *Eimeria meleagrimitis*, showing that differences in oocyst characters alone do not provide definitive identification of certain species of *Eimeria*. Despite the above limitations on the use of the general characters of oocysts as taxonomic tools, certain characters, for example, the sporopodia of *Eimeria funduli* (Duszynski et al., 1979) and the filament-bearing Stieda bodies of *Eimeria filamentifera* (Wacha and Christiansen, 1979) are so distinctive that they are particularly useful taxonomic tools.

Host and Site Specificity The characteristic host specificity and site within host preferences of different species of coccidia have long been recognized. These aspects of the biology of the coccidia are thoroughly reviewed by Joyner (Chapter 2, this volume). When considering the genetic aspects of host specificity, it should be noted that not only does the genetic constitution of the host influence susceptibility to the species normally occurring in the host (see above), but also the genetic background of the host may be important in determining whether cross-transmission will occur with species ordinarily foreign to it. Thus, Mayberry and Marquardt (1973) demonstrated cross-transmission of *E. separata* from the normal host, *Rattus*, to a particular inbred line of the mouse, *Mus musculus*. The ability of *E. separata* to complete development in the foreign host was associated with a specific allele (*a*) determining the agouti pattern (i.e., grizzled color of the fur resulting from alternating light and dark bands on the individual hairs) in certain inbred mouse strains (Mayberry et al., 1975). Furthermore, the development of *E. separata*, as measured by reproductive indices, was apparently greater in strains of mice carrying this allele in the doubly-recessive homozygous state (*aa*) than it was in those strains carrying a single allele (*a*) in the heterozygous state.

Pathogenicity The wide differences in pathogenicity existing between different species of coccidia are generally attributable to differences in the site of endogenous development. For example, among the coccidia of the domestic fowl, *Eimeria praecox* undergoes schizogony principally in epithelial cells in the superficial part of the intestinal villi and is practically innocuous (Long, 1967), while *E. necatrix* undergoes second generation schizogony deep in the lamina propria sometimes extending into the muscularis mucosa (Tyzzer, 1929; Stockdale and Fernando, 1975) and is, by contrast, so pathogenic that its species name was chosen to signify murderess (Johnson, 1938).

Despite such differences in the pathognomonic effects of different species, pathogenicity is generally a poor characteristic for differentiation of species, because differences in pathogenicity are also conditioned by such factors as infective dose, the age of the culture, the strain of the parasite (see below) and the age, immunological status, and genetic constitution of the host (Long, 1970a).

Prepatent Period The prepatent period, that is, the time between ingestion of oocysts by the host and the detection of a new generation of oocysts in the feces, is a species-specific characteristic, which is controlled largely by the parasite (see above). Because the prepatent period is only slightly modified by various environmental factors, it is relatively constant for each species and thus serves as a useful taxonomic tool for differentiating species of coccidia developing in the same host and having different developmental rates.

Immunological Variation It has long been recognized (and is now generally accepted) that each species of *Eimeria* stimulates a species-specific immunity (see also Rose, Chapter 8, this volume). For example, chickens immunized against six species still remain completely susceptible to infection with the seventh. This approach has been used by several authors in describing new species of avian coccidia. In the first description of *E. praecox*, Johnson (1930) found no cross-protection between it and *E. acervulina*, *E. maxima*, or *E. mitis*. Similar results were obtained by Levine (1942) and Edgar and Seibold (1964) in describing *E. brunetti* and *E. mivati*, respectively, while Tyzzer et al. (1932), in the second description of *E. necatrix*, found no cross-protection between it and *E. tenella*. None of these experiments was conducted on a quantitative basis, however, and an assessment of infection in specifically immunized hosts was based only on pathology and microscopic findings.

Detailed quantitative cross-immunity tests which measure oocyst production in immunized (experimental) and nonimmunized control chickens of an identical age are now preferred. Using such tests, Shirley (1979a) recently showed no evidence of cross-protection between *E. acervulina* and pure strains of *E. mivati*.

It is also possible to incorporate immunity studies in field surveys of the avian coccidia; by doing so Reid et al. (1965) were able to demonstrate the widespread occurrence of *E. mivati* in European poultry. Two techniques were used, one direct and the other indirect. For the direct method, 10 oocysts with dimensions within the range previously described for *E. mivati* were recovered from each fecal sample of interest, propagated, and used to immunize a group of chickens. These chickens were then challenged with a pure culture of *E. mivati*, as were previously nonimmunized chickens. For the indirect immunity challenge method, 10 chickens were removed from a commercial flock, subdivided into two groups of five, and either challenged with a pure culture of *E. mivati* or held as uninfected controls. Birds that gained weight after challenge were tentatively listed as immune, while those that lost weight or showed the presence of oocysts at postmortem were listed as susceptible.

A similar technique has recently been used by Karlsson and Reid (1977) to show the widespread distribution of *E. acervulina*, *E. brunetti*, *E. necatrix*, and *E. tenella* in the poultry flocks of northeast Georgia (U.S.A.).

Species of *Eimeria* from other hosts have been studied less extensively. Clarkson (1959), however, has shown that *Eimeria adenoeides* and *E. meleagridis* from the turkey are immunologically distinct; Becker and Hall (1933) made a similar observation with *E. miyairii* and *E. separata* from the rat.

Enzyme Variation The *Eimeria* are particularly suitable for enzyme electrophoresis. Large numbers of oocysts can be collected from the feces, and the removal of contaminating bacteria facilitates the easy interpretation of zymograms, the patterns of which are entirely of parasite origin.

Initial, independent studies by Rollinson (1975) and by Shirley (1975) indicated the value of enzyme electrophoresis as a taxonomic aid when it was found that most species from the chicken were characterized by enzymes with species-specific electrophoretic mobility profiles. Although some "common" variants were found, subsequent research has differentiated them, and each species from the chicken is now known to be characterized by enzymes with distinct migration profiles (see Shirley and Rollinson, 1979 and Figures 2 and 3). Of the seven species of *Eimeria* from the chicken, *E. brunetti*, for example, possesses variants of GPI and LDH, which show the greatest migration. This species is therefore easily recognized. Other enzymes subsequently investigated by Shirley (1975, 1978b) and Rollinson (1976) include aspartate aminotransferase (ASAT), glucose-6-phosphate dehydrogenase (G6PD), hexokinase (HK), leucine aminopeptidase (LAP), malate dehydrogenase (MDH), 6-phosphogluconate dehydrogenase (6PGD), and PGM. Each variant form of these enzymes has been arbitrarily codified and the extent of variation between the species from the chicken is shown in Table 1.

Characterization of variants of LDH by isoelectric focusing substantiated the results obtained by starch gel electrophoresis (Shirley and Lee, 1977; Figure 4).

Enzyme variants are stable markers and are useful, for example, in determining the purity of cultures passaged in vitro. Moreover, the finding that variants of at least GPI and LDH from *E. tenella* are stable throughout the ontogeny of the parasite (Shirley, 1975) could provide the basis for a study of other genera with an obligate or facultative two-host life cycle.

The objectivity of the enzyme markers has enabled a more critical study of the relationship between *E. acervulina* and *E. mivati*. (It had previously been suggested (Long, 1973a) that *E. mivati* was insufficiently distinct from *E. acervulina* to warrant its status as a species.) The recent findings of Shirley (1979a) showed no electrophoretic homology between pure cultures of these species (with an attendant absence of any cross-protection), and it was considered that *E. mivati* and *E. acervulina* must be regarded as separate species. However, a new problem has arisen which concerns the real possibility that *E. mivati* and *E. mitis* are synonyms. Pure cultures of *E. mivati* obtained by Shirley (1979a) seem not to produce gross lesions in chickens, although the weight gain of infected chickens can be seriously affected (Shirley, 1979b). These features are typical of *E. mitis* (Tyzzer, 1929; Joyner, 1958). Because the morphological appearances of oocysts of *E. mivati* and *E. mitis* are very similar, the relationship between these parasites needs to be investigated further.

Figure 2. Variants of glucose phosphate isomerase revealed by starch gel electrophoresis. 1,7 = *E. brunetti* (H); 2,8 = *E. maxima* (H); 3,9 = *E. tenella* (H); 4,10 = *E. mivati* (H); 5,11 = *E. acervulina* (H); 6,12 = *E. necatrix* (H).

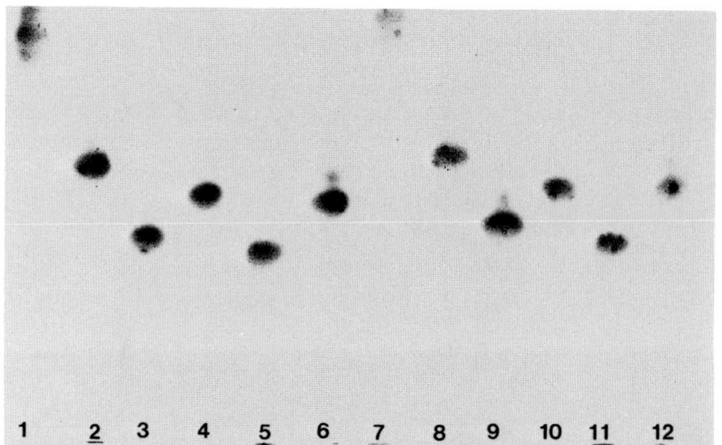

Table 1. Enzyme variation in *Eimeria* species from the chicken

	Enzyme and code number of variant								
Species / Strain	ASAT	G6PD	GPI	HK	LAP	LDH	MDH	6PGD	PGM
E. acervulina									
H[a]	3	3	7	5	3	2		5	2
W[b]	3	3	8	5	3	2		5	2
E. brunetti									
H			6			4			4
W	4	4	6	3	2	4		4	
E. maxima									
H	5	2	4	1		3		2	5
(*indentata*)		2	4	1		3			
E. mivati									
H[c]		3	2			1			3
MA		3	2			1			3
(Sharpsbridge)			2,10			1			
USA[c]			2,10			1			
(*diminuta*)		6	2			6			1
E. necatrix									
H			3			7	2		6
W	2	7		4	5	7		1	
E. praecox									
H		5	5			5			7
E. tenella									
H	1	1	1	2	6	8	1	3	8
W	1	1	9	2	4	8	1	3	9
Elberfeld			1			8			
TA		1	1			8	1		8

[a]H = Houghton strain.
[b]W = Weybridge strain.
[c]A pure culture obtained by a single passage in embryonated eggs.

Reprinted by permission from: Shirley, M. W., and Rollinson, D. Coccidia: The recognition and characterization of populations of *Eimeria*. In A. E. R. Taylor and R. Muller (eds.), Problems in the Identification of Parasites and Their Vectors. Symposia of the British Society for Parasitology. Oxford. Blackwell Scientific Publications, Ltd. Vol. 17, pp. 7–30 (1979).

Species of *Eimeria* from other hosts have also been examined by enzyme electrophoresis, albeit less comprehensively. Long and his colleagues studied two strains each of *E. meleagrimitis* (Long et al., 1977) and *E. dispersa* (Long and Millard, 1979a) from the turkey, and characterized GPI and LDH.

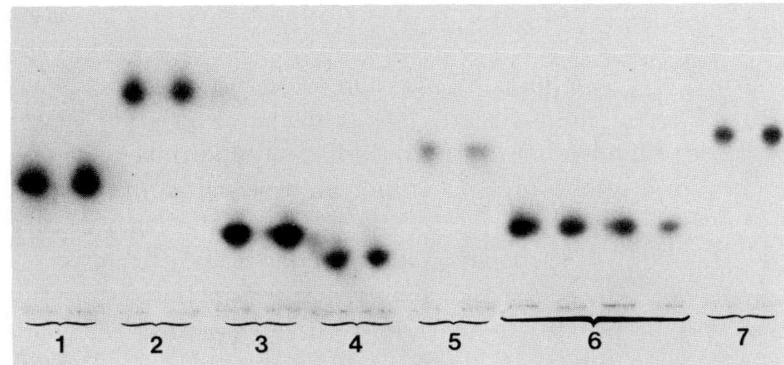

Figure 3. Variants of lactate dehydrogenase revealed by starch gel electrophoresis. 1 = *E. necatrix* (H); 2 = *E. brunetti* (H); 3 = *E. maxima* (H); 4 = *E. praecox* (H); 5 = *E. mivati* (H); 6 = *E. acervulina* (H); 7 = *E. tenella* (H).

Electrophoretic forms of these enzymes from *E. meleagrimitis* representative of a Yorkshire (U.K.) strain and a reference laboratory strain from Weybridge (Surrey, U.K.) differed. Contrarily, the Briston (U.K.) and U.S.A. strains of *E. dispersa* were indistinguishable. Rollinson (1977) confirmed the identity of a coccidium from the quail as a strain of *Eimeria bateri* by comparing its variants of ASAT, G6PD, GPI, HK, LAP, 6PGD, and PGM with those from the reference Weybridge strain.

Rollinson (1976, 1977) also characterized some species from the rabbit and sheep (Table 2). Different species were again characterized by unique enzyme variants. (The common variant of G6PD in *Eimeria ninakohlyakimovae* and *Eimeria ovina* requires further investigation in different buffer conditions.)

DNA Buoyant Density Characterization of coccidial DNA by buoyant density centrifugation in cesium chloride has been little investigated for use as a taxonomic aid. Lee and Fernando (1977) characterized DNA from unsporulated oocysts of *E. maxima*, *E. maxima* (*indentata*), and *E. tenella*. Differences were found between the two species but not between the strains of *E. maxima*. DNA from *E. maxima* and *E. maxima* (*indentata*) showed peaks of 1.707 and 1.682 g/cm^3 compared to a single peak of 1.711 g/cm^3 for *E. tenella*. These data were interpreted to confirm previous taxonomic studies; however, it is still necessary to demonstrate that all species of *Eimeria* from the chicken are characterized by DNA with species-specific buoyant density values. This information has not been reported, although the independent studies by Rollinson (unpublished) and Long and Newton (unpublished) have indicated that two avian species (*E. tenella* and *E. acervulina*) and *Eimeria coecicola* from the rabbit are all

Table 2. Enzyme variation in *Eimeria* species from the rabbit and sheep

Host	Parasite	Enzyme and code number of variant						
		LAP	G6PD	LDH	GPI	ASAT	HK	PGM
Rabbit								
	E. coecicola (W)[a]	2	2	3	1	2	2	
	E. intestinalis (W)	1	1	5	3	4		
	E. magna (W)	4	2	2	4	1	3	
	E. stiedai (W)	3	3	4	2	3	1	2
	E. irresidua (W)			1	5		4	1
Sheep								
	E. ninakohlyakimovae (W)		1	3	1	3		
	E. ovina (W)		1	1	2	1		
	E. weybridgensis (W)		2	2	3	2		

[a]W = Weybridge strain.

Reprinted by permission from: Shirley, M. W., and Rollinson, D. Coccidia: The recognition and characterization of populations of *Eimeria*. In A. E. R. Taylor and R. Muller (eds.), Problems in the Identification of Parasites and Their Vectors. Symposia of the British Society for Parasitology. Oxford. Blackwell Scientific Publications, Ltd. Vol. 17, pp. 7–30 (1979).

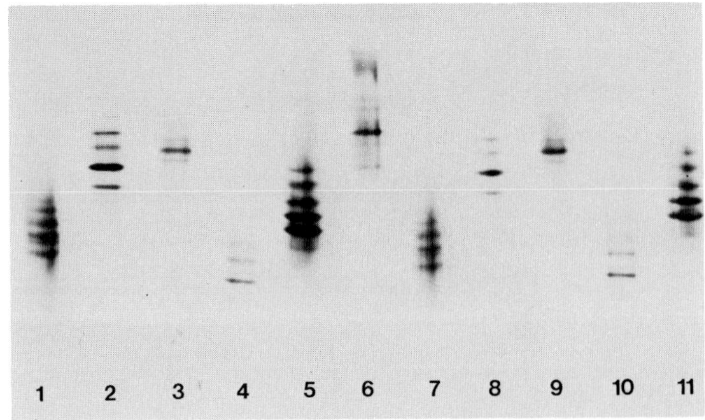

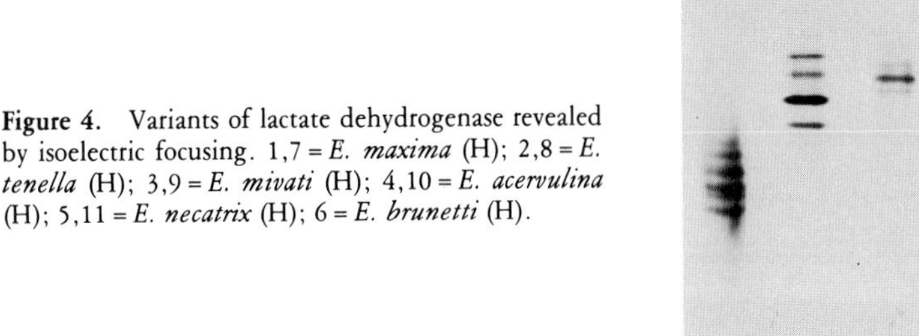
Figure 4. Variants of lactate dehydrogenase revealed by isoelectric focusing. 1,7 = *E. maxima* (H); 2,8 = *E. tenella* (H); 3,9 = *E. mivati* (H); 4,10 = *E. acervulina* (H); 5,11 = *E. necatrix* (H); 6 = *E. brunetti* (H).

characterized by DNA with a buoyant density of 1.711 ± 0.001 g/cm^3. The latter findings would seem to reduce the value of DNA buoyant density centrifugation as a taxonomic aid for the identification of species. However, it is possible that differences between genera may be more amenable to analysis by this technique.

Other Genera

Sarcocystis There remains much taxonomic confusion within the genus (see Levine, 1977; Levine, Chapter 1, this volume), and the relationship between *Sarcocystis* and *Frenkelia* has yet to be resolved. Cerná et al. (1978) recently suggested that sarcosporidians forming cysts in muscles belong to the subgenus *Sarcocystis* (i.e., *Sarcocystis* (*Sarcocystis*)) while sarcosporidians forming cysts in the brain belong to the subgenus *Frenkelia* (i.e., *Sarcocystis* (*Frenkelia*)).

Identification of different species of *Sarcocystis* relies almost entirely on traditional biosystematic approaches. Characters usually considered include the host involved, preferred site of development, duration of asexual stages in the intermediate host, length of patent period in the definitive host, timing of different stages, and morphology of parasites including ultrastructure of sarcocysts and microscopic appearance of sporocysts.

Some of the biological characters of species of *Sarcocystis* from domestic animals are shown in Table 3 (data compiled from Rommel and Geisel (1975), Dubey et al. (1977), and Levine (1977)).

Atkinson (1977) prepared zymograms of LDH, 6PGD, and PGM from three types of sarcocyst in bovine heart and diaphragm. They observed enzyme variants characteristic for each sarcocyst type. The specific identity of the sarcocysts was not known, but possibly correspond to *Sarcocystis cruzi*, *S. hirsuta*, and *S. hominis*. Clearly, the zymogram technique has much to offer taxonomic studies of *Sarcocystis* and, given the invariance of eimerian enzymes during development, an attempt to correlate enzyme types in sarcocysts and sporocysts could help to resolve some current problems in our knowledge of the life cycle of *Sarcocystis*.

Besnoitia Six species are recognized: *Besnoitia benetti*, *B. besnoiti*, *B. darlingi*, *B. jellisoni*, *B. tarandi*, and *B. wallacei*. They are differentiated principally on differences in intermediate hosts (Table 4), morphology, serology, and immunity. (The final host (cat) is known definitively for only *B. besnoiti* and *B. wallacei*.)

A study of the endogenous development of *B. wallacei* by Frenkel (1977) showed that this parasite, in contrast to *B. besnoiti*, *B. darlingi*, and *B. jellisoni* had only a limited period of tachyzoite multiplication. Frenkel (1977) also showed that homologous immunity developed in the two species, together with a slight cross-immunity in both directions.

Isospora Cross-transmission studies and morphological features both of oocysts and endogenous stages are still extensively relied upon for the identification and recognition of species. For example, until recently, *Isospora canis*, *I. felis*, and *I. rivolta* were recognized as the

3

Table 3. Characteristics of Sarcocystis from domestic animals

Species	Hosts: Prey	Hosts: Predator	Prepatent period (days)	Mean sporocyst size (μm)	Pathogenicity for prey host
S. cruzi	Cattle	Dog	9–10	16.3 × 10.8	Highly pathogenic
S. hirsuta	Cattle	Cat	7–9	12.5 × 7.8	Not or slightly pathogenic
S. hominis	Cattle	Man	9–10	14.7 × 9.3	Not or only slightly pathogenic
S. ovicanis	Sheep	Dog	8–9	14.8 × 9.9	Highly pathogenic
S. tenella	Sheep	Cat	11–14	12.4 × 8.1	Not pathogenic
S. miescheriana	Pig	Dog	9–10	12.6 × 9.6	Not known (presumed to be slight)
S. porcifelis	Pig	Cat	5–10	13–14 × 7–8	Pathogenic
S. bertrami	Horse	Dog		15.2 × 10.0	Thought not to be pathogenic
S. fayeri	Horse	Dog	12–27	12.0 × 7.9	Not known

Table 4. Characteristics of species of Besnoitia

Species	Geographic distribution	Hosts (natural)	Hosts (experimental)	Location of cysts
B. bennetti	Africa, Europe, Mexico	Horse, burro		Cutis, mucous membrane
B. besnoitia	Europe, Africa, China, Venezuela, U.S.S.R.	Cattle, goat, wildebeest, impala, kudu	Rabbit, sheep, goats, gerbils, hamsters, cell cultures	Cutis, subcutis, connective tissue, mucosal, serosal
B. darlingi	Panama	Lizards, opossum	Mice, marmoset, free-tailed bats	Myocardium, connective tissue
B. jellisoni	Peru, U.S.	Deer mice, kangaroo rats, opossums	White mice, rats, hamsters, guinea pigs, chicken embryos, cell cultures	Connective tissue, serosal
B. tarandi	Alaska	Reindeer, caribou		Fibrous connective tissue
B. wallacei	Hawaii	?	Rats, mice	Connective tissue

only species of *Isospora* from the cat and dog, with the latter species able to develop in both hosts. Dubey (1975a,b), however, confirmed the work of others including Mahrt (1966) and Dubey et al. (1970), in showing that oocysts of *I. rivolta* from cats and dogs, although morphologically similar, were in fact different species. The name *I. rivolta* was retained for the parasite from cats and the name *Isospora ohioensis* was introduced for the parasite from dogs. Dubey and Mahrt (1978) subsequently compared the endogenous development of the latter species with that of another "*I. rivolta*-like" species from the dog (Mahrt, 1967). Certain differences were found, and both parasites were considered to warrant separate status as species. That originally described by Mahrt (1966, 1967) as *I. rivolta* from dogs, but which did not develop in cats, was termed *Isospora neorivolta*. (No oocysts of this species are available for comparison with any other species.) Unfortunately, this rather limited approach to taxonomy cannot be regarded as definitive and, moreover, often relies upon the interpretation of data derived from relatively few observations.

Infraspecific Variation

Eimeria

Developmental Differences Alteration of the developmental rate of *E. tenella* following selection for precociousness was discussed above. The increase in the developmental rate of the selected strain (Wis-F) was accompanied by a loss of pathogenicity, which persisted as a stable trait through 25 generations of relaxed selection (Jeffers, 1975). The Wis-F strain and other attenuated strains derived from it elicit a protective immune response against challenge with pathogenic strains of *E. tenella* under both laboratory and simulated field conditions, thus making them potentially useful as vaccines (Jeffers, 1974b, 1976a; Johnson et al., 1979).

Serial passage in embryonated eggs of the Houghton strains of *E. mivati* and *E. tenella* enabled Long (1972a,b) to develop embryo-adapted lines of these species. Adaptation, particularly with *E. tenella*, was a gradual process requiring many consecutive passages in eggs before large numbers of oocysts were produced. Although Long (1972b) suggested that adaptation could have resulted from a selection of schizonts or merozoites capable of producing gametocytes and oocysts in the CAM, the mechanism would seem to be more complex in view of the number of passages required for complete adaptation.

Continued maintenance of *E. mivati* and *E. tenella* in embryonated eggs produced a further change in these lines; both were subsequently found to be markedly less pathogenic for chickens than their respective parent strains (Long, 1972a,b, 1973b, 1974b). The loss of pathogenicity is manifest through poorer development in the "normal" host (see also Shirley, 1979b), although with the embryo-adapted strain of *E. tenella* small second generation schizonts (whose growth is confined to within superficial epithelial cells) have replaced the larger forms so characteristic of the species (Long, 1973b). After 41 passages of *E. tenella* (H) in eggs, the loss of pathogenicity was substantially regained after one passage in chickens (Long, 1972a). However, a culture passaged 62 times in eggs followed by nine passages in chickens was still characterized by reduced pathogenicity (Long, 1974b).

Antigenicity has not been lost (Long, 1972a,b; Shirley, 1979b), and both embryo-adapted lines have been investigated for possible use as vaccines (Long and Millard, 1977, 1979b).

Immunological Variation Infraspecific immunological variation in the coccidia was first reported by Joyner (1969). He found that although the Houghton and Weybridge strains of *E. acervulina* were similar in several biological properties, complete immunity against one imparted only partial immunity to the other. Subsequently, immunological variation was reported among several laboratory and randomly chosen field strains of both *E. acervulina* and *E. maxima*, suggesting that extensive antigenic variation exists within each of these species (Long, 1974a; Norton and Hein, 1976; Jeffers, 1978). However, in most cases, the antigenic differences found were too small to be considered important limitations in the use of immunological techniques for the control of coccidiosis. When significant infraspecific immunological variation does exist, as has been found among certain strains of *E. maxima* by Long and Millard (1979c), adequate protection against both homologous and heterologous challenge can nevertheless be obtained if the oocysts used for immunization are obtained from the propagation of a mixture of several strains. Such an oocyst culture presumably contains both parental and hybrid progeny of the immunologically variable strains.

Pathogenicity Infraspecific variation in pathogenicity is difficult to measure precisely because this trait is so strongly influenced by environmental factors such as the infective dose, the age of the inoculum, and the age, immunological status, and genetic constitution of the host (Long, 1970a). Establishing the parameters that will most accurately measure differences in pathogenicity is also difficult. For example, while comparing pathogenicity and oocyst production in vivo and infectivity and oocyst production in vitro of the Beltsville, Weybridge, and Wisconsin strains of *E. tenella*, Doran et al. (1974) found that the ranking of the strains varied according to which parameter of measurement was considered. In comparisons in vivo, the Weybridge strain caused the highest mortality, while the Wisconsin strain caused the greatest decrease in weight gain. The three strains produced cecal lesions of equivalent severity. In vitro, the Wisconsin strain exhibited the highest percentage of infectivity, while the Beltsville strain produced the most oocysts.

A series of studies by McLoughlin and co-workers offer a unique opportunity to observe the pathogenicity of a strain of *E. tenella* over a long period of propagation under standardized laboratory conditions. This drug-sensitive strain provided a common base population from which they derived a number of drug-resistant strains in independent studies extending over a period of more than 12 yr. The published reports of several of these experiments provide observations on the pathogenicity of at least 89 different generations of the sensitive (parent) strain, based upon anticoccidial indices (McManus et al., 1968) obtained after inoculation of nonmedicated birds of similar age and genetic background with 10^5 oocysts. Despite the presumed maintenance of the integrity of this strain and a standardization of testing procedures, cultures derived from different generations of propagation were markedly different in their apparent pathogenicity (Table 5). The anticoccidial indices obtained ranged from 4–139 and the coefficient of variation among anticoccidial indices obtained for different generations of propagation within each study ranged from 14–58%. These observations suggest that although infraspecific variation in pathogenicity has been reported in *E. acervulina* (Jeffers, 1978), *E. maxima* (Norton and Hein, 1976), and *E. tenella* (Joyner and Norton, 1969; Shumard and Callender, 1970), the observed variation may be applicable only to a point in time in that the ranking of various strains in terms of pathogenicity may vary greatly from one generation to another. However, strains having reduced pathogenicity specifically associated with differences in endogenous development (see above) exhibit relatively stable responses during successive generations of propagation.

Toxoplasma Strains of *T. gondii* isolated from different naturally infected animals also differ in pathogenicity (Jacobs, 1974; Beverley, 1976). Again, critical characterization of genetic contributions to these differences has been difficult because of the important contribution of environmental variation to differences in virulence. For example, rapid propagation of the proliferative form of *T. gondii* in the mouse usually results in increased pathogenicity.

Other Characters

Enzyme Variation As yet, no major assessment of enzyme variation within the *Eimeria* has been undertaken, and the full extent of infraspecific variation is not known. However, as indicated above, some strain variation has been observed, particularly in those species from the chicken that have been studied more extensively.

Enzyme variation with *E. mivati* is well documented. Shirley et al. (1977) and Shirley (1978a) observed different variants of LDH and PGM from *E. mivati* (H) and *E. mivati* (*diminuta*); *E. mivati* (Sharpsbridge) was found to possess two variants of GPI, designated GPI-2 and GPI-10 (Shirley, 1979a). (The latter strain is the progeny of several oocysts and is not therefore a "pure" strain derived from a single oocyst.) All other strains of *E. mivati* so far examined are characterized by GPI-2, except a recently isolated strain which possesses GPI-10 (Shirley and Jeffers, unpublished). Rollinson et al. (1979) observed different variants of PGM in *E. maxima* (W) and *E. maxima* (*indentata*). Shirley and Rollinson (1979) also reported variation in the same enzyme in three of 13 isolates and strains of *E. maxima*. A total of 17 isolates

Table 5. Variation in the pathogenicity of different generations of the Beltsville strain of *E. tenella*[a]

Reference	Number of generations measured	Average anticoccidial index ± S.E.	Range of anticoccidial indices	Coefficient of variation (%)
McLoughlin and Gardiner (1967)	7	61 ± 13.3	15–116	58
McLoughlin and Gardiner (1968)	15	73 ± 10.2	4–129	54
McLoughlin and Chute (1968)	11	75 ± 11.3	4–120	50
McLoughlin (1970a)	5	88 ± 7.4	68–102	19
McLoughlin and Chute (1971)	5	102 ± 6.5	79–119	14
McLoughlin and Chute (1973a)	7	111 ± 7.6	84–132	18
McLoughlin and Chute (1973b)	5	73 ± 15.5	32–117	48
McLoughlin and Chute (1973c)	10	73 ± 4.4	56–95	19
McLoughlin and Chute (1974b)	4	68 ± 6.3	55–83	19
McLoughlin and Chute (1978)	9	79 ± 8.6	33–113	33
McLoughlin and Chute (1979)	11	81 ± 13.3	7–139	55

[a] Pathogenicity was measured by anticoccidial indices (McManus et al., 1968) obtained after inoculation of non-medicated 3-week-old chickens with 10^5 oocysts per bird.

and three strains of *E. maxima* examined by the latter authors showed no polymorphism in the enzymes GPI, LDH, and G6PD.

Fewer strains of the other avian species have been examined, although different variants of GPI have been found in the Houghton and Weybridge strains of both *E. acervulina* and *E. tenella* (Rollinson, 1975; Shirley, 1975).

Morphological Differences Norton and Joyner (1978) passaged the Weybridge strain of *E. maxima* in the presence of increasing levels of combinations of the anticoccidial drugs clopidol and methylbenzoquate and observed the appearance of oocysts containing two, three, and four sporocysts. Repeated passage of those oocysts with two sporocytes—each of which contained four sporozoites—raised their presence in the resulting yields of oocysts to about 80% after 10 to 14 passages. (Some normal oocysts were still produced.) The development of this line of *E. maxima* is of considerable interest because disporous, octozoic oocysts and a monoxenous life cycle are characteristics of the genus *Isospora*.

Electrophoresis of Whole Organism Proteins Bloomfield and Remington (1970) attempted to define genetic differences between three strains of *Toxoplasma* from comparisons of mobility profiles of proteins separated by electrophoresis in polyacrylamide gels. Lyophilized extracts of the parasites (and contaminating red blood cells) were used, but no differences in either the electrophoretic mobility of the proteins or the number of bands were found.

CONCLUDING REMARKS

A review of studies on the genetic basis for host response to the coccidia reveals that measurable variation in host susceptibility exists even in the absence of intentional selection. The effect of selection in exploiting this variation to further enhance host resistance has been demonstrated many times. The response of progeny from reciprocal crosses of host stocks having differential susceptibility to coccidia suggests that host resistance is conditioned by multiple genetic factors, lacking dominance and acting in an additive manner. Sex linkage, maternal effects, or cytoplasmic inheritance were not identified as important factors in determining resistance or susceptibility to coccidiosis. Nevertheless, the response of line-cross progeny from unselected inbred host lines suggests that heterosis may significantly influence the resistance to infection of progeny from specific line crosses. Despite the apparent importance of genetic variation in determining the host response to coccidia, we cannot suggest that exploitation of this variation through selection for increased host resistance is a practical way to control coccidiosis, or that it should supplant other methods (i.e., chemotherapeutic and immunological control measures) now shown to be effective.

Genetic studies on the parasite are still at a stage of infancy. Four genetic markers, that is drug resistance, developmental rate, enzyme variation, and temperature sensitivity, have been clearly delineated in the coccidia. Three of these markers have been useful in genetic recombination experiments, the results of which demonstrate that traits that may be used to definitively differentiate strains of coccidia are amenable to conventional genetic analysis. Although infraspecific variation has also been found in the pathogenicity, oocyst morphology, developmental characteristics, and antigenic make-up of coccidia, many basic questions regarding the genetic nature of this variation remain unanswered. The use of more refined methods for the detection and quantification of genetic variation should be useful in future studies. For example, isoallelic analyses based upon the electrophoretic mobility of specific enzymes through starch or acrylamide gel allows evaluation of genetic changes at single loci, a refinement much more difficult to achieve using other criteria. The applicability of these electrophoretic techniques for the identification and classification of different species and strains within species of coccidia has been confirmed in several recent studies. The continued application of genetic principles to the study of variation in the coccidia will contribute greatly to our knowledge of the biology of these organisms and may eventually lead to the development of more effective control measures.

3

LITERATURE CITED

Araujo, F. G., Williams, D. M., Grumet, F. C., and Remington, J. S. 1976. Strain-dependent differences in murine susceptibility to *Toxoplasma*. Infect. Immun. 13:1528–1530.

Atkinson, E. M. 1977. The use of electrophoresis in the identification of sarcocystis. Abstr. 421. Fifth International Congress of Protozoology, New York.

Becker, E. R., and Hall, P. R. 1933. Cross immunity and correlation of oocyst production during immunization between *E. miyairii* and *E. separata* in the rat. Am. J. Hyg. 18:220–223.

Becker, E. R., Jessen, R. J., Pattillo, W. H., and van Doorninck, W. M. 1956. A biometrical study of the oocyst of *Eimeria necatrix*, a parasite of the common fowl. J. Protozool. 3:126–131.

Becker, E. E., Zimmermann, W. J., and Pattillo, W. 1955. A biometrical study of the oocyst of *Eimeria brunetti*, a parasite of the common fowl. J. Protozool. 2:145–150.

Benacerraf, B., and McDevitt, H. O. 1972. Histocompatibility-linked immune response genes. Science 175:273–279.

Beverley, J. K. A. 1976. Toxoplasmosis in animals. Vet. Rec. 99:123–127.

Bloomfield, M. M., and Remington, J. S. 1970. Comparison of three strains of *Toxoplasma gondii* by polyacrylamide-gel electrophoresis. Trop. Geogr. Med. 22:367–370.

Buvanendran, V., and Kulasegaram, P. 1972. Resistance of breeds and breed crosses of chickens to experimental *Eimeria necatrix* infection. Br. Vet. J. 128:177–183.

Canning, E. U. 1962. Sexual differentiation of merozoites of *Barrouxia schneider*: (Butschli). Nature 195:720–721.

Canning, E. U. 1963. The use of histochemistry in the study of sexuality in the coccidia with particular reference to the Adeleidae. Progress in Protozoology, pp. 439–442. Academic Press, New York.

Canning, E. U., and Anwar, M. 1968. Studies on meiotic division in coccidial and malarial parasites. J. Protozool. 15:290–298.

Canning, E. U., and Morgan, K. 1975. DNA synthesis, reduction and elimination during life cycles of the eimeriine coccidian, *Eimeria tenella* and the hemogregarine, *Hepatozoon domerguei*. Exp. Parasitol. 38:217–227.

Carson, J. R. 1951. Exposure to disease agents of strains of chickens differing in resistance to leucosis. Poult. Sci. 30:213–230.

Catchpole, J., Norton, C. C., and Joyner, L. P. 1975. The occurrence of *Eimeria weybridgensis* and other species of coccidia in lambs of England and Wales. Br. Vet. J. 131:392–401.

Černá, Z., Kolarova, I., and Sulc, P. 1978. *Sarcocystis cernae* Levine, 1977, excystation, life-cycle and comparison with other heteroxenous coccidians from rodents and birds. Folia Parasitol. 25:291–297.

Challey, J. R. 1966. Changes in adrenal constituents and their relationship to corticosterone secretion in chickens selected for genetic resistance and susceptibility to cecal coccidiosis. J. Parasitol. 52:967–974.

Challey, J. R., Jeffers, T. K., and McGibbon, W. H. 1968. The response of commercial egg production stocks to experimental infection with *Eimeria tenella* and its relation to field mortality. Poult. Sci. 47:1197–1204.

Champion, L. R. 1951a. The inheritance of resistance to cecal coccidiosis in the domestic fowl

(*Gallus domesticus*). Ph.D. thesis, University of Wisconsin, Madison.

Champion, L. R. 1951b. The inheritance of resistance to *Eimeria tenella* in the domestic fowl. Poult. Sci. 31:911.

Champion, L. R. 1954. The inheritance of resistance to cecal coccidiosis in the domestic fowl. Poult. Sci. 33:670–681.

Chapman, D. 1978. Drug resistance in coccidia. In: P. L. Long, K. N. Boorman, and B. M. Freeman (eds.), Avian Coccidiosis, pp. 387–412. British Poultry Science, Ltd., Edinburgh.

Clarkson, M. J. 1959. The life history and pathogenicity of *Eimeria meleagridis* Tyzzer, 1927, in the turkey poult. Parasitology 49:519–528.

Dobell, C., and Jameson, A. P. 1915. The chromosome cycle in coccidia and gregarines. Proc. R. Soc. Lond. (Biol.) 89:83–94.

Doran, D. J., Vetterling, J. M., and Augustine, P. C. 1974. *Eimeria tenella*: An *in vivo* and *in vitro* comparison of the Wisconsin, Weybridge, and Beltsville strains. Proc. Helminth. Soc. Wash. 41:77–80.

Dubey, J. P. 1975a. Experimental *Isospora canis* and *Isospora felis* infection in mice, cats, and dogs. J. Protozool. 22:416–417.

Dubey, J. P. 1975b. *Isospora ohioensis* sp.n. proposed for *I. rivolta* of the dog. J. Parasitol. 61:462–465.

Dubey, J. P., and Mahrt, J. L. 1978. *Isospora neorivolta* sp.n. from the domestic dog. J. Parasitol. 64:1067–1073.

Dubey, J. P., Miller, N. L., and Frenkel, J. K. 1970. The *Toxoplasma gondii* oocysts from cat feces. J. Exp. Med. 132:636–662.

Dubey, J. P., Streitel, R. H., Stromberg, P. C., and Toussant, M. J. 1977. *Sarcocystis fayeri* sp.n. from the horse. J. Parasitol. 63:443–447.

Duszynski, D. W. 1971. Increase in size of *Eimeria separata* oocysts during patency. J. Parasitol. 57:948–952.

Duszynski, D. W., Solangi, M. A., and Overstreet, R. M. 1979. A new and unusual Eimerian (Protozoa: Eimeriidae) from the liver of the gulf killifish, *Fundulus grandis*. J. Wildl. Dis. 15:543–552.

Edgar, S. A., King, D. F., and Johnson, L. W. 1951. Control of avian coccidiosis through breeding or immunization. Poult. Sci. 30:911.

Edgar, S. A., and Seibold, C. T. 1964. A new coccidium of chickens, *Eimeria mivati* sp.n. (Protozoa: Eimeriidae) with details of its life history. J. Parasitol. 50:193–204.

Fernando, M. A., and Remmler, O. 1973a. *Eimeria diminuta* sp.n. from the Ceylon jungle fowl, *Gallus lafayettei*. J. Protozool. 20:357.

Fernando, M. A., and Remmler, O. 1973b. Four new species of *Eimeria* and one of *Tyzzeria* from the Ceylon jungle fowl *Gallus lafayettei*. J. Protozool. 20:43–45.

Fish, F. 1931. Quantitative and statistical analyses of infections with *Eimeria tenella* in the chicken. *Am. J. Hyg.* 14:560–576.

Fredeen, H. T. 1965. Genetic aspects of disease resistance. Anim. Breed. Abstr. 33:17–26.

Frenkel, J. K. 1977. *Besnoitia wallacei* of cats and rodents: With a reclassification of other cyst-forming isoporoid coccidia. J. Parasitol. 63:611–628.

Gavora, J. S., and Spencer, J. L. 1978. Breeding for genetic resistance to disease: Specific or general? World's Poult. Sci. J. 34:137–148.

Gowen, J. W. 1937. Contributions of genetics to understanding of animal diseases. J. Hered. 28:233–240.

Grant, V. 1971. Plant Speciation. Columbia University Press, New York.

Haberkorn, A. 1970. Die Entwicklung von *Eimeria falciformis* (Eimer 1870) in der weiben Maus (*Mus musculus*). Z. Parasitenkd. 34:49–67.

Hammond, D. M. 1973. Life cycles and development of coccidia. In: D. M. Hammond and P. L. Long (eds.), The Coccidia: *Eimeria*, *Isospora*, *Toxoplasma*, and Related Genera, pp. 45–79. University Park Press, Baltimore.

Hammond, D. M., Clark, W. N ., and Miner, M. L. 1961. Endogenous phase of the life cycle of *Eimeria auburnensis* in calves. J. Parasitol.

47:591–596.
Herrick, C. A. 1933–1934. The development of resistance to the protozoan parasite, *Eimeria tenella*, by the chicken. J. Parasitol. 20:329–330.
Hipke, H., and Hanson, E. D. 1974. Induction of temperature sensitivity in *Paramecium aurelia* by nitrosoguanidine. J. Protozool. 21:349–352.
Hutt, F. B. 1949. Genetics of the Fowl. McGraw-Hill, New York.
Hutt, F. B. 1958. Genetic Resistance to Disease in Domestic Animals. Comstock Publishing Associates, Ithaca, N.Y.
Hutt, F. B. 1964. Animal Genetics. Ronald Press, New York.
Jacobs, L. 1974. *Toxoplasma gondii*: Parasitology and transmission. Bull. N.Y. Acad. Med. 50:128–145.
Jeffers, T. K. 1968. Studies on genetic resistance to *Eimeria tenella* infection in the domestic fowl (*Gallus domesticus*). Ph.D. thesis, University of Wisconsin, Madison.
Jeffers, T. K. 1974a. Genetic transfer of anticoccidial drug resistance in *Eimeria tenella*. J. Parasitol. 60:900–904.
Jeffers, T. K. 1974b. Immunization against *Eimeria tenella* using an attenuated strain. Fifteenth World's Poultry Congress Proceedings, New Orleans. pp. 105–107.
Jeffers, T. K. 1975. Attenuation of *Eimeria tenella* through selection for precociousness. J. Parasitol. 61:1083–1090.
Jeffers, T. K. 1976a. Reduction of anticoccidial drug resistance by massive introduction of drug-sensitive coccidia. Avian Dis. 20: 649–653.
Jeffers, T. K. 1976b. Genetic recombination of precociousness and anticoccidial drug resistance in *Eimeria tenella*. Z. Parasitenkd. 50: 251–255.
Jeffers, T. K. 1978. Genetics of coccidia and the host response. In: P. L. Long, K. N. Boorman, and B. M. Freeman (eds.), Avian Coccidiosis, pp. 51–125. British Poultry Science Ltd., Edinburgh.
Jeffers, T. K., Challey, J. R., and McGibbon, W. H. 1969. Response to *Eimeria tenella* infection among lines of fowl selected for differential resistance to avian leukosis. Poult. Sci. 48:1604–1607.
Jeffers, T. K., Challey, J. R., and McGibbon, W. H. 1970. Response of several lines of fowl and their single cross progeny to experimental infection with *Eimeria tenella*. Avian Dis. 14:203–210.
Jeffers, T. K., and Wagenbach, G. E. 1969. Sex differences in embryonic response to *Eimeria tenella* infection. J. Parasitol. 55:949–951.
Jeffers, T. K., and Wagenbach, G. E. 1970. Embryonic response to *Eimeria tenella* infection. J. Parasitol. 56:656–662.
Johnson, J., Reid, W. M., and Jeffers, T. K. 1979. Practical immunization of chickens against coccidiosis using an attenuated strain of *Eimeria tenella*. Poult. Sci. 58:37–41.
Johnson, W. T. 1927. Immunity or resistance of the chicken to coccidial infection, p. 29. Oregon Agricultural Experiment Station Bulletin No. 230.
Johnson, W. T. 1930. Directors Biennial Report, 1928–1930, pp. 119–120. Oregon Agricultural Experiment Station.
Johnson, W. T. 1938. Coccidiosis of the chicken with special reference to the species, p. 5. Oregon Agricultural Experiment Station Bulletin No. 358.
Jones, E. E. 1932. Size as a species characteristic in coccidia: Variation under diverse conditions of infection. Arch. Protistenk. 76:130–170.
Joyner, L. P. 1958. Experimental *Eimeria mitis* infections in chickens. Parasitology 48:101–112.
Joyner, L. P. 1969. Immunological variation between two strains of *Eimeria acervulina*. Parasitology 59:725–732.
Joyner, L. P. 1970. Coccidiosis: Problems arising from the development of anticoccidial drug resistance. Exp. Parasitol. 28:122–128.
Joyner, L. P., Canning, E. U., Long, P. L., Rollinson, D., and Williams, R. B. 1978. A

suggested terminology for populations of coccidia (Eimeriorina), particularly of the genus *Eimeria* (Protozoa: Apicomplexa). Parasitology 77:27–31.

Joyner, L. P., and Norton, C. C. 1969. A comparison of two laboratory strains of *Eimeria tenella.* Parasitology 59:907–913.

Joyner, L. P., and Norton, C. C. 1975. Transferred drug resistance in *Eimeria maxima.* Parasitology 71:385–392.

Joyner, L. P., and Norton, C. C. 1976. Further observations on transferred drug-resistance in *Eimeria maxima.* Parasitology Part 2. 73:iii.

Joyner, L. P., and Norton, C. C. 1977. Further observations on the genetic transfer of drug resistance in *Eimeria maxima.* Parasitology 74:205–213.

Joyner, L. P., and Norton, C. C. 1978. The activity of methylbenzoquate and clopidol against *Eimeria maxima*: Synergy and drug resistance. Parasitology 76:369–377.

Kamei, K., Sato, K., and Tsunematsu, Y. 1976. A strain of mouse highly susceptible to *Toxoplasma.* J. Parasitol. 62:714.

Karlsson, T., and Reid, W. M. 1977. Prevalence of *Eimeria acervulina*, *Eimeria necatrix*, *Eimeria brunetti* and *Eimeria tenella* in Georgia (USA) as demonstrated by immunity challenge techniques. Br. Poult. Sci. 18:497–501.

Kheysin, Y. M. 1972. Morphophysiological properties of various stages of the life cycle of coccidia of the sub-order Eimeriidae. In: K. S. Todd (ed.), Life Cycles of Coccidia in Domestic Animals, pp. 23–101. University Park Press, Baltimore.

Klesius, P. H., and Hinds, S. E. 1979. Strain-dependent differences in murine susceptibility to coccidia. Infect. Immun. 26:1111–1115.

Klimes, B., and Orel, V. 1969. Investigation of genetic resistance to coccidiosis: I. Selection response of chickens to infection by *Eimeria tenella.* Acta Vet. (Brno) 38:51–57.

Klimes, B., and Orel, V. 1972. The influence of coccidiosis resistant, susceptible and unselected sires on progeny mortality after infection with *Eimeria tenella.* Avian Pathol. 1:65–67.

Klimes, B., Rootes, D. G., and Tanielian, Z. 1972. Sexual differentiation of merozoites of *Eimeria tenella.* Parasitology 65:131–136.

Lee, E-H., and Fernando, M. A. 1977. Characterization of coccidial DNA: An aid in species identification. Z. Parasitenkd. 53:129–131.

Lee, E-H., Remmler, O., and Fernando, M. A. 1977. Sexual differentiation in *Eimeria tenella* (Sporozoa: Coccidia). J. Parasitol. 63:155–156.

Levine, N. D. 1973. Introduction, history, and taxonomy. In: D. M. Hammond and P. L. Long (eds.), The Coccidia: *Eimeria*, *Isospora*, *Toxoplasma*, and Related Genera, pp. 1–22. University Park Press, Baltimore.

Levine, N. D. 1977. Nomenclature of *Sarcocystis* in the ox and sheep and of fecal coccidia of the dog and cat. J. Parasitol. 63:36–51.

Levine, P. P. 1940. The initiation of avian coccidial infection with merozoites. J. Parasitol. 26:337–343.

Levine, P. P. 1942. A new coccidium pathogenic for chickens, *Eimeria brunetti* n. sp. (Protozoa: Eimeriidae). Cornell Vet. 32:430–439.

Liburd, E. M. 1973. *Eimeria nieschulzi* infections in inbred and outbred rats: Infective dose, route of infection, and host resistance. Can. J. Zool. 51:273–279.

Long, P. L. 1959. A study of *Eimeria maxima* Tyzzer, 1929, a coccidium of the fowl (*Gallus gallus*). Ann. Trop. Med. Parasitol. 53: 325–333.

Long, P. L. 1965. Development of *Eimeria ten ella* in avian embryos. Nature 208:509–510.

Long, P. L. 1967. Studies on *Eimeria praecox* Johnson, 1930, in the chicken. Parasitology 57:351–361.

Long, P. L. 1968. The effect of breed of chickens on resistance to *Eimeria* infections. Br. Poult. Sci. 9:71–78.

Long, P. L. 1970a. Anticoccidial drugs: Factors affecting pathogenicity of avian coccidia. Exp. Parasitol. 28:4–10.

Long, P. L. 1970b. *Eimeria tenella*: Chemotherapeutic studies in chick embryos with a description of a new method (chorioallantoic membrane foci counts) for evaluating infections. Z. Parasitenkd. 33:329–338.

Long, P. L. 1972a. *Eimeria tenella*: Reproduction, pathogenicity and immunogenicity of a strain maintained in chick embryos by serial passage. J. Comp. Pathol. 82:429–437.

Long, P. L. 1972b. *Eimeria mivati:* Reproduction, pathogenicity and immunogenicity of a strain maintained in chick embryos by serial passage. J. Comp. Pathol. 82:439–445.

Long, P. L. 1973a. Studies on the relationship between *Eimeria acervulina* and *Eimeria mivati*. Parasitology 67:143–155.

Long, P. L. 1973b. Endogenous stages of a "chick embryo-adapted" strain of *Eimeria tenella*. Parasitology 66:55–62.

Long, P. L. 1973c. Pathology and pathogenicity of coccidial infections. In: D. M. Hammond and P. L. Long (eds.), The Coccidia: *Eimeria*, *Isospora*, *Toxoplasma*, and Related Genera, pp. 253–294. University Park Press, Baltimore.

Long, P. L. 1974a. Experimental infection of chickens with two species of *Eimeria* isolated from the Malaysian jungle fowl. Parasitology 69:337–347.

Long, P. L. 1974b. The growth of *Eimeria* in cultured cells and in chicken embryos: A review. In: Proceedings of the Symposium on Coccidia and Related Organisms, Guelph, Ontario, 1973, pp. 57–82. Murray Kelly Printing, London, Ontario.

Long, P. L., and Millard, B. J. 1977. *Eimeria*: Immunisation of young chickens kept in litter pens. Avian Pathol. 6:77–92.

Long, P. L., and Millard, B. J. 1979a. Studies on *Eimeria dispersa* Tyzzer, 1929 in turkeys. Parasitology 78:41–51.

Long, P. L., and Millard, B. J. 1979b. *Eimeria:* Further studies on the immunisation of young chickens kept in litter pens. Avian Pathol. 8:213–228.

Long, P. L., and Millard, B. J. 1979c. Immunological differences in *Eimeria maxima*: Effect of a mixed immunizing inoculum on heterologous challenge. Parasitology 79:451–457.

Long, P. L., Millard, B. J., and Shirley, M. W. 1977. Strain variation within *Eimeria meleagrimitis* from the turkey. Parasitology 75:177–182.

McDougald, L. R., and Jeffers, T. K. 1976a. *Eimeria tenella* (Sporozoa: Coccidia): Gametogony following a single asexual generation. Science 192:258–259.

McDougald, L. R., and Jeffers, T. K. 1976b. Comparative *in vitro* development of precocious and normal strains of *Eimeria tenella* (Coccidia). J. Protozool. 23:530–534.

McLoughlin, D. K. 1970a. Efficacy of buquinolate against ten strains of *Eimeria tenella* and the development of a resistant strain. Avian Dis. 14:126–130.

McLoughlin, D. K. 1970b. Coccidiosis: Experimental analysis of drug resistance. Exp. Parasitol. 28:129–136.

McLoughlin, D. K., and Chute, M. B. 1968. Drug resistance in *Eimeria tenella*: VII. Acriflavine-mediated loss of resistance to amprolium. J. Parasitol. 54:696–698.

McLoughlin, D. K., and Chute, M. B. 1971. Efficacy of decoquinate against eleven strains of *Eimeria tenella* and development of a decoquinate-resistant strain. Avian Dis. 15:342–345.

McLoughlin, D. K., and Chute, M. B. 1973a. Efficacy of clopidol against twelve strains of *Eimeria tenella*, and the development of a clopidol-resistant strain. Avian Dis. 17: 425–429.

McLoughlin, D K., and Chute, M. B. 1973b. Efficacy of Novastat against twelve strains of *Eimeria tenella* and the development of a Novastat-resistant strain. Avian Dis. 17:582–585.

McLoughlin, D. K., and Chute, M. B. 1973c. Efficacy of nequinate against thirteen strains of *Eimeria tenella* and the development of a nequinate-resistant strain. Avian Dis. 17:717–721.

McLoughlin, D. K., and Chute, M. B. 1974a. Drug resistance in *Eimeria tenella*: XI. Comparative response to amprolium of five isolants derived from a resistant strain. J. Parasitol. 60:835–837.

McLoughlin, D. K., and Chute, M. B. 1974b. The efficacy of monensin against one sensitive and thirteen drug resistant strains of *Eimeria tenella*. Poult. Sci. 53:770–772.

McLoughlin, D. K., and Chute, M. B. 1978. Robenidine resistance in *Eimeria tenella*. J. Parasitol. 64:874–877.

McLoughlin, D. K., and Chute, M. B. 1979. *Eimeria tenella* in chickens: Resistance to a mixture of sulfadimethoxine and ormetoprim (Rofenaid). Proc. Helminth. Soc. Wash. 46:265–269.

McLoughlin, D. K., and Gardiner, J. L. 1967. Drug resistance in *Eimeria tenella*: V. The experimental development of a nicarbazin-resistant strain. J. Parasitol. 53:930–932.

McLoughlin, D. K., and Gardiner, J. L. 1968. Drug resistance in *Eimeria tenella*: VI. The experimental development of an amprolium-resistant strain. J. Parasitol. 54:582–584.

McManus, E. C., Campbell, W. C., and Cuckler, A. C. 1968. Development of resistance to quinoline coccidiostats under field and laboratory conditions. J. Parasitol. 54:1190–1193.

Mahrt, J. L. 1966. Life cycle of *Isospora rivolta* (Grassi, 1879), Wenyon, 1923 in the dog. Ph.D. thesis, University of Illinois, Urbana.

Mahrt, J. L. 1967. Endogenous stages of the life cycle of *Isospora rivolta* in the dog. J. Protozool. 14:754–759.

Mayberry, L. F., and Marquardt, W. C. 1973. Transmission of *Eimeria separata* from the normal host, *Rattus*, to the mouse, *Mus musculus*. J. Parasitol. 59:198–199.

Mayberry, L. F., Plan, B., Nash, D. J., and Marquardt, W. C. 1975. Genetic dependence of coccidial transmission from rat to mouse. J. Protozool. 22(3):28a.

Mayhew, R. L. 1934. Studies on coccidiosis: VIII. Immunity or resistance to infection in chickens. J. Am. Vet. Med. Assoc. 38:729–734.

Mayr, E. 1963. Animal Species and Evolution. Harvard University Press, Cambridge, Mass.

Mayr, E. 1970. Populations, Species and Evolution. Harvard University Press, Cambridge, Mass.

Millen, T. W., Hill, J. F., and Arvidson, R. B. 1959. Inheritance of resistance to *Eimeria acervulina*. Poult. Sci. 38:1229.

Moultrie, F., Edgar, S. A., and King, D. F. 1953–1954. Breeding and immunizing chickens for resistance to coccidiosis: 1. Breeding chickens for resistance to coccidiosis, pp. 44–45. Alabama Agricultural Experiment Station 64th and 65th Annual Reports.

Norton, C. C., and Hein, H. E. 1976. *Eimeria maxima*: A comparison of two laboratory strains with a fresh isolate. Parasitology 72:345–354.

Norton, C. C., and Joyner, L. P. 1978. The appearance of bisporcystic oocysts of *Eimeria maxima* in drug-treated chicks. Parasitology 77:243–248.

Olisa, E. G., Herson, J., Headings, V. E., and Poindexter, H. A. 1977. *Toxoplasma gondii*: Survival time and variability in mouse host strains. Exp. Parasitol. 41:307–313.

Palafox, A. L., Alicata, J. E., and Kartman, L. 1949. Breeding chickens for resistance to cecal coccidiosis. World's Poult. Sci. J. 5:84–87.

Pellérdy, L. P., and Durr, U. 1970. Zum endogenen Entwicklungszyklus von *Eimeria stiedai* (Lindemann, 1865) Kisskalt und Hartmann, 1907. Acta Vet. Hung. 20:227–244.

Pfefferkorn, E. R., and Pfefferkorn, L. C. 1976. *Toxoplasma gondii*: Isolation and preliminary characterization of temperature-sensitive mutants. Exp. Parasitol. 39:365–376.

Pfefferkorn, E. R., and Pfefferkorn, L. C. 1979. Quantitative studies of the mutagenesis of *Toxoplasma gondii*. J. Parasitol. 65:364–370.

Pfefferkorn, E. R., Pfefferkorn, L. C., and Colby, E. D. 1977. Development of gametes and oocysts in cats fed cysts derived from cloned trophozoites of *Toxoplasma gondii*. J. Parasitol. 63:158–159.

Pfefferkorn, L. C. and Pfefferkorn, E. R. 1980. *Toxoplasma gondii*: Genetic recombination between drug resistant mutants. Exp. Parasitol. 50:305–316.

Reid, W. M., Friedhoff, K., Hilbrich, P., Johnson, J., and Edgar, S. A. 1965. The occurrence of the coccidium species *Eimeria mivati* in European poultry. Z. Parasitenkd. 25:303–308.
Reyer, W. 1937. Infektionsversuche mit *Barrouxia schneideri* an *Lithobius forficatus*, insbesondere zur Frage der Sexualität der Coccidiensporozoiten. Z. Parasitenkd. 9:478–522.
Rollinson, D. 1975. Electrophoretic variation of enzymes in chicken coccidia. Trans. R. Soc. Trop. Med. Hyg. 69:436–437.
Rollinson, D. 1976. Further electrophoretic studies on species of *Eimeria*. Parasitology Part 2. 73:iv.
Rollinson, D. 1977. Biochemical and biological characters of *Eimeria* species. Ph.D. thesis, University of London.
Rollinson, D., Joyner, L. P., and Norton, C. C. 1979. *Eimeria maxima*: The use of enzyme markers to detect the genetic transfer of drug-resistance between lines. Parasitology 78:361–367.
Rommel, M. 1970. Untersuchungen an experimentell infizierten Schweinen über die Beeinflussung der Morphologie der Oozysten von *Eimeria scabra* (Henry, 1931) und *E. polita* (Pellérdy, 1949) durch Ubervolkerung, Immunitat und Alter des Wirts. Z. Parasitenkd. 34:141–150.
Rommel, M., and Geisel, O. 1975. Untersuchungen über die Verbreitung und den Lebenszyklus einer Sarkosporidienart des Pferdes (*Sarcocystis equicanis* n. spec.). Berl. Muench. Tieraerztl. Wochenschr. 88:468–471.
Rose, M. E., Ogilvie, B. M., Hesketh, P., and Festing, M. F. W. 1979. Failure of nude (athymic) rats to become resistant to reinfection with the intestinal coccidian parasite *Eimeria nieschulzi* or the nematode *Nippostrongylus brasiliensis*. Parasite Immunol. 1:125–132.
Rosenberg, M. M. 1941. A study of the inheritance of resistance to *Eimeria tenella* in the domestic fowl. Poult. Sci. 20:472.
Rosenberg, M. M. 1948. Selection for resistance and susceptibility to cecal coccidiosis in the domestic fowl. Ph.D. thesis, University of Wisconsin, Madison.
Rosenberg, M. M., Alicata, J. E., and Palafox, A. L. 1954. Further evidence of hereditary resistance and susceptibility to cecal coccidiosis in chickens. Poult. Sci. 33:972–980.
Rosenberg, M. M., McGibbon, W. H., and Herrick, C. A. 1948. Selection for hereditary resistance and susceptibility to cecal coccidiosis in the domestic fowl. Proceedings of the 8th World's Poultry Congress, Copenhagen, pp. 745–751.
Roudabush, R. L. 1935. Merozoite infection in coccidiosis. J. Parasitol. 21:453–454.
Rutherford, R. L. 1943. The life cycle of four intestinal coccidia of the domestic rabbit. J. Parasitol. 29:10–32.
Ryley, J. F., and Hardman, L. 1978. Speciation studies with *Eimeria acervulina* and *Eimeria mivati*. J. Parasitol. 64:878–881.
Scholtyseck, E. 1953. Beitrag zur Kenntnis des Entwicklungsganges des Huhnercoccids *Eimeria tenella*. Arch. Protistenkd. 98:415–465.
Shirley, M. W. 1975. Enzyme variation in *Eimeria* species of the chicken. Parasitology 71:369–376.
Shirley, M. W. 1977. The development of some species and strains of *Eimeria* from single sporozoites and sporocysts. J. Protozool. 24:43A. No. 4.
Shirley, M. W. 1978a. Electrophoretic variation of enzymes: A further marker for genetic studies of the *Eimeria*. Z. Parasitenkd. 57:83–87.
Shirley, M. W. 1978b. Electrophoretic variants of phosphoglucomutase in different species and strains of *Eimeria* from the chicken. In: P. L. Long, K. N. Boorman, and B. M. Freeman (eds.), Avian Coccidiosis, pp. 127–134, British Poultry Science, Ltd., Edinburgh.
Shirley, M. W. 1979a. A reappraisal of the taxonomic status of *E. mivati* Edgar and Seibold, 1964, by enzyme electrophoresis and cross-immunity tests. Parasitology 78:221–237.
Shirley, M. W. 1979b. Studies on the pathogenicity of chicken-maintained (virulent) and

embryo-adapted (attenuated) strains of *Eimeria mivati*. Avian Pathol. 8:469–475.

Shirley, M. W. Maintenance of *Eimeria maxima* by serial passage of single sporocysts. J. Parasitol. In press.

Shirley, M. W., and Lee, D. L. 1977. Isoelectric focusing of coccidial enzymes. J. Parasitol. 63:390–392.

Shirley, M. W., and Millard, B. J. 1976. Some observations on the sexual differentiation of *Eimeria tenella* using single sporozoite infections in chicken embryos. Parasitology 73:337–341.

Shirley, M. W., Millard, B. J., and Long, P. L. 1977. Studies on the growth, chemotherapy and enzyme variation of *Eimeria acervulina* var. *diminuta* and *E. acervulina* var. *mivati*. Parasitology 75:165–176.

Shirley, M. W., and Rollinson, D. 1979. Coccidia: The recognition and characterization of populations of *Eimeria*. In: A. E. R. Taylor and R. Muller (eds.), Problems in the Identification of Parasites and Their Vectors. British Society of Parasitology Symposium, Vol. 17, pp. 7–30. Blackwell Scientific Publications, Ltd., Oxford.

Shumard, R. F., and Callender, M. E. 1970. Anticoccidial drugs: Screening methods. Exp. Parasitol. 28:13–24.

Stockdale, P. H. G., and Fernando, M. A. 1975. The development of the lesions caused by second generation schizonts of *Eimeria necatrix*. Res. Vet. Sci. 19:204–208.

Tyzzer, E. E. 1929. Coccidiosis in gallinaceous birds. Am. J. Hyg. 10:269–383.

Tyzzer, E. E., Theiler, H., and Jones, E. E. 1932. Coccidiosis in gallinaceous birds: II. A comparative study of species of *Eimeria* of the chicken. Am. J. Hyg. 15:319–393.

Vetterling, J. M. 1966. Endogenous cycle of the swine coccidium *Eimeria debliecki* Douwes, 1921. J. Protozool. 13:290–300.

Wacha, R. S., and Christiansen, J. L. 1979. *Eimeria filamentifera* sp. n. from the snapping turtle, *Chelydra serpentina* (Linne), in Iowa. J. Protozool. 26:353–354.

Wedekind, G. 1927. Zytologische Untersuchungen an *Barrouxia schneideri* (Gametembildung, Befruchtung und Sporogonie), zugleich ein Beitrag zum Reduktionsproblem. (Coccidienuntersuchungen 1). Z. Zellforsch. Mikr. Anat. 5:505–595.

Weppelman, R. M., Battaglia, J. A. and Wang, C. C. 1977. *Eimeria tenella*: The selection and frequency of drug-resistant mutants. Exp. Parasitol. 42:56–66.

Williams, D. M., Grumet, F. C., and Remington, J. S. 1978. Genetic control of murine resistance to *Toxoplasma gondii*. Infect. Immun. 19:416–420.

Yarwood, E. A. 1937. The life cycle of *Adelina crytocerci* sp. nov., a coccidian parasite of the roach *Crytocercus punctulatus*. Parasitology 29:370–390.

4

Ultrastructure

Bill Chobotar and Erich Scholtyseck

This work was supported in part by the Alexander von Humboldt-Stiftung and by a Faculty Development grant to Andrews University by the Merck Foundation, Rahway, N.J.

The contributions of electron microscopy to our understanding of the coccidia have been prodigious. Literally hundreds of publications dealing with ultrastructure have appeared in recent years. It was the electron microscope that revealed structural similarities of merozoites of *Sarcocystis* and *Toxoplasma* with their eimeriid counterparts, which suggested that these two genera were coccidia prior to the elucidation of the coccidian nature of their life cycle (Scholtyseck and Piekarski, 1965; Sénaud, 1967). The subphylum Apicomplexa was so named in recognition of the consistent presence in certain stages of an ultrastructural apical complex consisting of polar rings, rhoptries, conoid, micronemes, subpellicular microtubules, and micropores. Thus, electron microscope studies have provided details that serve as the basis of determining systematic relationships within the coccidia.

The electron microscope has also been an indispensable tool for analysis of developmental processes such as merogony, gamogony, fertilization, and sporogony.

In this chapter, the objective is to update our knowledge of the ultrastructure of the coccidia. However, because of the large volume of literature, it is necessary to restrict our discussion of the subject. Thus, we have concentrated largely on the classical coccidia found in the suborder Eimeririna, specifically, the better known members in the families Eimeriidae and Sarcocystidae. The reader is directed to a number of reviews that include ultrastructural information on different groups of coccidia: (Sénaud, 1967; Scholtyseck and Mehlhorn, 1970; Scholtyseck et al., 1970, 1971, 1972a; Vivier et al., 1970; Porchet-Henneré and Vivier, 1971; Porchet-Henneré, 1972; Hammond, 1973a,b; Scholtyseck, 1973a, 1979; Dubey, 1977; Mehlhorn and Heydorn, 1978; Beyer et al., 1978; and Speer, in press).

GENERAL ULTRASTRUCTURAL FEATURES OF THE MOTILE STAGES

The Pellicle

The pellicle consists of three unit membranes and is characteristic of all coccidian sporozoites and merozoites studied thus far (Figure 8). The outer membrane forms the plasmalemma and is a continuous structure enclosing the whole parasite. The middle and inner membranes are closely apposed and comprise the inner pellicular complex which has interruptions at the anterior and posterior polar rings and at the micropore. The plasmalemma and inner complex are separated by an electron-pale space of 15–20 nm. This description is based on studies of thin sections of motile stages in many species of coccidia. Recently, negative staining and freeze-fracture techniques have largely confirmed previous findings and have expanded our concept of pellicle morphology beyond the limits of ultrathin sections. In studies involving freeze-fracture techniques of motile stages of *Sarcocystis* species, *Eimeria nieschulzi,* and *Toxoplasma* species,

4

Porchet and Torpier (1977) and Dubremetz and Torpier (1978) have shown that three partition planes occur within the three pellicular membranes, resulting in the exposure of six cleavage faces (Figure 1). Under these conditions, the plasmalemma appears as in ultrathin sections, that is, as a continuous unit surrounding the whole parasite. A distinctive feature of the cleavage faces is the distribution of intramembranous particles (IMP) in mostly a random fashion within the plasmalemma, in contrast to the orderly pattern of IMP of the inner complex.

A departure from the random distribution of IMP in the plasmalemma is the arrangement of eight particles in a circle with a single central particle to form an apical rosette (Figure 2). Rosettes have been described from other Protozoa (Plattner et al. 1973; Satir et al., 1973) in which the membranes of specialized cytoplasmic vesicles (mucocysts and trichocysts) fuse with the plasmalemma to discharge their contents into the external medium. Because the apical region of invasive stages of coccidia is prominently involved in initial contact with host cells, the presence of a rosette there is significant. It has been suggested that the rosette area may be regarded as a "receptor-processor" system, which allows the parasite to recognize a potential host cell and to initiate processes involved in cell penetration, such as eventual rhoptry secretion (Porchet and Torpier, 1977; Dubremetz and Torpier, 1978).

In contrast to the rather uniform pattern of the plasmalemma, both fracture planes show that the inner membrane complex is made up of a series of rectangular contiguous strips or plaques aligned in longitudinal rows. Transverse sutures are present in the strips that are fitted together much like pieces in a puzzle (Figures 3 and 4). At the anterior end, the strips terminate on a truncated cone, the apical cap (Figure 3). The rows of strips are arranged in a helical pattern along the body and converge at the posterior end in a polar ring. The number of rows varies according to the species and coincides with half the number of subpellicular microtubules. Thus, in *Sarcocystis,* 11 rows and 22 microtubules occur (Porchet and Torpier, 1977), whereas in *E. nieschulzi,* the respective numbers are 13 and 26 (Dubremetz and Torpier, 1978). Treatment of motile stages by negative staining revealed a series of longitudinal lines radiating outward in a helical pattern from the posterior polar ring toward the anterior pole (Figure 5; Porchet-Henneré, 1975, 1976; D'Haese et al., 1977). These lines are especially prominent at the posterior end where they appear as a series of electron-dense pores or granules (Figure 6). By scanning electron microscopy, the lines are visible as a series of helical ribs on the surface of merozoites of *Sarcocystis tenella* (Mehlhorn and Scholtyseck, 1974). The number of lines and ribs present corresponds to the number of strips or plaques observed with the freeze-fracture method. On the P fracture faces of the inner complex, IMP are distributed in highly regular longitudinal alignments. Among these, each strip has two distinct lines made up of about twice the number of particles as the others (Figures 3, 4, and 7). These distinct lines correspond in number and arrangement with the underlying microtubules and suggest a functional interaction between the latter and the membranes (Porchet and Torpier, 1977; Dubremetz and Torpier, 1978).

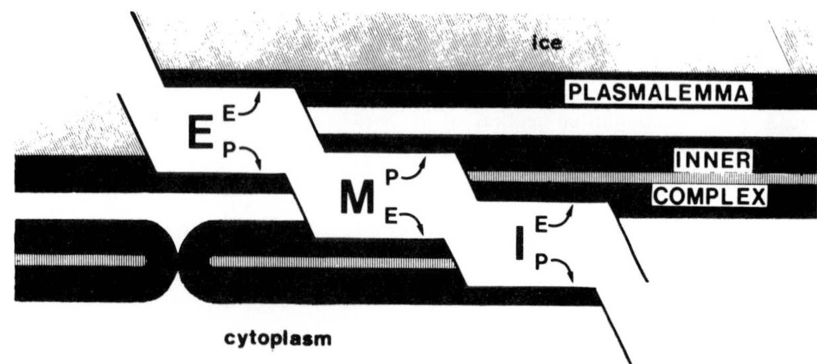

Figure 1. Schematic diagram of the fracture faces of a coccidian sporozoite or merozoite. The terminology is that of Branton et al. (1975) Reprinted by permission from: Dubremetz, J. F., and Torpier, G. Freeze fracture study of the pellicle of an eimerian sporozoite (Protozoa, Coccidia). *Journal of Ultrastructure Research* 62:94–109 (1978).

The structure and distribution of the rectangular strips observed by freeze-fracture methods lends strong support to an earlier view that the inner pellicular complex is composed of a series of flattened vesicles (Vivier and Petitprez, 1969; Figure 8). These vesicles develop during merogony, at which time they appear as crescent-shaped segments, each associated with a pair of subpellicular microtubules (Roberts et al., 1970a; Dubremetz, 1975; Ferguson et al., 1976; Dubremetz and Elsner, 1979). The segments become flattened and join under the plasmalemma of each developing merozoite.

Thus, by use of several available methods, the complex structural nature of the pellicle is revealed. However, many questions related to the functional significance of this organelle remain to be answered by future investigations. A diagrammatic representation of the present concept of the pellicle is given in Figure 9.

Polar Rings

Polar rings occur singly or in pairs at the apical pole of sporozoites and merozoites of coccidia. In some species, only one polar ring is present; it appears as an osmiophilic thickening at the anterior termination of the inner pellicular complex and serves to anchor the subpellicular microtubules (Colley, 1967; Ryley, 1969; Scholtyseck and Mehlhorn, 1970; Roberts and Hammond, 1970; Roberts et al., 1970b; Porchet-Henneré, 1975). In other species, these microtubules are attached to a second polar ring, located directly behind the first (Figure 20). In negatively stained specimens, these appear as an inner and outer ring, the latter having connections with the microtubules (Figure 10; D'Haese et al., 1977). This arrangement corresponds to that seen in longitudinal sections of motile stages, even where only a single polar ring has been reported (e.g., see Figure 3 in Roberts and Hammond (1970) and Figure 5b in Chobotar et al. (1975a)), suggesting that possibly two rings are present in most coccidia.

Microtubules

Microtubules are present in many stages of the life cycle of apicomplexans (Scholtyseck et al., 1972b). They are found in centrioles and in microgamete flagella, and form part of the mitotic apparatus during merogony. In sporozoites and merozoites, they make up the conoid and occur as subpellicular microtubules, which are arranged in regular intervals at the periphery directly beneath the inner pellicular complex. Each microtubule originates at an anterior polar ring and extends posteriorly; however, the length and number of these organelles varies with the species.

Twenty-two microtubules are consistently found in merozoites of the cyst-forming coccidia (*Frenkelia, Besnoitia, Sarcocystis, Toxoplasma,* and *Hammondia*); the number varies from 24–32 in the *Eimeria* but may reach 70 or more in *Klossia* and *Hemogregarina* (see reviews by Scholtyseck et al., 1970, 1972b; Scholtyseck, 1973a; Chobotar et al., 1975a).

4

Figures 2–4. Freeze-fracture replicas of *E. nieschulzi* sporozoites. Reprinted by permission from: Dubremetz, J. F., and Torpier, G. Freeze fracture study of the pellicle of an eimerian sporozoite (Protozoa, Coccidia). *Journal of Ultrastructure Research* 62:94–109 (1978). 2. Apical rosette of 8 + 1 IMP in the plasmalemma P face (*arrowhead*), surrounded by an irregular ring. × 100,000. 3. The apical cap is smooth except for channels and double ridges (*arrows*) aligned on the main lines of IMP on P_i × 60,000. 4. Fracture through the inner complex showing the alignments of IMP on the P_i face and the regularly spaced main lines (*arrows*). × 45,000.

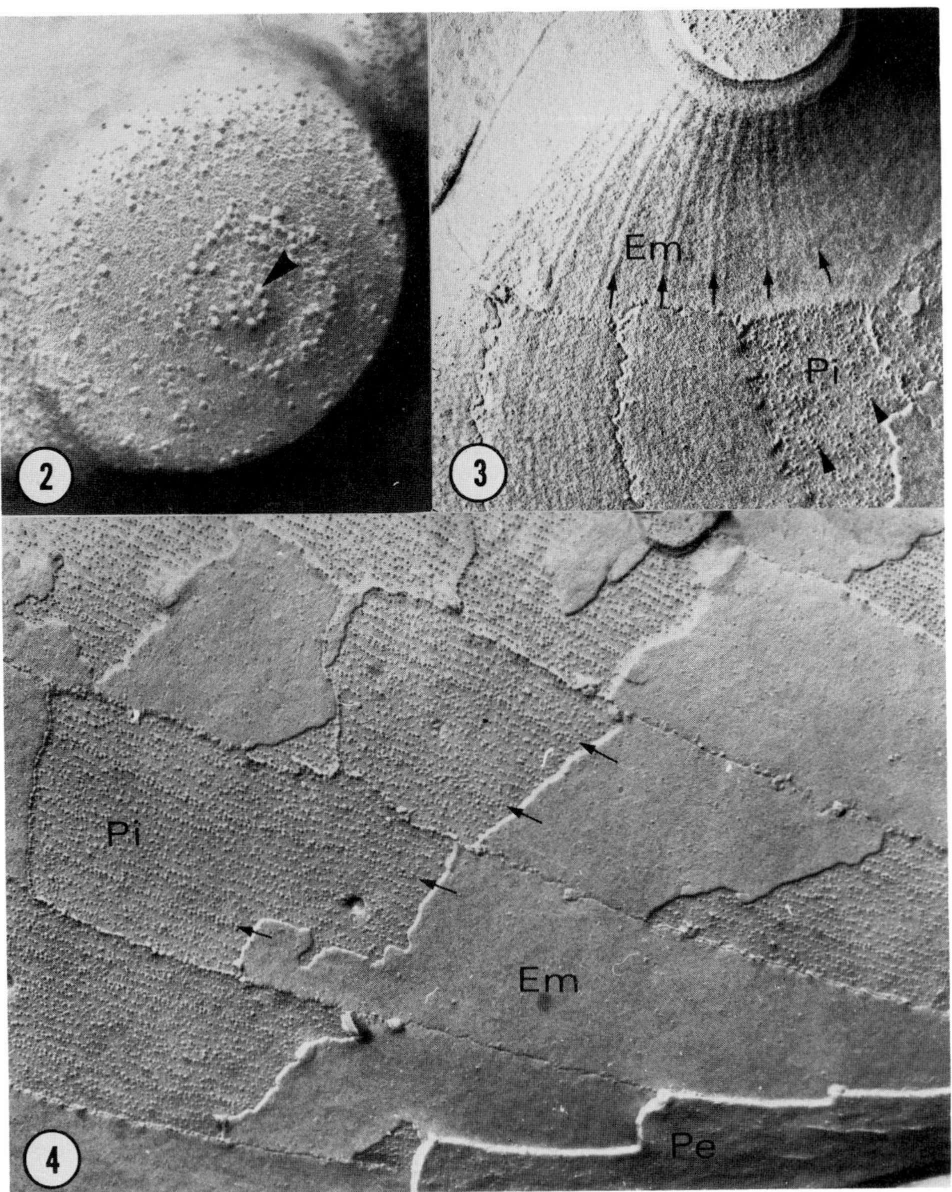

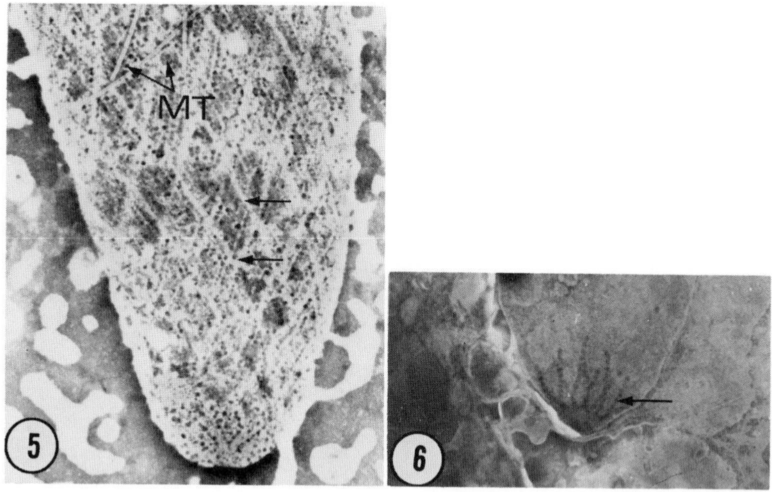

Figures 5 and 6. 5. Negatively stained merozoite of *Besnoitia jellisoni*. Ribs (*arrows*) represent margins of strips of the inner membrane complex of the pellicle. ×20,000. Reprinted by permission from: D'Haese, J., Mehlhorn, H. and Peters, W. Comparative electron microscope study of pellicular structures in Coccidia (*Sarcocystis*, *Besnoitia* and *Eimeria*). *International Journal for Parasitology* 7:505–518 (1977). 6. Negatively stained merozoite of *G. gilruthi*. The beaded lines are at edges of strips of the inner membrane complex. ×36,000. Reprinted by permission from: Porchet-Henneré, E. Structure du mérozoite de la Coccidie *Globidium gilruthi* de la caillete du mouton (d'après l'etude de coupes fines et de coloration négative). *Protistologica* 12:613–621 (1976).

The microtubules apparently do not extend to the posterior pole, despite some indication of this in thin-sectioned material. From studies of negatively stained preparations, the microtubules extend about halfway in sporozoites and merozoites of *Eimeria* species (Ryley, 1969; Robert and Hammond, 1970) and merozoites of *Sarcocystis* species (Porchet-Henneré, 1975; D'Haese et al., 1977), and about four-fifths of the way in merozoites of *Besnoitia* species (D'Haese et al., 1977).

The subpellicular microtubules of coccidia are similar to those found widely in cells of plants and animals. They are approximately 20 nm in diameter and are made up of about 12 longitudinally oriented protofilaments. The protofilaments appear to be composed of repeating subunits arranged in transverse or helical rows, giving the microtubules a striated appearance (Figures 10–12; Roberts and Hammond, 1970; Dubremetz, 1971b; Porchet-Henneré, 1976; D'Haese et al., 1977).

It is well known that microtubules are involved in cell movement, phagocytosis, and other cell membrane activities and that these activities can be reduced or stopped by use of microtubule inhibitors (Ward, 1971; Becker and Henson, 1973; Jensen and Edgar, 1976a). The flexing and gliding movements characteristic of sporozoites and merozoites have often been attributed to the presence of the subpellicular microtubules in close association with the pellicular complex. Heller and Scholtyseck (1971) observed arm-like extensions of the subpellicular microtubules in merozoites of *Eimeria stiedai,* which apparently connected the tubules to the inner pellicular complex. However, in merozoites of *Eimeria falciformis,* which were fixed in a flexed position, some microtubules were located deep in the cytoplasm in groups of two, three, or four (Scholtyseck et al., 1972b). Apparently, this arrangement was to avoid the folded area of the pellicle. On the side opposite the fold, the microtubules were close, but not connected to the pellicle. The microtubule-pellicle connections seen by Heller and Scholtyseck (1971) either are not common in the coccidia or may involve a cyclic interaction because dissociation from the pellicle can readily occur during flexing (D'Haese et al., 1977).

Studies of several members of the Apicomplexa in which microtubule inhibitors were used have produced inconclusive results. Jensen and Edgar (1976a) found that colcemid, vinblastine, and colchicine did not affect sporozoite (*Eimeria magna*) motility or prevent their penetration into cells, despite 6 hr of treatments. Conversely, sporozoites were immobilized and failed to enter cells in the presence of cytochalasin B, a chemical which interferes with microfilament contractility (Jensen and Edgar, 1976a). In contrast to these results, D'Haese et al. (1977) found that treatment with colchicine and urea stopped motility of merozoites of *Sarcocystis* and *Besnoitia* and sporozoites of *E. falciformis,* but did not destroy the microtubules. Motile stages of *Selenidium,* a gregarine with pellicular and microtubule arrangement similar to that of the coccidia, were immobilized in the presence of urea (which depolymerized the microtubules) and cytochalasin B, but not in the presence of colchicine (Schrével et al., 1974).

As a result, it has been suggested that subpellicular microtubules may not function in locomotion directly or that they may not be typical microtubules. Rather, they may be tubular

Figures 7 and 8. 7. Freeze-fracture replica of sporozoite of *E. nieschulzi*. Detail of the P_i face on inner complex strip showing the orderly distribution of IMP and a micropore. This strip has 23 lines, two of which (*arrowheads*) have twice as many IMP as the others. × 150,000. Reprinted by permission from: Dubremetz, J. F., and Torpier, G. Freeze fracture study of the pellicle of an eimerian sporozoite (Protozoa, Coccidia). *Journal of Ultrastructure Research* 62:94–109 (1978). 8. Merozoite of *T. gondii* showing anterior organelles. Note trimembranous pellicle of which the inner complex consists of two closely apposed membranes arranged in a series of flattened vesicles that adjoin at the *arrows*. × 65,000. Reprinted by permission from: Vivier, E., and Petitprez, A. Le complexe membranaire superficiel et son évolution lors de l'élaboration des individus-fils de *Toxoplasma gondii*. *Journal of Cell Biology* 43:329–342 (1969).

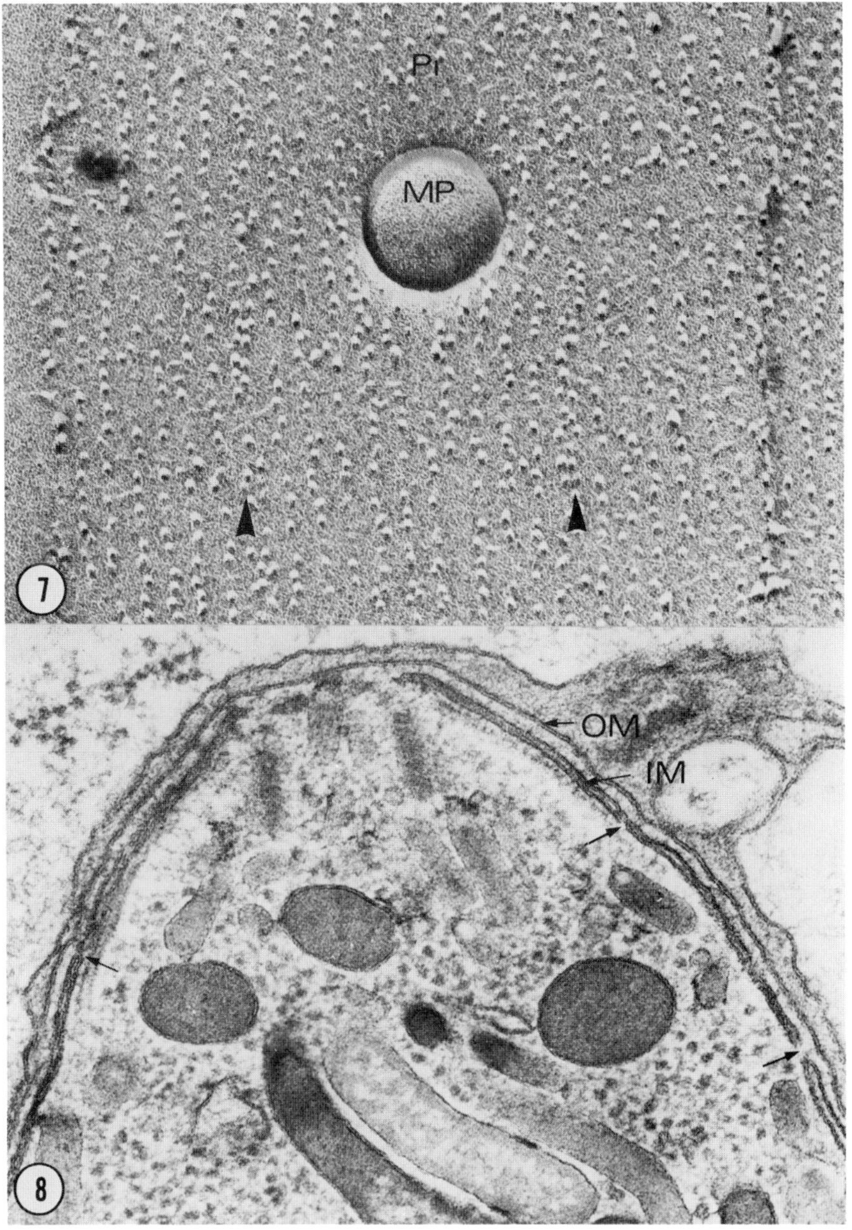

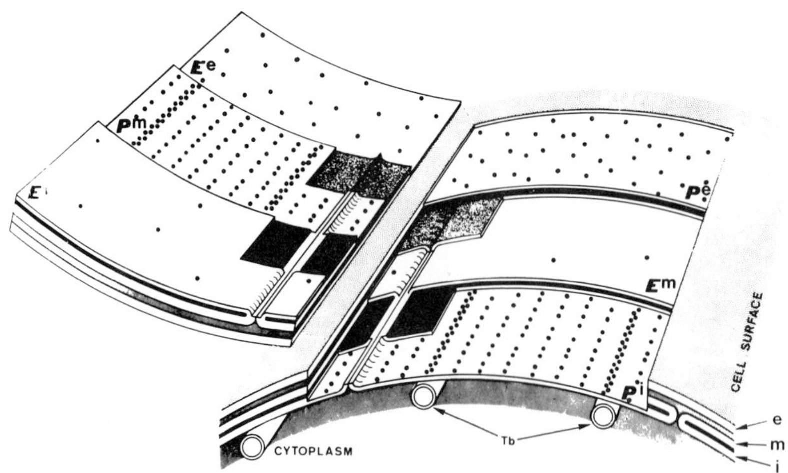

Figure 9. Diagram summarizing the coccidian pellicle in the strip area as revealed by freeze-fracture. The complementary fractures are shown, with their possible relationships with subpellicular microtubules. The terminology is that of Branton et al. (1975). Reprinted by permission from: Dubremetz, J. F., and Torpier, G. Freeze fracture study of the pellicle of an eimerian sporozoite (Protozoa, Coccidia). *Journal of Ultrastructure Research* 62:94–109 (1978).

aggregates of contractile elements which are inhibited by cytochalasin B (Jensen and Edgar, 1976a) or they may function indirectly in movement involving the ribbed pellicle and closely associated microtubules, making up a "cytoskeletal complex" which may act as an antagonist of contractile elements as yet undiscovered in the coccidia (D'Haese et al., 1977).

Conoid

The conoid was first described and named from merozoites of *Toxoplasma gondii* by Gustafson et al., (1954). Since then, it has been found in all coccidia and gregarines, but not in haemosporidians and piroplasms. The conoid is a hollow truncated cone of spirally arranged fibrillar structures from 25–30 nm in diameter (Figures 11 and 13*a,c*). The number of fibrillar elements in the conoid varies according to the species. For example, six to seven are found in *Eimeria callospermophili* (Scholtyseck et al., 1970), 18 in merozoites of *E. stiedai* (Heller, 1972), and 20 in merozoites of *S. tenella* (Porchet-Hennere, 1975).

In many species, two accessory structures, the conoidal rings (Scholtyseck et al., 1970; Scholtyseck, 1973a, 1979) or preconoidal annuli (Vivier and Petitprez, 1972), are found immediately anterior to the conoid (Figures 11 and 20). It is likely that these rings form an integral part of the conoid, because in negatively stained specimens (Porchet-Hennere, 1975; D'Haese et al., 1977) and in sectioned parasites (Scholtyseck et al., 1970; Scholtyseck, 1973a) connections occur between the rings and the conoidal cone. In the latter preparation, often a thin osmiophilic line appears to join the cone, with the rings forming a canopy-like membrane with a central pore at the apex (Figures 13*c* and 20).

Several authors have suggested that the fibrillar units making up the conoid are probably microtubules, because they closely resemble the subpellicular microtubules (Ryley, 1969; Porchet-Hennere, 1975, 1976; D'Haese et al., 1977). This has not been confirmed because of the inability, thus far, to isolate for study individual units making up the conoid.

Within the conoid are the ducts of the rhoptries and two microtubules that have the same diameter and arrangement of subunits as the subpellicular microtubules (Figure 11). They are usually eccentrically placed and extend a short distance into the cytoplasm (Ryley, 1969; Roberts and Hammond, 1970; Dubremetz, 1971b, 1975; Porchet-Hennere, 1975, 1976; D'Haese et al., 1977). These microtubules are apparently attached to the conoid, because they are seen in isolated conoids and may function to extend and retract the conoid.

The conoid protrudes through the anterior polar ring system, although no connections between the two occur. However, the relative position of the conoid varies with respect to the polar rings, so that at times either the anterior or the posterior margin of the conoid is opposite the rings. This observation led McLaren and Paget (1968) and others (Ryley, 1969; Roberts and Hammond, 1970; Scholtyseck et al., 1970; Scholtyseck, 1973a, 1979; Jensen and Edgar, 1978) to suggest that the conoid is a protrusible and retractable organelle. Variability of the apical portion of live invasive stages, including the formation of nipple-like projections, has been

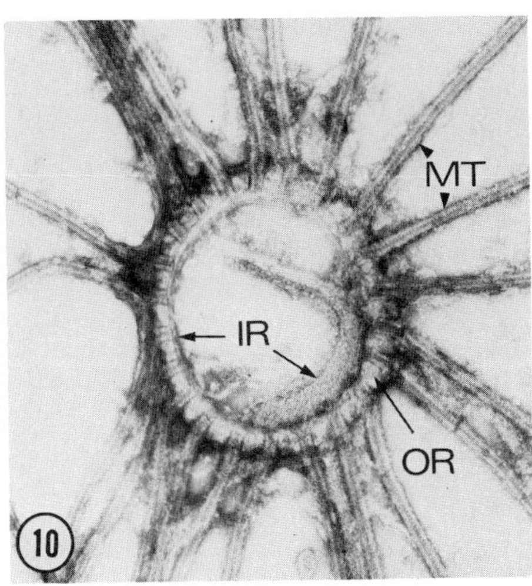

Figure 10. Negatively stained, isolated polar rings and subpellicular microtubules of *S. tenella* (= *S. ovifelis*). Note the presence of an inner (*IR*) and outer (*OR*) polar ring. The microtubules (*MT*) terminate in the outer ring. × 100,000. Reprinted by permission from: D'Haese, J., Mehlhorn, H., and Peters, W. Comparative electron microscope study of pellicular structures in Coccidia (*Sarcocystis, Besnoitia* and *Eimeria*). *International Journal for Parasitology* 7:505–518 (1977).

described in light microscope studies (Sénaud, 1967; Hammond et al., 1968; Speer, et al., 1971; Roberts et al., 1971; Fayer, 1972; Long and Speer, 1977).

The function most often attributed to the conoid is to aid in the penetration of host cells. Infective stages routinely enter cells apical end first; the idea of an extended conoid coupled with secretions from the rhoptries and micronemes to facilitate entrance has been an attractive theory for some time, although not yet confirmed (McLaren and Paget, 1968; Ryley, 1969; Scholtyseck et al., 1970; Roberts and Hammond, 1970; Vivier and Petitprez, 1972; Scholtyseck, 1973a, 1979).

Rhoptries and Micronemes

As described in many studies, rhoptries and micronemes are dense, osmiophilic structures usually found in the area between the nucleus and conoid of sporozoites and merozoites of Apicomplexa. A review of the early ultrastructural literature involving motile stages reveals a confusing variety of names for these organelles. The profusion of names can be attributed in part to publication of first-time descriptions of organisms that at that time were taxonomically widely separated.

In some studies, a distinction was not always made between rhoptries and micronemes, and in other studies, names for the two structures were reversed (Ryley, 1969; Roberts and Hammond, 1970; Porchet-Henneré and Vivier, 1971). Evidence for this is reflected in the considerable variation in the dimensions and morphology of these organelles (see the review by Scholtyseck and Mehlhorn, 1970), in contrast to the relative uniformity reported in later investigations. As subsequent studies revealed a relatively consistent morphology of sporozoites and merozoites, a more uniform terminology for these organelles emerged.

Several names have been used to describe the rhoptries. They were first identified in *Toxoplasma* and named "toxonemes" by Gustafson et al., (1954). In 1960, Garnham et al., described club-shaped structures in sporozoites of *Plasmodium gallinacea* as "paired organelles.' Jacobs (1967) and Ryley (1969) used the term "club-shaped organelles" because this seemed to best describe the organelles they observed. In his classic study of *Toxoplasma* and *Sarcocystis,* Sénaud (1967) introduced a new name, the "rhoptry," now used by most authors.

Rhoptries may vary considerably in shape, appearing as club-shaped, teardrop-shaped, elongate, and tortuous (Scholtyseck and Mehlhorn, 1970; Scholtyseck, 1973a, 1979). The anterior neck portions of rhoptries are narrow and duct-like and originate within the conoid. As they extend posteriorly, they gradually widen into a club or teardrop shape at their termination point (Figures 8, 13*a*, and 20). The club-shaped area sometimes has a dense homogeneous posterior part and a less dense, alveolar anterior part. In *Toxoplasma* merozoites, portions of the rhoptries often appear sponge-like (Figure 8).

Based largely on the study of thin sections, the number of rhoptries in motile stages varies from 2 to more than 20. In general, they occur in pairs in merozoites of *Isospora* and

Figures 11 and 12. Negatively stained specimens of *S. tenella* and *G. gilruthi*, respectively. Reprinted by permission from: Porchet-Henneré, E. Quelques précisions sur l'ultrastructure de *Sarcocystis tenella*. I. L'endozoite (après coloration négative). *Journal of Protozoology* 22:214–220 (1975) (Figure 11); and from: Porchet-Henneré, E. Structure du mérozoite de la Coccidie *Globidium gilruthi* de la caillete du mouton (d'après l'etude de coupes fines et de coloration négative). *Protistologica* 12:613–621 (1976) (Figure 12). 11. The apical structures of a merozoite including the preconoidal rings (R_1, R_2), the conoid (*C*), polar ring (*P*), and microtubules (*MT*). Note intraconoidal microtubules (*arrowheads*). × 148,000. 12. Detail of microtubules structure at high magnification. Note the cross-striations. × 150,000.

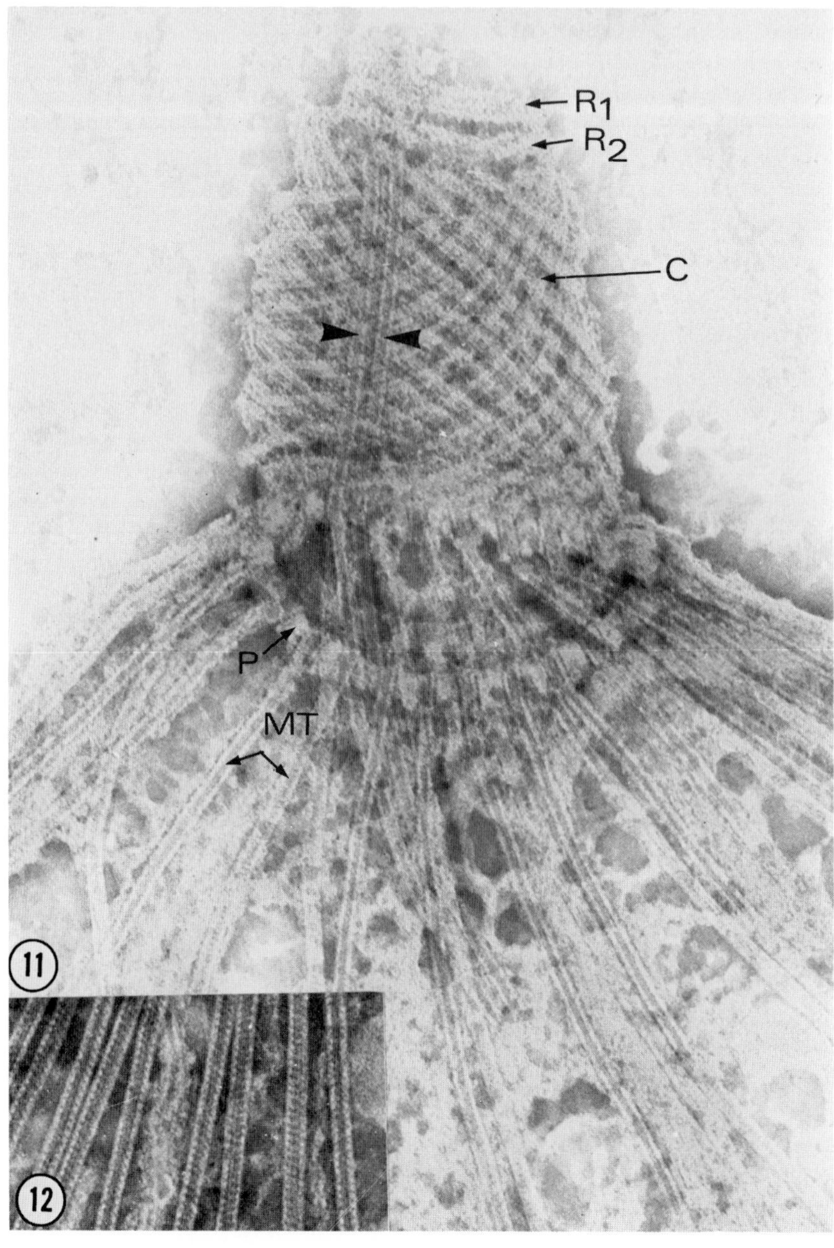

4

Figure 13. Coccidian merozoite showing the principal fine structural features. Reprinted by permission from: Scholtyseck, E. *Fine Structure of Parasitic Protozoa.* Heidelberg. Springer-Verlag (1979). *a,* longitudinal section of a merozoite; *b,* transverse and longitudinal section of a micropore; *c,* the conoid.

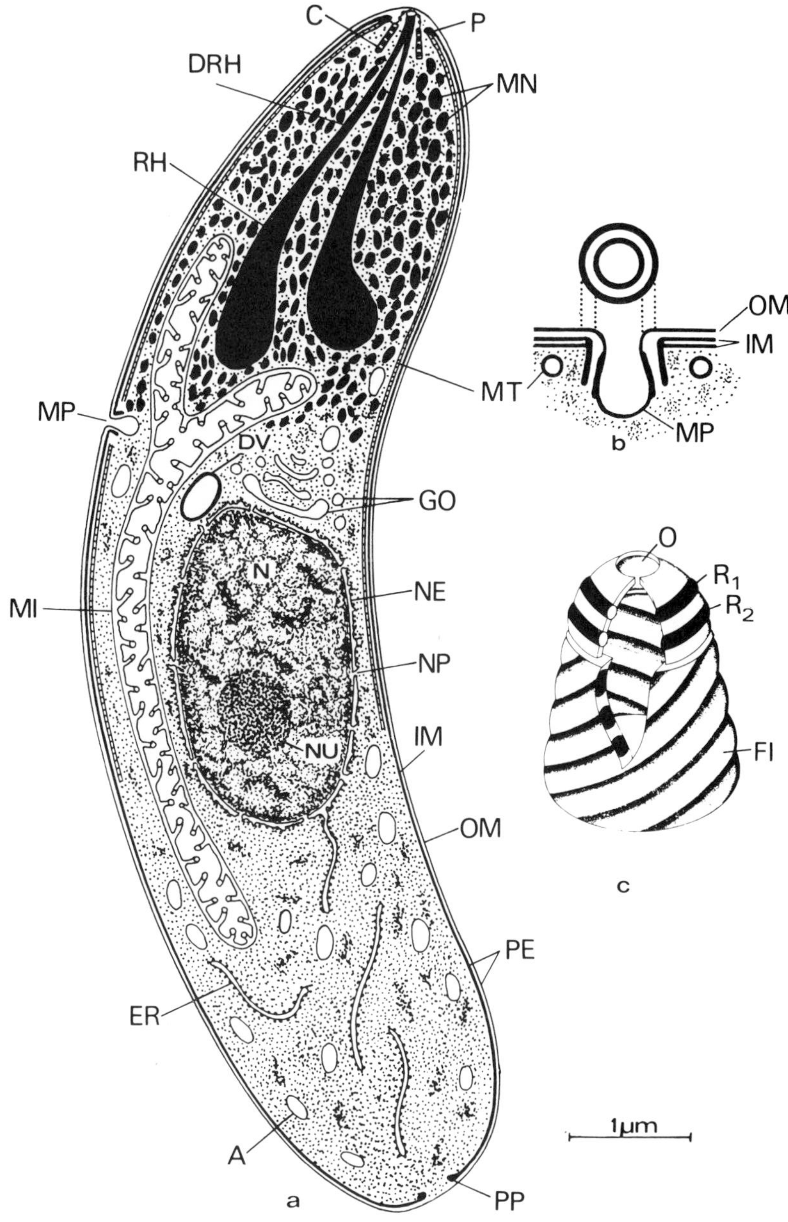

numerous *Eimeria* species, whereas more than two have been reported from *Frenkelia, Toxoplasma,* and *Sarcocystis* species. However, the number of rhoptries in these latter species must be considered in the light of one's interpretation of a single thin section which represents a minute portion of the whole cell. This is especially true in *Sarcocystis* in which the tortuosity of rhoptries may be increased among the large numbers of tightly packed micronemes (400 or more), making interpretation more difficult. Use of negative staining methods permits a more complete view of the organism; Porchet-Henneré (1975) has demonstrated two rhoptries in merozoites of *S. tenella,* originally thought to have several to many of these organelles.

Micronemes were first described from *Toxoplasma* by Gustafson et al. (1954). A proliferation of names followed, including sarconemes, lankesterellonemes, convoluted tubules, Cytoplasmastrange, and rod-shaped organelles. The nomenclature was considerably improved by use of the now widely accepted and less restrictive name "microneme" introduced by Jacobs (1967) to describe these small inclusions in *Toxoplasma.*

Micronemes have most often been described as tortuous or convoluted and arranged in bundles. In thin sections, the micronemes appear as dense osmiophilic structures, which usually occupy the anterior regions of sporozoites and merozoites. In cross-section, they appear round or slightly ovoid, and in longtitudinal section, they are elongate and rod-like with rounded ends (Figures 13*a,* 14, and 21), with dimensions of approximately 50–80 nm × 300–600 nm.

The origin and development of micronemes was studied in merozoites of cyst origin during endodyogeny in *S. tenella* (Mehlhorn et al., 1975d). In this species, the micronemes appeared to form as a result of confluence and densification of ribosome-like, spirally arranged granules within "vesicular formation zones," which originated from granular vacuoles located between the conoid and nucleus of the developing daughter cell.

Numerous attempts have been made to correlate the structure of rhoptries and micronemes with their function. It has often been suggested that these organelles constitute a single functional unit composed of larger (rhoptries) and smaller (micronemes) components. Scholtyseck and Mehlhorn (1970) have presented them as a common organelle connected by lateral ductules and Heller (1972) reported that rhoptries and micronemes of *E. stiedai* are formed by division of a single organelle and belong to the same system of ductules. Vivier and Petitprez (1972) reported that micronemes of *Toxoplasma* may represent stages in the development of rhoptries or sections of their stems. In sporozoites of *Isospora canis,* rhoptries and micronemes appeared to be larger and smaller branches respectively, of a single network (Jensen and Edgar, 1978). However, in other studies, some of which involved the use of negative staining and freeze-fracture techniques, the micronemes appear as distinctly separate bead-like units, apparently independent of the rhoptries (Ryley, 1969; Roberts and Hammond, 1970; Porchet-Hennéré, 1975, 1976; Figure 14).

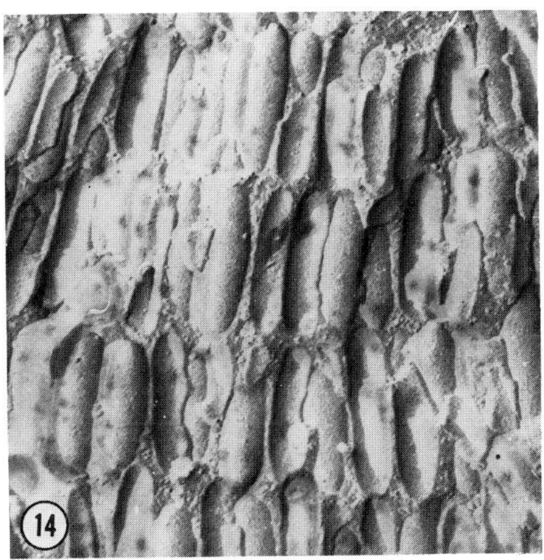

Figure 14. *S. tenella.* Freeze-fracture sample of merozoite micronemes. Note the uniform arrangement and apparent lack of interconnections. ×38,000. Reprinted by permission from: Porchet, E., and Torpier, G. Étude du germe infectieux de *Sarcocystis tenella* et *Toxoplasma gondii* par la technique du cryodécapage. *Zeitschrift für Parasitenkunde* 54:101–124 (1977).

It has been suggested that rhoptries may secrete substances used in the penetration of host cells (Scholtyseck and Mehlhorn, 1970; Vivier and Petitprez, 1972; Scholtyseck, 1973a, 1979; Hammond, 1973b; Jensen and Edgar, 1976b, 1978). Attempts to detect enzymes in rhoptries by cytochemical methods have produced contradictory results (Schrével, 1968; Vivier and Petitprez, 1972; Sénaud et al., 1972; Mehlhorn and Scholtyseck, 1973a).

Jensen and Edgar (1976b) reported that the entry of sporozoites of *E. magna* and subsequent membrane alterations were accompanied by the appearance of empty or partially empty membrane saccules, which were of the same general shape and location as the rhoptries. They also saw packets of osmiophilic material, analogous to rhoptry contents, aligned in channels which passed through the conoid. Similar vesicles were observed in the conoid area of merozoites of *Toxoplasma* by Vivier and Petitprez (1972). In another study, Jensen and Edgar (1978) described partially empty rhoptries in invading *I. canis* sporozoites concurrent with the appearance of extensive vesiculation of host cell cytoplasm next to the apical tip of the parasite. The vesicles and saccules were thought to represent secretory products of rhoptries which may aid in the penetration of host cells (Vivier and Petitprez, 1972; Jensen and Edgar, 1976b, 1978). Such secretions may be proteolytic (Lycke et al., 1965; Norrby et al., 1968) or may function to produce changes in the surface characteristics of the cell during the penetration process (Ladda et al., 1969; Bannister et al., 1975).

At present, existing evidence does not explain the role of rhoptries and micronemes in the coccidia. This is partly because of the difficulty in recording a dynamic process, such as secretion, with a static method, such as electron microscopy (Jensen and Edgar, 1976b).

The Micropore

The micropore is an ultrastructural organelle formed by an invagination of the limiting membrane, which continues uninterrupted through the invagination. In stages which have a complete pellicle, the inner pellicular complex also invaginates for a short distance, then abruptly terminates to form a thickened collar at the neck of the micropore. This collar is present whether the inner complex is present (as in merozoites and sporozoites) or not (as in microgamonts). Thus, in longitudinal section the micropore appears to be bounded by two lineal elements near the surface, whereas at the base only a single limiting membrane occurs (Figures 13*a,b,* 15–17, 19, 45, 46).

In cross-section, near the surface of the parasite, this organelle is composed of two concentric rings, which correspond to the two lineal elements of the micropore wall seen in longitudinal sections (Figure 13*b,* and 18).

Detailed reviews have been published which show that the basic structure and occurrence of micropores is relatively uniform in all representatives of the Apicomplexa where this

Figures 15–19. Micropores of *E. ferrisi* (Figures 15, 17, and 18) and *B. jellisoni* (Figures 16 and 19). Figure 15 is reprinted by permission from: Chobotar, B., Scholtyseck, E., Sénaud, J., and Ernst, J. V. A fine structural study of asexual stages of the murine coccidium *Eimeria ferrisi* Levine and Ivens, 1965. *Zeitschrift für Parasitenkunde* 45:291–306 (1975). Figures 16, 17, and 19 are reprinted by permission from: Sénaud, J., Chobotar, B., and Scholtyseck, E. Role of the micropore in nutrition of the Sporozoa: Ultrastructural Study of *Plasmodium cathemerium*, *Eimeria ferrisi*, *E. stiedai*, *Besnoitia jellisoni*, and *Frenkelia* sp. *Tropenmedizin und Parasitologie* 27:145–159 (1976). 15 and 17. Active micropores with adjacent vacuoles that have apparently pinched off from the micropore invagination. Note that the material in the vacuoles is similar to that of the parasitophorous vacuole. × 50,000 (15); × 25,000 (17). 16 and 19. active micropores with tortuous cavities. × 30,000 (16); × 48,000 (19). 18. cross-section of micropores near the orifice showing the inner and outer ring assembly, from a microgamont. × 28,000.

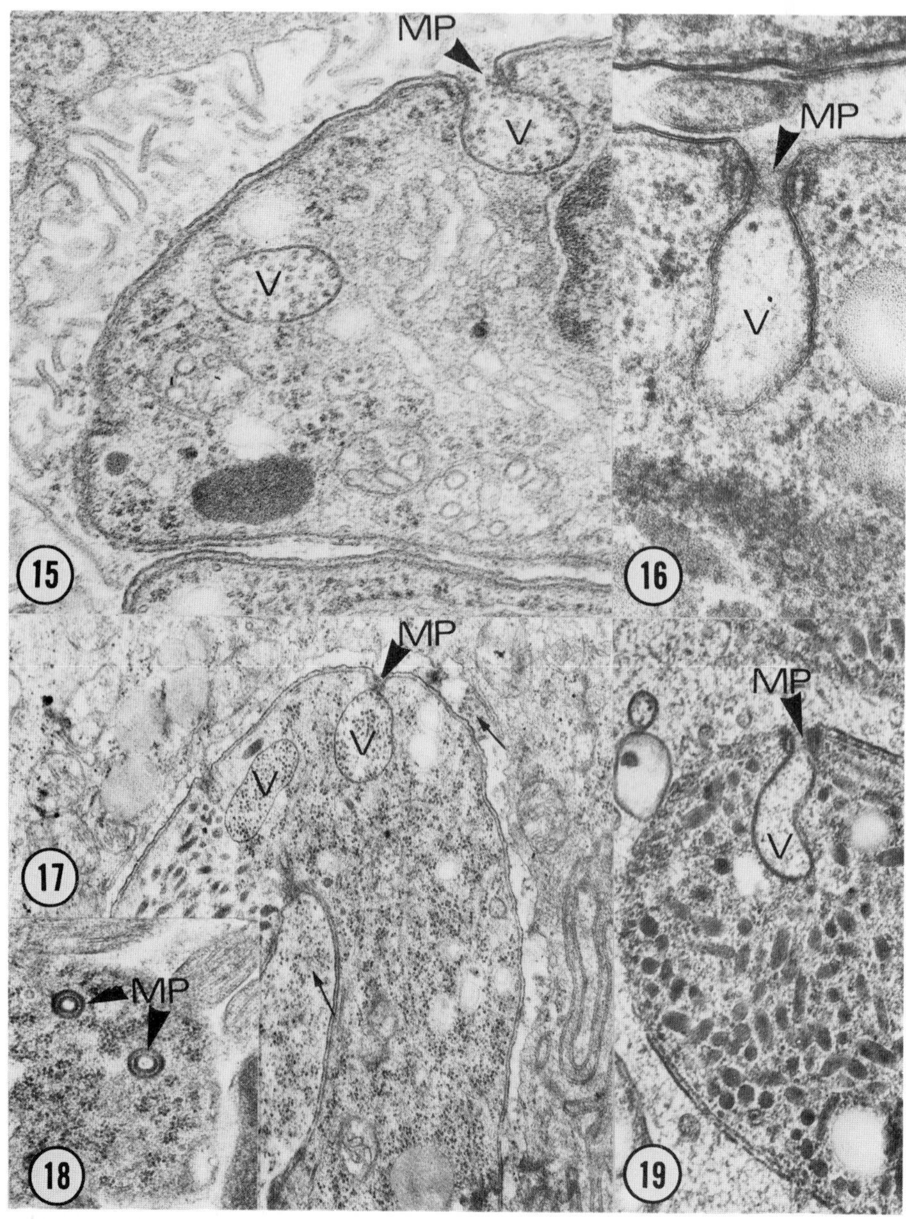

4

organelle has been found (Scholtyseck and Mehlhorn, 1970; Porchet-Henneré and Vivier, 1971; Scholtyseck, 1973a; Sénaud et al., 1976). In the coccidia, the micropore has been reported from every developmental stage, with the exception of the microgamete.

The micropore was first described from sporozoites of *Plasmodium falciparum* and was called a "micropyle" (Garnham et al., 1961). In 1965, Cheissin and Snigirevskaya reported the presence of a micropore in merozoites of *Eimeria intestinalis* and suggested that its function was possibly that of an "ultracytostome." In their study of trophozoites of *Coelotropha durchoni,* Vivier and Henneré (1965) suggested use of the term "micropore" for this structure. Shortly thereafter, Aikawa and his co-workers (reviewed by Aikawa, 1971) demonstrated that within the Haemosporidia this organelle was active in ingestion of host cell cytoplasm with the subsequent formation of food vacuoles and suggested a "functional" term, the "cytostome." In the intervening years, there has been a tendency to retain the term cytostome within the Haemosporidia and to use micropore for other members of the Apicomplexa.

In order to determine the exact number of micropores in a given stage, a complete set of serial sections would have to be examined; this has not been done to date. However, the available evidence suggests that the number of micropores increases with the size of the developmental stage (Scholtyseck and Mehlhorn, 1970; Mehlhorn, 1972a; Scholtyseck, 1973a; Michael, 1975; Sénaud et al., 1976; Ferguson et al., 1977d; Porchet-Henneré, 1977). For example, in merozoites and sporozoites of *Eimeria* and *Isospora* and in merozoites of *Toxoplasma, Sarcocystis, Frenkelia,* and *Besnoitia,* usually one or two micropores have been observed (Sénaud, 1967; Scholtyseck and Mehlhorn, 1970; Porchet-Henneré and Vivier, 1971; Roberts et al., 1972; Scholtyseck, 1973a; Chobotar et al., 1975a; Varghese, 1975; Ferguson et al., 1977d; Mehlhorn and Heydorn, 1978).

Larger numbers of micropores are reported from developing schizonts and gamonts. Four micropores were observed in an area of 2.5 μm^2, and 5 in 2.0 μm^2 in macrogamonts of *Grellia (Eucoccidium) dinophili* and *Eimeria maxima,* respectively (Bardele, 1966; Mehlhorn, 1972a). The largest number of micropores observed is nine, on the surface of a microgamont of *Eimeria auburnensis* (Hammond et al., 1969).

The increase in numbers of micropores has been associated with nutrition and growth (Scholtyseck and Mehlhorn, 1970), and with the surface area of the parasite (Ferguson et al., 1977d). Ferguson et al. (1977d) hypothesized that genetic control for synthesis of the plasmalemma might be related to a similar control for micropore formation so that when the surface area of the organism increases the number of micropores would also increase. This may explain the absence of micropores in microgametes and their presence in stages such as developing oocysts and sporocysts.

While evidence for participation of the micropore in the nurition of the Haemosporidia has been clearly established (Aikawa, 1971), a similar function has been more difficult to demonstrate in other apicomplexans. Nevertheless, data have been accumulating that suggest that in the coccidia the micropore may function as a cytostome with the formation of food vacuoles and that this activity is related to its structural appearance (Scholtyseck and Mehlhorn, 1970; Scholtyseck, 1973a; Michael, 1975; Sénaud et al., 1976; Ferguson et al., 1977d; Speer, in press).

There is considerable size variation of the micropore cavity among various apicomplexans, even within the same species. In descriptions of micropores from more than 20 species, Scholtyseck and Mehlhorn (1970) and Scholtyseck (1973a) reported that the width of the orifice ranged from 60–200 nm and the depth of the invagination from 50–350 nm. These variations in size have been interpreted as reflecting different degrees of activity. As a result, the terms "active" and "inactive" are used to describe functional and nonfunctional micropores (Sampson and Hammond, 1971; Scholtyseck, 1973a; Speer et al., 1973a; Michael, 1975; Sénaud et al., 1976; Ferguson et al., 1977d; Sibert and Speer, 1980).

Inactive micropores are characterized by a shallow depression with relatively parallel sides of the limiting membrane and an absence of food vacuoles in the cytoplasm (Figures 45 and 46). The majority of micropores observed appear to be inactive (Sampson and Hammond, 1971; Sénaud et al., 1976; Ferguson et al., 1977d; Speer, in press), but reliable interpretation from studies of single isolated thin sections is difficult.

Active micropores have enlarged cavities and often contain material of the same appearance as seen in the parasitophorous vacuole (Figures 15–17 and 19). Such micropores have been variously described: as opening into vacuoles in *Eimeria bovis* (Scholtyseck et al., 1966) or into cytoplasmic canals in *Sarcocystis* and *Besnoitia* merozoites (Sénaud, 1966, 1967); as having cavities and channels with varied sizes in *Besnoitia* (Sheffield, 1967) and *Lankesterella garnhami* (Büttner, 1968); as containing fluid (Snigirevskaya and Cheissin, 1969) or particulate matter of parasitophorous vacuole origin (Hammond et al., 1967; Sénaud and Černá, 1968; Michael, 1975; Sénaud et al., 1976). Thus far, *Eimeria alabamensis* is the only member of the coccidia reported to ingest unaltered host cell cytoplasm (Sampson and Hammond, 1971). Pinocytotic vesicles (Snigirevskaya and Cheissin, 1969) or vacuoles have often been observed near active micropores (Sampson and Hammond, 1971; Chobotar et al., 1975a; Sénaud et al., 1976). The contents of the vacuoles and that of the cavity of active micropores appear similar or identical. There is some evidence that host cell cytoplasm may become altered before entry into the parasite, as indicated by the dissimilarity of material in the parasitophorous vacuole, micropore cavity, and food vacuoles when compared to the host cell cytoplasm (Hammond et al., 1967; Strout and Scholtyseck, 1970; Sénaud et al., 1976). Recently, Sibert and Speer (1980) observed that active micropores of gamonts of *E. nieschulzi,* which contained material similar to that of the crescent bodies and which are frequently observed in eimerian infections, may be partially digested host cell matter. It is believed that other structures, such as intravacuolar tubules and invaginations of the parasite augment the micropore in gaining nutriments for the parasite (Michael, 1975; Varghese, 1976a).

4

From the foregoing, it is evident that much information is lacking concerning coccidian nutrition. Physiological and cytochemical methods are needed to reveal the mechanism of acquisition and delivery of nutritive elements for growth and maintenance of the parasite.

ULTRASTRUCTURE OF THE DEVELOPMENTAL STAGES

Sporozoites and Merozoites

The general features of sporozoites and merozoites appear to be consistent in the coccidia. Many organelles, especially those of the apical complex, have been discussed in detail above. Therefore, a general overview is given here, with an attempt to emphasize aspects not included previously.

Sporozoites and merozoites are elongate, fusiform motile stages capable of selecting and penetrating host cells and initiating intracellular development. They are enclosed with a trimembranous pellicle, which may contain one or more micropores. Twenty-two or more longitudinal subpellicular microtubules are distributed around the periphery, just beneath the inner membrane. The anterior of the body is packed with a series of organelles, including micronemes, two or more rhoptries, often at least one Golgi complex, one or two polar rings, and a conoid preceded by two preconoidal rings. Other cytoplasmic organelles include lipid bodies, amylopectin granules, mitochondria, endoplasmic reticulum, ribosomes, and a nucleus (Figures 13*a,* 20, and 21).

The most prominent and distinctive structures of sporozoites are the refractile bodies. These bodies are osmiophilic, electron-dense, homogeneous organelles with no perceptible substructure; usually two are present, one anterior and one posterior to the nucleus (Figure 21). They have been observed in many species of *Eimeria* (Colley, 1967; Roberts and Hammond, 1970; Vetterling et al., 1973b; Scholtyseck, 1973a, 1979), in *Lankesterella hylae* (Stehbens, 1966), and recently in a species of *Isospora* from sparrows (Milde, 1979). Sporozoites of other species of *Isospora* differ in that they have crystalloid bodies in the areas usually occupied by refractile bodies (Roberts et al., 1972; Mehlhorn and Markus, 1976; Dubey and Mehlhorn, 1978). Crystalloid bodies are generally irregularly shaped and not membrane-bound. They consist of dense or hollow spherical subunits with a range of 25–42 nm in the various species. Crystalloid bodies of similar size and appearance have been described from other apicomplexans, including *Aggregata* (Porchet-Henneré and Richard, 1971), *Grellia (Eucoccidium)* (Bardele, 1966), *Coelotropha* (Porchet-Henneré, 1971), *Sarcocystis* (Becker et al., 1979), and *Hammondia* (Sheffield et al., 1976). They are also widely found in the Haemosporidia (Garnham et al., 1962; Trefiak and Desser, 1973; Desser and Allison, 1979). In *Aggregata* (Porchet-Henneré and Richard, 1971), these structures are in the form of scattered granules in the oocyst

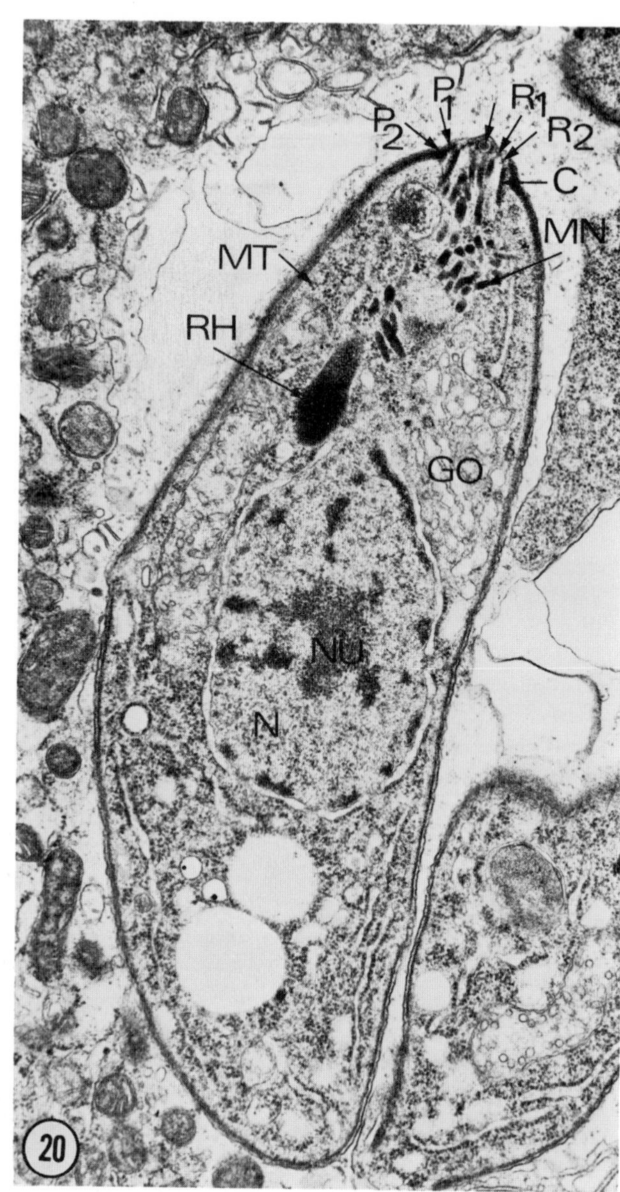

Figure 20. Merozoite of *E. ferrisi* showing typical fine structural features. Note the two polar rings and preconoidal rings and central pore-like structure at the apex of the conoidal area (*arrow*). × 18,000. Reprinted by permission from: Scholtyseck, E. *Fine Structure of Parasitic Protozoa*. Heidelberg. Springer-Verlag (1979). (Original, J. Sénaud).

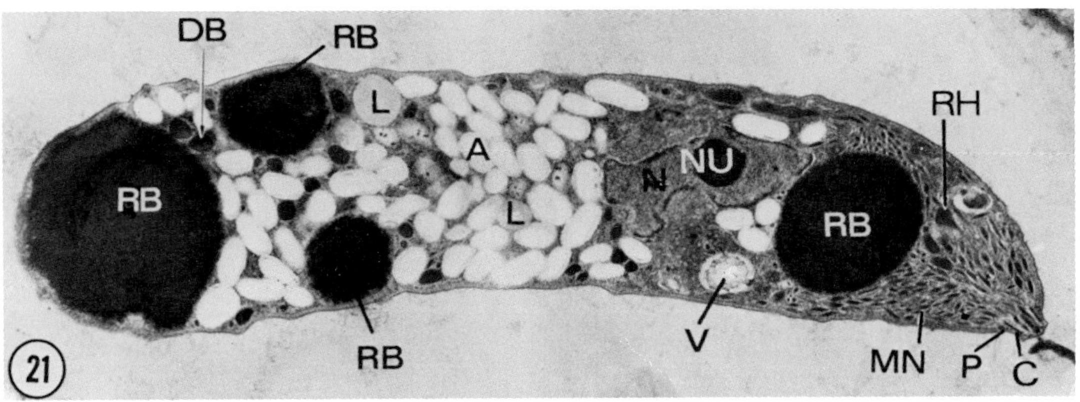

Figure 21. Longitudinal section of a sporozoite of *E. bovis* showing general ultrastructural features. Note the prominent refractile bodies (*RB*). ×12,000. Reprinted by permission from: Roberts, W. L., and Hammond, D. M. Ultrastructural and cytologic studies of the sporozoites of four *Eimeria* species. *Journal of Protozoology* 17:76–86 (1970).

and by aggregation become crystalline in the sporozoite. In an unusual finding, Milde (1979) described three forms of crystalloid bodies from three types of meronts in an *Isospora* species from sparrows. In one of the meront types, the crystalloid subunits appeared to be composed of hollow rods, most of which were parallel to each other forming short chains. Milde (1979) suggested that these inclusions may have originated from the homogeneous, electron-dense, refractile bodies of sporozoites. However, their significance in developing intracellular asexual stages is unknown.

The function of refractile and crystalloid bodies is unknown. Several authors have suggested that they constitute energy reserves for the parasite (Roberts and Hammond, 1970; Porchet-Henneré and Vivier, 1971; Trefiak and Desser, 1973; Hammond, 1973b; Vivier and Prouvost, 1977), but conclusive evidence to support this is lacking.

Host Cell Penetration and the Parasitophorous Vacuole

The mechanism of cell penetration by sporozoites and merozoites and the related question of whether the host cell membrane is interrupted at the point of entrance or remains intact has been of much interest and widely studied (reviewed by Doran, Chapter 6, this volume; Hammond, 1973b; Jensen and Hammond, 1975; Long and Speer, 1977).

From the available evidence, it is clear that the process of penetration is preceded by at least two events: motility of the parasites, including flexing and gliding, and the formation of an anterior stylet-like protuberance, the extended conoid, which can be thrust forward or retracted. Entrance is usually active and rapid (within seconds in many instances) and begins when the protuberance comes in contact with the cell surface. The area of the cell membrane which is in contact with the conoid region of the advancing parasite invaginates to form a cavity, while the membrane at the site of initial entry forms a ring causing a constriction of the parasite as it enters. As penetration is completed, the parasite is enclosed within a vacuole in the cell. Thus, for many species of coccidia, penetration is an active process initiated and carried out by the parasite.

Phagocytosis is another mechanism by which parasites enter cells. It has been a long-standing belief that sporozoites of chicken coccidia such as *Eimeria tenella* and *Eimeria acervulina* pass through the intestinal epithelium into the lamina propria where they are engulfed by macrophages which transport them to the glandular epithelium for further development. Eimerian sporozoites apparently have the ability to repeatedly invade and escape from macrophages, as well as survive within them for considerable periods of time, but further development is rare (El-Kasaby and Sykes, 1973; Rose, 1974; Michael, 1976; Long and Speer, 1977). Merozoites of *T. gondii* have been reported to enter cells by induced phagocytosis (Jones et al., 1972), but they also invade cells by an active process (Bommer et al., 1969).

Intracellular developmental stages of coccidia occupy a vacuolar space (the "parasitophorous vacuole") (Scholtyseck and Piekarski, 1965), which is usually bordered by a membrane. The nature of this membrane is apparently dependent on the mechanism of entry by the parasite. Several authors have shown that disruption of the host cell membrane occurred at the site of entry or after a brief invagination (Roberts et al., 1970b, 1971; Müller, 1975; Scholtyseck and Chobotar, 1975). In such cases, there are gaps in the membrane, or the parasite may be free in the host cell cytoplasm; however, membranes form within minutes to establish a continuous boundary of the parasitophorous vacuole (Roberts et al., 1970b; Müller, 1975).

On the other hand, sporozoites of *E. magna* and *I. canis* do not disrupt host cell membranes but produce a deep invagination, which is sealed off at the surface by short pseudopodia, leaving the parasite inside a completely membrane-bound vacuole. No interruption of host cell membranes occurs. However, the original invaginated host cell membrane breaks down quickly and is replaced by another membrane of host cell origin, which becomes the parasitophorous vacuole membrane (Jensen and Hammond, 1975; Jensen and Edgar, 1976b, 1978). This process is similar to that reported for *Plasmodium* and *Cryptosporidium* species except that no replacement of the original invaginated membrane occurred (Ladda et al., 1969; Vetterling et al., 1971; Bannister et al., 1975).

Thus, numerous studies on changes in the host cell membranes during penetration show obvious discrepancies. Interpretation is further complicated in that nearly all the reports involving cell penetration are based on in vitro studies in which environmental factors such as temperature, aging of media ingredients, variability and viability of cell types, and manipulation of cells during the fixation procedure may alter results significantly (Ladda et al., 1969; Vetterling et al., 1971; Jensen and Hammond, 1975). It is difficult to make valid comparisons with the in vivo situation in which the environment is much less variable and in which the rigidity of site and cell preference are well known (Long, 1976; Long and Speer, 1977).

The presence of a membrane-lined parasitophorous vacuole is characteristic of most species of coccidia, but there are exceptions. Extraintestinal meronts that occur prior to cyst formation in *Sarcocystis cruzi, Sarcocystis dispersa, Sarcocystis suihominis,* and *Frenkelia* species develop in free host cell cytoplasm of endothelial cells, not surrounded by a parasitophorous vacuole (Pacheco and Fayer, 1977; Sénaud and Černá, 1978; Heydorn and Mehlhorn, 1978; Göbel et al., 1978, Dubey et al., 1980). Host cell penetration was apparently not observed in these studies. Dubey et al. (1980) have suggested that if a parasitophorous vacuole does develop initially, the parasite might escape from it into the cytoplasm as *Trypanosoma cruzi* does, thereby avoiding destruction within phagolysosomes in macrophages, which this hemoflagellate often uses as a host cell. If among the endothelial cells, the precystic meronts also develop in reticuloendothelial cells, which are known to destroy ingested matter, then the ability to vacate the parasitophorous vacuole into the cytoplasm may be a means of survival within a hostile host cell (Dubey et al., 1980).

4

Development of Merozoites (Merogony)

A variety of terms have been used to describe the formation of merozoites in the coccidia. These include: schizogony, ectomerogony, endomerogony, exogenesis, endogenesis, endopolygeny, endodyogeny, multiple synchronous endopolygeny, and merogony. Schizogony, which is multiple fission, is the term previously used most frequently in the literature. However, Levine (1971) has pointed out that there are three types of multiple fission in the Apicomplexa—sporogony (formation of sporozoites), merogony (formation of merozoites), and gamogony (formation of gametes). Following the suggestions of Levine (1971) and Scholtyseck (1979), we will use the term merogony for any type of merozoite formation and consider the other terms as types of merogony. The different terms describe similar processes and differ chiefly in the dissimilar relationship between the nuclei and limiting membrane of the parasite and in the number of offspring produced.

After entering host cells and before nuclear division begins, sporozoites of some species of *Eimeria* may retain their shape and typical organelles from a few hours to several days before transforming into spheroid stages called uninucleate meronts (formerly called trophozoites). In other species, however, several nuclear divisions precede transformation to spheroid stages (reviewed by Speer, in press). Such multinucleate organisms which retain their elongate shape are called sporozoite-shaped meronts.

Transformation of uninucleate or multinucleate parasites to spheroid stages occurs by a gradual increase in width or by more rapid lateral outpocketings (Roberts et al., 1970b; Kelley and Hammond, 1972; Hammond, 1973b; Dubremetz and Elsner, 1979). Concurrent with the shape change is a gradual disappearance of the inner pellicular complex, subpellicular microtubules, and organelles of the apical complex (Colley, 1968; McLaren, 1969; Kelley and Hammond, 1972; Mehlhorn et al., 1973; Pacheco et al., 1975; Chobotar et al., 1975a; Dubremetz and Elsner, 1979). During transformation, the refractile bodies undergo changes in shape, size, and location. The anterior refractile body usually becomes smaller and disappears, whereas the posterior body undergoes fragmentation and gradually disappears (Roberts et al., 1970b; Kelley and Hammond, 1972; Hammond, 1973a,b; Dubremetz and Elsner, 1979) or becomes incorporated into merozoites (Hammond et al., 1970; Roberts et al., 1970b). Other ultrastructural changes include an increase in the size of the nucleus and nucleolus, an in crease in the amount of endoplasmic reticulum and ribosomes, an increase in the size and number of mitochondria, and a depletion of amylopectin granules (Hammond, 1973a,b; Pacheco et al., 1975).

The ultrastructural aspects of nuclear division preceding merozoite formation have been studied in *E. intestinalis, E. ovinoidalis (ninakohlyakimovae), E. necatrix, E. callospermophili, E. magna, E. falciformis, E. tenella, E. ferrisi, Globidium* species, and *E. bovis* (Snigirevskaya, 1969; Sénaud and Cerná, 1969; Roberts et al., 1970b; Dubremetz, 1971a, 1973, 1975; Danforth and Hammond, 1972; Kelley and Hammond, 1972, 1973; Hammond et al, 1973; Mehlhorn et al., 1973; Fernando and Stockdale, 1974; Hoppe, 1974, 1976; Pacheco et al., 1975; Chobotar et al., 1975a; Dubremetz and Elsner, 1979). This is a mitotic process associated with the appearance of kinetic centers (centrioles) and an eccentric or centrally placed intranuclear spindle.

The spindle apparatus consists of an undetermined number of microtubules (15 to 20 were found in *E. magna* (Hammond et al., 1973)) that originate at two dense poles, termed the centrocones (Sénaud, 1967). The centrocone of several species of *Eimeria* occurs within a protuberance or indentation of the nuclear envelope. One or two typical coccidian centrioles (nine single outer tubules surrounding a centrally located one) are present in the cytoplasm adjacent to the centrocone (Figure 22). In *E. magna, E. callospermophili* (Hammond et al., 1973), and *E. necatrix* (Dubremetz, 1971a, 1975), microtubules of the spindle originate at the apex of the centrocone and pass through openings in a two-layered membrane which represents an infolding of the nuclear envelope (Figure 22). Some spindle microtubules are continuous from pole to pole, whereas others are discontinuous and appear to terminate at dense structures, the kinetochores, which in *E. ovinoidalis (ninakohlyakimovae)* resemble the metaphase plate (Figure 23; Kelley and Hammond, 1973; Dubremetz, 1973; Hoppe, 1976; Dubremetz and Elsner, 1979).

As development proceeds, the chromosomes are drawn toward the centrocones and the nucleus becomes elongate (Figure 23). Division is finally completed by an infolding of the nuclear membrane at the midregion. The number of nuclear divisions is a species characteristic, as evidenced by the rather consistent number of merozoites produced by each meront generation in the host. It is a general phenomenon in coccidian development that merozoite formation is associated with the last nuclear mitosis (Porchet-Hennerė, 1972; Mehlhorn et al., 1973; Scholtyseck, 1973b; Dubremetz and Elsner, 1979).

There are two general processes of merozoite formation in the coccidia for which the terms exogenesis and endogenesis were proposed by Vivier (1970) and Porchet-Henneré (1972, 1977). Hammond (1973b) has suggested ectomerogony and endomerogony as analogous terms. We suggest use of the terms exogenesis and endogenesis because they are less restrictive and can be used for other phases (e.g., sporogony) of the life cycle (Ferguson et al., 1978b). Although in both processes merogenesis is associated with kinetic centers, the major difference lies in a dissimilar relationship between the mitotic poles and limiting membrane of the parasite. When the kinetic centers are associated with the limiting membrane during mitoses, as in the majority of the *Eimeria*, the merozoites form by exogenesis. When the merozoites form some distance from the limiting membrane as in *E. callospermophili* (Roberts et al., 1970a), *Toxoplasma* (Vivier, 1970; Sheffield, 1970), and *Sarcocystis* (Cerná and Sénaud, 1977), they form by endogenesis.

In exogenesis, merozoite formation is inititated by a thickening representing the inner membrane complex, and a conoid, immediately under the limiting membrane of the meront

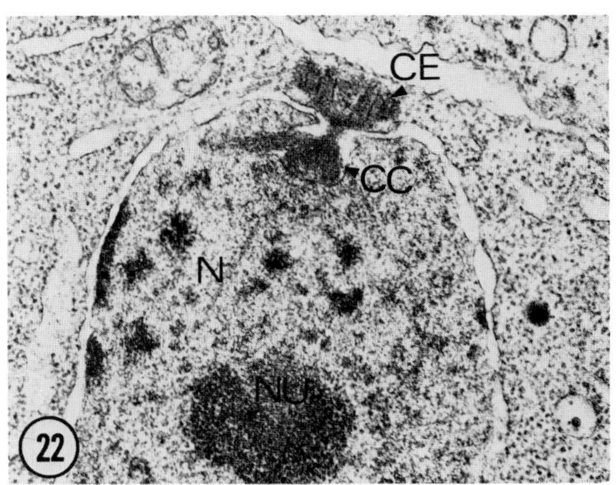

Figure 22. *Eimeria contorta.* Nucleus at the periphery of a meront just prior to initiation of merogony. Note the dense centrocone directed toward a pair of centrioles under the limiting membrane of the parasite (photograph supplied by B. E. G. Müller). ×12,000.

and adjacent to the centrioles and centrocone. This presumptive apical complex then becomes elevated at the surface of the meront; meanwhile, one or more osmiophilic bodies, the rhoptry anlagen, and a Golgi complex appear near the conoid (Figure 24). During elongation of the merozoite, subpellicular microtubules appear to be synthesized simultaneously with the inner membrane complex; the limiting membrane of the meront forms the outer membrane of the merozoite pellicle. In cross-sections, each pair of microtubules appears to be associated with a cresent-shaped section of the inner membrane complex (Roberts et al., 1970a; Dubremetz, 1975; Dubremetz and Elsner, 1979). In *E. necatrix* and *E. bovis,* an electron-dense fiber connects the conoid and the centrioles, suggesting that the latter plays a role in merozoite differentiation (Dubremetz, 1973; Dubremetz and Elsner, 1979). In the final stages of differentiation, typical rhoptries are formed, and the nucleus with the centrocone at its anterior end becomes incorporated into the merozoite, followed by entrance of the mitochondira and development of micronemes (Figure 25). Differentiated merozoites detach from the residual cytoplasm and appear in random or in tightly packed bundles.

Incorporation of two or more nuclei into merozoites has been reported for several species of *Eimeria* (Sénaud and Černá, 1969; Černá and Sénaud, 1971; Danforth and Hammond, 1972). The role of multinucleate merozoites in the life cycle is not known, but Danforth and Hammond (1972) suggested that they may represent stages in the formation of new merozoites.

With minor variations, the cytoplasmic changes just described occur during merogony in the following coccidia: *E. nieschulzi, E. tenella, E. magna, E. pragensis, E. ovinoidalis (ninakohlyakimovae), E. falciformis, E. necatrix, E. ferrisi, Globidium gilruthi,* and *E. bovis* (Colley, 1968; Sénaud and Černá, 1968, 1969; McLaren, 1969; Kelley and Hammond, 1972; Mehlhorn et al., 1973; Dubremetz, 1973, 1975; Fernando and Stockdale, 1974; Hoppe, 1974, 1976; Pacheco et al., 1975; Chobotar et al., 1975a; Porchet-Henneré, 1977; Dubremetz and Elsner, 1979).

In endogenesis, the merozoites originate internally in association with nuclei, centrocones, and centrioles. In later development, the merozoites assume a peripheral location and grow outward at the surface, with the limiting membrane of the mother meront forming the outer membrane of the merozoite, as in exogenesis.

Endodyogeny is characterized by the formation of two daughter organisms within the mother cell that is used up in the process (Figure 26). This type of reproduction is especially common in the cyst-forming coccidia. However, in recent years it has been found to occur in other coccidian genera.

The ultrastructural characteristics of endodyogeny are nearly identical in all the species studied. In *Toxoplasma* (Sheffield and Melton, 1968; Vivier, 1970), the process begins with the transformation of the parasite from a fusiform to an ovoid shape after entering the host cell. The Golgi apparatus divides into two parts, and two dome-shaped structures representing the inner membrane complex of the merozoite anlagen develop in the anterior of the mother cell.

Figures 23–25. Merogony by exogenesis in *E. bovis.* Reprinted by permission from: Dubremetz, J. F., and Elsner, Y. Y. Ultrastructural study of schizogony of *Eimeria bovis* in cell cultures. *Journal of Protozoology* 26:367–376 (1979). 23. Nuclear mitosis with a centrocone (and eventually a daughter nucleus) directed toward the merozoite anlagen. Note the spindle apparatus and structures resembling kinetochores. ×36,000. 24. Surface development of merozoites at a stage that corresponds to the one shown in Figure 23. ×5,700. 25. Part of a meront with merozoites in an advanced stage of development, although still attached to the residual cytoplasm. × 10,000.

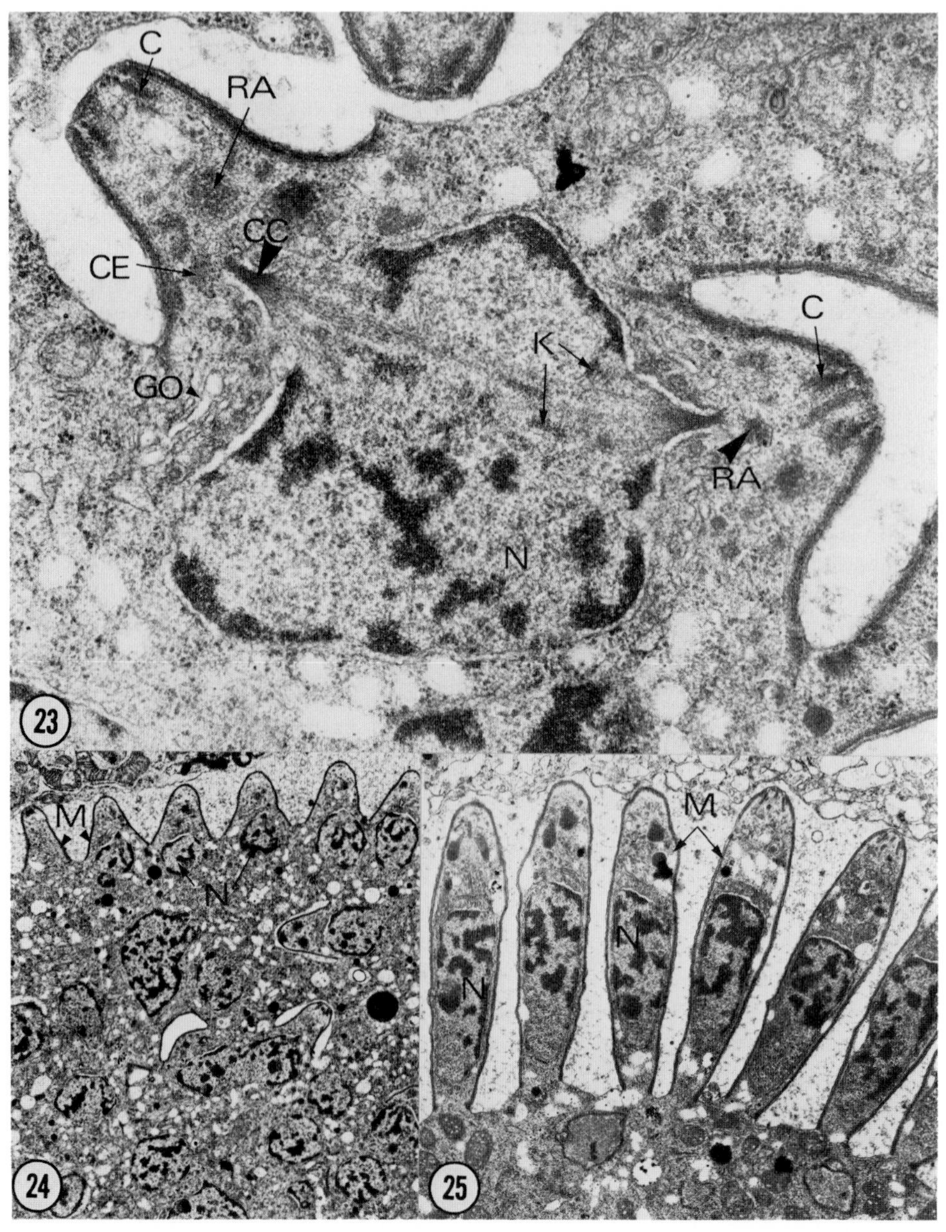

Nuclear spindles, centrocones, and centrioles are present (Kepka and Scholtyseck, 1970). Components of the apical complex appear as the inner membrane complexes grow posteriorly around each pole of the dividing nucleus. The nucleus then divides with each portion becoming incorporated into the daughter merozoite. As the merozoites reach the surface, the inner membrane complex of the mother cell disappears and its outer membrane joins the inner membrane complex of the daughter cell to complete formation of the pellicle.

In endopolygeny, two or more nuclear divisions occur before merozoite formation is initiated so that four or more nuclei are present (Hammond, 1973a,b). Except for the number of daughter merozoites formed, the process is similar to that of endodyogeny. Since merozoite development begins internally, both endodyogeny and endopolygeny can be considered as types of endogenesis.

An interesting variation of endopolygeny recently has been described in precystic extraintestinal merogony in *Sarcocystis* species (Černá and Sénaud, 1977; Pacheco and Fayer, 1977; Sénaud and Černá, 1978; Heydorn and Mehlhorn, 1978; Dubey et al., 1980) and in *Frenkelia* (Göbel et al., 1978). The meront, which maintains its trimembranous pellicle, develops in various organs, mainly in endothelial cells of small blood vessels, where it is situated immediately within the host cell cytoplasm. A parasitophorous vacuole is absent. Merogony is characterized by a marked enlargement and lobulation of the parasite nucleus, the appearance of a spindle apparatus and centrioles at each lobe, and the simultaneous formation of scores of daughter merozoites anlagen before nuclear division occurs (Figure 27). This process has been termed "synchronous multiple endopolygenesis" by some authors (Černá and Sénaud, 1977; Sénaud and Černá, 1978). The final stages of development, division and incorporation of nuclei, dedifferentiation of the inner membrane complex of the mother cell, and incorporation of its limiting membrane into the merozoite pellicle occurs as in endodyogeny.

Merozoites of the various species of coccidia vary somewhat in their size, the presence or absence of refractile bodies, and the number of subpellicular microtubules and of micronemes. Otherwise, there is a close structural resemblance among them. However, as studies of the different groups of coccidia have been published, a proliferation of names for merozoites have appeared, some of which include: zoite, endozoite, endodyocyte, cystozoite, schizozoite, tachyzoite, bradyzoite, proliferative forms, cyst forms, and metrocyte. Frenkel (1973, 1974) introduced the terms "tachyzoite" for rapidly multiplying precystic merozoites, which develop in "groups," and "bradyzoite" for the slowly multiplying organisms in cysts of the cyst-forming coccidia. We suggest some standardization of these terms for the sake of clarity and better communication both in the research laboratory and in the classroom. This would avoid, for example, the current and recent use in the literature of no less than five names (zoite, cystozoite, endodyocyte, endozoite, and bradyzoite) for the cyst merozoites of *Sarcocystis*. Therefore, we propose that the term "merozoite" be used as much as possible to describe the product of merogony. We would further suggest that the term "group" (Frenkel, 1973, 1974,

Figure 26. *T. gondii*. Development of two daughter merozoites by endodyogeny. Note that the inner membrane complex of the mother cell is absent in places. × 25,000. Reprinted by permission from: Vivier, E., and Petitprez, A. Le complexe membranaire superficiel et son évolution lors de l'élaboration des individus fils de *Toxoplasma gondii*. *Journal of Cell Biology* 43:329–342 (1969).

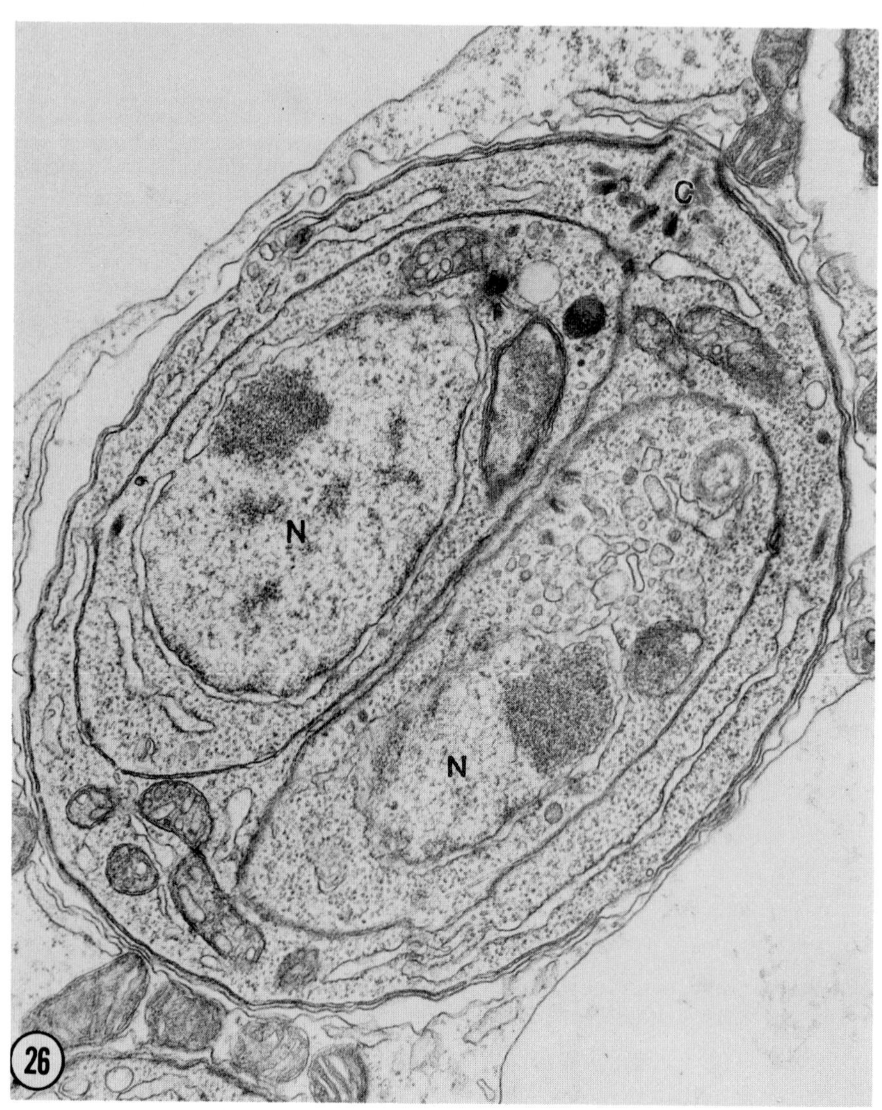

Figure 27. *S. dispersa*. A precystic meront in hepatic cells of *Mus musculus*. Note the merozoite anlagen developing at nuclear lobes of the enlarged nucleus (original, J. Sénaud and Ž. Černá). ×8,000.

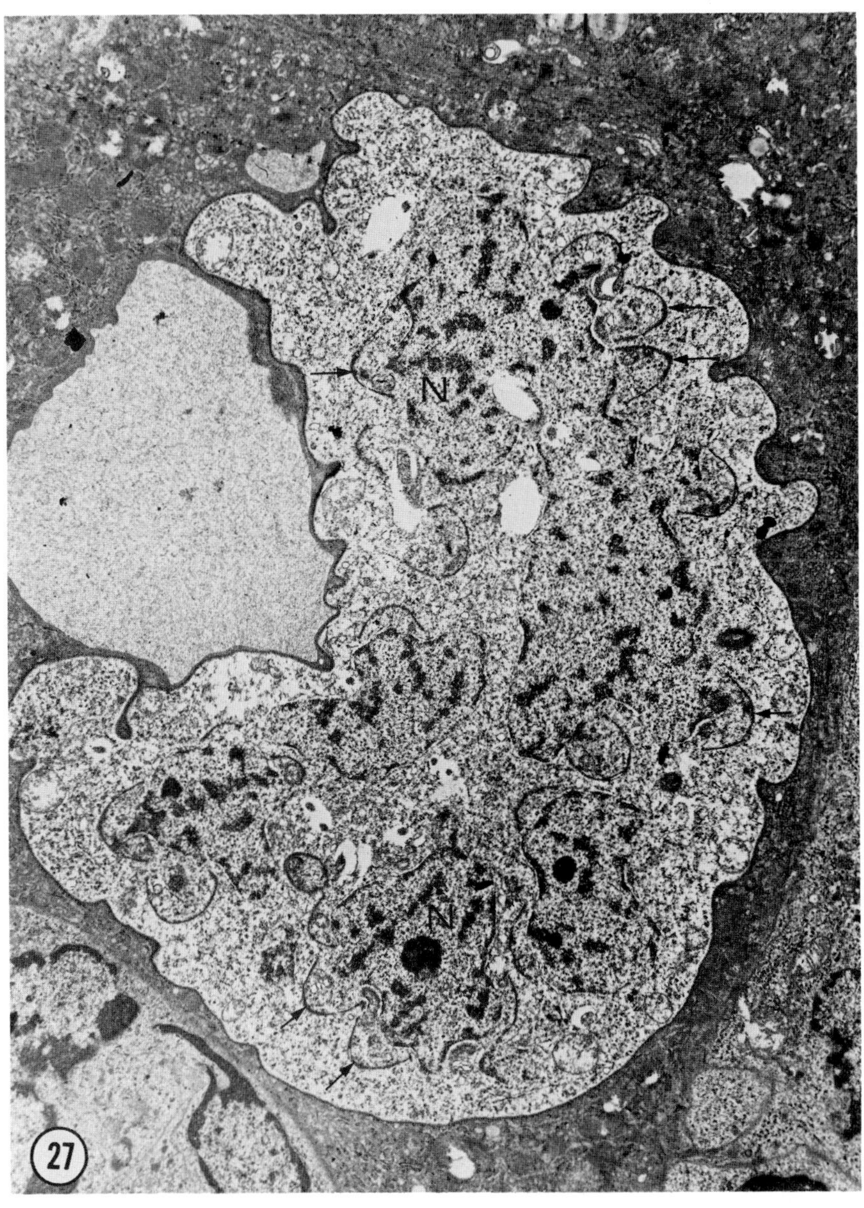

1977) is in reality a precystic proliferative meront which forms proliferative merozoites. The product of tissue cysts would then be cyst merozoites. The term "metrocyte" or mother cell (Sénaud, 1967) should be retained because it is a specialized merozoite (Mehlhorn and Heydorn, 1978; Scholtyseck, 1979) of tissue cysts whose structure and role are described below. Thus, the number of terms for the merozoites can be reduced from nine or more to essentially two, with a corresponding reduction in ambiguity and in the possibility of confusion.

Cyst Formation in the Coccidia

Members of the genera *Sarcocystis, Toxoplasma, Besnoitia, Frenkelia,* and *Hammondia* are known as the cyst-forming coccidia, because they produce characteristic tissue cysts in the intermediate host (usually an omnivore or herbivore) after ingesting sporulated oocysts or sporocysts from the final host (usually a carnivore). In all five genera, excysted sporozoites apparently leave the intestine and initiate one or more generations of precystic proliferative merogony-producing merozoites (tachyzoites), which invade new host cells, resulting in cyst formation. Members of the genus *Sarcocystis* produce cysts mostly in skeletal muscle, of *Hammondia* in skeletal and cardiac muscle, of *Toxoplasma* in the brain and in skeletal and cardiac muscle, of *Besnoitia* in various connective tissue cells (probably fibroblasts), and of the genus *Frenkelia* in the brain and spinal cord (summarized by Frenkel, 1977).

The ultrastructural aspects of cyst formation and mature cysts have been studied most extensively in *Sarcocystis* species (Scholtyseck et al., 1974; Mehlhorn et al., 1975a,b,c, 1976; Sheffield et al., 1977; Pacheco et al., 1978; Mehlhorn and Heydorn, 1978; Mehlhorn and Frenkel, 1980).

In *Sarcocystis* species, the parasite is situated within a typical parasitophorous vacuole bordered by a single unit membrane, which shortly becomes reinforced by a deposition of irregularly arranged underlying segments of osmiophilic material. This complex, termed the primary cyst wall ("Primärhülle," Mehlhorn and Scholtyseck, 1973b), is a characteristic of all the cyst-forming coccidia except *Besnoitia* (Sénaud, 1967; Kepka and Scholtyseck, 1970; Scholtyseck et al., 1974; Mehlhorn et al., 1976; Mehlhorn and Frenkel, 1980), and may reach a thickness of 20–100 nm early in the development of *Sarcocystis* species. The unreinforced areas of the cyst wall have a diameter of about 40 nm and form vesicle-like invaginations into the interior of the cyst (Figure 28 *a–c*). In some species, the reinforced areas develop characteristic protrusions which vary in length, shape, and micromorphology, according to the species. Thus, the walls of mature cysts may show protrusions, which are finger-like and tightly packed as in *Sarcocystis ovicanis* (Figures 28*d1* and 29), branched and cauliflower-like as in *S. tenella (S. ovifelis)* (Figure 28*d2*), or spherical blebs on short stalks as in *S. booliati* and *S. muris* (Mehlhorn et al., 1976; Kan and Dissanaike, 1976; Sheffield et al., 1977). Positioned immediately beneath the primary cyst wall in the interior of the parasitophorous vacuole is a band of amorphous ground substance, which may contain fibrillar and granular material. From the ground substance, numerous branching septa extend into the interior of the cyst, forming compartments which contain the parasites (Figure 29). Such compartments occur also in *Frenkelia* species whose cyst wall is very similar to that of *Sarcocystis* species, except that the surface in contact with host cell cytoplasm in the brain and spinal cord produces short irregular villosities and no prominent protrusions (Kepka and Scholtyseck, 1970; Chalupsky and Sénaud, 1973; Scholtyseck et al., 1974; Mehlhorn and Frenkel, 1980).

Cysts of *Sarcocystis* and *Frenkelia* species contain two types of parasites: metrocytes and merozoites (bradyzoites), whereas in cysts of *Hammondia, Besnoitia,* and *Toxoplasma* only merozoites are present. Both divide by endodyogeny in which two daughter cells are formed from a parent cell by a process described elsewhere in this chapter (Mehlhorn and Scholtyseck, 1973b; Scholtyseck, 1973a, 1979; Scholtyseck et al., 1974; Mehlhorn et al., 1975a,b,c; Rzepczyk and Scholtyseck, 1976; Sheffield et al., 1977; Pacheco et al., 1978). Metrocytes are globular to ovoid in shape with an irregular surface and are enclosed with a typical coccidian pellicle with underlying microtubules (Figure 31). One or more micropores are usually present. The cytoplasm is of a low electron density and contains ribosomes, mitochondria, endoplasmic reticulum, and a few amylopectin granules. The presence of a conoid has been reported, but rhoptries and micronemes are only occasionally found, indicating that these latter organelles are not typical of metrocytes.

Transformation from the fusiform-infecting merozoite to a metrocyte apparently occurs very early, as evidenced by the presence of one or two metrocytes in very young cysts of *S. cruzi (S. bovicanis)* (Pacheco et al., 1978). In early to intermediate stages of growth, the cyst is populated only by metrocytes. However, after an undetermined number of endodyogenies, merozoites are formed which accumulate in the central area of the cyst. By the time a cyst reaches maturity, it contains only merozoites and may be macroscopic, reaching a length of 10 mm or more (Mehlhorn et al., 1975c).

The structue of cyst merozoites is similar to that of coccidian sporozoites and merozoites described earlier, each possessing a trimembranous pellicle, micropores, preconoidal rings, conoid, polar rings, subpellicular microtubules, rhoptries, and micronemes. Cysts harboring only merozoites are highly infective to the final host and may survive in the intermediate host for years without further development. On the other hand, cysts containing only metrocytes, or metrocytes with small numbers of merozoites, are noninfective to the final host (Ruiz and Frenkel, 1976; Sheffield et al., 1977; Mehlhorn and Frenkel, 1980).

In a recent study by Mehlhorn and Frenkel (1980), it was observed that skeletal muscle cysts of *Hammondia hammondi* and *T. gondii* had a similar ultrastructure, except that merozoites in *H. hammondi* were smaller (4–5 μm) than in *T. gondii* (7–8 μm). *H. hammondi* has an obligate heteroxenous life cycle, in which tissue cysts from the intermediate host (prey) infect only the final predator host and sporulated oocysts infect only the intermediate host.

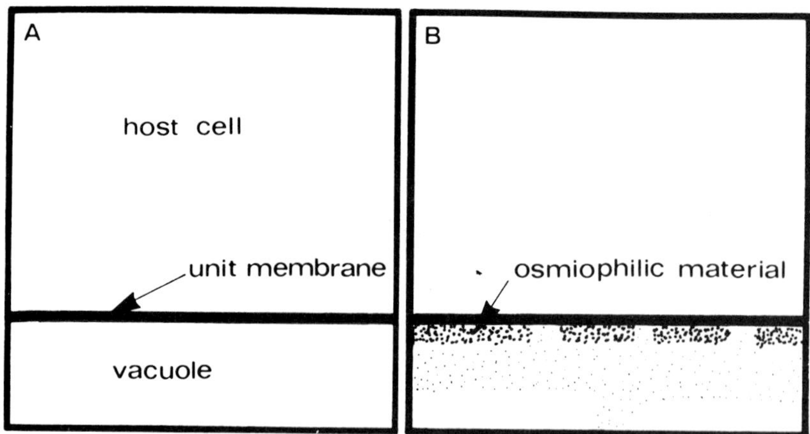

Cysts of *H. hammondi* and *T. gondii* were both limited by a relatively smooth but intensely folded primary cyst wall with an inner layer of ground substance. Young cysts of *H. hammondi* were in direct contact with sarconeme filaments, but older cysts were surrounded by an electron-pale zone, 0.5 μm thick, containing many small vesicles (Figure 30). In *T. gondii,* no such zone developed and the cyst wall remained directly adjacent to the muscle filaments. The cysts contained typical coccidian slender merozoites, which divide by endodyogeny. No compartments were found. The structure of *T. gondii* cysts in the brain is nearly identical to those found in muscle (Scholtyseck et al., 1974).

Several authors have described the cysts of *Besnoitia* species (Sheffield, 1968; Sénaud, 1969; Scholktyseck et al., 1974). The parasites lie in the parasitophorous vacuole of a hypertrophied host cell with one or more nuclei. The parasitophorous vacuole is lined with a unit membrane that possesses none of the features observed in *Sarcocystis, Frenkelia, Toxoplasma,* and *Hammondia* species; therefore, a primary cyst wall is absent in *Besnoitia* species. A ground substance and compartments are not present, but small elongate protrusions resembling microvilli and vesicles extend inward from the parasitophorus vacuole membrane. The outer surface of the host cell is greatly increased by the presence of irregularly shaped pseudopod-like extensions. This area is in direct contact with a thick (up to 15 μm) fibrous layer of host origin, termed the secondary cyst wall. This wall is a characteristic feature of *Besnoitia* species and of *S. tenella (S. ovifelis).*

Several species of *Isospora* have now been transmitted to the final host via experimentally infected rodents, making this heretofore perceived homoxenous parasite one that is facultatively heteroxenous (Frenkel and Dubey, 1972; Mehlhorn and Markus, 1976; Frenkel, 1977; Dubey and Mehlhorn, 1978). These investigations show that certain isosporan sporozoites can invade and persist for several months in extraintestinal organs of rodents. During this time, the parasite is viable and retains all structural characteristics of an isosporan sporozoite, including a prominent crystalloid body. The sporozoites invade lymphoreticular cells producing a unit membrane-lined parasitophorous vacuole. At first, the parasitophorus vacuole is electron-pale but later becomes filled with electron-dense granular and filamentous material, giving the appearance of a sheath or cyst wall by light microscopy. However, a primary cyst wall as seen in *Sarcocystis* species is not present. Frenkel (1977) has called these intracellular stages "monozoic cysts" and has proposed that those species possessing this trait be placed in a new genus, *Cystoisospora.* However, Levine (see Chapter 1 this volume) has synonymized *Cystoisospora* with *Isospora.* A comparison of tissue cysts produced by several coccidia is summarized in Figure 32.

Development of Gametes (Gamogony)

After a specific number of asexual generations, the merozoites produced develop into gamonts rather than meronts. The factors that determine sexual differentiation are not at present well

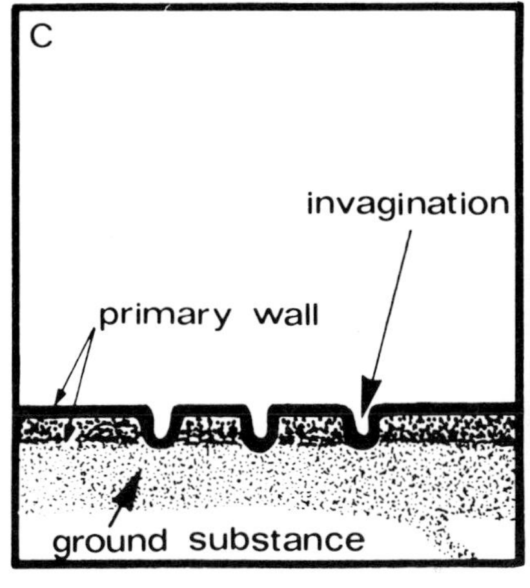

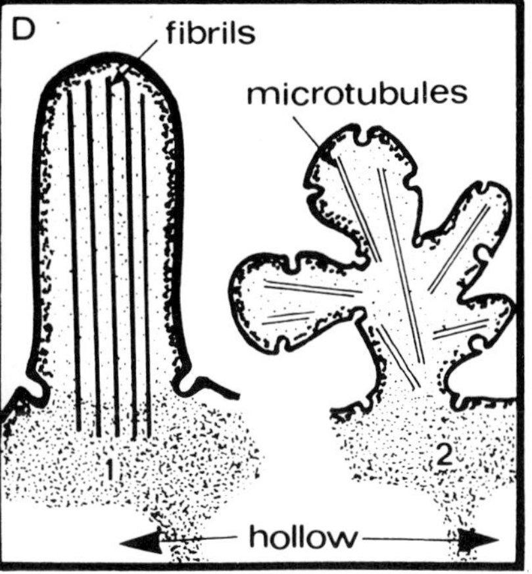

Figure 28. Diagrammatic representation of the formation of the primary cyst wall within a host cell by steps *A–D*. The complete primary wall may form protrusions (*D1* and *D2*) or not (C). Reprinted by permission from: Mehlhorn, H., Hartley, W. J., and Heydorn, A. O. A comparative ultrastructural study of the cyst wall of 13 Sarcocystis species. *Protistologica* 12:451–467 (1976).

understood. The consistency in the number of meront generations formed and in the timing of the whole cycle, including the appearance of sexual development, suggests that genotypic characteristics of the parasite are involved. However, interference with the hosts' immune system can significantly modify site and host specificity as well as the life cycle, indicating that these events are under partial control of the host (Long and Rose, 1970; Rose, 1973).

The early events of gamogony are similar to those already described for merogony. Merozoites from the final asexual generation penetrate host cells, where they reside in a parasitophorous vacuole and undergo transformation from a fusiform to a spheroid shape with the subsequent loss of the organelles of the apical complex.

Microgametogenesis There are many studies describing microgamete formation in the coccidia. Some of these have been reviewed by Scholtyseck et al. (1972a) and Scholtyseck (1973a). More recent reports include those by Scholtyseck et al. (1973, 1977); Vetterling et al. (1973a); Ferguson et al. (1974, 1977a, 1980a); Speer and Danforth (1976); Varghese (1976b); Madden and Vetterling (1977); Mehlhorn and Heydorn (1979); and Becker et al. (1979).

Microgametogenesis follows a similar pattern in the coccidia and can be divided into two phases: 1) the growth phase during which the gamont increases in size and nuclear division takes place similar to that observed in merogony; 2) the actual formation and differentiation of the microgametes. During the growth phase, it is usually difficult to identify microgamonts because of their similarity to meronts. In *Eimeria brunetti,* however, Ferguson et al. (1977a) found that the nuclei and cytoplasm of microgamonts differed in several ways from that of meronts. The most pronounced difference was in the microgamont nucleus which lacked a nucleolus and had chromatin which aggregated in dense patches at the periphery. There was a decrease in the size of the nucleus, which did not occur in the meront. Similar changes in nuclei of young microgamonts were observed in *E. magna* (Speer and Danforth, 1976).

Microgamete production is essentially a surface phenomenon in which the nuclei assume a peripheral location next to the limiting membrane of the parasite. In those species producing large numbers of microgametes, the surface area greatly increases by the development of fissures and invaginations into the interior to accommodate the developing gametes (Hammond et al., 1969; Mehlhorn, 1972c; Ferguson et al., 1977a). Other species which produce few microgametes per gamont show no such subdivisions.

Microgamete formation begins when the gamonts have peripheral nuclei over which develops a surface protrusion containing two centrioles (Figures 33 and 37*a,b*). Also present is an intranuclear spindle apparatus with one or two centrocones each directed to the centrioles (Figure 34). A mitochondrion is in close association with each nucleus. The centrioles develop into basal bodies and give rise to the flagella, which grow from the protrusion into the parasitophorous vacuole; each flagellum is covered by the limiting membrane of the gamont. Studies of microgametogenesis in *E. tenella* (Madden and Vetterling, 1977) by scanning electron

Figures 29 and 30. 29. Tissue cyst of *S. ovicanis*. Note the tightly packed protrusions of the primary cyst wall (*PCW*), the ground substance (*GS*) that forms the septa (*SE*) of the compartments, the merozoites (*M*), and metrocytes (*MC*). ×5,000. Reprinted by permission from: Mehlhorn, H., Heydorn, A. O., and Gestrich, R. Licht- und elektronenmikroskopische Untersuchungen an Cysten von *Sarcocystis ovicanis* Heydorn et al. (1975) in der Muskulatur von Schafen. *Zeitschrift für Parasitenkunde* 48:83–93 (1975). 30. Tissue cyst of *H. hammondi*. Note the relatively smooth primary cyst wall (*PCW*) and lack of compartments. ×10,000. Reprinted by permission from: Mehlhorn, H., and Frenkel, J. K. Ultrastructural comparison of cysts and zoites of *Toxoplasma gondii*, *Sarcocystis muris*, and *Hammondia hammondi* in skeletal muscle of mice. *Journal of Parasitology* 66:59–67 (1980).

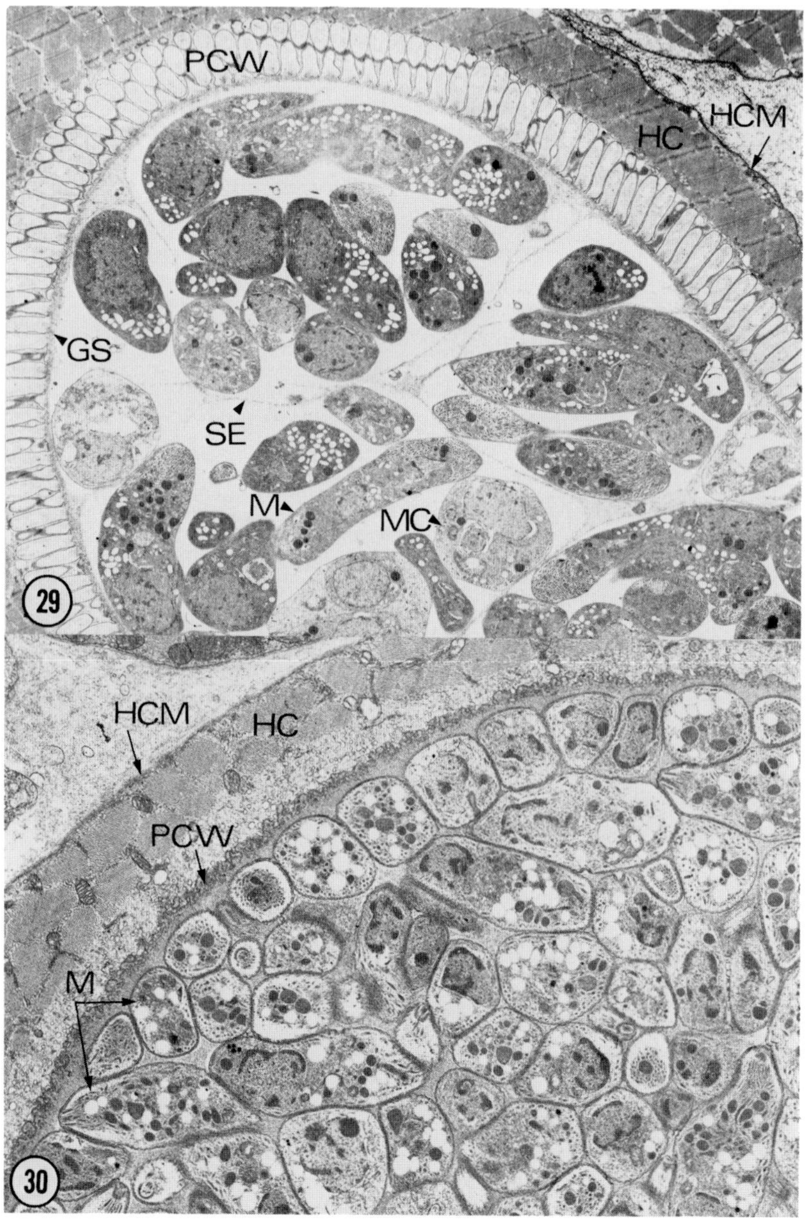

4

Figure 31. Diagrammatic representation of a metrocyte found in young to intermediate tissue cysts of *Sarcocystis* and *Frenkelia* species. Note the ovoidal shape in infolded pellicle. Reprinted by permission from: Scholtyseck, E. *Fine Structure of Parasitic Protozoa.* Heidelberg. Springer-Verlag (1979).

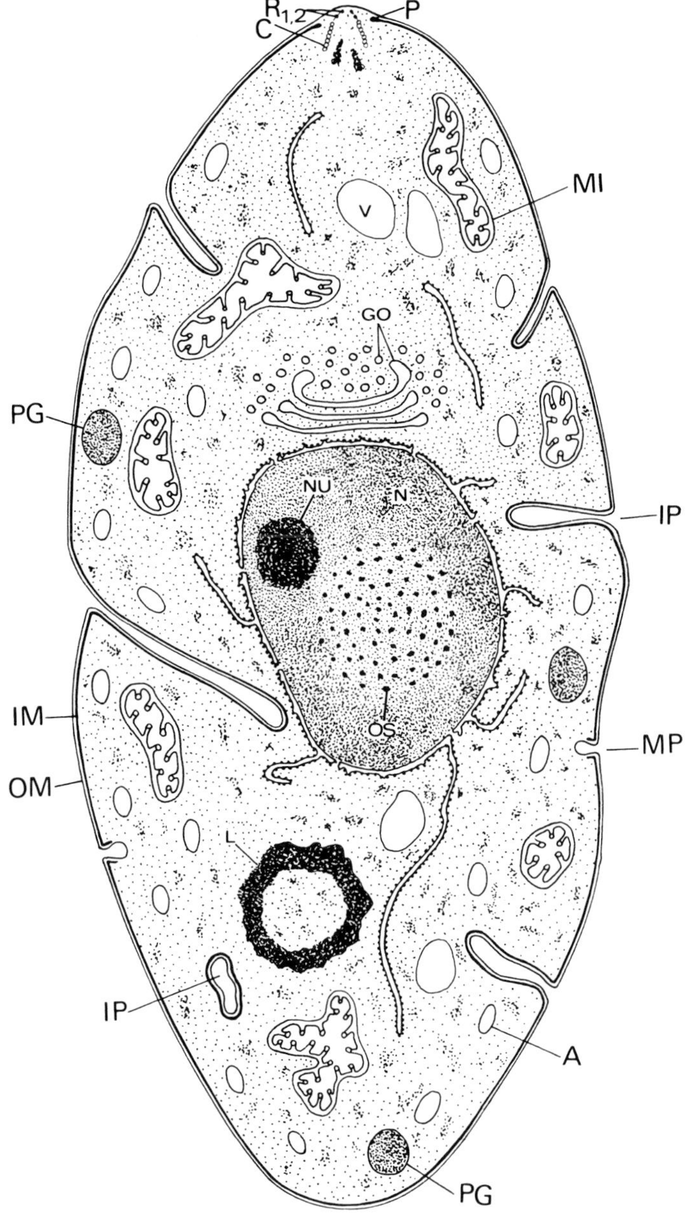

Figure 32. Diagrammatic representation of cysts in the different cyst-forming genera. *1.0*, a parasite in a parasitophorous vacuole (*PV*) bounded by a single membrane (*arrow*). *2.1*, in *Besnoitia* species cysts, the original PV is considerably enlarged and filled with large numbers of merozoites. Even in old cysts, the PV is lined by a single unreinforced membrane. The parasites divide by endodyogeny, but no metrocytes are present. The hypertrophied host cell becomes surrounded by a fibrous secondary cyst wall (*SCW*). *2.2*, young cysts of *Frenkelia* and *Sarcocystis* species (*2.2a*), and *Toxoplasma* and *Hammondia* species (*2.2b*). The membrane of the PV has become reinforced by osmiophilic material, thereby becoming a primary cyst wall (*PCW*). In cysts of *Frenkelia* and *Sarcocystis* (*2.2a*), metrocytes (*MC*) are present dividing by endodyogeny, whereas in *Toxoplasma* and *Hammondia*, the slender merozoites divide by endodyogeny. *3.1*, mature cysts of *Frenkelia* and *Sarcocystis* species are characterized by the presence of typical septa (*SE*) formed by the ground substance (*GS*). In *Frenkelia* and some *Sarcocystis* species, the PCW does not form long protrusions (*3.1a*), whereas in other sarcosporidia, typical protrusions are found (*3.1b*). With cysts of *S. tenella* (*S. ovifelis*), an SCW is present (*3.1c*). *3.2*, the PCW of *Toxoplasma* and *Hammondia* cysts remains relatively smooth, and internally are filled with slender merozoites. Septa are not present. Reprinted by permission from: Mehlhorn, H., and Frenkel, J. K. Ultrastructural comparison of cysts and zoites of *Toxoplasma gondii*, *Sarcocystis muris*, and *Hammondia hammondi* in skeletal muscle of mice. *Journal of Parasitology* 66:59–67 (1980).

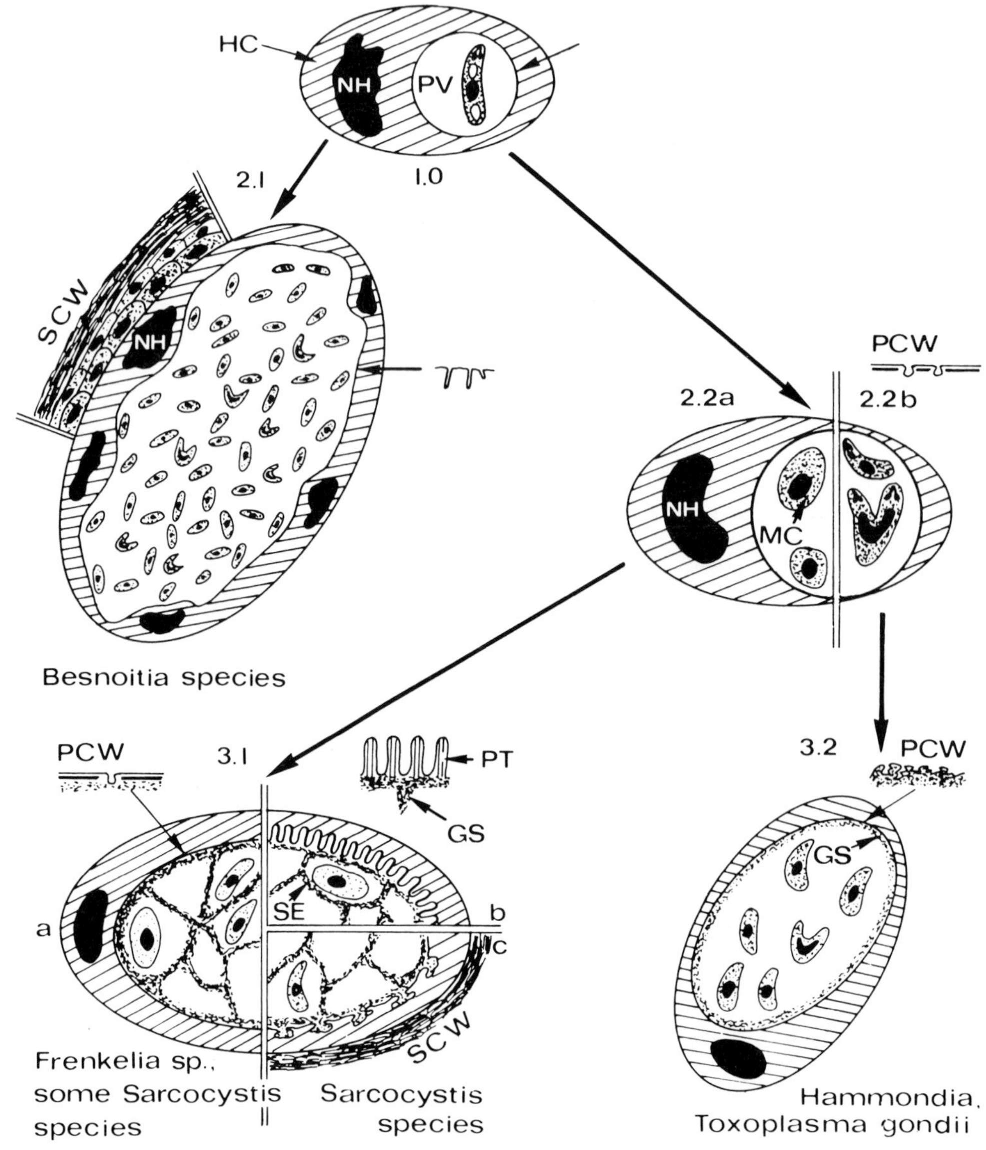

Figures 33 and 34. Microgamonts of *E. ferrisi*. Reprinted by permission from: Scholtyseck, E., Chobotar, B., Sénaud, J., and Ernst, J. V. Fine structure of microgametogenesis of *Eimeria ferrisi* Levine and Ivens, 1965 in *Mus musculus*. *Zeitschrift für Parasitenkunde* 51:229–240 (1977). 33. Intermediate microgamont with peripheral nuclei. ×11,000. 34. Peripheral nucleus showing a centrocone and adjacent centriole. × 24,000.

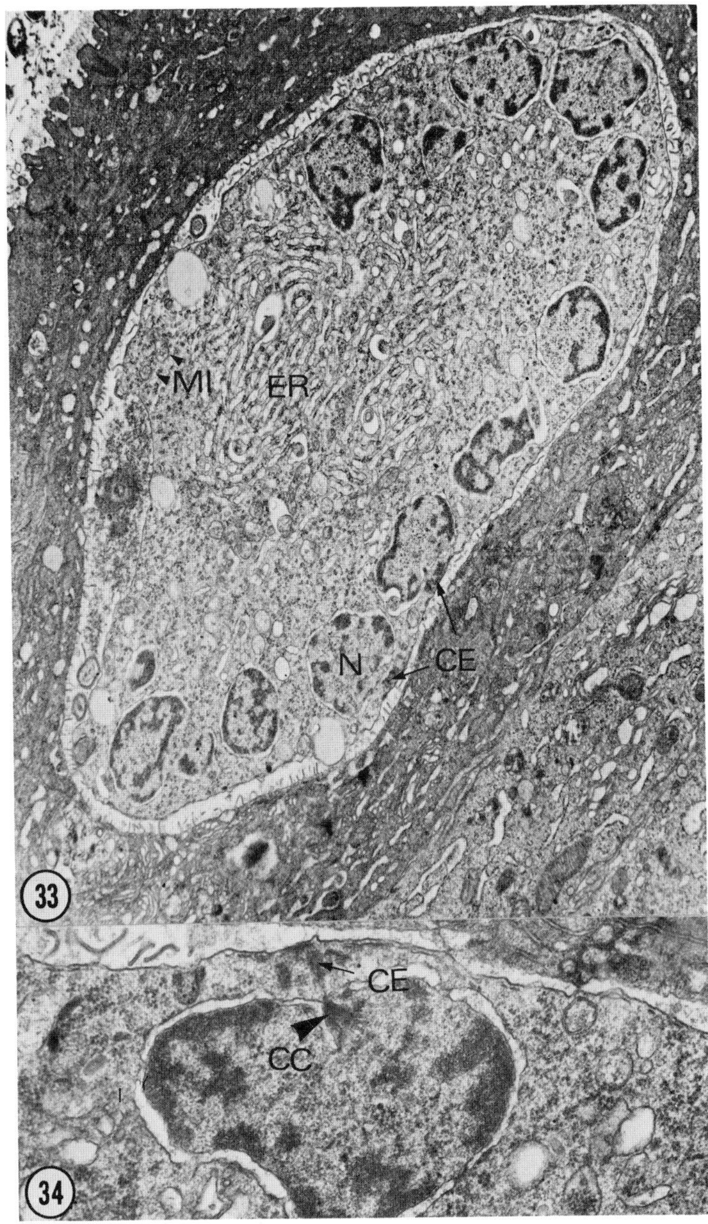

microscopy and in *Isospora* species (Milde, 1979) by transmission electron microscopy clearly show that the flagella emerge at the surface of the microgamont before emergence of the body of the microgamete. This is probably true for most species of coccidia, although verification is difficult from single thin sections. In some species, a dense membrane-like plaque, observed between the limiting membrane and the centrioles, may represent the anlage of the perforatorium (Hammond et al., 1969; Vetterling et al., 1973a; Speer and Danforth, 1976; Ferguson et al., 1977a). During flagellar growth, the nucleus becomes elongate and develops a dense osmiophilic area which contains the condensed chromatin, and a pale granular area (Figures 35 and 37*c,d*). The dense portion of the nucleus, accompanied by the mitochondrion, enters the protrusion to become the nucleus of the microgamete, and the pale portion remains in the cytoplasm (Figure 37*e*). A thickening of the pellicle forms a ring at the base of each protrusion so that the developing microgamete is attached to the gamont by a gradually narrowing stalk occupied by a channel connecting the dense and pale portions of the nucleus. Maturation of the microgamete is marked by elongation of the mitochondrion, nucleus, and flagella followed by separation from the residual cytoplasm in which the pale portion of the nucleus remains as a residual nucleus (Figures 36 and 37*e,f*). In some *Sarcocystis* species, the microgamont nucleus does not undergo subdivision prior to microgamete formation but becomes enlarged, forming lobes which reach to the periphery of the macrogamont (Mehlhorn and Heydorn, 1979; Becker et al., 1979). The chromatin undergoes condensation in each lobe, forming the presumptive microgamete nucleus. Twenty to 30 microgametes are formed simultaneously at the surface of the microgamont.

A mature microgamete has an elongate crescent-shaped body up to 7 μm long and is enclosed in a unit membrane. The dense nucleus occupies most of the body and is often pointed at its anterior end. This point is directed toward the perforatorium at the anterior pole of the microgamete. Immediately posterior to the perforatorium are the basal bodies; each flagellum is attached to a basal body and emerges near the anterior end of the macrogamete. Two or three flagella may be present, depending on the species. Regardless of whether two or three flagella occur, additional microtubules (four or more) are present; they appear to originate in the basal body zone and extend posteriorly along the body with some tubules terminating before others. The mitochondrion, which now possesses several rows of tubular cristae, is situated at the anterior end, usually in a groove of the nucleus (Figures 36, 37*f*, 38, and 39).

Macrogametogenesis The ultrastructural characteristics of macrogametogenesis have been described for various coccidia, including many *Eimeria* species (Lee and Millard, 1971; Scholtyseck et al., 1971; Mehlhorn, 1972b; Scholtyseck, 1973a, 1979; Speer et al., 1973b; Doens-Juteau and Sénaud, 1974; Varghese, 1975; Chobotar et al., 1975b, 1980; Wheat et al., 1976; Ferguson et al., 1977b; Pittilo and Ball, 1979), *Isospora* species (Pelster, 1973; Milde, 1979; Ferguson et al., 1980b), *T. gondii* (Pelster and Piekarski, 1972; Ferguson et al., 1975),

Figures 35 and 36. Microgametogenesis in *E. ferrisi.* Reprinted by permission from: Scholtyseck, E., Chobotar, B., Sénaud, J., and Ernst, J. V. Fine structure of microgametogenesis of *Eimeria ferrisi* Levine and Ivens, 1965 in *Mus musculus. Zeitschrift für Parasitenkunde* 51:229–240 (1977). 35. Early stages of microgamete differentiation showing dense and pale portions of the nucleus. ×24,000. 36. Free microgametes in the parasitophorous vacuole, one of which is sectioned longitudinally. ×13,000.

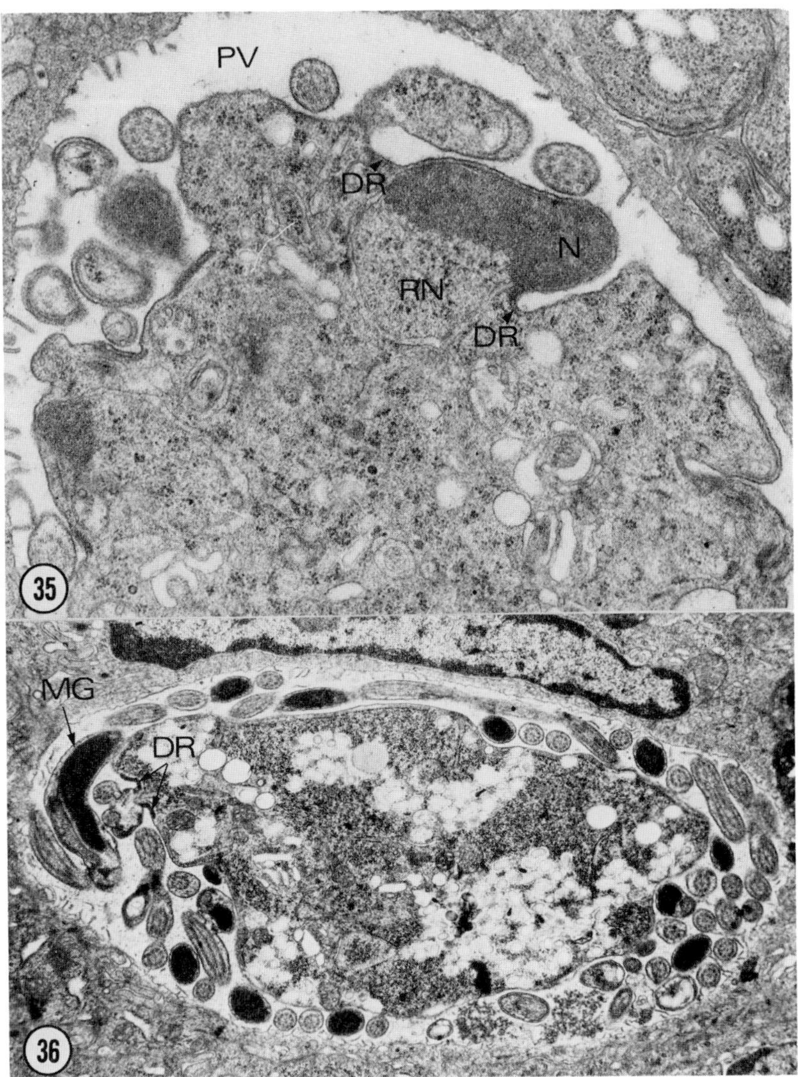

Figure 37. Diagrammatic representation of microgametogenesis. Reprinted by permission from: Scholtyseck, E. *Fine Structure of Parasitic Protozoa.* Heidelberg. Springer-Verlag (1979). See text for explanation.

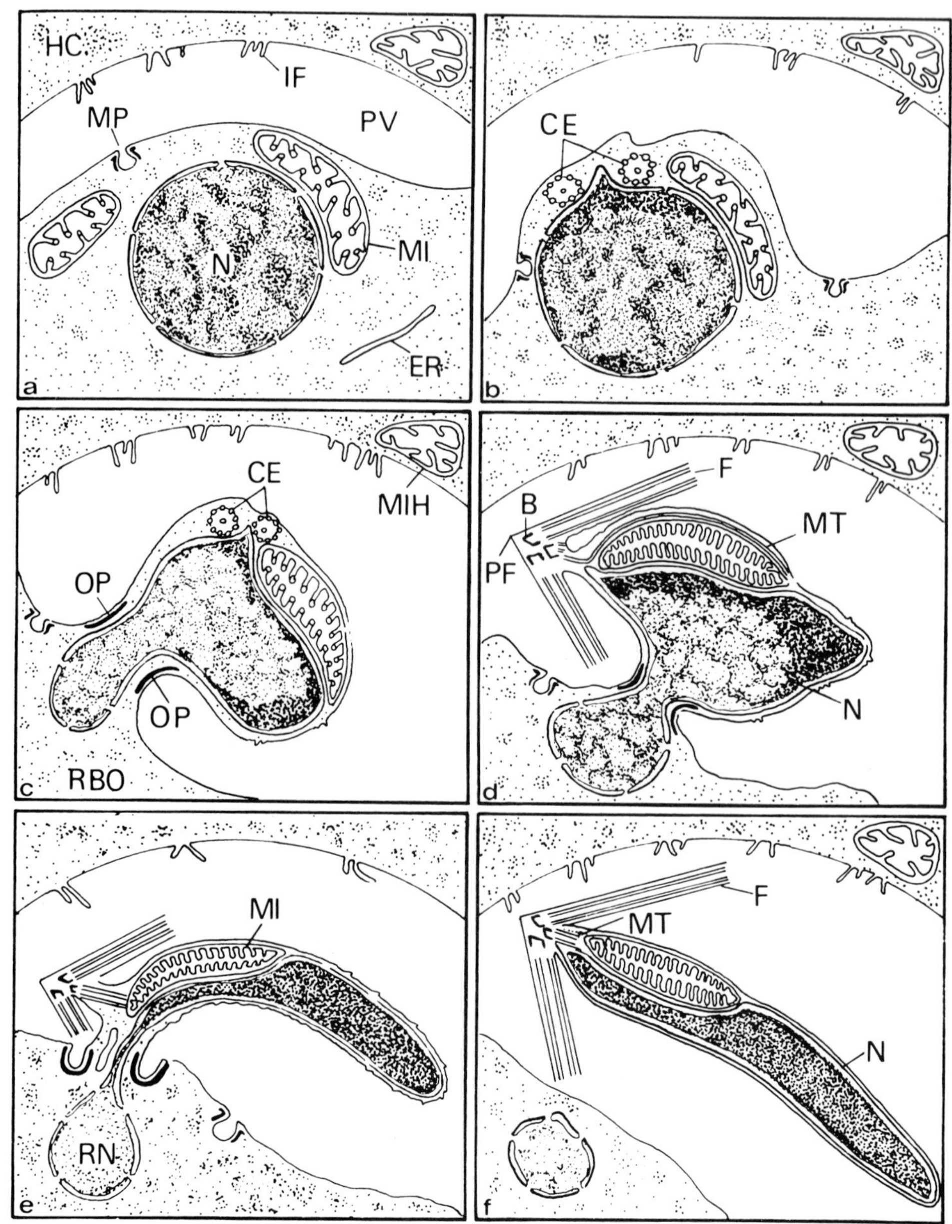

4

Figures 38 and 39. Longitudinal section of *E. maxima* and *E. magna*, respectively. 38. Note the prominent nucleus and three flagella (*F*). × 55,000. Reprinted by permission from: Mehlhorn, H. Elektronenmikroskopische Untersuchungen an Entwicklungsstadien von *Eimeria maxima* (Sporozoa, Coccidia): II. Die Feinstruktur der Mikrogameten. *Zeitschrift für Parasitenkunde* 40:151–163 (1972). 39. Microgamete with two flagella, one of which is attached for a short distance to the body. Note the perforatorium (*PF*) and auxiliary microtubules (*MT*). ×29,000. Reprinted by permission from: Speer, C. A., and Danforth, H. D. Fine-structural aspects of microgametogenesis of *Eimeria magna* in rabbits and in kidney cell cultures. *Journal of Protozoology* 23:109–115 (1976).

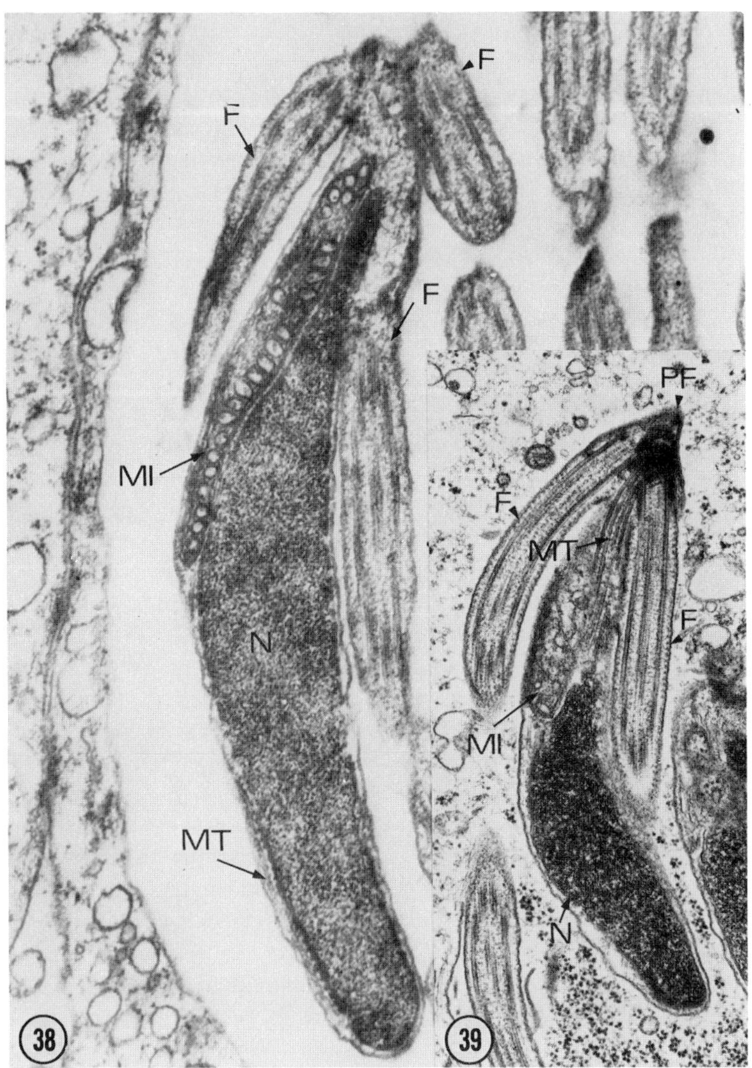

and *Sarcocystis* species (Vetterling et al., 1973a; Zaman and Colley, 1975; Mehlhorn and Heydorn, 1979; Becker et al., 1979).

These studies show that cytologic changes, which distinguish young macrogamonts, occur soon after host cell entry. After assuming an ellipsoid or spheroid shape during the dedifferentiation process, young macrogamonts can be differentiated from microgamonts because the former have an enlarged nucleus and a prominent osmiophilic nucleolus. Portions of the inner pellicular complex, some micronemes, and subpellicular microtubules may persist for a time as remnants of the former merozoite. There is usually a marked increase in the rough endoplasmic reticulum (rER), ribosomes, mitochondria, and Golgi complexes. Young macrogamonts have little or no lipid bodies or amylopectin granules, but these become increasingly numerous as development proceeds.

Most accounts describe a single limiting membrane in early stages, but the number of membranes often increases to two or three in later stages. In *E. maxima,* however, the outer two membranes of the merozoite pellicle persist (Mehlhorn, 1972b). In *T. gondii* (Ferguson et al., 1975) and in *Isospora felis* (Ferguson et al., 1980b), the trimembranous pellicle remains throughout macrogametogenesis. Some species retain a single limiting membrane through macrogamete development (see Chobotar et al., 1975b and Varghese, 1975). One to several micropores regularly occur in macrogamonts.

The most distinctive feature of macrogamonts is the presence of wall-forming (WF) bodies which begin to develop in relatively early stages, with those of type 1 (WF1) usually appearing later than those of type 2 (WF2). These bodies were originally described by Scholtyseck and Voigt (1964) and Scholtyseck et al. (1969), and were named because of their fate rather than the order of their appearance. Thus, the WF1 gave rise to the outer layer of the oocyst wall and the WF2 to the layer. Some species of *Sarcocystis* produce only one type of wall-forming body (WF1) (Vetterling et al., 1973a; Zaman and Colley, 1975).

Although the process of wall-forming body development is still incompletely understood, several recent reports show that the endoplasmic reticulum and Golgi complex are actively involved (Ferguson et al., 1975, 1980b; Wheat et al., 1976). The membranes of cisternae of the rER are heavily studded with ribosomes, except where they are adjacent to a Golgi complex. Small osmiophilic transfer vesicles bud off the surface of the smooth membrane and appear at the anterior face of the Golgi complex. These vesicles are thought to contain polypeptides and proteins that accumulate in the cisternae. These products are possibly complexed with other substances in the Golgi saccules and released in secretory vesicles at the posterior face of the Golgi in a manner similar to that described for protein synthesis in eukaryotes (Palade, 1975). It is believed that in macrogamonts, material from the secretory vesicles accumulates to form the WF1, because the newly formed WF1 nearby are very similar to the contents of the vesicles. Mature WF1 are usually homogeneous and strongly osmiophilic and sometimes bounded by a membrane.

The WF2 are synthesized within cisternae of the rER, often in close association with a Golgi complex. These bodies are usually not limited by a membrane; however, they remain folds consist mainly of two closely apposed unit membranes and are randomly distributed around the circumference of the vacuole. Some folds lose their connection from the host cell membrane and are believed to disintegrate, thus contributing to the amorphous and particulate material in the parasitophorous vacuole; such material may be involved in parasite nutrition (Hammond et al., 1967; Scholtyseck et al., 1971; Scholtyseck, 1973a; Michael, 1975).

Also within the parasitophorous vacuole, often associated with the surface of macrogamonts, are numerous intravacuolar tubules (IT). These were first described by Scholtyseck and Schäfer (1963) in *Eimeria perforans* and have been reported from many, but not all, species of coccidia studied (reviewed by Scholtyseck et al., 1971; Michael, 1975; and Ferguson et al., 1977b). In cross-section the IT resemble electron-clear vesicles, ranging from 50–110 nm in diameter, while in longitudinal sections they have transverse striations (Figures 41 and 42). Their length is difficult to determine because they undulate in and out of the plane of the section, but some portions measure up to 6 μm. Scholtyseck (1973a) suggested that the IT were of parasitic origin anchored at the parasite's surface with a direct structural connection to the host cell, and may serve for transport of materials between the host cell and the parasite. More recent reports have supported Scholtyseck's observations (Michael, 1975; Pittilo and Ball, 1979; Milde, 1979). Direct evidence for a transport role was demonstrated in *E. acervulina* in which some IT were connected to areas of the host cell containing pockets of ribosomes; some of the ribosomes were observed within the tubules themselves (Michael, 1975). In a species of *Isospora* within the cisternae during growth and are surrounded by an electron-pale space. Fully developed WF2 may vary in different species from a homogeneous osmiophilic appearance to a sponge-like labyrinthine nature (see Scholtyseck et al., 1971; Scholtyseck, 1973a). There is considerable variation in the size of the WF among the species, but both types may reach a diameter of 1.5 μm or more.

In young to intermediate macrogamonts, several membrane-bound vesicles with contents similar to that of the nucleus often occur at the periphery of the nucleus (Lee and Millard, 1971; Mehlhorn 1972a; Speer et al., 1973a; Varghese, 1975: Ferguson et al., 1975, 1977b, 1980b; Wheat et al., 1976). In some species, these vesicles appear to bud off the nuclear surface and have been termed Golgi adjuncts, multimembranous vacuoles, and nuclear detachment bodies. They persist throughout macrogamont development, but their function is unknown.

With further development, there is an increase in the size of the macrogamont and a corresponding increase in the size and number of organelles peculiar to this stage. At maturity, the parasite is no longer considered to be a macrogamont but a macrogamete. At this time, the macrogamete is bounded by one to three membranes, has WF usually near the periphery, with lipid bodies and amylopectin granules interior to the WF and a large nucleus and nucleolus (Figure 40).

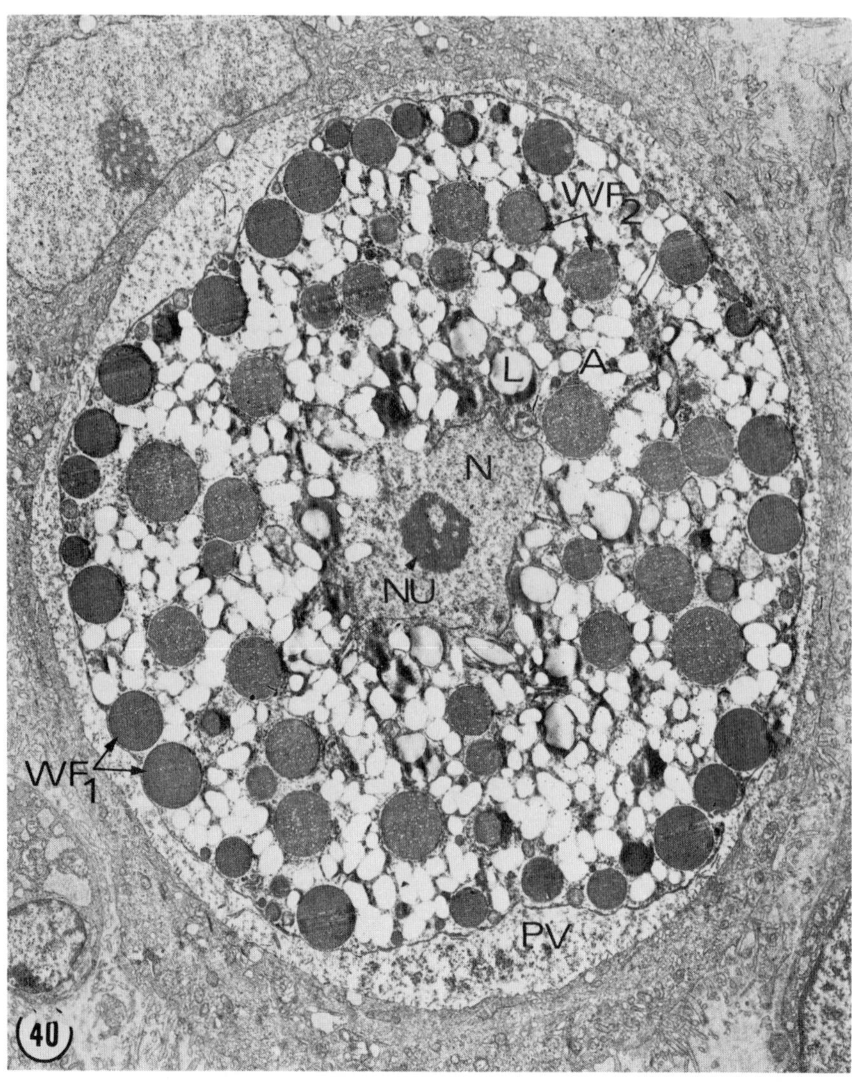

Figure 40. Macrogamete of *E. auburnensis.* Note the two types of wall-forming bodies (WF1 and WF2) with the WF1 at the periphery. ×7,500. Reprinted by permission from: Scholtyseck, E., Mehlhorn, H., and Hammond, D. M. Fine structure of macrogametes and oocysts of Coccidia and related organisms. *Zeitschrift für Parasitenkunde* 37:1–43 (1971).

In most species, the parasitophorous vacuole of the developing macrogamont is limited by a single membrane which often has numerous fine folds oriented toward the parasite. These from sparrows, the IT were formed by a differentiation of the limiting membrane of the parasite which at first formed a series of small elevations that grew outward to form an arc and eventually formed tubules (Milde, 1979; Figure 42).

Fertilization

Current opinion is that fertilization of a macrogamete by a microgamete resulting in the formation of a zygote occurs prior to oocyst wall formation. Many attempts have been made to record this act by electron microscopy, but only partial success has been achieved.

In 1970, Scholtyseck and Hammond saw longitudinal and cross-sections of a microgamete situated in a membrane-bound vacuole near the nucleus of a macrogamete of *E. bovis*. These authors believed that the microgamete actively enters the cytoplasm of the macrogamete and not a peripherally directed lobe of the nucleus, as proposed by earlier investigators. However, nuclear fusion was not observed by Scholtyseck and Hammond (1970).

In a scanning electron microscope study, Madden and Vetterling (1977) saw penetration of a macrogamete by the microgamete of *E. tenella*. In the same specimen, additional macrogametes adhered to the surface of the macrogamete; a similar observation was made by Marquardt (1966) in a light microscope study of *E. nieschulzi*.

In a more recent report (Sheffield and Fayer, 1980), fertilization in *S. cruzi (S. bovicanis)* differed from that of *E. bovis* and *E. tenella*. In *S. cruzi,* the body of the microgamete was positioned on the surface of the macrogamete between two membranes, one of host origin and the other of parasitic origin (Figure 43). The microgamete had a typical protozoan mitochondrion, several microtubules, and was invested by a single unit membrane. According to Sheffield and Fayer (1980), fertilization proceeds by fusion of the plasmalemmas of the two gametes, resulting in the formation of a narrow reinforced stalk through which the nucleoplasm of the microgamete passes into the macrogamete cytoplasm. A nearly identical process was recently observed by Schäffler (1979) in *Sarcocystis miescheriana (S. suicanis)* and by Entzeroth (1980) in a study of *Sarcocystis* species of roe deer. Fertilization in members of the Haemosporidia also occurs by membrane fusion and entry of microgamete contents into the macrogamete (Desser, 1972; Galucci, 1974).

As stated by Sheffield and Fayer (1980), the paucity of observed fertilizations makes interpretation of the differences seen in *Eimeria* and *Sarcocystis* difficult. Galucci (1974) noted that fertilization in *Haemoproteus columbae* is a rapid event, a fact that delayed electron microscope confirmation of the process for many years. Gamete to gamete contact and the changes subsequent to entrance of the microgamete or its contents may also be of short dura-

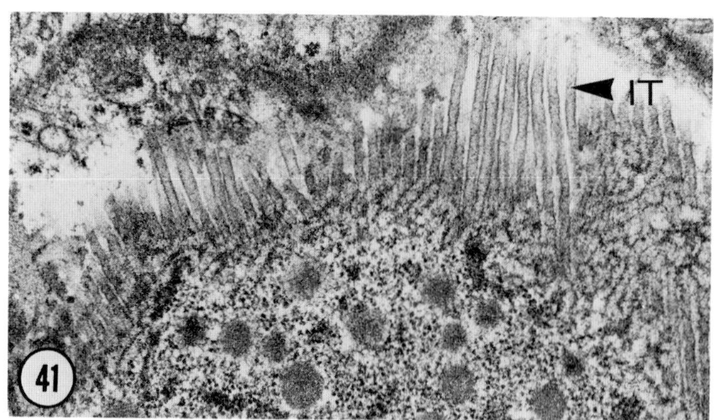

tion in the coccidia, thereby explaining, at least in part, the lack of observed fertilizations. Also, an ultrathin section represents only a minute portion of the relatively large macrogamete; this increases the possibility of an attached or penetrated microgamete being out of the plane of the section under observation.

Oocyst Wall Formation

Mature macrogametes and young zygotes are indistinguishable (Scholtyseck, 1973a), but cytoplasmic changes soon occur in the zygote, which signal the onset of wall formation. The ultrastructural details of oocyst wall development have been studied extensively by Scholtyseck et al. (1969, 1971); Lee and Millard (1971); Dubremetz and Yvore (1971); Mehlhorn (1972a); Scholtyseck (1973a); Vetterling et al. (1973a); Doens-Juteau and Sénaud (1974); Chobotar et al. (1975b, 1980); Ferguson et al. (1975, 1977c); Varghese (1975); Zaman and Colley (1975); Wheat et al. (1976); Michael (1978); Milde (1979); Becker et al. (1979); Mehlhorn and Heydorn (1979); and Sibert and Speer (1980).

Sibert and Speer (1980) have correctly pointed out that the majority of these studies describe incompletely formed walls. The young oocyst becomes gradually impermeable to fixatives and embedding media, thereby producing poor quality specimens for study. As a result, most of the observations deal with formation of the outer layer of the wall; however, the latter stages, immediately before oocyst release, have not been completely elucidated.

Initiation of wall formation has been consistently related to changes in the nature of the WF1 coupled with a proliferation of membranes at the surface of the zygote. The WF1 undergo disaggregation into smaller particles, which appear in the cytoplasm near the pellicle. Coincident with particulation of the WF1, additional membranes develop beneath the original zygote pellicle so that five or more membranes may eventually be present (Figure 45). The outer two or three membranes (numbers vary with the species) become elevated from the surface, providing a space in which particles of the WF1 are deposited forming the presumptive outer layer of the oocyst wall (Figure 46). As the material which forms the outer layer of the oocyst wall accumulates at the perimeter of the parasite, there is a corresponding decrease and eventual disappearance of the WF1.

In *Eimeria papillata,* no elevation of the peripheral membranes occurred initially. Rather, the WF1 particles accumulated in a layer beneath the pellicular membranes; the inner margin of this layer was continuous with the cytoplasm (Figure 44). Separation is accomplished later by membranes of the rER, which had accumulated below the granular layer (Chobotar et al., 1980).

Formation of the inner layer of the oocyst wall usually occurs before completion of the outer layer. The WF2 assume a more peripheral position and may undergo particulation and/or become markedly labyrinthine. The WF2, or portions of them, then move into intramembranous spaces beneath the developing outer layer and coalesce to form the inner layer

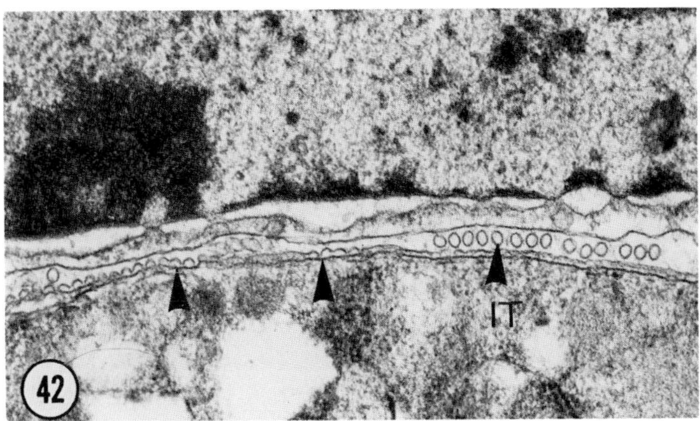

Figures 41 and 42. Intravacuolar tubules of a macrogamont in an *Isospora* species from sparrows. Reprinted by permission from: Milde, K. Light and electron microscope studies on isosporan parasites (Sporozoa) in sparrows (*Passer domesticus* L.). *Protistologica* 15:607–627 (1979). 41. Intravacuolar tubules (*IT*) originate in the macrogamont and appear to connect with the host cell. × 28,000. 42. Cross-sections of intravacuolar tubules (*IT*) which seem to be formed by the outer limiting membrane of the macrogamont. × 28,000.

(Figure 47) (see Ferguson et al., 1977c; Milde, 1979; Sibert and Speer, 1980). In several species of *Sarcocystis,* the outer layer of the oocyst wall seems to consist of material of WF1 origin and the inner layer of one (Vetterling et al., 1973a) to four (Zaman and Colley, 1975; Becker et al., 1979; Mehlhorn and Heydorn, 1979) closely applied membranes.

Thus, the several membranes formed at the periphery of the zygote and the WF, appear to play an important role in wall formation. However, little is known of the method by which wall material is transferred across the membranes. Sibert and Speer (1980) have suggested that in *E. nieschulzi* WF1 material possibly disaggregates into submicroscopic molecules for transfer and subsequently reaggregates in the intermembranous spaces. According to these authors, the WF2 material is probably transferred by exocytosis.

The cytoplasmic mass of a mature oocyst is termed the sporont. The sporont has a nucleus with a nucleolus, amylopectin granules, lipid bodies, mitochondria, endoplasmic reticulum, and ribosomes. Wall-forming bodies are absent.

At first, the newly formed layers are irregular in appearance and thickness (Figures 45–47). However, in subsequent development, they undergo condensation so that in mature oocysts of most coccidia the wall consists of an outer homogeneous electron-dense layer and an inner electron-lucent layer (Figures 48 and 49). Mature oocysts of some species of coccidia possess a micropyle. Sibert and Speer (1980) recently described a micropyle from intracellular oocysts of *E. nieschulzi.* Each micropyle was disc-shaped and consisted of 10 to 14 alternating electron-dense and -lucent bands, which were thought to have originated from membranous material during formation of the outer layer of the oocyst wall.

The fine structure of oocyst walls of coccidia has been reviewed by Speer and Duszynski (1975) and Speer et al. (1979). These reviews show that the mature coccidian oocyst wall is made up of from one to four layers. However, in most species studied, two prominent layers are present. The innermost layer is the most consistent in appearance (usually electron-lucent) and thickness. The outer layer is usually electron-dense, which in some species has a rough or uneven surface. Three distinct layers were described for *I. canis, I. canaria,* and *I. serini* (Speer et al., 1973b; Speer and Duszynski, 1975) and four layers for *I. lacazei* (Speer et al., 1979). In spite of the similarities in oocyst wall structure among the species of coccidia studied thus far, each species appear to have a unique wall structure. Based on ultrastructural differences in oocyst walls, a distinction was made by Speer and Duszynski (1975) between *I. canaria* and *I. serini,* which appear nearly identical by light microscopy. Such ultrastructural differences may be used as a basis for differentiating those species of coccidia (e.g., *Sarcocystis* species) that may be indistinguishable by light microscopy (Speer et al., 1979).

Development of Sporozoites (Sporogony)

Before an oocyst can be infective to a new host, the oocyst must undergo further development, that is, sporulation, which results in the formation of sporocysts and sporozoites. In most

Figure 43. Fertilization in *S. cruzi* (*S. bovicanis*). The microgamete at *upper right* lies inside the host layers and is fused with the macrogamete through a discontinuity at the raised parasite membrane layer. At the point of fusion, the gamete membranes are thickened (*arrows*). The microgamete nucleoplasm has passed through the area of fusion into the macrogamete cytoplasm. Note the mitochondrion (*MI*) and microtubules (*MT*) of the microgamete. ×41,000. Reprinted by permission from: Sheffield, H. G., and Fayer, R. Fertilization in the Coccidia: Fusion of *Sarcocystis bovicanis* gametes. *Proceedings of the Helminthological Society of Washington* 47:118–121 (1980).

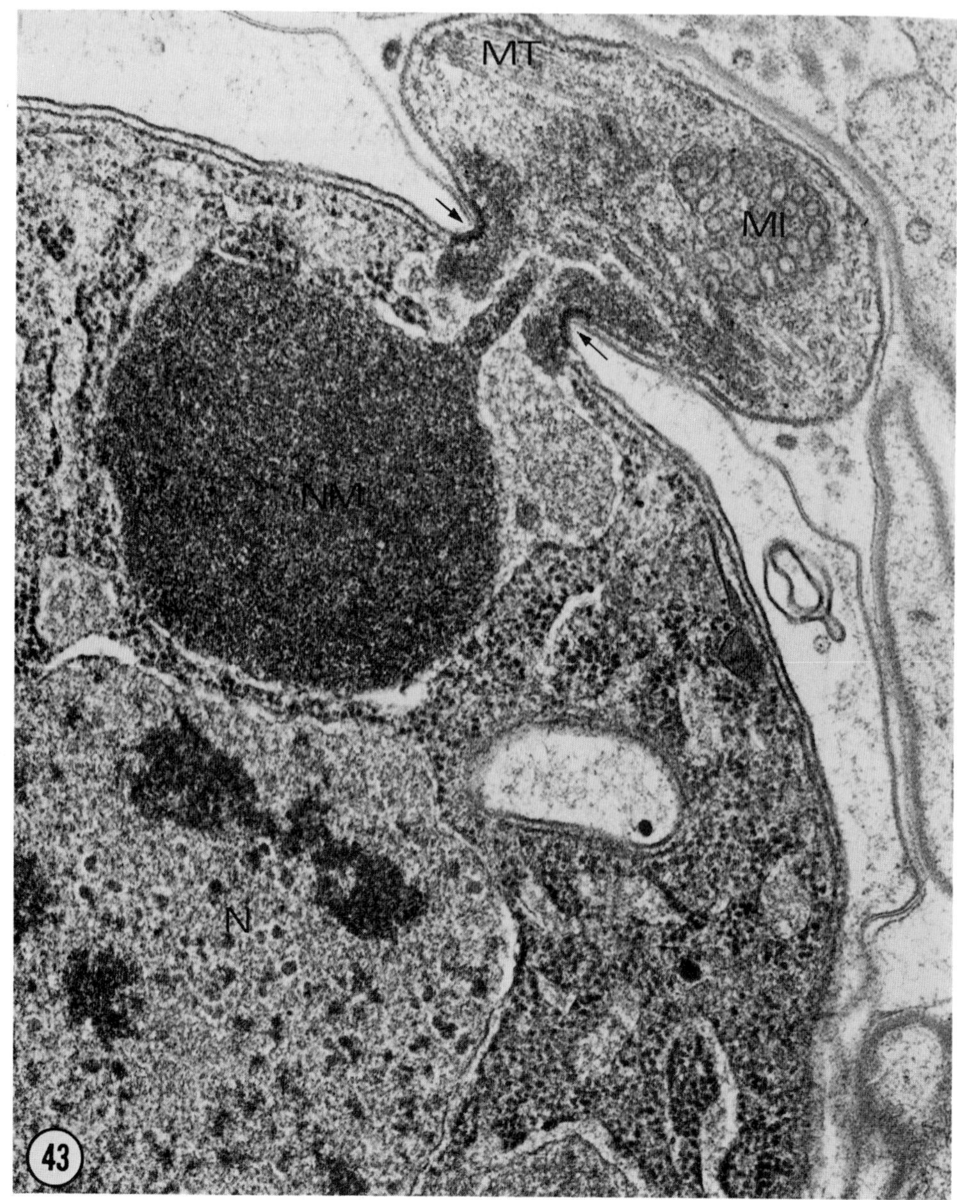

4

Figures 44 and 45. Young oocysts of *E. papillata*. Reprinted by permission from: Chobotar, B., Sénaud, J., Ernst, J. V., and Scholtyseck, E. Ultrastructure of macrogametogenesis and formation of the oocyst wall of *Eimeria papillata* in *Mus musculus*. *Protistologica* 16:115–124 (1980). 44. Portion of a parasite showing the WF1 material in a granular band at the periphery. Note the particulated units of the WF1 in small vesicles (*arrowhead*) which seem to be giving rise to the granular layer above. Membranes of the rough endoplasmic reticulum have initiated separation of the presumptive outer layer of the wall from the remainder of the cytoplasm (*arrow*). The WF2 have moved to a peripheral position and are spongy in appearance. × 23,000. 45. The outer layer of the oocyst wall is separated from the cytoplasm, and partial condensation of the granular material has occurred. Note the presence of five membranes and an inactive micropore. × 24,000.

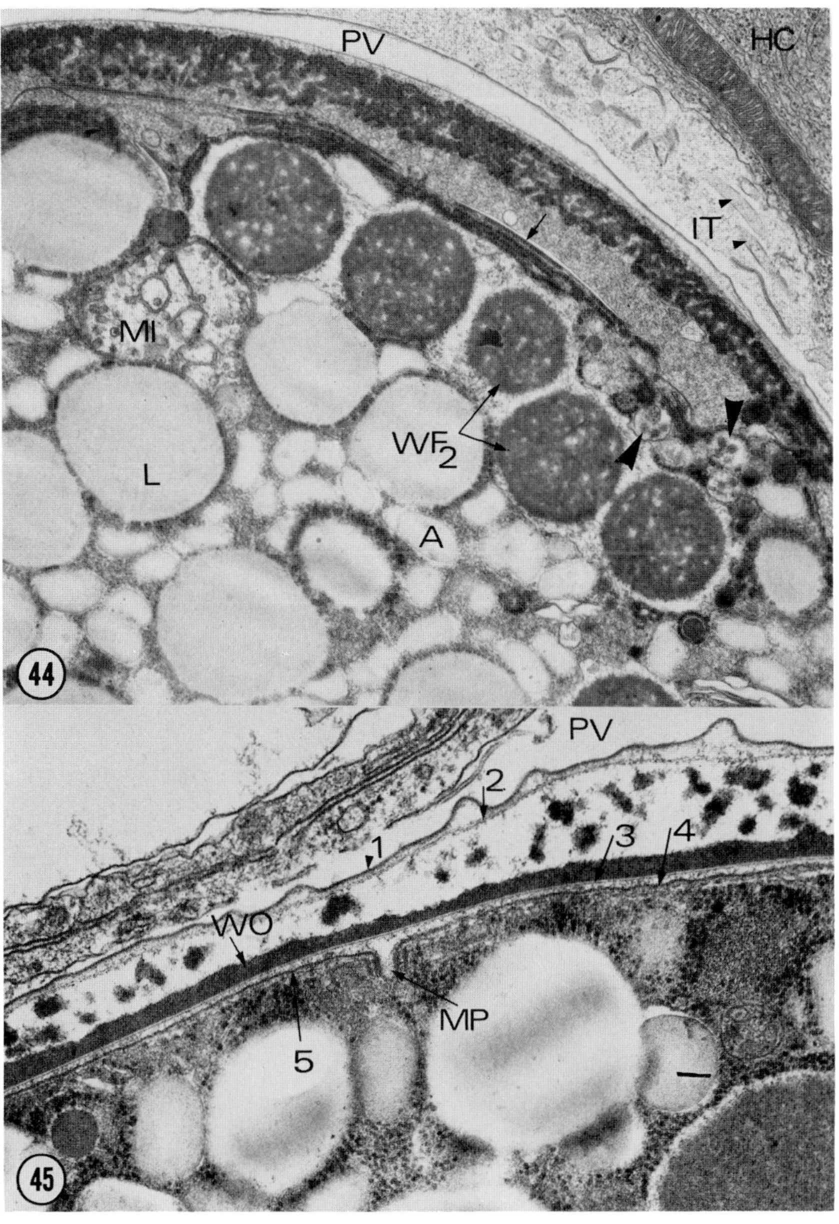

Figures 46–48. Reprinted by permission from: Sibert, G. J., and Speer, C. A. Fine structure of zygotes and oocysts of *Eimeria nieschulzi*. *Journal of Protozoology* 27:374–379 (1980). (Figures 46 and 48); and from Milde, K. Light and electron microscope studies on isosporan parasites (Sporozoa) in sparrows (*Passer domesticus* L.). *Protistologica* 15:607–627 (1979) (Figure 47).46. *E. nieschulzi*. Early stage in formation of the outer layer of the oocyst wall (*WO*) under elevated membranes of the zygote. × 24,600. 47. Formation of the inner layer of the oocyst wall in an *Isospora* species from sparrows. The particulated WF2 have fused together and are separated from the cytoplasm by a membrane (*arrowhead*). × 40,000. 48. *E. nieschulzi*. Fully formed oocyst wall consisting of an outer electron-dense layer and an inner electron-lucent layer. × 56,000.

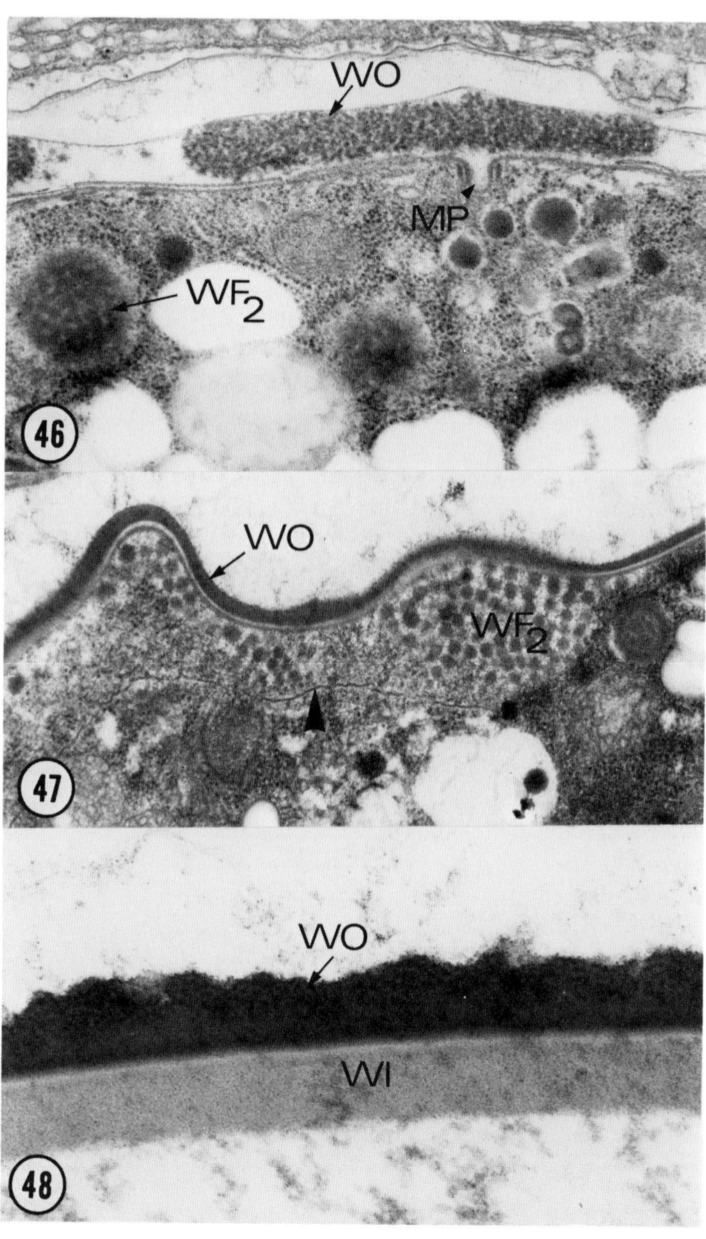

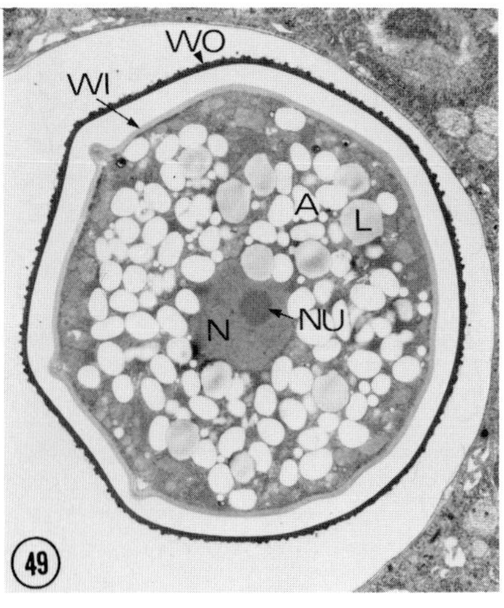

Figure 49. *E. nieschulzi.* Mature oocyst with a two-layered wall, amylopectin granules (*A*), and lipid bodies (*L*) in the cytoplasm. ×6,200. Reprinted by permission from: Sibert, G. J., and Speer, C. A. Fine structure of zygotes and oocysts of *Eimeria nieschulzi. Journal of Protozoology* 27:374–379 (1980).

Eimeria species, this process occurs externally, but in other species (e.g., *Sarcocystis* and *Frenkelia*), it occurs in the host.

The numerous attempts to study sporulation by electron microscopy in the Eimeriina have been largely unsuccessful, because the walls surrounding the sporont are impermeable to fixatives and embedding material. This problem has been recently overcome by use of a cryoprocessing method (Birch-Anderson et al., 1976). As a result, Ferguson and co-workers (1978a, 1979a,b) have now reported the details of sporulation of *E. brunetti* and *T. gondii* oocysts.

The early stages of sporulation in *T. gondii* and *E. brunetti* are similar. At the initiation of sporulation, there is an increase in the rER, ribosomes, and Golgi complexes, indicating an increase in protein synthesis. Two nuclear divisions produce four nuclei in each sporont by a process involving centrioles, intranuclear spindle fibers, and centrocones, as in merogony. Soon after nuclear division, invaginations appear in the limiting membrane, isolating the cytolasm into sporoblasts. Each sporoblast of *T. gondii* has two nuclei, whereas that of *E. brunetti* has one nucleus. A second unit membrane now envelops the parasite; the inner of these membranes is the limiting membrane and the outer one is associated with formation of the sporocyst wall.

In *E. brunetti,* each sporoblast becomes ellipsoid and the nucleus undergoes a final division, with a nucleus becoming oriented at each end. Sporozoite formation is initiated by the appearance of a dense plaque, the inner membrane complex, in the vicinity of each nucleus in close association with the limiting membrane of the parasite. The two sporozoites form at opposite ends of the sporoblast by posterior growth of the inner complex and invagination of the limiting membrane (Figure 50). The anterior end of the developing sporozoite has a conoid, rhoptry and microneme anlagen, subpellicular microtubules, and a centriole above the centrocone of the nucleus. With posterior progression of the pellicle, the refractile bodies, the nucleus, and amylopectin granules are incorporated into the body, and the sporozoites detach from the residual cytoplasm. Thus, sporogony in *E. brunetti* occurs by exogenesis, similar in nature to merogony in this species (Ferguson et al., 1976).

During this time, the Stieda body forms on the outer limiting membrane at one end of the sporocyst. This membrane loses its unit membrane character and develops into the sporocyst wall about 70 nm thick. The origin of the Stieda body and sporocyst wall is unknown, because no specific organelles appeared to be involved in their formation.

In *T. gondii,* sporocyst wall formation begins soon after the two sporoblasts are formed by the division of the zygote. Each sporoblast is at first limited by two membranes, but an additional membrane soon appears. The inner membrane represents the plasmalemma of the cytoplasmic mass and the two outer membranes give rise to the outer layers of the sporocyst wall. The inner layer is formed by an accumulation of osmiophilic material between the plasmalemma and the second membrane. During development of the inner layer, thickenings are observed which represent the precursors of specialized junctions of the plates forming this

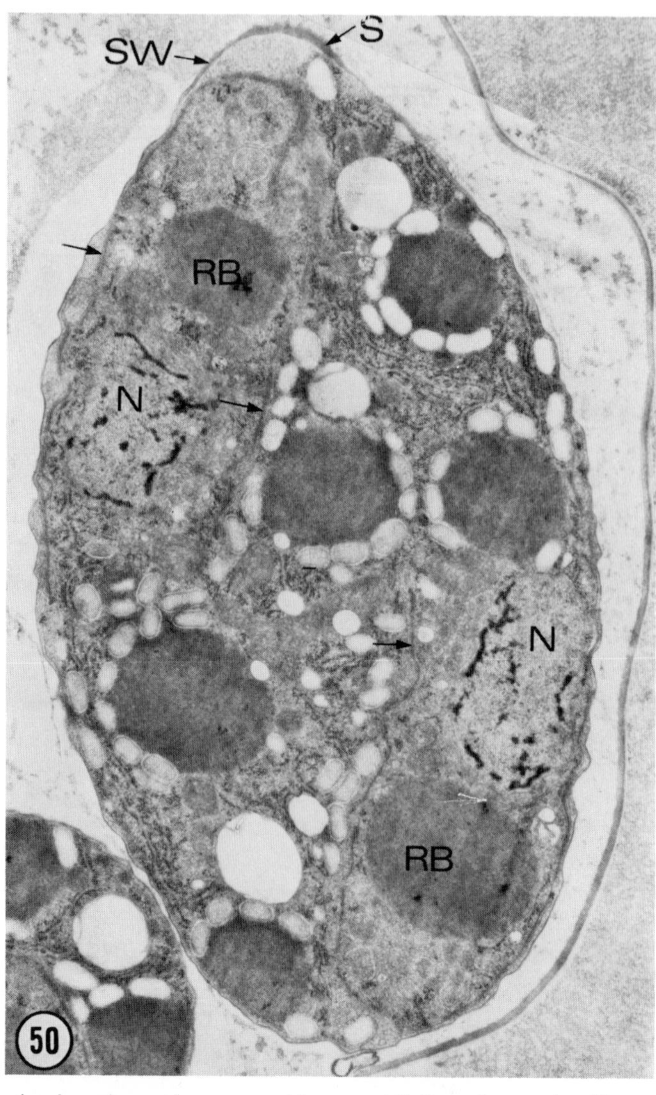

Figure 50. *E. brunetti.* Longitudinal section through an early sporocyst with two partially formed sporozoites. The extent of the invagination of the limiting membrane of the organism and growth of the inner membrane complex is marked (*arrows*). Note that the anterior refractile body and nucleus is enclosed at this stage. × 15,000. Reprinted by permission from: Ferguson, D. J. P., Birch-Andersen, A., Hutchinson, W. M., and Siim, J. Chr. Light and electron microscopy on the sporulation of the oocysts of *Eimeria brunetti*: II. Development into the sporocyst and formation of the sporozoite. *Acta Pathologica et Microbiologica Scandinavica. Section B: Microbiology* 86:13–24 (1978).

4

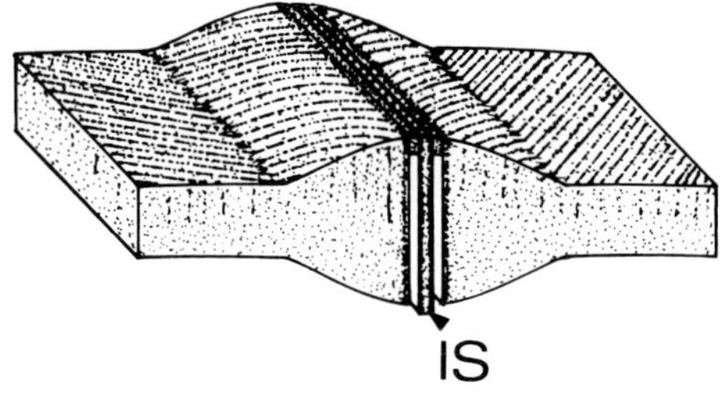

Figure 51. *T. gondii.* A 3-dimensional representation of the inner layer of the sporocyst wall at the junction between two plates showing the interposing strip (*IS*). Reprinted by permission from: Ferguson, D. J. P., Birch-Andersen, A., Siim, J. Chr., and Hutchinson, W. M. Ultrastructural studies on the sporulation of *Toxoplasma gondii*: II. Formation of the sporocyst and the sporocyst wall. *Acta Pathologica et Microbiologica Scandinavica. Section B: Microbiology* 87:183–190 (1979).

layer. The completed sporocyst wall has an outer layer of two closely apposed membranes and an inner layer made up of four curved plates. The margin of each plate has a lip-like thickening where it is joined to an adjacent plate by an interposing strip (Figures 51 and 52). A Stieda body does not develop in *T. gondii.*

A comparison of the sporocysts of *E. brunetti* and *T. gondii* reveals that the structural distinctiveness of each is representative of the two types of sporocysts widely found in the classical coccidia—that is, those with a Stieda body and those without one. On the basis of these differences, two distinct patterns of excystation are known to occur in the coccidia (Speer et al., 1973a, 1976; Speer and Duszynski, 1975; Duszynski and Speer, 1976; Christie et al., 1978; Ferguson et al., 1979c; Becker et al., 1979; Box et al., 1980). In sporocysts that have Stieda bodies, represented mainly by *Eimeria* species and some *Isospora* species, the sporozoites exit through an opening created by dissolution of the Stieda body by excysting fluid. In certain other species of *Isospora* and *Sarcocystis,* release of the sporozoites occurs randomly when the sporocyst walls collapse during the excystation process. The excysting fluid acts on the site of opposition between adjacent plates, causing the interposing strip to partially degenerate and separate from the margin of each plate, which leads to a rapid simultaneous collapse of the sporocyst (Figures 53 and 54).

ACKNOWLEDGMENTS

We hereby express our appreciation to Mrs. Edelgard Kirberg, Mrs. Ingeborg Schrauf, and Mrs. Brunhilde Zarbock for assistance in preparations of the illustrations and photographs; to Mrs. Angela Entzeroth for typing the manuscript; and to Dr. J. F. Dubremetz, Dr. E. Porchet, Prof. H. Mehlhorn, and Dr. Jean Sénaud for suggestions concerning the text.

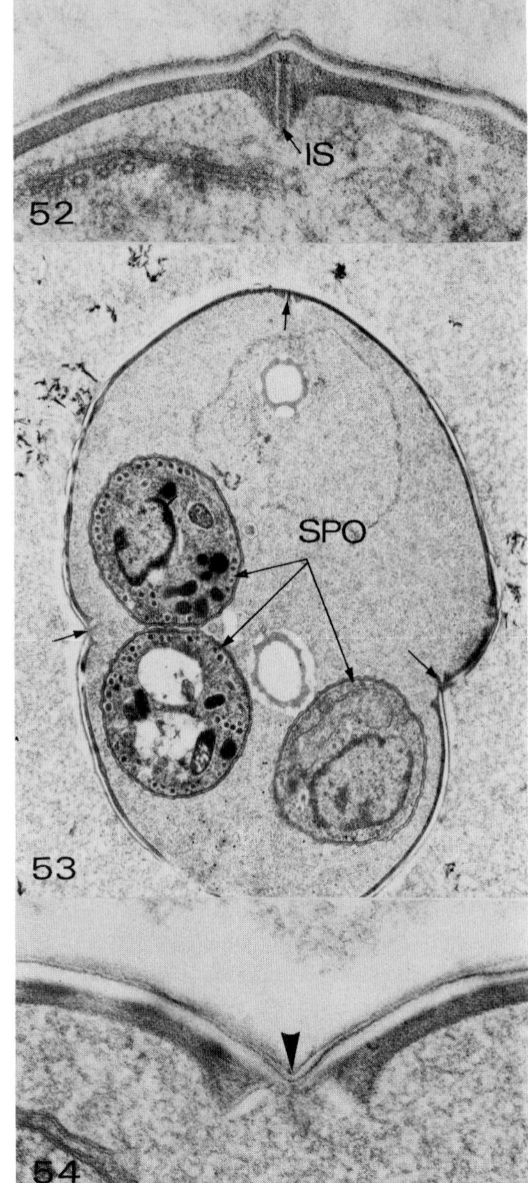

Figures 52–54. *T. gondii.* Reprinted by permission from: Ferguson, D. J. P., Birch-Andersen, A., Siim, J. Chr., and Hutchinson, W. M. An ultrastructural study on the excystation of the sporozoites of *Toxoplasma gondii. Acta Pathologica et Microbiologica Scandinavica. Section B: Microbiology* 87:277–283 (1979). 52. Cross-section showing the structure of the junction between plates of the inner layer of the sporocyst wall prior to treatment with excysting medium. ×90,000. 53. Section through a sporocyst at an early stage of excystation. The junctions of the plates forming the inner layer of the sporocyst wall can be seen (*arrows*). Note the beginning of an inward curling of the two lower junctions. ×15,000. 54. Cross-section showing the separation of the plates of the inner layer of the sporocyst wall. Note that the interposing strip has disappeared and that the outer wall has bent inward above the break. ×90,000.

4

ABBREVIATIONS

A	Amylopectin
C	Conoid
B	Basal body
CC	Centrocone
CE	Centriole
DB	Dark bodies
DR	Dense ring
DRH	Duct of rhoptries
ER	Endoplasmic reticulum
F	Flagellum
FI	Fibril
GO	Golgi apparatus
GS	Ground substance
HC	Host cell
HCM	Host cell membrane
IF	Intravacuolar folds
IM	Inner membranes of pellicle
IP	Invagination of pellicle
IR	Inner ring
IT	Intravacuolar tubules
IS	Interposing strip
K	Kinetochore
L	Lipid body
M	Merozoite
MC	Metrocyte
MG	Microgamete
MI	Mitochondrion
MIH	Mitochondrion of host cell
MN	Microneme
MP	Micropore
MT	Microtubules
N	Nucleus
NE	Nuclear envelope
NH	Nucleus of host cell
NM	Nucleus of microgamete

NP	Nuclear pore
NU	Nucleolus
OM	Outer membrane of pellicle
OP	Osmiophilic plaques
OR	Outer ring
P	Polar ring
PCW	Primary cyst wall
PE	Pellicle
PG	Protein granule
PT	Protrusions of primary cyst wall
PF	Perforatorium
PP	Posterior polar ring
PV	Parasitophorous vacuole
R_1R_2	Preconoidal rings
RA	Rhoptry anlagen
RB	Refractile body
RBO	Residual body
RH	Rhoptry
RN	Residual nucleus
SE	Septa
SCW	Secondary cyst wall
SPO	Sporozoite
V	Vacuole
WF1,2	Wall-forming bodies 1 and 2
WI	Inner layer of oocyst wall
WO	Outer layer of oocyst wall

4

LITERATURE CITED

Aikawa, M. 1971. Plasmodium: The fine structure of malarial parasites. Exp. Parasitol. 30:84–320.

Bannister, L. H., Butcher, G. A., Dennis, E. D., and Mitchell, G. H. 1975. Structure and invasive behaviour of *Plasmodium knowlesi* in merozoites in vitro. Parasitology 71:483–491.

Bardele, C. F. 1966. Elektronenmikroskopische Untersuchungen an dem Sporozoon *Eucoccidium dinophili* Grell. Z. Zellforsch. Mikrosk. Anat. 74:559–595.

Becker, E. L., and Henson, P. M. 1973. In vitro studies of immunologically induced secretion of mediators from cells and related phenomena. In: F. J. Dixon and H. G. Kunkel (eds.), Advances in Immunology, Vol. 16, pp. 93–193. Academic Press, New York.

Becker, B., Mehlhorn, H., and Heydorn, A. O. 1979. Light and electron microscopic studies on gamogony and sporogony of 5 *Sarcocystis* species *in vivo* and in tissue cultures. Zentralbl. Bakteriol. (Orig. A) 244:394–404.

Beyer, T. V., Shibalova, T. A., and Kostenko, L. A. 1978. Cytology of the Coccidia. (In Russian). Nauka, Leningrad.

Birch-Andersen, A., Ferguson, D. J. P., and Pontefract, R. D. 1976. A technique for obtaining thin sections of coccidian oocysts. Acta Pathol. Microbiol. Scand. 84:235–239.

Bommer, W., Heunert, H. H., and Milthaler, B. 1969. Kinematographische Studien über die Eigenbewegung von *Toxoplasma gondii.* Z. Tropenmed. Parasit. 20:450–458.

Box, E. D., Marchiondo, A. A., Duszynski, D. W., and Davis, P. C. 1980. Ultrastructure of *Sarcocystis* sporocysts from passerine birds and opossums: Comments on classification of the genus *Isospora.* J. Parasitol. 66:68–74.

Branton, D., Bullivant, S., Gilula, N.B., Moor, H., Mühlethaler, K., Northcote, D. H., Packer, L., Satir, B., Speth, V., Staehelin, L. A., Steere, R. L., and Weinstein, R. S. 1975. Freeze-etching nomenclature. Science 190:54–56.

Büttner, D. W. 1968. Das Cytostom von *Lankesterella garnhami.* Z. Zellforsch. Mikrosk. Anat. 88:126–137.

Černá, Ž., and Sénaud, J. 1971. Some peculiarities of the fine structure of merozoites of *Eimeria stiedai.* Folia Parasitologica 18:177–178.

Černá, Ž., and Sénaud, J. 1977. Sur un type nouveau de multiplication asexuée d'une Sarcosporidie, dans le foie de la souris. C. R. Acad. Sci., (D) (Paris) 285:347–349.

Chalupsky, J., and Sénaud, J. 1973. Premières observations des M. organismes (*Frenkelia* sp.), des petits mammifères sauvages dans le centre de la France (region de Besse-en-Chandesse) et observations faunistiques sur les petits rongeurs et insectivores. Ann. Biol. de Besse-en-Chandesse 7:147–174.

Cheissin, E. M., and Snigirevskaya, E. S. 1965. Some new data on the fine structure of the merozoites of *Eimeria intestinalis* (Sporozoa, Eimeriidea). Protistologica 1:121–126.

Chobotar, B., Scholtyseck, E., Sénaud, J., Ernst, J. V. 1975a. A fine structural study of asexual stages of the murine coccidium *Eimeria ferrisi* Levine and Ivens, 1965. Z. Parasitenkd. 45: 291–306.

Chobotar, B., Sénaud, J., Ernst, J. V., and Scholtyseck, E. 1975b. The ultrastructure of macrogametes of *Eimeria ferrisi* Levine and Ivens 1965 in *Mus musculus.* Z. Parasitenkd. 48:111–124.

Chobotar, B., Sénaud, J., Ernst, J. V., and Scholtyseck, E. 1980. Ultrastructure of macrogametogenesis and formation of the oocyst wall of *Eimeria papillata* in *Mus musculus.* Protistologica 16:115–124.

Christie, E., Pappas, P. W., and Dubey, J. P. 1978. Ultrastructure of excystment of *Toxoplasma gondii* oocysts. J. Protozool. 25:438–443.

Colley, F. C. 1967. Fine structure of sporozoites of *Eimeria nieschulzi*. J. Protozool. 14:217–220.

Colley, F. C. 1968. Fine structure of schizonts and merozoites of *Eimeria nieschulzi*. J. Protozool. 15:374–382.

Danforth, H. D., and Hammond, D. M. 1972. Stages of merogony in multinucleate merozoites of *Eimeria magna*. J. Protozool. 19:454–457.

Desser, S. S. 1972. Gametocyte maturation, exflagellation and fertilization in *Parahaemoproteus (= Haemoproteus) velans* (Coatney and Roudabush) (Haemosporina: Haemoproteidae): An ultrastructural study. J. Protozool. 19:287–296.

Desser, S. S., and Allison, F. 1979. Aspects of the sporogonic development of *Leucocytozoon tawaki* of the Fiordland Crested Penguin in its primary vector, *Austrosimulium ungulatum:* An ultrastructural study. J. Parasitol. 65:737–744.

D'Haese, J., Mehlhorn, H., and Peters, W. 1977. Comparative electron microscope study of pellicular structures in Coccidia *(Sarcocystis, Besnoitia* and *Eimeria).* Int. J. Parasitol. 7:505–518.

Doens-Juteau, O., and Sénaud, J. 1974. Étude ultrastructurale de la paroi du macrogamète et de l'oocyste chez *Eimeria tenella* Railliet et Lucet, 1891 Coccidie, Eimeriidae, parasite de l'intestine des poulets. Protistologica 10:261–270.

Dubey, J. P. 1977. *Toxoplasma, Hammondia, Besnoitia, Sarcocystis,* and other cyst-forming Coccidia of man and animals. In: J. P. Kreier (ed.), III. Gregarines, Haemogregarines, Coccidia, Plasmodia and Haemoproteids, pp. 101–237. Academic Press, New York.

Dubey, J. P., and Mehlhorn, H. 1978. Extraintestinal stages of *Isospora ohioensis* from dogs in mice. J. Parasitol. 64:689–695.

Dubey, J. P., Speer, C. A., and Douglass, T. G. 1980. Development and ultrastructure of first generation meronts of *Sarcocystis cruzi* in calves fed sporocysts from coyote species. J. Protozool. 27:380–387.

Dubremetz, J. F. 1971a. L'ultrastructure du centriole et du centrocone chez la coccidie *Eimeria necatrix*. Étude au course de la schizogonie. J. Microsc. 12:453–458.

Dubremetz, J. F. 1971b. Le conoide et les microtubules sous-pelliculaires du mérozoite d'*Eimeria necatrix* (Sporozoaire, Coccidiomorphe): Étude au microscope électronique. C. R. Acad. Sci. (D) (Paris) 272:600–603.

Dubremetz, J. F. 1973. Étude ultrastructurale de la mitose schizogonique chez la coccidie *Eimeria necatrix* (Johnson, 1930). J. Ultrastruct. Res. 42:354–376.

Dubremetz, J. F. 1975. La génèse des merozoites chez la coccidie *Eimeria necatrix:* Étude ultrastructurale. J. Protozool. 22:71–84.

Dubremetz, J. F., and Elsner, Y. Y. 1979. Ultrastructural study of schizogony of *Eimeria bovis* in cell cultures. J. Protozool. 26:367–376.

Dubremetz, J. F., and Torpier, G. 1978. Freeze fracture study of the pellicle of an eimerian sporozoite (Protozoa, Coccidia). J. Ultrastruct. Res. 62:94–109.

Dubremetz, J. F., and Yvore, P. 1971. Elaboration de la coque oocystique chez la coccidie *Eimeria necatrix* Johnson, 1930 (Sporozoaires, Coccidiomorphes). Étude au microscope électronique. C. R. Soc. Biol. (Paris) 165:862–866.

Duszynski, D. W., and Speer, C. A. 1976. Excystation of *Isospora arthopitheci* Rodhain, 1933 with notes on a similar process in *Isospora bigemina* (Stiles, 1891) Lühe, 1906. Z. Parasitenkd. 48:191–197.

El-Kasaby, A., and Sykes, A. H. 1973. The role of chicken macrophages in the parenteral excystation of *Eimeria acervulina.* Parasitology 66:231–239.

Entzeroth, R. 1980. Licht- und elektronenmikroskopische Untersuchungen zum Lebenszyklus von *Sarcocystis* sp. (Sporozoa, Coccidia) im Reh (*Capreolus capreolus*) und Hund (*Canis familiaris*). Doctoral dissertation, University of Bonn.

Fayer. R. 1972. Penetration of cultured cells by *Eimeria meleagriditis* and *E. tenella* sporozoites. J. Parasitol. 58:921–927.

Ferguson, D. J. P., Birch-Andersen, A., Hutchison, W. M., and Siim, J. Chr. 1976. Ultrastructural studies on the endogenous development of *Eimeria brunetti:* I. Schizogony. Acta Pathol. Microbiol. Scand. (B). 84:401–413.

Ferguson, D. J. P., Birch-Andersen, A., Hutchison, W. M., and Siim, J. Chr. 1977a. Ultrastructural studies on the endogenous development of *Eimeria brunetti*: II. Microgametogony and the microgamete. Acta Pathol. Microbiol. Scand. (B). 85:67–77.

Ferguson, D. J. P., Birch-Andersen, A., Hutchison, W. M., and Siim, J. Chr. 1977b. Ultrastructural studies on the endogenous development of *Eimeria brunetti*: III. Macrogametogony and the macrogamete. Acta Pathol. Microbiol. Scand. (B). 85:78–88.

Ferguson, D. J. P., Birch-Andersen, A., Hutchison, W. M., and Siim, J. Chr. 1977c. Ultrastructural studies on the endogenous development of *Eimeria brunetti*: IV. Formation and structure of the oocyst wall. Acta Pathol. Microbiol. Scand. (B). 85:201–221.

Ferguson, D. J. P., Birch-Andersen, A., Hutchison, W. M., and Siim, J. Chr. 1977d. The ultrastructure and distribution of micropores in the various developmental forms of *Eimeria brunetti.* Acta Pathol. Microbiol. Scand. (B). 85:363–373.

Ferguson, D. J. P., Birch-Andersen, A., Hutchison, W. M., and Siim, J. Chr. 1978a. Light and electron microscopy on the sporulation of the oocysts of *Eimeria brunetti*: I. Development of the zygote and formation of the sporoblasts. Acta Pathol. Microbiol. Scand. (B). 86:1–11.

Ferguson, D. J. P., Birch-Andersen, A., Hutchison, W. M., and Siim, J. Chr. 1978b. Light and electron microscopy on the sporulation of the oocysts of *Eimeria brunetti*: II. Development into the sporocyst and formation of the sporozoite. Acta Pathol. Microbiol. Scand. (B). 86:13–24.

Ferguson, D. J. P., Birch-Andersen, A., Hutchison, W. M., and Siim, J. Chr. 1980a. Ultrastructural observations on microgametogenesis and the structure of the microgamete of *Isospora felis.* Acta Pathol. Microbiol. Scand. (B). 88:151–159.

Ferguson, D. J. P., Birch-Andersen, A., Hutchison, W. M., and Siim, J. Chr. 1980b. Ultrastructural observations on macrogametogenesis and the structure of the macrogamete of *Isospora felis.* Acta Pathol. Microbiol. Scand. (B). 88:161–168.

Ferguson, D. J. P., Birch-Andersen, A., Siim, J. Chr., and Hutchison, W. M. 1979a. Ultrastructural studies on the sporulation of oocysts of *Toxoplasma gondii*: I. Development of the zygote and formation of the sporoblasts. Acta Pathol. Microbiol. Scand. (B). 87:171–181.

Ferguson, D. J. P., Birch-Andersen, A., Siim, J. Chr., and Hutchison, W. M. 1979b. Ultrastructural studies on the sporulation of *Toxoplasma gondii*: II. Formation of the sporocyst and the sporocyst wall. Acta Pathol. Microbiol. Scand. (B). 87:183–190.

Ferguson, D. J. P., Birch-Andersen, A., Siim, J. Chr., and Hutchison, W. M. 1979c. An ultrastructural study on the excystation of the sporozoites of *Toxoplasma gondii.* Acta Pathol. Microbiol. Scand. (B). 87:277–283.

Ferguson, D. J. P., Hutchison, W. M., Dunachie, J. F., and Siim, J. Chr. 1974. Ultrastructural study of early stages of asexual multiplication and microgametogony of *Toxoplasma gondii* in the small intestine of the cat. Acta Pathol. Microbiol. Scand. (B). 82:167–181.

Ferguson, D. J. P., Hutchison, W. M., and Siim, J. Chr. 1975. The ultrastructural development of the macrogamete and the formation of the oocyst wall of *Toxoplasma gondii.* Acta Pathol. Microbiol. Scand. (B). 83:491–505.

Fernando, M. A., and Stockdale, P. H. G. 1974. Fine structural changes associated with schizogony in *Eimeria necatrix.* Z. Parasitenkd. 43:105–114.

Frenkel, J. K. 1973. Toxoplasmosis: Parasite life cycle, pathology, and immunology. In: D. M. Hammond and P. L. Long (eds.), The Coccidia: *Eimeria, Isospora, Toxoplasma,* and Related Genera, pp. 343–410. University Park Press, Baltimore.

Frenkel, J. K. 1974. Advances in the biology of Sporozoa. Z. Parasitenkd. 45:125–162.

Frenkel, J. K. 1977. *Besnoitia wallacei* of cats and rodents: With a reclassification of other cyst-forming isosporoid coccidia. J. Parasitol. 63:611–628.

Frenkel, J. K., and Dubey, J. P. 1972. Rodents as vectors for feline coccidia, *Isospora felis* and *Isospora rivolta*. J. Infect. Dis. 125:69–72.

Galucci, B. 1974. Fine structure of *Haemoproteus columbae* Kruse during macrogametogenesis and fertilization. J. Protozool. 21:254–263.

Garnham, P. C. C., Bird, R. G., and Baker, J. R. 1960. Electron microscope study of motile stages of malaria parasites: I. The fine structure of the sporozoites of *Haemamoeba (Plasmodium) gallinacea*. Trans. R. Soc. Trop. Med. Hyg. 54:274–278.

Garnham, P. C. C., Bird, R. G., and Baker, J. R. 1962. Electron microscope studies of motile stages of malaria parasites: III. The ookinetes of *Haemamoeba* and *Plasmodium*. Trans. R. Soc. Trop. Med. Hyg. 56:116–120.

Garnham, P. C. C., Bird, R. G., Baker, J. R., and Bray, R. S. 1961. Electron microscope studies of motile stages of malaria parasites: II. The fine structure of the sporozoite of *Laverania (Plasmodium) falcipara*. Trans. R. Soc. Trop. Med. Hyg. 55:98–112.

Göbel, E., Rommel, M., and Krampitz, H. E. 1978. Ultrastrukturelle Untersuchungen zur ungeschlechtlichen Vermehrung von *Frenkelia* in der Leber der Rötelmaus. Z. Parasitenkd. 55:29–42.

Gustafson, P. V., Agar, H. D. and Cramer, D. I. 1954. An electron microscope study of *Toxoplasma*. Am. J. Trop. Med. Hyg. 3:1008–1022.

Hammond, D. M. 1973a. Life cycles and development of coccidia. In: D. M. Hammond and P. L. Long (eds.), The Coccidia: *Eimeria, Isospora, Toxoplasma,* and Related Genera, pp. 45–79. University Park Press, Baltimore.

Hammond, D. M. 1973b. Ultrastructure and development of coccidia. In: Proceedings of the Symposium on Coccidia and Related Organisms, pp. 11–43. University of Guelph, Guelph, Ontario.

Hammond, D. M., Chobotar, B., and Ernst, J. V. 1968. Cytological observations on sporozoites of *Eimeria bovis* and *E. auburnensis* and an *Eimeria* species from the Ord Kangaroo rat. J. Parasitol. 54:550–558.

Hammond, D. M., Roberts, W. L., Youssef, N. N., and Danforth, H. D. 1973. Fine structure of the intranuclear spindle poles of *Eimeria callospermophilli* and *E. magna*. J. Parasitol. 59:581–584.

Hammond, D. M., Scholtyseck, E., and Chobotar, B. 1967. Fine structure associated with nutrition of the intracellular parasite *Eimeria auburnensis*. J. Protozool. 14:678–683.

Hammond, D. M., Scholtyseck, E., and Chobotar, B. 1969. Fine structural study of microgametogenesis of *Eimeria auburnensis*. Z. Parasitenkd. 33:65–84.

Hammond, D. M., Speer, C. A., and Roberts, W. L. 1970. Occurrence of refractile bodies in merozoites of *Eimeria* species. J. Parasitol. 56:189–191.

Heller, G. 1972. Elektronenmikroskopische Untersuchungen zur Bildung und Struktur von Conoid, Rhoptrien, Mikronemen bei *Eimeria stiedae* (Sporozoa, Coccidia). Protistologica 8:43–51.

Heller, G., and Scholtyseck, E. 1971. Feinstrukturuntersuchungen zur Merozoitenbildung von *Eimeria stiedai*. Protistologica 7:451–460.

Heydorn, A. O., and Mehlhorn, H. 1978. Light and electron microscopic studies on *Sarcocystis suihominis:* 2. The schizogony preceding cyst formation. Zentrabl. Bakteriol. (Orig. A). 240:123–134.

Hoppe, G. 1974. La formation des merozoites chez la coccidie *Eimeria tenella* (Railliet et Lucet, 1891). Étude au microscope électronique. Protistologica 10:185–205.

Hoppe, G. 1976. The ultrastructure of early generation development and later schizonts of *Eimeria tenella* in the chicken. Protistologica 12:169–181.

Jacobs, L. 1967. Toxoplasma and toxoplasmosis. Adv. Parasitol. 5:1–45.

Jensen, J. B., and Edgar, S. A. 1976a. Effects of antiphagocytic agents on penetration of *Eimeria magna* sporozoites into cultured cells. J. Parasitol 62:203–206.

Jensen, J. B., and Edgar, S. A. 1976b. Possible secretory function of the rhoptries of *Eimeria magna* during preparation of cultured cells. J. Parasitol. 62:988–992.

Jensen, J. B., and Edgar, S. A. 1978. Fine structure of penetration of cultured cells by *Isospora canis* sporozoites. J. Protozool. 25:169–173.

Jensen, J. B., and Hammond, D. M. 1975. Ultrastructure of the invasion of *Eimeria magna* sporozoites into cultured cells. J. Protozool. 22:411–415.

Jones, T. C., Yeh, S., and Hirsch, J. G. 1972. The interaction between *Toxoplasma gondii* and mammalian cells: I. Mechanism of entry and intracellular fate of the parasite. J. Exp. Med. 136:1157–1172.

Kan, S. P., and Dissanaike, A. S. 1976. Ultrastructure of *Sarcocystis booliati* Dissanaike and Poopalachelvam, 1975 from the moon rat, *Echinosorex gymnurus*, in Malaysia. Int. J. Parasitol. 6:321–326.

Kelley, G. L., and Hammond, D. M. 1972. Fine structural aspects of early development of *Eimeria ninakohlyakimovae* in cultured cells. Z. Parasitenkd. 38:271–284.

Kepka, O., and Scholtyseck, E. 1970. Weitere Untersuchungen der Feinstruktur von *Frenkelia* spec. (= M-Organismus, Sporozoa). Protistologica 7:249–266.

Ladda, R., Aikawa, M. and Sprinz, H. 1969. Penetration of erythrocytes by merozoites of mammalian and avian malarial parasites. J. Parasitol. 55:633–644.

Lee, D. L., and Millard, B. J. 1971. The structure and development of the macrogamete and oocyst of *Eimeria acervulina*. Parasitology 62:31–34.

Levine, N. D. 1971. Uniform terminology for the protozoan Subphylum Apicomplexa. J. Protozool. 18:352–355.

Long, P. L. 1976. Intracellular cocccidia. In: C. R. Kennedy (ed.), Ecological Aspects of Parasitology, pp. 409–427. North Holland / Elsevier, Amsterdam.

Long, P. L., and Rose, M. E. 1970. Extended schizogony of *Eimeria mivati* in betamethasone treated chickens. Parasitology 60:147–155.

Long, P. L., and Speer, C. A. 1977. Invasion of host cells by Coccidia. In: Parasite Invasion, 15th Symposium of the British Society for Parasitology, pp. 1–26. Blackwell Scientific Publications, Ltd., Oxford.

Lycke, E., Lund, E., and Strannegard, Ö. 1965. Enhancement by lysozyme and hyaluronidase of the penetration by *Toxoplasma gondii* into cultured host cells. Br. J. Exp. Pathol. 46:189–199.

McLaren, D. J. 1969. Observations on the fine structural changes associated with schizogony and gametogony in *Eimeria tenella*. Parasitology 59:563–574.

McLaren, D. J., and Paget, G. E. 1968. A fine structural study on the merozoite of *Eimeria tenella* with special reference to the conoid apparatus. Parasitology 58:561–579.

Madden, P. A., and Vetterling, J. M. 1977. Scanning electron microscopy of *Eimeria tenella* microgametogenesis and fertilization. J. Parasitol. 63:607–610.

Marquardt, W. C. 1966. The living endogeneous stages of the rat coccidium, *Eimeria nieschulzi*. J. Protozool. 13:509–514.

Mehlhorn, H. 1972a. Elektronenmikroskopische Untersuchungen an Entwicklungsstadien von *Eimeria maxima* (Sporozoa, Coccidia): I. Die Feinstruktur der Makrogameten. Z. Parasitenkd. 39:161–182.

Mehlhorn, H. 1972b. Elektronenmikroskopische Untersuchungen an Entwicklungsstadien von *Eimeria maxima* (Sporozoa, Coccidia): II. Die Feinstruktur der Mikrogameten. Z. Parasitenkd. 40:151–163.

Mehlhorn, H. 1972c. Elektronenmikroskopische Untersuchungen an Entwicklungsstadien von *Eimeria maxima* aus dem Haushuhn: III. Der Differenzierungsprozess der Mikrogameten unter besonderer Berücksichtigung der Kernteilungen. Z. Parasitenkd. 40:243–260.

Mehlhorn, H., and Frenkel, J. K. 1980. Ultrastructural comparison of cysts and zoites of *Toxoplasma gondii, Sarcocystis muris,* and *Hammondia hammondi* in skeletal muscle of mice. J. Parasitol. 66:59–67.

Mehlhorn, H., Hartley, W. J., and Heydorn, A. O. 1976. A comparative ultrastructural study of the cyst wall of 13 *Sarcocystis* species. Protistologica 12:451–467.

Mehlhorn, H., and Heydorn, A. O. 1978. The Sarcosporidia (Protozoa, Sporozoa): Life cycle and fine structure. Adv. Parasitol. 16:43–93.

Mehlhorn H. and Heydorn, A. O. 1979. Electron microscopical study on gamogony of *Sarcocystis suihominis* in human tissue cultures. Z. Parasitenkd. 58:97–113.

Mehlhorn, H., Heydorn, A. O., and Gestrich, R. 1975a. Licht- und elektronenmikroskopische Untersuchungen an Cysten von *Sarcocystis ovicanis* Heydorn et al. (1975) in der Muskulatur von Schafen. Z. Parasitenkd. 48:83–93.

Mehlhorn, H., Heydorn, A. O., and Gestrich, R. 1975b. Licht- und elektronenmikroskopische Untersuchungen an Cysten von *Sarcocystis fusiformis* in der Muskulatur von Kälbern nach experimenteller Infektion mit Oocysten und Sporocysten von *Isospora hominis* Railliet et Lucet, 1891. Zentrabl. Bakteriol. (Orig. A). 231:301–322.

Mehlhorn, H., Heydorn, A. O., and Gestrich, R. 1975c. Licht- und elektronenmikroskopische Untersuchungen an Cysten von *Sarcocystis fusiformis* in der Muskulatur von Kälbern nach experimenteller Infektion mit Oocysten und Sporocysten der großen Form von *Isospora bigemina* des Hundes. Zentrabl. Bakteriol. (Orig. A). 232:392–409.

Mehlhorn, H., and Markus, M. B. 1976. Electron microscopy of stages of *Isospora felis* of the cat in the mesenteric lymph node of the mouse. Z. Parasitenkd. 51:15–24.

Mehlhorn, H., and Scholtyseck, E. 1973a. Cytochemistry of Toxoplasmatea *Sarcocystis, Frenkelia,* and *Besnoitia* at the ultrastructural level. In: P. de Puytorac and J. Grain (eds.), Progress in Protozool., 4th International Congress Protozool., UER Sciences, Clermont. p. 275.

Mehlhorn, H., and Scholtyseck, E. 1973b. Elektronenmikroskopische Untersuchungen an Cystenstadien von *Sarcocystis tenella* aus der Oesophagus-Muskulatur des Schafes. Z. Parasitenkd. 43:251–270.

Mehlhorn, H., and Scholtyseck, E. 1974. Die Parasit-Wirtsbeziehungen bei verschiedenen Gattungen der Sporozoen (*Eimeria, Toxoplasma, Sarcocystis, Frenkelia, Hepatozoon, Plasmodiun* und *Babesia*) unter Anwendung spezieller Verfahren. Microsc. Acta 75: 429–451.

Mehlhorn, H., Sénaud, J., Chobotar, B., and Scholtyseck, E. 1975d. Electron microscope studies of cyst stages of *Sarcocystis tenella:* The origin of micronemes and rhoptries. Z. Parasitenkd. 45:227–236.

Mehlhorn, H., Sénaud, J. and Scholtyseck, E. 1973. La schizogonie chez *Eimeria falciformis* (Eimer, 1870) Coccidie, Eimeriidae parasite de l'epithelium intestinal de la souris (*Mus musculus*): Étude au microscope électronique des mérozoites et de leur dévelopement au cours d'infections expérimentales. Protistologica 9:269–291.

Michael, E. 1975. Structure and mode of function of the organelles associated with nutrition of the macrogametes of *Eimeria acervulina.* Z. Parasitenkd. 45:347–361.

Michael, E. 1976. Sporozoites of *Eimeria acervulina* within intestinal macrophages in normal experimental infections. Z. Parasitenkd. 49:33–40.

Michael. E. 1978. The formation and final structure of the oocyst wall of *Eimeria acervulina:* A transmission and scanning electron microscope study. Z. Parasitenkd. 57:221–228.

Milde, K. 1979. Light and electron microscope studies on isosporan parasites (Sporozoa) in sparrows (*Passer domesticus* L.). Protistologica 15:607–627.

Müller, B. E. G. 1975. In vitro development from

sporozoites to first-generation merozoites in *Eimeria contorta* Haberkorn, 1971. Z. Parasitenkd. 47:23–34.

Norrby, R., Lindholm, L., and Lycke, E.1968. Lysosomes of *Toxoplasma gondii* and their possible relationship to the host cell penetration of *Toxoplasma* parasites. J. Bacteriol. 96:916–919.

Pacheco, N. D., and Fayer, R. 1977. Fine structure of *Sarcocystis cruzi* schizonts. J. Protozool. 24:382–388.

Pacheco, N. D., Sheffield, H. G., and Fayer, R. 1978. Fine structure of immature cysts of *Sarcocystis cruzi*. J. Parasitol. 64:320–325.

Pacheco, N. D., Vetterling, J. M., and Doran, D. J. 1975. Ultrastructure of cytoplasmic and nuclear changes in *Eimeria tenella* during first-generation schizogony in cell culture. J. Parasitol. 61:31–42.

Palade, G. 1975. Intracellular aspects of the process of protein synthesis. Science 189:347–358.

Pelster, B. 1973. Vergleichende elektronenmikroskopische Untersuchungen an den Makrogameten von *Isospora felis* und *I. rivolta*. Z. Parasitenkd. 41:29–46.

Pelster, B., and Piekarski, G. 1972. Untersuchungen zur Feinstruktur des Makrogameten von *Toxoplasma gondii*. Z. Parasitenkd. 39:225–232.

Pittilo, R. M., and Ball, S. J. 1979. The fine structure of the developing macrogamete of *Eimeria maxima*. Parasitology 79:259–265.

Plattner, H., Miller, F., and Bachmann, 1973. Membrane specializations in the form of regular membrane-to-membrane attachment sites in *Paramecium*, a correlated freeze-etching and ultrathin sectioning analysis. J. Cell Sci. 13:687–719.

Porchet-Henneré, E. 1971. La fécondation et la sporogenèse chez la coccidie *Coelotropha durchoni*: Étude en microscopie photonique et électronique. Z. Parasitenkd. 37:94–125.

Porchet-Henneré, E. 1972. Considérations générales sur les processus de schizogonie chez les Sporozoaires, à la lumière des données de la microscopie électronique. Année Biologique. 11:413–426.

Porchet-Henneré, E. 1975. Quelques précisions sur l'ultrastructure de *Sarcocystis tenella*: I. L'endozoite (après coloration négative). J. Protozool. 22:214–220.

Porchet-Henneré, E. 1976. Structure du mérozoite de la Coccidie *Globidium gilruthi* de la caillete du mouton (d'après l'étude de coupes fines et de coloration négative). Protistologica 12:613–621.

Porchet-Henneré, E. 1977. Étude ultrastructurale de la schizogonie chez la coccidie *Globidium gilruthi*. Protistologica 13:31–52.

Porchet-Henneré, E., and Richard, A. 1971. La sporogenèse chez la coccidie *Aggregata eberthi*: Étude en microscopie électronique. J. Protozool. 18:614–628.

Porchet, E., and Torpier, G. 1977. Étude du germe infectieux de *Sarcocystis tenella* et *Toxoplasma gondii* par la technique du cryodécapage. Z. Parasitenkd. 54:101–124.

Porchet-Henneré, E., and Vivier, E. 1971. Ultrastructure comparée des germes infectieux (sporozoites, mérozoites, schizozoites, endozoites etc) chez les Sporozoaires. Année Biologique. 10:1–2; 77–113.

Roberts, W. L., and Hammond, D. M. 1970. Ultrastructural and cytologic studies of the sporozoites of four *Eimeria* species. J. Protozool. 17:76–86.

Roberts, W. L., Hammond, D. M., Anderson, L. C., and Speer, C. A. 1970a. Ultrastructural study of schizogony in *Eimeria callospermophili*. J. Protozool. 17:584–592.

Roberts, W. L., Hammond, D. M., and Speer, C. A. 1970b. Ultrastructural study of the intra- and extracellular sporozoites of *Eimeria callospermophili*. J. Parasitol. 56:907–917.

Roberts, W. L., Mahrt, J. L., Hammond, D. M. 1972. Fine structure of the sporozoites of *Isospora canis*. Z. Parasitenkd. 40:183–194.

Roberts, W. L., Speer, C. A., and Hammond, D. M. 1971. Penetration of *Eimeria larimerensis* sporozoites into cultured cells as observed with the light and electron microscopes. J. Parasitol. 57:615–625.

Rose, M. E. 1973. Immunity. In: D. M. Hamnd and P. L. Long (eds.), The Coccidia: *Eimeria, Isospora, Toxoplasma,* and Related Genera, pp. 295–341. University Park Press, Baltimore.

Rose, M. E. 1974. Immune responses in infections with coccidia: Macrophage activity. Infect. Immun. 10:862–871.

Ruiz, A., and Frenkel, J. K. 1976. Recognition of cyclic transmission of *Sarcocystis muris* by cats. J. Infect. Dis. 133:409–418.

Ryley, J. F. 1969. Ultrastructural studies on the sporozoite of *Eimeria tenella.* Parasitology 59:67–72.

Rzepczyk, C., and Scholtyseck, E. 1976. Light and electron microscope studies on the *Sarcocystis* of *Rattus fuscipes,* an Australian rat. Z. Parasitenkd. 50:137–150.

Sampson, J. R., and Hammond, D. M. 1971. Ingestion of host cell cytoplasm by means of micropores in *Eimeria alabamensis.* J. Parasitol. 57:113–115.

Satir, B., Schooley, C., and Satir, P. 1973. Membrane fusion in a model system. Mucocyst secretion in *Tetrahymena.* J. Cell Biol. 56:153–176.

Schäffler, M. 1979. Licht- und elektronenmikroskopische Untersuchungen zur Gamogonie und Sporogonie von *Sarcocystis suicanis* im Dünndarm experimentell infizierter Hunde. Doctoral dissertation, University of Munich.

Scholtyseck, E. 1973a. Ultrastructure. In: D. M. Hammond and P. L. Long (eds.), the Coccidia: *Eimeria, Isospora, Toxoplasma,* and Related Genera, pp. 81–144. University Park Press, Baltimore.

Scholtyseck, E. 1973b. Die Deutung von Endodyogenie und Schizogonie bei Coccidien und anderen Sporozoen. Z. Parasitenkd. 42:87–104.

Scholtyseck, E. 1977. The significance of the fine structure of the Sporozoa. In: Proceedings of the First Japanese-German Co-operative Symposium on Protozoan Diseases. Vol. 1, pp. 75–89.

Scholtyseck, E. 1979. Fine Structure of Parasitic Protozoa. Springer-Verlag, Heidelberg.

Scholtyseck, E., and Chobotar, B. 1975. Electron microscope observations concerning the penetration of a host cell by *Eimeria ferrisi* in vivo. Z. Parasitenkd. 49:91–94.

Scholtyseck, E., Chobotar, B., Sénaud, J., and Ernst, J. V. 1977. Fine structure of microgametogenesis of *Eimeria ferrisi* Levine and Ivens, 1965 in *Mus musculus.* Z. Parasitenkd. 51:229–240.

Scholtyseck, E., and Hammond, D. M. 1970. Electron microscope studies of macrogametes and fertilization of *Eimeria bovis.* Z. Parasitenkd. 34:310–318.

Scholtyseck, E., Hammond, D. M. and Ernst, J. V. 1966. Fine structure of the macrogametes of *Eimeria perforans, E. stiedai, E. bovis,* and *E. auburnensis.* J. Parasitol. 52:975–987.

Scholtyseck, E., and Mehlhorn, H. 1970. Ultrastructural study of characteristic organelles (paired organelles, micronemes, micropores) of Sporozoa and related organisms. Z. Parasitenkd. 34:97–127.

Scholtyseck, E., Mehlhorn, H., and Friedhoff, 1970. The fine structure of the conoid of Sporozoa and related organisms. Z. Parasitenkd. 34:68–94.

Scholtyseck, E., Mehlhorn, H., and Hammond, D. M. 1971. Fine structure of macrogametes and oocysts of Coccidia and related organisms. Z. Parasitenkd. 37:1–43.

Scholtyseck, E., Mehlhorn, H., and Hammond, D. M. 1972a. Electron microscope studies of microgametogenesis in Coccidia and related groups. Z. Parasitenkd. 38:95–131.

Scholtyseck, E., Mehlhorn, H., and Müller, B. E. G. 1974. Feinstruktur der Cyste und Cystenwand von *Sarcocystis tenella, Besnoitia jellisoni, Frenkelia* sp. und *Toxoplasma gondii.* J. Protozool. 21:284–294.

Scholtyseck, E., Mehlhorn, H., and Sénaud, J. 1972b. Die subpellikulären Mikrotubuli in den Merozoiten von *Eimeria falciformis.* Z. Parasitenkd. 40:281–294.

Scholtyseck, E., Pellérdy, L., Mehlhorn, H., and

Haberkorn, A. 1973. Elektronenmikroskopische Untersuchungen über die Mikrogametenentwicklung des Mäusecoccids *Eimeria falciformis*. Acta Vet. Acad. Sci. Hung 23:61–73.

Scholtyseck, E., and Piekarski, G. 1965. Elektronenmikroskopische Untersuchungen über die Merozoiten von *Eimeria stiedai* und *E. perforans* und *Toxoplasma gondii:* Zur systematischen Stellung von *T. gondii.* Z. Parasitenkd. 26:91–115.

Scholtyseck, E., Rommel, A., and Heller, G. 1969. Licht- und elektronenmikroskopische Untersuchungen zur Bildung der Oocystenhülle bei Eimerien (*Eimeria perforans, E. stiedai* und *E. tenella*). Z. Parasitenkd. 31:289–298.

Scholtyseck, E., and Schäfer, 1963. Über schlauch-förmige Ausstülpungen an der Zellmembran der Makrogametocyten von *Eimeria perforans.* Z. Zellforsch. Mikrosk. Anat. 61: 214–219.

Scholtyseck, E., and Voigt, W. H. 1964. Die Bildung der Oocystenhülle bei *Eimeria perforans* (Sporozoa). Z. Zellforsch. Mikrosk. Anat. 62:279–292.

Schrével, J. 1968. L'ultrastructure de la région antérieure de la Grégarine *Selenidium* et son intérêt pour l'étude de la nutrition chez les Sporozoaires. J. Microsc. 7:391–410.

Schrével, J., Buissonet, S., and Metais, M. 1974. Action de l'urée sur la motilité et les microtubules sous-pelliculaires du protozoaire *Selenidium hollandei.* C. R. Acad. Sci. (D) (Paris) 278:2201–2204.

Sénaud, J. 1966. L'ultrastructure du micropyle des Toxoplasmida. C. R. Acad. Sci. (D) (Paris) 262:119–121.

Sénaud, J. 1967. Contribution á l'étude des Sarcosporidies et des Toxoplasmes. Protistologica 3:167–232.

Sénaud, J. 1969. Ultrastructure des formations kystiques de *Besnoitia jellisoni* (Frenkel, 1953), protozoaire, Toxoplasmea, parasite de la souris (*Mus musculus*). Protistologica 5:413–430.

Sénaud, J., and Černá, Ž. 1968. Étude en microscopie électronique des mérozoites et de la mérogonie chez *Eimeria pragensis* (Černá et Sénaud, 1968), Coccidie parasite de l'intestine de la souris (*Mus musculus*). Ann. Biol. de Besse-en-Chandesse 3:221–242.

Sénaud, J., and Černá, Ž. 1969. Étude ultrastructurale des mérozoites et de la schizongonie des Coccidies (Eimeriina): *Eimeria magna* (Perard, 1925) de l'intestine des lapins et *E. tenella* (Railliet et Lucet, 1891) des coecums des poulets. J. Protozool. 16:155–165.

Sénaud, J., and Černá, Ž. 1978. Le cycle de dévéloppement asexuée de *Sarcocystis dispersa* (Cerná, Kolarova et Sulc, 1977) chez la souris: Étude au microscope électronique. Protistologica 14:155–176.

Sénaud, J., Chobotar, B., and Scholtyseck, E. 1976. Role of the micropore in nutrition of the Sporozoa: Ultrastructural study of *Plasmodium cathemerium, Eimeria ferrisi, E. stiedai, Besnoitia jellisoni,* and *Frenkelia* sp. Tropenmed. Parasitol. 27:145–159.

Sénaud, J., Mehlhorn, H., and Scholtyseck, E. 1972. Cytochemische Untersuchungen an Cystenstadien von *Besnoitia jellisoni.* Z. Parasitenkd. 40:165–176.

Sheffield, H. G. 1967. The function of the micropyle in the cyst organism of *Besnoitia jellisoni.* J. Parasitol. 53:888–889.

Sheffield, H. G. 1968. Observations on the fine structure of the "cyst stage" of *Besnoitia jellisoni.* J. Protozool. 15:685–693.

Sheffield, H. G. 1970. Schizogony in *Toxoplasma gondii:* An electron microscope study. Proc. Helminth Soc. Wash. 37:237–242.

Sheffield, H. G., and Fayer, R. 1980. Fertilization in the Coccidia: Fusion of *Sarcocystis bovicanis* gametes. Proc. Helminth. Soc. Wash. 47:118–121.

Sheffield, H. G., Frenkel, J. K., and Ruiz, A. 1977. Ultrastructure of the cyst of *Sarcocystis muris.* J. Parasitol. 63:629–641.

Sheffield, H. G., and Melton, M. 1968. The fine structure and reproduction of *Toxoplasma gondii.* J. Parasitol. 54:209–226.

Sheffield, H. G., Melton, M. L., and Neva, F. A.

1976. Development of *Hammondia hammondi* in cell cultures. Proc. Helminth. Soc. Wash. 43:217–225.

Sibert, G. J., and Speer, C. A. 1980. Fine structure of zygotes and oocysts of *Eimeria nieschulzi.* J. Protozool. 27:374–379.

Snigirevskaya, E. S. 1969. Electron microscope study of the schizogony process in *Eimeria intestinalis.* Acta Protozool. 7:57–70.

Snigirevskaya, E. S., and Cheissin, E. M. 1969. The function of the micropore at endogenic developmental stages of *Eimeria intestinalis* (Sporozoa, Eimeriidea). Protistologica 5:209–214.

Speer, C. A. In vitro cultivation of protozoan parasites of man and domestic animals. In: J. B. Jensen (ed.), The Coccidia. CRC Press. In press.

Speer, C. A., and Danforth, H. D. 1976. Fine-structural aspects of microgametogenesis of *Eimeria magna* in rabbits and in kidney cell cultures. J. Protozool. 23:109–115.

Speer, C. A., Davis, L. R., and Hammond, D. M. 1971. Cinemicrographic observations on the development of *Eimeria larimerensis* in clutured bovine cells. J. Protozool. 18(suppl.):11.

Speer, C. A., and Duszynski, D. W. 1975. Fine structure of oocyst walls of *Isospora serini* and *Isospora canaria* and excystation of *Isospora serini* from the canary, *Serinus canarius* L. J. Protozool. 22:476–481.

Speer, C. A., Hammond, D. M., Mahrt, J. L., and Roberts, W. L. 1973a. Structure of the oocyst and sporocyst walls and excystation of sporozoites of *Isospora canis.* J. Parasitol. 59:35–40.

Speer, C. A., Hammond, D. M., Youssef, N. N., and Danforth, H. D. 1973b. Fine structural aspects of macrogametogenesis in *Eimeria magna.* J. Protozool. 20:274–281.

Speer, C. A., Marchiondo, A. A., Duszynski, D. W., and File, S. K. 1976. Ultrastructure of the sporocyst wall during excystation of *Isospora endocallimici.* J. Parasitol. 62:984–987.

Speer, C. A., Marchiondo, A. A., Müller, B., and Duszynski, D. W. 1979. Scanning and transmission electron microscopy of the oocyst wall of *Isospora lacazei.* Z. Parasitenkd. 59:219–225.

Stehbens, W. E. 1966. The ultrastructure of *Lankesterella hylae.* J. Protozool. 13:63–73.

Strout, R. G., and Scholtyseck, E. 1970. The ultrastructure of first generation development of *Eimeria tenella* (Railliet and Lucet, 1891) Fantham, 1909 in cell cultures. Z. Parasitenkd. 35:87–96.

Trefiak, W., and Desser, S. S. 1973. Crystalloid inclusions in species of *Leucocytozoon, Parahaemoproteus* and *Plasmodium.* J. Protozool. 20:73–80.

Varghese, T. 1975. The fine structure of the endogenous stages of *Eimeria labbeana:* 2. Mature macrogametes and young oocysts. Z. Parasitenkd. 46:43–51.

Varghese, T. 1976a. Fine structure of the endogenous stages of *Eimeria labbeana:* 3. Feeding organelles. Z. Parasitenkd. 49:25–32.

Varghese, T. 1976b. The fine structure of the endogenous stages of *Eimeria labbeana:* 4. Microgametogenesis. Z. Parasitenkd. 50:227–235.

Vetterling, J. M., Pacheco, N. D., and Fayer, R. 1973a. Fine structure of gametogony and oocyst formation in *Sarcocystis* sp. in cell culture. J. Protozool. 20:613–621.

Vetterling, J. M., Pacheco, N. D., and Madden, P. A. 1973b. Ultrastructure of dormant, activated and intracellular sporozoites of *Eimeria adenoides* and *E. tenella.* J. Parasitol. 59:15–27.

Vetterling, J. M., Takeuchi, A., and Madden, P. A. 1971. Ultrastructure of *Cryptosporidium wrairi* from the guinea pig. J. Protozool. 18:248–260.

Vivier, E. 1970. Observations nouvelles sur la reproduction asexuée de *Toxoplasma gondii* et considérations sur la notion d'endogenèse. C. R. Acad. Sci. (D) (Paris) 271:2121–2126.

Vivier, E., Devauchelle, G., Petitprez, A., Porchet-Henneré, E., Prensier, G., Schrével,

4

J., and Vinckier, D. 1970. Observations de cytologie comparée chez les Sporozoaires: I. Les structures superficielles chez les formes végétatives. Protistologica 6:127–150.

Vivier, E., and Henneré, E. 1965. Ultrastructure des stades végétatifs de la Coccidie *Coelotropha durchoni.* Protistologica 1:89–104.

Vivier, E., and Petitprez, A. 1969. Le complexe membranaire superficiel et son évolution lors de l'élaboration des individus-fils de *Toxoplasma gondii.* J. Cell Biol. 43:329–342.

Vivier, E., and Petitprez, A. 1972. Données ultrastructurales complémentaires, morphologiques et cytochimiques, sur *Toxoplasma gondii.* Protistologica 8:199–221.

Vivier, E., and Prouvost, J. 1977. Observations comparées sur des inclusions ordonnées minérales et organiques. Biol. Cellulaire 30: 159–164.

Ward, P. A. 1971. Leukotactic factors in health and disease. Am. J. Pathol. 64:521–530.

Wheat, B. E., Jensen, J. B., Ernst, J. V., and Chobotar, B. 1976. Ultrastructure of macrogametogenesis of *Eimeria mivati.* Z. Parasitenkd. 50:125–136.

Zaman, V., and Colley, F. C. 1975. Light and electron microscopic observations on the life cycle of *Sarcocystis orientalis* sp. n. in the rat (*Rattus norvegicus*) and the Malaysian reticulated python (*Python reticulatus*). Z. Parasitenkd. 47:169–185.

5

Biochemistry and Physiology of Coccidia

Ching Chung Wang

The actively growing phase of coccidia is inside a host cell. This fact underlies difficulties in isolating sufficient numbers of developing parasites for detailed quantitative biological investigations and, in turn, is reflected in limited biochemical understanding. In the few instances in which purified coccidial merozoites (Stotish and Wang, 1975) or schizonts (Fernando and Pasternak, 1977a) were obtained, they often consisted of a dying population incapable of further development. However, those who study coccidiosis have a major advantage over others who study related intracellular protozoan parasites such as *Plasmodium, Babesia, Theileria,* and *Anaplasma,* because coccidia do not require an intermediate host. Large numbers of coccidial zygotes, the unsporulated oocysts, can be purified easily in relatively synchronized form free from host tissues and other contaminants (Wang, 1976). Oocyst sporulation and excystation occur in vitro under well defined experimental conditions and are well suited to detailed biochemical investigations, which have been the major source of information in recent years.

In this chapter, biochemical knowledge of each phase of development of coccidia is discussed. Information on the various aspects of metabolism in the parasites, based largely on studies of the unsporulated oocysts, is examined critically. Differences in metabolism between parasites and their hosts are explored; in some cases, these provide an explanation for the mechanism of action of anticoccidial drugs. The discussion covers not only the *Eimeria* species but also *Toxoplasma gondii* which has been the subject of some elegant biochemical studies recently.

PHASES OF DEVELOPMENT

The development of coccidia through discrete phases of the life cycle is under specific genetic control. Unfortunately, little is known about the genetics of the parasites, and there is no information concerning signals triggering each phase of the development. Because of the relatively small size of the nuclei (Scholtyseck, 1973), there is not a clear indication of the chromosomal composition in the nuclei; it is generally accepted that coccidia exist as haploids during the asexual phase and as diploids at the stage of unsporulated oocysts (Canning and Morgan, 1975). Canning and Anwar (1968) noted the presence of two rows of five condensed chromosomes in the nucleus at metaphase from oocysts of *Eimeria tenella* and *Eimeria maxima.* This finding agrees with an earlier observation by Scholtyseck (1963) who had found five rod-like chromosomes in the nucleus of the *E. maxima* microgamete, but differs from the claim by Walton (1959) that there are only two chromosomes in *E. tenella* or *Eimeria bovis.* Definitive understanding is not yet available.

5

Genetic Homogeneity

Because coccidia are capable of massive asexual multiplication within relatively short periods of time, which is followed by sexual differentiation and zygote formation, they may have high frequencies of genetic variation. This has been amply demonstrated by the high incidence of drug resistance among the parasites. Thus, extreme caution must be exercised to ensure a genetically homogeneous population of parasites before further biochemical studies are carried out. Single oocyst cloning in chickens under isolated germ-free conditions has been widely used (Edgar and Seibold, 1964). The possibility of obtaining complete development with a single merozoite of *Eimeria falciformis* in mice was established by Haberkorn (1970a). Shirley and Millard (1976) and Lee et al. (1977) have introduced single sporozoites of *E. tenella* into chick embryos and produced viable oocysts. Shirley (1980) also maintained *E. maxima* by serial passages of single sporocysts in chickens. These techniques may provide, by a single passage, a homozygous population of oocysts, which can be preserved in liquid nitrogen in the form of sporocysts or sporozoites (Norton and Joyner, 1968). However, it can be argued that, because the oocysts undergo zygotic nuclear reduction during sporulation (Canning and Morgan, 1975), there may be two heterologous genotypes of sporocysts in one sporulated oocyst. A genetically homogeneous population thus may be obtained with confidence only by cloning single sporocysts or sporozoites.

There are a limited number of genetic markers identifiable in coccidia, mainly for drug resistance (Jeffers, 1974a; Joyner and Norton, 1975b). However, the electrophoretic profiles of various isozymes used to distinguish species of coccidia (Shirley, 1975) have also been shown to differ between various strains of the same species (Rollinson et al., 1979; Shirley and Rollinson, 1979). These markers are useful for identification of individual strains of coccidia.

T. gondii can be cloned in cell culture by the technique of plaque titration (Chaparas and Schlesinger, 1959). Its usefulness has been demonstrated in isolating temperature-sensitive and drug-resistant mutants, and estimating quantitatively the extent of mutagenesis induced by various means (Pfefferkorn and Pfefferkorn, 1976a,b, 1979).

Sporulation

Among the phases of development of various species of coccidia, the unsporulated oocysts of *E. tenella* are the most conveniently harvested. Because of their accumulation in the ceca of chickens 7–8 days after infection, the oocysts can be easily collected from the ceca. The host tissues may be digested with pepsin and the oocysts purified by a series of stepwise sucrose density differential centrifugations (Wang, 1976), with a final treatment with sodium hypochlorite (Wagenbach et al., 1966). The purified sample is free of biochemical or bacterial contamination, and the unsporulated oocysts remain unchanged at the initial stage of development when stored in 0.1 mM sodium dithionite at 0–4°C (Wang, 1976). The preservation lasts for at least 2 weeks without affecting appreciably the capability of oocyst sporulation, which proceeds with a synchrony of over 90% under optimal in vitro conditions (Wang, 1976). Centrifugal elutriation is another effective means of providing large numbers of purified oocysts (Stotish et al., 1977). Continuous-flow differential density flotation described by Vetterling (1969) is capable of isolating a large number of sporulated oocysts from fecal materials. However, the final preparation must be treated with sodium hypochlorite and further purified by sucrose density differential centrifugation before being used for biochemical investigations. The separation of unsporulated and sporulated oocysts is a more difficult task. Although the latter are known to have a slightly reduced dry weight (Wilson and Fairbairn, 1961), only partial separation from unsporulated oocysts can be achieved by centrifugation in saturated NaCl solution (Davis, 1973) or in 0.45 M sucrose solution (Wang, 1978a). However, repeated density centrifugation as described above can eventually provide a homogeneous sample of sporulated oocysts.

Morphological Changes The process of oocyst sporulation has been studied in considerable cytological detail in *E. tenella* and *E. maxima* (Canning and Anwar, 1968; Wagenbach and Burns, 1969). It is generally agreed that, following an early cytoplasmic contraction, the nucleus becomes spindle-shaped and moves to the surface membrane surrounding the cytoplasm. The first nuclear division then occurs, which may be a meiotic event because the DNA content in each daughter nucleus, estimated by Feulgen's stain, seems to be half of that of the parent nucleus (Canning and Morgan, 1975). The second division follows the first very closely; however, no apparent reduction in nuclear DNA content is observed in the newly formed nuclei. It thus may be a process of mitosis (see Table 1). Cytokinesis then commences, which leads to the formation of four sporoblasts and eventual sporocyst maturation. The last set of nuclear divisions has not been observed, but is thought to take place before the formation of sporozoites. It also may be a mitotic event, because studies of a haemogregarine *Hepatozoon domerguei* (Canning and Morgan, 1975) indicate that the nuclear DNA content of a sporozoite is about half of that in the zygote.

Respiration The optimal conditions for sporulation of most species of coccidial oocysts are aerobic at a temperature near 29°C (Dürr and Pellérdy, 1969). Frequently, oocysts attain the highest rate of sporulation in solutions of strong oxidizing agents such as dichromate or sulfuric acid (Wilson and Fairbairn, 1961). Smith and Herrick (1944) were the first to establish an association of respiratory activities of the oocyst with the process of sporulation. They noted that, after repeated washing, salt flotation, and antiformin treatment to reduce bacterial contamination, the unsporulated oocysts of *E. tenella* maintained a respiratory rate between 11.2 and 16.6 μl of $O_2/10^6$ oocysts/hr at 30°C in a Warburg apparatus for the initial 10 hr. The rate

Table 1. Range of DNA values of *E. tenella* nuclei throughout the life cycle

Nucleus	Number of readings	Range of DNA values in standard units	Arithmetic mean of values and standard errors[a]
Schizont	56	14.64–31.85	
Microgamete	22	13.25–15.11	13.95 ± 0.11
Zygote:			
Pre-first division	26	27.20–31.60	29.75 ± 9.17
Post-first division	52	14.20–32.00	
Post-second division	28	14.40–29.14	

[a]Adapted from Canning and Morgan, 1975.

nearly doubled during the 11th to 12th hr, but dropped to 0.4 μl O_2 / 10^6 oocysts / hr at the 48th hr when the sporulation was completed. Wilson and Fairbairn (1961) recorded respiration by *Eimeria acervulina* during sporulation at 30°C and found a lower initial rate of 3.5 μl of O_2 / 10^6 oocysts / hr, which gradually declined to 0.24 μl after 68 hr (Figure 1). Wang (1976) observed a rate of 6.0 ± 1.2 μl of O_2 / 10^6 oocysts / hr in an oxygen monitor during the initial 3 hr of *E. tenella* sporulation. The rate declined slowly and no sudden rise in respiratory activity was observed in the late stages of sporulation. Respiration and sporulation were both completely inhibited by 10 mM sodium cyanide, sodium azide, 0.1 mM sodium dithionite, or N_2 gas, but were restored when the inhibitors were removed. Respiration is thus essential for coccidial sporulation. Respiration also takes place at 0–4°C at a lower rate, but sporulation of *E. tenella* proceeds only to the stage of cytoplasmic contraction. Initiation of cytokinesis depends apparently on a biological event(s), which takes place only at higher temperatures (Wang, 1976). Wagenbach and Burns (1969) have studied the rate of respiration of *E. tenella* and *Eimeria stiedai* oocysts during sporulation and noticed a temporary decrease at the stage of spindle formation. The significance of this finding remains to be elucidated.

Carbohydrate and Lipid Metabolism Wilson and Fairbairn (1961) observed an almost linear decrease in polysaccharide storage in *E. acervulina* oocysts from 83 to 46 μg / 10^6 oocysts during the initial 10 hr of sporulation to the stage of sporoblast formation (Figure 1). Between the 10th and 25th hr of sporulation, the utilization of carbohydrate was diminished, whereas the lipid content dropped from 52 to 36 μg / 10^6 oocysts, while respiration continued. A net loss of 19% of the dry weight resulted upon completion of sporulation. A similar finding for sporulation of *E. tenella* was made by Wang et al. (1975b). There were amylopectin granules (Ryley et al., 1969) equivalent to 235 nmol of maltose in 10^6 unsporulated oocysts, but only those equivalent to 93 nmol of maltose were found in 10^6 sporulated oocysts. The granules were readily purified from unsporulated oocysts of *E. tenella* by digesting a crude oocyst homoge-

Figure 1. Metabolism of *E. acervulina* during sporulation (adapted from Wilson and Fairbairn, 1961). Reprinted by permission from: Ryley, J. F. Cytochemistry, physiology, and biochemistry. In D. M. Hammond and P. L. Long (eds.), The Coccidia: *Eimeria, Isospora, Toxoplasma,* and Related Genera. Baltimore. University Park Press, p. 161 (1973).

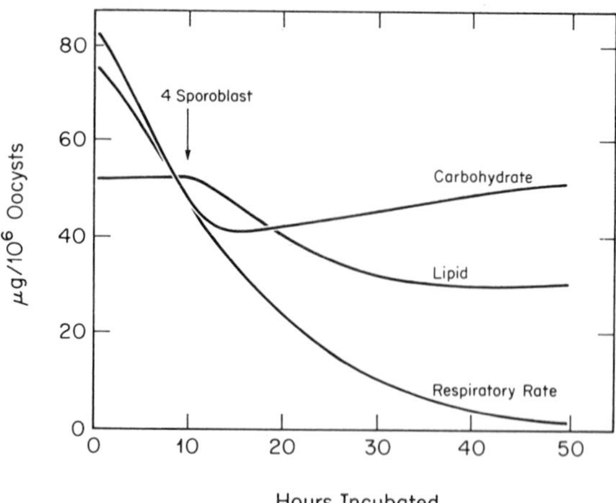

nate with sodium dodecyl sulfate (SDS) and pronase followed by centrifugation and washing (Wang et al., 1975b) or by repeated centrifugation of the crude extract at $3,000 \times g$ in 1.0 M sucrose (Stotish et al., 1978). α-Amylase treatment and chemical analysis revealed that the granules consist of glucose polymers (Wang et al., 1975b). It has been estimated that the granules contain a total of approximately 5×10^{-13} mol of glucose residues in each unsporulated oocyst of *E. tenella.* Assuming half of the glucose is fully utilized for energy production during sporulation, there are 9.5×10^{-12} mol of ATP generated in each oocyst.

The enzyme most likely responsible for channeling amylopectin into carbohydrate metabolism is amylopectin phosphorylase (Wang et al., 1975b). It has high activity (13 U/mg of protein) in the unsporulated oocyst of *E. tenella,* which causes the observed disappearance of amylopectin granules in 250 min under optimal conditions in vitro. The enzyme activity, however, declined at a linear rate during sporulation; sporulated oocysts contain less than 8% of the activity of unsporulated oocysts. No amylase activity was found in *E. tenella* oocysts.

The saponifiable lipids of *E. tenella* oocysts, analyzed by gas-liquid chromatography, decreased during sporulation from 91 to 47 μg per 10^6 oocysts (Weppelman et al., 1976). However, the nonsaponifiable lipids in the oocysts, which consist mainly of C_{24} and C_{26} primary fatty alcohols, identified by combined gas-liquid chromatography-mass spectrometry, increased from 16 to 44 μg per 10^6 oocysts. The added fatty alcohols were mainly associated with the cytoplasmic fraction.

During sporulation of *E. tenella,* CO_2 is the only detectable metabolite excreted into the medium (Wang, 1976), suggesting complete aerobic metabolism of carbohydrates and fatty acids in the oocysts.

Protein Metabolism Total oocyst nitrogen remains fairly constant during sporulation of *E. acervulina* (Wilson and Fairbairn, 1961). Total protein content in an *E. tenella* oocyst, estimated at 2.7×10^{-10} g, is changed very little by sporulation (Wang and Stotish, 1975b). However, 13% of the total protein in the soluble cytoplasm of *E. tenella* unsporulated oocysts is shifted to the $15,000 \times g$ pellet fraction in sporulated oocysts.

The Glycoprotein Stotish et al. (1976) demonstrated that the loss of cytoplasmic protein during sporulation reflects a single glycoprotein, which was identified readily in the cytoplasm of *E. tenella* unsporulated oocysts by polyacrylamide gel electrophoresis, but not in the cytoplasm of sporulated oocysts (Figure 2). The glycoprotein was purified to homogeneity by ammonium sulfate (25–60%) fractionation, DEAE-cellulose chromatography, and Sephadex G-200 filtration. The purity of the preparation was verified by SDS-gel electrophoresis, which also suggested a molecular weight of the glycoprotein of about 30,000. Amino acid and carbohydrate analysis by combined gas-liquid chromatography-mass spectrometry provided the composition of the glycoprotein, which consists of 141 amino acid residues, constituting about two-thirds of the molecular weight. The remainder of the molecule contains sugars, with

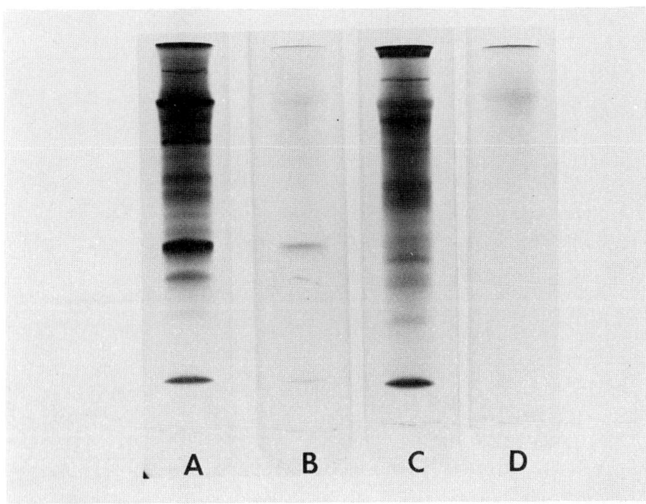

Figure 2. Comparison of unsporulated and sporulated soluble protein by polyacrylamide gel electrophoresis. Polyacrylamide gels (7.5%) stained with Coomassie brilliant blue (*A* and *C*) and by the periodic acid-Schiff technique (*B* and *D*). *A* and *B* contain 100 μg each of unsporulated oocysts-soluble protein; *C* and *D* contain 100 μg each of soluble protein from sporulated oocysts. Reprinted by permission of The American Society of Biological Chemists, Inc. from: Stotish, R. L., Wang, C. C., Hichens, M., Vanden Heuvel, W. J. A., and Gale, P. Studies of a glycoprotein in the oocysts of *Eimeria tenella*. *The Journal of Biological Chemistry* 251:302–307 (1976).

glucose and xylose as the major components. This glycoprotein is believed to be the first major protein purified from coccidia.

Stotish et al. (1976) induced an anti-glycoprotein serum in rabbits, and established a radioimmunoassay for the glycoprotein. The assay showed the presence of 144 μg of the glycoprotein in 1.0 mg of total cytoplasmic protein in *E. tenella* unsporulated oocysts. However, the glycoprotein virtually vanished from the cytoplasm in the late phase of sporulation when sporozoites were being formed. The material cross-reacting to the anti-glycoprotein serum was found on the membrane surface of *E. tenella* sporozoites by an immunofluorescence assay. This may mean that the glycoprotein becomes incorporated into the membranes of sporozoites during their formation. It is not known whether the glycoprotein remains intact while being incorporated into the membrane, nor is the biological function of the incorporated glycoprotein clear. Because it is an apparent surface antigen of the sporozoite, the possible presence of anti-glycoprotein antibody in chickens immunized against *E. tenella* infection was examined. Likewise, the possibility of passive immunization against *E. tenella* infection using rabbit anti-glycoprotein serum was tested. Neither approach was successful; the results were not surprising in view of the lack of protective humoral antibody response to *E. tenella* infection and the development of host immunity only during coccidial schizogony (Rose, 1973). Radial immunodiffusion and radioimmunoassay indicated the presence of a material partially cross-reacting to the anti-glycoprotein serum in the cytoplasm of *E. acervulina*. The observations made on *E. tenella* thus may apply to other species of coccidia.

Leucine Incorporation In spite of the restricted permeability of the oocyst wall (Ryley, 1973), some small molecules are capable of penetrating it to a limited extent. Wang and Stotish (1975b) observed incorporation of radiolabeled leucine into the trichloroacetic acid (TCA)-insoluble fraction of *E. tenella* oocysts during the initial 7 hr of sporulation (Figure 3). The incorporated radioactivity, which amounts to about 10^{-15} mol of leucine per oocyst, is concentrated in the cytoplasm. Most migrate in a narrow band in SDS-polyacrylamide gel electrophoresis at a rate corresponding to a molecular weight of about 50,000. The biological function of this minor protein(s) is not known, except that it may be located in the cytoplasm of sporozoites in the sporulated oocysts (Wang and Stotish, 1975b).

Polyribosomes Leucine incorporation during the early phase of *E. tenella* sporulation suggests active protein synthesis. Ribosomes were purified from the unsporulated oocysts of *E. tenella* at an average yield of 13.6 mg per 10^8 oocysts (Wang, 1978b). Electron microscopic examination indicated that most of the ribosomes, each about 20 nm in diameter, were densely beaded along strands of average length of 1.5 μm. These polysomes, which were likely in an active state of protein synthesis, were totally dissociated into monomeric ribosomes in the sporulated oocysts. Because single ribosomes suggest cessation of protein synthesis, the observation seems to reflect the dormant state of sporulated oocysts. Further analysis by sucrose density gradient sedimentation indicated that the content of *E. tenella* polysomes is largely dimin-

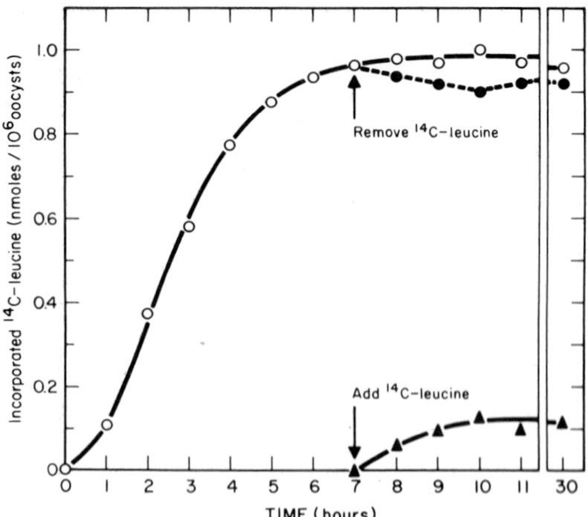

Figure 3. Incorporation of *l*-[^{14}C]leucine during sporulation. The labeled amino acid (4 μM (0.25 μCi/nmol)) was added to the unsporulated oocyst suspension at the beginning of incubation (○); at the beginning, but washed off with Ringer-phosphate buffer, pH 7.4, (by brief centrifugation), after 7 hr of incubation (●); or 7 hr after the beginning of sporulation (▲). Reprinted by permission from: Wang, C. C., and Stotish, R. L. Changes in nucleic acids and proteins in the oocysts of *Eimeria tenella* during sporulation. *The Journal of Protozoology* 22:438–443 (1975).

ished after 8 hr of sporulation, and turned into a distinct peak at about 80 S after 24 hr (Figure 4). This observation is in agreement with the suggestion, based on leucine incorporation data, that *E. tenella* oocysts synthesize protein only during the initial few hours of sporulation.

The 80 S ribosome of *E. tenella* oocysts consists of 2 subunits of about 40 S and 60 S, which are readily dissociated in high concentrations of KCl, and reassociated upon dialysis (Wang, 1978b). These subunits do not form hybrids with the corresponding ribosomal subunits of chicken liver. They have an electrophoretic pattern of proteins distinctively different from that of chicken liver ribosomal proteins, and an RNA profile of 5 S, 16 S, and 23 S in sucrose density gradient sedimentation. These data suggest that coccidial ribosomes may have several structural differences from host ribosomes. Other intracellular protozoan parasites seem to have similar aberrations in sizes of ribosomal RNA. Remington et al. (1970) fractionated the RNA of *T. gondii* by sedimentation and showed the presence of three major components of 24 S, 19 S, and 4–5 S. The ribosomal RNA of *Plasmodium knowlesi* was estimated to have components of 24.2 S, 17.4 S, and 3.8 S (Warhurst and Williamson, 1970), whereas ribosomal RNA of *Plasmodium berghei* consists of 25 S and 14.9 S species (Tokuyasu et al., 1969). These small, heterogeneous ribosomal RNAs may represent midway points of the hypothetical ribosomal evolution from prokaryotes to advanced eukaryotes (Reisner et al., 1968).

Enzyme Turnover A gradual decrease in the specific activities of some enzymes (e.g., dihydrofolate reductase and amylopectin phosphorylase) was observed in *E. tenella* oocysts during sporulation (Wang et al., 1975a, b). It is uncertain whether the decreases are due to inactivation or degradation of the enzymes. There is no detectable protease activity in crude extracts of unsporulated and sporulated oocysts of *E. tenella* at pH 7.0, but low levels of activity (∿15 μg of [methyl-^{14}C]glycinated hemoglobin hydrolyzed/mg of protein/30 min) were found in both extracts at pH 4.0. The protease has an apparent pI value of 6.3 in isoelectric focusing and is totally inhibited by 10^{-4} M phenylmethylsulfonyl fluoride, suggesting a serine-type protease (Wang and Stotish, 1978). This low activity implies limited protease degradation of proteins during coccidial sporulation.

However, leucine aminopeptidase activity was identified at equal levels (50 μmol of L-leucyl-*o*-nitroanilide/min/mg of protein) in unsporulated and sporulated oocysts of *E. tenella* (Wang and Stotish, 1978). Electrophoretic analysis showed three isozymes (I, II, and III) in unsporulated oocysts; they diminished during the late phase of sporulation with the simultaneous emergence of a new, fast-moving isozyme V (Figure 5).

The isozyme V, unlikely to have resulted from de novo protein synthesis because of its late emergence, is predominantly in the cytoplasm surrounding the sporocysts. It has slightly changed properties, and may perform a role during oocyst excystation. Isozymes I, II, and III may be involved in inactivating various enzymes during sporulation. Further studies are required.

5

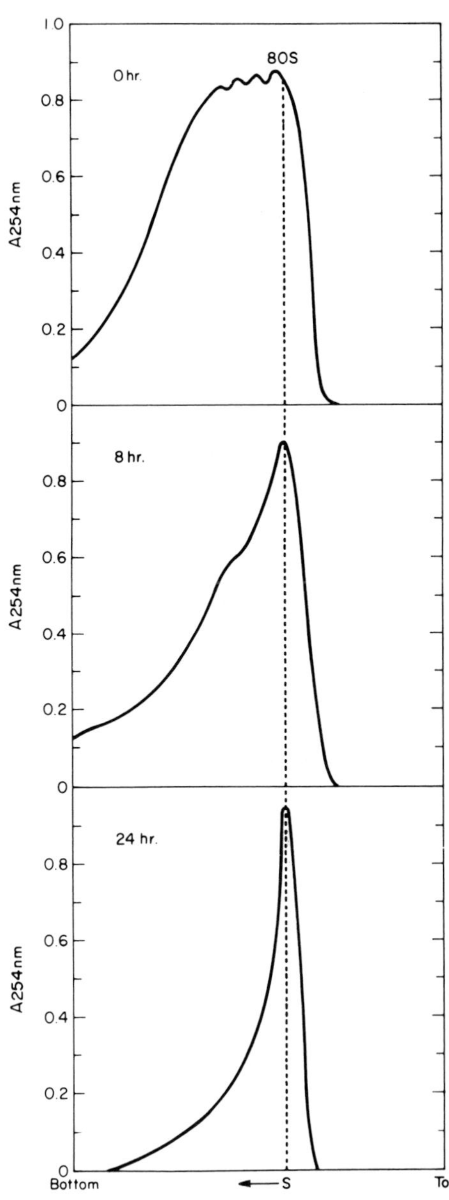

Figure 4. Sedimentation profiles of ribosomes purified from *E. tenella* oocysts. A suspension of ribosomes in 0.1 ml of Medium A (10–25 OD 260 nm units/ml) was layered on 25 ml of 15–30% linear sucrose density gradient prepared also in Medium A. The sample was centrifuged in a Spinco SW 65 rotor in a Spinco model L2-65 ultracentrifuge at 4°C and 60,000 rpm. The 80 S position in the profile was specificed by the chicken liver ribosomes preincubated in the in vitro protein synthesis reaction mixture. The number of hours demonstrated in each graph represents the duration of oocyst sporulation before harvest of ribosomes. Reprinted by permission from: Wang, C. C. The prokaryotic characteristics of *Eimeria tenella* ribosomes. *Comparative Biochemistry and Physiology. B: Comparative Biochemistry (Oxford)* 61:571–579 (1978).

Figure 5. Analysis of leucine aminopeptidase activities in *E. tenella* oocysts by polyacrylamide gel electrophoresis. Samples of crude cell-free extracts containing 0.10 mg of proteins were analyzed in slab gel electrophoresis using Ortec apparatus. Mixtures of *N*-benzoyl-DL-arginine, L-valine, DL-alanine, L-leucine, and glycine-β-naphthylamide at 0.2 mg/ml each were used as substrates. Bands of activity were then stained with black K salt. *A*, unsporulated oocysts; *B*, sporulated oocysts; *C*, cytoplasms of sporulated oocysts; *D*, sporocysts. Reprinted by permission from: Wang, C. C., and Stotish, R. L. Multiple leucine amino-peptidases in the oocysts of *Eimeria tenella* and their changes during sporulation. *Comparative Biochemistry and Physiology. B: Comparative Biochemistry (Oxford)* 61:307–313 (1978).

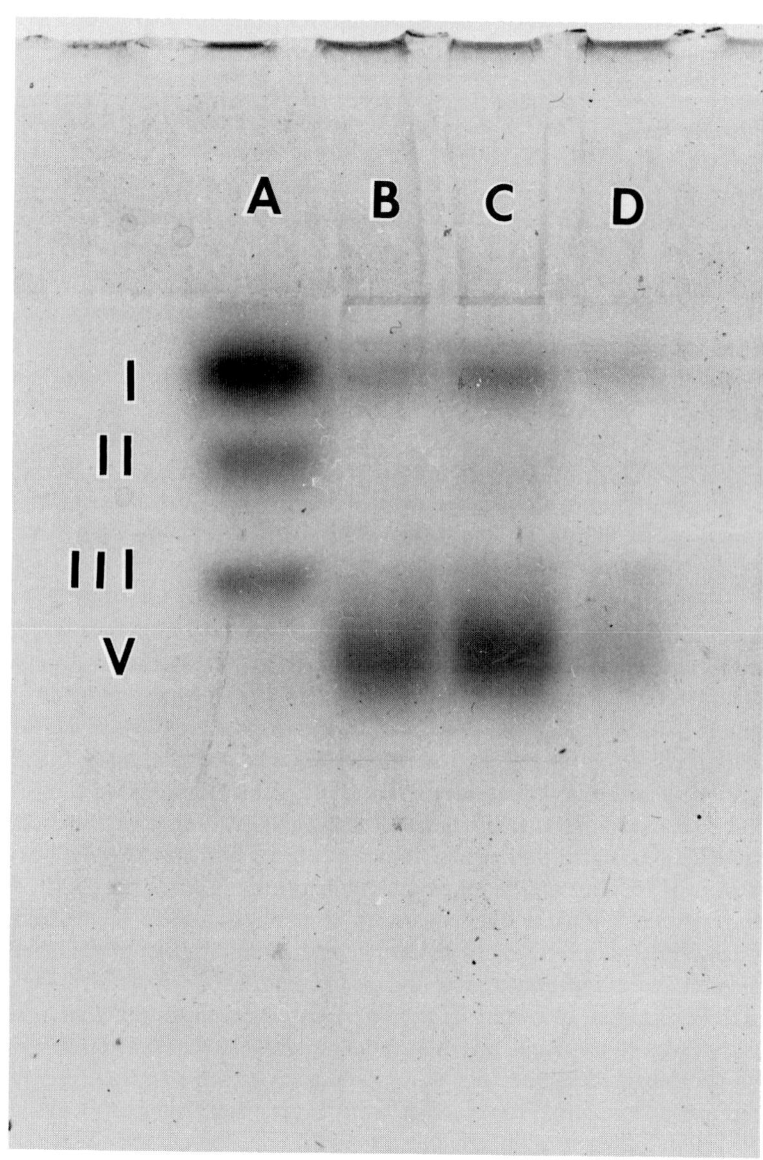

5

Nucleic Acid Metabolism Exogenous radiolabeled uridine is incorporated into the RNA fraction of nuclei and paranuclear bodies at 1.5×10^{-16} mol per *E. tenella* oocyst during the initial 6 hr of sporulation (Wang and Stotish, 1975b), suggesting limited turnover or synthesis of RNA. The bulk of RNA consisting mainly of the three ribosomal RNA species remains, however, relatively unchanged during sporulation (Wang, 1978b). No thymidine incorporation was detectable in the TCA-insoluble fraction, but the radiolabel in deoxyuridine was incorporated into RNA during *E. tenella* sporulation (Wang and Stotish, 1975b). This evidence strongly suggests a lack of DNA synthesis during *E. tenella* sporulation. Even if the failure to detect thymidine incorporation could be explained by low thymidine kinase in the parasite (Ouellette et al., 1974), the incorporation of the label from deoxyuridine into RNA indicates that the same nucleoside also should be incorporated into DNA if thymidylate synthetase is present and if DNA is being synthesized. Thymidylate synthetase has been demonstrated in the unsporulated oocysts of *E. tenella*, although the actual level of enzyme activity was not determined (Coles et al., 1979). This leaves the absence of DNA synthesis as the most reasonable explanation for the previous results. This conclusion is supported by measurements of the total DNA content in unsporulated and sporulated oocysts of *E. tenella*. With the samples spiked with radiolabeled *Escherichia coli* DNA, Wang and Stotish (1975b) found little change in cellular DNA content during *E. tenella* sporulation; there was about 5.8×10^{-12} g of DNA per oocyst before and after the process.

This conclusion seems to contradict the findings by Canning and Morgan (1975), which suggested similar levels of Feulgen-positive materials in each nucleus at the 2 and 4 nuclei stages of *E. tenella* sporulation, and thus could mean doubling of DNA content in the oocyst. However, since the *E. tenella* macrogamete nuclei were Feulgen-negative (Canning and Morgan, 1975), it is probable that only 50% of the total DNA was identifiable by Feulgen staining during the early phase of oocyst sporulation. Furthermore, the actual data show a wide range of variation in the contents of Feulgen-stainable materials among the post-first and post-second division nuclei; mostly intermediary values between that of a microgamete and pre-first division zygote were found (see Table 1). There is thus no firm evidence suggesting a net increase of DNA during sporulation. It is likely that all the DNA synthesis is completed prior to the maturation of unsporulated oocysts.

Among the enzymes involved in nucleic acid metabolism, thymidylate synthetase (Coles et al., 1979) and dihydrofolate reductase (Wang et al., 1975a) have been identified in unsporulated oocysts of *E. tenella*. There is little known about thymidylate synthetase, except that it has an apparent K_m of 6.54 μM for dUMP and 1.86 μM for N^5,N^{10}-methylene tetrahydrofolate. Dihydrofolate reductase was purified some 8,000-fold by methotrexate-aminoethyl-Sepharose affinity column chromatography. Its molecular weight was estimated as 240,000. It is present at 0.2 nmol of dihydrofolate reduced/min/mg of protein in the crude extracts of *E. tenella* unsporulated oocysts. Activity diminishes rapidly beyond detection during sporulation, which may be another indication of the lack of nucleic acid synthesis during sporulation.

The buoyant density of bulk DNA isolated from unsporulated oocysts of *E. tenella* was determined by CsCl equilibrium density gradient centrifugation: $\rho = 1.682$ g/cm^3 (Wang and Stotish, 1975b). Similar studies were carried out on DNA from *E. maxima* unsporulated oocysts, and two peaks were identified at 1.707 and 1.682 g/cm^3 (Lee and Fernando, 1977a). A Houghton strain of *E. tenella* has DNA with a buoyant density of 1.711 g/cm^3 (Shirley and Rollinson, 1979). It is not yet clear whether these differences in DNA buoyant densities can be used to distinguish different species or strains of coccidia.

Excystation

The sporulated oocysts of coccidia are capable of surviving at moderate temperature and humidity for extended periods of time, maintaining minimal rates of respiration (Wilson and Fairbairn, 1961; Wagenbach and Burns, 1969). This basal respiratory activity depends, at least in part, on consumption of the storage carbohydrate in the oocyst. Vetterling and Doran (1969) found that in *E. acervulina* sporulated oocysts stored at 4°C, the storage polysaccharide in sporozoites fell from 33.3 μg of glucose / 10^6 oocysts at 3 months to 21.3 μg at 1 yr, 7.8 μg at 2 yr, and 1.5 μg at 6 yr. The decrease in carbohydrate content was correlated with a decline in infectivity in chickens. Similar observations on carbohydrate utilization during storage were made by Kheysin (1959) who noted that reserves were longer-lasting in species of *Eimeria* than in *Isospora*. Kheysin also observed utilization of the carbohydrate in the residual body of the sporocyst as well as in the sporozoites themselves. Wagenbach and Burns (1969) noted that sporulated oocysts of *E. tenella* would survive for 4 days at 41.5°C in air, but only for 1 day at this temperature when incubated under N_2. It is not clear whether death reflects depletion of carbohydrate.

The Process of Excystation Excystation of coccidia is a process in which sporozoites are released from sporulated oocysts. It has been a focal point of interest and study for many years. A detailed account of previous efforts has been provided in a review by Ryley (1973). Briefly, the process involves removal of a round-shaped formation, that is, a micropyle, on the oocyst wall, by incubation at 41°C in solutions containing sodium dithionite, cysteine, or ascorbic acid under CO_2. In the presence of bile and pancreatic enzymes, the sporocystic plug, that is, the Stieda body, is removed, which allows the escape of sporozoites.

Although the involvement of CO_2 has led to the belief that it activates an enzyme precursor in the sporozoite (Hibbert and Hammond, 1968) or in the oocyst residual body (Ryley, 1973), there is little evidence supporting this theory. Instead, CO_2 was found replaceable by H_2S or NO (Jolley et al., 1976), which suggests that an anaerobic condition may be all that is required for removal of the micropyle. The freeing of sporocysts from oocysts is not a prerequisite for sporozoite release in nature (Lotze and Leek, 1968); bile and pancreatic enzymes may enter the oocyst after the micropyle is removed. Doran and Farr (1962) and Hibbert et al. (1969) noted that the use of bile could be replaced by taurocholate, glycocholate, deoxycholate, cholate, glycotaurocholate, or other surface-active agents. Wang and Stotish (1975a) found that *E. tenella* sporocysts could be pretreated in 2% taurocholate, washed, and then opened to release sporozoites upon incubation with pancreatic enzymes. The use of crude rather than purified preparations of pancreatic enzymes seems to be the best means of producing sporozoites in high yield; purified pancreatic trypsin or lipase are much less effective. α-Chymotrypsin has the best activity among purified pancreatic enzymes in catalyzing excystation of *E. acervulina* (Doran, 1966) and *E. tenella* (Wang and Stotish, 1975a). Use of specific inhibitors of trypsin and chymotrypsin suggests that the latter is one of the essential enzymes involved in releasing *E. tenella* sporozoites (Wang and Stotish, 1975a). Trypsin may be less important, but a combination of 0.01% purified chymotrypsin and 0.1% purified trypsin results in improved yield of *E. tenella* sporozoites (Chapman, 1978).

Purification of Sporocysts and Sporozoites The oocyst walls are highly susceptible to mechanical pressure and are ruptured easily by shearing forces. For avian coccidia, mechanical rupture of oocysts in the gizzard of the infected host is more likely to be the normal primary stage of excystation, rather than the removal of micropyles (Ryley, 1973). In laboratories, the best way of releasing sporocysts from oocysts is by mechanical shearing. Grinding in a Teflon-coated tissue grinder (Patton, 1965), shaking with grade 7 Ballotini glass beads (Ryley, 1973), and sonicating briefly (Stotish et al., 1976) all work satisfactorily for a high yield of sporocysts. Filtration through a column of glass beads has been widely adopted to purify the sporocysts (Wagenbach, 1969), but the yield is somewhat variable. Centrifugal elutriation may provide the most effective means of purifying large numbers of sporocysts at the rate of 10^8 cells per hr with an 80% yield (Stotish et al., 1977). The purified sporocysts can be stored in Ringer's phosphate buffer at 0–4°C for 1 month without significant loss of viability. The sporozoites harvested from incubating sporocysts with bile salts and pancreatic enzymes can also be purified through a glass bead column (Wagenbach, 1969). Yields and separation between sporocysts and sporozoites are poor. The centrifugal elutriation technique is the best method capable of providing $0.5–1.0 \times 10^8$ purified sporozoites with an average yield of 50% and minimal sporocyst contamination within a matter of 90 min (Stotish et al., 1977). Purified sporozoites may be stored in Ringer's phosphate buffer at 0–4°C for 3 weeks without detectable loss of infectivity (Long, 1973).

Metabolism in Sporozoites Respiration of coccidia during excystation was first observed by Vetterling (1968) on many different species of the parasite. Vigorous respiratory activity is initiated upon breakage of oocyst wall and proceeds at 41.5°C at a rate of 3.0±0.6 μl of O_2/hr/10^6 fresh *E. tenella* oocysts (Wang, 1975). The rate remains relatively constant for the first 3–4 hr, and gradually declines during the next few hours. The presence of taurocholate and pancreatic enzymes and subsequent freeing of sporozoites have no effect on the rate of respiration. Vetterling and Doran (1969) observed that during a 30-min period of excystation at 42.9°C, carbohydrate reserves in the sporozoites of *E. acervulina*, *Eimeria necatrix*, and *Eimeria meleagrimitis* fell from 33.3, 30.1, and 36.7 to 11.0, 9.4, and 13.3 μg of glucose per 10^6 oocysts, respectively. Ryley (1973) correlated the rates of respiration, given previously by Vetterling (1968), with the amounts of carbohydrate consumed and found that the ratios between moles of O_2 consumed and moles of glucose utilized were 6.35, 5.87, and 7.08 for the three parasites. These ratios suggest total breakdown of glucose molecules to CO_2 and H_2O during sporozoite respiration.

Excystation of coccidia, however, proceeds normally in the presence of 10 mM cyanide (Wagenbach and Burns, 1969), 0.1 mM dithionite, or N_2 (Wang, 1975). The sporozoites thus produced resumed respiration immediately upon removal of the anaerobic conditions and remained fully infective both in vivo and in chick embryos. There is thus no connection between the process of excystation and respiration by sporozoites. Indeed, *E. tenella* sporozoites can be released by the normal procedure of excystation from sporocysts pretreated with chloroform, ethyl acetate, or toluene at 20°C. Only when the sporocysts have been preheated at 55°C or higher, or pretreated with formaldehyde, does excystation no longer occur (Wang, 1975). These observations strongly suggest a passive nature of the coccidial excystation process, which may be merely dependent on the anaerobic environment, bile, pancreatic enzymes, optimal temperature, pH, and perhaps other factors existing in the intestinal tract of an infected host.

The emergence of sporozoites under anaerobic conditions is an interesting subject for study because it may mimic the events occurring in vivo. Ryley (1973) investigated the anaerobic endogenous metabolism of *E. tenella* sporozoites at 41°C and identified lactic acid as the major metabolite accounting for 74.5% of the consumed carbohydrate (see Table 2). The rest of the sugar metabolism leads to glycerol (14%) and CO_2 (8.4%). These results suggest glycolysis as the main pathway of glucose metabolism in sporozoites with a minimal contribution from the tricarboxylic acid cycle under anaerobic conditions (see under "Metabolism and Chemotherapy").

It has been generally accepted that energy derived from glucose metabolism in sporozoites may be needed for excystation and entry of sporozoites into host cells (Ryley, 1973). The notion implies that more amylopectin in sporozoites is utilized to meet the energy requirement by incomplete glucose metabolism under anaerobic conditions. This view is not supported by the evidence; data in Table 2 indicate a decrease of 30.05 μmol of glucose in the amylopectin reserve in *E. tenella* sporozoites containing 16 mg of N under anaerobic conditions over a period of 90 min. Because approximately 25 μg of N is found in 10^6 *E. acervulina* oocysts

5

Table 2. Anaerobic endogenous metabolism of *E. tenella* sporozoites[a]

	μmol	Mol/mol glucose
Amylopectin (as glucose)	−30.05	−1.00
Acid (from bicarbonate)	+50.25	+1.67
CO_2	+14.95	+0.50
Lactic acid	+45.50	+1.49
Pyruvic acid	0.00	0.00
Succinic acid	0.00	0.00
Glycerol	+8.60	+0.28

[A]Sporozoites of *E. tenella* (16mg of N in a total volume of 6 ml) were incubated in Ringer-bicarbonate medium for 90 min in Warburg flasks at 41 °C with gas phase of 5 % CO_2, 95 % N_2. Fermentation was stopped by tipping acid from a side bulb. Initial and final values are based on the combined contents of four manometer flasks. Values given are total amounts of substrate used and metabolites recovered, and also moles of metabolite formed per mol of substrate used (adapted from Ryley, 1973).

(Wilson and Fairbairn, 1961), the number of *E. tenella* sporozoites in the experiment is roughly equivalent to 6.4×10^8 *E. acervulina* oocysts. Vetterling (1968) indicated that 10^6 *E. acervulina* oocysts consume 22.3 μg (0.124 μmol) of glucose within 30 min of aerobic excystation, which represents a rate of glucose metabolism 7.9-fold higher than that observed under anaerobic conditions (Table 2). This observation, although requiring more experimental data for confirmation, may lead to an important new concept, that is, glucose metabolism in sporozoites may not serve any specific purpose, and the rate of metabolism may be dictated by the exogenous conditions. For example, addition of glucose to suspensions of *E. tenella* sporozoites accelerates the rate of glucose metabolism regardless of the amylopectin reserve in sporozoites (Ryley, 1973). The finding also raises the question whether an anaerobic environment, like that in vivo, for coccidial excystation and host cell invasion could improve the infectivity of sporozoites, because less amylopectin reserve in the sporozoite is consumed; this reserve may be needed for intracellular development of the parasite.

Oocyst Wall

It is generally accepted that the oocyst wall is derived from two distinct types of large granules which develop in the cytoplasm of macrogametocytes. By electron microscopic investigations, Scholtyseck and Voigt (1964) and Scholtyseck et al. (1969) demonstrated that the type II granules, which appear first in macrogametocytes, are slightly stained with osmium oxide but stained blue with mercuric bromphenol blue, suggesting the presence of protein. Type I granules appear subsequently and are homogeneously and densely stained by osmium oxide. Both types of granule migrate to the periphery of macrogametocyte during development and fuse to form the oocyst wall. The wall is known to consist of two layers; a thin, osmiophilic outer layer and a thicker, less electron-dense inner layer (Scholtyseck and Weissenfels, 1956; Nyberg and Knapp, 1970b). It is thus likely that the type I granules give rise to the outer layer and the type II granules give rise to the inner layer. For a more detailed description of the fine structure of oocyst wall, refer to Chapter 3 (Jeffers and Shirley, this volume).

Understanding of the nature of the oocyst wall has been complicated by occasional claims of the existence of a third outermost layer (Jackson, 1964; Ryley, 1980). This layer is thought to have a tan color, to represent about 20% of the wall on a dry weight basis (Ryley, 1980), and to be removable by sodium hypochlorite, which then gives the oocysts a white-colored appearance. There is, however, little evidence substantiating this belief other than a single light microscopic picture (Nyberg et al., 1968) showing a balloon formation around an *E. tenella* oocyst in the presence of sodium hypochlorite. This could be a film of contaminating materials caused by gases produced by the oxidizing power of sodium hypochlorite. Indeed, Nyberg and Knapp (1970a) observed, by scanning electron microscopy, a surface coating of granular debris over *E. tenella* oocysts removable by sodium hypochlorite and concluded that it could be host fecal materials. Ryley (1980) analyzed these materials and stated that "due to the large amount of inorganic matter chemical analysis was difficult. It did, however, appear to contain carbohydrate, and a protein characterized by a high proline content and the absence of basic amino acids." Unfortunately, no data or experimental details are available for further evaluation.

Stotish et al. (1978) used the unsporulated oocysts of *E. tenella*, isolated by the procedure of Wang (1976), which included a final step of 5% sodium hypochlorite treatment. These authors also purified oocyst wall fragments by sonification and sucrose density centrifugation. The purified wall fragments, shown in Figure 6, have an electron-dense outer layer about 10 nm thick and a lighter inner layer of 90 nm when examined by transmission electron microscopy (Figure 7*A*). They can be largely solubilized by incubation with 0.25 M dithiothreitol in 8.0 M urea at 100°C in vacuo for 15 hr, 0.1 N NaOH plus 0.5 M $NaBH_4$ at 37°C for 24 hr, 80% performic acid and 3% H_2O_2 at ᴮ10°C for 45 min, or 5% sodium hypochlorite at 25°C for 30 min. These wall fragments also leave thin, electron-dense strands in the pellet fraction, which are most likely the outer layer of the oocyst wall (Figure 7*B*). This isolated outer layer consists mainly of fatty alcohols, among which hexacosanol, the primary C_{26} alcohol, is the major component. It also contains some phospholipids and fatty acids, but no carbohydrate or protein. The solubilized inner layer has 80% of the total mass accounted for by a single glycoprotein band with an approximate molecular weight of 10,000 in SDS-polyacrylamide gel electrophoresis. This glycoprotein is the only nondialyzable component in the solubilized inner layer and contains nearly all the carbohydrates found in the oocyst wall. The carbohydrate composition of the glycoprotein is mannose, 0.005; galactose, 0.016; glucose, 0.207; and hexoseamine, 0.020

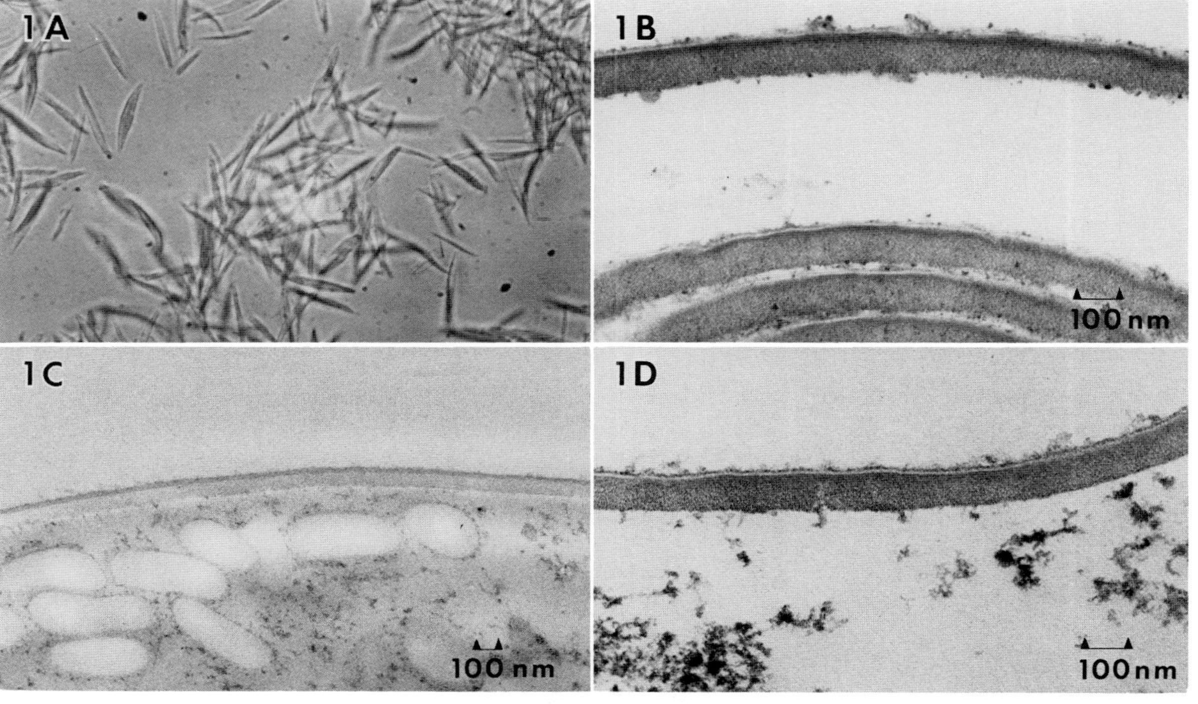

Figure 6. *E. tenella. A*, light microscopic photograph of aqueous suspension of purified oocysts walls of *E. tenella* unsporulated oocyst × 100. *B*, electron micrograph of purified oocysts wall preparation. Oocysts used for preparation had been purified as described in the text, including treatment with 5% NaOCl. *C*, electron micrograph of a section through an intact oocyst purified as described in the text. Comparison with *1B* illustrates conservation of structure during oocyst wall purification. *D*, electron micrograph of a section through an intact oocyst purified by centrifugal elutriation. Reprinted by permission from: Stotish, R. L., Wang, C. C., and Meyenhofer, M. Structure and composition of the oocyst wall of *Eimeria tenella*. *Journal of Parasitology* 64:1074–1081 (1978).

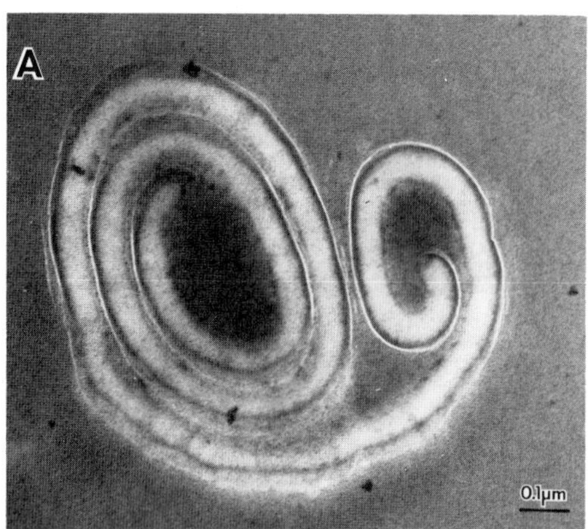

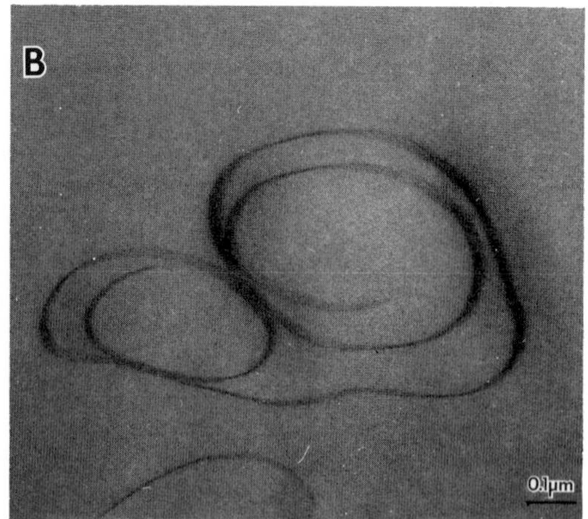

mg/mg of protein. Amino acid composition of the glycoprotein indicates relatively lower content of lysine and cysteine when compared with the total amino acid composition of purified oocyst wall. These two amino acids were found among small peptides in the solubilized inner layer. No significant amount of lipid was found in the solubilized inner layer.

Stotish et al. (1978) also isolated wall fragments from *E. tenella* unsporulated oocysts purified by repeated centrifugal elutriations (Stotish et al., 1977) without exposure to sodium hypochlorite. Their appearance is similar to that of those isolated from sodium hypochlorite-treated oocysts (Figure 6) under the electron microscope; no existence of a third outermost layer is apparent. Analysis of these untreated wall fragments gives similar compositions of lipids, carbohydrates, and proteins, and thus suggests no difference between the hypochlorite-treated and untreated oocyst wall structures.

These data have provided a simple picture of the structure of the *E. tenella* oocyst wall; a thin lipid outer layer covers a thick inner layer of small glycoprotein molecules linked together by a probable framework of disulfide linkages. Agents capable of breaking disulfide bonds by means of reduction or oxidation are capable of solubilizing the inner layer when it is exposed. This conclusion may provide an explanation for the mechanism of micropyle removal during excystation. The micropyle formed during oocyst sporulation is a small, circular, and indented area (Jolley et al., 1979), which could arise from the thinner outer layer of the oocyst wall. It may allow more penetration of small molecules, such as L-cysteine during in vitro excystation, to the inner layer to break disulfide bonds and thus destroy the wall structure beginning at the micropyle area. This theory is supported by the observation that the micropyles of the *Eimeria magna* oocysts can be removed by CO_2 plus L-cysteine as well as by sodium hypochlorite (Ryley and Robinson, 1976), which ruptures disulfide linkages by oxidation. After removal of the micropyle, L-cysteine increases sulfhydryl concentration in broken oocysts (Jolley et al., 1976) and causes thinning of the oocyst wall (Ryley, 1973). These findings cast doubt on the function of leucine aminopeptidase isozyme V in removing the micropyle (Wang and Stotish, 1978).

The oocyst wall is highly resistant to potassium dichromate, sodium hypochlorite, sulfuric acid, and sodium hydroxide, but is permeable to O_2, CO_2, NH_3, methyl bromide, carbon disulfide, phenolic compounds (El-Moukdad, 1976), and various organic solvents. Small, uncharged or hydrophobic molecules are usually capable of penetrating the oocyst wall of *E. tenella*. One example is 3-(4-nonylbenzoyl)acrylic acid which inhibits sporulation at 5 μM when the test is carried out in 0.05 N HCl (Wang, 1978a). In 0.05 N NaOH, this compound showed no inhibitory effect even at 10mM. One probable explanation is that only the undissociated acid form can pass through the oocyst wall. The inhibition of sporulation is probably due to a detergent-like activity of the compound.

The characteristic permeability of the oocyst wall is understandable in view of the predominant neutral fatty alcohols in the outer layer. The outer layer cannot be stripped off by sodium hypochlorite, but is partially removed by incubation with 1.0% *n*-decyl alcohol plus 0.5%

Figure 7. Electron micrographs of the purified wall fragments of unsporulated oocysts of *E. tenella*. Samples were fixed in glutaraldehyde (30 g/liter), post-fixed in OsO_4 (10 g/liter), dehydrated, embedded in Epon, and stained with uranyl magnesium acetate and lead citrate. *A*, the oocyst wall; *B*, the oocyst wall after treatment with 0.1 M dithiothreitol in 8 M urea and 0.01 M phosphate, pH 7.0, at 100°C for 12 hr. Reprinted by permission from: Wang, C. C. Biochemical and nutritional aspects of coccidia. In P. L. Long, K. N. Boorman, and B. M. Freeman (eds.), *Avian Coccidiosis*. Edinburgh. British Poultry Science, Ltd. pp. 135–184 (1978).

NP-40 at 29°C for 30 min (Stotish and Wang, unpublished). The treated oocysts have a 4-fold higher capacity for uptake of radiolabeled arginine and 2-deoxyglucose, and are lysed by sodium hypochlorite, an indication that the permeability is changed. The treated unsporulated oocysts of *E. tenella* lose the capability of sporulating, which may be due to penetration of *n*-decyl alcohol and NP-40 into the oocyst.

Entry into Host Cells

There has been considerable debate as to whether coccidia penetrate the host cells or enter by host invagination. Recent electron microscopy of *Eimeria* sporozoites in cell cultures seems to support the latter theory (Jensen and Hammond, 1975; Jensen and Edgar, 1978). There is no apparent rupture of host cell membrane during the sporozoite invasion, and the intracellular parasite ends up in a vacuole bounded by a membranous structure. The question of whether the entry represents an active process on the part of the parasite, the host, or both, is more difficult to answer. The conoid extrusion and the rhoptry-microneme complex observed among coccidial sporozoites and merozoites visually appear capable of serving secretory functions. Jensen and Edgar (1976b) presented electron micrographs showing entry of *E. magna* sporozoites into embryonic bovine tracheal cells and noted the appearance of empty membrane saccules in the apical region of the sporozoite. They suggested that rhoptry secretion aids the entry of the sporozoite by changing the surface characteristics of the host cell membrane. Similar microscopic observations have been reported on the entry of merozoites of *Plasmodium* species (Ladda et al., 1969; Bannister et al., 1975), *Babesia microti* (Rudzinska et al., 1975), and *Sarcocystis tenella* (Dubremetz and Porchet, 1978). Kilejian (1974) isolated a component of the cytoplasmic granules of *Plasmodium lophurae* trophozoite and identified it as a histidine-rich polypeptide with a molecular weight of 35,000–40,000. The protein was associated with rhoptries and micronemes of merozoites and, at very high concentrations, was capable of causing agglutination of erythrocytes and increasing their osmotic fragility (Kilejian, 1976), a finding that suggests a function in entry of merozoites into erythrocytes. De Souza and Souto-Padrón (1978) stained *T. gondii* with ethanolic phosphotungstic acid and showed, by electron microscopy, abundant presence of basic proteins in the rhoptries and micronemes. Lycke et al. (1975) isolated a penetration-enhancing factor among the proteins in *T. gondii* and found that it increased interaction between the parasite and its cultured host cells. In a well performed study, Dubremetz and Dissous (1980) used a French pressure cell to break the endozoites of *S. tenella*. They purified, by sucrose density gradient centrifugation, the pellicles, the micronemes, and the dense granules to apparent homogeneity. SDS-polyacrylamide gel electrophoresis of the pellicles showed a heterogeneous mixture of many proteins, but the purified micronemes contained only two major protein bands ($M_r = 20{,}000$ and 22,000) and the dense granules consisted of only one protein band of $M_r = 42{,}000$. Although the biological activities of these proteins are still not known, the finding represents major progress toward further understanding the biological functions of the organelles and their possible involvement in the entry of parasites into host cells.

The part likely to be played by the host cell during entry of coccidia also remains unclear. Jones et al. (1972) observed phagocytosis in macrophage, fibroblast, and HeLa cells during the entry of *T. gondii*; micropseudopods were extended by the cells to envelop the attached parasites in a typical phagocytical vacuole. That the usually nonphagocytic fibroblasts and HeLa cells should perform such a function, however, suggests stimulation from the parasite. Jensen and Edgar (1976a) tested various agents on Madin-Darby bovine kidney cells and studied subsequent entry by *E. magna* sporozoites. They found no effect from pretreating the host cells with sodium fluoride, iodoacetate, 2-deoxyglucose, colchicine, colcemid, or vinblastin on the frequency of parasite entry and thus concluded that host phagocytosis is not the mechanism for *E. magna* entry. This conclusion would have been more convincing if the anticipated drug effect on host cell metabolism were actually demonstrated in the experiments.

The basis of the remarkable host species specificity and strict site specificity in the in vivo development of coccidia is not well understood (for a review, see Joyner, Chapter 2, this volume). However, entry of the coccidian sporozoites into intestinal epithelial cells of foreign hosts has been repeatedly observed (Haberkorn, 1970b; Vetterling, 1976). The process of entry thus may not carry species specificity, although the specificity is exhibited when the parasites attempt to develop.

Intracellular Phase

After a coccidian sporozoite enters the host cell, it becomes difficult to follow by quantitative biochemical means. Cytochemical staining techniques yield limited information (reviewed by Ryley, 1973). The cultivation of coccidia in chick embryos (Long, 1965) and cell cultures (Patton, 1965) has provided some success with the study of parasite metabolism by autoradiography (Ouellette et al., 1973, 1974; Morgan and Canning, 1974) and by nutrient substitution (Ryley and Wilson, 1972; Doran and Augustine, 1978). The main avenue of understanding intracellular development of coccidia remains, however, heavily dependent on microscopic descriptions.

Asexual Phase The sporozoite contained in a vacuole of the wrong host cell cannot develop and survives for about 6–12 hr in vivo; for example, *E. maxima* in guinea fowls (Long and Millard, 1979). In the natural host, sporozoites are transformed into trophozoites, a process accompanied by the disappearance of polysaccharide granules in the parasites (Wagner and Foerster, 1964). Apparently, glucose metabolism is needed for this transformation. The trophozoite undergoes schizogony first by elongating the nucleus without eliminating the intact

5

nuclear membrane (Speer and Hammond, 1970). Elongated nucleolus and a spindle extending between opposite poles of a dividing nucleus were seen in early schizonts of *Eimeria ninakohlyakimovae* (Kelley and Hammond, 1972) and *Eimeria callospermophili* (Hammond, 1971). Relatively small spindles, lying eccentrically in the nucleus, also have been observed in early divisions of schizonts and merozoite formation (Danforth and Hammond, 1972). Cell division to form merozoites is accomplished by binary fission. According to Wagner and Foerster (1964), numerous polysaccharide granules were found in the newly formed merozoites, but they vanished again after maturation and development of the merozoites into trophozoites. It is not unlikely that a synthetic phase drawing energy from metabolizing the polysaccharides may be essential prior to cell division.

Merozoites have been isolated from infected tissues by many investigators, with little subsequent purification of the preparations (McLaren and Paget, 1968; Bedrnik, 1969; Shirley, 1975). Elsner et al. (1974) isolated schizonts and merozoites of *E. callospermophili* from cultured bovine cells, but the yield seemed to be relatively small and the viability of the isolated parasites was not verified. Stotish and Wang (1975) noted that *E. tenella* merozoites are relatively resistant to hyaluronidase and could be harvested in large numbers from infected cecal tissue or chorioallantoic membranes of embryonated eggs treated with the enzyme. Further purification by Ficoll density gradient centrifugation and glass bead filtration resulted in a 54% overall yield at 95% purity, which amounted to an average of 2×10^7 purified second generation merozoites per cecum or 1.5×10^6 per chorioallantoic membrane. The viability of these preparations was established by inoculating the merozoites into the ceca of chickens, resulting in oocyst production by 48 hr. Some preliminary biochemical studies on the purified *E. tenella* merozoites have been carried out and have provided information that may be compared to that obtained from the oocysts (Stotish et al., 1976; Wang and Stotish, 1978).

Recently, Fernando and Pasternak (1977a) succeeded in isolating chick intestinal cells infected with second generation schizonts of *E. necatrix*. The procedure of purification, which involves homogenizing mucosal scrapings in a Dounce homogenizer, sieving, and centrifugation, takes advantage of the larger size and higher resistance to shearing force and osmotic shock of the infected cells. The yield was high; approximately 3×10^7 infected cells could be obtained with 80% purity from the deep mucosal scrapings of four infected 10-week-old chickens. Further treatment of the purified infected cells with low concentrations of trypsin breaks the cell membranes and allows harvest of purified host nuclei (Fernando and Pasternak, 1977b) and the freed schizonts by sucrose density centrifugation (James, 1980a). James (1980a) used this technique to isolate schizonts of *E. necatrix* and *E. tenella* with a relatively poor yield (10–12% of the original population) but a seemingly high degree of purity under interference-contrast optics. Up to 77% of the purified schizonts were capable of excluding trypan blue. James (1980a) also was able to detect serine hydroxymethyl transferase activity in the crude extracts of these purified schizonts and to perform some interesting studies on thiamine uptake and the inhibitory effects of amprolium in *E. tenella* schizonts (James, 1980b). These findings are discussed under "Metabolism and Chemotherapy."

As new techniques become available for isolating coccidia from infected intestines during their intracellular phase, more investigators will undoubtedly use these purified specimens for further biochemical studies. A word of caution seems both suitable and necessary at this time. The intestines of chickens, mice, rats, etc., are loaded with bacteria that cannot be totally removed even under the heaviest regimens of antibiotics (Stotish and Wang, 1975). Any procedure of cell isolation, short of drastic steps, such as sodium hypochloride treatment, will not be able to remove all the bacteria. Bacterial contamination even at a relatively low level could introduce considerable error especially when uptake, metabolism, or incorporation of radiolabeled substrates in the parasite are studied. A routine check on bacterial counts in each sample thus seems in order. Microscopic examinations may not be the method with sufficient sensitivity; plating samples on nutrient agar and counting bacterial colonies after overnight incubation are recommended (Wang, 1976).

Sexual Differentiation Klimes et al. (1972) cultivated *E. tenella* in tissue cultures and noticed differences in intensities of periodic acid-Schiff (PAS) stains among the second or third generation schizonts. Because the PAS stain reflects mainly the polysaccharide content in merozoites, the authors postulated the existence of two types of second generation *E. tenella* merozoites; those densely stained by PAS may be the precursors of macrogametocytes, whereas those with light PAS stain may be destined to become microgametocytes. Although interesting, this hypothesis cannot be verified further by the staining technique. There is the other possibility, however, that the different intensities of PAS stain may represent different stages of maturation of the merozoites, because the polysaccharide granules are known to increase in merozoites during the process (Wagner and Foerster, 1964). After all, there seems no need to trigger sexual differentiation until a freed mature merozoite settles in a new host cell and begins to develop. McDougald and Jeffers (1976) were able to isolate a precocious strain of *E. tenella*, which produces viable oocysts after only one generation of schizogony. This strain proves that sexual differentiation is under genetic control of the parasite instead of the environment in which the parasite grows.

A sexual determination can be made simply by differential genetic expression. For example, a merozoite developing into a macrogametocyte may have, among other events, genetically amplified amylopectin synthetic activities, whereas for microgametocytes, extraordinarily active DNA synthesis and nuclear divisions must take place. Genes coding for these activities are apparently present in all the haploid merozoites or sporozoites; a selective expression may lead to the differentiation. The regulatory mechanism which controls the timing and the selection is totally unknown and remains a most interesting area to be explored.

The process of fusion between macrogamete and microgamete is poorly understood. There are microscopic pictures showing a macrogamete of *Eimeria nieschulzi* surrounded by

5

microgametes (Marquardt, 1966) or a microgamete at the nuclear membrane of a macrogamete of *E. maxima* (Scholtyseck, 1963). Scholtyseck and Hammond (1970) found a microgamete inside a macrogamete of *E. bovis* while the type II wall-forming bodies were undergoing fusion into larger units. The penetration of a macrogamete by a microgamete has been considered as a prerequisite for oocyst wall formation; this belief has not been supported by concrete evidence (Long, 1972). Following wall formation, the zygote continues to develop into the mature unsporulated oocyst, which can stay unchanged under anaerobic conditions for weeks (Wang, 1976), until the process of sporulation is triggered by oxygen. There is nothing known about the maturation process of the unsporulated oocyst, but it is believed to involve a vigorous synthetic phase to prepare the oocyst for later nuclear divisions during sporulation.

No method is yet available to separate coccidian macrogametes and microgametes or to isolate them in large enough numbers or pure enough forms for direct investigation. This failure contributes to the lack of understanding of the coccidial sexual phase to date and poses a severe limitation on knowledge of coccidian genetics, because no well controlled experiments on genetic recombination can be performed.

METABOLISM AND CHEMOTHERAPY

The accumulated biochemical information on many species of coccidia in different phases of development has made it possible to outline coccidial metabolism in general. Although many details are missing and numerous gaps in knowledge still exist, one can make an assertion from what has already been learned, that is, that coccidia go through complicated life cycles without significantly changing the basic pattern of metabolic activities. These activities involve: 1) consumption of polysaccharide storage prior to cell division, 2) cell division, 3) increase or preservation of polysaccharide storage, and 4) breaking out to invade a new host. The pattern repeats itself rhythmically during the asexual phase and sporulation. Only in sexual differentiation does the pattern break into two; the microgametocytes have activities 1 and 2 during their formation, whereas the macrogametocytes have mainly activity 3. The accumulation of polysaccharide granules in coccidia each time before they enter an extracellular environment is probably one of the most outstanding characteristics of coccidia among other sporozoan parasites, and may be attributed to the lack of an intermediary host.

In order to survive in the host intestinal tract during the extracellular phases, coccidia have also acquired the capability of withstanding pancreatic enzymes in their sporozoites (Ryley, 1973) and resistance to hyaluronidase in their merozoites (Stotish and Wang, 1975). These properties suggest a unique membrane structure of the parasite.

It has been found that other activities in coccidia, including purine and pyrimidine metabolism, mitochondrial function, and protein synthesis possess many unusual features. There are patterns of enzymic activities in the synthetic pathways of purines and pyrimidines of coccidia significantly different from those of the infected hosts. The mitochondria and ribosomes of coccidia have specific biochemical properties and have been among the targets for anticoccidial chemotherapy. Other studies on the mechanisms of action of some well known anticoccidial drugs also contributed to the further understanding of many unique aspects of coccidial metabolism. The differences in metabolism between coccidia and their hosts reflect the parasitic nature of the former and reflect the distance separating them along the route of evolution. They are the basis for chemotherapy against coccidiosis. Chemical structures of some of the established anticoccidial agents are presented in Chapter 9 (McDougald, this volume).

Carbohydrate Metabolism

Enzymes In addition to the identification of amylopectin in the oval-shaped polysaccharide granules (Ryley et al., 1969) and the discovery of amylopectin phosphorylase as the probable enzyme for producing glucose 1-phosphate from the granules (Wang et al., 1975b), many other enzymes involved in glucose metabolism have been found in large quantities in coccidia. Glucose-1-phosphate isomerase, glucose-6-phosphate mutase, hexokinase, glucose-6-phosphate dehydrogenase, 6-phosphogluconate dehydrogenase and lactic acid dehydrogenase are present in coccidian sporozoites, merozoites, and oocysts at such high levels that they are routinely used as electrophoretic markers for the purpose of coccidial species identification (Shirley and Rollinson, 1979). Frandsen (1976) identified glucose-6-phosphate dehydrogenase in crude extracts of *E. stiedae* unsporulated oocysts and calculated a specific enzyme activity of 0.37 μmol/min/mg of protein. The enzyme was purified about 50-fold by ammonium sulfate fractionation and DEAE-Sephadex column chromatography. Wang (1978a) found fructose-1,6-diphosphate aldolase in the unsporulated oocysts of *E. tenella* and partially characterized it to be of class I type, which is commonly found in animals (Mildvan et al., 1971). These data suggest that both glycolysis and pentose phosphate shunt may be actively functioning in coccidia.

There is ample evidence indicating a functioning tricarboxylic acid cycle in coccidia. Early histochemical studies suggested the presence of isocitrate dehydrogenase and succinate dehydrogenase in proliferative *T. gondii* (Capella and Kaufman, 1964). Isocitrate dehydrogenase and malate dehydrogenase were also identified in crude homogenates of *E. tenella* unsporulated oocysts with activities between 1 and 4 μmol/min/mg of protein (Wang, 1978a). Other related enzymes such as glutamate dehydrogenase (22.8 μmol/min/mg of protein of crude extract of *E. tenella* unsporulated oocysts; Wang et al., 1979a) and asparate aminotransferase (Shirley and Rollinson, 1979) are found in great abundance among coccidia, which suggests very active gluconeogenesis in the parasite. No study has been performed, however, on the synthetic enzyme(s) of amylopectin, which is apparently one of the most important enzymes in coccidia.

5

No isocitrate lyase or malate synthetase activity could be detected in crude extracts of *E. tenella* unsporulated oocysts (Wang, unpublished observation), excluding a possible glyoxylate pathway during this phase of *E. tenella* metabolism.

Because of the predominance of carbohydrate metabolism in coccidia, it is not difficult to exhaust the oxygen content in the cytoplasm of infected host cells during intracellular growth of the parasite. This may explain the massive production of lactic acid from the intestinal tissues infected with *E. acervulina* (van der Horst and Kouwenhoven, 1973) and the lowering of intestinal pH in the infected host (Stephens, 1965). Cytochemical data of Beyer (1970) also suggest fluctuations in succinate dehydrogenase and α-glycerophosphate dehydrogenase during intracellular development of rabbit coccidia and *E. tenella,* which suggests that aerobic glucose metabolism does not always reach its full capacity throughout the parasitic phase of coccidia.

Cofactors and Antimetabolites Heavy dependence upon semianaerobic carbohydrate metabolism is expected to make coccidia vulnerable to any disturbances in the glycolytic pathway or pentose phosphate shunt. This expectation is supported by results from a preliminary experiment performed by Warren (1968). In an effort to elucidate the vitamin requirements of the coccidia, Warren fed *E. acervulina*- or *E. tenella*-infected chickens with various single-vitamin-deficient diets and monitored the oocyst production. The results, presented in Table 3, indicate thiamine, biotin, and nicotinic acid as the three most needed vitamins; their deficiency led to over 90% suppression of oocyst production. Riboflavin deficiency also caused 75–89% inhibition of oocyst production. Thiamine, nicotinic acid, and riboflavin are precursors of thiamine pyrophosphate, NAD or NADP, and FAD, all of which are essential cofactors in carbohydrate metabolism. Many analogs of the three vitamins have long been known to have anticoccidial activities.

Amprolium and Thiamine Transport Amprolium (1-[(4-amino-2-propyl-5-pyrimidinyl)methyl]-2-picolinium chloride) is a chemical derivative of thiamine (Rogers, 1962) with anticoccidial activity and low toxicity (Cuckler et al., 1961). Its activity against coccidia can be reversed by thiamine in chickens (Rogers, 1962), as well as in cell culture (Ryley and Wilson, 1972). Because the drug lacks the hydroxymethyl group of thiamine on the thiazole moiety, it cannot be pyrophosphorylated and therefore cannot inhibit competitively the action of thiamine pyrophosphate as a coenzyme. It has a rather poor inhibitory effect on thiamine pyrophosphotransferase of rat brain cortex (Sharma and Quastel, 1965) or of the protozoan flagellate *Strigomonas oncopelti* (Bauchop and King, 1968), but it decreases the uptake of thiamine in the ligated intestinal loops of the fowl (Polin et al., 1963). Some amprolium-resistant mutants of *Lactobacillus fermenti* showed higher capacity and resistance to amprolium in their thiamine uptake (Kishi et al., 1971). Sharma and Quastel (1965) observed that transport of thiamine in rat brain cortex was Na^+-dependent, susceptible to ouabain and 2,4-dinitrophenol, and was apparently unrelated to the thiamine pyrophosphotransferase ac-

Table 3. Percentage of oocysts passed and percentage of weight gain of birds fed a single-vitamin-deficient diet compared with birds fed the "all vitamin" ration[a]

Diet deficient in	Percentage oocysts passed by birds infected with		Percentage weight gain in 12 days
	E. acervulina	*E. tenella*	
Thiamine	75 . 53.4		0.8
Biotin	1.8	6.8	93.9
Nicotinic acid	5.5	5.8	61.9
Riboflavin	10.9	24.9	57.5
Choline	26.7	40.3	75.3
Folic acid	64.1	38.2	75.5
Vitamin D	71.9	71.4	108.5
Calcium pantothenate	110.5	60.7	71.6
p-Aminobenzoic acid	84.6	83.0	99.6
Vitamin C	33.1	154.6	85.7
Pyridoxine	115.8	94.5	50.7
Vitamin E	212.9	49.1	110.8
Vitamin K	130.8	95.7	114.6
Lipoic (thioctic) acid	72.4	161.3	89.7
Hesperidin	204.4	94.2	114.4
Inositol	124.5	156.2	104.6
Vitamin A	204.2	153.2	81.1
Linolenic acid	133.2	160.6	83.7
Cyanocobalamin	307.3	164.2	102.7

[a]Adapted from Warren, 1968.

tivity in the same tissue. Amprolium (10 μM), although not inhibitory to the transferase, completely inhibited this carrier-mediated active transport of thiamine. This potent inhibition has been considered to be the mode of anticoccidial action of amprolium. However, the high therapeutic index of amprolium remains unexplained by the mode of action, unless it is assumed that either the intracellular coccidia are very sensitive to inhibition of thiamine transport, or that the parasite has its own thiamine transport system which may be much more susceptible to amprolium than that of the host cell.

In a recent study, James (1980b) assayed the uptake of thiamine by isolated second generation schizonts of *E. tenella* and by chicken intestinal epithelial cells. He found active transport of thiamine in both cell preparations. The parasite has a higher substrate affinity ($K_m = 0.07$ μM) than the host cell ($K_m = 0.36$ μM) (see Figure 8). Both thiamine transport systems were competitively inhibited by amprolium, but the inhibitory potencies differed widely. The drug has a K_I value of 7.6 μM for *E. tenella* thiamine transport and a K_I of 362.5 μM for that of intestinal epithelial cells (Figure 9). When thiamine transport in the schizonts of an *E. tenella* line

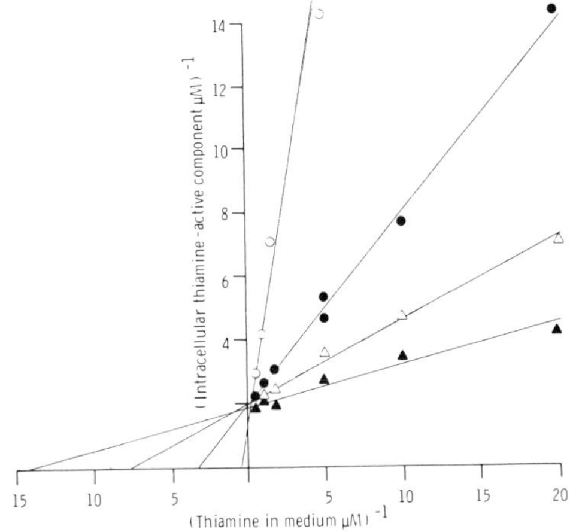

Figure 8. *E. tenella* schizont uptake of thiamine. Lineweaver-Burk plot of initial rate of active thiamine uptake by schizonts, in the absence of amprolium (▲) and with amprolium at 100 μM (△), 250 μM (●) and 1,000 μM (○). Reprinted by permission from: James, S. Thiamine uptake in isolated schizonts of *Eimeria tenella* and the inhibitory effects of amprolium. *Parasitology* 80:313–322 (1980).

resistant to amprolium was investigated, the substrate had a K_m of 0.132 μM and amprolium showed a much higher K_I of 115 μM. This well performed experiment confirmed the long-postulated mode of action of amprolium and accounted for the high therapeutic index. Thiamine transport in the parasite is 50 times more susceptible to amprolium than that in the host. When coccidia become resistant to the drug, it is, at least in one instance, due to a partial loss (15-fold) of susceptibility of the thiamine transport to amprolium.

Nicotinamide Derivatives There have been several derivatives of nicotinamide found active against coccidiosis. Ball et al. (1965) observed that pyridine-3-sulfonamide markedly reduced oocyst output in *E. acervulina* infections, when included in the diet at 60–250 ppm. 6-Aminonicotinamide was active against *E. tenella* and *E. necatrix* infections. Both drugs, however, have rather narrow margins between activity and toxicity and both could be nullified by equal amounts of nicotinamide. Ryley and Wilson (1972) found 6-aminonicotinamide inhibitory to the growth of *E. tenella* in cell cultures at 0.02 ppm; the activity was completely reversed by 0.2 ppm of nicotinamide. A recent patent from Japan also showed *N*-chloroacetyl-2-methyl-5-nitronicotinamide to be an effective drug against *E. tenella* infection in chickens (Morisawa et al., 1976).

6-Aminonicotinamide has been known to function by replacing the nicotinamide moiety in NAD or NADP through the action of NADase (Dietrich et al., 1958). The 6-aminonicotinamide analogs of NAD and NADP are inactive as cofactors for various dehydrogenases. In vivo administration of 6-aminonicotinamide to C-57 black mice bearing the 755 tumor resulted in marked inhibition of 3-phosphoglyceraldehyde dehydrogenase, β-hydroxybutyrate dehydrogenase, and α-ketoglutarate dehydrogenase of the tumor. The lack of anticoccidial therapeutic value in this drug is likely due to this type of nondiscriminatory inhibition of dehydrogenases in many cells and tissues.

Riboflavin Analogs Partial control of *E. acervulina* infection was achieved by administration of riboflavin analogs 10-ethyl- and 10-*n*-pentyl-7,8-dimethyl-isoalloxazines at 500–1,000 ppm (Ball and Warren, 1967). The compounds were also toxic; both activity and toxicity could be reversed by one-tenth the amount of riboflavin. More recently, it has been found that four distinct types of riboflavin antagonists possess substantial, broad-spectrum anticoccidial activities (Graham et al., 1977): 7-chloro-10-D-ribotylisoalloxazine, 8-methylamino-8-nor-riboflavin, 9-azariboflavin, and 5-deazariboflavin. The most potent derivative, 5-deazariboflavin, protects chickens against *E. tenella* infection at a feed level of only 5 ppm when the diet contains 4.8 ppm of riboflavin. The anticoccidial effect of 10 ppm of the drug can be abolished when riboflavin is increased to 16.8 ppm in the diet.

Extensive biochemical studies of the properties of 5-deazariboflavin have been done in recent years. Spencer et al. (1976) synthesized 5-deaza-FMN and 5-deaza-FAD from 5-deazariboflavin with a partially purified FAD synthetase complex from *Brevibacterium ammoniagenes*, containing both phosphorylating and adenylating activities. The C-5 position is the

Figure 9. Dixon plot of action of amprolium on intestinal cells (---) and *E. tenella* amprolium-sensitive schizonts (——). *Verticle lines* indicate changes in active uptake of thiamine in the presence of amprolium at 250 μM compared with uptake in the absence of drug. Thiamine concentration is 0.3 μM. Reprinted by permission from James, S. Thiamine uptake in isolated schizonts of *Eimeria tenella* and the inhibitory effects of amprolium. *Parasitology* 80:313–322 (1978).

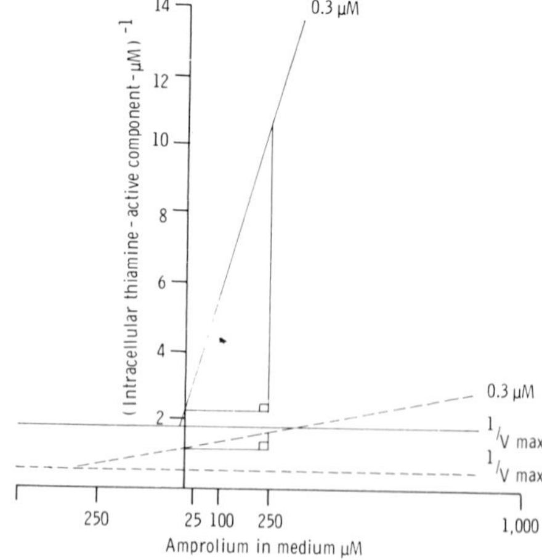

locus for hydrogen transfer in deazaflavin redox reactions. Fisher et al. (1976) examined the reduction of 5-deazariboflavin by NADH in the presence of the *Beneckea harveyi* NADH:oxidoreductase and obtained a very low E_0 of -0.310 V. Deaza-FMN-reconstituted yeast NADPH:oxidoreductase, deaza-FAD-reconstituted pig kidney D-amino acid:O_2 oxidoreductase, and deaza-FAD-reconstituted *Aspergillus niger* D-glucose:O_2 oxidoreductase were reduced by their substrates at rates 100,000-fold lower than those of the holoenzymes; the reduced deaza FAD could not be reoxidized by oxygen. These interesting observations may provide the basis of the anticoccidial activity and the toxicity of 5-deazariboflavin.

Lipid Metabolism

There is little knowledge of the lipid metabolism of coccidia. Charney et al. (1971) examined the effects of an essential fatty acid-deficient diet on coccidiosis in chickens, and found that the lesions and mortality caused by *E. tenella* and *E. mivati* infection were significantly reduced. The severity of infection could be restored by 5% corn oil supplement but not by 5% hydrogenated coconut oil. The results suggest an incapability of coccidia to synthesize some of the essential unsaturated fatty acids. Gas-liquid chromatographic analysis of fatty acids in *E. tenella* oocysts identified oleic acid as the most predominant component constituting 68–75% of the total fatty acids (Weppelman et al., 1976); palmitic acid and stearic acid each made up about 10%. No apparent change of fatty acid composition in the oocysts took place during sporulation.

Cholesterol was identified in *E. tenella* at about 10 μg per 10^6 oocysts (Weppelman et al., 1976). Phospholipids and fatty alcohols were found in large amounts in the outer layer of the oocyst wall; the alcohols were also synthesized in the cytoplasm during sporulation. However, none of the lipid synthetic pathways has been carefully examined in coccidia. There is no anticoccidial drug known to interfere specifically with lipid metabolism. *t*-Butylaminoethanol is an anti-*E. tenella* drug; its activity in infected chickens can be totally reversed by choline or dimethylaminoethanol at ratios of 1:200 or 1:20, respectively (McManus et al., 1979). Its toxicity in chickens is also overcome by the two compounds at ratios of 1:10 and 1:7. Because choline is required by coccidia (see Table 3) and is a part of phosphatidylcholine, *t*-butylaminoethanol may act against the parasite by inhibiting phospholipid synthesis. However, a moderate synergistic interaction between the drug and sulfaquinoxaline argues for a possible inhibition of methyl group transfer from choline or dimethylaminoethanol by the drug as the mode of anticoccidial action (McManus and Rogers, 1979). The conclusion is far from being clear and cannot explain why *t*-butylaminoethanol is only active against *E. tenella*.

Biotin, the essential cofactor in biosynthesis of fatty acids and fatty alcohols, is also an essential vitamin for the growth of coccidia (see Table 3). Warren and Ball (1967) were able to inhibit oocyst production with *E. acervulina*, *E. maxima*, and *E. tenella* by feeding diets containing 30–50% dried egg white. These effects could be reversed by daily doses of biotin, which presumably overcome the biotin deficiency caused by avidin in the egg white. A therapeutically useful biotin antagonist has not yet been found for coccidiosis.

Mitochondrial Function

Large numbers of mitochondria have been observed in coccidia during every phase of their development (Scholtyseck, 1963). From the vigorous respiratory activities observed in coccidial sporulation and excystation and the calculated ratios of 6 between oxygen reduced and glucose consumed (Ryley, 1973) by sporozoites under aerobic conditions, there is no doubt that coccidia have fully functional mitochondria during their extracellular phase. Because the mitochondria have the same detailed, internal tubular ultrastructures throughout the life cycle (Scholtyseck, 1973), they may remain relatively unchanged and retain full activity during the intracellular phase.

Mitochondrial Structure Ryley (1973) treated the sporozoites of *E. tenella* with diaminobenzidine and observed by electron microscopy dense granules on the tubular cristae of the mitochondria which could be either cytochrome oxidase or peroxidase activity. A similar observation was also made on the mitochondria of *T. gondii* by Akao (1971). Superoxide dismutase activity was found in the mitochondrial fraction of *E. tenella* unsporulated oocysts (Wang, 1978a). It consists of a single protein band in polyacrylamide gel electrophoresis and has an estimated molecular weight of 40,000 in Sephadex G-100 gel filtration. The enzyme is insensitive to KCN but is inactivated by chloroform-ethanol treatment (McCord and Fridovich, 1969; Tsuchihashi, 1973). Diethyldithiocarbamate, a copper-chelating agent known to inactivate the Cu^{2+}-Zn^{2+} superoxide dismutase of mouse liver (Heikkila et al., 1976), has no effect on the enzyme from *E. tenella*. Dialysis against 8-hydroxquinoline and guanidine, a method developed by Ose and Fridovich (1976) to remove manganese from *E. coli* Mn^{2+}-superoxide dismutase, inactivates the enzyme from *E. tenella* unsporulated oocysts. The enzyme, however, could be reactivated by further dialysis against 0.01 mM $MnCl_2$. These results suggest that the *E. tenella* superoxide dismutase is a manganese-activated enzyme as is the superoxide dismutase of *E. coli* (Keele et al., 1970), the superoxide dismutase in the cytoplasm of *Tritrichomonas foetus* (Lindmark and Muller, 1974), and the superoxide dismutase in the mitochondrial matrix of chicken liver (Weisiger and Fridovich, 1973). But the Cu^{2+}-Zn^{2+} enzyme, found in the cytoplasm of all the higher eukaryotes as the primary superoxide dismutase (Fridovich, 1974), is absent from the *E. tenella* oocyst. This finding would classify *E. tenella* with the luminous fungi *Pleurotus olearius* and *Neurospora crassa* (Lavelle et al., 1974), and group *E. tenella* with the most primitive eukaryotes (Lumsden and Hall, 1975).

Intact mitochondria were isolated from *E. tenella* unsporulated oocysts by gentle homogenization of the oocysts in 0.21 M mannitol, 0.07 M sucrose, 0.01 mM EDTA, and 0.01 M Tris-phosphate, pH 7.6, and by centrifugation (Wang, 1975). Amylopectin granules were the

Figure 10. Inhibition of respiration in mitochondria from *(left) E. tenella* wild type and *(right) E. tenella* amquinate-resistant mutant. Succinate (10mM) was the substrate, and each assay contained 4.0 mg of protein per ml. The inhibitors: amquinate, ○——○; 2-hydroxy-3-(4-phenoxyphenyl)propyl1,4-naphthoquinone, ●——●; antimycin A, ∇——∇; and 2-heptyl-4-hydroxyquinoline-*N*-oxide, ▼——▼. Reprinted by permission from: Wang, C. C. Studies of the mitochondria from *Eimeria tenella* and inhibition of electron transport by quinoline coccidiostats. *Biochimica et Biophysia Acta (Amsterdam)* 396:210–219 (1975).

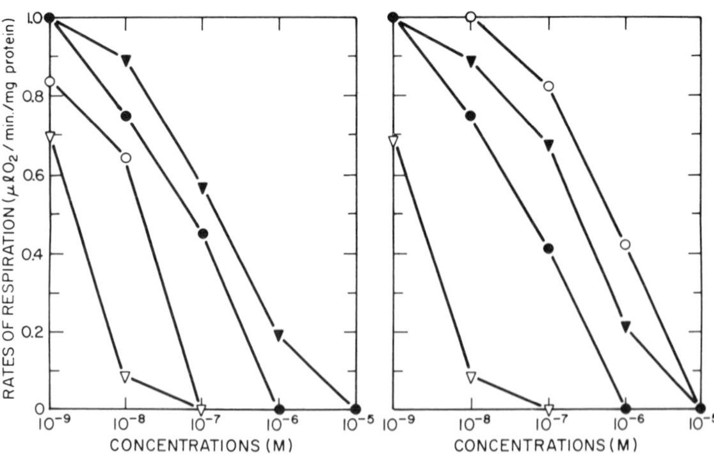

major contaminant in the final preparation, but could be largely removed by α-amylase digestion. The isolated mitochondria have distinctive tubular cristae structures and respire in response to succinate, malate plus pyruvate, and L-ascorbate at rates of 1.00, 0.40, and 0.25 μl of O_2/min/mg of protein, respectively. Spectrophotometric analysis of the cytochromes in mitochondria indicated strong absorption bands at 563 and 605 nm in difference spectra, which suggest the presence of *b*-type and *a*-type cytochromes. But absorption maxima characteristic of *c*-type cytochrome is absent. Further studies with CO treatment of the respiratory pigments demonstrated a major band at 415 nm and two minor bands at 534 and 570 nm, coinciding well with the α-, γ-bands of CO cytochrome *o* (Kamen and Horio, 1970), which suggests that cytochrome oxidase from *E. tenella* mitochondria may be the *o*-type. The mitochondrial respiration is inhibited by cyanide, azide, carbon monoxide, antimycin A, and 2-heptyl-4-hydroxyquinoline-*N*-oxide, but is relatively resistant to rotenone and amytal.

Anticoccidial Quinolones There is a family of quinolone derivatives (e.g., buquinolate, amquinate, methyl benzoquate, and decoquinate) known as effective anticoccidial agents (Ryley and Betts, 1973). They have, in general, rather low toxicities, but high frequencies of resistance have been developed against these drugs among coccidia (McManus et al., 1968). Coccidia gaining resistance to one quinolone also become resistant to other quinolones, which suggests that all the anticoccidial quinolones may have the same mechanism of action.

Wang (1976) observed that the quinolones inhibit respiration of *E. tenella* with ID_{50} values of $1–2 \times 10^{-5}$ M during sporulation and 3×10^{-6} M during excystation. The inhibition of respiration, which was fully reversible, led to failure of sporulation but did not affect excystation. The respiration of an amquinate-resistant strain of *E. tenella* is much less subject to inhibition by the quinolones. When the drugs were tested on succinate or malate plus pyruvate-supported respiration in isolated *E. tenella* mitochondria (Wang, 1975), 50% inhibition was achieved at 3 pmol of drug per mg of mitochondrial protein (see Figure 10). The inhibition could not be reversed by coenzyme Q_6 or Q_{10} and did not affect L-ascorbate-supported respiration or submitochondrial succinate dehydrogenase or NADH dehydrogenase. The site of action of the quinolones was thus tentatively identified near cytochrome *b* in *E. tenella* mitochondria. Mitochondria from the amquinate-resistant strain were 100-fold less susceptible to the quinolones (Figure 10), which agreed with the observed resistance in chickens. The results suggest that the mechanism of anticoccidial action of the quinolones is by inhibiting mitochondrial respiration in coccidia. None of the quinolones exhibited any inhibitory effect on chicken liver mitochondrial respiration, which may explain the lack of drug toxicity.

Respiration by submitochondrial particles of *E. tenella* resistant to amquinate is also less susceptible to the quinolones (Wang, 1975), which rules out changes in mitochondrial membrane drug permeability as the possible cause of drug resistance. The reason for high frequency of resistance to the quinolones among coccidia was attributed to a possible non-Mendelian genetic control in mitochondria of the parasite, but further studies are needed to verify this hypothesis.

5

Anticoccidial Hydroxynaphthoquinones Many derivatives of 2-hydroxynaphthoquinone have long been known to have antimalarial (Fieser et al., 1948), anticoccidial (Rogers, 1967), and anti-*Theileria* (McHardy, 1978) activities. They inhibit mitochondrial succinate dehydrogenase and NADH dehydrogenase activities of different sources and the inhibition can be reversed by coenzyme Q (Catlin et al., 1968; Skelton et al., 1971). 2-Hydroxy-3-(4-*trans*-cyclohexylcyclohexyl)-1,4-naphthoquinone and 2-hydroxy-3-(4-phenoxyphenyl)propyl-1,4-naphthoquinone, two toxic antimalarial and anticoccidial agents, strongly inhibited the respiration and sporulation of *E. tenella* (Wang, 1976). They also blocked succinate-supported respiration in isolated *E. tenella* mitochondria, and both the wild type and the amquinate-resistant mitochondria were equally susceptible to these compounds (see Figure 10), which suggest different modes of anticoccidial action between the hydroxynaphthoquinones and the quinolones. The two hydroxynaphthoquinones were potent inhibitors of succinate dehydrogenase-CoQ reductase and NADH dehydrogenase-CoQ reductase from mitochondria of both *E. tenella* and chicken liver (Wang, 1975). These inhibitory effects, partially reversible by CoQ, may be the basis of both the activity and toxicity of the two drugs.

Clopidol A pyridone anticoccidial drug, otherwise known as Coyden or meticlorpindol, has a chemical structure related to that of quinolone, but does not have an obvious effect on the mitochondrial respiration of *E. tenella* (Wang, 1975). However, synergism between clopidol and methyl benzoquate in anticoccidial activities often has been observed in vivo (Ryley, 1975; Greuel et al., 1975). Furthermore, collateral sensitivities to the quinolones were observed in two laboratory strains of *E. acervulina* resistant to clopidol (Jeffers and Challey, 1973), but decoquinate resistance acquired by coccidia was not accompanied by increased sensitivity to clopidol. Joyner and Norton (1978) isolated drug-resistant strains of *E. maxima* by serial passages through chickens under drug pressure. They observed that resistance to methyl benzoquate was readily acquired by a strain already resistant to clopidol, whereas resistance to clopidol in a methyl benzoquate-resistant strain requires many passages. They also made the interesting observation that an *E. maxima* strain made resistant to both drugs produced oocysts of which 80% were bisporocystic, that is, an oocyst consisted of only two sporocysts, but each sporocyst had four sporozoites (Norton and Joyner, 1978). When a strain of *E. tenella* resistant to clopidol and a methyl benzoquate-resistant strain were passaged together in chickens, no progeny acquiring resistance to both drugs could be identified (Joyner and Norton, 1975b). The result suggests that the two genetic markers coding for resistance to the two drugs may be too close together to allow any detectable frequency of genetic recombination.

All the observations seem to tie the mode of anticoccidial action of clopidol closely with that of the quinolones. Because the latter are known to block electron transport down the cytochrome chain in coccidial mitochondria, Wang (1978a) proposed the probable presence of an alternative pathway of electron transport in coccidial mitochondria which might be inhibited specifically by clopidol. In the clopidol-resistant strains, this alternative pathway may be spared so that the normal function of the cytochrome chain becomes vital for survival of the parasites; any perturbation by the quinolones may arrest their development. In contrast, resistance to the quinolones is achieved by decreased susceptibility of the cytochrome chain to the drugs (Wang, 1975). Thus, a functioning cytochrome chain should be present in all drug-resistant strains. This hypothesis is plausible, but proof is lacking.

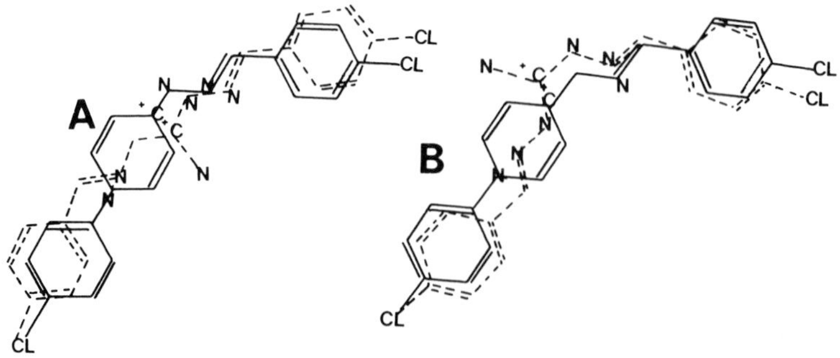

Oxidative Phosphorylation Uncouplers The isolated mitochondria of *E. tenella* are not coupled well enough for one to study the process of oxidative phosphorylation in the parasite (Wang, 1975). However, a few established anticoccidial drugs have been shown to affect oxidative phosphorylation in mammalian mitochondria. Robenidine (1,3-bis-[(*p*-chlorobenzylidene)amino]guanidine) (Kantor et al., 1970) was found to inhibit oxidative phosphorylation in rat liver mitochondria at a rather elevated concentration of 20 mM (Wong et al., 1972). At a level of 16 nmol per mg of mitochondrial protein, the drug also demonstrated an oligomycin-type activity by inhibiting Cl-CCP–induced ATPase of the mitochondria. These moderate inhibitory activities of robenidine are also shared by many other guanidine derivatives known to have no anticoccidial action (Pressman, 1963) and are thus unlikely to be the mechanism of activity against the parasites. There have been numerous publications reporting the frequent development of resistance to robenidine among many species of coccidia in recent years (Jeffers, 1974b; Norton and Joyner, 1975; Chapman, 1976; Lee and Fernando, 1977b). The mechanism of resistance development remains unknown. Joyner and Norton (1975a) described an interesting robenidine-resistant strain of *E. maxima* which required 132 ppm of robenidine in the chicken diet for normal development. The robenidine-dependence may bear resemblance to the mutants of *E. coli*, which are dependent on streptomycin, a guanidinoglycoside antibiotic capable of binding to the 30 S bacterial ribosomal subunit (Birge and Kurland, 1969). However, robenidine exerted no inhibitory effect on *E. tenella* wild type ribosome-mediated in vitro protein synthesis (Wang, 1978b). It is not known whether robenidine may cause miscoding in protein synthesis in the parasite.

Douglas et al. (1977) recently synthesized a family of 1-substituted 4(1H)-pyridinone hydrazones and found many of them highly effective against *E. tenella* and *E. acervulina* infections in chickens. One derivative, 4-chlorobenzaldehyde-1-(4-chlorophenyl)-4(1H)-pyridinylidene hydrazone fluorosulfonate, showing very high in vivo anticoccidial activities, was completely inactive against a robenidine-resistant strain of *E. tenella*. Conformational analysis of the two compounds in their bioactive forms by ^{13}C-nuclear magnetic resonance spectroscopy indicates that the former compound in its A conformation bears a qualitative resemblance to the robenidine cation in conformations W and S (see Figure 11). The average deviation data indicate a slight preference for superposition with the S conformation of the robenidine cation. These structural and conformational similarities can explain the cross-resistance of the two drugs and may also suggest similar modes of anticoccidial action.

5

Figure 11. Superposition of 4-chlorobenzaldehyde-1-(4-chlorophenyl-4(1H)-pyridinylidene hydrazone fluorosulfonate (*solid lines*) with the robenidine cation in *A* conformation W; the average distance deviation was 0.73 Å for superposition of 22 atoms. *B* conformation S; the average distance deviation was 0.58 Å for 22 atoms superposed. Reprinted by permission from: Douglas, A. W., Fisher, M. H., Fishinger, J. J., Gund, P., Harris, E. E., Olson, G., Patchett, A. A., and Ruyle, W. V. Anticoccidial 1-substituted 4(1H)-pyridinone hydrazones. *Journal of Medicinal Chemistry* 20:939–943 (1977).

4,4-Dinitrocarbanilide, which is the active ingredient of the anticoccidial drug nicarbazin (Cuckler et al., 1955), was shown to inhibit succinate-linked NAD reduction in beef heart mitochondria (Dougherty, 1974). The compound is very potent; the concentration required for half-maximum inhibition is 5×10^{-8} M. The compound also inhibits energy-dependent transhydrogenase and the accumulation of Ca^{2+} by rat liver mitochondria in the presence of ATP. These results suggest an interaction between the carbanilide and the mitochondrial energy transduction chain. A similar observation was also made on other carbanilide derivatives; Hamilton (1970) studied the mode of antibacterial activities of the oxidative phosphorylation uncoupler trichlorocarbanilide and found that it affected the bacterial plasma membrane, rendering it permeable to Cl^-. Wang (1978a) tested both robenidine and 4,4′-dinitrocarbanilide on chicken erythrocytes and observed that both drugs caused massive leakage of K^+ from the cells. The K^+ efflux, caused by drug concentrations ranging from 10^{-5} to 10^{-4} M, was not coupled with an influx of H^+. Wang (1978a) also observed that 4,4′-dinitrocarbanilide, virtually insoluble in water, could be solubilized by adding bovine serum albumin (BSA) to the drug suspension. A linear relationship between the logarithm of BSA concentration and the concentration of solubilized drug, monitored spectrophotometrically, suggests drug binding to the protein.

Because robenidine and 4,4′-dinitrocarbanilide are guanidino and ureido compounds, they should have the capability of binding to proteins. This protein binding may cause uncoupled mitochondria and leaky membranes. It should also make the discovery of true mechanisms of anticoccidial actions of the two drugs very difficult. It could be assumed, however, that the protein in coccidia with the lowest dissociation constant in binding to the drug is the true therapeutic target. It might be isolated by drug affinity column chromatography.

Membrane Transport

Previous discussions have indicated unique properties of coccidial cell membranes because of the resistance of sporozoites and merozoites to digestive enzymes. Once the sporozoite or merozoite enters a host cell, the internal host cell membrane forming a vacuole around the parasite quickly disappears, leaving the latter in direct contact with the cytoplasm of host cell (Jensen and Hammond, 1975; Rudzinska et al., 1975). It is not known how nutrients are transported across the parasite membrane. At least in theory, one would expect that mere passive diffusion should satisfy all the nutritional requirements, because cytoplasm is present on both sides of the parasite membrane. Harold and Van Brunt (1977) have found that *Streptococcus faecalis* loses the electrical potential across the cytoplasmic membrane as well as the concentration gradients of H^+, K^+, and Na^+ when treated with ionophorous antibiotics such as gramicidin D, gramicidin A, and valinomycin plus nigericin. The bacteria can grow normally in the presence of these ionophores, however, when the medium is enriched with a high concentration of K^+ (0.28 M), a low concentration of Na^+ (0.01 M), and a slightly alkaline pH (7.7), a condition mimicking that in cellular cytoplasm.

On closer examination, however, coccidia must have acquired some capability to actively transport nutrients toward the late phase of schizogony or sexual differentiation, because the host cell plasma membrane has become so deteriorated by then that it seems to accumulate gel-phase lipid and is highly leaky (Thompson et al., 1979). The host cells become stainable with trypan blue, which suggests that most of the nutrients originally concentrated in the host cell cytoplasm may have been lost. The coccidia must be able to take up necessary nutrients against a concentration gradient to support their vigorous growth and multiplication during these phases of development. The dinitrophenol-inhibitable thiamine transport observed in the second generation schizonts of *E. tenella* provides a good example (James, 1980b).

This conclusion is also supported by the discoveries during the past decade of numerous polyether monocarboxylate ionophores as effective anticoccidial drugs. Anticoccidial agents nigericin (Gorman and Hamill, 1971), dianemycin (Hamill et al., 1969), monensin (Shumard and Callender, 1967), lasalocid (Mitrovic and Schildknecht, 1973), salinomycin (Danforth et al., 1977), narasin (Berg and Hamill, 1978), and lonomycin (Cruthers et al., 1978) all belong to the same family of antibiotics which are known to destroy cross-membrane gradients of H^+, K^+, Na^+, and Ca^{2+} (Pressman et al., 1967). According to the Mitchell hypothesis (Mitchell, 1976), the active oxidation of electron donors is accompanied by expulsion of protons across the cell membrane, leading to an electrochemical gradient of protons ($\Delta\ \mu_{H}{}^{+}$), which is represented by the following equation:

$$\Delta\ \mu_{H}{}^{+} = \Delta\Psi - \frac{2.3\ RT}{F}\Delta pH$$

where $\Delta\Psi$ represents the electrical potential across the membrane and ΔpH is the chemical difference in proton concentrations across the membrane. It is $\Delta\ \mu_{H}{}^{+}$ or one of its two components that is postulated to be the immediate driving force for the inward movement of transport substrates. Ramos and Kaback (1977a) studied membrane vesicles isolated from *E. coli* by flow dialysis and used the distribution of weak acids to measure ΔpH and the distribution of a lipophilic cation triphenylmethyl phosphonium to measure $\Delta\Psi$. They noted that nigericin, one of the polyether monocarboxylate ionophores, reduces the difference in proton concentrations across the membrane (ΔpH). But the decrease in ΔpH was accompanied by an increase in transmembrane electrical potential ($\Delta\Psi$), which resulted in no change in $\Delta\ \mu_{H}{}^{+}$. Further titration studies with nigericin and valinomycin led Ramos and Kaback (1977b) to conclude that there are two classes of transport systems; those that are driven primarily by $\Delta\ \mu_{H}{}^{+}$ (lactose, proline, serine, glycine, tyrosine, glutamate, leucine, lysine, cysteine, and succinate) and those that are driven primarily by ΔpH (glucose 6-phosphate, D-lactate, glucuronate, and gluconate). However, both transport systems can be driven by $\Delta\Psi$ alone when there is no longer a differ-

5

ence in proton concentration across the membrane. Therefore, nigericin inhibits transport of glucose 6-phosphate, D-lactate, glucuronate, and gluconate only when the external pH is lower than the internal pH.

These findings lead to interesting speculations on the probable mode of action of anticoccidial ionophores. Assuming that they act against the parasites by inhibiting their carbohydrate transport like that observed on *E. coli* membrane vesicles, the pH value in the host cytoplasm will have to be lower than that in the coccidial cytoplasm. This is apparently achieved toward the later stages of intracellular coccidial development because 1) large amounts of lactic acid were produced by the parasites (van der Horst and Kouwenhoven, 1973), which lowered the pH in the intestine (Stephens, 1965), and 2) the host cell membranes were so deteriorated that they were no longer a barrier between the host cell cytoplasm and the intestinal tract (Thompson et al., 1979). However, during the early intracellular phase when coccidia have not yet caused any appreciable damage, the pH values on both sides of the parasite plasma membrane should be nearly equal and the host cell membrane should be relatively intact. The orally administered ionophore may act primarily on the host cell membrane. Because the pH in the intestinal tract is on the average of 6–7 (Ruff and Reid, 1975), it is not difficult to visualize inhibition of intestinal cell carbohydrate transport by the ionophores. The effect may deprive carbohydrate supply from coccidia and thus kill the parasite. When the ionophores are used for prophylactic purposes, blocked host carbohydrate transport may be the primary mode of anticoccidial action. This indirect means of controlling coccidiosis may also explain the narrow margins of drug safety (Weppelman et al., 1977) and the lack of resistance development to the drugs among coccidia in chickens (Jeffers, 1978).

The increase in transmembrane electrical potential ($\Delta\Psi$) caused by polyether monocarboxylate ionophores (Ramos and Kaback, 1977a) was further investigated by Lichtshtein et al. (1979) with mouse neuroblastoma-rat glioma hybrid NG 108-15 cells. They observed that monensin, known to catalyze transmembrane exchange of H^+ for Na^+ (Wang, 1978a), increases the $\Delta\Psi$ by 20–30 mV in media containing high Na^+ but not in media of high K^+. This hyperpolarizing effect was completely blocked by ouabain, suggesting that the ionophore-induced Na^+ influx may have stimulated the electrogenic activity of the Na^+,K^+-ATPase which catalyzes simultaneous movement of 3 eq of Na^+ out and 2 eq of K^+ into the cell (Baker, 1965). This same effect is, however, unlikely to exist on an intracellular coccidium even if some monensin molecules can penetrate the host to reach the parasite plasma membrane, because the host cytoplasm surrounding the parasite contains high K^+ and low Na^+. Thus, when the host cell remains relatively intact, the coccidia inside it have relatively small ΔpH across the membrane and unchanged $\Delta\Psi$ under direct action of monensin; that is, the $\Delta \mu_H{}^+$ of the coccidial plasma membrane is not directly affected by the ionophore. This conclusion supports the previous notion that anticoccidial ionophores may exert indirect effect on the parasite by interacting with the host plasma membrane.

There has been some recent interest in the possible accumulation of monensin or lasalocid in *E. tenella* sporozoites by in vitro incubation and in the possible curtailing effect on the subsequent intracellular development of the treated sporozoites. Negative results were obtained from a few preliminary experiments (McDougald and Galloway, 1976). Smith and Strout (1979) incubated *E. tenella* sporozoites with radiolabeled lasalocid or narasin and found only residual drug uptake, which had no apparent dependence on the cell number or the incubation temperature. The sporozoite-associated radioactivity was steadily removed with each successive wash, which suggests lack of high affinity, specific binding of the ionophores to cell membrane. It is apparent that extreme caution must be exercised both in carrying out the experiments and in interpreting the results.

Nucleic Acid Metabolism

Although there is very little metabolism of nucleic acids observed during the extracellular phase of coccidial development, the enormously high rates of cellular multiplication during intracellular growth are undoubtedly accompanied by vigorous nucleic acid synthesis. The large collection of sulfa drugs, antifolates, and purine and pyrimidine derivatives known to be effective anticoccidial agents not only suggests the need for an unperturbed nucleic acid metabolism for coccidial growth, but also points out ample differences between the host and parasite in the metabolism to allow selective inhibition. This area of metabolism and chemotherapy of coccidia requires some detailed discussion.

Folate Metabolism The sulfonamides were among the first compounds found to have anticoccidial activities (Levine, 1938; Horton-Smith and Boyland, 1946; Cuckler and Ott, 1947). The activities are antagonized by *p*-aminobenzoic acid (PABA) (Horton-Smith and Boyland, 1946), but are synergized with the antifolate pyrimethamine in controlling coccidiosis (Kendall, 1956). These findings have led to the belief that coccidia synthesize their own folic acid, and the sulfa drugs act against the parasites by inhibiting their folic acid synthesis. This conclusion is supported by Warren's finding (1968) that deficiency of folic acid in infected chickens has little suppressive effect on the development of coccidia (see Table 3). Joyner (1960) observed that 25 mg/kg of folic acid daily reversed the growth inhibition on chickens caused by 0.004% pyrimethamine plus 0.05% sulfadimidine in the diet, without reducing therapeutic efficacy against *E. tenella* infections in chickens. Apparently, the parasite is incapable of taking up exogenous folic acid. Although biosynthesis of folic acid in coccidia has not yet been studied, dihydropteroate synthetase was identified, isolated, and studied in *P. berghei* (McCullough and Maren, 1974) and *Plasmodium chabaudi* (Walter and Königk, 1974). A good correlation between the in vitro inhibition of dihydropteroate synthetase and in vivo antimalarial activities of some sulfa drugs support the suggested mode of drug action.

5

Rogers et al. (1964) described a series of 2-substituted PABA and found them active against certain intestinal species of coccidia in chickens. These compounds, exemplified by 2-ethoxy-PABA, which was later derivatized to form 4-acetamido-2-ethoxybenzoic acid methyl ester (ethopabate) as a prodrug for commercial anticoccidial use in vivo, potentiate the anticoccidial action of pyrimethamine. Their anticoccidial activity is antagonized by the simultaneous administration of an equal weight of PABA and is thus most likely acting like the sulfa drugs. However, 2-ethoxy-PABA does not inhibit the growth of sulfa-sensitive bacteria, nor does it act against *E. tenella* in infected chickens (Rogers et al., 1964). The 2-ethoxy-PABA also lacks activity against *E. tenella* when grown in cultured embryonic chick kidney epithelial cells (Wang, unpublished observation), which suggests that the lack of activity of the drug against cecal coccidia in vivo is not due to problems in drug distribution. A simple explanation could be that the dihydropteroate synthetase in intestinal coccidia may have a substrate specificity differing from that of cecal coccidia and bacteria, and can recognize 2-substituted PABA as substrate.

The investigation of dihydrofolate reductase in *E. tenella* unsporulated oocysts (Wang et al., 1975a) not only demonstrated a high molecular weight (240,000) of the enzyme, a characteristic shared by the dihydrofolate reductase of other protozoan parasites (Ferone et al., 1969; Gutteridge and Trigg, 1971), but also showed a high susceptibility of the coccidial enzyme to pyrimethamine. The kinetic constants presented in Table 4 indicate similar K_I values of methotrexate on dihydrofolate reductase of *E. tenella* as well as chicken liver. But pyrimethamine has a K_I of 3 nM on the parasite enzyme and a 12-fold higher K_I of 36 nM on chicken liver dihydrofolate reductase. This moderate difference between K_I values may explain the anticoccidial activity of pyrimethamine and its relatively small therapeutic index of 1.5 (Joyner, 1960; McManus et al., 1967).

One other enzyme, serine hydroxymethyl transferase, which catalyzes the transfer of the hydroxymethyl group from serine to tetrahydrofolate to form glycine and N^5, N^{10}-methylene tetrahydrofolate, has been identified in crude extracts of purified *E. necatrix* second generation schizonts (James, 1980a). An average enzyme activity of 0.035 nmol of product formed/min/mg of protein was found in the extract. The parasite enzyme has a relatively high pH optimum of 8.25. In *P. lophurae*, Platzer (1972) observed that N^{10}-formyltetrahydrofolate synthetase and N^5,N^{10}-methylene tetrahydrofolate dehydrogenase were both absent, whereas serine hydroxymethyl transferase was present in abundance (2.5 nmol of product formed/min/mg of protein) in the cytosolic fraction of the parasite (Platzer, 1977). Smith et al. (1976) noted that among many compounds tested, L-serine was the only methyl group donor for *P. knowlesi* in rhesus monkey erythrocytes. The methyl group from serine was about evenly distributed between methionine and thymidylate formed in the parasite. It was postulated that serine hydroxymethyl transferase may be the pivotal enzyme for the 1-carbon pool in *Plasmodia*; inhibition of the enzyme may demonstrate in vivo antimalarial activity. It is not clear

Table 4. Kinetic constants for dihydrofolate reductase from *E. tenella* and from chicken liver

Kinetic constants	Dihydrofolate reductase	
	E. tenella	Chicken liver
K_m values[a] (M)		
NADPH	$2.9 \pm 0.2 \times 10^{-6}$	$2.3 \pm 9.2 \times 10^{-6}$
Dihydrofolate	$4.7 \pm 0.5 \times 10^{-7}$	$9.6 \pm 1.0 \times 10^{-7}$
Maximum rate[b]		
Dihydrofolate	325 ± 25	275 ± 25
Folate	65 ± 5	ND[c]
K_I values (M)		
Methotrexate	$1.7 \pm 0.8 \times 10^{-9}$	$0.25 \pm 0.1 \times 10^{-9}$
Pyrimethamine	$3.0 \pm 1.0 \times 10^{-9}$	$3.6 \pm 0.5 \times 10^{-8}$

[a]Measured at 90 μM concentration of the other substrate.

[b]Measured in units of moles of substrate reduced/min/mol of methotrexate binding sites. Assayed with 90 μM NADPH and 90 μM substrate (adapted from Wang et al., 1975a).

[c]ND = not done.

whether the 1-carbon metabolism in coccidia is similar to that in *Plasmodia.* However, the relatively low activity of serine hydroxymethyl transferase in *E. necatrix* schizonts suggests possible sources of methyl groups other than serine for the organism.

Purine Metabolism All the parasitic protozoa examined to date seem to be unable to synthesize the purine ring de novo, because of the failure of radiolabeled glycine and formate in minimal defined media to label the nucleic acid purines of the parasites. *Trypanosoma cruzi* (Gutteridge and Gaborak, 1979), *Leishmania braziliensis* (Marr et al., 1978b), *P. lophurae* (Walsh and Sherman, 1968), and *Trypanosoma megna* (Bone and Steinert, 1956) are among the best documented parasites incapable of synthesizing purines de novo and dependent on purine salvage. Hypoxanthine, adenine, and adenosine have been regarded as the major sources of purines for the parasites (Königk, 1977). However, recent studies on *Plasmodia* (Van Dyke et al., 1977) and the discoveries of purine nucleoside hydrolase in *Trypanosoma gambiense* (Schmidt et al., 1975) and adenine aminohydrolase in *Leishmania* (Marr et al., 1978b) increasingly favor the hypothesis that hypoxanthine may be the main, if not the only, supply of purines to these parasites.

There has been an increasing number of studies on the purine metabolism of coccidia in recent years. All the evidence accumulated to date suggests a very similar pattern of activities in this family of parasites.

Lack of de Novo Purine Nucleotide Synthesis Perrotto et al. (1971) incubated trophozoites of *T. gondii* with high concentrations of [^{14}C]formate and [^{14}C]glycine for 2 hr and found little incorporation of radioactivity into the parasite DNA. By pulse-labeling for 2 hr and by autoradiography, Wang (unpublished) observed little incorporation of [^{14}C]formate into TCA-insoluble fractions of *E. tenella* schizonts in embryonic chick kidney epithelial cells. No activity of amidophosphoribosyl transferase, which is the first enzyme on the de novo purine nucleotide synthetic pathway and a primary site of its feedback regulation (Wyngaarden and Ashton, 1959), could be detected in the crude extracts of *E. tenella* unsporulated oocysts (Wang and Simashkevich, in press). When [^{14}C]glycine or [^{14}C]formate is incubated with crude extracts of *E. tenella* unsporulated oocysts in the presence of all the necessary cofactors and substrates for de novo purine nucleotide synthesis, no incorporation of radioactivity into nucleotides or IMP could be detected by polyethyleneimine (PEI)-cellulose adsorption or high pressure liquid chromatographic analysis (see Table 5). However, when a crude extract of chicken liver was examined in the same experiments, substantial incorporation of [^{14}C]glycine into IMP and its precursors and similarly significant incorporation of [^{14}C]formate into IMP were observed (Table 5). Glycine incorporation was dependent on the presence of phosphoribosylpyrophosphate (PRPP), whereas formate incorporation was stimulated by adding 5-amino-1-ribosyl-4-imidazole-carboxamide 5′-phosphate (AICAR) to the reaction mixture, as expected from the known de novo purine nucleotide synthetic pathway.

Table 5. Incorporation of radiolabeled substrates into purine nucleotides

Substrates	Reaction mixtures	Substrates incorporated into purine nucleotides (nmol)	
		Chick liver	*E. tenella*
[C]Glycine (1.2 μmol)	Complete	332.5	2.2
	−PRPP	7.7	0.5
	Boiled extract	0	1.0
[^{14}C]Formate (4.0 μmol)	Complete	139.0	0
	+AICAR (4.0 μmol)	364.5	0
	Boiled extract	0	1.7
[^{14}C]Hypoxanthine (3.3 μmol)	Complete	43.6	402.2
	−PRPP	0	0

The complete reaction mixture contained in 2.0 ml (in micromoles): aspartate, 4.0; glutamine, 4.0; ATP, 5.0; $MgCl_2$, 10.0; $KHCO_3$, 20.0; KCl, 100; Tris-Cl, pH 7.8, 100; PRPP, 3.0; N^5,N^{10}-methyl tetrahydrofolate, 3.0; phosphoenolpyruvate, 30; pyruvate kinase, 28.6 U; crude extracts, 10 mg of protein. The incubation was carried out at 37 °C for 20 min. Purine nucleotides were trapped on PEI-cellulose and washed and radioactivities were counted (adapted from Wang and Simashkevich, in press).

These results have indicated clearly that *E. tenella*, and probably other species of coccidia, are incapable of de novo synthesis of the purine ring.

Purine Salvage The same *E. tenella* extract, while incapable of de novo purine formation, demonstrated very high activity of hypoxanthine-guanine phosphorribosyl transferase (HGPRT). The specific activity is approximately 10-fold higher than that in the chicken liver extract (Table 5) and has an average value of 1.5 nmol of hypoxanthine converted/min/mg of protein when assayed under optimal conditions. It was partially purified by GMP-agarose affinity column chromatography and seemed to be a single enzyme recognizing both hypoxanthine and guanine as substrates (Wang and Simashkevich, in press). The enzyme differs from chicken liver HGPRT by an apparently higher pI and heat lability. It is inhibited by allopurinol with an ID_{50} of 1.7×10^{-4} M, whereas the chicken liver enzyme is not affected by this drug. The observations resemble those made on *T. cruzi* (Marr et al., 1978a) and *Leishmania* (Marr and Berens, 1977). Allopurinol also has long been known to act against *E. tenella* in cell cultures at 100 ppm (Ryley and Wilson, 1972). Other enzymes involved in purine salvage, adenine phosphoribosyl transferase (APRT) and adenosine kinase, were also found at substantial levels in the crude extract of *E. tenella* unsporulated oocysts: APRT, 170 pmol of substrate converted/min/mg of protein; adenosine kinase, 900 pmol of substrate converted/min/mg of protein (Wang and Simashkevich, in press). It thus seems that *E. tenella* has the capability of salvaging hypoxanthine, guanine, adenine, and adenosine.

Autoradiographic studies indicate moderate incorporation of [^{3}H]adenosine into the nucleic acids of *E. tenella* schizonts grown in chick embryos (Morgan and Canning, 1974). However, because the incubation with radioligand in the experiment lasted for 5 days, which resulted in labeling both the host and parasite, it is uncertain whether adenosine itself or some metabolic product(s) of adenosine was the immediate substrate salvaged by *E. tenella*. Wang et al. (1979a) pulsed-labeled *E. tenella* schizonts grown in cell cultures with radiolabeled hypoxanthine, adenine, or adenosine for 2 hr. They noted that only hypoxanthine was incorporated into parasite nucleic acids to a significant extent when examined by autoradiography. A similar autoradiographic study (Wang and Simashkevich, 1980) showed considerable accumulation of radioactivity in *E. tenella* schizonts by a 10-min pulse treatment of the infected cell cultures with [^{14}C]hypoxanthine when the washed sample was not treated with 5% TCA. The radiolabel, largely absent from the host cells, is extractable with 5% TCA, and thus may be mostly nucleotide derivatives of hypoxanthine. Pfefferkorn and Pfefferkorn (1977a) observed that Lesch-Nyhan human skin fibroblasts, which are defective in HGPRT, support the intracellular growth of *T. gondii*. Autoradiography established that radioactive hypoxanthine and guanine supplied to the infected cells were incorporated specifically into the perchloric acid-insoluble fraction of the parasite (see Figure 12). The RNA and DNA from infected and uninfected Lesch-Nyhan cells were isolated after incubation with [^{3}H]hypoxanthine for varying

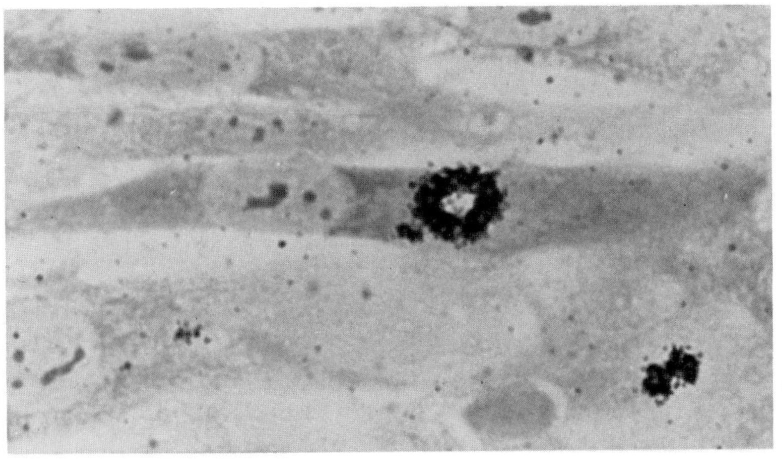

Figure 12. Autoradiographic localization of labeled hypoxanthine incorporated into Lesch-Nyhan cells infected by *T. gondii*. The cells were exposed to medium containing 10 μCi/ml during the interval of 16–24 hr after infection and then processed for autoradiography. The time of exposure was 2 days. The infected cell was doubly infected and contains a group of eight parasites. Reprinted by permission from: Pfefferkorn, E. R., and Pfefferkorn, L. C. *Toxoplasma gondii:* Specific labeling of nucleic acids of intracellular parasites in Lesch-Nyhan cells. *Experimental Parasitology* 41:95–104 (1977).

lengths of time; only the RNA and DNA from infected cells were significantly labeled (Figure 13). Further analysis of the labeled RNA indicated that 71% of the radioactivity was in adenine, whereas 29% was associated with guanine, suggesting predominant conversion of IMP to AMP in *T. gondii*. The [^{3}H]guanine-labeled RNA still had 93% of the total radioactivity in guanine, which reflects a lack of interconversion between GMP and AMP in the parasite.

In another investigation, Pfefferkorn and Pfefferkorn (1976b) noted that the growth of *T. gondii* in human fibroblast cultures was inhibited by adenine arabinoside (ara-A) at low concentrations that had no effect on the host cell. The specific target of this inhibition seemed to be synthesis of DNA by the parasite. By chemical mutagenesis, a mutant of *T. gondii* (ara-A-1) was isolated which was 50-fold more resistant to ara-A. The wild type and mutant *T. gondii* were isolated from cell cultures, purified in large numbers, and disrupted to prepare crude extracts for enzyme assays (Pfefferkorn and Pfefferkorn, 1978).

The results presented in Table 6 indicate that wild type *T. gondii* has significant activities of adenosine deaminase, deoxyadenosine deaminase, and adenosine kinase, but very little deoxyadenosine kinase. *T. gondii* (ara-A-1) has similar activities in all these enzymes except adenosine kinase, which is essentially absent from the mutant. This lack of adenosine kinase may be the basis of resistance to ara-A, because an adenosine kinase-less mutant (FR5) of 3T6 cells was notably more resistant to ara-A than the parent cell. Presumably, the *T. gondii* adenosine kinase serves to phosphorylate ara-A; without this enzyme the formation of ara-A triphosphate, which is the inhibitor of DNA synthesis, is blocked. The absence of adenosine kinase from *T. gondii*, however, has no apparent effect on the development of the parasite; both the wild type and mutant *T. gondii* grown in FR5 cells were equally efficiently labeled by [^{3}H]adenosine. This lack of difference suggests an alternative pathway of salvaging adenosine as the really functional means; adenosine can be deaminated to form inosine, which upon

Table 6. Specific enzyme activities involved in purine metabolism in extracts of *T. gondii*

	pmol nucleotide/min/mg protein[a]			
Extract assayed	Adenosine deaminase	Deoxyadenosine deaminase	Adenosine kinase	Deoxyadenosine kinase
Human fibroblast (HF)	650	580	2,100	1.20
Wild type *T. gondii* from HF	160	190	940	0.30
Ara-A-1 *T. gondii* from HF	120	130	36	0.34
Ara-A-1 *T. gondii* from FR5[b]	ND[c]	ND	<0.1	ND

[a]Averages of two or more independent determinations.
[b]A mutant of 3T6 cells lacking adenosine kinase (adapted from Pfefferkorn and Pfefferkorn, 1978).
[c]ND = not done.

Figure 13. Kinetics of incorporation of [^{3}H]hypoxanthine into the RNA and DNA of infected and uninfected Lesch-Nyhan cells. One-half of a series of replicate cultures was infected with a high multiplicity of *T. gondii.* Microscopic examination later in the course of infection showed that approximately 65% of the cells were infected. The calculated incorporation per cell was based on the total cell number, not on the number of infected cells. Duplicate infected and uninfected cultures were labeled at 8-hr intervals with medium containing 5.0 μCi/ml of [^{3}H]hypoxanthine. At the end of the labeling period, the incorporation into RNA and DNA was measured. The data from successive 8-hr periods of labeling were summed to give a kinetic picture of nucleic acid synthesis. RNA, infected cells ●——●; RNA, uninfected cells, ○---○; DNA, infected cells, ▲——▲; DNA, uninfected cells, △- - -△. Reprinted by permission from: Pfefferkorn, E. R., and Pfefferkorn, L. C. *Toxoplasma gondii:* Specific labeling of nucleic acids of intracellular parasites in Lesch-Nyhan cells. *Experimental Parasitology* 41:95–104 (1977).

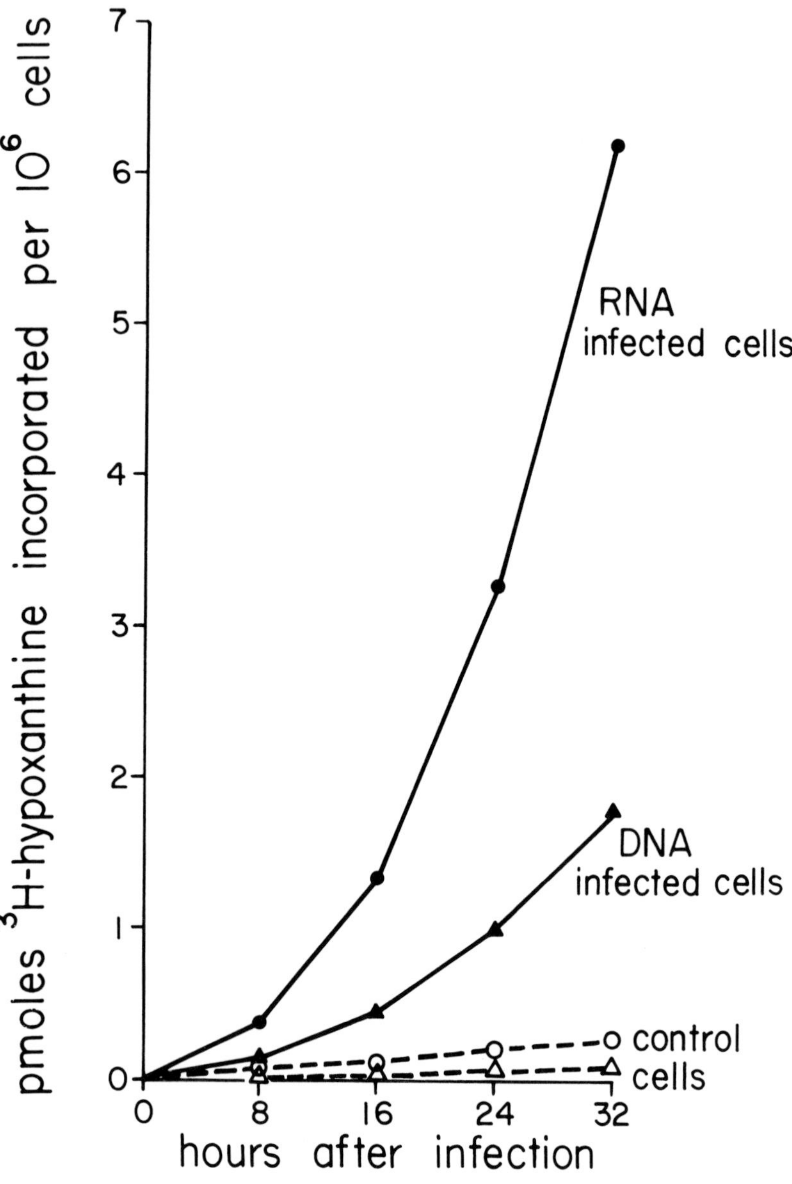

phosphorylic cleavage yields hypoxanthine, which can then enter the nucleotide pool through HGPRT.

In the intestinal mucosa, the site of coccidial development, there also is no active de novo purine synthesis (Mackinnon and Deller, 1973). It has high activities of HGPRT and APRT and extremely high levels of purine nucleoside phosphorylase (Yamada, 1961) and alkaline phosphatase (Kunitz, 1960). These enzymes may be essential for purine salvage by the intestinal mucosa. Their presence also suggests that hypoxanthine may be the major component in the pool of purines and purine nucleosides of intestinal mucosa. This then implies that hypoxanthine salvage could be the main route of purine supply for coccidia.

Arprinocid The compound, 9-(2-chloro-6-fluorobenzyl)adenine, is an anticoccidial agent (Miller et al., 1977). It is readily absorbed by medicated chickens and is excreted in the urine mainly as arprinocid and arprinocid-1-*N*-oxide (Jacob et al., 1978). These two compounds also have been identified as the major products after incubation of either arprinocid or arprinocid-1-*N*-oxide with chicken liver microsomes (Wolf et al., 1978). Control of coccidiosis can be achieved by feeding chickens with either arprinocid or the 1-*N*-oxide; the two drugs are equally active by oral administration (Lire et al., 1976).

The biochemical activities of arprinocid have been extensively examined (Wang et al., 1979a). It functions as an inhibitor of dihydrofolate reductase of *E. tenella* by competing with NADPH at a K_I value of 3×10^{-6} M, but this inhibition contributes little to anticoccidial action. It also exerts specific inhibition on the transport of hypoxanthine and guanine in HeLa cells (Wang et al., 1979b) and Chinese hamster lung fibroblasts (Slaughter and Barnes, 1979). The data presented in Table 7 indicate competitive arprinocid inhibition of hypoxanthine and guanine uptake with K_I values of 33 and 79 μM, respectively. Little inhibition of adenine transport in HeLa cells by arprinocid is observed. There is a reasonably good correlation between inhibitory effects on hypoxanthine transport in HeLa cells and anticoccidial activities among the analogs of arprinocid. In the presence of 67 μM arprinocid in *E. tenella*-infected embryonic chick kidney epithelial cell cultures, uptake of [^{3}H]hypoxanthine by the parasite is totally inhibited. This same concentration of arprinocid, equivalent to 20 ppm, is also the ED_{50} value

Table 7. Inhibition of purine transport in HeLa cells by arprinocid

Substrates	K_m (M)	V_{max} (pmol/min/10^6 cells)	Arprinocid[a] K_I (M)
Hypoxanthine	8.3×10^{-5}	1,000	3.3×10^{-5}
Guanine	2.5×10^{-4}	1,670	7.9×10^{-5}
Adenine	4.0×10^{-5}	500	4.4×10^{-3}

[a]Arprinocid acts as competitive inhibitor with each substrate (adapted from Wang et al., 1979b).

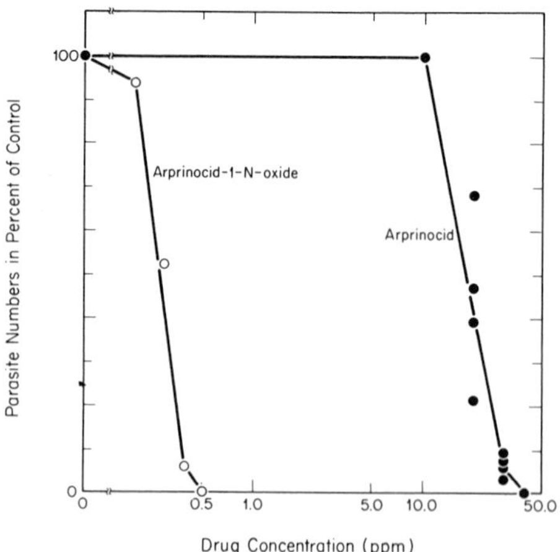

Figure 14. The activities of arprinocid and arprinocid-1-*N*-oxide against *E. tenella* in embryonic chick kidney epithelial cell cultures. Reprinted by permission from: Wang, C. C., and Simashkevich, P. M. A comparative study of the biological activities of arprinocid and arprinocid-1-*N*-oxide. *Molecular and Biochemical Parasitology* 1:137–149 (1980).

of the drug activity against *E. tenella* in cell cultures (see Figure 14). The in vitro anticoccidial activity of arprinocid is partially reversible by hypoxanthine and is thus in good agreement with the conclusion that arprinocid acts against coccidia by inhibiting their hypoxanthine transport.

Arprinocid-1-*N*-oxide, on the other hand, showed little activity in all the tested biochemical assays (Wang et al., 1979a) and is 50% as active as arprinocid in inhibiting hypoxanthine transport. It has, however, extremely potent anticoccidial activity in cell cultures; the ED_{50} value is about 0.30 ppm (see Figure 14), and the activity is not affected by hypoxanthine. A similar observation also has been made by Latter and Wilson (1979). They observed good activity of arprinocid against *E. tenella* in chick embryo liver cell cultures, but not in chick embryo kidney cell cultures. The published work by Wolf et al. (1978) showing chicken liver microsomal conversion of arprinocid to its 1-*N*-oxide led these authors to test the latter compound in chick embryo kidney cell cultures and to find good activity.

Wang and Simashkevich (1980) analyzed the tissue levels of arprinocid and its 1-*N*-oxide in chickens fed with 70 ppm of radiolabeled arprinocid in the diet; they found 0.64 ppm of arprinocid and 0.33 ppm of arprinocid-1-*N*-oxide in the liver and intestine. While this tissue level of arprinocid (1.4 μM) seems much too low when compared with its K_I on hypoxanthine transport and ED_{50} of in vitro anticoccidial activity, the level of arprinocid-1-*N*-oxide is nearly the same as the in vitro ED_{50} against *E. tenella*. It is thus likely that the 1-*N*-oxide is the main active entity in vivo.

The mode of anticoccidial action of arprinocid-1-*N*-oxide seems to be dilation of rough endoplasmic reticulum in coccidia (Wang et al., 1981). The electron micrographs in Figure 15 show that exposure of *E. tenella* second generation schizonts grown in cell culture to 67 μM arprinocid-1-*N*-oxide at 39°C for 3 hr results in formation of giant vacuoles in *E. tenella* merozoites (Figure 15*C*). There are membranous structures around the vacuoles and numerous spherical masses of cytoplasm of varying sizes inside the vacuoles. The vacuole formation can be prevented by the presence of SKF-525A, a known inhibitor of microsomal metabolism (Cooper et al., 1954), which suggests that further microsomal metabolism may be needed for the anticoccidial action of arprinocid-1-*N*-oxide. There was no vacuole found in *E. tenella* merozoites which underwent similar treatment with arprinocid (see Figure 15*B*). Arprinocid-1-*N*-oxide demonstrated a high affinity type II binding (K_s = 20.4 μM) to the heme moiety of microsomal cytochrome P-450 in studies of difference spectra (Wang et al., 1981). In the electron paramagnetic resonance investigation, both the low and the high spin signals of the heme iron in cytochrome P-450 were drastically altered by the 1-*N*-oxide, but unaffected by arprinocid. The different properties of the two compounds may be attributed to the metal ion chelating capability of the 1-*N*-oxide.

A similar observation was made on the effects of the adenosine derivatives nucleocidin, cordycepin, and puromycin and its aminonucleoside on the fine structure of *Trypanosoma*

5

Figure 15. Electron micrographs of drug-treated *E. tenella* merozoites in cultured embryonic chick kidney epithelial cells. *E. tenella* developed to the second generation schizonts in cell cultures were incubated with drugs at 39°C for 3 hr. *A*, no-drug control; *B*, 67 μM arprinocid; *C*, 67 μM arprinocid-1-*N*-oxide; *D*, 67 μM arprinocid-1-*N*-oxide plus 5.0 ppm of SKF-525A. Reprinted by permission from: Wang, C. C., Simashkevich, P. M., and Fan, S. S. The mechanism of anticoccidial action of arprinocid-1-*N*-oxide. *Journal of Parasitology* 67:137–149 (1981).

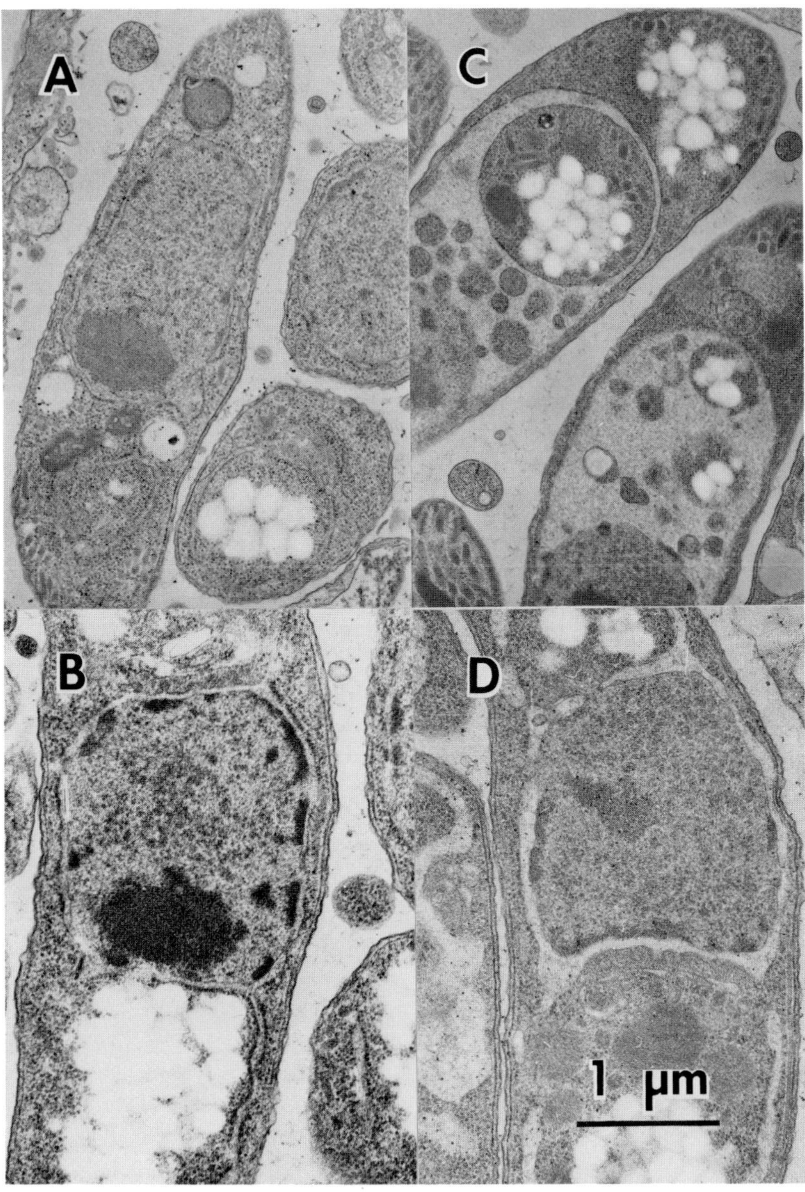

rhodesience (Williamson and Macadam, 1976). All four drugs induced electron-lucent cytoplasmic clefts in the cytoplasm; the clefts had an intimate relationship with the rough endoplasmic reticulum and produced excessive lysosomal vacuolation. Williamson and McLaren (1974) also noted that these drug effects could not be reproduced in vitro. The 1-*N*-oxide derivatives of these four drugs may be the active entity in vivo. Thus, many pharmacologically active purine derivatives may have their metabolites as the active agents and may not necessarily act on the purine metabolism.

Pyrimidine Metabolism

Salvage of Pyrimidines Since Roberts et al. (1970) discovered the lack of incorporation of [^{3}H]thymidine into *E. callospermophili* in cell cultures, similar observations have been made with *E. tenella* in cell cultures (Ouellette et al., 1973, 1974) and in embryonated eggs (Morgan and Canning, 1974). It seems that coccidia may have rather weak thymidine kinase activity. On the other hand, however, uridine is readily incorporated into *E. tenella* (Ouellette et al., 1973), and uridine and deoxyuridine are incorporated into *T. gondii* at substantial levels in cultured cells (Pfefferkorn and Pfefferkorn, 1976b). Deoxyuridine was proved to be a general precursor for both RNA and DNA in *T. gondii*, which suggests possible conversion of deoxyuridine to uracil and then uridylic acid in the parasite. This possibility is supported by the observation that a *T. gondii* mutant resistant to 5-fluorodeoxyuridine (FUDR) was also resistant to fluorouracil and fluorouridine (Pfefferkorn and Pfefferkorn, 1977b). Autoradiographic analysis showed that the FUDR-resistant mutant failed to incorporate either deoxyuridine, uracil, or uridine. Further studies by Pfefferkorn (1978) indicated that both the wild type and the mutant *T. gondii* had little or no uridine kinase, but uridine and deoxyuridine phosphorylases were found in both parasites (Table 8). The critical difference between the mutant and wild type parasites proved to be a 100-fold lower concentration of uracil phosphoribosyl transferase

Table 8. Specific enzyme activities involved in pyrimidine metabolism in extracts of *T. gondii*[a]

	pmol product/mg protein/min		
Enzymes	Human fibroblasts	Wild type *T. gondii*	FUDR-1 *T. gondii*
Uridine kinase	2.4	0.06	0.05
Deoxyuridine phosphorylase	1	80	60
Uridine phosphorylase	20	2,300	1,800
Uracil phosphoribosyl transferase	<0.01	14.5±0.6	<0.01

[a]Adapted from Pfefferkorn, 1978.

5

in the FUDR-resistant mutant. A back mutant of the resistant strain, selected for its ability to utilize uracil, simultaneously regained uracil phosphoribosyl transferase and sensitivity to FUDR. Thus, by a series of elegant experiments, Pfefferkorn and Pfefferkorn (1976b, 1977b) and Pfefferkorn (1978) have demonstrated a mandatory phosphorolysis of uridine and deoxyuridine to produce uracil before conversion to uridylic acid by uracil phosphoribosyl transferase and incorporation into *T. gondii* nucleic acids. FUDR, fluorouridine, and fluorouracil apparently act against the parasite by the same route. Additional supporting evidence for this conclusion is the significant incorporation of radiolabeled uracil into the nucleic acids of *T. gondii* grown in human fibroblast cells when no incorporation of uracil into the latter can be detected (Pfefferkorn and Pfefferkorn, 1977c). A similar observation also was made on uracil incorporation into *E. tenella* grown in embryonic chick kidney epithelial cells (Wang and Simashkevich, unpublished).

Uracil phosphoribosyl transferase thus must be a unique enzyme in coccidia and apparently the only identified enzyme in pyrimidine and pyrimidine nucleoside salvage. The ability of *T. gondii* to survive without this enzyme, however, points out the lack of need for pyrimidine salvage by the parasite. This fact also has been reflected by the relatively high frequency of resistance to FUDR among *T. gondii* (Pfefferkorn and Pfefferkorn, 1979). Other anticoccidial agents, such as emimycin (Patchett and Wang, 1976), seem to act in the same way as 5-fluorouracil and are known to produce drug-resistant populations easily among bacteria (De Zeeuw and Tynan, 1969). Glycarbylamide (Cuckler et al., 1958) is another anticoccidial agent which bears a close structural resemblance to 5-formamidoimidazole-4-carboxamide. The ribonucleotide derivative of the latter is an intermediate in de novo purine synthesis. Since it is not possible for the drug to act against the parasites by blocking de novo purine synthesis because coccidia do not synthesize purine, an alternative explanation is that it might be taken by a certain phosphoribosyl transferase of the parasite as substrate and converted to a ribonucleotide. Due to the relatively high frequency of resistance development in coccidia against glycarbylamide (Gardiner and McLoughlin, 1963), there is a possibility that uracil phosphoribosyl transferase may be the enzyme converting the drug to its ribonucleotide. Recent evidence, however, indicated that it is not the case (Wang and Simashkevich, unpublished).

De Novo Pyrimidine Synthesis The independence of coccidia from pyrimidine salvage suggests an adequately functioning pyrimidine synthetic machinery in the parasite. Although little information is available on the subject among coccidia, the conversion of dihydro-orotate to orotate has been studied in Kinetoplastida and *Plasmodium* by Gutteridge et al. (1979). The mechanism of the reaction in the two groups of protozoa was quite distinct. In the Kinetoplastida, the enzyme is an hydroxylase, which occurs in the soluble fraction of the cell and probably requires tetrahydrobiopterin for activity. In contrast, in *Plasmodium*, the enzyme is a dehydrogenase, which is probably mitochondrial, positioned at the ubiquinone level, and which can be inhibited by menoctone. Both enzymes are distinct from the mammalian enzyme. Should the coccidial enzyme belong to either category, there is a good possibility of finding an inhibitor specifically blocking the parasite pyrimidine synthesis. Because it is not known whether pyrimidine salvage alone is capable of supporting the growth of coccidia, a combination of the inhibitor with 5-fluorouracil may provide an effective control of the parasites.

At least in theory, a specific inhibitor of coccidial thymidylate synthetase should be able to block both synthetic and salvage pathways for thymidylate. This enzyme already has been detected in the crude extracts of *E. tenella* oocysts (Coles et al., 1979); further characterization of it may suggest specific ways of inhibition. However, because most of the effective thymidylate synthetase inhibitors have been analogs of pyrimidine nucleotides (Reyes and Heidelberger, 1965), to reach the target enzyme in vivo may require the action of coccidial uracil phosphoribosyl transferase on the pyrimidine analog. The inhibitor thus would have the same resistance problem as that of 5-fluorouracil.

6-Azauracil is another uracil analog having moderate anticoccidial activity (Miller et al., 1979). Its mode of action is presumably first to be converted by the coccidial orotate phosphoribosyl transferase to 6-azauridylate (Wang and Simashkevich, unpublished), which inhibits orotidylate decarboxylase (Handschumacher, 1960; Rubin et al., 1962). 6-Azauridylate (1 to 4 μM) is known to exert 93–100% inhibition (Mylari et al., 1977). However, many derivatives of 6-azauracil, each having some aryl group attached to the N^1-position, have shown exceedingly potent anticoccidial activities (Chappel et al., 1974; Mylari et al., 1977; Miller et al., 1979). The benzyl derivative had a minimal effective dose in vivo of 60 mg/kg against *E. tenella*, showed no sign of degradation in the dosed animals, yet indicated weaker inhibitory effect on yeast orotidylate decarboxylase; 30 μM of 1-benzyl-6-azauracil gave 100% inhibition (Mylari et al., 1977). It is thus unlikely that 1-benzyl-6-azauracil, 1-(3′,5′-dichlorophenyl)-6-azauracil, or 2,3-chloro-4-(4-chlorobenzoyl)phenyl-as-triazine-3,5(2H,4H)-dione could have inhibition of orotidylate decarboxylase as their mode of action. The real mode of action remains to be discovered.

Protein Synthesis

There are many antibiotics known to bind to bacterial ribosomes to inhibit protein synthesis (Pestka, 1971) and also found active against coccidiosis. They include tetracycline, chloramphenicol (Ryley and Wilson, 1976), lincomycin (Arakawa and Todd, 1968), spiramycin (Cuckler and Malanga, 1957), tylosin (Reid et al., 1969), and rosamycin (Panitz, 1974). The purified ribosomes from *E. tenella* oocysts were coupled with chicken liver activating enzymes and GTP and ATP generating systems and were found actively incorporating radioactive amino acids into TCA-precipitable materials (Wang, 1978b). This in vitro protein synthesis was utilized to test many ribosome-binding antibiotics, of which tetracycline was found to be the most potent inhibitor. Chloramphenicol also inhibited the protein synthesis. An examination of the binding between radiolabeled chloramphenicol and *E. tenella* ribosomes by Scatchard analysis revealed that approximately 1 chloramphenicol molecule was bound to one ribosome with a dissociation constant of 10^{-5} M, which is 16 times higher than that between *E. coli* ribosome and chloramphenicol (Fernandez-Münoz et al., 1971), but much lower than that with mammalian ribosomes (Pestka, 1971). The actions of these antibiotics on coccidia thus may be by inhibiting protein synthesis in the parasite. Similar observations have been made on ribosomes of *P. knowlesi* and *P. lophurae*, both of which are sensitive to chlortetracycline, whereas clindamycin, doxymycin, tetracycline, and spiramycin show causal prophylactic activity at the tissue stage of *P. berghei* development (Hill, 1975; Sherman, 1976; Sherman and Jones, 1976). These parasites may have pharmacological properties in their ribosomes resembling those of bacterial ribosomes and thus may provide a suitable target for chemotherapy. Unfortunately, none of the known ribosome-binding antibiotics possess very potent anticoccidial activities.

CONCLUSION

In this chapter, most of the pertinent biochemical information on coccidia known to date has been reviewed. The volume of information gathered has far exceeded that anticipated; the technological progress alone made during the last few years will undoubtedly dictate rapid advancement in our understanding of the parasite within the next few years.

Purified *E. tenella* oocysts can now develop in synchronized steps throughout sporulation and excystation in a test tube through a variety of experimental conditions. Oocysts can provide a vast source of material for detailed studies on metabolic pathways, individual enzymes, mitochondria, ribosomes, etc. The purified schizonts may be useful for investigating the membrane and transport properties of coccidia. The micronemes, dense granules, and pellicles purified from *S. tenella* may provide a basis of understanding the mechanism of coccidial invasion of host cells. Plaque formation by *T. gondii* in cultured transformed cell lines may become increasingly useful as a technique for isolating coccidial mutants. There are also many attainable tasks waiting to be fulfilled, such as the isolation of nuclei and the identification of chromosomal compositions, the isolation of plasma membranes and identification of surface components, the isolation of specific messenger RNAs from the polyribosomes, and the investigation of the mechanisms of protein synthesis in coccidia.

There are also important long-term goals which may not be so easy to achieve in the near future. There must be ways of separating macrogametocytes from microgametocytes and, if possible, continue their propagation. Well controlled in vitro genetic recombination must be made possible to gain further understanding of the genetics of coccidia. The mechanisms of genetic triggering of each developmental phase in the life cycle of coccidia require intensive study.

On the side of chemotherapeutic treatment of coccidiosis, many effective anticoccidial agents have been discovered by empirical screens during the past decades. The emphasis on practical orientation in these efforts in the past has led to a lack of understanding on how or why a drug works against the parasites or how the coccidia become drug-resistant. Some of the preliminary studies on these subjects reviewed in this chapter have pointed out specific areas of parasite metabolism, which may differ significantly from the host metabolism to become targets of effective chemotherapy in the future. Knowledge of the mechanism(s) of drug resistance may help to prevent resistance by combining two different drugs from the very beginning of drug application. With the diminishing chances of finding new anticoccidial drugs by empirical means, because numerous drugs already have been found, rational approaches may be the preferable way to design more specific and more sensitive methods to find better and new drugs in the future.

LITERATURE CITED

Akao, S. 1971. *Toxoplasma gondii:* Localization of peroxidase activity. Exp. Parasitol. 29: 250–254.

Arakawa, A., and Todd, A. C. 1968. Cellular response of calves to first-generation schizonts of *Eimeria bovis* after treatment of calves with sulfamethazine and lincomycin hydrochloride. Am. J. Vet. Res. 29:1549–1559.

Baker, P. F. 1965. Phosphorous metabolism of intact crab nerve and its relation to the active transport of ICNS. J. Physiol. 180:383–423.

Ball, S. J., and Warren, E. W. 1967. Activity of riboflavin analogues against *Eimeria acervulina.* Vet. Rec. 80:581–582.

Ball, S. J., Warren, E. W. and Parnell, E. W. 1965. Anticoccidial activity of nicotinamide antagonists. Nature (Lond.) 208:397.

Bannister, L. H., Butcher, G. A., Dennis, E. D., and Mitchell, G. H. 1975. Structure and invasive behaviour of *Plasmodium knowlesi* merozoites *in vitro.* Parasitology 71:483–491.

Bauchop, T., and King, L. 1968. Amprolium and thiamine pyrophosphotransferase. Appl. Microbiol. 16:961–962.

Bedrnik, P. 1969. Cultivation of *Eimeria tenella* in tissue cultures: I. Further development of second generation merozoites in tissue cultures. Acta Protozool. 7:87–98.

Berg, D. H., and Hamill, R. L. 1978. The isolation and characterization of narasin, a new polyether antibiotic. J. Antibiot. (Tokyo) 31:1–6.

Beyer, T. V. 1970. Coccidia of domestic animals. Some metabolic peculiarities of particular stages of the life cycle. J. Parasitol. 56(Sect. 11):28–29.

Birge, E. A., and Kurland, C. G. 1969. Altered ribosomal protein in streptomycin-dependent *Escherichia coli.* Science 166:1282–1284.

Bone, G. J., and Steinert, M. 1956. Isotopes incorporated in the nucleic acids of *Trypanosoma megna.* Nature (Lond.) 178:308–309.

Canning, E. U., and Anwar, M. 1968. Studies on meiotic division in coccidial and malarial parasites. J. Protozool. 15:290–298.

Canning, E. U., and Morgan, K. 1975. DNA synthesis, reduction and elimination during life cycles of the *Eimeriine coccidian, Eimeria tenella* and the *Haemogregarine, Hepatozoon domerguei.* Exp. Parasitol. 38:217–227.

Capella, J. A., and Kaufman, H. E. 1964. Enzyme histochemistry of *Toxoplasma gondii.* Am. J. Trop. Med. Hyg. 13:664–666.

Catlin, J. C., Pardini, R. S., Daves, G. D., Heidker, J. C., and Folkers, K. 1968. New hydroxyquinones, apparent inhibitors of coenzyme Q enzyme systems. J. Am. Chem. Soc. 90:3572–3574.

Chaparas, S. D., and Schlesinger, R. W. 1959. Plague assay of Toxoplasma on monolayers of

5

chick embryo fibroblasts. Proc. Soc. Exp. Biol. Med. 102:431–437.

Chapman, H. D. 1976. *Eimeria tenella:* Experimental studies on the development of resistance to robenidine. Parasitology 73:265–273.

Chapman, H. D. 1978. Studies on the excystation of different species of *Eimeria in vitro.* Z. Parasitenkd. 56:115–121.

Chappel, L. R., Howes, H. L., and Lynch, J. E. 1974. The site of action of a broad-spectrum aryltriazine anticoccidial, CP-25,415. J. Parasitol. 60:415–420.

Charney, M.Z., Reid, W. M., McDougald, L. R., and Johnson, J. 1971. Effects of essential fatty acid deficiency on coccidiosis in the domestic fowl. Poult. Sci. L(6):1801–1805.

Coles, A. M., Swoboda, B. E. P., and Ryley, J. F. 1979. Thymidylate synthetase as a chemotherapeutic target in the treatment of avian coccidiosis. Proceedings for the British Society for Parasitology 79:li.

Cooper, J. R., Axelrod, J., and Brodie, B. B. 1954. Inhibitory effects of β-diethylaminoethyl diphenylpropylacetate on a variety of drug metabolic pathways *in vitro.* J. Pharmacol. Exp. Ther. 112:55–63.

Cruthers, L. R., Szanto, J., Linkenheimer, W. H., Maplesden, D. C., and Brown, W. E. 1978. Anticoccidial activity of lonomycin (SQ 12,525) in chicks. Poult. Sci. 57:1227–1233.

Cuckler, A. C., Chapin, L. R., Malanga, C. M., Rogers, E. F., Becker, H. J., Clark, R. L., Leanza, W. J., Pessolano, A. A., Shen, T. Y., and Sarett, L. H. 1958. Antiparasitic drugs: II. Anticoccidial activity of 4,5-imidazoledicarboxamide and related compounds. Proc. Soc. Exp. Biol. Med. 98:167–170.

Cuckler, A. C., Cobb, W. R., McManus, E.C., and Ott, W. H. 1961. Amprolium: 6. Efficacy for turkey coccidiosis. Poult. Sci. 40:1392.

Cuckler, A. C., and Malanga, C. M. 1957. The effects of antibiotics on avian coccidiosis and enterohepatitis. Antibiot. Annual 1956–1957 New York, Medical Encyclopedia Inc., pp. 592–595.

Cuckler, A. C., Malanga, C. M., Basso, A. J., and O'Neill, R. C. 1955. Antiparasitic activity of substituted carbanilide complexes. Science. 122:244–245.

Cuckler, A. C., and Ott, W. H. 1947. The effect of sulfaquinoxaline on the development stages of *Eimeria tenella.* J. Parasitol. 33(suppl.):10–11.

Danforth, H. D., and Hammond, D. M. 1972. Merogony in multinucleate merozoites of *Eimeria magna* Perard, 1925. J. Protozool. 19:454–457.

Danforth, H. D., Ruff, M. D., Reid, W. M., and Miller, R. L. 1977. Anticoccidial activity of salinomycin in battery raised broiler chickens. Poult. Sci. 56:926–932.

Davis, L. R. 1973. Techniques. In: D. M. Hammond and P. L. Long (eds.), The Coccidia: *Eimeria, Isospora, Toxoplasma,* and Related Genera, pp. 411–458. University Park Press, Baltimore.

De Souza, W., and Souto-Padrón, T. 1978. Ultrastructural localization of basic proteins on the conoid, rhoptries and micronemes of *Toxoplasma gondii.* Z. Parasitenkd. 56:123–129.

De Zeeuw, J. R., and Tynan, E. J. 1969. Pyrimidine reversal of emimycin inhibition of *Escherichia coli.* J. Antibiot. (Tokyo) 22:386–387.

Dietrich, L. S., Friedland, I. M., and Kaplan, L. A. 1958. Pyridine nucleotide metabolism: Mechanism of action of the niacin antagonist, 6-aminonicotinamide. J. Biol. Chem. 233: 964–968.

Doran, D. J. 1966. Pancreatic enzymes initiating excystation of *Eimeria acervulina* sporozoites. Proc. Helminth. Soc. Wash. 33:42–43.

Doran, D. J., and Augustine, P. C. 1978. *Eimeria tenella:* Vitamin requirements for development in primary cultures of chicken kidney cells. J. Protozool. 25:544–546.

Doran, D. J., and Farr, M. M. 1962. Excystation of the poultry coccidium, *Eimeria acervulina.* J. Protozool. 9:154–161.

Dougherty, H. W. 1974. Inhibition of mitochondrial energy transduction of carbanilides. Fed. Proc. 33:1657.

Douglas, A. W., Fisher, M. H., Fishinger, J. J.,

Gund, P., Harris, E. E., Olson, G., Patchett, A. A., and Ruyle, W. V. 1977. Anticoccidial 1-substituted 4(1H)-pyridinone hydrazones. J. Med. Chem. 20:939–943.

Dubremetz, J. F., and Dissous, C. 1980. Characteristic proteins of micronemes and dense granules from *Sarcocystis tenella* endozoites (Protozoa, Coccidia). Mol. Biochem. Parasitol. 1:279–289.

Dubremetz, J. F., and Porchet, E. 1978. *Sarcocystis tenella:* Pénétration des endozites dans la cellule hôte *in vitro*. J. Protozool. 25(Abstr. 159):525A.

Dürr, U., and Pellérdy, L. 1969. Zum sauerstoff verbrauch der Kokzidienoocysten während der sponilation. Acta. Vet. Hung. 19:307–310.

Edgar, S. A., and Seibold, C. T. 1964. A new coccidium of chickens *Eimeria mivati* sp. n. (Protozoa: Eimeriidae) with details of its life history. J. Parasitol. 50:193–204.

El-Moukdad, A. R. 1976. Über die Wirkung verschiedener Desinfektionsmettel auf präparasitäre Entwicklungsstadien. Wien. Tierärztl. Mschr. 63:399–405.

Elsner, Y. Y., Roberts, W. L., Shigematsu, A., and Hammond, D. M. 1974. Isolation of schizonts and merozoites of *Eimeria callospermophili* from cultured bovine intestinal cells. J. Parasitol. 60:531–532.

Fernandez-Münoz, R., Monroe, R. E., Torres-Pinedo, R., and Vasquez, D. 1971. Substrate and antibiotic-binding sites at the peptidyltransferase center of *Escherichia coli* ribosomes. Eur. J. Biochem. 23:185–193.

Fernando, M. A., and Pasternak, J. 1977a. Isolation of chick intestinal cells infected with second-generation schizonts of *Eimeria necatrix*. Parasitology 74:19–26.

Fernando, M. A., and Pasternak, J. 1977b. Isolation of host-cell nuclei from chick intestinal cells infected with second-generation schizonts of *Eimeria necatrix*. Parasitology 74:27–32.

Ferone, R., Burchall, J. J., and Hitchings, G. H. 1969. *Plasmodium berghei* dihydrofolate reductase isolation, properties and inhibition by antifolates. Mol. Pharamcol. 5:49–59.

Fieser, L. F., Berlinger, E., Bondhus, F. J., Chang, F. C., Dauben, W. G., Ettlinger, M. G., Fawaz, G., Fields, M., Fieser, M., Heidelberger, C., Heymann, H., Seligman, A. M., Vaughan, W. R., Wilson, A. G., Wilson, E., Wu, M-I., Leffler, M. T., Hamlin, K. E., Hathaway, R. J., Matson, E. J., Moore, E. E., Moore, M. B., Rapala, R. T., and Zaugg, H. E. 1948. Naphthoquinone antimalarials: I. General survey. J. Am. Chem. Soc. 70:3151.

Fisher, J., Spencer, R., and Walsh, C. 1976. Enzyme-catalyzed redox reactions with the flavin analogues 5-deazariboflavin, 5-deazariboflavin-5′-phosphate, and 5-deazariboflavin-5′-diphosphate, 5′→5′-adenosine ester. Biochemistry 15:1054–1064.

Frandsen, J. C. 1976. Partial purification and some properties of glucose-6-phosphate dehydrogenase from *Eimeria stiedae* (Lindemann, 1865) Kisskalt & Hartmann, 1907 (Protozoa: coccidia). Comp. Biochem. Physiol. 54B:537–541.

Fridovich, I. 1974. Evidence for the symbiotic origin of mitochondria. Life Sci. 14:819–826.

Gardiner, J. L., and McLoughlin, D. K. 1963. Drug resistance in *Eimeria tenella:* III. Stabilities of resistance to glycarbylamide. J. Parasitol. 49:657–659.

Gorman, M., and Hamill, R. L. Nigericin for treating coccidiosis. U. S. Patent 3,555,150. 1971 January 12.

Graham, D. W., Brown, J. E., Ashton, W. T., Brown, R. D., and Rogers, E. F. 1977. Anticoccidial riboflavine antagonists. Experientia 33:1274–1276.

Greuel, E., Kuil, H., and Robl, R. 1975. Synergism between metichlorpindol and methylbenzoquate against *E. acervulina*. Z. Parasitenkd. 46:163–165.

Gutteridge, W. E., Dave, D., and Richards, W. H. G. 1979. Conversion of dihydroorotate to orotate in parasitic protozoa. Biochim. Biophys. Acta 582:390–401.

Gutteridge, W. E., and Gaborak, M. 1979. A re-

5

examination of purine and pyrimidine synthesis in the three main forms of *Trypanosoma cruzi.* Int. J. Biochem. 10:415–422.

Gutteridge, W. E., and Trigg, P. I. 1971. Action of pyrimethamine and related drugs against *Plasmodium knowlesi in vitro.* Parasitology 62:431–444.

Haberkorn, A. 1970a. Die Entwicklung von *Eimeria falciformis* (Eimer, 1870) in der weissen Maus (Mus musculus). Z. Parasitenkd. 34:49–67.

Haberkorn, A. 1970b. Zur Empfänglickeit nicht spezifischer wirte für Schizogonie-Stadien verschiedener *Eimeria* arten. Z. Parasitenkd. 35:156–161.

Hamill, R. L., Hoehn, M. M., Pittenger, G. E., Chamberlin, J., and Gorman, M. 1969. Dianemycin, an antibiotic of the group affecting ion transport. J. Antibiot. (Tokyo) 22:161–164.

Hamilton, W. A. 1970. The mode of action of membrane-active antibacterials. FEBS Symp. 20:71–79.

Hammond, D. M. 1971. The development and ecology of coccidia and related intracellular parasites. In: A. M. Fallis (ed.), Ecology and Physiology of Parasites, a Symposium. pp. 3–19. Toronto University Press, Toronto.

Handschumacher, R. E. 1960. Orotidylic acid decarboxylase: Inhibition studies with azauri dine-5′-phosphate. J. Biol. Chem. 235: 2917–2919.

Harold, F. M., and Van Brunt, J. 1977. Circulation of H^+ and K^+ across the plasma membrane is not obligatory for bacterial growth. Science 197:372–373.

Heikkila, R., Cabbot, F., and Cohen, G. 1976. *In vivo* inhibition of superoxide dismutase in mice by diethyldithiocarbamate. J. Biol. Chem. 251:2182–2185.

Hibbert, L. E., and Hammond, D. M. 1968. Effects of temperature on *in vitro* excystation of various *Eimeria* species. Exp. Parasitol. 23:161–170.

Hibbert, L. E., Hammond, D. M., and Simmons, J. R. 1969. The effects of pH, buffers, bile and bile acids on excystation of sporozoites of various *Eimeria* species. J. Protozool. 16: 441–444.

Hill, J. 1975. The activity of some antibiotics and long-acting compounds against the tissue stages of *Plasmodium berghei.* Ann. Trop. Med. Parasitol. 69:421–427.

Horton-Smith, C., and Boyland, E. 1946. Sulphonamides in the treatment of caecal coccidiosis of chickens. Br. J. Pharmacol. 1:139–152.

Jackson, A. R. B. 1964. The isolation of viable coccidial sporozoites. Parasitology 54:87–93.

Jacob, T. A., Buhs, R. P., Rosegay, A., Carlin, J., Vanden Heuvel, W. J. A., and Wolf, F. J. 1978. Identification of 6-amino-9-(2-chloro-6-fluorobenzyl) purine-1-*N*-oxide, a urinary metabolite of 6-amino-9-(2-chloro-6-fluorobenzyl) purine, MK-302, arprinocid. Fed. Proc. 37:813.

James, S. 1980a. Isolation of second-generation schizonts of avian coccidia and their use in biochemical investigations. Parasitology 80: 301–312.

James, S. 1980b. Thiamine uptake in isolated schizonts of *Eimeria tenella* and the inhibitory effects of amprolium. Parasitology 80: 313–322.

Jeffers, T. K. 1974a. Genetic transfer of anticoccidial drug resistance in *Eimeria tenella.* J. Parasitol. 60:900–904.

Jeffers, T. K. 1974b. *Eimeria acervulina* and *E. maxima:* Incidence and anticoccidial drug resistance of isolants in major broiler-producing areas. Avian Dis. 18:331–342.

Jeffers, T. K. 1978. *Eimeria tenella:* Sensitivity of recent field isolants to monensin. Avian Dis. 22:157–161.

Jeffers, T. K., and Challey, J. R. 1973. Collateral sensitivity to 4-hydroxyquinolines in *Eimeria acervulina* strains resistant to meticlorpindol. J. Parasitol. 59:624–630.

Jensen, J. B., and Edgar, S. A. 1976a. Effects of antiphagocytic agents on penetration of *Eimeria magna* sporozoites into cultured cells.

J. Parasitol. 62:203–206.
Jensen, J. B., and Edgar, S. A. 1976b. Possible secretory function of the rhoptries of *Eimeria magna* during penetration of cultured cells. J. Parasitol. 62:988–992.
Jensen, J. B., and Edgar, S. A. 1978. Fine structure of penetration of cultured cells by *Isospora canis* sporozoites. J. Protozool. 25:169–173.
Jensen, J. B., and Hammond, D. M. 1975. Ultrastructure of the invasion of *Eimeria magna* sporozoites into cultured cells. J. Protozool. 22:411–415.
Jolley, W. R., Allen, J. V., and Nyberg, P. A. 1979. Micropyle and oocyst wall changes associated with chemically mediated *in vitro* excystation of *Eimeria stiedae* and *Eimeria tenella.* Int. J. Parasitol. 9:199–204.
Jolley, W. R., Burton, S. D., Nyberg, P. A., and Jensen, J. B. 1976. Formation of sulfhydryl groups in the walls of *Eimeria stiedae* and *E. tenella* oocysts subjected to *in vitro* excystation. J. Parasitol. 62:199–202.
Jones, T. C., Yeh, S., and Hirsch, J. G. 1972. The interaction between *Toxoplasma gondii* and mammalian cells: I. Mechanism of entry and intracellular fate of the parasite. J. Exp. Med. 136:1157–1172.
Joyner, L. P. 1960. The relationship between toxicity and coccidiostatic efficacy of pyrimethamine and sulphonamides and their relative reversal by folic acid. Res. Vet. Sci. 1:2–9.
Joyner, L. P., and Norton, C. C. 1975a. Robenidine-dependence in a strain of *Eimeria maxima.* Parasitology 70:47–51.
Joyner, L. P., and Norton, C. C. 1975b. Transferred drug-resistance in *Eimeria maxima.* Parasitology 71:385–392.
Joyner, L. P., and Norton, C. C. 1978. The activity of methyl benzoquate and clopidol against *Eimeria maxima:* Synergy and drug resistance. Parasitology 76:369–377.
Kamen, M. D., and Horio, T. 1970. Bacterial cytochromes: I. Structural aspects. Annu. Rev. Biochem. 39:673–700.
Kantor, S., Kennett, R. L., Jr., Waletzky, E., and Tomcufcik, A. S. 1970. 1,3-Bis-(*p*-chlorobenzilidene-amino)guanidine hydrochloride, (Robenzidene): New poultry anticoccidial agent Science 168:373–375.
Keele, B. B., McCord, J. M., and Fridovich, I. 1970. Superoxide dismutase from *Escherichia coli* B. A new manganese-containing enzyme. J. Biol. Chem. 245:6176–6181.
Kelley, G. L., and Hammond, D. M. 1972. Fine structural aspects of early development of *Eimeria ninakohlyakimovae* in cultured cells. Z. Parasitenkd. 38:271–284.
Kendall, S. B. 1956. Synergy between pyrimethamine and sulphonamides used in the control of *Eimeria tenella.* Proc. R. Soc. Med. 49:874–877.
Kheysin, E. M. 1959. Cytochemical investigations of different stages of the life cycle of coccidia of the rabbit. Proceedings of the XVth International Congress on Zoology, London, 1958, pp. 713–716.
Kilejian, A. 1974. A unique histidine-rich polypeptide from the malaria parasite *Plasmodium lophurae.* J. Biol. Chem. 249:4650–4655.
Kilejian, A. 1976. Does a histidine-rich protein from *Plasmodium lophurae* have a function in merozoite penetration? J. Protozool. 23:272–277.
Kishi, H., Okumoto, C., and Hiraoka, E. 1971. Isolation and properties of mutants of *Lactobacillus fermenti* resistant to amprolium. J. Vitam. 17:59–63.
Klimes, B., Rootes, D. G., and Tanielian, Z. 1972. Sexual differentiation of merozoites of *Eimeria tenella.* Parasitology 65:131–136.
Königk, E. 1977. Salvage syntheses and their relationship to nucleic acid metabolism. Bull. WHO 55:249–252.
Kunitz, M. 1960. Intestinal alkaline phosphatase: I. The kinetics and thermodynamics of reversible inactivation. II. Reactivation by zinc ions. J. Gen. Physiol. 43:1149–1169.
Ladda, R. L., Aikawa, M., and Sprinz, H. 1969. Penetration of erythrocytes by merozoites of mammalian and avian malarial parasites. J.

Parasitol. 55:633–644.
Latter, V. S., and Wilson, R. G. 1979. Factors influencing the assessment of anticoccidial activity in cell culture. Parasitology 79:169–175.
Lavelle, F., Durosay, P., and Michelson, A. M. 1974. Purification et proprietes de la superoxyde dismutase du champignon *Pleurotus olearius.* Biochimie 56:451–458.
Lee, E.-H., and Fernando, M. A. 1977a. Characterization of coccidial DNA, an aid in species identification. Z. Parasitenkd. 53:129–131.
Lee, E.-H., and Fernando, M. A. 1977b. Drug resistance in coccidia: A robenidine-resistant strain of *Eimeria tenella.* Can. J. Comp. Med. 41:466–470.
Lee, E.-H., Remmler, O., and Fernando, M. A. 1977. Sexual differentiation in *Eimeria tenella* (Sporozoa: Coccidia). J. Parasitol. 63:155–156.
Levine, P. P. 1938. The effect of sulfanilamide on the course of experimental avian coccidiosis. Cornell Vet. 29:309.
Lichtshtein, D., Dunlop, K., Kaback, H. R., and Blume, A. J. 1979. Mechanism of monensin-induced hyperpolarization of neuroblastoma-glioma hybrid NG 108-15. Proc. Natl. Acad. Sci. U. S. A. 76:2580–2584.
Lindmark, D. G., and Müller, M. 1974. Superoxide dismutase in the anaerobic flagellates, *Tritrichomonas foetus* and *Monocercomonas sp.* J. Biol. Chem. 249:4634–4637.
Lire, E. P., Barker, W. M., and McCrae, R. C. Use of 6-amino-9-(substituted benzyl) purines and their corresponding *N'*-oxides as coccidiostats. U. S. Patent 3,953,597. 1976 April 27.
Long, P. L. 1965. Development of *Eimeria tenella* in avian embryos. Nature (Lond.) 208:509–510.
Long, P.L. 1972. Observations on the oocyst production and viability of *Eimeria mivati* and *E. tenella* in the chorioallantois of chicken embryos incubated at different temperatures. Z. Parasitenkd. 39:27–37.
Long, P. L. 1973. The growth of *Eimeria* in cultured cells and in chicken embryos: A review. In: M. A. Fernando, B. M. McGraw, and H. C. Carlson (eds.), Coccidia and Related Organisms. University of Guelph, Guelph, Ontario.
Long, P. L., and Millard, B. J. 1979. Rejection of *Eimeria* by foreign hosts. Parasitology 78:239–247.
Lotze, J. C., and Leek, R. G. 1968. Excystation of the sporozoites of *Eimeria tenella* in apparently unbroken oocysts in the chicken. J. Protozool. 15:693–697.
Lumsden, J., and Hall, D. O. 1975. Superoxide dismutase in photosynthetic organisms provides an evolutionary hypothesis. Nature 257:670–672.
Lycke, E., Carlsberg, N., and Norrby, R. 1975. Interactions between *Toxoplasma gondii* and its host cells: Function of the penetration-enhancing factor of *Toxoplasma.* Infect. Immun. 11:853–861.
McCord, J. M., and Fridovich, I. 1969. Superoxide dismutase: An enzymic function for erythrocuprein (hemocuprein). J. Biol. Chem. 244:6049–6055.
McCullough, J. L., and Maren, T. H. 1974. Dihydropteroate synthetase from *Plasmodium berghei:* Isolation, properties, and inhibition by dapsone and sulfadiazine. Mol. Pharmacol. 10:140–145.
McDougald, L. R., and Galloway, R. B. 1976. Anticoccidial drugs: Effects on infectivity and survival intracellularly of *Eimeria tenella* sporozoites. Exp. Parasitol. 40:314–319.
McDougald, L. R., and Jeffers, T. K. 1976. *Eimeria tenella* (Sporozoa: Coccidia): Gametogony following a single asexual generation. Science 192:258–259.
McHardy, N. 1978. *In vitro* studies on the action of menoctone and other compounds on *Theileria parva* and *T. annulata.* Ann. Trop. Med. Parasitol. 72:501–511.
Mackinnon, A. M., and Deller, D. J. 1973. Purine nucleotide biosynthesis in gastrointestinal mucosa. Biochim. Biophys. Acta 319:1–4.

McLaren, D. J., and Paget, G. E. 1968. A fine structural study on the merozoite of *Eimeria tenella* with special reference to the conoid apparatus. Parasitology 58:561–571.

McManus, E. C., Campbell, W. C., and Cuckler, A. C. 1968. Development of resistance to quinoline coccidiostats under field and laboratory conditions. J. Parasitol. 54:1190–1193.

McManus, E. C., Oberdick, M. T., and Cuckler, A. C. 1967. Response of six strains of *Eimeria brunetti* to two antagonists of para-aminobenzoic acid. J. Protozool. 14:379–381.

McManus, E. C., and Rogers, E. F. 1979. *Eimeria tenella:* Synergistic interaction of sulfaquinoxaline and *t*-butylaminoethanol in the chicken. Exp. Parasitol. 48:235–238.

McManus, E. C., Rogers, E. F., Miller, B. M., Judith, F. R., Schleim, K. D., and Olson, G. 1979. *Eimeria tenella:* Specific reversal of *t*-butylaminoethanol toxicity for parasite and host by choline and dimethylaminoethanol. Exp. Parasitol. 47:13–23.

Marquardt, W. C. 1966. The living endogenous stages of the rat coccidium *Eimeria nieschulzi.* J. Protozool. 13:509–514.

Marr, J. J., and Berens, R. L. 1977. Antileishmanial effect of allopurinol: II. Relationship of adenine metabolism in *Leishmania* species to the action of allopurinol. J. Infect. Dis. 136:724–732.

Marr, J. J., Berens, R. L., and Nelson, D. J. 1978a. Antitrypanosomal effect of allopurinol: Conversion *in vivo* to aminopyrazolopyrimidine nucleotides by *Trypanosoma cruzi.* Science 201:1018–1020.

Marr, J. J., Berens, R. L., and Nelson, D. J. 1978b. Purine metabolism in *Leishmania donovani* and *Leishmania braziliensis.* Biochim. Biophys. Acta 544:360–371.

Mildvan, A. S., Kobes, R. D., and Rutter, W. J. 1971. Magnetic resonance studies of the role of the divalent cation in the mechanism of yeast aldolase. Biochemistry 10:1191–1204.

Miller, B. M., McManus, E. C., Olson, G., Schleim, K. D., Van Iderstine, A. A., Graham, D. W., Brown, J. E., and Rogers, E. F. 1977. Anticoccidial and tolerance studies in the chicken with two 6-amino-9-(substituted benzyl) purines. Poult. Sci. 56: 2039–2044.

Miller, M. W., Mylari, B. L., Howes, H. L., Jr., Lynch, M. J., Lynch, J. E., and Koch, R. C. 1979. Anticoccidial derivatives of 6-azauracil: 2. High potency and long plasma life of N1-phenyl structures. J. Med. Chem. 22: 1483–1487.

Mitchell, P. 1976. Vectorial chemistry and molecular mechanics of chemiosmotic coupling: Power transmission by proticity. (The Ninth CIBA Medial Lecture) Biochem. Soc. Trans. 4:399–430.

Mitrovic, M., and Schildknecht, E. G. 1973. Anticoccidial activity of antibiotic X-537A in chickens. Poult. Sci. 52:2065.

Morgan, K., and Canning, E. U. 1974. Incorporation of ^{3}H-thymidine and ^{3}H-adenosine by *Eimeria tenella* grown in chick embryos. J. Parasitol. 60:364–367.

Morisawa, Y., Kataoka, M., Kitano, N., and Matsuzawa, T. 1976. Sankyo Co., Ltd. Nitronicotinamide derivatives for control of coccidiosis. Japan. Kokai 76,148,032 (Cl A61K31/455) 1976 Dec. 18. Appl. 75/71,058, June 12, 1975; 7.

Mylari, B. L., Miller, M. W., Howes, H. L., Jr., Figdor, S. K., Lynch, J. E., and Koch, R. C. 1977. Anticoccidial derivatives of 6-azauracil: I. Enhancement of activity by benzylation of nitrogen-1. Observations on the design of nucleotide analogs in chemotherapy. J. Med. Chem. 20:475–483.

Norton, C. C., and Joyner, L. P. 1968. The freeze preservation of coccidia. Res. Vet. Sci. 9:598–600.

Norton, C. C., and Joyner, L. P. 1975. The development of drug-resistant strains of *Eimeria maxima* in the laboratory. Parasitology 71:153–165.

Norton, C. C., and Joyner, L. P. 1978. The appearance of bisporocystic oocysts of *Eimeria*

5

maxima in drug-treated chicks. Parasitology 77:243–248.

Nyberg, P. A., Bauer, D. H., and Knapp, S. E. 1968. Carbon dioxide as the initial stimulus for excystation of *Eimeria tenella* oocysts. J. Protozool. 15:144–148.

Nyberg, P. A., and Knapp, S. E. 1970a. Scanning electron microscopy of *Eimeria tenella* oocysts. Proc. Helminth. Soc. Wash. 37:29–32.

Nyberg, P. A., and Knapp, S. E. 1970b. Effect of sodium hypochlorite on the oocyst wall of *Eimeria tenella* as shown by electron microscopy. Proc. Helminth. Soc. Wash. 37:32–36.

Ose, D., and Fridovich, I. 1976. Superoxide dismutase: Reversible removal of manganese and its substitution by cobalt, nickel or zinc. J. Biol. Chem. 251:1217–1218.

Ouellette, C. A., Strout, R. G., and McDougald, L. R. 1973. Incorporation of radioactive pyrimidine nucleosides into DNA and RNA of *Eimeria tenella* (Coccidia) cultures *in vitro*. J. Protozool. 20:150–153.

Ouellette, C. A., Strout, R. G., and McDougald, L. R. 1974. Thymidylic acid synthesis in *Eimeria tenella* (Coccidia) cultured *in vitro*. J. Protozool. 21:398–400.

Panitz, F. 1974. Anticoccidial activity of rosamycin. J. Parasitol. 60:530–531.

Patchett, A. A., and Wang, C. C. Methods for treating coccidiosis with emimycin and its derivatives. U. S. Patent No. 3,991,185. 1976 November 9.

Patton, W. H. 1965. *Eimeria tenella:* Cultivation of the asexual stages in cultured animal cells. Science 150:767–769.

Perrotto, J., Keister, D. B., and Gelderman, A. H. 1971. Incorporation of precursors into *Toxoplasma* DNA. J. Protozool. 18:470–473.

Pestka, S. 1971. Inhibitors of ribosome functions. Ann. Rev. Biochem. 40:697–710.

Pfefferkorn, E. R. 1978. *Toxoplasma gondii:* The enzymic defect of a mutant resistant to 5-fluorodeoxyuridine. Exp. Parasitol. 44:26–35.

Pfefferkorn, E. R., and Pfefferkorn, L. C. 1976a. *Toxoplasma gondii:* Isolation and preliminary characterization of temperature-sensitive mutants. Exp. Parasitol. 39:365–476.

Pfefferkorn, E. R., and Pfefferkorn, L. C. 1976b. Arabinosyl nucleosides inhibit *Toxoplasma gondii* and allow the selection of resistant mutants. J. Parasitol. 62:993–999.

Pfefferkorn, E. R., and Pfefferkorn, L. C. 1977a. *Toxoplasma gondii:* Specific labeling of nucleic acids of intracellular parasites in Lesch-Nyhan cells. Exp. Parasitol. 41:95–104.

Pfefferkorn, E. R., and Pfefferkorn, L. C. 1977b. *Toxoplasma gondii:* Characterization of a mutant resistant to 5-fluorodeoxyuridine. Exp. Parasitol. 42:44–55.

Pfefferkorn, E. R., and Pfefferkorn, L. C. 1977c. Specific labeling of intracellular *Toxoplasma gondii* with uracil. J. Protozool. 24:449–453.

Pfefferkorn, E. R., and Pfefferkorn, L. C. 1978. The biochemical basis for resistance to adenine arabinoside in a mutant of *Toxoplasma gondii.* J. Parasitol. 64:486–492.

Pfefferkorn, E. R., and Pfefferkorn, L. C. 1979. Quantitative studies of the mutagenesis of *Toxoplasma gondii.* J. Parasitol. 65:364–370.

Platzer, E. G. 1972. Metabolism of tetrahydrofolate in *Plasmodium lophurae* and duckling erythrocytes. Trans. N. Y. Acad. Sci. 34:200–208.

Platzer, E. G. 1977. Subcellular distribution of serine hydroxymethyltransferase in *Plasmodium lophurae.* Life Sci. 20:1417–1424.

Polin, D., Wynosky, E. R., and Porter, C. C. 1963. *In vivo* absorption of amprolium and its competition with thiamine. Proc. Soc. Exp. Biol. Med. 114:273–277.

Pressman, B. C. 1963. The effects of guanidine and alkylguanidines on the energy transfer reactions of mitochondria. J. Biol. Chem. 238:401–409.

Pressman, B. C., Harris, E. J., Jagger, W. S., and Johnson, J. H. 1967. Antibiotic-mediated transport of alkali ions across lipid barriers. Proc. Natl. Acad. Sci. U. S. A. 58:1949–1956.

Ramos, S., and Kaback, H. R. 1977a. The electrochemical proton gradient in *Escherichia coli* membrane vesicles. Biochemistry 16:848–854.

Ramos, S., and Kaback, H. R. 1977b. The relationship between the electrochemical proton gradient and active transport in *Escherichia coli* membrane vesicles. Biochemistry 16:854–859.

Reid, W. M., Taylor, E. M., and Johnson, J. 1969. A technique for demonstration of coccidiostatic activity of anticoccidial agents. Trans. Am. Microsc. Soc. 88:148–159.

Reisner, A. H., Rowe, J., and Macindoe, H. M. 1968. Structural studies on the ribosomes of *Paramecium:* Evidence for a "primitive" animal ribosome. J. Mol. Biol. 32:587–610.

Remington, J. S., Bloomfield, M. M., Russel, E., and Robinson, W. S. 1970. The RNA of *Toxoplasma gondii.* Proc. Soc. Exp. Biol. Med. 133:623–626.

Reyes, P., and Heidelberger, C. 1965. Fluorinated pyrimidines: XXVI. Mammalian thymidylate synthetase: Its mechanism of action and inhibition by fluorinated nucleotides. Mol. Pharmacol. 1:14–30.

Roberts, W. L., Elsner, Y. Y., Shigematsu, A., and Hammond, D. M. 1970. Lack of incorporation of ^{3}H-thymidine into *Eimeria callospermophili* in cell cultures. J. Parasitol. 56:833–834.

Rogers, E. F. 1962. Thiamine antagonists. Ann. N. Y. Acad. Sci. 98:412–429.

Rogers, E. F. 2-(4-Cyclohexylcyclohexyl)-3-hydroxy-1,4-naphthoquinone in controlling coccidiosis in poultry. U. S. Patent 3,347,472. 1967 Oct. 17.

Rogers, E. F., Clark, R. L., Becker, H. J., Pessolano, A. A., Leanza, W. J., McManus, E. C., Andriuli, F. J., and Cuckler, A. C. 1964. Antiparasitic drugs: V. Anticoccidial activity of 4-amino-2-ethoxybenzoic acid and related compounds (29616). Proc. Soc. Exp. Biol. Med. 117:488–492.

Rollinson, D., Joyner, L. P., and Norton, C. C. 1979. *Eimeria maxima:* The use of enzyme markers to detect the genetic transfer of drug resistance between lines. Parasitology 78:361–367.

Rose, M. E. 1973. Immunity. In: D. M. Hammond, and P. L. Long (eds.), The Coccidia: *Eimeria, Isospora, Toxoplasma,* and Related Genera, pp. 295–341. University Park Press, Baltimore.

Rubin, R. J., Jaffe, J. J., and Handschumacher, R. E. 1962. Qualitative differences in the pyrimidine metabolism of *Trypanosoma equiperdum* and mammals as characterized by 6-azauracil and 6-azauridine. Biochem. Pharmacol. 11:563–572.

Rudzinska, M. A., Trager, W., Lewengrub, S., and Gubert, E. 1975. Invasion of *Babesia microti* into erythrocytes. J. Protozool. 22(suppl.):28A.

Ruff, M. D., and Reid, W. M. 1975. Coccidiosis and intestinal pH in chickens. Avian Dis. 19:52–58.

Ryley, J. F. 1973. Cytochemistry, physiology and biochemistry. In: D. M. Hammond and P. L. Long (eds.), The Coccidia: *Eimeria, Isospora, Toxoplasma,* and Related Genera, pp. 145–181. University Park Press, Baltimore.

Ryley, J. F. 1975. Lerbek, a synergistic mixture of methyl benzoquate and clopidol for the prevention of chicken coccidiosis. Parasitology 70:377–384.

Ryley, J. F. 1980. Recent developments in coccidian biology; Where do we go from here? Parasitology 80:189–209.

Ryley, J. F., Bentley, M., Manners, D. J., and Stark, J. R. 1969. Amylopectin, the storage polysaccharide of the coccidia *Eimeria brunetti* and *E. tenella.* J. Parasitol. 55:839–845.

Ryley, J. F., and Betts, M. J. 1973. Chemotherapy of chicken coccidiosis. Adv. Pharmacol. Chemother. 11:221–293.

Ryley, J. F., and Robinson, T. E. 1976. Life cycle studies with *Eimeria magna* Perard, 1925. Z. Parasitenkd. 50:257–275.

Ryley, J. F., and Wilson, R. G. 1972. Growth factor antagonism studies with coccidia in tissue culture. Z. Parasitenkd. 40:31–34.

Ryley, J. F., and Wilson, R. G. 1976. Laboratory studies with some older anticoccidials. Para-

5

sitology 73:287–309.

Schmidt, G., Walter, R. D., and Königk, E. 1975. A purine nucleoside hydrolase from *Trypanosoma gambiense:* Purification and properties. Tropenmed. Parasitol. 26:19–26.

Scholtyseck, E. 1963. Untersuchungen über den Kernverhaltnisse und das Wachstum bei Coccidiomorphen unter besonderer Berucksichtigung von *Eimeria maxima.* Z. Parasitenkd. 22:428–474.

Scholtyseck, E. 1973. Ultrastructure. In: D. M. Hammond and P. L. Long (eds.), The Coccidia: *Eimeria, Isospora, Toxoplasma,* and Related Genera, pp. 81–144. University Park Press, Baltimore.

Scholtyseck, E., and Hammond, D. M. 1970. Electron microscope studies of macrogametes and fertilization in *Eimeria bovis.* Z. Parasitenkd. 34:310–318.

Scholtyseck, E., Rommel, A., and Heller, G. 1969. Licht- und elektroenmikroskopische Untersuchungen zur Bildung der Oocystenhulle bei Eimerien *(Eimeria perforans, E. stiedae* und *E. tenella.* Z. Parasitenkd. 31:289–298.

Scholtyseck, E., and Voigt, W.-H. 1964. Die Bildung der Oocystenhulle bei *Eimeria perforans* (Sporozoa). Z. Zellforsch. Mikrosk. Anat. 62:279–292.

Scholtyseck, E., and Weissenfels, N. 1956. Elektronenmikroskopische Untersuchungen von Sporozoen: I. Die Oocystenmembran des Huhnercoccids *Eimeria tenella.* Arch. Protistenk. 101:215–222.

Sharma, S. K., and Quastel, J. H. 1965.Transport and metabolism of thiamine in rat brain cortex *in vitro.* Biochem. J. 94:790–800.

Sherman, I. W. 1976. The ribosomes of the simian malaria *Plasmodium knowlesi:* II. A cell-free protein synthesizing system. Comp. Biochem. Physiol. 53B:447–450.

Sherman, I. W., and Jones, L. A. 1976. Protein synthesis by a cell-free preparation from the bird malaria *Plasmodium lophurae.* J. Protozool. 23:277–281.

Shirley, M. W. 1975. Enzyme variation in *Eimeria* species of the chicken. Parasitology 71:369–376.

Shirley, M. W. 1980. Maintenance of *Eimeria maxima* by serial passage of single sporocysts. J. Parasitol. 66:172–173.

Shirley, M. W., and Millard, B. J. 1976. Some observations on the sexual differentiation of *Eimeria tenella* using single sporozoite infections in chicken embryos. Parasitology 73:337–341.

Shirley, M. W., and Rollinson, D. 1979. Coccidia: The recognition and characterization of populations of *Eimeria.* In: A. E. R. Taylor and R. Miller (eds.), Problems in the Identification of Parasites and Their Vectors. 17th Symposium of the British Society on Parasitology, pp. 7–30. Blackwell Scientific Publications, Ltd., Oxford.

Shumard, R. F., and Callender, M. E. 1967. Monensin, a new biologically active compound: VI. Anticoccidial activity. Antimicrob. Agents Chemother.-1967. 369–377.

Skelton, F. S., Bowman, C. M., Porter, T. H., and Folkers, K. 1971. New quinolinequinone inhibitors of mitochondrial reductase systems and reversal by coenzyme Q. Biochem. Biophys. Res. Commun. 43:102–107.

Slaughter, R. S., and Barnes, E. M., Jr., 1979. Hypoxanthine transport by Chinese hamster lung fibroblasts: Kinetics and inhibition by nucleosides. Arch. Biochem. Biophys. 197:349–355.

Smith, B. F., and Herrick, C. A. 1944. The respiration of the protozoan parasite, *Eimeria tenella.* J. Parasitol. 30:295–302.

Smith, C. C., McCormick, G. J., and Canfield, C. J. 1976. *Plasmodium knowlesi: In vitro* biosynthesis of methionine. Exp. Parasitol. 40:432–437.

Smith, C. K., II, and Strout, R. G. 1979. *Eimeria tenella:* Accumulation and retention of anticoccidial ionophores by extracellular sporozoites. Exp. Parasitol. 48:325–330.

Speer, C. A., and Hammond, D. M. 1970. Nuclear divisions and refractile-body changes in sporozoites and schizonts of *Eimeria callospermophili* in cultured cells. J. Parasitol. 56:461–467.

Spencer, R., Fisher, J., and Walsh, C. 1976. Preparation, characterization and chemical properties of the flavin coenzyme analogs 5-deazariboflavin, 5-deazariboflavin-5′-phosphate, and 5-deazariboflavin-5′-phosphate, 5′→5-adenosine ester. Biochemistry 15: 1043–1053.

Stephens, J. P. 1965. Some physiological effects of coccidiosis caused by *Eimeria necatrix* in the chicken. J. Parasitol. 51:331–335.

Stotish, R. L., Simashkevich, P. M., and Wang, C. C. 1977. Separation of sporozoites, sporocysts and oocysts of *Eimeria tenella* by centrifugal elutriation. J. Parasitol. 63: 1124–1126.

Stotish, R. L., and Wang, C. C. 1975. Preparation and purification of merozoites of *Eimeria tenella*. J. Parasitol. 61:700–703.

Stotish, R. L., Wang, C. C., Hichens, M., Vanden Heuvel, W. J. A., and Gale, P. 1976. Studies of a glycoprotein in the oocysts of *Eimeria tenella*. J. Biol. Chem. 251:302–307.

Stotish, R. L., Wang, C. C., and Meyenhofer, M. 1978. Structure and composition of the oocyst wall of *Eimeria tenella*. J. Parasitol. 64:1074–1081.

Thompson, J. E., Fernando, M. A., and Pasternak, J. 1979. Induction of gel-phase lipid in plasma membrane of chick intestinal cells after coccidial infection. Biochim. Biophys. Acta 555:472–484.

Tokuyasu, K., Ilan, J., and Ilan, J. 1969. Biogenesis of ribosomes in *Plasmodium berghei*. Milit. Med. 134(suppl.):1032–1038.

Tsuchihashi, M. 1973. Zur kenntnis der blutkatalase. Biochem. Z. 140:63–112.

van der Horst, C. T. G., and Kouwenhoven, B. 1973. Biochemical investigation with regard to infection and immunity of *Eimeria acervulina* in the fowl. Z. Parasitenkd. 42:23–38.

Van Dyke, K., Trush, M. A., Wilson, M. E., and Stealey, P. K. 1977. Isolation and analysis of nucleotides from erythrocyte-free malarial parasites *(Plasmodium berghei)* and potential relevance to malaria chemotherapy. Bull. WHO 55:253–264.

Vetterling, J. M. 1968. Oxygen consumption of coccidial sporozoites during *in vitro* excystation. J. Protozool. 15:520–522.

Vetterling, J. M. 1969. Continuous-flow differential density flotation of coccidial oocysts and a comparison with other methods. J. Parasitol. 55:412–417.

Vetterling, J. M. 1976. *Eimeria tenella:* Host specificity in gallinaceous birds. J. Protozool. 23:155–158.

Vetterling, J. M., and Doran, D. J. 1969. Storage polysaccharide in coccidial sporozoites after excystation and penetration of cells. J. Protozool. 16:772–775.

Wagenbach, G. E. 1969. Purification of *Eimeria tenella* sporozoites with glass bead columns. J. Parasitol. 55:833–838.

Wagenbach, G. E., and Burns, W. C. 1969. Structure and respiration of sporulating *Eimeria stiedae* and *E. tenella* oocysts. J. Protozool. 16:257–263.

Wagenbach, G. E., Challey, J. R., and Burns, W. C. 1966. A method for purifying coccidian oocysts employing clorox and sulfuric acid-dichromate solution. J. Parasitol. 52:1222.

Wagner, W. H., and Foerster, O. 1964. Die PAS-AO Methode, eine Spezialfarburg fur Coccidien in Gewebe. Z. Parasitenkd. 25:28–48.

Walsh, C. J., and Sherman, I. W. 1968. Purine and pyrimidine synthesis by the avian malaria parasite, *Plasmodium lophurae*. J. Protozool. 15:763–770.

Walter, R. D., and Königk, E. 1974. Purification and properties of the 7,8-dihydropteroate-synthesizing enzyme from *Plasmodium chabaudi*. Hoppe-Seylers Z. Physiol. Chem. 355:431–437.

Walton, A. C. 1959. Some parasites and their chromosomes. J. Parasitol. 45:1–20.
Wang, C. C. 1975. Studies of the mitochondria from *Eimeria tenella* and inhibition of electron transport by quinolone coccidiostats. Biochim. Biophys. Acta 396:210–219.
Wang, C. C. 1976. Inhibition of the respiration of *Eimeria tenella* by quinolone coccidiostats. Biochem. Pharamacol. 25:343–349.
Wang, C. C. 1978a. Biochemical and nutritional aspects of coccidia. In: P. L. Long, K. N. Boorman, and B. M. Freeman (eds.), Avian Coccidiosis, pp. 135–184. British Poultry Science, Ltd., Edinburgh.
Wang, C. C. 1978b. The prokaryotic characteristics of *Eimeria tenella* ribosomes. Comp. Biochem. Physiol. 61B:571–579.
Wang, C. C., and Simashkevich, P. M. 1980. A comparative study of the biological activities of arprinocid and arprinocid-1-*N*-oxide. Mol. Biochem. Parasitol. 1:335–345.
Wang, C. C., and Simashkevich, P. M. Purine-metabolism in a protozoan parasite *Eimeria tenella.* Proc. Natl. Acad. Sci. U.S.A. In press.
Wang, C. C., Simashkevich, P. M., and Fan, S. S. 1981. The mechanism of anticoccidial action of arprinocid-1-*N*-oxide. J. Parasitol. 67:137–149.
Wang, C. C., Simashkevich, P. M., and Stotish, R. L. 1979a. Mode of anticoccidial action of arprinocid. Biochem. Pharmacol. 28:2241–2248.
Wang, C. C., and Stotish, R. L. 1975a. Pancreatic chymotrypsin as the essential enzyme for excystation of *Eimeria tenella.* J. Parasitol. 61:923–927.
Wang, C. C., and Stotish, R. L. 1975b. Changes in nucleic acids and proteins in the oocysts of *Eimeria tenella* during sporulation. J. Protozool. 22:438–443.
Wang, C. C., and Stotish, R. L. 1978. Multiple leucine amino-peptidases in the oocysts of *Eimeria tenella* and their changes during sporulation. Comp. Biochem. Physiol. 61B:307–313.
Wang, C. C., Stotish, R. L., and Poe, M. 1975a. Dihydrofolate reductase from *Eimeria tenella:* Rationalization of chemotherapeutic efficacy of pyrimethamine. J. Protozool. 22:564–568.
Wang, C. C.,Tolman, R. L., Simashkevich, P. M., and Stotish, R. L. 1979b. Arprinocid, an inhibitor of hypoxanthine-guanine transport. Biochem. Pharmacol. 28:2249–2260.
Wang, C. C., Weppelman, R. M., and Lopez-Ramos, B. 1975b. Isolation of amylopectin granules and identification of amylopectin phosphorylase in the oocysts of *Eimeria tenella.* J. Protozool. 22:560–564.
Warhurst, D. C., and Williamson, J. 1970. Ribonucleic acid from *Plasmodium knowlesi* before and after chloroquine treatment. Chem. Biol. Interact. 2:89–106.
Warren, E. W. 1968. Vitamin requirements of the coccidia of the chicken. Parasitology 58:137–148.
Warren, E. W., and Ball, S. J. 1967. Anticoccidial activity of egg white and its counteraction by biotin. Vet. Rec. 80:578–579.
Weisiger, R. A., and Fridovich, I. 1973. Mitochondrial superoxide dismutase site of synthesis and intramitochondrial localization. J. Biol. Chem. 248:4793–4796.
Weppelman, R. M., Olson, G., Smith, D. A., Tamas, T., and Van Iderstine, A. 1977. Comparison of anticoccidial efficacy, resistance and tolerance of narasin, monensin and lasalocid in chicken battery trails. Poult. Sci. 56:1550–1559.
Weppelman, R. M., Vanden Heuvel, W. J. A., and Wang, C. C. 1976. Mass spectrometric analysis of the fatty acids and non-saponifiable lipids of *Eimeria tenella* oocysts. Lipids 11:209–215.
Williamson, L., and Macadam, R. F. 1976. Drug effects on the fine structure of *Trypanosoma rhodesiense:* Puromycin and its amino-nucleoside, cordycepin and nucleocidin.

Trans. R. Soc. Trop. Med. Hyg. 70:130–137.
Williamson, J., and McLaren, D. J. 1974. Fatty acid metabolism and drug-induced cytoplasmic clefts in trypanosomes. Trans. R. Soc. Trop. Med. Hyg. 68:263.
Wilson, P. A. G., and Fairbairn, D. 1961. Biochemistry of sporulation in oocysts of *Eimeria acervulina*. J. Protozool. 8:410–416.
Wolf, F. J., Steffens, J. J., Alvaro, R. F., and Jacob, T. A. 1978. Microsomal conversion of MK-302, arprinocid [6-amino-9-(2-chloro-6-fluorobenzyl) purine] to 6-amino-9-(2-chloro-6-fluorobenzyl) purine-1-*N*-oxide by liver microsomes from the chicken and the dog and to 2-chloro-6-fluorobenzyl alcohol by liver microsomes from the rat and mouse. Fed. Proc. 37:814.
Wong, D. T., Horng, J-S., and Wilkinson, J. R. 1972. Robenzidene, an inhibitor of phosphorylation. Biochem. Biophys. Res. Commun. 46:621–627.
Wyngaarden, J. B., and Ashton, D. M. 1959. The regulation of activity of phosphoribosylpyrophosphate amidotransferase by purine ribonucleotides: A potential feedback control of purine biosynthesis. J. Biol. Chem. 234:1492–1496.
Yamada, E. W. 1961. The phosphorolysis of nucleosides by rabbit bone marrow. J. Biol. Chem. 236:3043–3046.

6

Behavior of Coccidia in Vitro

David J. Doran

Why do we cultivate coccidia? The main reasons are to study their behavior and development and the antiparasitic effects. Embryo infections are valuable because they offer an environment at the tissue level and may allow development of the whole endogenous cycle. In cell culture, the parasite is visible and the immediate environment can be altered. Host-parasite relationships can be observed at the cellular level, and the behavior of the parasite can be studied under a variety of conditions.

There have been numerous reviews of work with *Eimeria* and *Toxoplasma* in cell culture and avian embryos. The review by Doran (1973) covered work with these two genera and with *Besnoitia, Isospora,* and *Sarcocystis.* In that review, data concerning *Toxoplasma gondii* came from 192 publications of original research since 1929, and data pertaining to *Eimeria* and *Isospora* came from 82 publications of original research since 1965. In preparing the present review, 54 papers on *T. gondii* and 100 papers on *Eimeria* and *Isospora* published since early 1972 were found, which involved use of either cell culture or avian embryos. Although the pace of research on cultivation of coccidia has certainly not diminished, there has been a rather pronounced shift in the type of research. Before 1972, most work (especially with *Eimeria*) was descriptive, pertaining mostly to morphology and mechanics of cultivation; after 1972, most work (especially with cell culture) was concerned with drug action, cytochemistry and biochemistry, ultrastructure, genetics, and immunity. Studies such as these before 1972 were briefly mentioned in a previous review (Doran, 1973), in a section entitled "Embryo and Cell Culture Applications." It is more appropriate that this work be covered in other chapters in this volume; therefore it is omitted from the present review.

Species in the genera *Besnoitia, Eimeria, Isospora,* and *Toxoplasma* have been cultivated in both cell culture and avian embryos, whereas *Hammondia, Hepatozoon, Klossia,* and *Sarcocystis* have been cultivated in cell culture only. Much of the work with *Besnoitia, Eimeria, Isospora,* and *Sarcocystis* in cell culture was carried out by the late Dr. Datus Hammond and/or his former graduate students. This chapter is dedicated to Dr. Hammond—an excellent teacher, a fine gentleman, and a respected colleague for many years.

BEHAVIOR IN CELL CULTURE

Extracellular Movement

Besnoitia jellisoni and *Sarcocystis* species zoites and *Hepatozoon* sporozoites flex, glide, and pivot (Fayer et al., 1969; Fayer, 1970; Hendricks and Fayer, 1973). These same movements, along with undulating, probing, and helical movements, have been reported for *Eimeria* sporozoites. Of these six types of motion, only gliding and undulating have been reported for

6

T. gondii zoites. However, when other terms used to describe the movement of *T. gondii* outside of cells (see Doran, 1973) are analyzed and compared with those used to describe movement of *Besnoitia, Eimeria, Hepatozoon,* and *Sarcocystis,* it is apparent that organisms of all five genera perform similar movements in cell culture.

Movement of *Eimeria* sporozoites is limited in tissue culture medium without cells, but when sporozoites are placed near cultured cells, the sporozoites actively move and quickly enter the cells (Long and Speer, 1977).

T. gondii zoites are most active immediately after release from a cell (Jacobs, 1953). Zoites can move up to 20 μm in one direction and maintain this motion for about 15 min (Pulvertaft et al., 1954). Reported speeds have ranged from one to two body lengths/s (Bommer et al., 1969a) to as much as 400 times the body length/s (Lund et al., 1961).

Entering and Leaving Cells

Active Penetration Numerous workers have observed *Eimeria* sporozoites actively entering and leaving cells. According to Speer and Hammond (1970b), the process of leaving is similar to that of entering (Figures 1–6). Roberts et al., (1971) observed that *Eimeria larimerensis* sporozoites thrust a slender anterior protuberance into a cell to a depth of 1–2 μm. The protuberance is sometimes retracted and inserted at different locations near the original insertion point. Usually, the protuberance moves laterally within the cytoplasm for 1–2 s. It then seems to swell, increase in width, and decrease in length. Within 3 s, the remainder of the sporozoite enters the cell. Intervals up to 1 min have been reported for complete entry of several other eimerians into cells (Fayer and Hammond, 1967; Clark and Hammond, 1969; Kelley and Hammond, 1970; Speer and Hammond, 1970b; Speer et al., 1970a). *Eimeria alabamensis* sporozoites, which usually enter nuclei of intestinal cells in the natural host, take 20–30 min to enter cells in culture (Sampson et al., 1971). Sporozoites may leave cells immediately after entry or may remain 5–7 days from the time of inoculation (Clark and Hammond, 1969). Escape of host cell cytoplasm is frequently seen after a sporozoite leaves a cell, but rarely after it enters (Speer et al., 1971). The act of leaving a cell could be associated with the artificial environmental conditions, the type of cell, or both (Fayer and Hammond, 1967).

T. gondii zoites have also been observed in the process of actively entering and leaving cells. As with *Eimeria,* the process of leaving is the same as that of entering (Hirai et al., 1966; Bommer et al., 1969a). Cook and Jacobs (1958) and Hirai et al. (1966) could not confirm the observation by Pulvertaft et al. (1954) that active entry kills cells. The description by Pulvertaft et al. (1954) of entry into cells is interesting. The zoite (*a*) selects a point on the "circumference" of a cell for invasion, applies itself to the surface, and then slides around until it reaches a selected point, and (*b*) always enters a cell from below at a point where the cytoplasm underlies the "concavity near the nucleus." Hirai et al. (1966) observed that free and attached

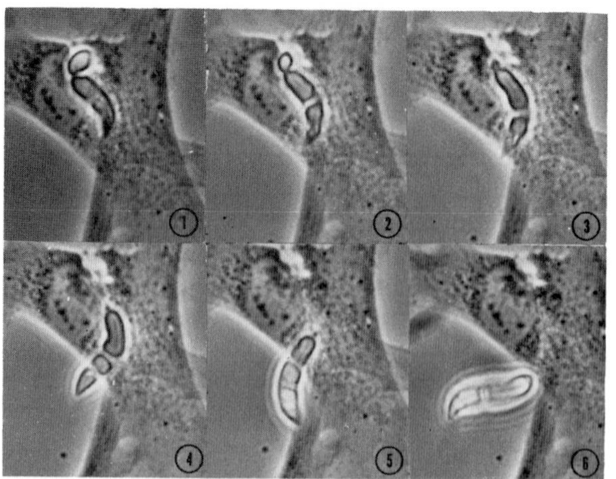

parasites do not behave in the same manner. Those attached never penetrate through the cell membrane in the immediate area of attachment, but move to another part of the cell or even to another cell before penetrating. Those present in the medium show no particular movement. They suddenly become active and penetrate directly into cells. A zoite extends its anterior end and makes a small hole in the cell membrane (Bommer, 1969; Bommer et al., 1969a). This hole through which the parasite enters is 1.5 μm wide and appears within a few seconds after initial contact (Hirai et al., 1966; Bommer et al., 1969a). The zoite constricts, becomes gourd-shaped (Hirai et al., 1966), and undergoes "amoeboid deformation" (Bommer, 1969) while pulling its way through the hole (Bommer et al. 1969a). Complete entry has been reported to take 15–30 s (Bommer et al., 1969a), an average of 40 s (Hirai et al., 1966), and 30–60 min (Schmidt-Hoensdorf and Holz, 1953).

Of considerable interest is whether *T. gondii* and the eimerian parasites secrete a substance that aids active penetration of cells. Studies concerning (*a*) the penetration-promoting effects of lysozyme, hyaluronidase, and lysed parasites (Lycke et al., 1965); (*b*) the extraction of a penetration-enhancing factor (PEF) from lysed zoites, and the effect it produces in cell culture (Lycke and Norrby, 1966; Norrby and Lycke, 1967; Lycke et al., 1968); (*c*) the demonstration of lysosomes in the parasite before and after inoculation into culture (Norrby, 1970b); and (*d*) the fractionation, purification, and localization of PEF (Norrby, 1970a) contribute strong circumstantial evidence that *T. gondii* does secrete a substance aiding penetration. On the other hand, Fayer et al. (1970), working with *Eimeria adenoeides* sporozoites, found that hyaluronidase does not increase the number of intracellular sporozoites and that both chondroitin sulfate and hyaluronic acid do not reduce the number of intracellular sporozoites when these agents are included in the medium at the time of inoculation and the infected cells are fixed 1 hr later.

Passive Ingestion (Phagocytosis) Phagocytosis has been suggested as a means by which *Eimeria* sporozoites gain entrance into cells (Strout et al., 1965; Doran and Vetterling, 1967a; Jensen and Hammond, 1974). Jensen and Edgar (1976) treated cells of an established bovine kidney cell line (Madin-Darby) with antiphagocytic agents (reagents blocking glycolysis, thus reducing phagocytosis) and found that entrance of sporozoites into cells was not inhibited. These results suggest that phagocytosis is not the method by which *Eimeria magna* sporozoites enter cells. Cultured macrophages, however, seem to phagocytize *E. tenella* sporozoites, and the sporozoites may survive several days within these cells (Rose and Long, 1976).

Pulvertaft et al. (1954) state that they observed phagocytosis of *T. gondii* zoites by murine macrophages. On the other hand, Holz and Albrecht (1953) reported that macrophages show no tendency to flow around the parasite and engulf it, but move in the opposite direction from the parasite. Zoites pretreated with formalin have been found within HeLa cells (Lycke et al., 1965; Norrby and Lycke, 1967), suggesting that phagocytosis occurs. However, Hirai et al.

6

Figures 1–6. *E. larimerensis* sporozoite penetrating and then leaving a sixth passage embryonic bovine kidney cell immediately after inoculation. Living specimens; phase-contrast microscopy. × 1,150 Reprinted by permission from: Speer, C. A., and Hammond, D. M. Development of *Eimeria larimerensis* from the Uinta ground-squirrel in cell culture. *Zeitschrift für Parasitenkunde (Berlin)* 35:105–118 (1970). 1. Sporozoite penetrating cell, showing constriction in the posterior third of sporozoite body at site of entrance. 2. Constriction in the posterior fifth of sporozoite body. 3. Intracellular sporozoite immediately after completion of penetration. 4. Early stage in exit from the host cell, with constriction in the anterior quarter of sporozoite body at the site of exit. 5. Constriction in the posterior third of sporozoite body. 6. Extracellular sporozoite immediately after leaving the host cell.

(1966) did not observe this phenomenon in a cinematographic study using formalin-killed parasites and swine kidney cells. Jones et al. (1972) studied the entry of zoites into HeLa cells, mouse fibroblasts, and mouse macrophages and concluded that phagocytosis takes place in all three cell types. They regarded the process as "induced phagocytosis" in the case of fibroblasts and HeLa cells, because these cells are not normally phagocytic.

Time and Rate of Entry into Cells

B. jellisoni zoites (obtained from exudate) enter cells 20–30 min after inoculation (Doby and Akinshina, 1968; Akinshina and Doby, 1969b). According to Doby and Akinshina (1968), it takes 45 min for all zoites to penetrate, providing that the exudate is used within 1 hr after it is obtained.

Eimeria sporozoites are intracellular within 1 hr after inoculation. *E. alabamensis* sporozoites (Sampson et al., 1971) and *Eimeria vermiformis* sporozoites (Kelley and Youssef, 1977) enter within 5 min. Cell type may influence time of entry because *Eimeria bovis* entered two cell types after 3 min and others only after 30 min (Fayer and Hammond, 1967). Entry rates have not been quantitated. However, it has been reported for various species that (*a*) the number of intracellular sporozoites increases for up to 3 hr, remains constant up to 6 hr, and then decreases (Strout et al., 1965); (*b*) all that enter do so by 2.5–3 hr (Doran and Vetterling, 1967b); (*c*) entry continues for 24 hr (Strout and Ouellette, 1968, 1970); and (*d*) the maximum number in different cell types is reached in 1–8 hr (*Eimeria callospermophili*) and in 24–48 hr (*Eimeria bilamellata*)(Speer et al., 1970a). Species and strains vary with regard to infectivity. *Eimeria dispersa* sporozoites penetrate cells more quickly than *Eimeria gallopavonis;* 4 hr after inoculation, 67–98% of *E. dispersa* and 23–56% of *E. gallopavonis* had penetrated (Doran and Augustine, 1977). The percentages of penetration 4 hr after inoculation with the Wisconsin, Weybridge, and Beltsville strains of *Eimeria tenella* were 36%, 19%, and 23%, respectively (Doran et al., 1974).

T. gondii zoites are also usually intracellular within 1 hr of inoculation. Occasionally, they can be found as early as 15 min (Lund et al., 1961; Hansson and Sourander, 1965). The rate of entry, expressed as percentage of infected cells, increases in proportion to the time of incubation from 3–8 hr (Cook and Jacobs, 1958; Kaufman et al., 1958) before decreasing. No increase after 20 min (Matsubayashi and Akao, 1963) and an increase up to 24 hr (Arai et al., 1958) have also been reported. Kaufman et al. (1958) stated that the decrease in penetration rate with time is caused by a loss of viability of parasites that remain extracellular and by a decrease in the total number in the supernatant fluid. With zoites obtained from exudate, the highly virulent RH strain is more invasive than the Beverley strain (Matsubayashi and Akao, 1963; Kusunoki, 1977), the 113-CE strain (Kaufman et al., 1958), and the MF, JQ, and 113-CE strains (Hogan et al., 1961). The work of Kusunoki (1977) indicated that invasiveness may depend on the source of the organisms. Using the Beverley strain, he found that tachyzoites (from exudate), sporozoites (excysted from occysts), and bradyzoites (from cysts) give RNIU (parasites/100 cells) values of 100, 75, and 55, respectively.

Intracellular Movement and Location

After entry into cells, *Eimeria* sporozoites and *Besnoitia, Sarcocystis,* and *Toxoplasma* zoites move through the cytoplasm and become situated close to the cell nucleus. Some *Eimeria* sporozoites even indent the nuclear membrane (Strout et al., 1965; Doran and Vetterling, 1967a; Roberts et al., 1971; Sampson et al., 1971). Fayer (1970) found one to five *Sarcocystis* zoites close to a nucleus. He stated that before coming to rest they can move backwards and forwards for 1 μm in each direction. According to Hirai et al. (1966), *T. gondii* zoites stop immediately after passing through the cell membrane; when movement is resumed, it takes 10–90 s from time of entry to reach the nucleus, not always by the shortest route. While moving to the cell nucleus, *E. alabamensis* and *E. larimerensis* sporozoites leave "trails" or pathways in the cytoplasm (Roberts et al., 1971; Sampson et al., 1971; Speer et al., 1971). Intracellular sporozo ites flex (Clark and Hammond, 1969), probe (Kelley and Hammond, 1970), and move laterally (Speer and Hammond, 1970b) or change direction (Roberts et al., 1971). When movement stops, sporozoites immediately become blunt or rounded at the anterior end and assume the shape characteristic of intracellular forms (Clark and Hammond, 1969).

Besnoitia, Hammondia, Isospora, Sarcocystis, and *Toxoplasma* zoites and *Eimeria* and *Hepatozoon* sporozoites become situated within a cytoplasmic vacuole (parasitophorous vacuole). With *T. gondii,* the vacuole is formed when the zoite starts to increase in size (Lund et al., 1961; Hansson and Sourander, 1965). Several investigators have found that the vacuole is not always present in *T. gondii*-infected cells (Akinshina, 1959; Sourander et al., 1960) and in cells infected with *Eimeria* (Doran and Vetterling, 1967a; Doerr and Höhn, 1972). It is likely that in all of these cases the parasite had just entered and the vacuole had not formed by the time the cultures were fixed and examined.

E. alabamensis usually develops within the nuclei of intestinal cells in the host. However, in cell culture only 1% of those found intracellularly were also intranuclear (Sampson et al., 1971). Other species have been observed to pass through nuclei (Speer and Hammond, 1970b; Speer et al., 1971) or to be situated in fixed preparations in such a way as to give the impression that they are intranuclear (Doran and Vetterling, 1967a). Roberts et al. (1971) observed that (*a*) the parasites pass through the nuclear membrane in much the same way that they pass through the cell membrane, (*b*) sporozoites usually leave a nucleus immediately after entering, but some remain for as long as 30 min and occasionally move around within the nucleus, and (*c*) sporozoites leaving a nucleus usually leave the cell. Occasionally, *T. gondii* zoites are found that seem to be intranuclear. However, rather than being attributed to active penetration, their location is considered to be caused by folding of the nucleus about the cytoplasmic vacuole con-

6

taining the parasite (Sourander et al., 1960) or to a "layering" of the parasite upon the nuclear membrane (Remington et al., 1970). *B. jellisoni* zoites are also found within nuclei (Akinshina and Doby, 1969a,b; Fayer et al., 1969). Figure 7 shows a zoite that seems to have just left a nucleus.

Considering the closeness of coccidial stages to cell nuclei, the question naturally arises as to whether the parasite obtains something necessary for development from the nucleus. The results obtained by Jones (1973) with *T. gondii* and enucleated cells are of interest. Enucleated fibroblast cells were obtained by centrifugation of glass-adherent cells in media containing cytochalasin B. Zoites entered the enucleated cells and multiplied in a manner identical to that seen in normal fibroblasts and cytochalasin B-treated nucleated fibroblasts, suggesting that the cellular factor(s) required for multiplication is not derived from the nucleus. The concentration of parasites near the nucleus of cells in culture may be entirely a result of the volume of cytoplasm in this area being much greater than that in the processes extending outward.

Development

Besnoitia

Cells Supporting Development There are no reported failures to obtain development of two of the species in this genus since the initial report by Bigalke (1962) with *Besnoitia besnoiti*. This species develops in (*a*) primary bovine kidney, lamb kidney, and sheep thyroid, and (*b*) cell line hamster kidney, human cervical carcinoma (HeLa), and green marmoset kidney (Vero). *B. jellisoni* develops in (*a*) primary embryonic chicken fibroblasts and hamster kidney fibroblasts, human fibroblasts, mouse macrophages, and swine kidney, (*b*) cell line embryonic bovine spleen and trachea, embryonic human lung, and rabbit kidney, and (*c*) established cell line Madin-Darby bovine kidney (MDBK) and Madin-Darby cat kidney (MDCK).

Mode and Rate of Multiplication After *B. jellisoni* zoites enter cells, they increase in size, become ovoid, and multiply within 3–5 hr after inoculation (Doby and Akinshina, 1968). According to Bigalke (1962), *B. besnoiti* multiplies in cell culture by synchronous binary fission. Akinshina and Doby (1969a) and Fayer et al. (1969) found stages indicating that *B. jellisoni* multiplies by endodyogeny (Figure 8). They also found stages suggesting that binary fission and even schizogony might take place.

The two immediate descendants from one zoite constitute a generation. The generation time, which is the interval between two generations, is frequently used in work with *Besnoitia, Hammondia,* and *Toxoplasma* to express the rate of multiplication. The generation time of *B. jellisoni* in primary hamster fibroblasts is 6.3 hr, whereas in macrophages from mice that are nonimmune, immune to *Besnoitia,* and immune to *Toxoplasma* it is 12.8 hr, 38.5 hr, and 12.8 hr, respectively (Hoff and Frenkel, 1974).

Aggregates and Cytopathological Effects Aggregates of zoites are formed by repeated multiplication. They are either rosette-shaped (Bigalke, 1962; Doby and Akinshina, 1968;

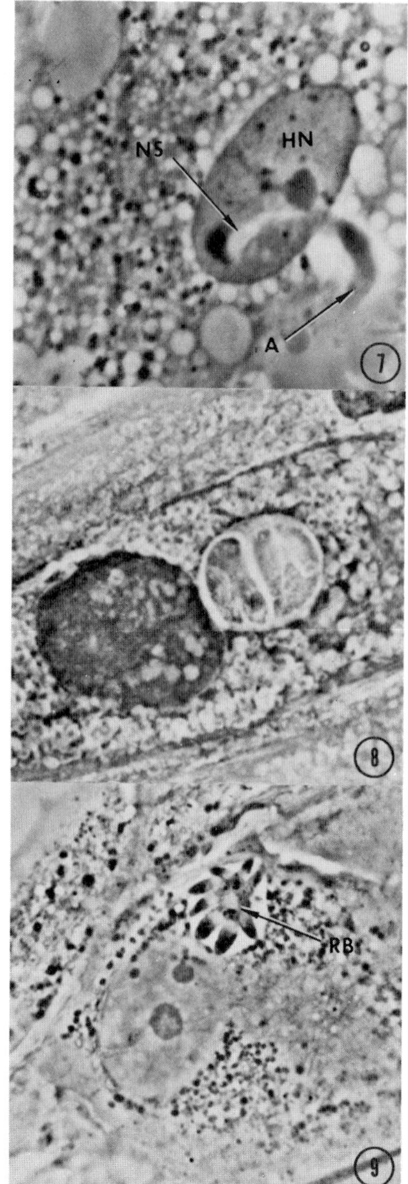

Figures 7–9. *B. jellisoni* in embryonic bovine spleen cells. Reprinted by permission from: Fayer, R., Hammond, D. M., Chobotar, B., and Elsner, Y. Y. Cultivation of *Besnoitia jellisoni* in bovine cell cultures. *Journal of Parasitology* 55:645–653 (1969). 7. A zoite leaving a cell nucleus. *NS*, area in nucleus where nucleoplasm has been displaced; *HN*, host cell nucleus; *A*, anterior tip of zoite; 5 days. × 2,500. 8. A pair of zoites in a single vacuole, each zoite containing two daughters; 4 days. × 2,500. 9. A rosette-shaped aggregate with eight zoites and a round body (*RB*); 4 days. × 1,200.

Fayer et al., 1969), shaped like a bunch of bananas (Bigalke, 1962; Fayer et al., 1969), or lacking any obvious organization (Fayer et al., 1969). A few rosette-shaped aggregates (Figure 9) and aggregates shaped like bananas contain structures interpreted as residual bodies (Fayer et al., 1969). Fayer et al. (1969) observed that rosette-shaped aggregates contained from three to eight organisms 2 days after inoculation. They also found one aggregate with no obvious organization in a vacuole within an embryonic bovine spleen cell 9 days after inoculation.

The cytoplasm of infected cells becomes vacuolated to the extent that the monolayer looks foamy (Fayer et al., 1969). The nucleus enlarges (Akinshina and Doby, 1969a), contains an enlarged nucleolus and relatively small particles of chromatin (Fayer et al., 1969), and eventually becomes pyknotic (Akinshina and Doby, 1969a).

Release of Organisms, Reinvasion of Cells, and "Histopathological" Effects Extracellular organisms can be found 30–40 hr after inoculation (Doby and Akinshina, 1968). Fayer et al. (1969) found maximum numbers of intracellular organisms after 3 and 4 days and large numbers of extracellular organisms after 5 and 6 days. In two of their experiments, they found a second peak in intracellular organisms at 8 and 9 days. This indicates that organisms leave the cells in which they develop, invade new cells, and go through a second "cycle" of development. However, in seven other experiments, the numbers of intracellular organisms declined after the initial peak. The authors offered two explanations for this decline: 1) pathological changes may have made the cells unsuitable as potential host cells; and 2) the destruction of large numbers of cells may have reduced their numbers below a level that would support further development.

Besnoitia resembles *Toxoplasma* in that infection results in a loss of most of the cell monolayer (Bigalke, 1962; Fayer et al., 1969). However, *Besnoitia* may differ from *Toxoplasma* in the manner in which organisms leave cells. Fayer et al. (1969) frequently observed organisms passing through cell membrane. Rupture of the membrane, enabling organisms within the cell to be passively expelled, has not been reported.

Serial Passage Bigalke (1962) first showed that *B. besnoiti* could be serially passaged in cell culture. However, the parasites were lost after four passages. Neuman (1974) serially passaged this species in primary cultures of sheep thyroid cells every 5–10 days and maintained it for 4½ yr without loss of virulence. He found the peak numbers of organisms 4–6 days after inoculation, serially passaged this species six times in HeLa cells and cell line hamster kidney cells, but found that the growth rate was much slower in these cells than in the sheep thyroid cells.

Eimeria

Cells Used and Extent of Development Gavrilov and Cowez (1941) established cultures with explants from rabbit intestine parasitized with various developmental stages of *Eimeria perforans*. Twelve days later they found gametocytes in cells away from the explant. If these were new cells, as the workers claimed, the right to be considered the first to obtain in vitro development of an eimerian belongs to them. If the cells were merely cells from the explant that had moved, then Patton (1965) was the first to obtain development using sporozoites. Patton obtained development to mature first generation schizonts with *E. tenella* in established cell line bovine kidney (Madin-Darby) cells and cell line embryonic quail fibroblasts. Since Patton's work, attempts have been made to cultivate 29 species of *Eimeria.* The best results, judged by the completion of the life cycle, with 27 of the species are shown in Tables 1 and 2. Two other species, *Eimeria maxima* and *Eimeria praecox,* failed to develop (Doran, 1971b). When sporozoites were used as the inoculum, oocysts were obtained only with *E. tenella;* when merozoites from the host were used, oocysts were obtained with *Eimeria acervulina, E. bovis, E. magna, Eimeria meleagrimitis,* and *E. tenella.* There are also three reports of the development of either gametocytes or oocysts after cultivation of parasitized tissue. Doerr and Höhn (1972) cultivated rabbit liver infected with *Eimeria stiedae* sporozoites and schizonts and obtained gametocytes; Long (1969) cultured cells from the chorioallantoic membrane (CAM) containing second generation merozoites and gametocytes of *E. tenella* and obtained oocysts; and Onaga et al. (1974) cultured cells from the CAM infected with *E. tenella* sporozoites and obtained oocysts.

Attempts have also been made to obtain development with sporozoites and merozoites produced in culture. Speer (1979) reported development of second generation *E. magna* merozoites when sporozoites excysted from oocysts produced in vitro were inoculated into culture. Bedrnik (1969a) transferred *E. tenella* merozoites from one culture to another and claimed to have obtained not only a third generation, but also a fourth and a fifth generation of schizonts. However, Doran (1974) inoculated cultures with first and second generation *E. tenella* merozoites and found that, although some merozoites entered cells, there was no development. Hammond et al. (1969) observed that first generation *E. bovis* merozoites entered mouse macrophages and, if inoculated with sterile fluid from the cecum of a calf, they also entered embryonic bovine tracheal cells. However, in both cell types there was no development. *Eimeria zuernii* merozoites do not enter cells (Speer et al., 1973a).

Most of the species listed in Tables 1 and 2 have been reported either to not develop or to develop but not as far in a variety of other cell types. An analysis of all reports since 1965, when sporozoites were used as the inoculum, clearly shows that more of the life cycle in cell culture is more likely to be seen if tissues from the natural host are used; the best development of 21 of the species was in cultures derived from natural host tissue (Table 1). Attempts have been made with 19 of these species to obtain development in cells from nonhost animals. With eight of these, development proceeded as far as it did in cells of host origin; with the other 11, it stopped prematurely. There have been no reported attempts using sporozoites to obtain development of *E. adenoeides, E. bilamellata, Eimeria contorta, E. magna,* and *Eimeria nieschulzi* in cultures established from tissue of the host.

6

Table 1. Development of *Eimeria* in cell cultures when sporozoites from in vivo-produced oocysts were used as the inoculum

Species[a]	Host	Cells[b]	Type of culture[c]	Best development[d]	First report
E. acervulina	Chicken	Chicken kidney	P	msch (4)	Naciri-Bontemps (1976)
E. adenoeides	Turkey	Bovine kidney (E)	CL	msch (1)	Doran (1969b)
E. alabamensis	Cow	Bovine kidney	ECL	tr (2)	Sampson et al. (1971)
E. auburnensis	Cow	Bovine kidney	ECL	msch (1)	Clark and Hammond (1969)
		Bovine spleen (E)	CL	msch (1)	Clark and Hammond (1969)
		Bovine trachea (E)	CL	msch (1)	Clark and Hammond (1969)
E. bilamellata	Ground squirrel	Bovine kidney	ECL	msch (1)	Speer et al. (1970a)
		Bovine kidney (E)	CL	msch (1)	Speer et al. (1970a)
		Bovine trachea (E)	CL	msch (1)	Speer et al. (1970a)
		Bovine synovia (E)	CL	msch (1)	Speer et al. (1970a)
		Hamster kidney	CL	msch (1)	Speer et al. (1970a)
E. bovis	Cow	Bovine trachea (E)	CL	isch (2)	Hammond and Fayer (1968)
E. brunetti	Chicken	Chicken kidney	P	msch (3)	Ryley and Wilson (1972)
E. callospermophili	Ground squirrel	Ground squirrel embryo (W)	P	isch (2)	Speer et al. (1970a)
E. canadensis	Cow	Bovine kidney (E)	ECL	msch (1)	Müller et al. (1973a)
		Bovine kidney (E)	CL	msch (1)	Müller et al. (1973a)
		Bovine kidney	P	msch (1)	Müller et al. (1973a)
		Bovine liver (E)	CL	msch (1)	Müller et al. (1973a)
E. contorta	Rat	Bovine kidney (E)	ECL	msch (2)	Müller et al. (1973b)
		Bovine kidney (E)	CL	msch (2)	Müller et al. (1973b)
E. crandallis	Sheep	Bovine liver (E)	CL	msch (1)	de Vos and Hammond (1971)
		Ovine thyroid (E)	CL	msch (1)	de Vos and Hammond (1971)
		Ovine trachea (E)	CL	msch (1)	de Vos and Hammond (1971)
E. diminuta (= *E. mivati*)(EA)	Jungle fowl	Chicken trachea	P	msch (2)	Shirley et al. (1977)
E. dispersa (T)	Turkey	Turkey kidney	P	msch (2)	Doran and Augustine (1977)
		Chicken kidney	P	msch (2)	Doran and Augustine (1977)
E. ellipsoidalis	Cow	Bovine kidney	ECL	msch (1)	Speer and Hammond (1971a)
		Bovine spleen (E)	CL	msch (1)	Speer and Hammond (1971a)
		Bovine trachea (E)	CL	msch (1)	Speer and Hammond (1971a)
E. gallopavonis	Turkey	Turkey kidney	P	mac	Doran and Augustine (1977)
E. larimerensis	Ground squirrel	Bovine kidney	ECL	msch (2)	Speer et al. (1973c)
E. magna	Rabbit	Bovine kidney	ECL	msch (2)	Speer and Hammond (1971b)
		Bovine liver (E)	?	msch (2)	Speer and Hammond (1971b)
E. meleagrimitis	Turkey	Turkey kidney	P	mac	Augustine and Doran (1978)
E. mivati (EA)	Chicken	Chicken kidney	P	msch (2)	Long (1974a)
E. necatrix	Chicken	Chicken kidney	P	msch (2)	Doran (1971b)
E. nieschulzi	Rat	Bovine intestine (E)	CL	msch (1)	Speer et al. (1970b)
		Bovine kidney (E)	CL	msch (1)	Speer et al. (1970b)
		Bovine liver (E)	CL	msch (1)	Speer et al. (1970b)
E. ninakohlyakimovae	Sheep	Bovine intestine (E)	CL	msch (1)	Kelley and Hammond (1970)
		Bovine kidney	ECL	msch (1)	Kelley and Hammond (1970)
		Ovine kidney (E)	CL	msch (1)	Kelley and Hammond (1970)
		Ovine thymus (E)	CL	msch (1)	Kelley and Hammond (1970)
		Ovine thyroid (E)	CL	msch (1)	Kelley and Hammond (1970)
		Ovine trachea (E)	CL	msch (1)	Kelley and Hammond (1970)

—continued

Table 1. Development of *Eimeria* in cell cultures when sporozoites from in vivo-produced oocysts were used as the inoculum

Species[a]	Host	Cells[b]	Type of culture[c]	Best development[d]	First report
E. stiedae	Rabbit	Rabbit liver	P	msch (?)	Doerr and Höhn (1972)
		Goat fibroblasts	P	msch (?)	Doerr and Höhn (1972)
E. tenella	Chicken	Chicken kidney	P	Oocyst	Doran (1970c)
		Guinea fowl	P	Oocyst	Doran and Augustine (1973)
		Partridge kidney	P	Oocyst	Doran and Augustine (1973)
		Pheasant kidney	P	Oocyst	Doran (1971c)
		Quail kidney	P	Oocyst	Doran and Augustine (1973)
		Turkey kidney	P	Oocyst	Doran and Augustine (1973)
E. tsunodae	Quail	Chicken embryo (W)	P	msch (1)	Ogimoto et al. (1976)
		Chicken kidney	P	msch (1)	Ogimoto et al. (1976)
		Hamster kidney	P	msch (1)	Ogimoto et al. (1976)
		Quail embryo (W)	P	msch (1)	Ogimoto et al. (1976)
E. uzura	Quail	Chicken embryo (W)	P	msch (1)	Ogimoto et al. (1976)
		Chicken kidney	P	msch (1)	Ogimoto et al. (1976)
		Hamster kidney	P	msch (1)	Ogimoto et al. (1976)
		Quail embryo (W)	P	msch (1)	Ogimoto et al. (1976)
E. vermiformis	Mouse	Mouse embryo (W)	P	msch (1)	Kelley and Youssef (1977)
		Bovine kidney	CL	msch (1)	Kelley and Youssef (1977)
		Bovine kidney	ECL	msch (1)	Kelley and Youssef (1977)
E. zuernii	Cow	Bovine kidney	P	msch (1)	Speer et al. (1973a)
		Bovine kidney (E)	CL	msch (1)	Speer et al. (1973a)
		Bovine kidney (E)	ECL	msch (1)	Speer et al. (1973a)
		Bovine trachea (E)	CL	msch (1)	Speer et al. (1973a)

[a]EA = embryo adapted strain; T = turkey strain.

[b]E = embryonic; W = freshly trypsinized whole embryo.

[c]P = primary; CL = cell line; ECL = established cell line (Madin-Darby); ? = not stated.

[d]msch = mature schizont; tr = trophozoites; isch = immature schizont; mac = macrogamete. Numbers in parentheses indicate generation; question mark indicates generation not stated.

It is interesting how frequently the established cell line MDBK appears in Tables 1 and 2. With sporozoites as the inoculum (Table 1), this line is either the best or one of the best for 11 of the 16 non-avian species, including five of the six species from cattle. With merozoites as the inoculum (Table 2), MDBK is more favorable for two of the four non-avian species. Of interest is the absence of results using intestinal cells in both the tables, especially because almost all species that have been cultivated parasitize the intestine. Cultivation of nine species using intestinal cells has been attempted, but, unfortunately, intestinal cells from the natural host were used with only three species. *E. alabamensis* (Sampson et al., 1971), *E. bovis* (Fayer and Hammond, 1967), and *E. meleagrimitis* (Doran and Vetterling, 1968) either did not develop or developed less than is shown in Table 1.

Asexual Development Eimerian sporozoites contain a large refractile body (RB) posterior to the nucleus. This RB (eosinophilic or clear globule) diminishes in size during first generation schizogony. In addition to the posterior RB, freshly excysted sporozoites of seven species have been reported to contain several smaller RBs anterior and/or posterior to the

Table 2. Development of *Eimeria* in cell cultures when merozoites from the host were used as the inoculum

Species	Cells[a]	Type of culture[b]	Merozoite generation[c]	Best development[d]	First report
E. acervulina	Chicken kidney	P	4	Oocyst	Naciri-Bontemps (1976)
E. bovis	Bovine kidney	P	1	Oocyst	Speer and Hammond (1973)
E. brunetti	Chicken fibroblasts (E)	P	2	gam	Shibalova (1970)
	Quail fibroblasts (E)	P	2	gam	Shibalova (1970)
E. larimerensis	Bovine kidney	ECL	2	gam	Speer et al. (1973c)
E. magna	Bovine kidney	ECL	?	Oocyst	Speer and Hammond (1972a)
E. meleagrimitis	Turkey kidney	P	1	msch (2)	Augustine and Doran (1978)
	Turkey kidney	P	2	Oocyst	Augustine and Doran (1978)
	Turkey kidney	P	3	Oocyst	Augustine and Doran (1978)
E. stiedae	Rabbit kidney	P	?	msch (4)	Coudert and Provôt (1974)
E. tenella	Chicken cecum	P	2 (C)	Oocyst	Bedrnik (1969b)
	Chicken embryo (W)	P	2	Oocyst	Bedrnik (1969b)
	Chicken embryo (Wc)	P	2	Oocyst	Bedrnik (1969b)
	Chicken fibroblasts (E)	P	2	Oocyst	Bedrnik (1967b)
	Chicken liver	P	2	Oocyst	Bedrnik (1969b)
	Chicken liver and cecum	P	2	Oocyst	Bedrnik (1969b)

[a]E = embryonic; W = freshly trypsinized whole embryo; Wc = freshly trypsinized whole embryo minus the cecum.
[b]P = primary; ECL = established cell line (Madin-Darby).
[c]? = not stated; C = along with cells at the time of inoculation.
[d]Gam = gametocyte; msch = mature schizont. Numbers in parentheses indicate generation.

nucleus. These gradually disappear after a sporozoite enters a cell. Details concerning numbers and rates of disappearance were summarized by Doran (1973). Doran and Augustine (1977) found that in two species the additional anterior RB does not disappear as fast as it does in other species. At 4 hr after inoculation into cell culture, all intracellular *E. dispersa* sporozoites and 97% of the intracellular *E. gallopavonis* sporozoites contained two RBs; at 48 hr, 90% and 95% of intracellular *E. dispersa* and *E. gallopavonis* sporozoites, respectively, also contained the anterior RB. Two RBs have also been found in immature schizonts (Fayer and Hammond, 1967; Clark and Hammond, 1969; Ryley and Wilson, 1972; Doran and Augustine, 1977; Kelley and Youssef, 1977) and in merozoites (Hammond et al., 1970; Speer and Hammond 1970b, 1971b) of various species. Doran and Augustine (1977) found that 50% of immature *E. dispersa* schizonts at 48 hr contained two RBs. The function and significance of the RBs are unknown. The reason for the retention of more than one RB by some species during schizogony is also unknown. It probably is not merely a characteristic of growth in culture, because Doran (1978) found numerous immature *E. dispersa* schizonts with two RBs in tissue from the turkey.

Fayer and Hammond (1967) described the nuclear changes in *E. bovis* sporozoites before the parasites assumed the round shape characteristic of trophozoites (Figure 10, *a–d*). Types A and B sporozoites of Doran and Vetterling (1968) correspond to those shown in Figure 10 *a* and *d*. Change from a vesicular to a compact nucleus has also been described for other species (Clark and Hammond, 1969; Speer et al., 1970a; Sampson et al., 1971). According to Kelley and

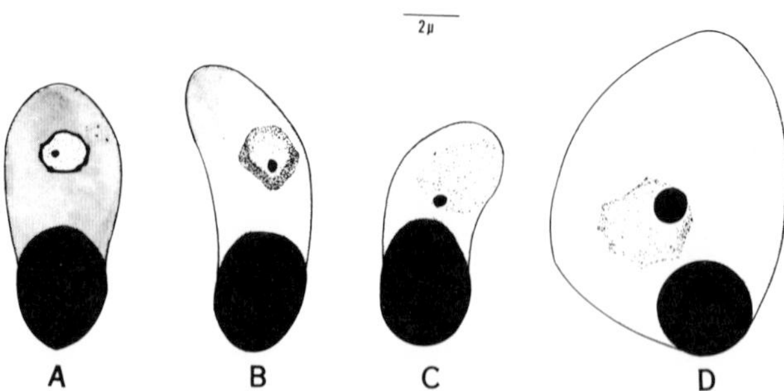

Hammond (1970), *Eimeria ninakohlyakimovae* is transformed into a trophozoite by either a gradual increase in width of the sporozoite more pronounced at the anterior end or by a lateral outpocketing of the sporozoite, usually in the posterior portion of the body. In both instances, the body increases in size as it becomes spherical. Transformation of *Eimeria crandallis* sporozoites to spherical trophozoites is by an increase in width of organisms and/or formation of a lateral outpocketing (de Vos and Hammond, 1971). Also, some sporozoites lengthen into elongate immature schizonts (sporozoite-shaped schizonts) that later become ellipsoidal. In *E. alabamensis,* outpocketing occurs at the anterior end and extends posteriorly along one side (Sampson et al., 1971). Doran and Vetterling (1968) found type B sporozoites and binucleate schizonts of *E. meleagrimitis* with lateral outpocketings. They thought that the sporozoites were probably being transformed to trophozoites, but considered the outpocketing to be a form of abnormal development.

Speer and Hammond (1970a) described nuclear division in detail. In *E. callospermophili,* the first division usually occurs within 8–10 hr after inoculation (Speer and Hammond, 1970a); in *E. larimerensis,* it occurs after 15 hr and takes 2 hr to complete (Speer and Hammond, 1970b). Among various species, immature first generation schizonts develop from either spherical trophozoites or elongated sporozoites. Sporozoite-shaped schizonts, which later transform into spherical schizonts after lateral outpocketing and enlargement in size, have been reported for *E. alabamensis, Eimeria auburnensis, E. bilamellata, Eimeria canadensis, E. callospermophili, E. larimerensis,* and *E. magna.* Transformation of *E. magna* sporozoite-shaped schizonts to spherical schizonts takes 41 min at 37°C and occurs after the formation of eight nuclei. Sporozoite-shaped schizonts flex while intracellular (Speer and Hammond, 1969, 1970b; Speer et al., 1970a; Speer et al., 1971). Some even penetrate new cells (Speer and Hammond, 1969; Speer et al., 1970a) in a manner similar to that of sporozoites (Speer et al., 1970a).

Merozoite formation in 13 species of *Eimeria* has been studied in cell culture. Merozoites form by a radial budding process at the periphery either of the intact immature schizont (Figure 11) or of subdivisions of the schizont called blastophores (Figure 12). All species studied (*E. alabamensis, E. auburnensis, E. bilamellata, E. bovis, E. callospermophili, E. canadensis,* E. contorta, E. crandallis, E. larimerensis, E. magna, E. ninakohlyakimovae, E. tenella, and *E. zuernii*) form merozoites from intact schizonts; seven of them *(E. auburnensis, E. bovis, E. canadensis, E. crandallis, E. ninakohlyakimovae, E. tenella* and *E. zuernii*) also form merozoites from blastophores. De Vos et al. (1972) observed that some *E. crandallis* schizonts became lobulated. The lobules separated to form individual schizonts, and formation of merozoites from these was like that from schizonts that have not undergone subdivision. According to Strout and Ouellette (1970), *E. tenella* merozoites of both the first and second generation that develop from intact schizonts are usually arranged in the shape of a rosette. In some instances, they are simply detached from the central mass irregularly without evidence of

Figure 10. Semidiagrammatic representation of nuclear changes associated with transformation of *E. bovis* sporozoites into trophozoites; sketches of intracellular specimens from spleen cell culture 8 days after inoculation. Reprinted by permission from: Fayer, R., and Hammond, D. M. Development of first-generation schizonts of *Eimeria bovis* in cultured bovine cells. *Journal of Protozoology* 14:764–772 (1967). *A*, sporozoite with vesicular nucleus; well defined layer of peripheral chromatin; small, somewhat eccentric nucleolus; deeply stained, ellipsoidal, posterior refractile body. *B*, sporozoite with thicker, paler, and less distinct peripheral layer of chromatin; nucleus and nucleolus somewhat enlarged. *C*, sporozoite with relatively large nucleolus; peripheral chromatin has become dispersed and nucleus is no longer vesicular. *D*, trophozoite with nucleus similar to that of sporozoite in *C*, except that nucleolus is larger; posterior refractile body has become spherical.

symmetry or pattern. Long (1969) found four different types of mature *E. tenella* schizonts, and some of the rosette-shaped type 2 schizonts contained two rows of merozoites. Merozoite formation was completed in *E. magna* in 56 min (Speer and Hammond, 1971b); in *E. larimerensis*, it took 60–90 min (Speer and Hammond, 1970b).

Merozoites usually contain a single nucleus. However, Speer and Hammond (1971b) found *E. magna* merozoites that were multinucleate. Seventy-two hours after inoculation, 30% of the mature schizonts had four (a range of two to eight) multinucleate merozoites. No further development was observed.

According to Strout and Ouellette (1970), after *E. tenella* first and second generation schizonts are mature, the merozoites within become spontaneously activated. The schizont becomes spherical and nearly detached from the cell. Activated merozoites are not released from the schizont simultaneously. Some penetrate the extremely elastic membrane; others, after 2 or 3 hr of activity, are released when the schizont membrane eventually tears and disintegrates. Occasionally, a few merozoites become inactive within the collapsed membrane. According to Sampson et al. (1971), when a first generation *E. alabamensis* merozoite leaves a cell, another immediately follows. There is apparently no constriction of the body while leaving and 10–20 min elapse before other merozoites leave the schizont; however, one to four merozoites remain in the vacuole after others have escaped. Merozoites of *E. callospermophili* do undergo constriction as they leave (Roberts et al., 1970).

Hammond et al. (1969) and Doran (1974) give the only data concerning numbers of merozoites produced in culture. They inoculated groups of cell cultures with known numbers of sporozoites and, at daily intervals after inoculation, changed the medium and counted the number of free merozoites in the medium removed. With an inoculum of 270,000 *E. bovis* sporozoites, Hammond et al. (1969) recovered 15 million first generation merozoites between the 10th and 30th days after inoculation. This is only about 55 merozoites for every sporozoite placed into culture. However, the yields obtained by Doran (1974) were lower. With 85,000 *E. tenella* sporozoites, maximum yield of first generation merozoites between the 48th and 82nd hr after inoculation was 470,000; the maximum yield of second generation merozoites between the 95th and 156th hr was 1,820,000. This is only about 5 and 21 merozoites/sporozoite. Low yields of extracellular merozoites in culture may be attributed to (*a*) only small numbers of sporozoites entering cells, surviving, and developing; (*b*) some merozoites remaining intracellular; and (*c*) death of merozoites in the culture medium during the 24 hr between counts.

Several investigators have added bile, bile salts, and trypsin to media to stimulate a release of merozoites from schizonts and penetration into cells. Hammond et al. (1969) found that *E. bovis* merozoites do not enter cells if preincubated for 10 min in a medium containing 0.2% trypsin and 5% bile. A medium containing 0.5% trypsin and 2% bovine bile increases the motility of intra- and extracellular *E. callospermophili* merozoites and increases the number of

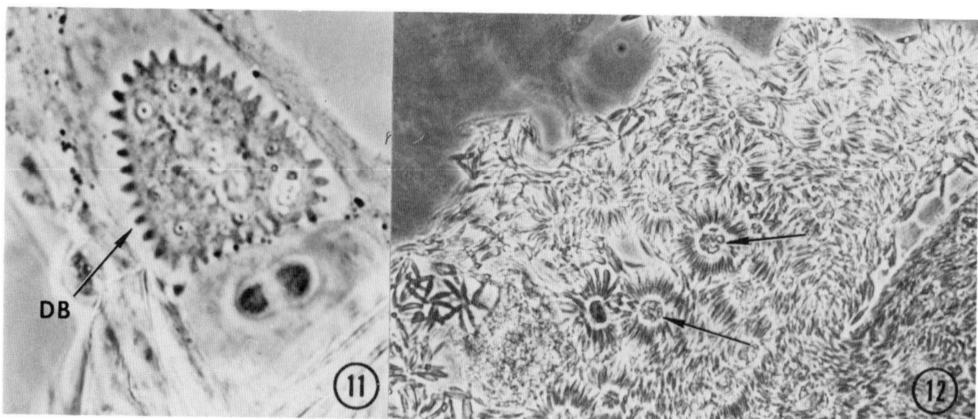

merozoites leaving a schizont (Speer et al., 1970a). In media containing 4% bile or 0.5% of a bile salt, or one of these in combination with 0.2% trypsin, extracellular merozoites of *E. bilamellata, E. callospermophili, E. larimerensis, E. nieschulzi,* and *E. ninakohlyakimovae* are markedly motile, and merozoites within schizonts, except for those of *E. ninakohlyakimovae,* become active and leave the cell (Speer et al., 1970b). In such media, merozoites move for longer distances than normal, and those of *E. callospermophili* penetrate new host cells (Speer et al., 1970a). Speer et al. (1971) found that 1% bovine bile or sodium taurocholate alone increases the activity of *E. larimerensis* merozoites that are extracellular or within schizonts.

First generation merozoites penetrate new cells shortly after leaving schizonts. Those of *E. callospermophili* take 5–10 s to enter and then move toward the cell nucleus (Speer et al., 1970a). Mature second generation schizonts of *Eimeria brunetti* (Ryley and Wilson, 1972), *E. contorta* (Müller et al., 1973b), *Eimeria mivati (diminuta)* (Shirley et al., 1977), *E. meleagrimitis* (Augustine and Doran, 1978), *Eimeria necatrix* (Doran, 1971b), and *E. tenella* (Doran, 1970c; Strout and Ouellette, 1970; Itagaki et al., 1974; McDougald and Jeffers, 1976) are most frequently found in clusters (Figure 13). Consequently, it is assumed that first generation merozoites invade cells in the immediate vicinity of the schizont in which they develop. Single cells frequently are found to contain between 15 and 20 schizonts (Figure 14).

Strout and Ouellette (1970) observed that, after leaving a cell, second generation *E. tenella* merozoites occasionally form a complete circle either by the bending of one or by apparent fusion of two merozoites. These workers also found intracellular T-shaped merozoites capable of movement in either direction. This author frequently has seen bent and nearly circular *E. tenella* merozoites, but has thought them to be degenerate forms because internal structures other than a nucleus were not visible. Intracellular merozoites, especially those of the second generation, change shape and move slowly through the cytoplasm, giving the appearance of amoeboid movement.

E. tenella probably undergoes a third asexual generation in cell culture. Strout and Ouellette (1970) thought that small, mature schizonts with merozoites 10 μm long that consistently appeared at the time of gametogony in embryonic chick kidney cells belonged to the third generation. Doran (1970c) found schizonts 12–20 μm in diameter with 5 to 17 merozoites 6 and 7 days after placing sporozoites into nonembryonic chick kidney cell cultures. These schizonts, thought to belong to a third generation, were usually situated close to a cluster of immature or mature second generation schizonts (Figure 15).

Bedrnik (1967a,b, 1969b) inoculated cell cultures with second generation *E. tenella* merozoites obtained from the cecum of chickens. Merozoites entered cells within 5 min. Trophozoites appeared after 1–2 hr and mature third generation schizonts appeared at 14 hr (Bedrnik, 1969b) and 24 hr (Bedrnik, 1967a,b). Merozoites invaded cells in cultures prepared from trypsinized whole chick embryo rather unevenly, sometimes accumulating two or more within a cell, giving the impression that merozoites seek out certain cells (Bedrnik, 1969b).

Figures 11 and 12. 11. *E. bovis* schizont in which budding of merozoites has begun. Dark bodies (*DB*) at tips of some merozoites are probably developing paired organelles. A nucleus is visible at the base of merozoites in a portion of the schizont. Embryonic bovine tracheal cells; 9 days. ×1,100. Reprinted by permission from: Hammond, D. M., and Fayer, R. Cultivation of *Eimeria bovis* in three established cell lines and in bovine tracheal cell line cultures. *Journal of Parasitology* 54:559–568 (1968). 12. Large, ruptured *E. ninakohlyakimovae* schizont, with released merozoites and residual bodies of blastophores (*arrows*) surrounded by almost completely formed merozoites. Embryonic lamb tracheal cells; 16 days. ×200. Reprinted by permission from: Kelley, G. L., and Hammond, D. M. Development of *Eimeria ninakohlyakimovae* from sheep in cell cultures. *Journal of Protozoology* 17:340–349 (1970).

Schizonts developed in HeLa cells, but the merozoites were not released (Bedrnik, 1967a). In chick embryo cell cultures, merozoites were released and quickly invaded new cells (Bedrnik, 1969b). Additional small, mature schizonts appeared 24–48 hr later. On the basis of size and time of appearance, these were assumed to constitute a fourth generation. In other work, Bedrnik (1969a) mentioned a fifth generation. Only three generations are known to exist in the host.

Eimeria may multiply in cell culture by a process other than schizogony. Speer and Hammond (1971b) and Speer et al. (1973c) found second generation schizonts of *E. magna* and *E. larimerensis*, respectively, that contained only two or four merozoites. Figures 16 and 17 suggest that endodyogeny, or a process very close to it, takes place. Although they did not discuss the formation of such schizonts, Doran and Augustine (1977) and Augustine and Doran (1978) also found first generation *E. dispersa* schizonts and third generation *E. meleagrimitis* schizonts that contained only two merozoites.

Sexual development Reports concerning the development of *E. tenella* with sporozoites from the host as the inoculum are limited to only two cell types from the definitive host in addition to primary kidney cells. In primary fibroblasts, Shibalova (1970) obtained gametocytes, whereas Patton (1965), Shibalova (1968, 1969a,b), and Bedrnik (1969a) obtained only schizonts; in primary chorioallantoic membrane cells, Long (1969) obtained no development. With second generation merozoites as the inoculum, Bedrnik (1969b) obtained oocysts in primary cecal cells. Our repeated attempts (unpublished) to obtain oocysts by inoculating sporozoites into primary cultures of embryonic chick cecum have failed. However, the monolayers at 41°C invariably overgrew, and lasted only 4–5 days, although they contained only small numbers of cells (10–15% confluency) in a medium with a low (2–5%) serum content.

Gametocytes and oocysts of *E. tenella* appear 6 days after inoculation with sporozoites (Doran, 1970c) and 2–3 days after inoculation with second generation merozoites (Bedrnik, 1969b). McDougald and Jeffers (1976) found that sporozoites of an *E. tenella* strain that had been selected for precociousness developed to oocysts in 88–96 hr. Immature macrogametes were present as early as 64 hr. These authors concluded that the reason for the earlier appearance of oocysts with this strain is that gametogony is initiated by first generation merozoites.

Strains of *E. tenella* vary considerably in cell culture. Doran et al. (1974) compared development of the Beltsville, Weybridge, and Wisconsin strains and found that the percentage of infection at 4 hr after inoculation was higher with the Wisconsin strain than with the other two strains, but that oocyst production at 6, 7, and 8 days was much lower. The Beltsville strain produced the most oocysts. In addition to there being fewer Wisconsin strain oocysts, at 6 days there were also noticeably fewer macrogametes and, especially, second generation schizonts. We suggested that perhaps some time before 6 days, schizonts of the Wisconsin strain destroyed more of the monolayer of cells than other strains.

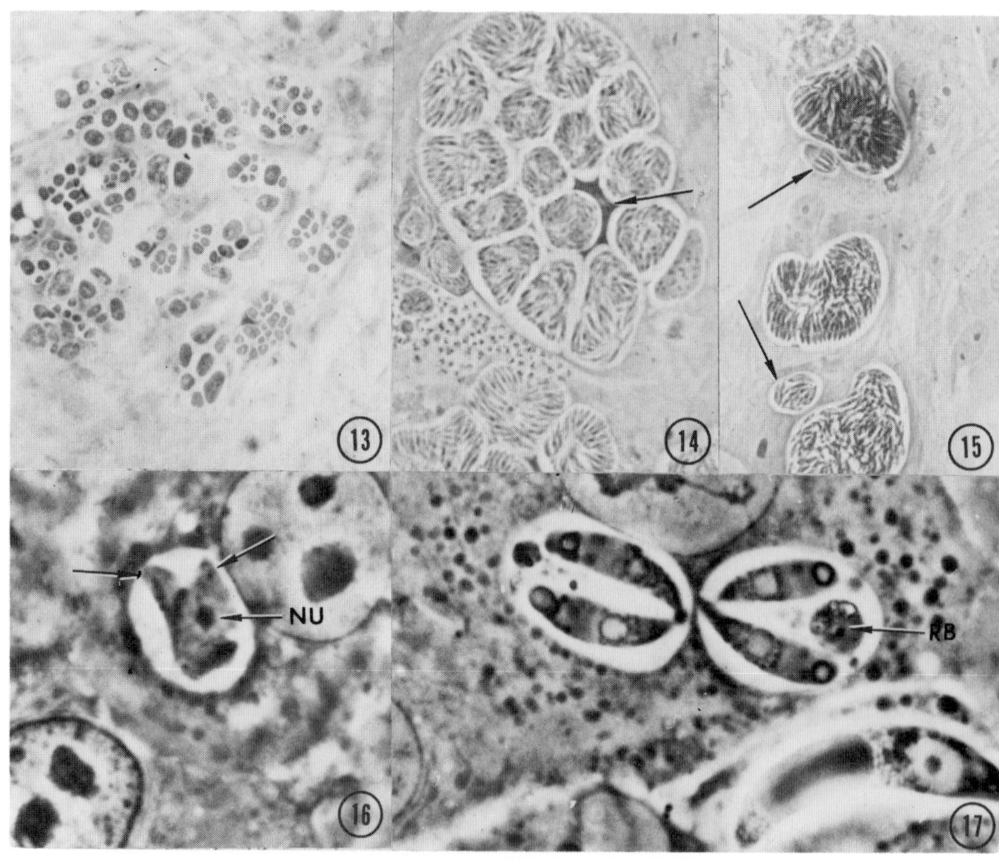

Figures 13–17. 13. Cluster of mature second generation *E. tenella* schizonts in chicken kidney cells; 4 days. × 250. (Doran, original.) 14. A single chicken kidney cell containing 15 mature *E. tenella* schizonts; 4 days. × 480. *Arrow* indicates host cell nucleus. (Doran, original.) 15. Mature third generation *E. tenella* schizonts (*arrows*) in chicken kidney cells; 6 days. × 520. (Doran, original.) 16. Early stage in the formation of two second generation *E. larimerensis* merozoites (*arrows*). *NU,* nucleus. Madin-Darby bovine kidney cells, phase-contrast microscopy; 32 hr. × 2,000. Reprinted by permission from: Speer, C. A., Hammond, D. M., and Elsner, Y. Y. Schizonts and immature gamonts of *Eimeria larimerensis* in cultured cells. *Proceedings of the Helminthological Society of Washington* 40:147–153 (1973). 17. Two second generation *E. larimerensis* schizonts, each with two merozoites and a residual body (*RB*). Madin-Darby bovine kidney cells, phase-contrast microscopy; 48 hr. × 2,000. Reprinted by permission from: Speer, C. A., Hammond, D. M., and Elsner, Y. Y. Schizonts and immature gamonts of *Eimeria larimerensis* in cultured cells. *Proceedings of the Helminthological Society of Washington* 40:147–153 (1973).

6

Figures 18–22. 18. *E. tenella* oocysts in chicken kidney cells. Phase-contrast microscopy; 7 days. ×940. (Doran, original). 19. *E. tenella* macrogametes (*MAC*) and microgametocytes (*MIC*) in chicken kidney cells; 6 days. × 1,000. (Doran, original.) 20. *E. tenella* microgametocytes in chicken kidney cells; 6 days. ×600. (Doran, original.) 21. Two chicken kidney cells completely filled with *E. tenella* oocysts, macrogametes, and microgametocytes; 6 days. × 380. *HCN, host cell nucleus. Reprinted by permission from: Doran, D. J. Eimeria tenella:* From sporozoites to oocysts in cell culture. *Proceedings of the Helminthological Society of Washington* 37:84–92 (1970). 22. *E. tenella* elongate macrogamete (*A*), elongate oocyst (*B*), and normal oocysts (*C*) in chicken kidney cells; 7 days. ×900. (Doran, original.)

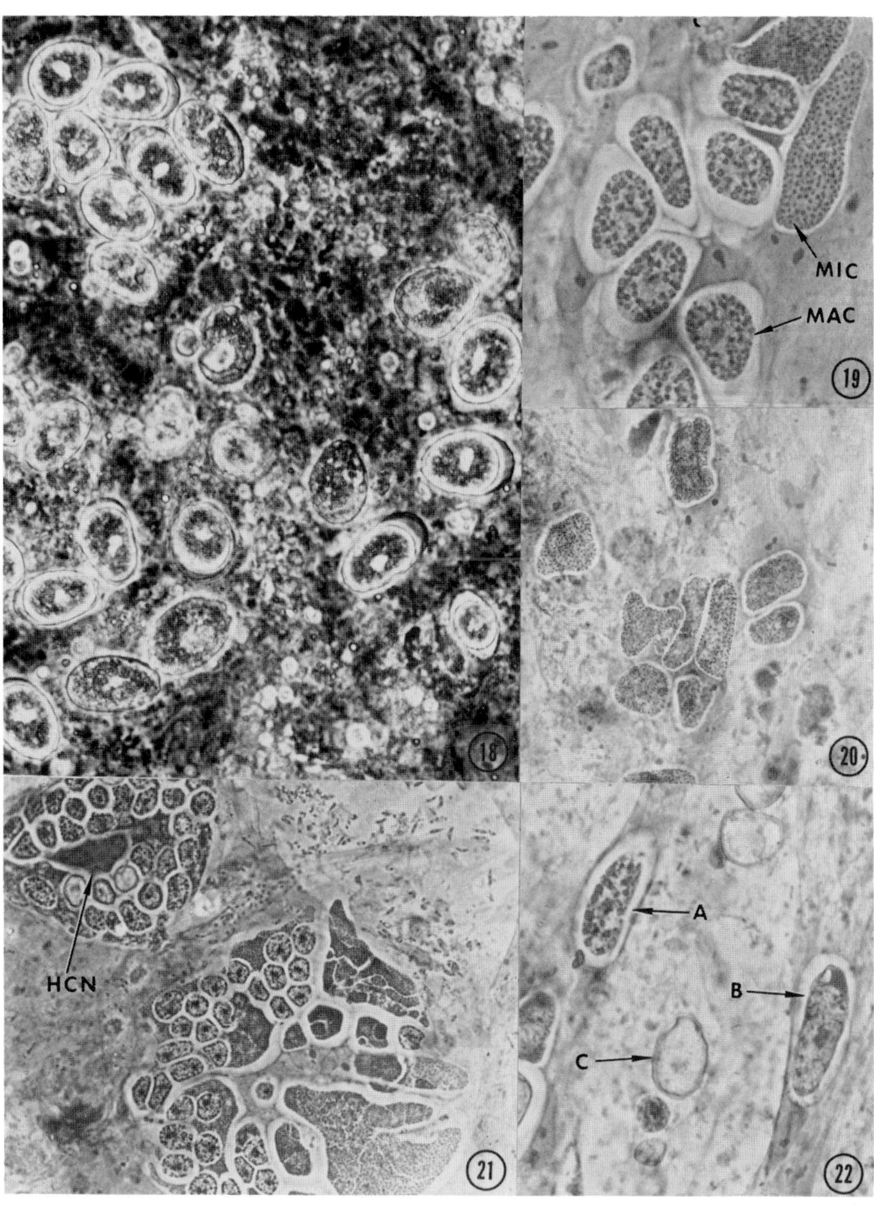

A single *E. tenella* gametocyte or oocyst is rarely seen. They usually occur in clusters (Figure 18) within islets or patches of cells (Bedrnik, 1967b, 1969b; Doran, 1970c; Strout and Ouellette, 1968, 1970). Before the formation of oocysts, macrogametes and microgametocytes are usually situated together (Figure 19). Within a cluster of gametocytes, the ratio of macrogametes to microgametocytes is usually more than 10:1. However, quite frequently each sexual stage is found by itself in groups of 5 to 18 (Figure 20) and occasionally one can scan several consecutive microscopic fields under low magnification (× 380) and observe only one of the sexual stages. Single cells are often found that are completely filled with gametocytes and oocysts (Figure 21). Macrogametes are usually ovoid or egg-shaped (Figure 19). Occasionally, they become greatly elongated, and oocysts develop from these which are nearly twice the length of a normal oocyst (Figure 22).

Speer and Hammond (1972b) showed for the first time that macrogametes can be motile. In a cinemicrographic study, they observed that *E. magna* macrogametes move through the cytoplasm of cells in an amoeboid manner. They also found that macrogametes leave clear pathways in the cytoplasm and occasionally pass through the cell membrane into an adjacent cell. Macrogametes usually contain a single nucleus. However, Doran and Augustine (1973) and Speer (1979), working with *E. tenella* and *E. magna*, respectively, found macrogametes that were multinucleate. These workers believed such forms were abnormalities caused by in vitro cultivation and have no other significance.

The number of *E. tenella* gametocytes developing in relation to the number of asexual schizonts is extremely small (Bedrnik, 1967b; Strout and Ouellette, 1968, 1970). Oocysts are produced only sporadically 1–2 days after inoculation with second generation merozoites (Bedrnik, 1967b). When cultures are established with parasitized tissue from the CAM, oocysts are produced, but the number is small in comparison with the number of gametocytes (Long, 1969). With sporozoites as the inoculum, oocysts have been consistently produced in this laboratory, when primary chicken kidney cells were used. However, the yields of oocysts have been very disappointing. The best yield yet obtained (Doran, unpublished) in a Leighton tube was 17,000. This yield seems good considering the small area of cells for development, but is poor considering the 50,000 sporozoites required to produce it.

Oocysts produced in culture, which are found intra- and extracellularly (Bedrnik, 1969b; Doran, 1971a), sporulate when removed and kept at room temperature and produce infection when fed to their natural host (Long, 1969; Doran, 1970c; McDougald and Jeffers, 1976; Augustine and Doran, 1978; Speer, 1979).

Pathological Effects The nucleus of a parasitized cell either decreases (Speer et al., 1970a) or increases in size (Fayer and Hammond, 1967; Clark and Hammond, 1969; Sampson et al., 1971; de Vos et al., 1972). When *E. alabamensis* develops intranuclearly, the nucleus becomes larger than if development is in the cytoplasm (Sampson et al., 1971). The nucleus may also become vacuolated and irregularly shaped (Speer et al., 1970a) and cells may become binucleate (Clark and Hammond, 1969; Speer et al., 1970a; Sampson et al., 1971) or multinucleate (Clark and Hammond, 1969). The nucleolus enlarges (Fayer and Hammond, 1967; Hammond and Fayer, 1968; Kelley and Hammond, 1970; de Vos et al., 1972) and the chromatin either remains unchanged (Clark and Hammond, 1969) or becomes finely granular or no longer visible (Fayer and Hammond, 1967; Hammond and Fayer, 1968; Kelley and Hammond, 1970).

Cytoplasm of a parasitized cell becomes granular (Fayer and Hammond, 1967; Hammond and Fayer, 1968), vacuolated and granular (Clark and Hammond, 1969; Kelley and Hammond, 1970; Speer et al., 1970a), or only vacuolated (Doran and Vetterling, 1967a; Bedrnik, 1969b; de Vos et al., 1972). Cells enlarge (Patton,1965; Doran and Vetterling, 1967a; Clark and Hammond, 1969; Speer et al., 1970a); cells infected with large *E. meleagrimitis* schizonts become from two to three times larger than normal (Doran and Vetterling, 1967a), and cells with *E. tenella* schizonts become from eight to ten times larger (Patton, 1965).

Cytopathological changes are most noticeable either soon after mature schizonts are formed or when merozoites are leaving a cell. Clark and Hammond (1969) observed that vacuolization and granulation are at first limited to a cell parasitized with a schizont, but then spread to neighboring nonparasitized cells and eventually affect the entire monolayer. They stated that the timing of degenerative changes in relation to the appearance of mature schizonts suggests that this stage may be toxic to cells.

Like *Toxoplasma, Eimeria* also destroys cells. The host cell membrane is destroyed as merozoites escape (Sampson et al., 1971; Kelley and Youssef, 1977). Cells containing free merozoites or schizonts with merozoites become detached from the cell layer (Doran and Vetterling, 1967b, 1968). Unlike *Toxoplasma* zoites, *Eimeria* merozoites seldom reinvade cells. Consequently they produce nothing like circular foci or plaques. However, eimerians do affect the cell monolayer. ''Breaks'' appear in the monolayer after inoculation with 1 million or more second generation merozoites (Bedrnik, 1969b). With sporozoites, the effect on the monolayer depends on the size of the inoculum. Doran (1971a) inoculated cell cultures with doses of up to 1 million sporozoites that had been freed of debris (oocyst and sporocyst walls). Retardation of cell growth increased with increased dosage. Cultures inoculated with 250,000 sporozoites or fewer were from 95%–100% confluent at 7 days, whereas those given more were progressively less confluent. In cultures given 1 million sporozoites, the cell monolayers, which were only 40% confluent at 7 days, contained a large amount of cellular debris and many vacuolated cells.

Cell Culture and Host Comparisons Most of the reports of development in cell culture compare it with development in the natural host. From these comparisons, it is evident that schizonts of nearly all of the species that develop in cell culture (*a*) are smaller and contain fewer merozoites than schizonts of the same generation in the host and (*b*) first appear in cell culture and in the natural host at about the same time. There are exceptions, however. Niciri-

Bontemps (1976) observed that *E. acervulina* schizonts of all four generations in culture were larger than those in the chicken. Speer et al. (1973a) found mature first generation *E. zuernii* schizonts that were several times larger than those from the host; the cultured schizonts contained 500 to 1,000 merozoites, and those from the host contained from 24 to 36. Consequently, these authors thought that Davis and Bowman (1957), who described development in the host, might have missed the first generation. With *E. bovis*, the time that first generation schizonts appear varies in different cell types; in most of the cells used, it was about the same as in the host, but in some it was several days earlier (Fayer and Hammond, 1967; Hammond and Fayer, 1968). First generation schizonts of *Eimeria ellipsoidalis* matured in 5–6 days in cell culture (Speer and Hammond, 1971a), whereas mature schizonts were found in calves 10–14 days after inoculation (Hammond et al., 1963). Because of this great time difference, the former workers believed that the schizonts in calves, which were described as belonging to an unknown generation, were those of a generation later than the first. Mature first generation schizonts of two ground squirrel species, *E. callospermophili* and *E. larimerensis*, also appear in culture in about one-half the time needed for development in the host (Speer and Hammond, 1970b; Speer et al., 1970a).

Oocysts of *E. acervulina* and *E. magna* from culture are similar in size, shape, and appearance to those obtained from chickens and rabbits, respectively (Niciri-Bontemps, 1976; Speer and Hammond, 1972a). Cultured *E. meleagrimitis* oocysts are smaller than those from the turkey (Augustine and Doran, 1978). In primary chick kidney cells, oocysts of *E. tenella* are mostly similar in size to those from chickens, but, as previously mentioned, a few are considerably longer (Doran, 1970c); in primary guinea fowl, partridge, pheasant, quail, and turkey kidney cells, oocysts are smaller than those from the chicken (Doran and Augustine, 1973).

Hammondia Sheffield et al. (1976) cultivated *Hammondia hammondi* in primary whole mouse embryo, primary monkey kidney, and cell line human diploid fibroblasts (WI-38). Sporozoites penetrated cells in a manner similar to that of *T. gondii.* Multiplication was exclusively by endodyogeny. Rosettes containing eight parasites were present on the 4th day after inoculation; after 5–8 days, parasitophorous vacuoles contained many parasites (Figures 23–28). The generation time was 24 hr. Unlike *T. gondii,* long-term cultivation was not successful; no parasites were found 4–6 weeks after inoculation. Subculture to a second culture (serial passage) was also unsuccessful. The authors stated that failure to maintain the parasite in culture may indicate differences between its growth requirements and those of *T. gondii.*

Hepatozoon Chao and Ball (1972) grew the invertebrate stages of *Hepatozoon* species and *Hepatozoon rarefaciens,* two reptilian haemogregarines, in cultures of Grace's cell line of *Aedes aegypti* (later found to be that of a moth, *Aedes eucalypti*). Uninucleate oocysts, re-

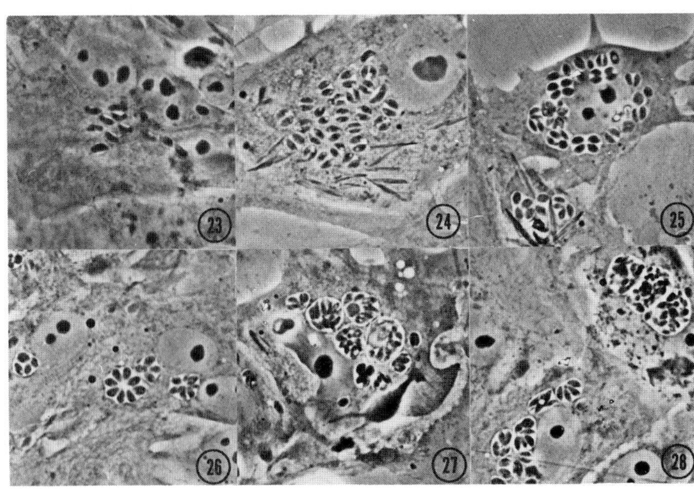

Figures 23–28. *H. hammondi* in monkey kidney cells. Phase-contrast microscopy. ×560. 23. Fixed one day after inoculation. 24. Two days. 25. Three days. 26. Four days. 27. Six days. 28. Eight days. Reprinted by permission from: Sheffield, H. G., Melton, M. L., and Neva, F. A. Development of *Hammondia hammondi* in cell cultures. *Proceedings of the Helminthological Society of Washington* 43:217–225 (1976).

moved from the hemocoels of *Culex tarsalis* 6 days after the mosquitos had bitten, developed to produce infective sporozoites. Ball and Chao (1973) reported that all vector stages of *H. rarefaciens* could be grown in a *Culex pipiens* cell line culture, starting with gametocytes that were obtained from blood of an infected gopher snake. Oocysts formed and sporogony was completed in 20 days at 23–24°C.

Hendricks and Fayer (1973) studied the vertebrate phase of development of a member of this genus. They obtained *Hepatozoon griseisciuri* oocysts from mites that had been given blood of an infected squirrel (*Sciurus carolinensis*). Sporozoites, excysted from sporocysts released from the oocysts, were placed in cultures of primary adult squirrel kidney (PSK) and primary neonatal squirrel heart, kidney, and spleen. Trophozoites were present in all four cell types at 72 hr. Mature schizonts, which developed only in PSK, were present after 5 days.

Isospora Successful attempts to obtain development of isosporans when sporozoites excysted from oocysts were inoculated into cell cultures are shown in Table 3. There are apparently only two reported failures, both of which were with *Isospora felis* and *Isospora rivolta.* Sheffield and Melton (1970a) obtained no development in primary cultures of monkey kidney cells, and Shibalova and Petrenko (1972) found no development in the seven cell types used for *Isospora bigemina.*

According to Fayer (1972b), Fayer and Mahrt (1972), and Fayer and Thompson (1974), isosporans multiply by endodyogeny. Organisms are arranged in pairs within one parasitophorous vacuole and frequently seem to be attached at one end (Figure 29). They are shorter than sporozoites, stain more intensely, and contain a larger nucleus. The report of Turner and Box (1970) concerning development of *Isospora lacazei* and / or *Isospora chloridis* by schizogony was surprising. Photomicrographs showing early development of the first generation indicate schizogony, but other micrographs, especially the one of infected canary cells at 4 days, show groups of organisms ("schizonts") that do not have the arrangement that normally follows schizogony. A few groups have four "pairs" of organisms in somewhat of a rosette, which strongly suggests that endodyogeny had taken place. Because Turner and Box (1970) obtained the oocysts for their work from the droppings of trapped English sparrows, the possibility that a few *Eimeria* sporozoites contaminated their inoculum cannot be ruled out.

Fayer and Mahrt (1972) did not find paired *Isospora canis* zoites until the 3rd day after inoculation. With *Isospora felis,* zoites were found 1 day after inoculation and were more abundant after 4–7 days (Fayer and Thompson, 1974). Shibalova and Petrenko (1972) did not state when they found paired *I. bigemina,* but did say that rosettes were present at 60 hr. They used an inoculum of 100,000 sporozoites and found that it was much too large. Cell monolayers were almost completely disintegrated by 111 hr.

Sheffield and Melton (1970b) fed brains containing *T. gondii* oocysts to cats and obtained *Isospora*-like oocysts in the feces. These oocysts measured approximately 10 μm. Sheffield and

Table 3. Development of *Isospora* in cell cultures when sporozoites were used as the inoculum

Species	Cells[a]	Type of culture[b]	Best development[c]	First report
I. bigemina (small form)	Chicken fibroblasts	P	Rosettes	Shibalova and Petrenko (1972)
	Human amnion	CL	Rosettes	Shibalova and Petrenko (1972)
	Marmoset kidney	CL	Rosettes	Shibalova and Petrenko (1972)
	Monkey kidney	CL	Rosettes	Shibalova and Petrenko (1972)
	Quail fibroblasts	P	Rosettes	Shibalova and Petrenko (1972)
	Rabbit kidney	CL	Rosettes	Shibalova and Petrenko (1972)
	Sheep kidney (E)	CL	Rosettes	Shibalova and Petrenko (1972)
I. canis	Bovine kidney (E)	CL	pz	Fayer and Mahrt (1972)
	Bovine trachea (E)	CL	pz	Fayer and Mahrt (1972)
	Canine intestine (E)	P	pz	Fayer and Mahrt (1972)
	Canine kidney (E)	ECL	pz	Fayer and Mahrt (1972)
I. felis	Bovine trachea (E)	CL	pz	Fayer and Thompson (1974)
	Canine kidney	P	pz	Fayer and Thompson (1974)
	Canine kidney (E)	ECL	pz	Fayer and Thompson (1974)
	Human amnion (E)	CL	pz	Fayer and Thompson (1974)
	Human cervical carcinoma	CL	pz	Fayer and Thompson (1974)
	Human esophagus (E)	CL	pz	Fayer and Thompson (1974)
	Human intestine (E)	CL	pz	Fayer and Thompson (1974)
	Human lung	CL	pz	Fayer and Thompson (1974)
I. lacazei and/or *I. chloridis*	Canary fibroblasts (E)	P	msch (2)	Turner and Box (1970)
	Chicken fibroblasts (E)	P	msch (2)	Turner and Box (1970)
I. rivolta	Bovine kidney (E)	CL	pz	Fayer (1972b)
	Canine kidney	CL	pz	Fayer (1972b)
	Canine kidney	ECL	pz	Fayer (1972b)

[a]E = embryonic.
[b]P = primary; CL = cell line; ECL = established cell line (Madin-Darby).
[c]Pz = paired zoites; msch (2) = mature schizonts of the second generation.

Melton excysted the sporozoites and inoculated them into primary monkey kidney cell cultures. They found paired organisms and groups of four to eight. Multiplication was by endodyogeny, and the growth pattern in culture was similar to that of *T. gondii.* Shibalova and Petrenko (1972) obtained oocysts by experimental infection of cats. Their *I. bigemina* oocysts measured 12–18 μm—slightly larger than those obtained by Sheffield and Melton (1970b), but still similar to *Toxoplasma* oocysts. Shibalova and Petrenko (1972) called attention to the similarity in development of *I. bigemina* with *Toxoplasma.* It is possible that they were working with *T. gondii* rather than *I. bigemina.*

Klossia Moltmann (1978) first reported development of a member of this genus. He cultivated explants of snail (*Copea nemoralis*) kidney infected with *Klossia helicina.* After 3

6

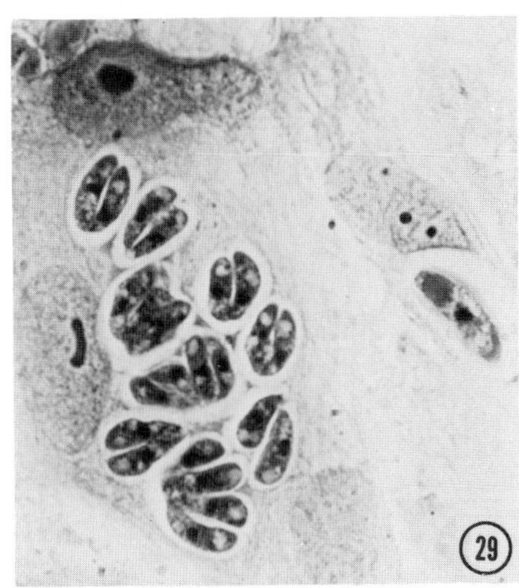

Figure 29. Paired *I. canis* organisms in embryonic canine kidney cells; a sporozoite is in an adjacent cell; 9 days. Reprinted by permission from: Fayer, R., and Mahrt, J. L. Development of *Isospora canis* (Protozoa; Sporozoa) in cell culture. *Zeitschrift für Parasitenkunde* 38:313–318 (1972).

days at room temperature, fibroblast- and epithelial-like cells had grown outward from the explant. Motile parasites, coming from the explant, infected the new cells and developed to schizonts and macrogamonts. Sporogony did not occur.

Sarcocystis Fayer (1970) was the first to cultivate a member of this genus in cell culture. *Sarcocystis* species zoites obtained from cysts found in leg or breast muscle of purple grackles (*Quiscalus quiscula*) developed to macrogametes (Figure 30), microgametocytes (Figures 31–33), and oocysts (Figure 34) in cell line embryonic bovine kidney (Fayer, 1970, 1972a; Vetterling et al., 1973) and cell line embryonic bovine trachea (Fayer, 1970, 1972a; Fayer and Thompson, 1975). Development proceeded to gametocytes only in established cell line Madin-Darby canine kidney cells and in primary embryonic chicken and turkey kidney cells (Fayer, 1970). Asexual development evidently did not take place in cell culture; banana-shaped zoites from muscle cysts transformed directly into macrogametes and microgametocytes. Unfortunately, all attempts to obtain sporulation of the oocysts produced in culture were unsuccessful (Fayer, 1972a). Further insight into the relationships between *Sarcocystis* and other coccidia might be obtained if we knew the mode of development of *Sarcocystis* sporozoites.

Dubremetz et al. (1975) inoculated primary cultures of embryonic sheep kidney cells with *Sarcocystis tenella* zoites obtained from the esophagus of sheep. They found oblong stages that corresponded to the macrogametes described by Fayer (1970, 1972a), but did not find microgametocytes or oocysts.

Toxoplasma

Cells Supporting Development *T. gondii,* which was first cultivated by Levaditi et al. (1929), is easier to cultivate in cell culture than any other intracellular protozoan parasite. It develops in a wide variety of cells from 11 warm-blooded animals. The following primary cultures support development: (*a*) human amnion, epithelium (skin, nasal, pharyngeal, and testicular), fibroblasts, kidney, leukocytes, liver, macrophages, myometrium, muscle, retinoblast cells, epithelial-like cells from tonsils, uterus; (*b*) cow kidney, skeletal muscle; (*c*) embryonic chicken blood monocytes, cartilage, freshly trypsinized whole embryo, fibroblasts, heart, intestine, liver, lung, muscle, nerve ganglion, spleen, subcutaneous tissue; (*d*) dog kidney; (*e*) guinea pig bone marrow, freshly trypsinized whole embryo, kidney, macrophages; (*f*) hamster kidney, fibroblasts, and macrophages; (*g*) monkey fibroblasts, heart, kidney, spleen, testicular epithelium; (*h*) mouse freshly trypsinized whole embryo, fibroblasts, heart, intestine, kidney, lung, macrophages, muscle; (*i*) rabbit corneal endothelium, kidney, liver, macrophages, spleen; (*j*) rat freshly trypsinized whole embryo, heart, kidney, leukocytes (buffy coat), macrophages, retinal cells (ganglion cells, bipolar nerve cells, rods, neuroglial cells, retinal pigment epithelium, retinal blood vessel); and (*k*) swine kidney. Cell line cultures supporting development are: (*a*) human amnion (FL), bone marrow (Detroit-6 and -98 of Berman and

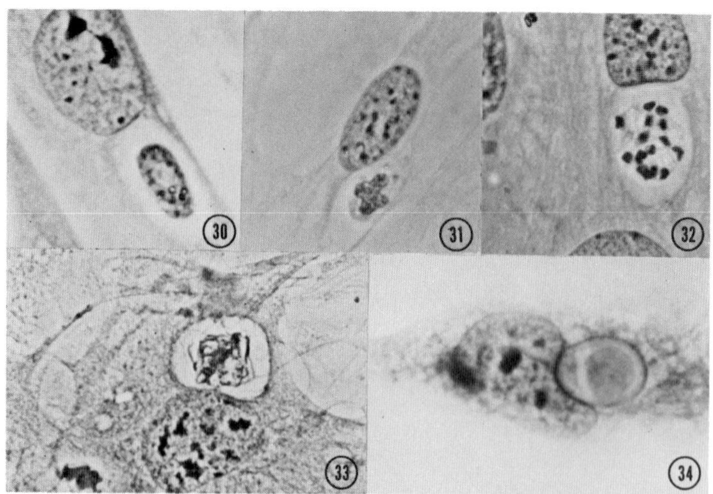

Stulberg), carcinoma of the cervix (HeLa), carcinoma of the larnyx (Hep 1 and 2 of Moore), carcinoma of the mouth (KB of Eagle), conjunctiva (Chang), fibroblasts (MRF), malignant foreskin (Detroit-89 of L), intestine (Henle), leukemial monocytes (J-111 of Osgood), liver (Chang), fetal lung (RU-1, WI-38), pleura (Detroit-116 of Berman and Stulberg), retinoblastoma (RB); (*b*) cow kidney (MDBK); (*c*) hamster epithelial cells (SV-40); (*d*) marmoset kidney (Vero); and (*e*) mouse whole embryo (Balb/3), fibroblasts (L-cell of Earle, L-929).

Two attempts to obtain development of *T. gondii* in cells from cold-blooded animals have been reported. Stuer (1972) obtained no development in primary gonadal and skin epithelial cells from rainbow trout and fathead minnows. He did not say whether the organisms entered the cells. In mosquito cell lines (*Aedes albopictes, A. aegypti, Aedes w-albus*), there was no penetration at 35°C (Buckley, 1973). Both of these studies were probably undertaken with the idea of determining whether there might be reservoirs or vectors for the parasite.

Mode and Rate of Multiplication Tachyzoites multiply very soon after entering cells. The earliest times reported are from 2–5 hr with the RH strain (Lund et al., 1961; Bommer, 1969) and within 3 hr with the Beverley strain (Kusunoki, 1977). Before multiplication begins, zoites become ovoid (Sheffield and Melton, 1968) or round (Schmidt-Hoensdorf and Holz, 1953) and increase in size. Time-lapsed photography studies in cell culture (Hansson and Sourander, 1965; Bommer, 1969; Bommer et al., 1969b) confirmed an earlier observation by Goldman et al. (1958) that multiplication is by endodyogeny. Sporozoites also multiply by endodyogeny (Kusunoki, 1977). However, it is possible that endodyogeny may not be the only method by which this species multiplies. Longitudinal fission and "some stages" of endodyogeny have been reported by Akinshina and Zassuchin (1965). From a cinemicrographic study of infected mouse macrophages, Meyer and de Oliveira Musacchio (1974) believed that multiplication is by binary fission. The best evidence for multiplication by a method other than endodyogeny was provided by Azab and Beverley (1974). They inoculated zoites of the Rh strain and zoites of a low virulence strain (Beverley) into embryonic lung cells and obtained some large, multinucleate forms highly suggestive of schizogony. The authors provided good evidence, including fluorescent-antibody staining, that they were dealing with a pure *T. gondii* infection. According to Bommer (1969) and Bommer et al. (1969b), in endodyogeny the two daughter zoites that are formed leave together or in quick succession by active penetration of the maternal membrane. The pair of new zoites remain connected to the remnants of the membrane for a long time, forming a V-shaped "butterfly" figure. Later, the membrane is sloughed off. Although he did not refer to multiplication as endodyogeny, Lund et al. (1961) observed that it takes 15 min for the nucleus to divide and 30–45 min for the parasite to divide. Division of the parasite has also been reported to take 60–90 min (Hansson and Sourander, 1965; Bommer, 1970).

Tachyzoites of the more virulent strains multiply faster than those of the less virulent strains. The generation time was 4.85 hr for the RH strain and 6.35 and 7.46 hr for the less virulent S-7 and M-7741 strains, respectively (Kaufman and Maloney, 1962). In an earlier

Figures 30–34. 30. *Sarcocystis* sp. macrogamete. Embryonic bovine tracheal (EBTr) cells; 24 hr. ×1,800. (Fayer, original.) 31. Young *Sarcocystis* sp. microgametocyte with a 4-lobed nucleus. EBTr cells; 24 hr. × 1,800. Reprinted by permission from: Fayer, R., and Thompson, D. E. Cytochemical and cytological observations on *Sarcocystis* sp. propagated in cell culture. *Journal of Parasitology* 61:466–475 (1975). 32. Young *Sarcocystis* sp. microgametocyte with 14 nuclei. EBTr cells; 30 hr. × 1,800. Reprinted by permission from: Fayer, R., and Thompson, D. E. Cytochemical and cytological observations on *Sarcocystis* sp. propagated in cell culture. *Journal of Parasitology* 61:466–475 (1975.) 33. *Sarcocystis* sp. microgametocyte with microgametes. EBTr cells; 30 hr. × 1,500. Reprinted by permission from: Fayer, R. Gametogony of *Sarcocystis* sp. in cell culture. *Science* 175:65–66 (1972). Copyright 1972 by the American Association for the Advancement of Science. 34. *Sarcocystis* sp. oocyst. EBTr cells; 30 hr. × 2,400. Reprinted by permission from: Fayer, R., and Thompson, D. E. Cytochemical and cytological observations on *Sarcocystis* sp. propagated in cell culture. *Journal of Parasitology* 61:466–475 (1975).

study, the generation time was 7.00 hr for the RH strain and 7.42 and 15.67 hr for the less virulent S-5 and 113-CE strains, respectively (Kaufman et al., 1958). Within a strain, tachyzoites reproduce faster than bradyzoites and sporozoites. With the Beverley strain, the generation times of tachyzoites, bradyzoites, and sporozoites were 10.0, 15.5, and 12.5 hr, respectively (Kusunoki, 1977). Jones et al. (1972) found that the average generation time of the RH strain tachyzoites was 5.5 days in mouse fibroblasts, 8 days in mouse macrophages, and 9 days in HeLa cells. It was later found that the division time was prolonged if the macrophages were obtained from immunized mice (Jones et al., 1975).

Aggregates and Cytopathological Effects An aggregate is formed by repeated synchronous division of zoites. In cell culture work, this aggregation, which is usually rosette-shaped, has also been referred to as a clone or terminal colony. Occasionally, rosettes do not develop. Hogan et al. (1961) obtained rosettes with the RH strain in six cell types, but with the MF strain the rosettes developed in only five of the cell types, and with the 113-CE strain in none. Lund et al. (1961) stated that the "characteristic" rosette-like appearances often disappear as multiplication proceeds but are reestablished by newly formed organisms in other cells. Rosettes have been found as early as 4 hr after inoculation (Balducci and Tyrrell, 1956). Kusunoki (1977) found rosettes of the Beverley strain 36 hr after inoculation of cultures with sporozoites. The cytoplasm of the cell is completely filled with growing RH strain rosettes 30 hr after inoculation (Lund et al., 1963b). There is apparently no limit to the number of rosettes, provided the host cell remains intact. A single cell can contain 200 organisms (Lund et al., 1961), and as many as 32 organisms can be in one rosette (Beverley, 1969). Single rosettes of the RH strain contain more organisms than those of the less virulent 113-CE strain (Kaufman et al., 1958).

An infected cell divides normally (Meyer and de Oliveira, 1942; Lund et al., 1961; Lund et al., 1963b) or it may become multinucleate (Akinshina, 1959). Before rupture of a cell, the nucleus becomes pyknotic (Akinshina, 1959; Bickford and Burnstein, 1966) and frequently is pushed toward the periphery of the cell by a developing rosette (Vischer and Suter, 1954; Akinshina, 1959). Mouse macrophage nuclei, but not those of HeLa cells and mouse fibroblasts, swell and the nucleolus increases in density (Jones et al., 1972). It has been reported that the cytoplasm becomes vacuolated and "frothy" and then disappears. However, the absence of any appreciable cytopathological changes before rupture of the cell membrane also has been reported (Chernin and Weller, 1957).

Release of Organisms, Reinvasion of Cells, and Histopathological Effects Schmidt-Hoensdorf and Holz (1953) suggested that zoites leave a cell by active screw-like movements, leaving a persistent intact cell membrane with multiple perforations. The more generally accepted view is that zoites are passively expelled after rupture of the cell membrane. Rupture of a cell does not depend on the presence of any fixed number of organisms, but occurs when the elastic capacity of the cell membrane is exhausted as a result of the multiplication of parasites (Lund et al., 1961).

Zoites invade adjoining cells within a few seconds after destroying the cell in which they develop (Lund et al., 1961). Cells remote from the original areas of infection remain uninfected (Hogan et al., 1961). Repetition of the developmental cycle results in degenerative changes in the layer of cells. Plaques appear as white, irregular (Foley and Remington 1969), or circular (Chaparas and Schlesinger, 1959; Akinshina and Zassuchin, 1969) areas against a pink background of viable cells (Foley and Remington, 1969). A plaque consists of a peripheral ring of infected cells surrounding a central mass of degenerated cells with pyknotic nuclei (Bickford and Burnstein, 1966). The central area is later sloughed off leaving a clear area (Akinshina, 1965; Bickford and Burnstein, 1966).

Degenerative changes in the cell layer coincide with the liberation of large numbers of parasites (Chernin and Weller, 1954a). More virulent strains multiply faster than those less virulent and, consequently, are released into the medium earlier. According to Lund et al. (1963a), organisms of the RH strain are released from HeLa cells beginning at 24 hr. The number in the medium reaches a maximum at 5 days, about the time that plaques or necrotic lesions appear (Lycke and Lund, 1964a; Foley and Remington, 1969). The speed with which cell layers are destroyed depends on the size of the inoculum and the virulence of the strain. According to Kaufman et al. (1959), the RH strain destroys cells more rapidly than the avirulent 113-CE strain, and cultures infected with avirulent strains sometimes are not destroyed but develop a "chronic" or prolonged infection.

According to Jacobs (1956), the number of organisms present at any one time in the medium is small compared to the number that can be recovered from exudates of animals. Also, there is a high proportion of nonviable forms. Under optimal growth conditions, yields in monolayer cultures of from 10 (Akinshina and Gracheva, 1964) to 18 (Shimizu, 1961) million organisms/ml of culture medium have been obtained. Treatment of the monolayer with trypsin can substantially increase the yield above that normally released (Csóka and Kulcsár, 1968b, 1970). Using HeLa cells in suspension culture, Valkoln and Cinatl (1978) obtained a yield of 20 million organisms/ml of culture medium. This was at 7 days after inoculation when all cells were completely destroyed.

Cysts (pseudocysts) have frequently been found in cultures. Work carried out in cell culture indicates that the thick, rigid cyst wall is derived either from only the host cell (Hansson and Sourander, 1965; Bommer, 1969), only the parasite (Matsubayashi and Akao, 1962), or both the host cell and the parasite (Matsubayashi and Akao, 1963, 1965). Cyst formation depends on rate of multiplication. Cysts are found earlier (Hogan, 1961) and more frequently (Hogan et al., 1961; Matsubayashi and Akao, 1963) with the slow-growing, less virulent strains than with the faster growing, more virulent strains. The MF and 113-CE strains produce cysts at 72 hr (Hogan et al., 1961), whereas the RH strain may require 7 days (Hoff et al., 1977) but more often requires 2–3 weeks (Hogan et al., 1961).

Isolation, Long-Term Maintenance, and Serial Passage The parasite can be isolated from macerated bits of tissue by inoculation into cell cultures (Jacobs et al., 1954; Osaki et al., 1964;

6

Akinshina, 1965; Chang et al., 1972; Hayes et al., 1973). However, Abbas (1967b) found that cell culture was about 300 times less sensitive than mice for isolation from acutely infected tissue and was of no value for isolation from subacutely or chronically infected vertebrate tissue.

Long-term maintenance in cultures without serial passage has been reported by numerous workers. At 26–28°C, zoites survive 44 days with no medium change and 80 days with a change every 21 days (Bell, 1961). At 4°C with no medium change, zoites are viable after 90 days and, when glycerin is added to the medium before refrigeration, after 120 days (Stewart and Feldman, 1965). However, survival time is usually longer when cultures are kept close to 37°C. With a medium change every 3 or 4 days, viable zoites can be maintained for as long as 33 months (Schuhová, 1960). A chronic or prolonged infection occurs (Schuhová, 1957; Bickford and Burnstein, 1966). A few islets of cells remain after initial cell destruction (Schuhová, 1960; Bickford and Burnstein, 1966). These grow and form a new sheet of cells after 17 days (Schuhová, 1960) or from 26–30 days (Bickford and Burnstein, 1966). Cell destruction again takes place, but is not as severe as before (Bickford and Burnstein, 1966). Eventually, an equilibrium is established between cell destruction and cell growth (Schuhová, 1960; Lund et al., 1963a; Bickford and Burnstein, 1966). Lund et al. (1963a) believed that this equilibrium was established only when less than one-half of the cells were initially infected.

The parasite has been serially propagated in cell culture for long intervals of time. This has been carried out in hanging drop (Carrel) cultures (Meyer and de Oliveira, 1943, 1945), roller tubes (Chernin and Weller, 1954a,b, 1957), and stationary tubes or other culture vessels (Sabin and Olitsky, 1937; Cook and Jacobs, 1958; Lycke and Lund, 1964b; Akinshina, 1965; Stewart and Feldman, 1965; Csóka and Kulcsár, 1968b). Meyer and de Oliveira (1943, 1945) incubated freshly inoculated cultures at 38–39°C for 24 hr and then kept them at ambient temperatures (close to 28°C). After the cultures had become adapted to ambient temperatures, they were transferred and new tissue was added every 7–13 days; viable parasites remained for 3 yr. Cultures have also been kept at 22°C (Cook and Jacobs, 1958; Akinshina, 1965) and 4°C (Akinshina, 1964, 1965). Storage at these lower temperatures prolongs the interval between transfers. At 22°C, only three transfers were made during 3 months (Cook and Jacobs, 1958); at 4°C, transfers were made every 3 or 4 weeks (Akinshina, 1964). Csóka and Kulcsár (1968b) used two temperatures between transfers of the RH strain. They incubated newly inoculated cultures at 37°C for 3 days and then at 4°C for 2 days. Forty-five transfers were made during the 8 months that the zoites survived.

Factors Other Than Strain Differences Influencing Penetration, Survival, and Development

Parasitological

Age of the Parasite and Time When Obtained from the Host Doran and Vetterling (1969) examined sporozoites obtained from oocysts of *E. meleagrimitis* of different ages and showed that with older parasites penetration and development was poorer. In a similar study with previously frozen sporozoites, development of *E. adenoeides* and *E. tenella* was greater at 48 hr with sporozoites frozen when 3 or 4 weeks old than with sporozoites frozen when 36 and 54 weeks old (Doran, 1970a). The use of sporozoites that have been frozen and stored in liquid nitrogen vapor has advantages for many studies. Methods for storage have been described (Doran, 1969a,b).

Second generation *E. tenella* merozoites taken from the cecal mucosa of chickens 6 days after infection always produce sexual stages when inoculated into cell culture, whereas those taken on the 5th day rarely do so (Bedrnik, 1969a,b, 1970). Merozoites of *E. magna* taken from the rabbit 3.0 days after infection undergo only further asexual development in culture; those taken at 3.5–4.0 days produce a few sexual stages, but those taken later, at up to 5.5 days, produce progressively more sexual stages (Speer et al., 1973b). Oocysts of *H. rarefaciens,* taken from hemocoels of *C. tarsalis* 6 days after the mosquitos have bitten, develop to produce infective sporozoites; 3-day-old oocysts reach the sporocyst stage but do not develop further (Chao and Ball, 1972). Zoites of *T. gondii* taken from peritoneal exudate earlier than 4 days after inoculation produce a heavier infection in culture than those taken from exudate at 4 days or later (Shimizu, 1961; Hirai et al., 1966). Cultures are frequently rinsed several times after inoculation; several investigators (Lund et al., 1961; Shimizu, 1961; Hirai et al., 1966) state that the reason for rinsing is to remove toxins.

Size of the Inoculum Strout et al. (1969b) inoculated cultures with different quantities of *E. tenella* sporozoites ranging from 10,000 to 1 million and compared infection and development 96 hr later. Their data represent only what was present at 96 hr and give no indication of the relationship between dosage and the number of parasites entering cells. However, they did find more asexual development in cultures inoculated with 10,000 than in those inoculated with 1 million sporozoites. Doran (1971a) compared infection at 4 hr and oocyst production at 6 and 7 days in cultures inoculated with 25,000 to 1 million *E. tenella* sporozoites freed of debris (oocyst and sporocyst walls). The number of intracellular sporozoites at 4 hr is roughly proportional to the size of the inoculum up to 500,000; with 750,000 and 1 million sporozoites, the number is similar to that with 500,000. The number of oocysts produced was greatest with an inoculum of 100,000 sporozoites, but rapidly declined with inocula of 250,000 and more.

The size of the *T. gondii* inoculum is usually expressed as a dilution of pure peritoneal exudate and not as an actual parasite count. Regardless of method, the size of the inoculum influences the number of parasites that enter cells. The number entering is proportional to the size of the inoculum (Holz and Albrecht, 1953; Lycke and Lund, 1964a; Norrby, 1970b). The inoculum of the RH strain for maximal invasion is one that results in five parasites per cell (Hirai et al., 1966). Size of the inoculum also influences the amount (Hogan et al., 1961) and time (Balducci and Tyrrell, 1956) of initial cell destruction, the plaque size (Akinshina and Zasuchina, 1966), the time when large numbers of organisms appear in the culture media

6

(Balducci and Tyrrell, 1956; Chernin and Weller, 1957), and the time when degeneration of the cell layer is complete (Cook and Jacobs, 1958).

Treatment before Inoculation Freeze thawing and cleaning sporozoite suspensions of oocyst and sporocyst walls by passage over glass beads affect the number of *Eimeria* sporozoites that enter cells. Frozen and thawed sporozoites are less invasive than fresh sporozoites (Doran, 1969a, 1970a). Sporozoites, cleaned by passage over glass beads, are less invasive than untreated sporozoites (Doran, 1970b, 1971a). The difference between treated and untreated sporozoites increases with dosage (Figure 35*A*). Freeze thawing and passage over beads seems to alter the sporozoite and reduce its ability to penetrate cells or survive for 4 or 5 hr in a cell. However, it is more probable that some of the sporozoites in the inoculum that are morphologically viable are actually dead. Passage over glass beads affects the survival of parasites between 5 and 48 hr (Figure 35*B*). Freeze thawing of sporozoites has no effect on development; the percentages of development at 48 hr are similar for frozen and unfrozen sporozoites (Doran, 1969a, 1970a). However, cleaning by passage over beads has a decided effect on development. The number of schizonts is directly proportional to the size of the inoculum in cultures inoculated with cleaned suspensions but not in cultures inoculated with uncleaned suspensions (Figure 35*C*). The decrease in relative yield with uncleaned inocula of more than 250,000 sporozoites is probably caused by the effect of debris on the monolayer. The yield of *E. tenella* oocysts at 7 days is three times greater with cleaned than with uncleaned suspensions (Doran, 1971a).

Results with *E. auburnensis* suggest that pretreatment of oocysts with CO_2 before grinding to release sporocysts may adversely affect development (Clark and Hammond, 1969).

Irradiation of *E. tenella* oocysts with gamma rays (7,000 to 21,000 rad) does not kill sporozoites or diminish their ability to penetrate cells (Klimes et al., 1972). However, sporozoites from oocysts irradiated with 10–15 rad develop only to mature first generation schizonts; those from nonirradiated oocysts develop to oocysts (Sokolic et al., 1973).

Environmental

Culture Chamber *T. gondii* zoites inoculated into roller tubes (round tubes in a revolving drum) infect fewer cells than those inoculated into stationary tube cultures (Kaufman et al., 1958, 1959). With the RH strain, the percentage of infected cells in roller tubes is about one-fourth that in stationary cultures; the comparable figure with strain 113-CE is about one-third (Kaufman et al., 1959). These investigators indicated that the greater cell penetration in the stationary cultures is probably the result of more prolonged contact between the organisms and cells. As measured by lysis time of infected cells instead of direct counts, differences between infection in roller and stationary tubes are not detectable (Cook and Jacobs, 1958).

Temperature Patton (1965) found mature *E. tenella* schizonts at 41°C but not at 37°C. In the only detailed study of the effects of different temperatures on development, Strout et al.

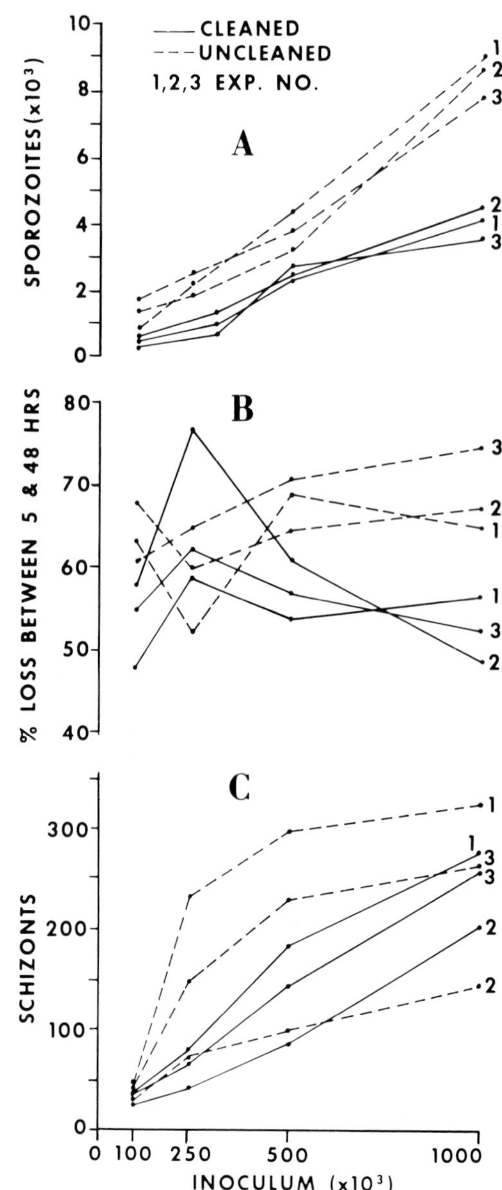

Figure 35. Effect of debris (oocyst and sporocyst walls) on behavior of *E. adenoeides* in embryonic bovine kidney cells. Reprinted by permission from: Doran, D. J. Survival and development of *Eimeria adenoeides* in cell cultures inoculated with sporozoites from cleaned and uncleaned suspensions. *Proceedings of the Helminthological Society of Washington* 37:45–48 (1970). *A*, number of intracellular sporozoites 5 hr after inoculation; *B*, percentage loss of intracellular stages between 5 and 48 hr; *C*, number of schizonts at 48 hr.

6

(1969a) observed that *E. tenella* underwent schizogony at temperatures from 35–43°C, that mature schizonts were present at 36–42°C, and that the optimum temperature for schizogony was 41°C. Itagaki et al. (1974) found that although more *E. tenella* sporozoites penetrated cells at 37°C than at 40°C, development was better at 41°C than at 37°C. They also found that *E. brunetti* did not develop at 37°C. However, *E. acervulina* completes four asexual generations at 37°C (Naciri-Bontemps, 1976). Mature *I. lacazei* and/or *I. chloridis* schizonts develop at 37°, 41°, and 43°C, but fewer developing parasites are found in cells grown at 37°C than in those grown at higher temperatures (Turner and Box, 1970).

The optimal temperature for multiplication of *T. gondii* is 37°C (Akinshina, 1965). Plaques are formed at 42°C, but they are fewer at 42°C than at 37°C (Akinshina and Zasuchina, 1966, 1967). The rate of multiplication decreases as the temperature is lowered (Cook and Jacobs, 1958; Kaufman, 1961; Maloney and Kaufman, 1964). The number of zoites per cell after 24 hr at 38°, 29°, and 4°C was 8.80, 2.23, and 1.44, respectively (Kaufman, 1961). Balducci and Tyrrell (1956) placed heavily inoculated cultures at 4°C and found that zoites were gradually released into the medium and could be detected there for 4 weeks. However, Cook and Jacobs (1958) kept cultures at 4°C for 30 days and found no degeneration caused by proliferation of organisms. They also found that the culture material was noninfective to mice by intraperitoneal inoculation.

Debris (Oocysts and Sporocyst walls) and Toxins After inoculation of cell cultures with either *Eimeria* or *T. gondii,* most investigators rinse their cells one or more times and then add fresh medium. With *Eimeria,* it is assumed that rinsing is to remove debris. Patton (1965) stated that debris is toxic to cells. According to Fayer and Hammond (1967), debris sometimes hinders observations, but does not seem to have any appreciable toxic effect on cells. The latter workers prepared an extract of *E. bovis* oocysts (debris removed) and found that it killed cells within 1 hr. The lack of proportionality between size of uncleaned inocula and numbers of schizonts shown in Figure 35*c* can hardly be due to toxins carried over with the oocysts. Because the uncleaned suspensions were thoroughly washed before inoculation, the lack of proportionality is probably entirely due to debris.

As previously mentioned, cultures infected with *Toxoplasma* are rinsed to remove toxins carried over with the exudate. However, opinions differ as to whether a "toxotoxin" produces an effect in culture. Some have claimed that there is an effect (Lund et al., 1961; Shimizu, 1961; Lycke and Lund, 1964b; Hirai et al., 1966); others have said that there is no evidence that a toxin causes degenerative changes in culture (Chernin and Weller, 1954a, 1957; Balducci and Tyrrell, 1956; Cook and Jacobs, 1958; Hansson and Sourander, 1965).

Medium and Medium Changes According to Turner and Box (1970), more *I. lacazei* and/or *I. chloridis* sporozoites penetrate cells when serum is omitted from the inoculation medium. There is no good comparative study to determine whether the same is true of any eimerian species. However, the type of serum may be of importance for both penetration and survival of *Eimeria* species. More *E. tenella* sporozoites penetrate cells when suspended in a medium composed of 90% Hanks' balanced salt solution (HBSS), 5% lactalbumin hydrolysate (LAH; 2.5% solution in HBSS), and 5% fetal calf serum rather than one composed of Medium 199 and 5% chicken serum (Doran, 1971a). The use of sera from chickens immune to *E. tenella* and *E. maxima* reduces the number of intracellular stages between 24 and 48 hr in primary chicken kidney cells, but has little effect on established infections in primary chorioallantoic membrane cells (Long and Rose, 1972). Serum may also affect development. Itagaki et al. (1974) reported that development of *E. brunetti* is reduced by the addition of serum. However, with second generation merozoites as the inoculum, Bedrnik (1970) found no differences in development of mature third generation schizonts in cells maintained in phosphate-buffered saline (PBS), PBS and 10% calf serum, or PBS and various components of Medium 199. Strout (1975), in abstract only (with no details), mentioned using a chemically defined serum-free medium for cultivating *E. tenella.* He reported that development paralleled that in media with sera and extracts, but quantitatively exceeded that in Eagle's Basal Medium and Eagle's Minimum Essential Medium with serum. However, with the chemically defined medium, oocysts were not found until 7 days after inoculation, 1 day later than usual (Doran, 1970c, 1971a).

Numerous media (with serum) have been used for cultivating *E. tenella.* Some are better than others and, with sporozoites as the inoculum, the differences are apparent only during the latter part of the life cycle. Patton (1965) found no differences in development of first generation schizonts in several different media. However, after 4–5 days, differences become pronounced (Doran, 1971a). In a medium (HEBM) containing Eagle's Basal Medium (with HBSS) and either 5% or 10% fetal calf serum, second generation schizonts are scarce, macrogametes contain either no wall-forming bodies or a few small ones, and the number of oocysts are small and incompletely formed. However, in a medium containing 90% HBSS, 5%LAH, and 5% fetal calf serum, there are more sexual stages, macrogametes contain larger granules, and the oocysts are more numerous and have completely formed walls. These differences are probably due to differences in cell maintenance. Monolayers of cultures maintained in both media are in good condition at 4 days, but later monolayers in HEBM and calf serum are not in as good condition and contain few of the elevated areas or patches of cells essential for development of sexual stages.

In most of the work with *E. tenella* in this laboratory, primary chick kidney cells were grown in a medium composed of 80% HBSS, 10% LAH, and 10% fetal calf serum (Medium A) and maintained after inoculation in 80% HBSS, 5% LAH, and 5% fetal calf serum (Medium B). Medium A has not been used for maintenance after inoculation because growth is too fast and cell layers overgrow and roll off the cover slip by 7 days (Doran 1970c). As previously mentioned, the yield of oocysts in this system of Medium A before inoculation and Medium B after inoculation (A/B) has been very disappointing. In hopes of increasing the yield, 15 other systems were tested (Doran and Augustine, 1976). Oocyst yields from 26–57%

6

greater than those in A/B were obtained when cells were maintained before inoculation in Medium A and after inoculation in a medium composed of (*a*) 90% of either Eagle's Basal Medium, Eagle's Minimum Essential Medium, or Medium 199 with HBSS (HEBM, HMEM, or H199); (*b*) 5% LAH; and (*c*) 5% fetal calf serum. In systems involving either HEBM, HMEM, or H199, oocysts yield was lowest when LAH was absent both before and after inoculation. Yield increased when LAH was present after inoculation, and exceeded that with A/B when LAH was present both before and after inoculation. Elevated areas or patches of cells believed necessary for gametogony were abundant in each of the seven systems supporting the higher yields; they were sparse in systems involving only HEBM and HMEM. Lactalbumin hydrolysate, which was present in six of the seven systems producing the higher yields, may provide a better physical environment by supporting greater retention of elevated areas or patches of cells. The reason for the three yields higher than in A/B is probably because LAH contains one or more growth factors which, in combination with those in HEBM, HMEM, and H199, provide a better nutritional environment than that provided in system A/B.

Ťos-Luty et al. (1971) compared multiplication of *T. gondii* in media containing different sera. Multiplication was higher with human serum negative in the complement-fixation test (CFT) for toxoplasmosis than with rat and horse sera negative in the CFT or with human, rabbit, and rat sera positive in the CFT. Lund et al. (1963a) found that more zoites enter cells when the medium contains no serum. With 1% human serum, about one-half as many zoites were intracellular at 3 hr than when there is no serum; with 10% and 20% human serum, the number decreased further. These workers also found that with all concentrations of human serum there is a 2-day delay in the liberation of parasites into the medium. Somewhat different results were obtained by Shimizu (1961, 1963); the rate of multiplication increased rather than decreased when calf serum was added (Shimizu, 1961, 1963).

Penetration and development of *Eimeria* at different pH values have not be compared. However, most workers have generally tended to keep the inoculation and maintenance media at about pH 7.2. A pH of 7.4 is beneficial for development of *I. lacazei* and/or *I. chloridis* (Turner and Box, 1970). Penetration by *T. gondii* zoites is good when the pH is 6.8–8.0; with higher and lower values, cells are inferior and multiplication is poor (Shimizu, 1963). The optimal pH is reported to be either 7.0 (Shimizu, 1961, 1963) or 7.5 (Akinshina and Zasuchina, 1967). Upon gassing their cultures and using a controlled environmental culture system, Dvorak and Howe (1977) found that there was no relation between penetration of cells and pH in the range of 6.9–7.8 and that increasing the CO_2 from 0.5–3.7 mM and using bicarbonate ion concentrations lower or higher than 36.25 mM inhibit penetration.

More *E. tenella* oocysts develop if a total volume of 10 ml rather than 2 ml is used for maintaining cells in Leighton tubes (Doran, 1971a). With 2 ml, medium changes are necessary on days 4, 5, and 6 after inoculation because of a drop in pH from 7.0–7.2 to about 6.4. If 10 ml is used, the pH changes very little, and a medium change is unnecessary. The higher oocyst yield with 10 ml is probably a result of better regulation of pH as well as a result of retention of second generation merozoites free in the medium on days 4–6 that would otherwise be removed from culture by a medium change. Routine medium changes are necessary in order to obtain high yields of *T. gondii* (Chernin and Weller, 1954b, 1957; Shimizu, 1961, 1963). Changes every 3 or 4 days gave yields as high as 2.7 million organisms/ml of culture medium; with no medium change, about one-fifth that number is obtained (Chernin and Weller, 1954b, 1957).

Cellular

Cell Type Some cell types are better than others for cultivating *Eimeria;* differences in survival and development have been found. Between 5 and 48 hr, the loss of *E. meleagrimitis* in primary bovine kidney cells is 41%, whereas it is about 81% in primary turkey intestinal cells (Doran and Vetterling, 1968). With eimerians that do not complete their life cycle from a sporozoite, the best cell type might be considered the one in which the life cycle proceeds the farthest. As previously mentioned, this is usually one from the definitive host. Some species develop to mature first generation schizonts in several different cell types, but differ in the time of appearance of schizonts (Doran and Vetterling, 1967a; Hammond and Fayer, 1968; Clark and Hammond, 1969; Kelley and Hammond, 1970; Doran, 1971a; de Vos and Hammond, 1971; Ogimoto et al., 1976) and percentage development (Hammond and Fayer, 1968; Clark and Hammond, 1969; Speer and Hammond, 1970b; Speer et al., 1970a; Sampson et al., 1971). With *E. tenella,* the best cell type is the one in which the most oocysts are produced. More are found in primary cultures of nonembryonic chicken kidney cells (PCK) than in kidney cells from the embryo (Doran, 1970c; Itagaki et al., 1974). Oocyst production is greater in PCK than in primary kidney cells from the guinea fowl, partridge, pheasant, quail, and turkey (Doran and Augustine, 1973).

Cell type also influences development of *Besnoitia, Isospora,* and *Sarcocystis.* Although confluency of the cells at the time of inoculation was not considered, the data of Fayer et al. (1969) suggested that embryonic bovine spleen and tracheal cells are more favorable than an established cell line of MDBK for development of *B. jellisoni.* Although they did not present quantitative data, Akinshina and Doby (1969a) reported that 1) fibroblast-like cells are better than epithelial-like cells for development of *B. jellisoni,* 2) fibroblast-like cells from swine and rabbit kidney are better than embryonic chick fibroblasts, and 3) chick embryo fibroblasts are better than epithelial-like cells from swine and rabbit kidney. According to Turner and Box (1970), *I. lacazei* and/or *I. chloridis* progress to mature second generation schizonts in primary cultures of both canary and chicken kidney, but many more develop in canary cells. Fayer (1970) found that *Sarcocystis* species developed in cell line bovine kidney and trachea, established cell line MDCK, and primary chicken and turkey kidney cells, but did not penetrate embryonic chicken muscle cells. This finding is surprising considering that *Sarcocystis* is a parasite of muscle tissue.

6

Penetration and development of *T. gondii* also depend on cell type. There are differences in the rates of penetration (Vischer and Suter, 1954; Csóka and Kulcsár, 1968a), rates of multiplication (Vischer and Suter, 1954; Hogan et al., 1960; Lund et al., 1963a; Matsubayashi and Akao, 1963), and timing of cytopathogenic effects (Cook and Jacobs, 1958; Hogan et al., 1960; Akinshina and Zassuchin, 1965; Hansson and Sourander, 1965). Cook and Jacobs (1958) found that lysis was slower in conjunctival epithelium (Chang) and human intestine (Henle) than in primary monkey kidney cells. The differences in timing, which were found not to be due to differences in rates of penetration, were thought to be caused either by metabolic factors, by physical differences (e.g., continuity of cell sheets), or by the resistance of fibroblast-like cells that continued to proliferate after epithelial-like cells were destroyed. In the only comparative study of infection in a single cell type derived from different hosts, Vischer and Suter (1954) observed slower intracellular development in macrophages from rabbits, guinea pigs, and rats than in those from mice. The authors stated that the rate of development was correlated with the susceptibility of the four species of animals to infection. Human cervical carcinoma (HeLa) cells are the most frequently used. However, in seven comparisons of HeLa with other cell types (Hogan et al., 1960; Hogan, 1961; Lund et al., 1963a; Matsubayashi and Akao, 1963; Csóka and Kulcsár, 1968a; Bommer, 1969; Jones et al., 1972), HeLa was found to be the inferior cell type in three (Hogan, 1961; Lund et al., 1963a; Csóka and Kulcsár, 1968a). The best indication that cell preference is not a matter of differences in cell maintenance or some other cultural factor is in the work of Csóka and Kulcsár (1968a). They grew HeLa cells and primary human amnion cells together and found that far more parasites entered the amnion cells.

Passage Number The number of times cells are serially passed influences the development of several species of *Eimeria*. Neither *E. meleagrimitis* nor *E. necatrix* develops in primary cultures or in the first and second passages of embryonic bovine kidney; development starts in cells of the third (*E. necatrix*) or fourth *(E. meleagrimitis)* passages and increases through the 20th passage (Doran and Vetterling, 1967a). The opposite seems to be true for several nonavian species in established cell line MDBK. *E. larimerensis* develops to mature second generation schizonts in the 112th to 114th passages (Speer et al., 1973c) but only to immature second generation schizonts in the 180th passage (Speer and Hammond, 1970b). In the 123rd passage, *E. zuernii* develops to mature first generation schizonts; in the 224th passage it proceeds only to immature first generation schizonts (Speer et al., 1973a). *E. canadensis* develops to mature first generation schizonts in the 113th passage; in the 220th to 230th passages, there is no development (Müller et al., 1973a).

In the only report concerning passage number with *T. gondii,* Dvorak and Howe (1977) found that more zoites penetrated embryonic bovine skeletal muscle cells when the cultures were primary rather than secondary.

Age and Number of Cells Opinions differ concerning the effect of age of cells (interval in culture before inoculation) on multiplication of *T. gondii.* Human amnion cells 7 and 41 days old are equally susceptible (Csóka and Kulcsár, 1968a), and there is no difference in timing of lysis in monkey kidney cells 6, 14, 21, and 28 days old (Cook and Jacobs, 1958). With HeLa cells, the number of organisms produced 8 days after inoculation in cultures 2 days old is about four times that in those 5 days old (Shimizu, 1961).

The number of cells in a culture influences development of both *E. tenella* and *T. gondii.* Maintenance of cultures of chicken kidney cells long enough for *E. tenella* oocysts to develop requires the use of a cell suspension that will result in less than 35% confluency of cells at the time of inoculation with sporozoites (Doran, 1971a). In cultures seeded with 25,000 and 50,000 cells, 1 week elapsed before one-half of the cells infected with *T. gondii* lysed; in cultures with 5,000 cells, lysis occurred in 24 hr (Lund et al., 1963a).

Trypsinization of Tissue The degree to which kidney tissue is trypsinized before establishment of primary cultures is important in the development of *E. tenella.* Doran (1970c) found that gametocytes and oocysts did not develop if tissue was trypsinized to a point where the cell suspension from which the cultures were prepared contained mostly single cells and few aggregates. Two days after inoculation of such cultures with sporozoites, very few patches of epithelial-like cells are seen; most cells are widely dispersed and growing in a fibroblast-like manner. However, gametocytes and oocysts are found when the suspension is composed mostly of cell aggregates. Two days after inoculation of cultures derived from this type of suspension, epithelial-like cells are abundant and most of them contain parasites. After 5 days, however, there are only a few epithelial-like patches; instead, there are dense areas of growth containing mostly fibroblast-like cells, and these contain the parasites. Bedrnik (1967b) reported that sexual stages were not found in fibroblasts, but were strictly limited to the islands of epithelial-like cells. He used second generation merozoites as the inoculum, and the interval between inoculation and appearance of sexual stages was only from 1–2 days. It is believed by this author that sexual stages and oocysts of *E. tenella* develop only from sporozoites that enter elevated areas or patches of cells, epithelial-like at first, that are produced from cell aggregates. Because development from the first generation on takes place primarily in clusters, merozoites probably do not migrate very far before entering cells. Those that undergo further development enter cells adjacent to the cell in which the schizont develops. Bedrnik (1967b) believes that a certain type of cell is required for gametocyte development. However, it is probable that cells in depth provide certain physiological and/or physical conditions necessary for development. Several explant cultures (1-mm square pieces) of chicken kidney were inoculated with 500 to 700 sporozoites. The growth medium was removed and sporozoites were placed directly onto the tissue; maintenance medium was added 15 min later. After 7 days, more than one-half of the cells in the stained sections contained gametocytes and oocysts (Doran, unpublished).

It seems that primary aggregates may be favorable for only *E. tenella.* No development of *E. acervulina, E. praecox, E. maxima,* or *E. mivati* and only second generation schizonts of

6

E. necatrix occurred when aggregates were used (Doran, 1971b). Müller et al. (1973a) obtained no development of *E. canadensis* in aggregates of bovine intestine and only development of mature first generation schizonts in those of bovine kidney. Using *E. magna* merozoites from the host as the inoculum, Speer et al. (1973b) found little difference in development between primary rabbit kidney cultures with and without aggregates, and Speer and Hammond (1973) found no difference in the quantity of oocysts in primary bovine kidney cultures with and without aggregates.

BEHAVIOR IN AVIAN EMBRYOS

Development and Pathological Effects

Besnoitia Frenkel (1953) first cultivated this parasite. He obtained *B. jellisoni* zoites from cysts found in white-footed (deer) mice and inoculated them into chick embryos. Infections developed that were fatal in 4–12 days. Frenkel later (1965) stated that *B. jellisoni* produces a general, usually fatal infection in chick embryos.

Pathology is similar to that produced by *T. gondii*. Bigalke (1962) inoculated *B. besnoiti* zoites into the chorioallantoic cavity and yolk sac of chick embryos. Circumscribed, yellow-white foci that measured 0.5–4 mm in diameter were found on the chorioallantoic and yolk sac membranes. The lesions consisted of necrotic material with cellular infiltration with numerous intra- and extracellular parasites nearby.

Bigalke (1962) maintained *B. besnoiti* by serial passage in chick embryos. Organisms were passaged six times. During these passages, the virulence of the organisms apparently increased. Embryos began to die on the 19th day after the initial passage, but after passages 2, 3, and 4, and 5 and 6, the mortality began on days 11, 7, and 5, respectively.

Eimeria Long (1965) first reported the complete development of an eimerian species (*E. tenella)* in chick embryos. Since then, other species have undergone development. Table 4 lists the species that developed when either sporozoites or second generation merozoites were inoculated into the allantoic cavity of chicken embryos. The species from birds complete their life cycles, but *E. stiedai* from the rabbit develops only to gametocytes. Attempts to obtain development of *E. maxima* (Long, 1966; Shibalova et al., 1969) and a turkey strain of *E. dispersa* (Long and Millard, 1979) were unsuccessful. These are the only species tried that did not develop. There are a few unsuccessful attempts to obtain complete development of species listed in Table 4. The following are noteworthy: 1) Ryley (1968), using the same strain of *E. brunetti* used by Long (1966), reported no development; 2) Long (1974b) stated that

Table 4. Development of *Eimeria* in embryonated chicken eggs after inoculation of sporozoites or second-generation merozoites into the allantoic cavity

Species	Inoculum	Best development	First report
E. acervulina	Sporozoites	Oocysts	Long (1973b)
E. brunetti	Sporozoites	Oocysts	Long (1966)
	Merozoites	Oocysts	Shibalova (1969a)
E. mivati (diminuta)	Sporozoites	Oocysts	Long (1974d)
E. grenieri	Sporozoites	Oocysts	Long and Millard (1978)
E. mitis	Sporozoites	Oocysts	Shibalova (1970)
E. mivati	Sporozoites	Oocysts	Long and Tanielian (1965)
E. necatrix	Sporozoites	Oocysts	Shibalova (1970)
E. praecox	Sporozoites	Oocysts	Shibalova (1970)
E. stiedai	Sporozoites	Gametocytes	Fitzgerald (1970)
E. tenella	Sporozoites	Oocysts	Long (1965)
	Merozoites	Oocysts	Long (1966)

numerous attempts to obtain development of *E. praecox* and *E. acervulina* in his laboratory had failed; and 3) Long (1966) obtained no development of *E. necatrix* with second generation merozoites and only mature second generation schizonts with sporozoites.

The report of Long (1973b) concerning production of oocysts by *E. acervulina* is surprising. All previous attempts (Long and Tanielian, 1965; Long, 1966; Shibalova et al., 1969; Shibalova, 1970; Itagaki et al., 1972) were unsuccessful. Long (1973b) inoculated embryos with *E. acervulina*. Oocysts were produced when 1×10^7 sporozoites were inoculated, but no oocysts were produced with smaller inocula. It was subsequently shown that the *E. acervulina* culture contained an extremely small proportion of *E. mivati* and it was this parasite which was grown (Shirley, 1979).

Attempts have been made to obtain development by routes other than the allantoic cavity. In his initial work, Long (1965) did not obtain infection after inoculating *E. tenella* sporozoites into the amniotic cavity, via the yolk sac, onto the CAM, and intravenously. Later, Long (1974b) reported development when *E. tenella* and *E. mivati* sporozoites were inoculated onto a "dropped" CAM. The CAM was dropped by drilling holes in both the air space end of the shell and in a position over the CAM; the membrane was then dropped by applying negative pressure to the hole in the air space end. Sporozoites were inoculated onto the surface of the CAM (chorion) through the hole over the CAM. Only 2 of 24 embryos inoculated with *E. tenella* became infected and only 7 of 34 embryos inoculated with *E. mivati* became infected. Far fewer oocysts can be harvested after sporozoites are placed onto the CAM than after they are inoculated by way of the allantoic cavity. Long stated that the more carefully the membrane is dropped, the less likely it is that the embryo will be infected. He also stated that damage to the membranes by this technique is probably unavoidable and suggested that the CAM becomes infected only when the chorion is damaged and sporozoites can enter the allantoic cavity by way of the damaged site. Ishii and Onaga (1971) inoculated *E. tenella* sporozoites into the yolk sac and found large numbers of mature schizonts, but no gametocytes or oocysts, in tissue of the yolk stalk 5 and 8 days after inoculation. Long (1974b) also mentions that development occurs on the yolk membrane when *E. tenella* is inoculated via the yolk sac. Long (1971a) again attempted to obtain infection by the intravenous route with large numbers of sporozoites cleaned of debris (oocyst and sporocyst hulls). Schizogony and gametogony took place in the bile duct epithelium and in cells lining the sinusoids of the liver, but only a few gametocytes were found. Treatment of embryos with dexamethasone before inoculation increased the number of gametocytes. Inoculation of *E. praecox* and *E. maxima* intravenously into embryos, some of which were pretreated with dexamethasone, failed to produce infection (Long, 1974b). When embryos were inoculated intravenously with *E. brunetti, E. mivati,* and *E. necatrix* sporozoites, infection occurred in the liver of some embryos of each group; development did not occur elsewhere (Long and Millard, 1973).

Attempts have also been made to obtain development in different species of embryonated eggs from sporozoites inoculated into the allantoic cavity. Long (1965) found that *E. tenella* would not develop in quail and turkey embryos. However, Rollinson (1976) found small numbers of oocysts of this species on the CAM of quail embryos. *E. dispersa* does not develop in quail and turkey embryos (Long and Millard, 1979). An embryo-adapted strain of *E. tenella* developed to oocysts in duck and quail embryos, but *E. brunetti, E. acervulina,* a regular pathogenic strain of *E. tenella,* and an embryo-adapted strain of *E. mivati* did not (Long and Millard, 1975). In goose embryos pretreated with dexamethasone, *E. tenella* developed to second generation schizonts; in untreated embryos, there was no development (Long, 1971a).

There have been two recent reports of the development of oocysts after inoculation of single sporozoites into the allantoic cavity of chick embryos. Lee et al. (1976), using single *E. tenella* and *E. diminuta* (from the Malaysian jungle fowl, *Gallus lafayetti)* sporozoites, found gametocytes and oocysts on the CAM 7–10 days after inoculation. Shirley and Millard (1976), using single sporozoites of an embryo-adapted strain of *E. tenella* that had been serially passaged from 66 to 78 times, obtained yields of 1,800 to 2,000 oocysts/sporozoite. These two studies are important in that they prove the bisexual nature of a coccidial sporozoite.

Long (1971b) showed for the first time that an eimerian species can be maintained by serial passage in fertile eggs. He recovered unsporulated *E. tenella* oocysts from the CAM, sporulated them, excysted the sporozoites from sporocysts released from the oocysts, and inoculated the sporozoites into the allantoic cavity of uninfected embryos. By this method, he maintained the parasite through 16 passages. Since this first successful attempt, Long and his colleagues have made considerable progress with the serial passage of several species of *Eimeria.* Long (1972c) reported that a Houghton strain of *E. tenella* had undergone 42 embryo passages.

During this time, the oocyst yield increased dramatically—in essence, the strain had become embryo-adapted. It had also lost its pathogenicity to both embryos and chickens, but regained it during a second passage in chickens. Long (1973a) observed that the strain had lost its ability to produce characteristically large, mature second generation schizonts in both embryos and chickens. However, after two passages in chickens, the large schizonts reappeared in both embryo and chicken. Later, Long (1974c) reported that the embryo-adapted strain had been passaged 62 times over a 2½ yr period and that it no longer regained its pathogenicity after nine passages in chickens. A 75th passage yielded 7,500,000 oocysts from an inoculum of only 5,000 sporozoites (Long and Millard, 1975). *E. mivati* has been serially passaged 22 times (Long, 1972b). With this species, there is also an increased oocyst production in embryos and a decreased pathogenicity for chicks. Shirley et al. (1977) serially passaged *E. mivati (diminuta)* 12 times in embryos. By the 10th passage, more oocysts were produced with an inoculum of 20,000 sporozoites than with an inoculum of 80,000 sporozoites, a result that suggests that this parasite, too, was becoming embryo-adapted.

In *E. tenella* infections, maturation of colonies of second generation schizonts produces lesions on the CAM. Long (1970a) described two types of lesions; one type was large (0.3–1.0 mm in diameter) and often occurred along the walls of the blood vessels, whereas the other was small (0.05–0.1 mm in diameter) and scattered more evenly through the membrane (Figure 36). In infections with the embryo-adapted strain of this species, Long (1973a) found more small lesions and fewer large lesions along the blood vessels of the CAM (Figure 37). Long (1966) found that (*a*) the size of *E. tenella* schizonts varies more in the CAM than in the cecum of chicks; (*b*) many second generation *E. tenella* merozoites are not released from the CAM; (*c*) *E. necatrix* schizonts are located in groups, well below the epithelium, that seem to be surrounded by a membranous layer; and (*d*) merozoites of *E. brunetti, E. mivati,* and *E. tenella,* but not of *E. necatrix,* are found in the allantoic cavity. In *E. tenella* infections, hemorrhaging occurs (Baldelli et al., 1968; Ryley, 1968; Long, 1966, 1972c, 1974c) and, if the dose is large enough, the embryo dies (Long, 1970b, 1972c, 1974c). In infections with *E. mivati,* there is no hemorrhage, death is rare, and pathological changes are restricted to loss of epithelial tissue from the CAM into the allantoic cavity and a few gross lesions caused by host cell reactions on the CAM (Long, 1972b). Jeffers and Wagenbach (1970) found that infection with *E. tenella* did not affect the weight of the egg or the hatched chick. They also found that mortality was higher when oocysts were given to chicks hatched from eggs previously inoculated with sporozoites than with chicks hatched from noninoculated eggs.

Isospora There are apparently only two reports of infection of chick embryos with isosporan parasites. Baldelli et al. (1971) obtained isosporan oocysts from a cat, excysted the sporozoites, and inoculated the sporozoites into the chorioallantoic cavity. There was no development. However, Arcay de Peraza (1976) mentioned (in abstract only) that sporozoites

6

of *Isospora frenkeli,* a new species she found in cats, underwent schizogony and developed to oocysts in cells of the CAM.

Toxoplasma Levaditi et al. (1929) first reported cultivation of *T. gondii* in the embryonated chicken egg. Since then, all reported attempts to obtain infection in avian embryos have been successful. The parasite can be isolated from macerated or excised bits of tissue following inoculation into embryonated eggs (Warren and Russ, 1948; Anwar and Oureshi, 1954; Jacobs et al., 1954; Jacobs, 1956). Zoites can be inoculated intravenously, into the yolk sac, or into the embryonic membranes or cavities. A comparative study of all three routes has not been made, but the yolk sac and allantoic cavity routes have been compared. Resultant infections have been either similar (Abbas, 1967b), less intense in the yolk sac (Nockiewicz, 1972), or more intense in the yolk sac (Olisa, 1976).

Embryos are not as sensitive to infection with small numbers of organisms as mice (Weinman, 1944; Eichenwald, 1956; Abbas, 1967b). Eichenwald (1956) stated that, of the three systems for isolation (chick embryo, cell culture, and animal), chick embryo is the least sensitive. Abbas (1967b) compared the three methods for isolating the RH strain from acute, subacute, and chronic infections. For isolation from acutely infected tissue, mice inoculated intraperitoneally were at least 16 times as sensitive as embryos; for subacutely and chronically infected tissues, mice were, respectively, 50 and 10 times as sensitive as embryos. Abbas (1967b) stated that infections were detected earlier in embryos than in mice, although the latter were more sensitive. He also stated that zoites could be found within embryos in 10 days, whereas, except for the highly virulent strains, they did not appear in mice for 6 weeks.

Wolfson (1942) showed for the first time that *T. gondii* can be maintained by serial passage in fertile eggs. The parasite is not always demonstrable after one or more early passages, but it can be readily demonstrated after repeated subinoculation (Jacobs and Melton, 1954). The virulence of a strain of *T. gondii* can be maintained in chick embryos at its preisolation level (Jacobs, 1953; Jacobs and Melton, 1954). A strain is occasionally found that is difficult to establish. However, once established, it can be maintained readily by weekly passages (Jacobs, 1956). MacFarlane and Ruckman (1948) passed the RH strain 27 times in 10- and 11-day-old embryos. The LD-50 titers for these passages and for passages between embryos and mice inoculated intracerebrally were about the same. Jacobs (1953) stated that the RH strain was maintained 2–3 yr without loss of virulence. Jacobs and Melton (1954) maintained the 113-CE strain by weekly passages in embryos for 1 yr.

Macroscopic yellow-white lesions (abscess-like elevations or islands, colonies, pocks, nodules, plaques) appear on the chorioallantoic membrane (Wolfson, 1941, 1942; Weinman, 1944; Anwar and Oureshi, 1954; Lund et al., 1963b) and amniotic membrane (MacFarlane and Ruckman, 1948). Lesions can usually be found 1 week after inoculation or earlier (Jacobs, 1953). They range in diameter from 0.5 mm (MacFarlane and Ruckman, 1948) to 20 mm (An-

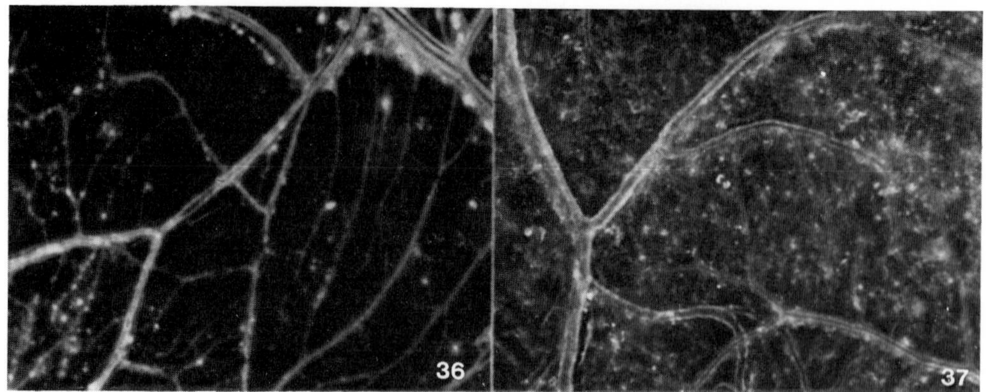

war and Oureshi, 1954) and are frequently visible through the unbroken shell on transillumination (Warren and Russ, 1948). The lesion or area around it is necrotic (Warren and Russ, 1948; Kuwahara, 1959) or thickened (MacFarlane and Ruckman, 1948) and contains numerous zoites (Wolfson, 1941; MacFarlane and Ruckman, 1948; Warren and Russ, 1948). Jögiste and Ukhov (1969) reported the presence of extracellular colonies on the CAM, the transformation of proliferating mesenchymal cells into macrophages, and the cystic form of the parasite within the macrophages.

Zoites can also be found in tissues within the embryo. The parasite is disseminated through the blood stream (Jacobs, 1953) and has been reported to reproduce within blood cells (Zassuchin, 1963; Novinskaya, 1965; Galuzo et al., 1969). Wolfson (1941) found zoites in erythrocytes, monocytes, and plasma of embryonic duck blood, but not in other tissues of the embryo. Anwar and Oureshi (1954) found, in descending order of prevalence: (*a*) large lesions containing zoites in the CAM, kidney, bone marrow, and brain; and (*b*) "micro-lesions" in the cerebellum, heart, liver, spleen, and intestine. Cysts have also been reported in a variety of tissues (Abbas, 1967a; Jögiste and Ukhov, 1969).

Olisa (1976) inoculated 12 freshly laid fertile eggs with 5,000 zoites. Three of the chicks hatched, but their heads continually rotated to the right, they were unable to stand, and they had respiratory difficulties.

Factors Influencing Development and Severity of Pathological Effects

Strain, Size, and Age of the Inoculum As previously mentioned, strains of *E. tenella* and *E. mivati* that become embryo-adapted by serial passage in embryos produce more oocysts and are less pathogenic than the parent strain that has not been passaged. The following information pertains to strains and species that have not been serially passaged. Long (1972b) compared inocula of *E. mivati* sporozoites ranging from 5,000 to 180,000 and found that, although the reproductive rate was highest with 10,000 sporozoites, more oocysts were produced with an inoculum of 40,000 sporozoites; beyond 40,000 there was no further increase. Inocula of 11,000 to 56,000 *E. necatrix* sporozoites and 17,000 to 85,000 *E. tenella* sporozoites produced good infections in chick embryos, and inocula higher than 85,000 *E. tenella* sporozoites resulted in early deaths but no visible parasites (Long, 1966). Ryley (1968), also working with *E. tenella*, found that 10,000 sporozoites killed 30% of the embryos, whereas 100,000 killed all embryos on the 5th day. Long (1970b) compared hemorrhagic mortality of the Weybridge and Houghton strains of *E. tenella*. With inocula of 5,000 to 15,000, and 45,000 sporozoites of the virulent Weybridge strain, the mortality rates were 70%, 92%, and 100%, respectively; with similar inocula of the less virulent Houghton strain, the mortality rates were only 29%, 68%, and 92%, respectively. With both strains, deaths caused by an inoculum of 45,000 began at 89 hr, but with the other inocula they did not begin until 85–92 hr. Jeffers and Wagenbach (1970) also found that embryo death during the hemorrhagic phase of the infection (4th–9th days)

6

Figures 36 and 37. Chorioallantoic membrane (CAM; fresh preparation) showing lesions associated with *E. tenella* schizogony. Approximately ×5. 36. CAM of embryo 118 hr after inoculation with sporozoites not passaged in embryos. Note the larger discrete lesions along the blood vessels. Reprinted by permission from: Long, P. L. *Eimeria tenella:* Chemotherapeutic studies in chick embryos with a description of a new method (chorioallantoic membrane foci counts) for evaluating infections: *Zeitschrift für Parasitenkunde* 33:329–338 (1970). 37. CAM of embryo 118 hr after inoculation with 40,000 sporozoites of an *E. tenella* strain serially passaged 41 times in embryos. Note numerous very small lesions. Reprinted by permission from: Long, P. L. Endogenous stages of a "chick embryo-adapted" strain of *Eimeria tenella. Parasitology* 66:55–62 (1973).

was highly dependent on inoculum level. Deaths occur earlier and slightly more frequently among embryos inoculated with large doses of sporozoites. Although mortality during the hatching period (18th through 21st days) is less dose-dependent than hemorrhagic mortality, inoculations even at a low level reduce the hatchability of embryos. Long (1970c) inoculated embryos with similar numbers of *E. tenella* sporozoites obtained from oocysts 15, 65, 122, and 162 days old. The percentages of mortality were 83.3, 16.6., 5.5, and 7.6, respectively. The average numbers of focal lesions per CAM were 140.0, 15.4, 3.2, and 3.5, respectively.

Jacobs and Melton (1954) inoculated 7-day-old embryos with 100, 1,000, 10,000, or 100,000 *T. gondii* zoites of strain 113-CE. All embryos died, but those given the smaller inocula survived the longest. The survival time with 100 zoites was about 14 days, whereas with 100,000 zoites, it was only 8 days. The size of lesions and the time that they appear depend on the size of the inoculum (Wolfson, 1942; Weinman, 1944). Jõgiste and Ukhov (1969) claimed that the RH strain produced changes on the CAM similar to those produced by an avirulent strain but produced no necrotic foci. Unfortunately, they used 1,500 zoites of the RH strain and 65,000 zoites of the avirulent strain.

Temperature of Incubation Development of *E. tenella* is more rapid at the temperature of the chicken (41°C) than at slightly lower temperatures. At 38°C, mature second generation schizonts appear at 4.0–4.2 days (Long, 1965, 1970b; Ryley, 1968) and gametocytes and oocysts appear at 7.0 days (Long, 1965; Ryley, 1968). At 39°C, mature second generation schizonts appear at 3.4 days (Long, 1970b) and gametocytes and oocysts appear at 5.0 days (Long, 1970b; Shibalova, 1970). Mortality of embryos is also higher and more rapid at 41°C (Long, 1970b). At 38°C, an inoculum of 30,000 sporozoites causes no mortality, but at 39° and 41°C the embryo mortality is 78% and 95%, respectively. Long (1972a) found that oocyst production of *E. mivati* and *E. tenella* was reduced and delayed in embryos incubated at temperatures below 41°C. He studied the sporulation in oocysts of both species recovered from embryos incubated at temperatures between 37° and 39°C at night. Sporulation rates were low, especially in oocysts from embryos infected with *E. mivati.* When infected embryos are incubated at 41°C, oocysts must be harvested at a certain time. Long (1972b) determined that the best time to harvest the embryo-adapted strain of *E. mivati* from the infected CAM was at 130 hr. There were more oocysts at 144 hr, but sporulation was best in oocysts recovered from 130–132 hr. Oocysts harvested at 168 hr failed to sporulate because they had been subjected to 41°C for nearly 36–48 hr.

There has been no detailed study on the development of *T. gondii* or *Besnoitia* species at different temperatures. However, Warren and Russ (1948) mentioned that *T. gondii* multiplies equally well when eggs are incubated at 35° and 37.5°C.

Age, Strain, and Sex of the Embryo Long (1970b) found that mortality caused by *E. tenella* occurs earlier and is higher in 9-day-old than in 11-day-old embryos. He also compared infection in three strains of embryos (S, C, and K) that had been bred selectively for susceptibility or resistance to Marek's disease. The S strain, which was bred for susceptibility, was much more susceptible to infection than the other two strains. However, 3-week-old chicks of the S strain were less susceptible than embryos of the C and K strains. Long (1970b) stated that the difference in susceptibility of chicks and embryos of the S strain is due to host resistance factors that develop within the first few weeks after hatching. Long (1971b) found Rhode Island Red embryos more susceptible to infection than Brown Leghorn embryos. However, later (1972b) he found the reverse to be true with *E. mivati;* Brown Leghorns produced three times as many oocysts as Rhode Island Reds. Long also (1974d) found that Brown Leghorn embryos produced more *E. mivati (diminuta)* oocysts than Light Sussex embryos. Using nine different mating types and different dosages, Jeffers and Wagenbach (1969) determined that mortality during the hemorrhagic period in *E. tenella* infections is significantly higher in female than in male embryos.

T. gondii (RH strain) multiplies equally well in eggs 6–12 days old (MacFarlane and Ruckman, 1948; Warren and Russ, 1948), but deaths occur over a longer period, that is, from 4–8 days, in the younger embryos (MacFarlane and Ruckman, 1948). Calcified plaques in the CAM are found only occasionally in 7-day-old embryos but more frequently in 12-day-old embryos.

ACKNOWLEDGMENTS

I hereby express my appreciation to M. Barry Chute and Robert B. Ewing for assistance in preparation of the illustrations; to Gary L. Kelley, Michael D. Ruff, Harley G. Sheffield, and Clarence A. Speer for reviewing the manuscript; and to Susan Charbonneau and Ellen Mendonca for secretarial assistance.

LITERATURE CITED

Abbas, A. M. A. 1967a. Toxoplasmosis of chickens experimentally infected during embryonic life. Trans. R. Soc. Trop. Med. Hyg., 61:514–516.

Abbas, A. M. A. 1967b. Comparative study of methods used for isolation of *Toxoplasma gondii.* Bull. WHO 36:344–346.

Akinshina, G. T. 1959. Cultivation of *Toxoplasma gondii* in tissue cultures. Bull. Exp. Biol. Med. 47:47–50.

Akinshina, G. T. 1964. Prolonged preservation of *Toxoplasma* in tissue cultures. (In Russian; English summary). Byull. Eksp. Biol. Med. 58:98–100.

Akinshina, G. T. 1965. Use of the methods of tissue cultures for diagnostics of toxoplasmosis and for maintenance of *Toxoplasma's* strains. (In Russian). In: I. G. Galuzo (ed.), Toxoplasmosis of Animals, pp. 386–390. Academy of Science of the Kazakh SSR, Alma-ata.

Akinshina, G. T., and Doby, J. M. 1969a. Multiplication de *Besnoitia jellisoni* Frenkel,

6

1953 (Protozoaires Toxoplasmatea) en cultures de cellules de tissus d'origines differentes. Protistologica 5:249–253.

Akinshina, G. T., and Doby, J. M. 1969b. Étude comparée de la multiplication de *Toxoplasma gondii* et de *Besnoitia jellisoni* dans les cultures de cellules. In: Progress in Protozoology. Proceedings of the Third International Congress on Protozoology, p. 222.

Akinshina, G. T., and Gracheva, L. I. 1964. Production of toxoplasmosis antigen by tissue culture methods. (In Russian; English summary). Med. Parazit. Bolezni 33:661–665.

Akinshina, G. T., and Zassuchin, D. N. 1965. Multiplication of *Toxoplasma* in tissue culture. In: Progress in Protozoology. Proceedings of the Second International Congress on Protozoology, International Congress Series 91, pp. 99–100. Excerpta Medica Foundation, The Hague.

Akinshina, G. T., and Zassuchin, D. N. 1969. Contribution a l'étude de la variabilite de *Toxoplasma gondii*: II. Obtention de clônes resistants aux medicaments in vitro. In: Progress in Protozoology. Proceedings of the Third International Congress on Protozoology, p. 223.

Akinshina, G. T., and Zasuchina, G. D. 1966. Method of investigating mutations in protozoa *(Toxoplasma gondii)*. (In Russian). Genetika 2:71–75.

Akinshina, G. T., and Zasuchina, G. D. 1967. A new method of indicating the presence of *Toxoplasma gondii* in tissue cultures. Bull. Exp. Biol. Med. 63:443–445.

Anwar, A., and Oureshi, M. 1954. Experimental toxoplasmosis in chick embryos. In: Proceedings of the Sixth Pakistan Science Conference, Karachi, 1954. III. (Abstracts), pp. 225–226.

Arai, H., Saito, H., and Nomura, T. 1958. Invasion and multiplication of *Toxoplasma gondii* in HeLa cells. Jpn. J. Med. Prog. 45:663–669.

Arcay de Peraza, L. 1976. Desarrollo en la memerana corio-alantoidea de embrion de pollo (MCA) de un nuevo coccidia de gato: *Isospora frenkeli* sp. nov. In: Resumenes de Trabajos Libres, Congreso Fourth Latinoamericano de Parasitologia, San Jose, Costa Rica, 7–11 Dec., p. 26.

Augustine, P. C., and Doran, D. J. 1978. Development of *Eimeria meleagrimitis* Tyzzer from sporozoites and merozoites in turkey kidney cell cultures. J. Protozool. 25:82–86.

Azab, M., and Beverley, J. K. A. 1974. Schizogony of *Toxoplasma gondii* in tissue cultures. Z. Parasitenkd. 44:33–41.

Baldelli, B., Frescura, T., Ambrosi, M., Polidori, A., and Mughetti, L. 1971. Experimental toxoplasmosis in cats and cultivation of *Toxoplasma* oocysts in chicken embryos. Bull. Soc. Ital. Biol. Exp. 47:416–418.

Baldelli, B., Frescura, T., and Asdrubali, G. 1968. Cultivation of *Eimeria tenella* in chick embryos. Vet. Ital. 19:101–107.

Balducci, D., and Tyrrell, D. 1956. Quantitative studies of *Toxoplasma gondii* in culture of trypsin-dispersed mammalian cells. Br. J. Exp. Pathol. 37:168–175.

Ball, G. H., and Chao, J. 1973. The complete development of the sporogonous stages of *Hepatozoon rarefaciens* cultured in a *Culex pipiens* cell line. J. Parasitol. 59:513–515.

Bedrnik, P. 1967a. Further development of the second generation of *Eimeria tenella* merozoites in tissue cultures. Folia Parasitol. 14:361–363.

Bedrnik, P. 1967b. Development of sexual stages and oocysts from the 2nd generation of *Eimeria tenella* merozoites in tissue cultures. Folia Parasitol. 14:364.

Bedrnik, P. 1969a. Some results and problems of cultivation of *Eimeria tenella* in tissue cultures. Acta Vet. (Brno.) 38:31–35.

Bedrnik, P. 1969b. Cultivation of *Eimeria tenella* in tissue cultures: I. Further development of second generation merozoites in tissue cultures. Acta Protozool. Warszawa 7:87–98.

Bedrnik, P. 1970. Cultivation of *Eimeria tenella* in tissue cultures: II. Factors influencing a further development of second generation merozoites in tissue culture. Acta Protozool.

Warszawa 7:253–261.
Bell, J. B. 1961. *Toxoplasma gondii* infection of tissue culture cells and animals. Va. J. Sci. 12:163.
Beverley, J. K. A. 1969. The biology of *Toxoplasma* infections. In: Seventh Symposium of the British Society on Parasitology, London, pp. 43–49.
Bickford, A. A., and Burnstein, T. 1966. Maintenance of *Toxoplasma gondii* in monolayer cultures of human epithelial (H. Ep. 2) cells. Am. J. Vet. Res. 27:319–325.
Bigalke, R. D. 1962. Preliminary communication on the cultivation of *Besnoitia besnoiti* (Marotel, 1912) in tissue culture and embryonated eggs. J. S. Afr. Vet. Med. Assoc. 33:523–532.
Bommer, W. 1969. The life cycle of virulent *Toxoplasma* in cell culture. Aust. J. Exp. Biol. Med. Sci. 47:505–512.
Bommer, W. 1970. The life cycle of *Toxoplasma gondii*. J. Parasitol. 56(4; Sect. II, Part I):31.
Bommer, W., Heunert, H. H., and Milthaler, B. 1969a. Kinematographische Studien über die Eigenbewegung von *Toxoplasma gondii*. Z. Tropenmed. Parasitol. 20:450–458.
Bommer, W., Hofling, K. H., and Heunert, H. H. 1969b. Multiplication of *Toxoplasma gondii* in cell cultures. Ger. Med. Monthly 14:399–405.
Buckley, S. M. 1973. Survival of *Toxoplasma gondii* in mosquito cell lines and establishment of continuous infection in Veno cell cultures. Exp. Parasitol. 33:23–26.
Chang, Chung-Ho, Stulberg, C., Bollinger, R. O., Walker, R., and Brough, A. J. 1972. Isolation of *Toxoplasma gondii* in tissue culture. J. Pediatr. 81:790–791.
Chao, J., and Ball, G. H. 1972. *In vitro* culture of the vector phase of snake haemogregarines in mosquito cell lines. J. Parasitol. 58:148–152.
Chaparas, S. D., and Schlesinger, R. W. 1959. Plaque assay of *Toxoplasma* on monolayers of chick embryo fibroblasts. Proc. Soc. Exp. Biol. Med. 102:431–437.
Chernin, E., and Weller, T. H. 1954a. Serial propagation of *Toxoplasma gondii* in roller tube cultures of mouse and of human tissues. Proc. Soc. Exp. Biol. Med. 85:68–72.
Chernin, E., and Weller, T. H. 1954b. Further observations on the growth of *Toxoplasma gondii* in roller tube tissue cultures. J. Parasitol. 40(5; Sect. 2):21.
Chernin, E., and Weller, T. H. 1957. Further observations on the growth of *Toxoplasma gondii* in roller tube cultures of mouse and primate tissues. J. Parasitol. 43:33–39.
Clark, W. N., and Hammond, D. M. 1969. Development of *Eimeria auburnensis* in cell cultures. J. Protozool. 16:646–654.
Cook, M. K., and Jacobs, L. 1958. Cultivation of *Toxoplasma gondii* in tissue cultures of various derivations. J. Parasitol. 44:172–182.
Coudert, P., and Provôt, F. 1974. Developpement interne et schizogonie en cultures cellulares d' *Eimeria stiedie* (Lindmann, 1865) Kissalt et Hartmann, 1907. C. R. Acad. Sci. (D) (Paris) 279:911–913.
Csóka, R., and Kulcsár, G. 1968a. Cultivation of *Toxoplasma gondii* in primary human amniotic cell culture. Acta Microbiol. Acad. Sci. Hung. 15:11–15.
Csóka, R., and Kulcsár, G. 1968b. Maintenance of *Toxoplasma gondii* in primary human amniotic cell culture. Acta Microbiol. Acad. Sci. Hung. 15:357–360.
Csóka, R., and Kulcsár, G. 1970. Comparative study on sensitivity to *Toxoplasma gondii* of human primary amniotic cell culture and of mice. Acta Microbiol. Acad. Sci. Hung. 17:85–89.
Davis, L. R., and Bowman, G. W. 1957. The endogenous development of *Eimeria zuernii* a pathogenic coccidium of cattle. Am. J. Vet. Res. 18:569–574.
de Vos, A. J., and Hammond, D. M. 1971. Development of *Eimeria crandallis* from sheep in cell cultures. J. Protozool. 18(suppl.):11.
de Vos, A. J., Hammond, D. M., and Speer, C. A. 1972. Development of *Eimeria crandallis* from sheep in cultured cells. J. Protozool. 19:335–343.

6

Doby, J. M., and Akinshina, F. T. 1968. Possibilitiés de dévelopment de *Besnoitia jellisoni* (Protozoaire parasite Toxoplasmatea) en culture de cellules. Quelques aspects de son comportement en fibroblasts d'embryon de poulet. C. R. Soc. Biol. (Paris) 162:1207–1210.

Doerr, U., and Höhn, H. 1972. Zur in vitro-Kultur von *Eimeria stiedae*. Z. Parasitenkd. 3:56–57.

Doran, D. J. 1969a. Freezing excysted coccidial sporozoites. J. Parasitol. 55:1229–1233.

Doran, D. J. 1969b. Cultivation and freezing of poultry coccidia. Acta Vet. (Brno.) 38:25–30.

Doran, D. J. 1970a. Effect of age and freezing on development of *Eimeria adenoeides* and *E. tenella* sporozoites in cell culture. J. Parasitol. 56:27–29.

Doran, D. J. 1970b. Survival and development of *Eimeria adenoeides* in cell cultures inoculated with sporozoites from cleaned and uncleaned suspensions. Proc. Helminth. Soc. Wash. 37:45–48.

Doran, D. J. 1970c. *Eimeria tenella:* From sporozoites to oocysts in cell culture. Proc. Helminth. Soc. Wash. 37:84–92.

Doran, D. J. 1971a. Increasing the yield of *Eimeria tenella* oocysts in cell culture. J. Parasitol. 57:891–900.

Doran, D. J. 1971b. Survival and development of 5 species of chicken coccidia in primary chicken kidney cell cultures. J. Parasitol. 57:1135–1137.

Doran, D. J. 1971c. Comparative development of *Eimeria tenella* in primary cultures of kidney cells from the chicken, pheasant, partridge, and turkey. J. Parasitol. 57:1376–1377.

Doran, D. J. 1973. Cultivation of coccidia in avian embryos and cell culture. In: D. M. Hammond and P. L. Long (eds.), The Coccidia: *Eimeria, Isospora, Toxoplasma,* and Related Genera, pp. 183–252. University Park Press, Baltimore.

Doran, D. J. 1974. *Eimeria tenella:* Merozoite production in cultured cells and attempts to obtain development of culture-produced merozoites. Proc. Helminth. Soc. Wash. 41: 169–173.

Doran, D. J. 1978. The life cycle of *Eimeria dispersa* Tyzzer, 1929, in turkeys. J. Protozool. 25:293–297.

Doran, D. J., and Augustine, P. C. 1973. Comparative development of *Eimeria tenella* from sporozoites to oocysts in primary kidney cell cultures from gallinaceous birds. J. Protozool. 20:658–661.

Doran, D. J., and Augustine, P. C. 1976. *Eimeria tenella:* Comparative oocysts production in primary cultures of chicken kidney cells maintained in various media systems. Proc. Helminth. Soc. Wash. 43:126–128.

Doran, D. J., and Augustine, P. C. 1977. *Eimeria dispersa* and *Eimeria gallopavonis:* Infectivity, survival, and development in primary chicken and turkey kidney cell cultures. J. Protozool. 24:172–176.

Doran, D. J., and Vetterling, J. M. 1967a. Comparative cultivation of poultry coccidia in mammalian kidney cell cultures. J. Protozool. 14:657–662.

Doran, D. J., and Vetterling, J. M. 1967b. Cultivation of the turkey coccidium *Eimeria meleagrimitis* Tyzzer, 1929, in mammalian kidney cell cultures. Proc. Helminth. Soc. Wash. 34:59–65.

Doran, D. J., and Vetterling, J. M. 1968. Survival and development of *Eimeria meleagrimitis* Tyzzer, 1929, in bovine kidney and turkey intestine cell cultures. J. Protozool. 15:796–802.

Doran, D. J., and Vetterling, J. M. 1969. Influence of storage period on excystation and development in cell culture of sporozoites of *Eimeria meleagrimitis* Tyzzer, 1929. Proc. Helminth. Soc. Wash. 36:33–35.

Doran, D. J., Vetterling, J. M., and Augustine, P. C. 1974. *Eimeria tenella:* An in vivo and an in vitro comparison of the Wisconsin, Weybridge, and Beltsville strains. Proc. Helminth. Soc. Wash. 41:77–80.

Dubremetz, J. F., Porchet-Henneré, E., and Parenty, M. D. 1975. Croissance de *Sarcocystis* tenella en culture cellulaire. C. R. Acad. Sci. (D) (Paris) 280:1793–1795.

Dvorak, J. A., and Howe, C. L. 1977. *Toxoplasma gondii*-vertebrate cell interactions: I. The influence of bicarbonate ion, CO-2, pH and host cell culture age on the invasion of vertebrate cells in vitro. J. Protozool. 24: 416–419.

Eichenwald, H. 1956. The laboratory diagnosis of toxoplasmosis. Ann. N. Y. Acad. Sci. 64:207–214.

Fayer, R. 1970. *Sarcocystis:* Development in cultured avian and mammalian cells. Science 168:1104–1105.

Fayer, R. 1972a. Gametogony of *Sarcocystis* sp. in cell culture. Science 175:65–66.

Fayer, R. 1972b. Cultivation of feline *Isospora rivolta* in mammalian cells. J. Parasitol. 58:1207–1208.

Fayer, R., and Hammond, D. M. 1967. Development of first-generation schizonts of *Eimeria bovis* in cultured bovine cells. J. Protozool. 14:764–772.

Fayer, R., Hammond, D. M., Chobotar, B., and Elsner, Y. Y. 1969. Cultivation of *Besnoitia jellisoni* in bovine cell cultures. J. Parasitol. 55:645–653.

Fayer, R., and Mahrt, J. L. 1972. Development of *Isospora canis* (Protozoa; Sporozoa) in cell culture. Z. Parasitenkd. 38:313–318.

Fayer, R., Romanowski, R. D., and Vetterling, J. M. 1970. The influence of hyaluronidase and hyaluronidase substrates on penetration of cultured cells by eimerian sporozoites. J. Protozool. 16:432–436.

Fayer, R., and Thompson, D. E. 1974. *Isospora felis:* Development in cultured cells with some cytological observations. J. Parasitol. 60:160–168.

Fayer, R., and Thompson, D. E. 1975. Cytochemical and cytological observations on *Sarcocystis* sp. propagated in cell culture. J. Parasitol. 61:466–475.

Fitzgerald, P. R. 1970. Development of *Eimeria stiedae* in avian embryos. J. Parasitol. 56:1252–1253.

Foley, V. L., and Remington, J. S. 1969. Plaquing of *Toxoplasma gondii* in secondary cultures of chick embryo fibroblasts. J. Bacteriol. 98:1–3.

Frenkel, J. K. 1953. Infections with organisms resembling *Toxoplasma,* together with the description of a new organism: *Besnoitia jellisoni.* In: Atti del Sixth Congresso Internationale di Microbiologia 5:426–434.

Frenkel, J. K. 1965. The development of the cyst of *Besnoitia jellisoni:* Usefulness of this infection as a biologic model. In: Progress in Protozoology. Proceedings of the Second International Congress on Protozoology, International Congress Series 91, pp. 187–188. Excerpta Medica Foundation, The Hague.

Galuzo, I. G., Konovalova, S. I., and Krivkova, A. M. 1969. The behavior of avirulent strains of *Toxoplasma* in tissue cultures and in chick embryos. In: Progress in Protozoology. Proceedings of the Third International Congress on Protozoology, pp. 229–230.

Gavrilov, W., and Cowez, S. 1941. Essai de culture in vitro de tissus de moustiques et d'intestins de lapins adultes infectes. Ann. Parasitol. Hum. Comp. 18:180–186.

Goldman, M., Carver, R. K., and Sulzer, A. J. 1958. Reproduction of *Toxoplasma gondii* by internal budding. J. Parasitol. 44:161–171.

Hammond, D. M., and Fayer, R. 1968. Cultivation of *Eimeria bovis* in three established cell lines and in bovine tracheal cell line cultures. J. Parasitol. 54:559–568.

Hammond, D. M., Fayer, R., and Miner, M. L. 1969. Further studies on *in vitro* development of *Eimeria bovis* and attempts to obtain second-generation schizonts. J. Protozool. 16:298–302.

Hammond, D. M., Sayin, F., and Miner, M. L. 1963. Uber den Entwicklungszyklus und die Pathogenität von *Eimeria ellipsoidalis* Becker and Frye, 1929, in Kalbern. Berl. Muench. Tieraerztl. Wochenschr. 76:331–333.

Hammond, D. M., Speer, C. A., and Roberts, W. 1970. Occurrence of refractile bodies in merozoites of *Eimeria* species. J. Parasitol. 56:189–191.

Hansson, H. A., and Sourander, P. 1965. *Tox-*

6

oplasma gondii in cell cultures from rat retina. Virchows Arch. (Pathol. Anat.) 338:224–236.

Hayes, K., Billson, F. A., Jack, I., Sanderson, L., Rogers, J. G., Greer, C. H., and Booth, L. 1973. Cell culture isolation of *Toxoplasma gondii* from an infant with unusual ocular features. Med. J. Aust. 60-61:1297–1299.

Hendricks, L. D., and Fayer, R. 1973. Development of *Hepatozoon griseisciuri* in cultured squirrel cells. J. Protozool. 20:550–554.

Hirai, K., Hirato, K., and Yanagawa, R. 1966. A cinematographic study of the penetration of cultured cells by *Toxoplasma gondii.* Jpn. J. Vet. Res. 14:81–90.

Hoff, R. L., Dubey, J. P., Behbehani, A. N., and Frenkel, J. K. 1977. *Toxoplasma gondii* cysts in cell culture: New biological evidence. J. Parasitol. 63:1121–1124.

Hoff, R. L. and Frenkel, J. K. 1974. Cell mediated immunity against *Besnoitia* and *Toxoplasma* in specifically and cross-immunized hamsters and in cultures. J. Exp. Med. 139:560–580.

Hogan, M. J. 1961. Discussion in the symposium on toxoplasmosis. Surv. Ophthalmol. 6:734.

Hogan, M. J., Yoneda, C., Feeney, L., Zweigart, P., and Lewis, A. 1960. Morphology and culture of *Toxoplasma.* Arch. Ophthalmol. 67:655–667.

Hogan, M. J., Yoneda, C., and Zweigart, O. 1961. Growth of toxoplasma strains in tissue culture. Am. J. Ophthalmol. 51:920–930.

Holz, A., and Albrecht, M. 1953. Die Zuchtung von *Toxoplasma gondii* in Zellkulturen. Z. Hyg. Infekt. 136:605–609.

Ishii, T., and Onaga, H. 1971. Development of *Eimeria tenella* and *E. brunetti* in chick embryos. (In Japanese; English summary) Jpn. J. Parasitol. 20:45–51.

Itagaki, K., Hifayama, N., Tsubokura, M., and Otsuki, K. 1974. Development of *Eimeria tenella, E. brunetti,* and *E. acervulina* in cell cultures. Jpn. J. Vet. Sci. 36:467–482.

Itagaki, K., Tsubukura, M., and Taira, Y. 1972. Basic biological studies on avian coccidium development of *Eimeria tenella, E. brunetti,* and *E. acervulina* in chick embryos. (In Japanese; English summary) Jpn. J. Vet. Sci. 34:143–149.

Jacobs, L. 1953. The biology of *Toxoplasma.* Am. J. Trop. Med. Hyg. 2:365–389.

Jacobs, L. 1956. Propagation, morphology, and biology of *Toxoplasma.* Ann. N. Y. Acad. Sci. 64:154–179.

Jacobs, L., Fair, J. R., and Bickerton, J. H. 1954. Adult ocular toxoplasmosis. Report of a parasitologically proved case. Arch. Ophthalmol. 53:63–71.

Jacobs, L., and Melton, M. L. 1954. Modifications in virulence of a strain of *Toxoplasma gondii* by passage in various hosts. Am. J. Trop. Med. Hyg. 3:447–457.

Jeffers, T. K., and Wagenbach, G. E. 1969. Sex differences in embryonic response to *Eimeria tenella* infection. J. Parasitol. 55:949–941.

Jeffers, T. K., and Wagenbach, G. E. 1970. Embryonic response to *Eimeria tenella* infection. J. Parasitol. 56:656–662.

Jensen, J. B., and Edgar, S. A. 1976. Effects of antiphagocytic agents on penetration of *Eimeria magna* sporozoites into cultured cells. J. Parasitol. 62:203–206.

Jensen, J. B., and Hammond, D. M. 1974. Penetration of *Eimeria magna* into cultured cells. Presented at the 49th Annual Meeting, American Society of Parasitologists, August 4–9, Kansas City, Mo.

Jögiste, A. K., and Ukhov, J. I. 1969. Host-parasite relationships in chick embryos infected with *Toxoplasma gondii.* In: Progress in Protozoology. Proceedings of the Third International Congress on Protozoology, pp. 230–231.

Jones, T. C. 1973. Multiplication of toxoplasma in enucleate fibroblasts. Proc. Soc. Exp. Biol. Med. 42:1268–1271.

Jones, T. C., Len, L., and Hirsch, J. G. 1975. Assessment in vitro of immunity against *Toxoplasma gondii.* J. Exp. Med. 141:466–482.

Jones, T. C., Yeh, S., and Hirsch, J. G. 1972. The interaction between *Toxoplasma gondii* and mammalian cells: I. Mechanism of entry and intracellular fate of the parasite. J. Exp.

Med. 136:1157–1172.

Kaufman, H. E. 1961. Discussion in the symposium on toxoplasmosis. Surv. Ophthalmol. 6:734.

Kaufman, H. E., and Maloney, E. D. 1962. Multiplication of *Toxoplasma gondii* in tissue culture. J. Parasitol. 48:358–361.

Kaufman, H. E., Melton, M. L., Remington, J. S., and Jacobs, L. 1959. Strain differences of *Toxoplasma gondii.* J. Parasitol. 45:189–190.

Kaufman, H. E., Remington, J. S., and Jacobs, L. 1958. Toxoplasmosis: The nature of virulence. Am. J. Ophthalmol. 46(Nov., Part 2):255–261.

Kelley, G. L., and Hammond, D. M. 1970. Development of *Eimeria ninakohlyakimovae* from sheep in cell cultures. J. Protozool. 17:340–349.

Kelley, G. L., and Youssef, N. N. 1977. Development in cell cultures of *Eimeria vermiformis* Ernst, Chobotar and Hammond, 1971. Z. Parasitenkd. 53:23–29.

Klimes, B., Tanielian, Z., and Ali, N. A. 1972. Excystation and development in cell culture of irradiated oocysts of *Eimeria tenella.* J. Protozool. 19:500–504.

Kusunoki, Y. 1977. In vitro study on the invasion and multiplication of *Toxoplasma gondii* sporozoites in cultured cells. (In Japanese; English summary) Jpn. J. Parasitol. 26:6–16.

Kuwahara, Ch. 1959. Observations on the culture of *Toxoplasma* in embryonated chick eggs and a histological study on chick embryo. (In Japanese; English summary) J. Osaka City Med. Cent. 8:907–925.

Lee, E.-H., Remmler, O., and Fernando, M. A. 1976. Oocysts from single sporozoite infections of *Eimeria tenella* in chick embryos. Presented at the 51st Annual Meeting of the American Society of Parasitologists, August 22–25, San Antonio, Texas.

Levaditi, C., Sanchis-Bayarri, V., Lepine, P., and Schoen, R. 1929. Étude sur l'encephalomyelite provoquee par le *Toxoplasma cuniculi.* Ann. Inst. Pasteur, Paris 43:673–736.

Long, P. L. 1965. Development of *Eimeria tenella* in avian embryos. Nature 208:509–510.

Long, P. L. 1966. The growth of some species of *Eimeria* in avian embryos. J. Parasitol. 56:575–581.

Long, P. L. 1969. Observations on the growth of *Eimeria tenella* in cultured cells from the parasitized chorioallantoic membranes of the developing chick embryo. Parasitology 59:757–765.

Long, P. L. 1970a. *Eimeria tenella:* Chemotherapeutic studies in chick embryos with a description of a new method (chorioallantoic membrane foci counts) for evaluating infections. Z. Parasitenkd. 33:329–338.

Long, P. L. 1970b. Some factors affecting the severity of infection with *Eimeria tenella* in chicken embryos. Parasitology 60:435–447.

Long, P. L. 1970c. Studies on the viability of sporozoites of *Eimeria tenella.* Z. Parasitenkd. 35:1–6.

Long, P. L. 1971a. Schizogony and gametogony of *Eimeria tenella* in the liver of chick embryos. J. Protozool. 18:17–20.

Long, P. L. 1971b. Maintenance of intestinal protozoa *in vivo* with particular reference to *Eimeria* and *Histomonas* In: A. E. R. Taylor and R. Muller (eds.), Isolation and Maintenance of Parasites in Vivo, pp. 65–75, Blackwell Scientific Publications, Ltd., Oxford.

Long, P. L. 1972a. Observations on the oocyst production and viability of *Eimeria mivati* and *E. tenella* in the chorioallantois of chicken embryos incubated at different temperatures. Z. Parasitenkd. 39:27–37.

Long, P. L. 1972b. *Eimeria mivati:* Reproduction, pathogenicity and immunogenicity of a strain maintained in chick embryos by serial passage. J. Comp. Pathol. 82:439–445.

Long, P. L. 1972c. *Eimeria tenella:* Reproduction, pathogenicity and immunogenicity of a strain maintained in chick embryos by serial passage. J. Comp. Pathol. 82:429–437.

Long, P. L. 1973a. Endogenous stages of a "chick embryo-adapted" strain of *Eimeria tenella.* Parasitology 66:55–62.

Long, P. L. 1973b. Studies on the relationship between *Eimeria acervulina* and *Eimeria mivati.* Parasitology 67:143–155.

Long, P. L. 1974a. The growth of embryo-adapted strains of *Eimeria* in cell cultures. In: Proceedings of the 3rd International Congress of Parasitology, Munich. Vol. 1. p. 121.

Long, P. L. 1974b. The growth of *Eimeria* in cultured cells and in chicken embryos: A review. In: Proceedings of the Symposium on Coccidia and Related Organisms, pp. 57–82. University of Guelph Press, Guelph, Ontario.

Long, P. L. 1974c. Further studies on the pathogenicity and immunogenicity of an embryo-adapted strain of *Eimeria tenella.* Avian Pathol. 3:255–268.

Long, P. L. 1974d. Experimental infection of chickens with two species of *Eimeria* isolated from the Malaysian jungle fowl. Parasitology 69:337–347.

Long, P. L., and Millard, B. J. 1973. *Eimeria* infection of chick embryos: The effect of known anticoccidial drugs against *E. tenella* and *E. mivati.* Avian Pathol. 2:111–125.

Long, P. L., and Millard, B. J. 1975. Pathogenicity, immunogenicity and site specificity of an attenuated strain of *Eimeria tenella.* J. Protozool. 22:53A.

Long, P. L., and Millard, B. J. 1978. Studies on *Eimeria grenieri* in the guinea fowl *(Numida meleagris).* Parasitology 76:1–9.

Long, P. L., and Millard, B. J. 1979. Studies on *Eimeria dispersa* Tyzzer, 1929, in turkeys. Parasitology 78:41–51.

Long, P. L., and Rose, M. E. 1972. Immunity to coccidiosis: Effect of serum antibodies on cell invasion by sporozoites of *Eimeria* in vitro. Parasitology 65:437–445.

Long, P. L., and Speer, C. A. 1977. Invasion of host cells by Coccidia. Symposium, Proceedings of the British Society of Parasitology 15:1–26.

Long, P. L., and Tanielian, Z. 1965. The isolation of *Eimeria mivati* in Lebanon during the course of a survey of *Eimeria* spp. in chickens. "Magon" Scientific Series, Lebanon Agr. Res. Inst. 6:1–18.

Lund, E., Lycke, E., and Sourander, P. 1961. A cinematographic study of *Toxoplasma gondii* in cell cultures. Br. J. Exp. Pathol. 42:357–362.

Lund, E., Lycke, E., and Sourander, P. 1963a. Some aspects of cultivation of *Toxoplasma gondii* in cell cultures. Acta Pathol. Microbiol. Scand. 57:199–210.

Lund, E., Lycke, E., and Sourander, P. 1963b. Study on cultured cells infected with *Toxoplasma gondii.* In: Progress in Protozoology. Proceedings of the First International Congress on Protozoology, p. 365.

Lycke, E., and Lund, E. 1964a. A tissue culture method for titration of infectivity and determination of growth rate of *Toxoplasma gondii.* Acta Pathol. Microbiol. Scand. 60:221–233.

Lycke, E., and Lund, E. 1964b. A tissue culture method for titration of infectivity and determination of growth rate of *Toxoplasma gondii.* Acta Pathol. Microbiol. Scand. 60:209–220.

Lycke, E., Lund, E., and Strannegärd, Ö. 1965. Enhancement by lysosome and hyaluronidase of the penetration by *Toxoplasma gondii* into cultured host cells. Br. J. Exp. Pathol. 46:189–199.

Lycke, E., and Norrby, R. 1966. Demonstration of a factor of *Toxoplasma* parasites into cultured host cells. Br. J. Exp. Pathol. 47:248–256.

Lycke, E., Norrby, R., and Remington, J. 1968. Penetration-enhancing factor extracted from *Toxoplasma gondii* which increases its virulence for mice. J. Bacteriol. 96:785–788.

McDougald, L. R., and Jeffers, T. K. 1976. Comparative in vitro development of precocious and normal strains of *Eimeria tenella* (Coccidia). J. Protozool. 23:530–534.

MacFarlane, J. O., and Ruckman, I. 1948. Cultivation of *Toxoplasma gondii* in the developing chick embryo. Pro. Soc. Exp. Biol. Med. 67:1–4.

Maloney, E. D., and Kaufman, H. E. 1964. Multiplication and therapy of *Toxoplasma gondii* in tissue culture. J. Bacteriol. 88:319–321.

Matsubayashi, H., and Akao, S. 1962. Morphological studies on the development of *Tox-*

oplasma cysts and a comment on the mechanism of cyst production. (In Japanese; English summary) Jpn. J. Parasitol. 11(4):13.

Matsubayashi, H., and Akao, S. 1963. Morphological studies on the development of the *Toxoplasma* cyst. Am. J. Trop. Med. Hyg. 12:321–333.

Matsubayashi, H., and Akao, S. 1965. The application of immunoelectron microscopy to the study of cyst development in toxoplasmosis. (In Japanese; English summary) Jpn. J. Parasitol. 14(4):55.

Meyer, H., and de Oliveira, M. X. 1942. Observacoes sobre divisoes mitoticas em celulas parasitadas. Ann. Acad. Bras. Cienc. 14:289–292.

Meyer, H., and de Oliveira, M. X. 1943. Conservacao de protozoaires em bulteras de tecido mantidas a temperatura ambiente. Rev. Bras. Biol. 3:341–343.

Meyer, H., and de Oliveira, M. X. 1945. Resultados de 3 anos de observacao de cultivo de "Toxoplasma" (Nicolle e Manceaux, 1909) em cultura de tecido. Rev. Bras. Biol. 5:145–146.

Meyer, H., and de Oliveira Musacchio, M. 1974. *Toxoplasma gondii* in tissue cultures; A microcinematographic study in phase contrast. Rev. Inst. Med. Trop. Sao Paulo 16:7–9.

Moltmann, V. G. 1978. Die Entwicklung von *Klossia Lelicina* (Coccidia, Adeleidea) in Schneckennieren-Gewebekulturen. Methode und erste Ergebnisse. Zentralbl. Bakteriol. (B) p. 257.

Müller, B. E. G., de Vos, A. J., and Hammond, D. M. 1973a. In vitro development of first-generation schizonts of *Eimeria canadensis* (Bruce, 1921). J. Protozool. 20:293–297.

Müller, B. E. G., Hammond, D. M., and Scholtyseck, E. 1973b. In vitro development of first and second generation schizonts of *Eimeria contorta* Haberkorn, 1971. (Coccidia, Sporozoa). Z. Parasitenkd. 41:173–185.

Naciri-Bontemps, M. 1976. Reproduction of the cycle of coccidia *Eimeria acervulina* (Tyzzer, 1929) in cell cultures of chicken kidneys. Ann. Rech. Vet. 7:223–230.

Neuman, M. 1974. Cultivation of *Besnoitia besnoiti* Marotel, 1912, in cell culture. Tropenmed. Parasitol. 25:243–249.

Nockiewicz, A. 1972. Biological properties of *Toxoplasma gondii* cultivated in the chicken embryo. Acta Biol. Cracov. (Ser. Zool.). 15:29–33.

Norrby, R. 1970a. An immunological study on the host cell penetration factor of *Toxoplasma gondii*. p. 79. Elanders Boktryckeri Aktiebolag, Goteborg. Sweden.

Norrby, R. 1970b. Host cell penetration of *Toxoplasma gondii*. Infect. Immun. 3:250–255.

Norrby, R., and Lycke, E. 1967. Factors enhancing the host cell penetration of *Toxoplasma gondii*. J. Bacteriol. 93:53–58.

Novinskaya, V. F. 1965. Use of the chick embryos in diagnostics of toxoplasmosis. (In Russian) In: I. G. Galuzo (ed.), Toxoplasmosis of Animals, pp. 391–393, Academy of Science of the Kazakh SSR, Alma-ata.

Ogimoto, K., Komatsu, M., Tsunoda, K., and Tanaka, Y. 1976. Development of *Eimeria* from Japanese quail in cell cultures. (In Japanese) Jpn. J. Parasitol. 25(suppl.):23.

Olisa, E. G. 1976. Experimental observation on the adaptation and survival of *Toxoplasma* gondii to hen's eggs. Lab. Invest. 34:341.

Onaga, H., Ishii, T., and Koyama, T. 1974. Development of *Eimeria tenella* in cultured cells from the parasitized chorioallantoic membrane. (In Japanese; English summary) Jpn. J. Vet. Sci. 36:73–80.

Osaki, H., Oka, Y., Yamamoto, K., and Matsuo, N. 1964. Isolation of *Toxoplasma* by tissue culture. (In Japanese; English summary) Jpn. J. Parasitol. 13(4):284.

Patton, W. H. 1965. *Eimeria tenella:* Cultivation of the asexual stages in cultured animal cells. Science 150:767–769.

Pulvertaft, R. J., Valentine, J. C., and Lane, W. F. 1954. The behavior of *Toxoplasma gondii* on serum-agar culture. Parasitology 44:478–485.

Remington, J. S., Earle, P., and Yagura, T. 1970.

Toxoplasma in nucleus. J. Parasitol. 56:390–391.

Roberts, W. L., Hammond, D. M., and Speer, C. A. 1970. Ultrastructural study of the intra- and extracellular sporozoites of *Eimeria callospermophili*. J. Parasitol. 56:907–917.

Roberts, W. L., Speer, C. A., and Hammond, D. M. 1971. Penetration of *Eimeria larimerensis* sporozoites into cultured cells as observed with the light and electron microscopes. J. Parasitol. 57:615–625.

Rollinson, D. 1976. Development of *Eimeria tenella* in quail embryos. Trans. R. Soc. Trop. Med. Hyg. 70:21.

Rose, M. E., and Long, P. L. 1976. Immunity to coccidiosis; Interactions in vitro between *Eimeria tenella* and chicken phagocytic cells. In: H. Van den Bossche (ed.), Biochemistry of Parasites and Host-Parasite Relationships, pp. 449–455. Elsevier/North Holland Biomedical Press, Amsterdam.

Ryley, J. F. 1968. Chick embryo infections for the evaluation of anticoccidial drugs. Parasitology 58:215–220.

Ryley, J. F., and Wilson, R. G. 1972. The development of *Eimeria brunetti* in tissue culture. J. Parasitol. 58:660–663.

Sabin, A. B., and Olitsky, P. K. 1937. *Toxoplasma* and obligate intracellular parasitism. Science 85:336–338.

Sampson, J. R., Hammond, D. M., and Ernst, J. V. 1971. Development of *Eimeria alabamensis* from cattle in mammalian cell cultures. J. Protozool. 18:120–128.

Schmidt-Hoensdorf, F., and Holz, J. 1953. Zur Biologie und Morphologie des *Toxoplasma gondii*. Z. Hyg. Infekt. 136:601–604.

Schuhová, V. 1957. Langfristige Kulturen des *Toxoplasma gondii* in He-La-Zellen. Zentralbl. Bakteriol. (B) 168:631–636.

Schuhová, V. 1960. Long-term culture of *Toxoplasma gondii* in HeLa cells. J. Hyg. Epidemiol. Microbiol. Immunol. 4:131–132.

Sheffield, H. G., and Melton, M. L. 1968. The fine structure and reproduction of *Toxoplasma gondii*. J. Parasitol. 54:209–226.

Sheffield, H. G., and Melton, M. L. 1970a. *Toxoplasma gondii:* The oocyst, sporozoite and infection of cultured cells. Science 167:892–893.

Sheffield, H. G., and Melton, M. L. 1970b. Observations on the sporozoites of *Toxoplasma gondii* and their behavior in cultured cells. J. Parasitol. 56(4; Sect. II, Part I):315.

Sheffield, H. G., Melton, M. L., and Neva, F. A. 1976. Development of *Hammondia hammondi* in cell cultures. Proc. Helminth. Soc. Wash. 43:217–225.

Shibalova, T. A. 1968. Cultivation of *Eimeria tenella* on chicken embryos and in tissue cultures. (In Russian) Parazitologiia, Leningrad 2:483–484.

Shibalova, T. A. 1969a. Cultivation of coccidian endogenous stages in chicken embryos and tissue cultures. In: Progress in Protozoology. Proceedings of the Third International Congress on Protozoology, p. 355–356.

Shibalova, T. A. 1969b. Cultivation of the asexual stages of *Eimeria tenella* in cultured tissue cells. (In Russian) Tsitologiia 11:707–713.

Shibalova, T. A. 1970. Cultivation of the endogenous stages of chicken coccidia in embryos and tissue cultures. J. Parasitol. 56(4; Sect. II, Part I):315–316.

Shibalova, T. A., Korolev, A. M., and Sobchak, I. A. 1969. Cultivation of chicken coccidia in chick embryos. (In Russian) Veterinariia 11:68–71.

Shibalova, T. A., and Petrenko, V. I. 1972. *Isospora bigemina:* Sporozoite-induced development of the life cycle in tissue culture. (In Russian; English summary) Akad. Nauk. USSR 6:201–205.

Shimizu, K. 1961. Studies on toxoplasmosis: III. Observations on the tissue culture method of *Toxoplasma gondii*. (In Japanese; English summary) Jpn. J. Vet. Sci. 23:33–44.

Shimizu, K. 1963. Studies on toxoplasmosis: V. Complemental observations on the tissue culture method, especially the effect of the nutrient fluid upon the invasion and multiplication of the organisms. Jpn. J. Vet. Res. 11:1–11.

Shirley, M. W. 1979. A reappraisal of the taxonomic status of *Eimeria mivati* Edgar and Siebold, 1964, by enzyme electrophoresis and cross immunity tests. Parasitology 78:221–237.

Shirley, M. W., and Millard, B. J. 1976. Some observations on the sexual differentiation of *Eimeria tenella* using single sporozoite infections in chicken embryos. Parasitology 73:337–341.

Shirley, M. W., Millard, B. J., and Long, P. L. 1977. Studies on the growth, chemotherapy and enzyme variation of *Eimeria acervulina* var *diminuta* and *E. acervulina* var *mivati.* Parasitology 75:165–182.

Sokolic, A., Tanielian, Z., and Ali, N. A. 1973. Studies in cell culture of development and antigenicity of (60Co) irradiated *Eimeria tenella.* "Magon" Scientific Series, Lebanon Agr. Res. Inst. 49:1–9.

Sourander, P., Lycke, E., and Lund, E. 1960. Observations on living cells infected with *Toxoplasma gondii.* Br. J. Exp. Pathol. 41:176–178.

Speer, C. A. 1979. Further studies on the development of gamonts and oocysts of *Eimeria magna* in cultured cells. J. Parasitol. 65:591–598.

Speer, C. A., Davis, L. R., and Hammond, D. M. 1971. Cinemicrographic observations on the development of *Eimeria larimerensis* in cultured bovine cells. J. Protozool. 18(suppl.):11.

Speer, C. A., de Vos, A. J., and Hammond, D. M. 1973a. Development of *Eimeria zuernii* in cell cultures. Proc. Helminth. Soc. Wash. 40:160–163.

Speer, C. A., and Hammond, D. M. 1969. Cinemicrographic observations on the development of *Eimeria callospermophili* in cultured cells. J. Protozool. 16(suppl.):16.

Speer, C. A., and Hammond, D. M. 1970a. Nuclear divisions and refractile body changes in sporozoites and schizonts of *Eimeria callospermophili* in cultured cells. J. Parasitol. 56:461–467.

Speer, C. A., and Hammond, D. M. 1970b. Development of *Eimeria larimerensis* from the Uinta ground-squirrel in cell culture. Z. Parasitenkd. 35:105–118.

Speer, C. A., and Hammond, D. M. 1971a. Development of *Eimeria ellipsoidalis* from cattle in cultured bovine cells. J. Parasitol. 57:675–677.

Speer, C. A., and Hammond, D. M. 1971b. Development of first- and second-generation schizonts of *Eimeria magna* from rabbits in cell cultures. Z. Parasitenkd. 37:336–353.

Speer, C. A., and Hammond, D. M. 1972a. Development of gametocytes and oocysts of *Eimeria magna* from rabbits in cell culture. Proc. Helminth. Soc. Wash. 39:114–118.

Speer, C. A., and Hammond, D. M. 1972b. Motility of macrogamonts of *Eimeria magna* coccidia in cell culture. Science 178:763–765.

Speer, C. A., and Hammond, D. M. 1973. Development of second-generation schizonts, gamonts and oocysts of *Eimeria bovis* in bovine kidney cells. Z. Parasitenkd. 42:105–113.

Speer, C. A., Hammond, D. M., and Anderson, L. C. 1970a. Development of *Eimeria callospermophili* and *E. bilamellata* from the Uinta ground squirrel *Spermophilus armatus* in cultured cells. J. Protozool. 17:274–284.

Speer, C. A., Hammond, D. M., and Elsner, Y. Y. 1973b. Further asexual development of *Eimeria magna* merozoites in cell cultures. J. Parasitol. 59:613–623.

Speer, C. A., Hammond, D. M., and Elsner, Y. Y. 1973c. Schizonts and immature gamonts of *Eimeria larimerensis* in cultured cells. Proc. Helminth. Soc. Wash. 40:147–153.

Speer, C. A., Hammond, D. M., and Kelley, G. L. 1970b. Stimulation of motility in merozoites of five *Eimeria* species by bile salts. J. Parasitol. 56:927–929.

Stewart, G. L., and Feldman, H. A. 1965. Use of tissue culture cultivated *Toxoplasma* in the dye test and for storage. Proc. Soc. Exp. Biol. Med. 188:542–546.

Strout, R. G. 1975. *Eimeria tenella* (Coccidia): Development in cells cultured in a chemically defined medium. Presented at the 50th Annual Meeting of the American Society of Para-

sitologists, November 10–15, New Orleans.
Strout, R. G., and Ouellette, C. A. 1968. Gametogony of *Eimeria tenella* (Coccidia) in cell cultures. Science 163:695–696.
Strout, R. G., and Ouellette, C. A. 1970. Schizogony and gametogony of *Eimeria tenella* in cell culture. Am. J. Vet. Res. 31:911–918.
Strout, R. G., Ouellette, C. A., and Gangi, D. P. 1969a. *Eimeria tenella:* Temperature and asexual development in cell culture. Exp. Parasitol. 25:324–328.
Strout, R. G., Ouellette, C. A., and Gangi, D. P. 1969b. Effect of inoculum size on development of *Eimeria tenella* in cell cultures. J. Parasitol. 55:406–411.
Strout, R. G., Solis, J., Smith, S. C., and Dunlop, W. R. 1965. In vitro cultivation of *Eimeria acervulina* (Coccidia). Exp. Parasitol. 17:241–246.
Stuer, D. 1972. Toxoplasminfektion bein Fischgeweben und Fischen in vitro. Z. Parasitenkd. 38:58.
Tos-Luty, S., Dutkiewicz, J., and Uminski, J. 1971. Influence of different sera on the in vitro reproduction of *Toxoplasma gondii.* Acta Parasitol. Pol. 19:227–236.
Turner, M. B. F., and Box, E. D. 1970. Cell culture of *Isospora* from the English sparrow, *Passer domesticus domesticus.* J. Parasitol. 56:1218–1223.
Valkoln, A., and Cinatl, J. 1978. Cultivation of *Toxoplasma gondii* in suspension cultures. J. Protozool. 25:40A.
Vetterling, J. M., Pacheco, N. D., and Fayer, R. 1973. Fine structure of gametogony and oocyst formation in *Sarcocystis* sp. in cell culture. J. Protozool. 20:613–621.
Vischer, W. A., and Suter, E. 1954. Intracellular multiplication of *Toxoplasma* in adult mammalian macrophages cultivated *in vitro*. Proc. Soc. Exp. Biol. Med. 86:413–419.
Warren, J., and Russ, S. B. 1948. Cultivation of *Toxoplasma* in embryonated eggs: An antigen derived from chorioallantoic membrane. Proc. Soc. Exp. Bio. Med. 67:85–89.
Weinman, D. 1944. Human toxoplasma. Puerto Rico J. Publ. Hlth. Trop. Med. 20:125–193.
Wolfson, F. 1941. Mammalian toxoplasma in erythrocytes of canaries, ducks, and duck embryos. Am. J. Trop. Med. 21:653–658.
Wolfson, F. 1942. Maintenance of human *Toxoplasma* in chicken embryos. J. Parasitol. 28(suppl.):16–17.
Zassuchin, D. N. 1963. Observations on the biology of *Toxoplasma.* In: Progress in Protozoology. Proceedings of the First International Congress on Protozoology, pp. 366–368.

7

Pathology and Pathogenicity

M. A. Fernando

The pathology and pathogenicity of coccidial infections were reviewed by Long in 1973. The present discussion of this subject is essentially an update of the previous review with the addition of a section on the pathology and pathogenicity of *Toxoplasma*, *Sarcocystis*, and related genera. Therefore, an historical review is not included and much of the earlier work is not discussed.

The more recent work on factors affecting the pathogenicity of coccidia, the general host response to these organisms, and the pathological changes induced are discussed, as well as the nature of the pathogenetic mechanisms involved and possible lines for future research.

FACTORS AFFECTING THE PATHOGENICITY OF COCCIDIA

It is generally accepted that the pathogenicity of a given species of coccidia is variable. Factors inherent within the particular isolate or strain, environmental factors acting on the parasite, both outside and within the host, and the genetic makeup and immune status of the host itself tend to influence the severity of the infection and the outcome of the disease. These factors, the manner in which they influence coccidial infections, and some of the more practical implications are discussed below. Long (1973) has reviewed much of the older literature on this subject; the present discussion, therefore, is mostly limited to recent work with excerpts from Long's (1973) review.

Dose of Oocysts

It has long been considered that the reaction of a host to coccidial infection depends on the numbers of sporulated oocysts ingested. Because coccidia have a self-limiting life cycle terminating in the formation of oocysts, this assumption has, within limits, been proved correct.

Long (1973, 1978) reviewed the published work on the relationship of dose of oocysts to both the numbers of new oocysts produced by the infected host and the severity of the pathological changes induced. Although increasing the number of oocysts ingested gives rise to a proportional increase in the oocyst output, several workers have noted a "crowding" effect when excessive numbers of oocysts are given, resulting in fewer oocysts being produced per oocyst ingested (Krassner, 1963; Lotz and Leek, 1970; Williams, 1973). Long (1973) offered several possible explanations for this observation. 1) Interferon or an interferon-like substance may be produced in response to a coccidial infection; interferon has been shown by Fayer and Baron (1971) to inhibit the intracellular development of *Eimeria tenella* in vitro. Long and Milne (1971) could not detect any antiviral substance in the allantoic fluid of chick embryos infected with *E. tenella* and found only small amounts in the sera of three of the large numbers of

E. tenella-infected chickens examined. However, these small amounts may be significant locally, at the site of infection. 2) The exposure of the host to massive numbers of early stages of the parasite may stimulate immune mechanisms effective against later stages in the life cycle. This kind of an effect has been shown by Lotz and Leek (1970) in sheep infected with large numbers of *Eimeria intricata* oocysts. These authors found degeneration of second generation schizonts 17 days after heavy infections and suggested that immunity stimulated by the early stages led to this reaction. 3) In some species, the "crowding factor" may operate in a simpler way, that is, large numbers of early stages destroy tissue so that merozoites fail to invade new cells.

An increase in the number of oocysts ingested by the host is usually accompanied by an increase in the severity of the disease (Hein, 1968, 1969, 1971, 1974; Long, 1973). However, Leathem and Burns (1968) noted that very heavy doses of oocysts produced lower mortality in cecal coccidiosis of chickens. It is possible that the invasion of very large numbers of sporozoites and/or the development of the early stages produce a host reaction resulting in loss of some invasive stages (Rose et al., 1975).

Viability and Virulence of Oocysts

As noted earlier, intraspecific variations and environmental factors influence the virulence and viability of oocysts. Long (1973) discussed the evidence presented by various workers to show that isolates or strains of the same species may differ in their pathogenicity. Norton and Hein (1976) compared two laboratory strains and a fresh isolate of *Eimeria maxima*. They found that one of the laboratory strains was less pathogenic but produced more oocysts than the other two strains examined. Dikovskata (1974) investigated the pathogenicity of 13 strains of *E. tenella* from different areas of the USSR. The mortality of infected birds varied from 12.5–80% between these strains. Lee (E. H. Lee, personal communication) noted that a drug-resistant strain of *E. tenella*, isolated from a field outbreak, was more pathogenic than other field strains and a laboratory-maintained strain. It may, therefore, be possible to select strains of a given species of coccidia for high or low virulence. For example, one could conceivably alter a laboratory strain by always collecting the oocysts passed by the host during a selected, short, time interval. Other parasites, particularly nematodes, have been selected for certain traits in this manner (Le Jambre, 1977; Michel et al., 1979).

The viability of oocysts and their ability to infect a host are affected by many environmental factors. Oocysts must undergo sporulation before they are infective; this process is dependent on temperature, humidity, and oxygen tension. The optimum temperature has been noted as 29° and 30°C (Marquardt et al., 1960; Long, 1973).

Exposure to either high or very low temperatures prior to sporulation is detrimental to a high proportion of ooocysts. Long (1973) noted that, of unsporulated oocysts held at 4°C for 14 weeks, 46% sporulated when moved to 28°C for 48 hr, but oocysts held at 4°C for 26 weeks failed to sporulate when placed at 28°C. However, Marquardt et al. (1960) found that unsporulated oocysts of *Eimeria zuernii* were unharmed by freezing in water at −7° to −8°C for as long as 2 months. After 2 weeks at this temperature, 92% sporulated, and after 2 months, 74% sporulated when transferred to room temperature. These results represent differences between the two species of Eimerian oocysts; *E. zuernii* is more adapted to survive at lower temperatures in the unsporulated state than the coccidia of chickens. It would be interesting to test the survival of coccidial oocysts from species of birds that inhabit colder climates, for example, Canada geese. Skene et al. (1981) had evidence to suggest that *Eimeria hermani* oocysts could survive a Canadian winter, but it is not known whether they over-wintered in the sporulated or unsporulated state.

Sporulated oocysts of most species survive for long periods of time at low temperatures. Helle (1970) had strong evidence to suggest that oocysts from ovine coccidia over-wintered in Norway to provide the main source of infection for lambs in spring. The environment in question has a permanent snow cover and subzero temperatures from December to March or April. This author believes that oocysts survive in the soil and that lambs become infected because they eat soil in the spring. Skene et al. (1981) found that Canada geese in a waterfowl sanctuary in Southern Ontario passed small numbers of oocysts throughout the winter months and that goslings hatched in the spring began passing *E. hermani* oocysts when 7 days old. Kheysin (1972) reported the survival of a few oocysts of *Eimeria acervulina* and *E. tenella* exposed to a Siberian winter. The ability of sporulated oocysts to survive even the harshest of winters plays an important part in the infection of susceptible animals in the spring. Although not important in poultry, this aspect is exceedingly important in the epidemiology of coccidiosis in cattle, sheep, and wildlife.

High temperatures are more detrimental to oocysts. Platz (1977) studied the survival of *E. tenella* oocysts after heat decay of coccidia containing chicken manure. When peak temperatures were 65–66°C, none of the oocysts survived. However, at 60°C, some oocysts did survive and were able to sporulate and infect 1-day-old chicks.

Marquardt et al. (1960) studied the effects of centrifugation, relative humidity, solar radiation, oxygen, pH, and certain chemical agents on the sporulation of oocysts of *E. zuernii*. At 10^{-6} M $HgCl_2$, only 3% sporulated, but various other chemicals, including copper sulfate and ammonium chloride, had no effect. Unsporulated oocysts exposed to sunlight for 8 hr did not survive. Centrifugation in water at 6,000 × *g* for 2 hr did not kill the parasite.

Flotation in a saturated solution of sodium chloride is the method most frequently used to separate oocysts from fecal debris. It is therefore important to assess the effects on oocysts of prolonged contact with this solution. Ryley and Ryley (1978) studied these effects in 10 species of chicken, rabbit, and cattle coccidia. Even though appreciable deformation and collapse of the oocysts were noted after 1–2 days of contact, these effects were reversible on washing, and subsequent sporulation was not affected. *Eimeria stiedai* was found to be the least sensitive and *E. tenella* the most sensitive.

Site of Development within the Host

Most coccidial species develop within the intestinal tract of their hosts. Those that develop in other organs, for example, *Eimeria truncata* in goose kidney and *E. stiedae* in rabbit biliary epithelium, tend to cause more damage. Within the intestinal tract itself, the species that develop superficially, within villous epithelial cells, are less pathogenic than those that develop deeper, within the lamina propria. Long (1973) discusses this subject in detail.

Age and Sex of Host

Younger animals are generally assumed to be more susceptible to infectious disease than their older counterparts. However, it has been shown, both in chickens and mammals, that older animals raised coccidia-free are as susceptible or more susceptible than very young ones to similar doses of oocysts (see Long, 1973, for review). Visco et al. (1978) found no difference in the prevalence of coccidial infections in cats of different ages. Hein (1971, 1974) found that growth retardation was more severe in 2-week-old than in 6-week-old chickens given similar doses of *Eimeria necatrix* or *Eimeria brunetti*. At high dose levels, however, the oocyst production was higher in the older birds. Krassner (1963) made similar observations using *E. acervulina*. Rose (1967) found that excystation of *E. tenella* sporozoites was more rapid in chicks aged 4, 5, and 6 weeks than in those 0, 1, 2, and 3 weeks of age. Also, in birds 0–1 week of age, a greater proportion of sporulated oocysts were discharged in the feces a few hours after inoculation.

Strain or Breed of Host

Most of the work on genetic resistance or susceptibility to coccidial infections has been done with chickens. This aspect is reviewed in detail by Chobotar and Scholtyseck (Chapter 4, this volume) and by Jeffers (1978). Suffice it to say that genetic resistance to death from coccidiosis need not necessarily be accompanied by a decrease in other pathogenic effects such as weight loss or blood loss.

Other Factors

It is often found that the environment within the intestine, the bacterial flora, and the nutritional status of the host affect the pathogenicity of coccidial parasites. For example, several workers have suggested a synergism between *E. tenella* and the normal intestinal flora of chickens. Visco and Burns (1972a) found that 14-day-old and 40-day-old bacteria-free chicks were more resistant than conventional controls to cecal coccidiosis, as assessed by mortality, gross cecal pathology, and weight gain. Oocyst production was not affected. Visco and Burns (1972b), using monoflora and diflora chicks, observed that *Clostridium perfringens* monoflora chicks developed the most severe clinical disease when infected with *E. tenella*. Other workers have also associated *C. perfringens* with clinical disease caused by *E. tenella* and *E. brunetti* infections (Hein and Timms, 1972; Bradley and Radhakrishnan, 1973). Nagey and Mathey (1972) incriminated pathogenic *Escherichia coli* in the deaths of chicks infected with a non-lethal dose of *E. brunetti*. Owen (1975) was unable to infect gnotobiotic mice with *Eimeria falciformis*, even though the oocysts hatched within the intestine. Visco and Burns (1972c) observed that highly susceptible strains of birds remained so, even in the absence of the normal bacterial flora. They did not find a difference in the pathogenicity of *E. tenella* in susceptible strains of bacteria-free and conventional chickens with the doses they used. They stated, however, that if the dose of oocysts was lowered, a critical point may be reached when the pathogenic effects would be more severe in the conventional birds. Therefore, it seems that *E. tenella*, at least, is able to cause disease in both bacteria-free and conventional chicks and that the bacterial flora tends to increase the pathogenic effects of the parasite.

Similarly, it has been shown that aflatoxins in the diet increase the severity of *E. tenella* infections (Wyatt et al., 1975) and *E. acervulina* infections (Ruff and Wyatt, 1978).

The nutritional status of the host is known both to increase and decrease the severity of coccidiosis. Sharma et al. (1973) showed that chicks infected with *E. tenella* and fed 24% crude protein had a higher mortality rate than those fed 16 or 20% crude protein. However, in *E. acervulina* infections, the higher crude protein diet was protective against weight loss. Britton et al. (1964) also found that birds on a high protein diet were more susceptible to infection. Pout and Catchpole (1974) found that lambs on a lower plane of nutrition had a shorter period of weight loss and recovered, after coccidial infection, faster than better nourished lambs.

PATHOLOGY AND PATHOGENESIS WITH CLINICAL CORRELATION

Under natural conditions, most birds and mammals pass small numbers of coccidial oocysts in their feces, without apparent ill effects. Coccidiosis becomes important as a disease when animals live, or are reared, under conditions that permit the build-up of infective oocysts in the environment. The intensive rearing of domesticated animals and the concentration of wild birds and mammals in waterfowl sanctuaries and game farms seem to provide these conditions. Here, susceptible hosts are exposed to heavy infections leading to clinical disease.

7

Published work on the pathology and pathogenesis of coccidial infections can be divided into two categories: observations on naturally occurring outbreaks of coccidiosis and descriptions of experimentally induced disease. In the latter, animals are usually infected with known doses of one or more species of coccidia. This discussion includes both types of studies; however, the numerous papers published on the subject are not reviewed.

General Response of the Intestinal Tract

Because the endogenous development of most species of coccidia occurs within the intestines of their hosts, a discussion of the general response of the intestine to coccidial infection is warranted. Both asexual and sexual stages are site-specific and are most frequently found within villous epithelial cells. Less frequently, they infect crypt epithelial cells and cells within the lamina propria.

The development of species within villous epithelial cells of the small intestine is often accompanied by villous atrophy (Figures 1 and 2), which has been reported in intestinal coccidiosis of lambs, chickens, and man caused by *Eimeria crandallis*, *E. acervulina*, and *Isospora belli*, respectively (French et al., 1964; Pout, 1967a,b, 1974a; Brandborg, 1971; Fernando and McCraw, 1973; Trier et al., 1974).

This change in the villous structure to blunt, greatly shortened, mucosal projections with abnormal epithelial differentiation occurs in a wide variety of etiologically unrelated intestinal diseases, lending support to the hypothesis that the intestinal mucosa responds nonspecifically to a wide variety of stimuli (Sprinz, 1962). Hence, altered morphology seems to reflect a response inherent to the mucosa and nonspecific to the damaging agent.

More recently, Ferguson and Jarrett (1975) had evidence to indicate the thymus dependence of experimental villous atrophy in rats, produced by the nematode parasite *Nippostrongylus brasiliensis*. They consider villous atrophy, at least in the rat, to be a hypersensitivity reaction in the small intestine. Manson-Smith et al. (1979) made similar observations in mice infected with *Trichinella spiralis*.

More severe damage to the intestinal mucosa accompanies the development of coccidia that parasitize the lamina propria and beyond. For example, hemorrhage and extensive destruction of tissue occur within the intestine during the development of second generation schizonts of *E. necatrix*. The villi, however, appear in structure unchanged (Figure 2).

The intestines of animals heavily infected with coccidia may appear flaccid and distended. Oikawa and Kawaguchi (1974) studied the effect of coccidial infection on acetylcholine-induced contraction of the digestive tract of chickens. They found that *E. tenella* infection caused a depression of cecal and distal small intestinal response and that *E. acervulina* infection

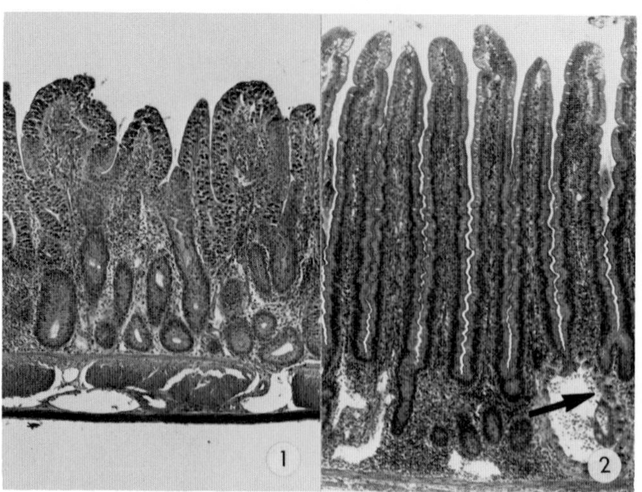

also caused a depression of distal small intestinal response but an enhancement of the response in the proximal (infected) part of the small intestine.

Infections in Birds

The Domestic Fowl Commercial chickens are reared under conditions ideal for the propagation of coccidial infections. The disease itself is not seen as frequently as it might be, because of the continuous use of anticoccidial drugs. However, outbreaks do occur, especially during the emergence of drug-resistant strains.

Fairly accurate descriptions of disease caused by the various species of coccidia after experimental infections are available (see Long, 1973, for review). The clinical disease is an enteritis, which may or may not be accompanied by hemorrhage.

Hemorrhage, and in heavy infections, anemia and death occur in *E. tenella* and *E. necatrix* infections. Heavy *E. necatrix* infections are more severe, perhaps because almost the entire small intestine is parasitized. These two species are the most pathogenic and, in both, lesions are associated with the development of second generation schizonts in cells within the lamina propria. Stockdale and Fernando (1975) described the development of the lesions associated with *E. necatrix* schizonts. First generation merozoites were found in crypt epithelial cells, probably those around the ruptured first generation schizont (Figure 4). As development proceeded, infected cells were found in the lamina propria, around the crypts they initially parasitized. It has, however, not been proved with any certainty whether the infected crypt cells themselves migrate out into the lamina propria or whether the trophozoites re-enter or are phagocytized by other cells found within the lamina propria, for example, macrophages or fibroblasts. Knowledge of what actually happens would help us understand the pathogenic mechanisms involved in both *E. necatrix* and *E. tenella* infections.

Contrary to the belief that lesions, particularly hemorrhage, occur during the maturation of colonies of large second generation schizonts, it has been found in this laboratory that hemorrhage into the lamina propria begins to appear as early as 72 hr post-infection. Hein (1971) also noted "lesions" as early as day 3, in chickens infected with over 80,000 oocysts of *E. necatrix*. At this time, the schizonts are immature and the infected cells are not enlarged. A mechanism other than mechanical disruption by large, infected host cells must be sought. The main lesions seem to be a progressive destruction of uninfected tissue, including blood vessels and the muscularis mucosa, and their replacement by an inflammatory infiltrate and hemorrhage (Figure 5). The lesions are most severe about 5 days post-infection. Heterophils are found as early as 65 hr post-infection, migrate into crypt lumina immediately after the rupture of first generation schizonts, and are seen up to 5 days post-infection (Figures 3 and 4). Lymphocytes arrive later in the infection, during the maturation of second generation schizonts.

7

Figures 1 and 2. 1. Partial villous atrophy in chicken duodenum, 5 days post-infection with *E. acervulina*. × 70. 2. Small intestine of chicken, 4 days post-infection with *E. necatrix*. Note morphologically normal villi with destruction of tissue deep in the lamina propria. *Arrow* indicates developing second generation schizonts. × 70.

Hein (1971) studied the effects of various doses of *E. necatrix* in 2- and 6-week-old chickens. Severe hemorrhagic enteritis was seen after infections with 20,000 or more oocysts; there were no survivors when the dose was 80,000 or more. Death tended to be sudden and occurred earlier as the dose of oocysts was increased. Birds were often found dead before any marked change was noted in either mean weight or packed cell volume and before blood was found in the feces.

Michael and Hodges (1972) compared the pathogenic effects of single (20,000) and repeated (5,000 × 4 oocysts) doses of *E. necatrix* infections in 6-week-old chicks. The single dose was more pathogenic and caused mortality. Lesions typical of severe hemorrhagic enteritis were evident in both types of infections, but were detected earlier with the single dose. The repeated infections may have produced some immunity. Sudden death, similar to that described by Hein (1971), was also observed by Michael and Hodges (1972).

Heavy infections of other species of chicken *Eimeria* (*E. acervulina*, *E. brunetti*, and *E. maxima*) produce lesions, suppression or loss of weight, and lowered egg production (Hein, 1968, 1974, 1976; Michael and Hodges, 1971; Long, 1973; Sharma et al., 1973).

With *E. acervulina*, the main lesions seen in the upper small intestine are associated with villous atrophy and increased cellularity of the lamina propria (Pout, 1967a,b; Fernando and McCraw, 1973). The lesions are accompanied by an increased rate of epithelial cell turnover and shortened crypt cell generation cycle (Fernando and McCraw, 1973, 1977). In an ultrastructural study of infected villous epithelial cells, Humphrey and Turk (1974) observed centrioles in epithelial cells of the middle third of the villi during the recovery period, indicating rapid epithelial cell turnover even during this period. Other ultrastructural abnormalities observed at 5 days post-infection included nonspecific alterations of cell membranes and organelles similar to those seen in other types of mucosal injury. The mitochondrial alterations observed were similar to those that have been noted after cell injury by compounds that uncouple oxidative phosphorylation. Humphrey and Turk (1974) suggested that the observed mitochondrial abnormalities in *E. acervulina* infection may indicate that an oxidative phosphorylation uncoupling toxin is released as a result of the infection. They and Sharma and Fernando (1975) found vesicles that have the appearance of lipid accumulations in infected cells. These vesicles are perhaps an indication of the malfunctioning of intracellular transformations of absorbed fatty acids, monoglycerides, and diglycerides.

In *E. maxima* infections, the lesions are seen mainly during the development of the large male and female gamonts in villous epithelial cells. The intestinal wall is thickened, and the villi are shortened and infiltrated with mononuclear cells. A thick mucoid discharge is seen in the lumen.

The lesions seen in *E. brunetti* infections have been described by Davies (1963) and Hein (1974). Typical lesions of hemorrhagic enteritis were observed in the distal part of the small intestine, 4–5 days post-infection. Normal intestinal contents were scanty and the mucous mem-

Figures 3–6. 3 and 4. Chicken intestinal crypts, 67 hr post-infection with *E. necatrix*. × 770. Note large numbers of heterophils in distended crypt (Figure 3) and early stages of the second asexual generation within crypt epithelial cells (*arrow*, Figure 4). Heterophils are seen surrounding the crypt and within the crypt lumen in Figure 4. 5. Heavily infected small intestine of chicken, 5 days post-infection. Note the absence of tissue in large areas of the lamina propria. Mature second generation schizonts are seen migrating towards the intestinal lumen. The lumen contains blood, mucus, and tissue debris. × 70. 6. Goose kidney harobring gametocytes of *E. truncata*. × 190.

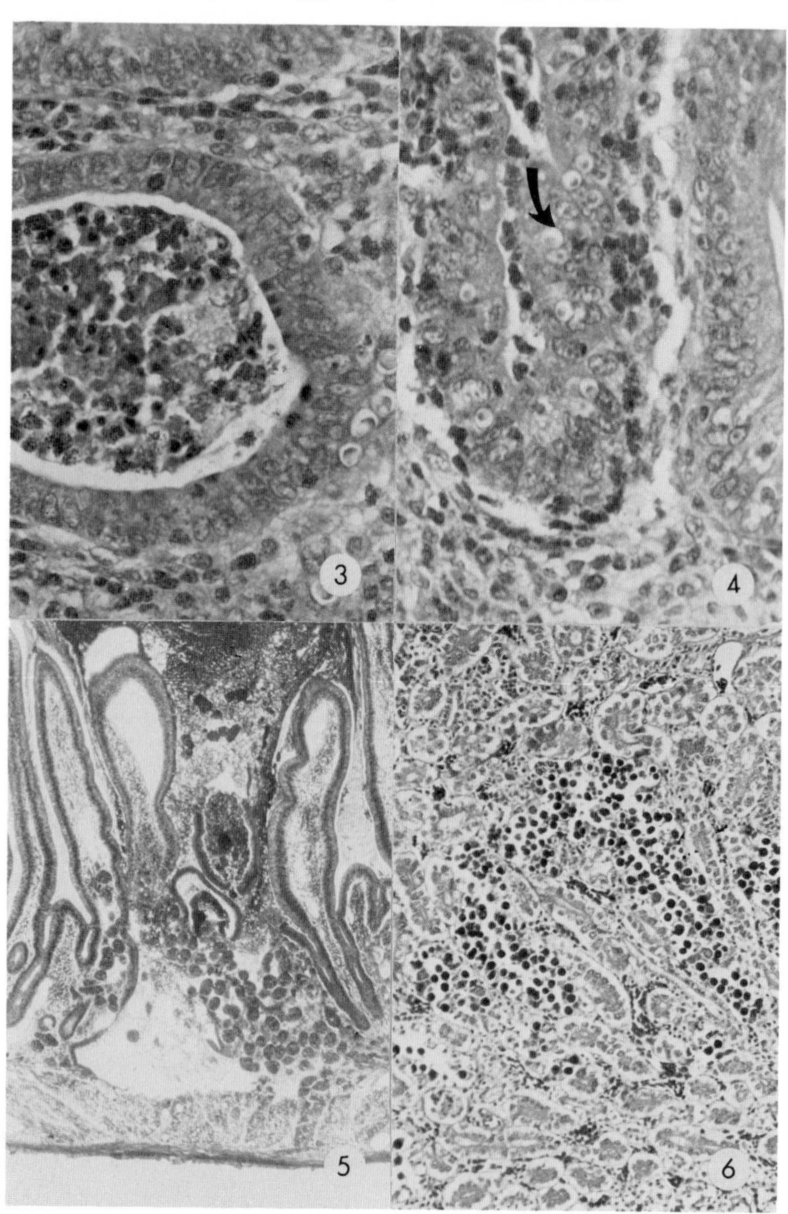

brane was covered with a reddish mucoid exudate, which persisted during the 5th and 6th days. By the 7th day, the birds were beginning to recover.

Turkeys, Ducks, and Geese In turkey coccidia, the sexual stages produce disease and give rise to marked lesions. *Eimeria adenoeides* lesions occur in the lower small intestine and ceca, *Eimeria meleagrimitis* shows a necrotic lesion in the duodenum and upper small intestine, and *Eimeria meleagridis* infections are associated with yellow caseous material localized in the ceca (Clarkson, 1960). *E. adenoeides* and *E. meleagrimitis* are known to be very pathogenic.

Hein (1969) described the pathogenic effects of *E. adenoeides* and *E. meleagrimitis* in turkey poults. Small doses of oocysts led to depressed weight gains and the severity of the disease increased progressively as the dose increased. Mortality coincided with schizogony when high doses of oocysts were given and with gametogony when low doses of oocysts were given.

In turkey poults experimentally infected with *E. meleagrimitis*, edema, lymphocytic infiltration, and mucoid degeneration of the epithelial cells at the tips of the villi occurred 4 days post-infection (Hawkins, 1952). Very little hemorrhage was noted in the infected intestine.

In ducks, *Tyzzeria perniciosa* seems to be the most pathogenic species, producing lesions similar to those of *E. necatrix* in chickens (Long, 1973). Recently, Nation and Wobeser (1977) reported that 10 of 12 species of wild ducks examined in central Saskatchewan, Canada, were infected with renal coccidia. The greatest prevalence of infection occurred in female and juvenile birds and no gross lesions attributable to coccidia were found. Microscopic lesions were focal in nature. These authors did not give a precise diagnosis of the species involved but stated that it most closely resembled the published descriptions of *Eimeria boschadis*, the species found in white mallards in Sweden. Wobeser (1974) also reported renal coccidiosis in three of 45 mallards and two of seven pintail ducks collected in Saskatchewan. He found a mild ureteritis and interstitial nephritis associated with the coccidial infection and suggests that the parasite could be pathogenic.

In geese, two forms of coccidiosis are recognized: a renal and an intestinal form. *E. truncata* is responsible for renal coccidiosis (Figure 6), whereas 13 species undergo development in the intestine.

Renal coccidiosis has been described as a common disease of domestic geese in Europe and in North America. Critcher (1950) first reported the disease in Canada geese at the Pea Island Migratory Waterfowl Refuge, North Carolina. Hanson et al. (1957) cited further winter losses of Canada geese at Pea Island, attributed to renal coccidiosis. While these and other authors (Levine et al., 1950; Farr, 1954) presented evidence of the pathogenicity of *E. truncata* to both domestic and wild geese, Klimes (1963) supported by Pellérdy (1974) believed that it is harmless, or only mildly pathogenic and rarely causes death.

Of the intestinal species of goose coccidia, *Eimeria anseris* and *Eimeria nocens* have been described as causing acute disease in domestic goslings (Klimes, 1963). Klimes (1963) also noted that the pathogenicity of the above species was greater when they occurred together. Randall and Norton (1973) reported acute intestinal coccidiosis in a flock of 200 geese with about 50 deaths and in a flock of 40 birds with eight deaths. *E. nocens* was the only coccidial parasite identified.

Infections in Mammals

Cattle Levine and Ivens (1970) reported that most of the coccidia of cattle produce some pathogenic effects on their hosts. *E. zuernii* and *Eimeria bovis* (Figures 7–9) are considered to be the most pathogenic. Hemorrhagic enteritis occurs in infections with both these species. In severe clinical cases, there is marked diarrhea, with feces containing stringy masses of mucus and clotted blood, often accompanied by a loss of appetite, dehydration, and general weakness (Fox, 1978).

The disease caused by *E. zuernii* has been difficult to reproduce experimentally. Stockdale and Niilo (1976) were able to reproduce it in calves by treating them with dexamethasone, 12, 15, and 16 days post-infection. Untreated calves developed diarrhea 19–23 days post-inoculation. In infected and dexamethasone-treated calves, however, the diarrhea changed to dysentery on days 20 and 21 post-infection. If the calves survived long enough, the dysentery reverted back to diarrhea. These results confirm the earlier observations of Niilo (1970b).

Outbreaks of coccidiosis in cattle during the cold winter months is common in the Prairie Provinces of Canada and in the northwestern states of the U. S. The disease, called "winter coccidiosis," usually affects weaned calves, 6 months of age or older. The mechanism of infection is unknown and *E. zuernii* is mostly incriminated. Various predisposing factors, such as stress due to cold, have been associated with the disease (Niilo, 1970a).

Stockdale (1977) described the pathogenesis of the lesions produced by *E. zuernii* in calves. The second asexual generation and the gametogonic stages seemed to be the pathogenic stages. Lesions were seen later than 17 days post-infection, and the proximal 30 cm of the spiral colon and the whole of the cecum were affected. In most of the calves, the epithelium was completely absent from this area, and the lamina propria was covered with a diphtheritic membrane. The intestine was hyperemic, dilated, and flaccid. Calves began to recover after 26 days post-infection.

A similar diphtheritic membrane was observed by Hammond et al. (1944) in calves infected with *E. bovis*. Fitzgerald and Mansfield (1972) studied the effects of a coccidial infection with 98% *E. bovis* oocysts in Holstein-Friesian calves. Animals surviving severe infections weighed an average of 22–27 kg less than controls after 10 months. Feed consumption was reduced for 13 weeks and water for 4–5 weeks post-infection.

Figures 7–9. Sections of the small and large intestine of calves after inoculation of *E. bovis* oocysts. 7. Eighteen days after inoculation. Note large first generation schizont in central lacteal, which has caused gross distension of the villus. × 350. 8. Sixteen days after inoculation. Note young gametocytes (*g*) in the eipthelial cell of a crypt in the large intestine. The parasitized cell has an enlarged nucleus. × 750. 9*a* and *b*. Eighteen days after inoculation. Note large number of macrogametocytes in *a* and oocysts in *b* which have destroyed the epithelium lining the crypts. × 400. (Figures 7–9 are kindly supplied by P. L. Long from photographs provided by the late D. M. Hammond.)

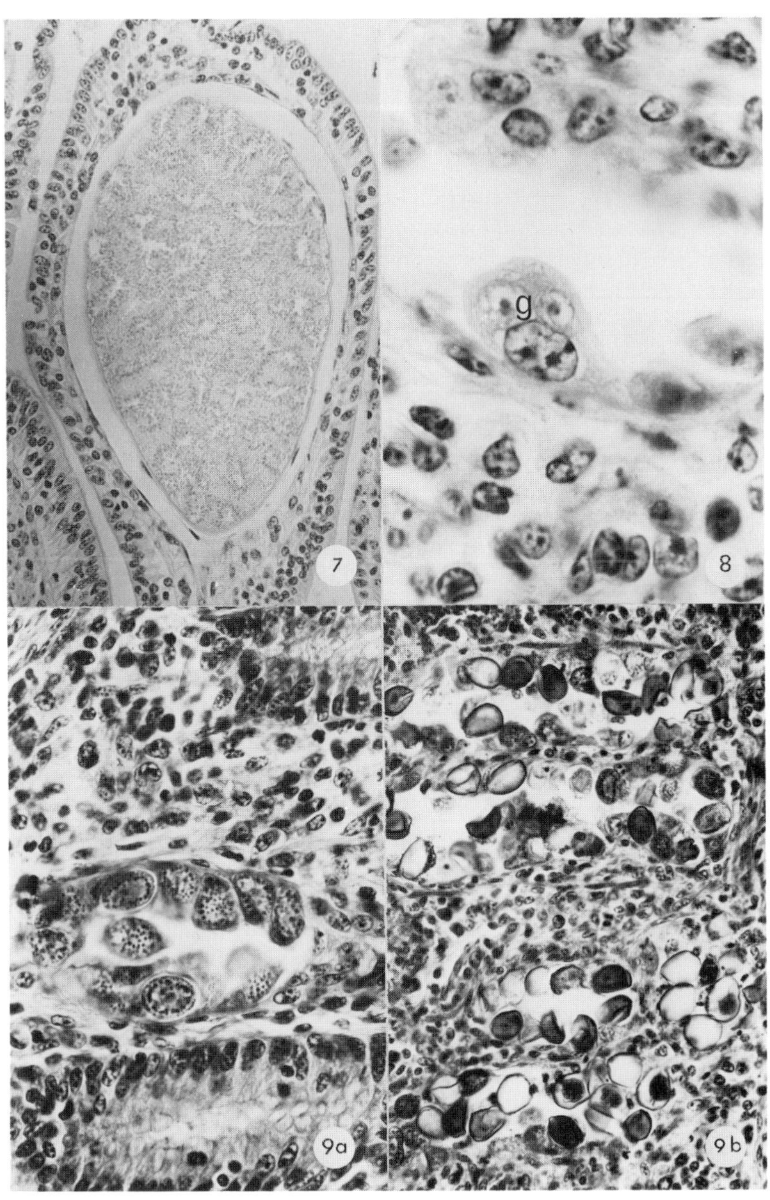

Bovine coccidiosis may be accompanied by nervous signs that include opisthotonos, medial strabismus, tetanic spasms, and convulsions (Julian et al., 1976; Jolley and Bergstrom, 1977). In the two outbreaks reported by Julian et al. (1976), the affected animals died in convulsions after an illness of 1 to several days, during which time they showed periodic nervous signs. At necropsy, the blood vessels in the brain were found to be dilated and congested, with occasional hemorrhage into the Virchow-Robin spaces. The spiral colon and cecum were heavily infected with coccidia, and oocysts were found in scrapings. The etiology of the nervous signs remains obscure. Julian et al. (1976) suggested that toxins produced by the coccidia may be the cause and should perhaps be considered together with hypoglycemic tetany.

Bovine cryptosporidiosis is being increasingly reported in outbreaks of diarrhea in calves (Meuten et al., 1974; Schmitz and Smith, 1975; Powell et al., 1976; Pohlenz et al., 1978). Pohlenz et al. (1978) do not consider it possible to draw conclusions as to whether Cryptosporidia were responsible for the diarrhea in the calves. However, these authors recommend that Cryptosporidia be regarded as a common enteric pathogen of calves.

Sheep Large numbers of oocysts may be found in the feces of clinically normal sheep. In young lambs, clinical disease in the form of diarrhea and failure to gain weight is diagnosed as coccidiosis because of the presence of a large number of coccidial oocysts in fecal samples. Pout (1976) believes that coccidiosis is only a contributory factor to the diarrhea, which may be caused primarily by a nutritional imbalance or a bacterial infection. *Eimeria ovina* is the commonest species found in sheep and is considered to be the most pathogenic.

In a series of papers, Pout and his co-workers discussed several aspects of the pathogenicity of sheep coccidia and the pathology and clinical aspects of coccidial infections in lambs (Pout, 1973, 1974a,b; Pout et al., 1973; Pout and Catchpole, 1974). They confirmed that lambs acquired a high oocyst burden at 2–4 weeks of age with a progressive reduction thereafter. Experimental infection of coccidia-free lambs with pure cultures of *Eimeria arloingi* (*E. ovina*) and *E. crandallis* gave rise to a lesion similar to that of the "flat mucosa" in areas of the small intestine with a high density of parasites (Pout, 1974a). Pout and co-workers also noted a reduction in alkaline phosphatase activity judged by histochemical techniques and a diminution of the width of the brush border even in areas that were not parasitized. Pout and Catchpole (1974) found that diarrhea occurred at about the beginning of patency in all animals.

Pigs Coccidial parasites are not considered a major cause of disease in pigs and many animals are asymptomatic carriers. Recent studies have shown, however, that disease does occur in young pigs (Morin et al., 1980). Stuart et al. (1978) reported that scours associated with coccidia was diagnosed in baby pigs from 54 swine farms by pathologists at the University of Georgia. Previously healthy pigs developed scours at 5–10 days of age and lesions were characterized by villous atrophy and various degrees of erosion of the mucosa of the jejunum

7

and ileum. Although morbidity was high, mortality was usually low to moderate. *Isospora suis* was the only species of coccidia diagnosed. Rommel (1970) studied the course of *Eimeria scabra* and *Eimeria polita* infection in 80 piglets and weaner pigs reared coccidia-free. Infection with 200 or more oocysts of either species gave rise to temporary constipation or diarrhea in weaners. The infection was milder in suckling pigs.

Dogs and Cats At least 13 species of coccidia are found in canine feces; however, little is known about clinical coccidiosis in dogs (Dubey, 1976, 1978).

Dubey et al. (1978) presented circumstantial evidence to attribute clinical disease and pathological changes in a 10-week-old pup, which died after a history of weight loss, to coccidiosis. These authors believe that the species involved is an unidentified, *Isospora ohioensis*-like coccidial parasite pathogenic to dogs. In a subsequent study of the effects of experimental *I. ohioensis* infection in young pups, Dubey (1978) found that five of 21 newborn pups infected became ill, but none of the 26 older pups showed any clinical signs. Villous atrophy and necrosis of the tips of the villi were seen at autopsy.

The pathogenicity of the other species of dog coccidia is not known. Under experimental conditions, none of the six parasite-free dogs had diarrhea after ingesting 100,000 sporulated *Isospora canis* oocysts (Dubey, 1976).

With the exception of *Toxoplasma*, feline coccidia are probably nonpathogenic to cats under experimental conditions (Dubey, 1976). Whether clinical coccidiosis occurs under natural conditions is not known for certain and needs further investigation.

Cats are infected with *Toxoplasma* by ingesting infected animals or sporulated oocysts. Simultaneous to the cycle in the intestine, *Toxoplasma* invades the extraintestinal organs of the cat. Toxoplasmosis in cats is generally asymptomatic, but the acute infection may be accompanied by fever, hepatitis, mesenteric adenitis, and pneumonia with dyspnea (Frenkel, 1978). The subacute or chronic disease may be accompanied by anemia, leukopenia, chorioretinitis, iritis, and encephalitis (see below). The pathology of feline toxoplasmosis has been described by Hirth and Nielson (1969).

Rabbits and Rodents Several species of coccidia produce pathogenic effects in rabbits. The biliary coccidia *E. stiedai* and the intestinal species *Eimeria magna* and *Eimeria irrisidua* are known to be the most pathogenic (see Long, 1973, for review).

The pathogenicity of rodent coccidia is not well understood. Recently, Mesfin et al. (1977) described the pathological changes caused by *E. falciformis* var. *pragensis* in mice. Depression, anorexia, weight loss, diarrhea or dysentery, and dehydration were most pronounced, 8–10 days post-infection. Thirty-one percent mortality occurred in mice infected with 20,000 oocysts; however, none of the mice infected with 500 oocysts died.

The enteric lesions were restricted to the large intestine and consisted of crypt and villous epithelial cell destruction and submucosal edema. Hemorrhage was sometimes seen, 7–11 days post-infection. The most significant finding was the necrotic enteritis seen 8–12 days after infection in mice with 5,000 to 20,000 oocysts.

Man Human infection with coccidia has been considered both rare and usually subclinical. However, acute or chronic diarrhea associated with coccidiosis is being reported with increasing frequency (Brandborg et al., 1970; Trier et al., 1974; Syrkis et al., 1975; Meisel et al., 1976; Ravenel et al., 1976).

Documented complications include steatorrhea and malabsorption (French et al., 1964; Brandborg et al., 1970). Trier et al. (1974) described a case of chronic intestinal coccidiosis with intermittent diarrhea for more than 20 yr and malabsorption for more than 7 yr. The coccidial infection was documented for 10 months by biopsy examination. The mucosal lesion was characterized by shortened villi, hypertrophied crypts, and infiltration of the lamina propria with eosinophils and other leukocytes. Various stages of *I. belli* were identified in the small intestine by light and electron microscopy.

Ravenel et al. (1976) reported a case of 26-yr duration diagnosed as *I. belli* infection. It is difficult to explain the prolonged infection in these two cases without reinfection. Trier et al. (1974) suggested the occurrence of numerous generations of schizogony over the years or sporulation of the oocysts within the duodenal lumen and reinfection by excysted sporozoites. These authors did observe sporulated oocysts within the duodenal lumen of their patient.

TOXOPLASMA, SARCOCYSTIS, AND RELATED GENERA

Toxoplasma

Since the discovery of *Toxoplasma gondii* in a North African rodent, *Ctenodactylus gundi*, by Nicolle and Manceaux (1908), it has been reported from a wide range of vertebrate hosts in most regions of the world. The voluminous literature that has since accumulated on the clinical and pathological aspects of toxoplasmosis is not reviewed here. However, several excellent reviews are available (Beverley, 1969; Frenkel, 1971, 1973a; Quinn and McCraw, 1972; Turner, 1978). More recent work on the pathogenicity of the organism and the comparative aspects of the response of different host species to *T. gondii* infection are discussed.

Factors Affecting Pathogenicity Strains isolated from both man and animals vary greatly in virulence. This variation, whether the infective forms are cysts in tissues or oocysts from cat feces, is an important factor affecting the host response to infection (Work et al., 1970; Bever-

7

ley et al., 1971b; Sasaki et al., 1974). Most naturally occurring strains are of low virulence, but an increase in virulence may occur after infection in an unusually susceptible individual host animal, or as a result of gametogony and oocyst production in cats and repassage of the asexual cycle in the original host (Beverley, 1976).

There is evidence to suggest that the virulence of a strain of *T. gondii* may also be influenced by the immunological competence of the host. Thus, a strain of low virulence for older pigs caused more severe lesions in the tissue when piglets were infected at 1 day of age (Beverley et al., 1978). Similar results were obtained by Kulasiri (1962) in newborn rats using the virulent RH strain. Beverley et al. (1978) suggested that their results reflect the immaturity of the reticuloendothelial system of the pig in the first few days of life.

Acquired immunity in the host is another factor influencing the pathogenicity of the parasite. This is particularly true of congenital toxoplasmosis and of infections acquired during the first few days of life. In the congenital form, availability of maternal antibodies at the time of fetal infection can alter the course of the disease (Beverley, 1969; Frenkel, 1971). Colostral antibodies may similarly affect the disease in the newborn and in very young animals.

Although most mammals and birds are susceptible to infection with *T. gondii*, certain host species are found to be genetically more resistant to the effects of primary infection. For example, cattle are more resistant than sheep (Munday, 1978) and rats are genetically highly resistant (Jacobs, 1956). This variation is a factor to be taken into consideration when assessing the pathogenicity of a given strain of *T. gondii*. It may also prove to be important in the epidemiology and control of toxoplasmosis. Munday (1978) suggested that the apparent resistance of cattle to Tasmanian strains of *T. gondii* is important in both animal health and public health, because cattle, rather than sheep, may conceivably be used to graze heavily infected pastures.

Pathology and Clinical Correlation A diversity of pathological manifestations has been attributed to infection with *T. gondii*. This is not surprising, considering its wide host range and great variation in the virulence of different strains. In addition, the organism does not seem to show a special affinity for any particular cell type (Koestner and Cole, 1960b; Frenkel, 1971). Toxoplasmosis as a disease, however, is less common than serological surveys of *Toxoplasma* antibodies suggest (Jones, 1973; Van der Wagen et al., 1974; Reimann et al., 1975; Waldeland, 1976b). This indicates that asymptomatic or subclinical toxoplasmosis is the commoner form (Frenkel, 1971, 1973a; Feldman, 1974; Turner, 1978; Krick and Remington, 1978). A distinction should therefore be made between clinical disease and subacute infection. Krick and Remington (1978) stated that ingestion of the organism, parasitemia, invasion of multiple organs, and chronic infection may occur without apparent clinical signs, or clinical disease may appear at any time during or after parasitemia. In some instances, the active infection may be brief, with immunity developing before any significant lesions are produced (Frenkel, 1973a).

In man and in most domestic animals, toxoplasmosis occurs either as a congenital or an acquired infection (Frenkel, 1971, 1973a; Turner, 1978). Congenital toxoplasmosis was the earliest form to be recognized in man and causes the greatest concern (Beverley, 1969; Desmonts and Couvreur, 1974). It is reported that 30–40% of women who first became infected during pregnancy transmitted the infection to their babies, 5–24% of whom became ill and died in the newborn period (Frenkel, 1973b). Infections early in pregnancy may cause fetal death and abortion, miscarriage, or stillbirth. Infections contracted later may result in advanced congenital toxoplasmosis postnatally or in the development of disease within the first few days of life (Beverley, 1969). Reports on the relationship between chronic toxoplasmosis and spontaneous abortion are often contradictory. Some workers have found an association (Jones et al., 1969; Kimball et al., 1971), while others have not (Southern, 1972; Stray-Pederson and Lorentzen-Styr, 1977). Frenkel (1971) supported the concept that toxoplasmosis of recent origin results in an occasional abortion but that chronic infections rarely do. In a recent retrospective study of 523 pregnant women, Johnson et al. (1979) found that the percentage of women with *Toxoplasma* antibody and a past history of spontaneous abortion was not statistically different from the percentage of women without antibody and a past history of spontaneous abortion. This aspect of transplacental transmission of *Toxoplasma* infection requires further investigation.

In sheep, toxoplasmosis is recognized as a common cause of abortion and neonatal death, occurring more commonly than in man or in any other domestic animal (Beverley and Watson, 1962; Beverley and Mackay, 1962; Hartley and Boyes, 1964; Waldeland, 1976a). As many as 74% of abortions in sheep in New Zealand (Hartley and Boyes, 1964) and 54% in Norway (Waldeland, 1976a) are caused by *Toxoplasma*. As in man, it is generally considered that congenital *T. gondii* infections of lambs occur only when their dams acquired an infection for the first time during pregnancy. Munday (1972) failed to detect toxoplasmosis among 178 lambs born to ewes believed to be chronically infected with *T. gondii*.

Toxoplasmosis has been associated with stillbirth in cattle in South America (Watson, 1972), and Sanger et al. (1953) were able to confirm congenital toxoplasmosis in a newborn calf by isolating the organism. Munday (1978), however, could not induce abortion or fetal infection in 4 pregnant cows with experimental infections of up to 7×10^6 tissue cysts.

Sanger and Cole (1955) isolated *T. gondii* from pigs collected aseptically as they emerged from the birth canal. They also isolated organisms from the placenta, milk, and heart of a sow which was consistently asymptomatic. Work et al. (1970) were able to produce experimentally maternal disease and fetal infections in swine. Fetal mortality ranged from 50–100% in the RH strain-infected litters and 30% in the porcine strain-infected litter.

Congenital toxoplasmosis has also been reported in dogs (Smit, 1961). A recent report described the presumed occurrence of a fatal congenital infection in a newborn seal captured off the coast of Alaska about 1 hr after birth (Van Pelt and Dieterich, 1973). Evidently, toxo-

7

plasmosis is not common in seals, and how they acquire it is unknown. Feldman (1974) found that 50 seals he examined were serologically negative.

The pathology of the fetus in congenital toxoplasmosis has been studied best in human and ovine infections. The lesions in the human fetus have been reviewed by Beverley (1969) and Frenkel (1971). The manifestations of fully developed human congenital toxoplasmosis are those of multiple organ damage and dysfunction, and *T. gondii* may be present in virtually every organ (Larsen, 1977). Generalized lymphadenopathy, edema, blood-stained exudates in the body cavities, and hepatosplenomegaly have been reported in the visceral stage of the disease. If this stage is survived or bypassed, the infection becomes localized in the central nervous system (CNS). The choroid and retina are the most frequently affected. Less often there is a focal necrotizing encephalitis, which may be accompanied by calcification and, in rare instances, obstructive hydrocephalus may occur.

The pathology of the fetus in ovine abortion due to toxoplasmosis has been described by Beverley et al. (1971b). These workers found that the changes seen in the fetus were not sufficiently extensive or severe to suggest that they had been the cause of fetal death. These changes were also much less frequent and very much less severe than those seen in the placenta. Beverley et al. (1971a) described the pathology of the ovine placenta and suggested that placental damage is perhaps the primary cause of fetal death in toxoplasmosis of the ovine (Figures 12 and 13). Significant placental lesions, although reported, are not generally associated with toxoplasmosis in man (Frenkel, 1971).

Although necrotizing retinochoroiditis is reported as the commonest clinical manifestation of congenital toxoplasmosis in man (Beverley, 1969), no macroscopic evidence of pathological change and no necrosis was found in the uveal tract of congenitally infected ovine fetuses (Beverley et al., 1971b). No calcification, hydrocephalus, or fits were noted as part of the ovine syndrome. Beverley et al. (1971b) suggested that these differences in lesions between human and ovine congenital toxoplasmosis may be due to differences in strains infecting man and sheep; strains from ovine infections were of low virulence and nonfatal to mice, while those isolated from human congenital infections have been so virulent that they were fatal to mice at the first inoculation.

In sheep, leukoencephalomalacia was present in 80% of the cases of intrauterine *Toxoplasma* infection and seems to be characteristic of the disease in this host (Figure 14). It has not been seen in other types of intrauterine infections in sheep or in aborted piglets and calves (Hartley and Kater, 1963).

In acquired toxoplasmosis, the general pattern of infection and cyst formation occurs in all host species and, as in the congenital form, the typical reaction of all tissues to proliferating organisms is necrosis. The extent of cellular necrosis and hypersensitivity reactions, if any, will determine the severity of the infection (Turner, 1978). The organs and tissues most affected are similar to those affected in congenital toxoplasmosis, namely, brain, myocardium, lymph

Figures 10–13. 10 and 11. Fundus photographs of human eyes showing acute (Figure 10) and healed (Figure 11) retinitis. The healed lesion shows scar tissue with proliferation of retinal pigment. Reprinted by permission and through the courtesy of Dr. Michael Easterbrook, Dept. of Ophthalmology, Toronto General Hospital. 12 and 13. Material used in Figures 12 and 13 obtained through the courtesy of Dr. N. C. Palmer, Pathologist, Veterinary Services Branch, Ontario Ministry of Agriculture and Food. Ovine placenta from abortion induced by a *Toxoplasma* infection. 12. Cotyledons. 13. Section showing *Toxoplasma* cyst (*arrow*) and mineralization (*double arrow*). × 770.

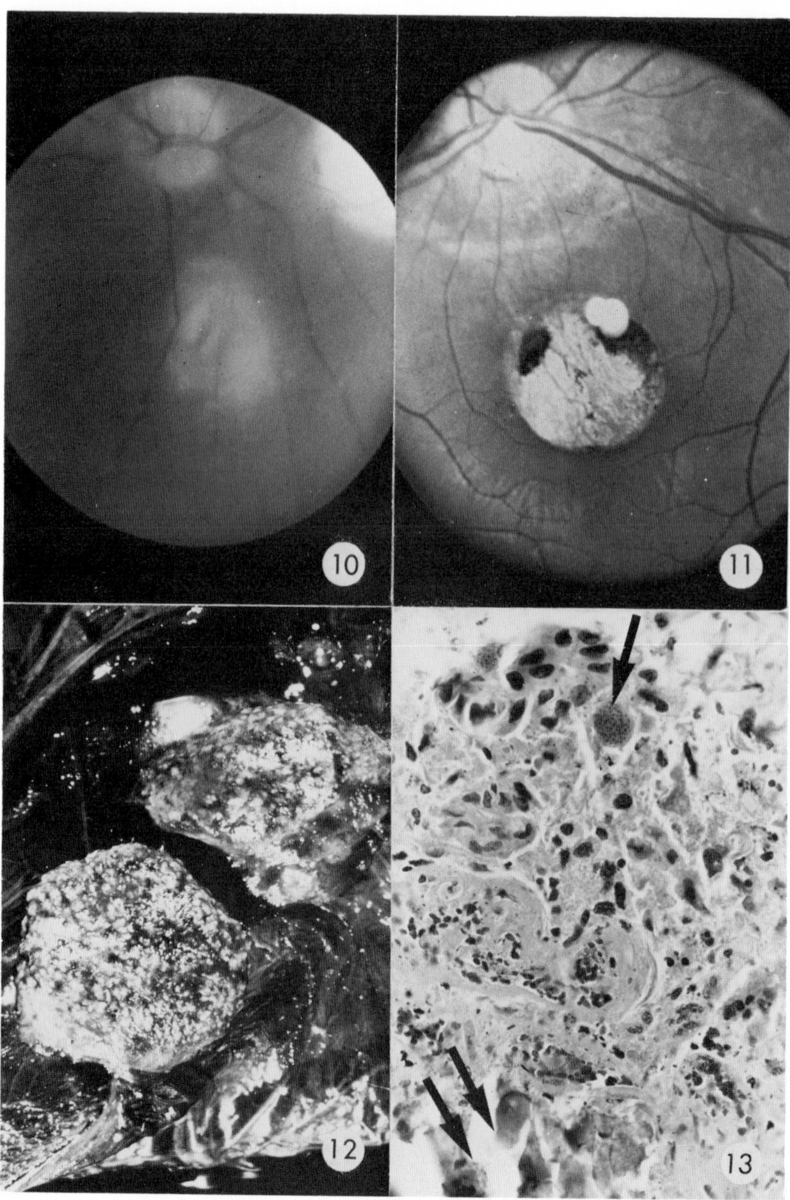

7

nodes, intestines, pancreas, and liver (Figures 15–17). Even in subclinical infections, *T. gondii* may be diffusely distributed throughout the body. Consequently, organisms may persist in cysts in all tissues, probably for the life of the host. Therefore, it is not surprising to see a report of transfer of toxoplasmosis following an organ transplant. Reynolds et al. (1966) reported the death of a patient from acute disseminated toxoplasmosis following transplantation of a kidney from his mother. The patient received immunosuppressive therapy before and after the transplant.

In man, the most common clinical manifestation is lymphadenopathy. The cervical nodes are most frequently involved and may be accompanied by fever, malaise, and peripheral mononucleosis. This lymphoid hyperplasia has been interpreted as an immune reaction in lymphoreticular tissue (Frenkel, 1971). Acute enteritis and mesenteric adenitis have been seen in cats and pigs after ingestion of infective forms (Frenkel et al., 1969; Dubey et al., 1979), but have not yet been reported in man. In generalized toxoplasmosis in man, pneumonia, myocarditis, hepatitis, encephalitis, and retinochoroiditis have been reported. A typhus-type rash has also been associated with acute toxoplasmosis in man. Teutsch et al. (1979) reported several of the above symptoms in an outbreak of toxoplasmosis in 37 patrons of a riding stable in Atlanta, Georgia. Fever, headache, and lymphadenopathy were present in over 80% of the patients. This report is also the first to associate clinical disease in man with ingestion of *Toxoplasma* oocysts.

Toxoplasmic encephalitis, myocarditis, and pneumonia can be fatal in man and enteritis can be fatal in animals (Frenkel, 1971). Usually, myocarditis itself may not be suspected antemortem, being overshadowed by symptoms of general infection and CNS involvement (Theologides and Kennedy, 1969). In isolated toxoplasmic myocarditis, the usual clinical feature is that of chronic cardiomyopathy, and at autopsy, the heart may show hypertrophy and dilatation. Eight cases of pericarditis associated with toxoplasmosis have been reported (Miranda et al., 1976). In most of these cases, pericarditis was the most visible symptom and, except in two, recovery was complete.

There are only a few reported cases of clinical toxoplasmosis in domestic animals other than the epidemic outbreaks of abortion in sheep, as discussed earlier. In experimental infections, Koestner and Cole (1961) found cerebral lesions in 74% of infected sheep. Perivascular cuffing and *Toxoplasma* cysts associated with "circling disease" in sheep was reported as early as 1950 by Wickham and Carne. McErlean (1974) reported the findings in two sheep showing progressive paralysis. There were no macroscopic lesions, although extensive perivascular cuffing in the spinal cord associated with *Toxoplasma* cysts were observed. This and other similar reports are now considered to describe acute sarcocystosis (Dubey, 1976).

In pigs, organisms are frequently isolated without obvious lesions or clinical disease. *T. gondii* has been isolated from striated muscle, brain, lung, stomach, and large intestine

Figures 14–17. 14. Leukoencephalomalacia in brain of aborted ovine fetus. × 190. Material used in Figure 14 obtained through the courtesy of Dr. N. C. Palmer, Pathologist, Veterinary Services Branch, Ontario Ministry of Agriculture and Food. 15. Heart of a 4-month-old dog that died of generalized toxoplasmosis. Note *Toxoplasma* cyst (*arrow*) and degeneration and fibrosis of heart muscle. Clinical signs included lameness of hind legs. Lesions and *Toxoplasma* were seen in skeletal and cardiac muscle, brain, spinal cord, and liver. × 510. 16. Pancreas of an 8-year-old cat that died of generalized toxoplasmosis. *Arrow* indicates *Toxoplasma* cyst. Interstitial pancreatitis was seen in other areas of pancreas. *Toxoplasma* was also found in spleen, liver, brain, meninges, lungs, kidney, and intestine. × 510. Material used in Figures 15 and 16 is from the collection of the Department of Pathology, Ontario Veterinary College. 17. Fetal calf heart showing *Toxoplasma* cyst (*arrow*). Calf aborted at 5 months pregnancy. × 770. Material used in Figure 17 obtained through the courtesy of Dr. N. C. Palmer, Pathologist, Veterinary Services Branch, Ontario Ministry of Agriculture and Food.

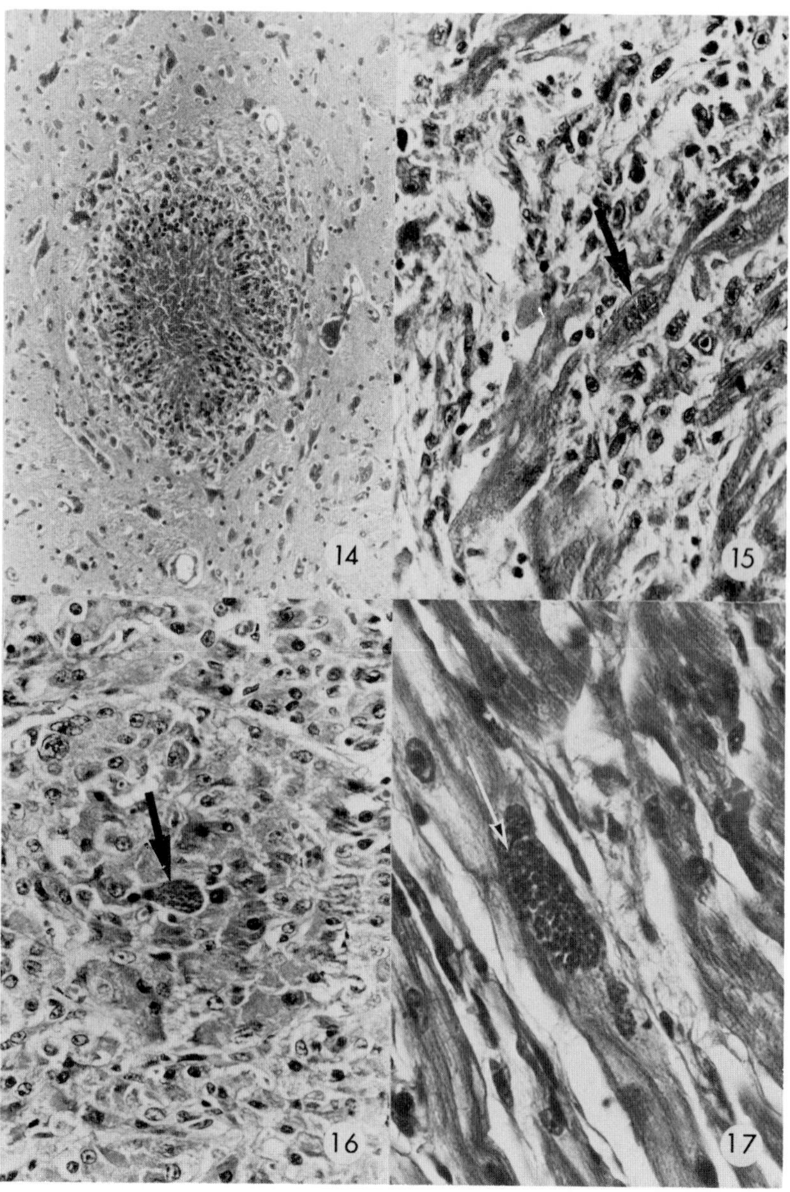

(Katsube et al., 1975). Toxoplasmosis was diagnosed in a 4-week-old pig from a litter of 16 born to a clinically normal sow (Dubey et al., 1979). The affected pig and seven other littermates died after diarrhea had developed within 1–2 weeks of birth. Multifocal necrosis of blood vessels and other tissues were found in the small intestine, mesenteric lymph nodes, liver, lungs, and brain. Numerous tachyzoites, ultrastructurally indistinguishable from *T. gondii*, were found in the lesions. Dubey et al. (1979) believed that the most likely source of infection was oocysts from cats on the farm. Thus, they suggested that toxoplasmosis be considered in the differential diagnosis of diarrhea in young pigs.

Overt clinical toxoplasmosis in cattle is rare and most reported cases involve young animals. Sanger et al. (1953) reported toxoplasmosis in four herds in Ohio, and Corner et al. (1963) described the pathology of a disease of adult cattle in Ontario, presumed to be toxoplasmosis, which they called "Dalmeny disease." However, Dalmeny disease now is considered synonymous with acute sarcocystosis (Dubey, 1976). More recently, Ferguson (1979) described the clinical and pathological findings in a 5-month-old Friesian cross calf, one of three calves that died. Each had had a similar clinical history of unthriftiness, weakness, and difficulty in breathing over several weeks. The major pathological changes were restricted to the lungs and kidneys and comprised interstitial pneumonia with multifocal necrosis, and glomerular necrosis. Many *Toxoplasma* cysts were found in both organs. This case should be reassessed, taking into consideration the etiology of Dalmeny disease.

In horses, antibodies to *T. gondii* have been detected; however, little is known about clinical toxoplasmosis. There have been three reports of encephalomyelitis caused by a protozoan resembling *T. gondii* (Cusick et al., 1974; Dubey et al., 1974; Beech and Dodd, 1974). Dubey (1976) reexamined these lesions and is of the opinion that the disease is related to a *Sarcocystis*-like organism and not to *Toxoplasma*. Al-Khalidi and Dubey (1979) surveyed 500 horses, 1–18 years of age, slaughtered at a meat packing plant in Creston, Ohio, for the prevalence of *Toxoplasma* infection. By feeding equine tissues to specific pathogen-free (SPF) cats and by inoculation into mice, *T. gondii* was isolated from at least seven horses. One anomalous finding in this study was the isolation of *T. gondii* from three of five cats fed pools of tissue from 128 horses thought to be serologically negative. The authors concede that more work needs to be done regarding the persistence of antibody and organism in horses, perhaps by using experimentally infected animals.

Skeletal muscle atrophy has been associated with a case of clinical canine toxoplasmosis (Holliday et al., 1963). These authors described *Toxoplasma*-induced encephalomyelitis resulting in damage to the motor neurons of the spinal cord, leading to atrophy of the skeletal muscles. The encephalomyelitis was a widely disseminated, nonsuppurative form involving primarily the gray matter. In a similar case reported earlier, mild to moderate focal disseminated myositis was seen in addition to the lesions described above (Hartley et al., 1958). Myositis, but not atrophy, has also been reported in man. Taking into consideration the prevalence of widely disseminated CNS lesions, Holliday et al. (1963) suggested that skeletal muscle atrophy may occur frequently and that toxoplasmosis should be considered in the differential diagnosis of dogs showing atrophy of skeletal muscles.

The neuropathology associated with porcine, canine, ovine, and bovine toxoplasmosis has been described (Koestner and Cole, 1960a,b, 1961). In naturally and experimentally infected animals, these authors found cerebral lesions in 53% of the pigs, 74% of the dogs, 75% of the sheep, and 47% of the cattle. Lesions in the CNS in cattle were mild and no predilection site was found in any of the animal species studied. In all of these species, toxoplasmosis was characterized by focal necrosis and vascular damage in acute infections and by glial nodules, repair, and scar formation in chronic infections. In sheep and cattle, chronic infections were also associated with vascular mineralization. Organisms, either free or in cysts, were demonstrated in the majority of cases and, in chronic infections, cysts were most frequently found in the cerebral cortex.

Lesions in porcine toxoplasmosis were strictly focal, the microglial response came quickly, and well circumscribed glial nodules were seen concurrently with foci of early necrosis on the 6th day post-inoculation. These lesions were also relatively small, averaging 150 μm in diameter. In contrast, Koestner and Cole (1960b, 1961) found extensive areas of necrosis, gliosis, and demyelination in dogs and sheep.

Cerebral calcification, common to chronic toxoplasmosis in children, seems to be rare in animals. Also, subependymal lesions, common in congenital toxoplasmosis in children, were seen only in 3 of the 63 dogs studied by Koestner and Cole (1960b). The lesions in the dogs were also extremely mild compared to the extensive areas of periventricular necrosis seen in infants.

Piper et al. (1970) compared the ocular lesions caused by *T. gondii* in 60 dogs, 11 cats, 10 pigs, 18 sheep, and 10 cattle. Lesions were found most frequently in the iris, ciliary body, and retina and were characterized by infiltration of large mononuclear cells, histiocytes, and occasional lymphocytes. Focal necrosis, especially in the retina, was frequently observed. Ovine and bovine lesions differed from those of other species by the predominance of epithelioid cells in sheep and plasmacytes in cattle. Masses of organisms, severe necrosis, and suppurative myositis occurred in the extraocular muscles of dogs, cats, and sheep. In man, retinochoroiditis is the most frequent lesion associated with chronic toxoplasmosis (Frenkel, 1971). Here, ocular lesions originate in the retina with necrosis of parasitized cells followed by inflammation and formation of glial nodules (Figure 10). These may destroy the retina and lead to scar formation (Figure 11) (Frenkel, 1971). In the adjacent choroid and sclera, granulomatous inflammation and necrosis are found in chronic lesions. Piper et al. (1970), however, found that it was uncommon for retinal and choroidal lesions to occur together in any of the animal species they studied (described above), and the underlying sclera was never involved.

Hypersensitivity is mainly responsible for the immunopathology associated with toxoplasmosis. Frenkel (1971) speculated that the granulomatous response in the choroid and sclera in man is a hypersensitivity reaction to the prolonged toxoplasmic proliferation in the retina with continued antigen release. If this were so, it is interesting to speculate further as to why such a reaction in the choroid is rare in other animal species, even though organisms proliferate in the retina. Necrotizing lesions after cyst rupture are thought to result from delayed hypersensitivity, whereas the periventricular lesion in human neonatal toxoplasmosis is said to be a consequence of antigen-antibody reaction of the immediate hypersensitivity or Arthus type.

As stated earlier, *T. gondii* persists in cysts, in the tissues of an infected host, perhaps for life. Recrudescent toxoplasmosis or relapse may therefore occur in chronically infected hosts. In man, this is increasingly recognized in immunosuppressed patients (Cheever et al., 1965; Reynolds et al., 1966; Cohen, 1970). Frenkel (1957) studied the role of immunosuppression in hamsters and found that cortisone was able to produce relapse and radiation was able to potentiate this effect, whereas nitrogen mustard neither produced relapse nor potentiated the effect of cortisone.

Sarcocystis

Historically, *Sarcocystis* has been considered a nonpathogenic organism incidentally encountered at autopsy in animal tissue. However, it is now known that several species of *Sarcocystis* are pathogenic to their intermediate hosts (Dubey, 1976). Dubey (1976) reviewed the species that occur in cattle, sheep, swine, horse, man, and mule deer, and discussed their pathogenicity to the host animals. *Sarcocystis* in man has recently been reviewed by Beaver et al. (1979).

Pathogenicity *Sarcocystis* species are mainly pathogenic to their intermediate hosts. This is not surprising, because merozoites from the intermediate host develop into gamonts in the intestine of the definitive host.

More than one species of Sarcocystis may parasitize a single intermediate host; these may vary in their pathogenicity to that host. For example, *Sarcocystis cruzi* is more pathogenic to cattle than the other two species that infect cattle.

Within the intermediate host itself, the schizont stages that develop primarily in vascular endothelial cells are considered more pathogenic than the cyst stage found in skeletal and cardiac muscle and in neural tissue. Regardless of the stage, the number of sarcocysts observed in tissues bears no relationship to the severity of the disease (Frelier et al., 1979).

Pathology and Clinical Correlation The pathological changes seen in several intermediate host species infected with various species of *Sarcocystis* have been described (Munday et al., 1975, 1977; Dubey, 1976; Koller et al., 1977; Leek et al., 1977; Frelier et al., 1979; Beaver et al., 1979). The best studied is bovine sarcocystosis.

Bovine sarcocystosis has emerged as an important disease, although cattle are relatively resistant to toxoplasmosis. "Dalmeny disease" (Corner et al., 1963) was subsequently described in a group of 12 calves from a farm near Peterborough, Ontario (Meads, 1976), five of which died. At necropsy, schizonts, structurally identical to *S. cruzi*, were found in one of them. Similar clinical findings were described by Frelier et al. (1977) in a group of eight yearling Holstein-Friesian heifers from a dairy farm in Seneca County, New York State.

In a subsequent paper, Frelier et al. (1979) discussed the clinical disease and pathological features observed in the Seneca County outbreak. The clinical disease was characterized by cachexia, peripheral lymph node enlargement, and anemia. The two heifers necropsied had different developmental stages of the parasite, and each stage was characterized by specific histopathological findings. Vascular endothelial schizonts, seen in the one heifer, were associated with mild, mononuclear cell infiltration, alveolar capillary fibrinous thrombi, and multifocal splenic necrosis. Young *S. cruzi* cysts, seen in the cardiac and skeletal muscles of the second heifer, were associated with a multifocal, degenerative myositis.

Johnson et al. (1975) discussed the pathogenesis of sarcocystosis associated with the different developmental stages in experimentally infected cattle. The acute febrile stage of the disease occurred during the development of endothelial cell-associated schizonts. This stage lasted up to about 33 days following infection, and the most severe pathological changes occurred between 26 and 33 days, when schizonts were found throughout the body. The second phase of the disease began with the development of immature cysts, containing only metrocytes, in myofibers. These cysts were found in cardiac and skeletal muscles, later than 33 days. When schizonts were distributed throughout the body, Johnson et al. (1975) observed a severe inflammatory response, marked by a diffuse mononuclear cell infiltration, hemorrhage, and edema of the parasitized organs. There was also an active proliferation of lymphoid cells in the lymphoreticular tissue. After the disappearance of the schizonts and the appearance of cysts, the inflammatory response changed to a subacute state. Hemorrhage and edema were no longer present. The changes in the central nervous system during schizogony were characterized by petechial hemorrhage and mild, perivascular mononuclear cell infiltration of cerebellum, brain stem, cerebrum, and spinal cord, as well as mild mononuclear cell infiltration of the meninges.

Mahrt and Fayer (1975) found that the anemia seen between the 4th and 5th weeks after experimental infection of calves was accompanied by an increase in serum lactate dehydrogenase, aspartate aminotransferase, and creatine phosphokinase.

Fayer et al. (1976) found that cows became ill and aborted after experimental *Sarcocystis* infection. However, there were no apparent parasites or lesions in tissues of the aborted fetuses and viable calves examined in this study. These authors suggested that intrauterine transmission of *Sarcocystis* either does not occur or is quite infrequent. The cause of fetal death and abortion is at present not clear. The other two species, *Sarcocystis hirsuta* and *Sarcocystis hominis*, that infect cattle are considered nonpathogenic (Dubey, 1976).

In sheep, *Sarcocystis* infection is common throughout the world. Two species infect sheep, *Sarcocystis ovicanis*, involving a dog-sheep cycle, and *Sarcocystis tenella*, involving a cat-sheep cycle. *S. tenella* is considered nonpathogenic, while *S. ovicanis* is known to be highly pathogenic to lambs and causes abortion in pregnant ewes (Gestrich et al., 1974; Leek et al., 1977; Leek and Fayer, 1978).

Experimental infection of lambs with *S. ovicanis* sporocysts resulted in an acute, severe disease characterized by anemia, inappetence, weight loss, fever, reduced serum protein, and death (Leek et al., 1977). Leek et al. (1977) suggested that the decrease in serum protein levels is indicative of a possible loss of kidney function. This is substantiated by their histopathological findings of glomerulonephritis. The most prominent gross and histopathological lesions were in the cardiac and skeletal muscles. Gestrich et al. (1974) found that the most prominent lesions were lymph node swelling, myositis, and petechial hemorrhages; Munday et al. (1975) noted myositis and encephalitis in fatal infections. It seems, therefore, that the clinical and pathological features of *S. ovicanis* infections in sheep resemble those of *S. cruzi* infections in cattle. Gestrich et al. (1974), Leek et al. (1977), and Heydorn and Gestrich (1976) found schizonts in vascular endothelial cells throughout the body 24–29 days after experimental infections with *S. ovicanis*; however, Munday et al. (1975) found larger schizonts in many organs after 15 days.

Leek and Fayer (1978) studied experimental *Sarcocystis*-induced abortion in sheep. No evidence of intrauterine transmission was obtained in this study, although 8 of 11 ewes either aborted, died, or became moribund before term. All inoculated ewes became clinically ill, and the severity of the illness seemed to be directly proportional to the numbers of sporocysts administered. However, only three schizonts of *Sarcocystis* were identified in capillaries in the fetal membranes of two ewes. Organisms were not found in any fetus, placenta, uterine tissues, or newborn lambs. Therefore, as in cattle, the mechanism causing death and abortion is not clear. Anemia and the absence of intrauterine transmission tend to distinguish sarcocystosis from toxoplasmosis in sheep.

Hartley and Blakemore (1974) reported encephalomyelitis with myelomalacia in two young sheep associated with an unidentified sporozoan parasite. This and two other reports of similar illness in sheep (Olafson and Monlux, 1942; McErlean, 1974) are now recognized as further manifestations of sarcocystosis (Dubey, 1976).

Sarcocystis infection is considered common in pigs; there seem to be three species, *Sarcocystis miescheriana* (pig-dog cycle), *Sarcocystis suihominis* (pig-man cycle), and *Sarcocystis porcifelis* (pig-cat cycle). *S. suihominis* and *S. porcifelis* have been reported to be pathogenic (Golubkovan and Kisliakova, 1974; Heydorn, 1977).

Golubkovan and Kisliakova (1974) reported diarrhea, lameness, and myositis in pigs after the ingestion of *S. porcifelis* sporocysts from cat feces. Heydorn (1977) infected 4- to 6-week-old piglets with 50,000 to 5×10^6 sporocysts of *S. suihominis*. The animals became ill 12 days post-infection, and about one-half of the piglets that received 1 million or more sporocysts died between 14 and 17 days post-infection.

Koller et al. (1977) studied the pathology of experimental *Sarcocystis hemionilatrantis* infection in mule deer fawns. Clinical signs were first noticed at 18 days; fawns died 27–63 days post-infection. Schizonts were not seen in endothelial cells, as in cattle and sheep, but were seen within macrophages in tissue, and within perivascular spaces in skeletal and cardiac muscle. However, fawns were not examined before 27 days post-infection, although clinical signs were first noticed at 18 days. Early endothelial schizont stages therefore may have been missed. An important finding in this study was bilateral adrenal cortical hemorrhage in infected fawns.

Beaver et al. (1979) reviewed *Sarcocystis* infections in man. They suggested that, because nowhere are people regularly eaten by predators, an infection with *Sarcocystis* in human muscle is always zoonotic. These authors reported five cases of *Sarcocystis* infection in man with clinical manifestations, all of which were diagnosed by biopsy examination. They reviewed the 35 cases previously reported and stated that seven morphological types were recognized, each representing one of several different species, all of which are zoonotic and none of which can be designated *Sarcocystis lindemanni*. Conclusive evidence of pathogenicity in the mature cyst in any of the 40 cases reported was not found.

It therefore seems that, at least in domestic animals, the schizont stages of some species of *Sarcocystis* are very pathogenic and able, in heavy infections, to cause clinical disease, abortion, and death. The most consistent clinical finding was anemia and lowered serum protein levels. These findings distinguish sarcocystosis from toxoplasmosis. Although abortion was induced in experimentally infected pregnant cattle and ewes, there was no evidence to suggest intrauterine transmission of *Sarcocystis*.

Other Related Genera

Besnoitia, *Hammondia*, and *Frenkelia* are other coccidial genera with asexual stages in intermediate hosts. Frenkel (1977) discussed the interrelationships between these and *Toxoplasma*, *Sarcocystis*, *Cystoisospora*, and *Isospora*.

At least five species of *Besnoitia* have been described to date, and at least *Besnoitia wallacei* has been transmitted from cats to mice and back again to cats (Frenkel, 1977). Transmission of other species through carnivores has not been successful.

In the intermediate hosts, *Besnoitia* species form thick-walled cysts in connective tissue visible to the naked eye. Adrenal infection of hamsters by *Besnoitia jellisoni* has been reported and offered as a model for adrenal necrosis to Addison's disease in man (Frenkel, 1977). *B. jellisoni* cysts have also been reported in hamster eyes (Frenkel, 1961). *Besnoitia besnoiti*

7

infection in cows gives rise to lesions on the skin, referred to as "elephant skin," and to orchitis (Pols, 1960).

Frenkel (1977) found cysts of *B. jellisoni*, 30–60 days following the injection of oocysts into mice or rats. The most heavily infected organ was the heart, with some cysts in liver and lung.

Hammondia hammondi oocysts fed to mice multiplied initially in the gut wall and mesenteric lymph nodes and, after 11–16 days, cysts were seen in skeletal and cardiac muscle (Frenkel and Dubey, 1975). This was accompanied by necrosis, myositis, and myocarditis.

PATHOGENIC MECHANISMS

The mechanisms by which the various species of coccidia exert their pathogenic effects on the host have been the subject of a vast array of published research. The major effort has been directed to coccidial infections in chickens, mainly because this host is both easy to work with and economically important. However, many of the mechanisms responsible for the pathogenesis of the morphological and biochemical lesions seen in coccidial infections are still poorly understood.

Pathophysiological Changes

Because most species of coccidia parasitize the host's intestinal tract, changes in the physiological functions of the intestine contribute a great deal to the disease produced.

Several authors have documented a lowering of the intestinal pH during coccidial infections (Stephens et al., 1974; Ruff et al., 1974a; Ruff and Reid, 1975; Anderson et al., 1977). Ruff et al. (1974a) observed a decrease in the pH of the intestinal contents of *E. acervulina*-infected, conventional and gnotobiotic chickens measured at 7 days post-infection. The decrease was unaffected by a change in the dose of oocysts or the severity of the infection. Reduced feed intake or increased gizzard acidity was found not to be the cause. The authors conceded that their data give no indication of the physiological and/or biochemical cause for the decrease in pH. Ruff and Reid (1975) observed a similar increase in intestinal acidity in chickens infected with *Eimeria mivati*, *E. maxima*, or *E. necatrix*, 7–9 days post-infection. The greatest decrease in pH was in the heaviest infected areas of the intestine. Infection with *E. tenella* did not alter the intestinal pH.

Anderson et al. (1977) found that *E. meleagrimitis* or *Eimeria gallopavonis*, but not *E. adenoeides*, significantly decreased the pH in the parasitized region of the turkey intestine.

Kouwenhoven and Van der Horst (1969, 1972) presented evidence that the increased acidity of the coccidia-infected intestine adversely affected its ability to absorb vitamin A and xanthophyll. As a result, the carotene content of the blood decreased rapidly in experimental birds infected with *E. acervulina*. Plasma vitamin A levels were, however, maintained by the liver. Ruff et al. (1974b) found lowered blood carotenoid levels in chickens infected with *E. mivati*, *E. acervulina*, *E. maxima*, *E. necatrix*, *E. brunetti*, or *E. tenella*. Yvore and Mainguy (1972) suggest that the loss of plasma xanthophylls in coccidia-infected chickens is related to modifications in their transport mechanism rather than to difficulties in absorption.

Anorexia and a depression in body weight are common clinical signs observed during coccidial infections. Several workers have shown that the anorexia is only partially responsible for the body weight depression, because normal animals starved to the same extent did not lose as much weight (Preston-Mafham and Sykes, 1970; Michael and Hodges, 1972; Allen et al., 1973; Takhar and Farrell, 1979a).

Therefore, one has to look at possible changes in the digestion and/or absorption of nutrients in infected animals. A reduction in the activity of digestive enzymes, such as disaccharidases, associated with the brush border has been demonstrated (Enigk and Dey-Hazra, 1976; Major and Ruff, 1978a). Major and Ruff (1978b) noted a loss in pancreatic weight and an overall decrease in amylolytic activity in the pancreas of chickens infected with *E. acervulina*, *E. maxima*, or *E. necatrix*. Amylolytic activity of the surface mucosa was reduced in *E. acervulina* and *E. maxima* infections. These findings indicate that coccidial infections can reduce the digestive capacity of the host. Turk (1972) suggested, on the basis of indirect evidence, that protein digestion is also impaired by coccidial infections.

The effect of coccidial infections upon the absorption of several nutrients has been studied. A decrease in the absorption of both amino acids and glucose has been noted. Ruff (1974) and Ruff et al. (1976) observed that methionine absorption was decreased in *E. necatrix*- and in *E. acervulina*-infected chickens but not in *E. tenella*-infected chickens. The low pH was not considered a factor in the decreased absorptive capacity of the upper small intestine. Giese et al. (1971) studied the absorption of [^{14}C]glucose in chickens infected with four species of coccidia, 6 days post-infection. The depression of glucose absorption was greatest in *E. necatrix*-infected birds and less so in *E. maxima*, *E. acervulina*, and *E. brunetti*, in that order. Stein and Marquardt (1973) found a decrease in the absorption of [^{14}C]glucose in rats, 8 days after *Eimeria nieschulzi* infection. This decrease was noted both in the intestinal tissues and hepatic portal plasma and seemed to coincide with the appearance of clinical signs and growth rate changes. The low glucose levels in the portal plasma indicate that the host has no compensatory absorption of glucose by other parts of the small intestine.

Zinc and oleic acid absorption was decreased by *E. acervulina* and *E. necatrix* infections but not by *E. tenella* (Turk and Stephens, 1970). Calcium absorption was increased on the 1st day and decreased on the 6th day after infection with *E. acervulina*, *E. necatrix*, and *E. brunetti* (Turk, 1973).

In addition to a decrease in the capacity to absorb nutrients, several workers have provided evidence to indicate an increase in the permeability of intestinal mucosa for the passage of

7

plasma proteins into the intestinal lumen (Preston-Mafham and Sykes, 1967; Rose and Long, 1969; Enigk et al., 1970). The exact mechanism, morphological or otherwise, by which this change is effected is not clearly understood. Altered morphology, such as disrupted tight junctions between epithelial cells, has been associated with increased mucosal permeability in nematode infections (Murray et al., 1971).

Taking both the decrease in absorptive capacity and the increase in intestinal permeability into consideration, Sharma and Fernando (1975) studied the effect of *E. acervulina* infection on the retention of nutrients during the acute (4–8 days post-infection) and recovery (9–12 days post-infection) phases of the disease. They found a decreased protein retention during the acute phase but not during the recovery phase. A greater rate of protein anabolism during the recovery phase was indicated by their results, coinciding with the compensatory growth observed during this period (Sharma et al., 1973). A reduction in the retention of percent gross energy with a concomitant increase in the ether extract and gross energy of the excreta was noted in the acute phase of *E. acervulina* infection (Sharma and Fernando, 1975). Similarly, in *E. acervulina*-infected birds, Takhar and Farrell (1979a) found that the efficiency of utilization of metabolizable energy was lowered from 0.73 in control birds to 0.43 during 0–8 days post-infection and 0.52 during 9–16 days post-infection. These authors also had evidence to suggest that anorexia and the concomitant reduction in water intake were the primary causes of the reduction in weight gain. Of the total weight gain, only 7.5 g/kg was in the form of fat and 213 g/kg was in the form of protein compared with 45 and 210 g/kg, respectively, for noninfected birds.

Takhar and Farrell (1979b) had experimental results to show that a single infection of 16-day-old chickens with 1.1×10^6 oocysts of *E. acervulina* per bird will provide virtually complete protection against the adverse effect on energy and nitrogen metabolism of secondary infections of similar magnitude.

Coccidiosis can result in diarrhea of varying severity and consequent changes in the concentration of plasma proteins and electrolytes. Allen et al. (1973) noted a decrease in protein, sodium, and chloride concentrations, and an increase in the potassium concentration, in the plasma of *E. brunetti*-infected chickens. The fall in sodium and chloride was coincident with the appearance of mucoid diarrhea and a rise in plasma potassium. The severe hypoproteinemia, the electrolyte disturbance, and the drastic reduction in extracellular fluid may result in an irreversible shock situation causing death. This may explain the sudden deaths encountered by Hein (1971) in *E. necatrix* infections.

It seems, therefore, that anorexia and the concomitant reduction in water intake, the decreased digestive and absorptive capacity, and the increased permeability of the intestinal mucosa are collectively responsible for the weight depression observed in coccidiosis. The diarrhea, with its associated plasma electrolyte changes and dehydration, may be the cause of death in at least some coccidial infections.

Other Contributory Factors

As stated earlier, there is evidence to indicate that villous atrophy is a T-cell–mediated hypersensitivity reaction (Ferguson and Jarrett, 1975; Manson-Smith et al., 1979). Villous atrophy occurs in intestinal coccidiosis of chickens, lambs, pigs, and man (Pout, 1967a,b, 1974a; Brandborg, 1971; Fernando and McCraw, 1973; Stuart et al., 1978) and is at least partially responsible for malabsorption and steatorrhea. The increased epithelial cell turnover associated with villous atrophy allows immature cells to appear on the villi. These cells may lack properly formed cell junctions, contributing to the increased mucosal permeability observed in coccidial infections. These immature cells also lack the complement of enzymes necessary for optimal microvillar digestion and transport of absorbed nutrients across the cell. Thus, villous atrophy and the increased epithelial cell turnover associated with it may play a significant role in maldigestion, malabsorption, and increased mucosal permeability occurring in intestinal coccidiosis.

Rose and Long (1969) and Rose et al. (1975) found that sporozoite invasion of the intestinal epithelial cells was accompanied by an increased vascular permeability of the mucosa. The effect is enhanced by immunity, and these authors suggested that release of histamine may be involved. No data are as yet available on the contribution of mast cells to the pathogenesis of the lesions produced by coccidial parasites.

Because coccidia are intracellular parasites, the response and the behavior of the parasitized host cell would certainly influence the pathogenicity of the parasite and the pathogenesis of the lesions induced. Small schizonts of *E. maxima*, developing in villous epithelial cells, cause little pathology during a primary infection. Large schizonts of *E. necatrix*, in cells within the lamina propria, are associated with cellular destruction and an acute inflammatory reaction. Is it the size or is it the location of the parasite that makes the difference? Is it perhaps the behavior or a change in the behavior of the infected cell that is the cause? We do not have answers to these questions at present. When we do, we may be a little closer to understanding the pathogenetic mechanisms operating in coccidial infections.

LITERATURE CITED

Al-Khalidi, N. W., and Dubey, J. P. 1979. Prevalence of *Toxoplasma gondii* in horses. J. Parasitol. 65:331–334.

Allen, W. M., Berrett, S., Hein, H., and Hebert, C. N. 1973. Some physiopathological changes associated with experimental *Eimeria brunetti* infection in the chicken. J. Comp. Pathol. 83:369–375.

Anderson, W. I., Ruff, M. D., Reid, W. M., and Johnson, J. K. 1977. Effects of turkey coccidiosis on intestinal pH. Avian Pathol. 6:125–130.

Beaver, P. C., Gadgil, R. K., and Morera, P. 1979. Sarcocystosis in man: A review and report of five cases. Am. J. Trop. Med. Hyg. 28:819–844.

7

Beech, J., and Dodd, D. C. 1974. *Toxoplasma*-like encephalomyelitis in the horse. Vet. Pathol. 11:87–96.

Beverley, J. K. A. 1969. Toxoplasmosis in man. Br. J. Hosp. Med. 2:645–653.

Beverley, J. K. A. 1976. Toxoplasmosis in animals. Vet. Rec. 99:123–127.

Beverley, J. K. A., Henry, L., and Hunter, D. 1978. Experimental toxoplasmosis in young piglets. Res. Vet. Sci. 24:139–146.

Beverley, J. K. A., and Mackay, R. R. 1962. Ovine abortion and toxoplasmosis in the East Midlands. Vet. Rec. 74:499–501.

Beverley, J. K. A., and Watson, W. A. 1962. Further studies on toxoplasmosis and ovine abortion in Yorkshire. Vet. Rec. 74:548–552.

Beverley, J. K. A., Watson, W. A., and Payne, J. M. 1971a. The pathology of the placenta in ovine abortion due to toxoplasmosis. Vet. Rec. 88:124–128.

Beverley, J. K. A., Watson, W. A., and Spence, J. B. 1971b. The pathology of the foetus in ovine abortion due to toxoplasmosis. Vet. Rec. 88:174–177.

Bradley, R. E., and Radhakrishnan, C. V. 1973. Coccidiosis in chickens: Obligate relationship between *Eimeria tenella* and certain species of cecal microflora in the pathogenesis of the disease. Avian Dis. 17:461–476.

Brandborg, L. L. 1971. Structure and function of the small intestine in some parasitic diseases. Am. J. Clin. Nutr. 24:124–132.

Brandborg, L. L., Goldberg, S. B., and Breidenbach, W. C. 1970. Human coccidiosis—a possible cause of malabsorption. N. Engl. J. Med. 283:1306–1313.

Britton, W. M., Hill, C. H., and Barber, C. W. 1964. A mechanism of interaction between dietary protein levels and coccidiosis in chicks. J. Nutr. 82:306–310.

Cheever, A. W., Valsamis, M. P., and Rabson, A. S. 1965. Necrotizing toxoplasmic encephalitis and herpetic pneumonia complicating treated Hodgkin's disease. N. Engl. J. Med. 272:26–29.

Clarkson, M. J. 1960. The coccidia of the turkey. Ann. Trop. Med. Parasitol. 54:253–257.

Cohen, S. N. 1970. Toxoplasmosis in patients receiving immunosuppressive therapy. J. Am. Med. Assoc. 211:657–660.

Corner, A. H., Mitchell, D., Meads, E. B., and Taylor, P. A. 1963. Dalmeny disease, an infection of cattle presumed to be caused by an unidentified protozoan. Can. Vet. J. 4:252–264.

Critcher, S. 1950. Renal coccidiosis in Pea Island Canada geese. Wildl. N. Carolina. 14:14–15.

Cusick, P. K., Sells, D. M., Hamilton, D. P., and Hardenbrook, H. J. 1974. Toxoplasmosis in two horses. J. Am. Vet. Med. Assoc. 164: 77–80.

Davies, S. F. M. 1963. *Eimeria brunetti*, additional cause of intestinal coccidiosis in the domestic fowl in Britain. Vet. Rec. 75:1–4.

Desmonts, G., and Couvreur, J. 1974. Congenital toxoplasmosis: A prospective study of 378 pregnancies. N. Engl. J. Med. 290: 1110–1116.

Dikovskata, V. E. 1974. Intraspecies variability in *Eimeria tenella* in the chicken. (In Russian) Parazitologiia 8:548–552.

Dubey, J. P. 1976. A review of *Sarcocystis* of domestic animals and of the coccidia of cats and dogs. J. Am. Vet. Med. Assoc. 169: 1061–1078.

Dubey, J. P. 1978. Pathogenicity of *Isospora ohioensis* infection in dogs. J. Am. Vet. Med. Assoc. 173:192–197.

Dubey, J. P., Davis, G. W., Koestner, A., and Kiryer, K. 1974. Equine encephalomyelitis due to a protozoan parasite resembling *Toxoplasma gondii*. J. Am. Vet. Med. Assoc. 165:249–255.

Dubey, J. P., Weisbrode, S. E., and Rogers, W. A. 1978. Canine coccidiosis attributed to an *Isospora ohioensis*-like organism: A case report. J. Am. Vet. Med. Assoc. 173:185–191.

Dubey, J. P., Weisbrode, S. E., Sharma, S. P., Al-Khalidi, N. W., Zimmerman, J. L., and Gaafar, S. M. 1979. Porcine toxoplasmosis in Indiana. J. Am. Vet. Med. Assoc. 173: 604–609.

Enigk, K., and Dey-Hazra, A. 1976. Activity of disaccharidases of the intestinal mucosa of the chicken during infection with *Eimeria necatrix*. Vet. Parasitol. 2:117–185.

Enigk, K., Schanzel, E., Scupin, E., and Dey-Hazra, A. 1970. Der enterale plasmaproteinverlust bei der coccidiose des Huhnes. Zent. Veterinarmed. Reihe B. 17:522–526.

Farr, M. M. 1954. Renal coccidiosis of Canada geese. J. Parasitol. 40(suppl.):46.

Fayer, R., and Baron, S. 1971. Activity of interferon and an inducer against development of *Eimeria tenella* in cell culture. J. Protozool. 18(suppl.):21.

Fayer, R., Johnson, A. J., and Lunde, M. 1976. Abortion and other signs of disease in cows experimentally infected with *Sarcocystis fusiformis* from dogs. J. Infect. Dis. 134:624–628.

Feldman, H. A. 1974. Toxoplasmosis: An overview. Bull. N. Y. Acad. Med. 50:110–127.

Ferguson, H. W. 1979. Toxoplasmosis in a calf. Vet. Rec. 104:392–393.

Ferguson, A., and Jarrett, E. E. E. 1975. Hypersensitivity reactions in the small intestine: I. Thymus dependence of experimental partial villous atrophy. Gut 16:114–117.

Fernando, M. A., and McCraw, B. M. 1973. Mucosal morphology and cellular renewal in the intestine of chickens following a single infection of *Eimeria acervulina*. J. Parasitol. 59:493–501.

Fernando, M. A., and McCraw, B. M. 1977. Changes in the generation cycle of duodenal crypt cells in chickens infected with *Eimeria acervulina*. Z. Parasitenkd. 52:213–218.

Fitzgerald, P. R., and Mansfield, M. E. 1972. Effects of bovine coccidiosis on certain blood components, feed consumption, and body weight changes of calves. Am. J. Vet. Res. 33:1391–1397.

Fox, J. E. 1978. Bovine coccidiosis: A review, including safety studies with decoquinate for prevention. Vet. Pract. 59:599–603.

Frelier, P., Mayhew, I. G., Fayer, R., and Lunde, M. N. 1977. Sarcocystosis: A clinical outbreak in dairy calves. Science 195:1341–1342.

Frelier, P. F., Mayhew, I. G., and Pollock, R. 1979. Bovine sarcocystosis: Pathologic features of naturally occurring infection with *Sarcocystis cruzi*. Am. J. Vet. Res. 40:651–657.

French, J. M., Whitby, J. L., and Whitfield, A. G. W. 1964. Steatorrhea in man infected with coccidiosis (*Isospora belli*). Gastroenterology 47:642–648.

Frenkel, J. K. 1957. Effects of cortisone, total body irradiation and nitrogen mustard on chronic latent toxoplasmosis. Am. J. Pathol. 33:618–619.

Frenkel, J. K. 1961. Pathogenesis of toxoplasmosis with a consideration of cyst rupture in infection. Surv. Ophthalmol. 6:799–825.

Frenkel, J. K. 1971. Toxoplasmosis. In: R. A. Marcial-Rojas (ed.), Pathology of Protozoal and Helminthic Diseases, pp. 254–290. Williams & Wilkins Company, Baltimore.

Frenkel, J. K. 1973a. Toxoplasmosis: Parasite life cycle, pathology and immunology. In: D. M. Hammond and P. L. Long (eds.), The Coccidia: *Eimeria*, *Isospora*, *Toxoplasma*, and Related Genera, pp. 343–410. University Park Press, Baltimore.

Frenkel, J. K. 1973b. *Toxoplasma* in and around us. Bioscience 23:343–352.

Frenkel, J. K. 1977. *Besnoitia wallacei* of cats and rodents: With a reclassification of other cyst-forming isosporoid coccidia. J. Parasitol. 63:611–628.

Frenkel, J. K. 1978. Toxoplasmosis in cats: Diagnosis, treatment and prevention. Comp. Immunol. Microbiol. Infect. Dis. 1:15–20.

Frenkel, J. K., and Dubey, J. P. 1975. *Hammondia hammondi:* A new coccidium of cats producing cysts in muscles of other mammals. Science 189:222–224.

Frenkel, J. K., Dubey, J. P., and Miller, N. L. 1969. *Toxoplasma gondii*: Fecal forms separated from eggs of the nematode *Toxocara cati.* Science 164:432–433.

Gestrich, R., Schmitt, M., and Heydorn, A. O. 1974. Pathogenität von Hunden für Lämmer. Berl. Muench. Tieraerztl. Wochenschr. 87: 362–363.

Giese, W., Stoll, U., Dey-Hazra, A., and Enigk, K. 1971. Der Einfluss vershiedener *Eimeria-*

Arten auf absorption und Stoffwechsel von 14C-Glucose bei Hühnerküken. Exp. Parasitol. 29:440–450.

Golubkovan, D. I., and Kisliakova, Z. I. 1974. The source of infection for swine *Sarcocystis*. (In Russian) Veterinariia 11:85–86.

Hammond, D. M., Davis, L. R., and Bowman, G. W. 1944. Experimental infections with *Eimeria bovis* in calves. Am. J. Vet. Res. 5:303–311.

Hanson, H. C., Levine, N. D., and Ivens, V. 1957. Coccidia (Protozoa: Eimeriidae) of North American wild geese and swans. Can. J. Zool. 35:715–733.

Hartley, W. J., and Blakemore, W. F. 1974. An unidentified sporozoan encephalomyelitis in sheep. Vet. Pathol. 11:1–12.

Hartley, W. J., and Boyes, B. W. 1964. Incidence of ovine perinatal mortality in New Zealand with particular reference to intra-uterine infections. N. Z. Vet. J. 12:33–36.

Hartley, W. J., and Kater, J. C. 1963. The pathology of *Toxoplasma* infection in the pregnant ewe. Res. Vet. Sci. 4:326–332.

Hartley, W. J., Lindsay, A. B., and Mackinnon, M. M. 1958. *Toxoplasma* meningo-encephalomyelitis and myositis in a dog. N. Z. Vet. J. 6:124–127.

Hawkins, P. A. 1952. Coccidiosis in Turkeys. Technical Bull. Vol. 52. Michigan State University, Agricultural Experimental Station, East Lansing.

Hein, H. 1968. The pathogenic effects of *Eimeria acervulina* in young chicks. Exp. Parasitol. 22:1–11.

Hein, H. 1969. *Eimeria adenoides* and *E. meleagrimitis*: Pathogenic effect in turkey poults. Exp. Parasitol. 24:163–170.

Hein, H. 1971. Pathogenic effects of *Eimeria necatrix* in young chickens. Exp. Parasitol. 30:321–330.

Hein, H. 1974. *Eimeria brunetti*: Pathogenic effects in young chickens. Exp. Parasitol. 36:333–341.

Hein, H. E. 1976. *Eimeria acervulina*, *E. brunetti*, and *E. maxima*: Pathogenic effects of single or mixed infections with low doses of oocysts in chickens. Exp. Parasitol. 39:415–421.

Hein, H. E., and Timms, L. 1972. Bacterial flora in the alimentary tract of chickens infected with *Eimeria brunetti* and in chickens immunized with *Eimeria maxima* and cross-infected with *Eimeria brunetti*. Exp. Parasitol. 31:188–193.

Helle, O. 1970. Winter resistant oocysts in the pasture as a source of coccidial infection in lambs. Acta Vet. Scand. 11:545–564.

Heydorn, A. O. 1977. Beiträge zum Lebenszyklus der Sarkosporidien: IX. Entwicklungszuklus von *Sarcocystis suihominis* n. spec. Berl. Muench. Tieraerztl. Wochenschr. 90: 218–224.

Heydorn, A. O., and Gestrich, R. 1976. Beiträge zum Lebenszyklus der Sarkosporidien: VII. Entwicklungsstadien von *Sarcocystis ovicanis* im Schaf. Berl. Muench. Tieraerztl. Wochenschr. 89:1–5.

Hirth, R. S., and Nielson, S. W. 1969. Pathology of feline toxoplasmosis. J. Small Anim. Pract. 10:213–221.

Holliday, T. A., Olander, H. J., and Wind, A. P. 1963. Skeletal muscle atrophy associated with canine toxoplasmosis: A case report. Cornell Vet. 53:288–301.

Humphrey, C. D., and Turk, D. E. 1974. The ultrastructure of chick intestinal absorptive cells during *Eimeria acervulina* infection. Poult. Sci. 53:1001–1008.

Jacobs, L. 1956. Propagation, morphology and biology of *Toxoplasma*. Ann. N. Y. Acad. Sci. 64:154–179.

Jeffers, T. K. 1978. Genetics of coccidia and the host response. In: P. L. Long, K. N. Boorman, and B. M. Freeman (eds.), Avian Coccidiosis, pp. 51–125. British Poultry Science, Ltd., Edinburgh.

Johnson, A. J., Hildebrandt, P. K., and Fayer, R. 1975. Experimentally induced *Sarcocystis* infection in calves: Pathology. Am. J. Vet. Res. 36:995–999.

Johnson, A. M., Roberts, H., Wetherall, B.,

McDonald, P. J., and Need, J. A. 1979. Relationship between spontaneous abortion and presence of antibody to *Toxoplasma gondii*. Med. J. Aust. 1:579–580.

Jolley, W. R., and Bergstrom, R. C. 1977. Summer coccidiosis in Wyoming calves. Vet. Med. Small Anim. Clin. 72:218–219.

Jones, S. R. 1973. Toxoplasmosis: A review. J. Am. Vet. Med. Assoc. 163:1038–1042.

Jones, M. H., Sever, J. L., Baker, T. H., Hallett, J. H., Goldberg, E. D., Justus, K. M., and Gilkson, M. R. 1969. Toxoplasmosis and abortion. Am. J. Obstet. Gynecol. 104:919–920.

Julian, R. J., Harrison, K. B., and Richardson, J. A. 1976. Nervous signs in bovine coccidiosis. Mod. Vet. Pract. 57:711–718.

Katsube, Y., Hagiwara, T., and Imaizumi, K. 1975. Latent infection of *Toxoplasma* in swine. Jpn. J. Vet. Sci. 37:245–252.

Kheysin, Y. M. 1972. Life Cycles of Coccidia of Domestic Animals. (English translation, F. K. Plous, Jr.). University Park Press, Baltimore.

Kimball, A. C., Kean, B. H., and Fuchs, F. 1971. The role of toxoplasmosis in abortion. Am. J. Obstet. Gynecol. 111:219–226.

Klimes, B. 1963. Coccidia of the domestic goose. Zentralbl. Vet. Med. B. 10:427–448.

Koestner, A., and Cole, C. R. 1960a. Neuropathology of porcine toxoplasmosis. Cornell Vet. 50:362–384.

Koestner, A., and Cole, C. R. 1960b. Neuropathology of canine toxoplasmosis. Am. J. Vet. Res. 21:831–844.

Koestner, A., and Cole, C. R. 1961. Neuropathology of ovine and bovine toxoplasmosis. Am. J. Vet. Res. 22:53–66.

Koller, R. D., Kistner, T. P., and Hudkins, G. G. 1977. Histopathologic study of experimental *Sarcocystis hemionilatrantis* infection in fawns. Am. J. Vet. Res. 38:1205–1209.

Kouwenhoven, B., and Van der Horst, C. J. G. 1969. Strongly acid intestinal content and lowered protein, carotene and Vitamin A blood levels in *Eimeria acervulina* infected chickens. Z. Parasitenkd. 32:347–353.

Kouwenhoven, B., and Van der Horst, C. J. G. 1972. Disturbed intestinal absorption of Vitamin A and carotenes and the effect of a low pH during *Eimeria acervulina* infection in the domestic fowl (*Gallus domesticus*). Z. Parasitenkd. 38:152–161.

Krassner, S. M. 1963. Factors in host susceptibility and oocyst infectivity in *Eimeria acervulina* infections. J. Protozool. 10:327–332.

Krick, J. A., and Remington, J. S. 1978. Toxoplasmosis in the adult: An overview. N. Engl. J. Med. 298:550–553.

Kulasiri, C. 1962. The behaviour of suckling rats to oral and intraperitoneal infection with a virulent strain of *Toxoplasma gondii*. Parasitology 52:193–198.

Larsen, J. W. 1977. Congenital toxoplasmosis. Teratology 15:213–217.

Leathem, W. D., and Burns, W. C. 1968. Duration of acquired immunity of the chicken to *Eimeria tenella* infection. J. Parasitol. 54:227–232.

Leek, R. G., and Fayer, R. 1978. Sheep experimentally infected with *Sarcocystis* from dogs: II. Abortion and disease in ewes. Cornell Vet. 68:108–123.

Leek, R. G., Fayer, R., and Johnson, A. J. 1977. Sheep experimentally infected with *Sarcocystis* from dogs: I. Disease in young lambs. J. Parasitol. 63:642–650.

Le Jambre, L. F. 1977. Genetics of vulvar morph types in *Haemonchus contortus: Haemonchus contortus cayugensis* from the Finger Lakes region of New York. Int. J. Parasitol. 7:9–14.

Levine, N. D., and Ivens, V. 1970. The Coccidian Parasites (Protozoa, Sporozoa) of Ruminants. Illinois Biological Monographs 44, University of Illinois Press, Urbana.

Levine, N. D., Morrill, C. C., and Schmittle, S. C. 1950. Renal coccidiosis in an Illinois gosling. N. Am. Vet. 31:738–739.

Long, P. L. 1973. Pathology and pathogenicity of coccidial infections. In: D. M. Hammond and P. L. Long (eds.), The Coccidia: *Eimeria*, *Isospora*, *Toxoplasma*, and Related Genera, pp. 253–294. University Park Press, Baltimore.

Long, P. L. 1978. The problem of coccidiosis:

General considerations. In: P. L. Long, K. N. Boorman, and B. M. Freeman (eds.), Avian Coccidiosis, pp. 3–28. British Poultry Science, Ltd., Edinburgh.

Long, P. L., and Milne, B. S. 1971. The effect of a interferon inducer on *Eimeria maxima* in the chicken. Parasitology 62:295–302.

Lotz, J. C., and Leek, R. G. 1970. Failure of development of the sexual phase *Eimeria intricata* in heavily inoculated sheep. J. Protozool. 17:414–417.

McErlean, B. A. 1974. Ovine paralysis associated with spinal lesions of toxoplasmosis. Vet. Rec. 94:264–266.

Major, J. R., Jr., and Ruff, M. D. 1978a. Disaccharidase activity in the intestinal tissue of broilers infected with coccidia. J. Parasitol. 64:706–711.

Major, J. R., Jr., and Ruff, M. D. 1978b. *Eimeria spp*: Influence of coccidia on digestion (amylolytic activity) in broiler chickens. Exp. Parasitol. 45:234–240.

Manson-Smith, D. F., Bruce, R. G., and Parrott, D. M. V. 1979. Villous atrophy and expulsion of intestinal *Trichinella spiralis* are mediated by T cells. Cell. Immunol. 47:285–292.

Mahrt, H., and Fayer, R. 1975. Hematologic and serologic changes in calves experimentally infected with *Sarcocystis fusiformis*. J. Parasitol. 61:967–969.

Marquardt, W. C., Senger, C. M., and Seghetti, L. 1960. The effect of physical and chemical agents on the oocyst of *Eimeria zuernii* (Protozoa, Coccidia). J. Protozool. 7:186–189.

Meads, E. B. 1976. Dalmeny disease—another outbreak—probably sarcocystis. Can. Vet. J. 17:271.

Meisel, J. L., Perera, D. R., Meligro, C., and Rubin, C. E. 1976. Overwhelming watery diarrhea associated with a *Cryptosporidium* in an immunosuppressed patient. Gastroenterology 70:1156–1160.

Mesfin, G. M., Bellamy, J. E. C., and Stockdale, P. H. G. 1977. The pathological changes caused by *Eimeria falciformis* var. *pragensis* in mice. Can. J. Comp. Med. 42:496–510.

Meuten, D. J., Van Kruiningen, H. J., and Lein, D. H. 1974. Cryptosporidiosis in a calf. J. Am. Vet. Med. Assoc. 165:914–917.

Michael, E., and Hodges, R. D. 1971. The pathogenic effects of *Eimeria acervulina*: A comparison of single and repeated infections. Vet. Rec. 89:329–333.

Michael, E., and Hodges, R. D. 1972. The pathogenic effects of *Eimeria necatrix*: A comparison of single and repeated infections. Vet. Rec. 91:958–962.

Michel, J. F., Lancaster, M. B., and Hong, C. 1979. The effect of age, acquired resistance, pregnancy and lactation on some reactions of cattle to infection with *Ostertagia ostertagi*. Parasitology 79:157–168.

Miranda, A., Rubiés-Prat, J., Cerdá, E., and Foz, M. 1976. Pericarditis associated to acquired toxoplasmosis in a child. Cardiology 61: 303–306.

Morin, M., Robinson, Y., and Turgeon, D. 1980. Intestinal coccidiosis in baby pigs. Can. Vet. J. 21:65.

Munday, B. L. 1972. Transmission of Toxoplasma infection from chronically infected ewes to their lambs. Br. Vet. J. 128:71–72.

Munday, B. L. 1978. Bovine toxoplasmosis: Experimental infections. Int. J. Parasitol. 8:285–288.

Munday, B. L., Barker, I. K., and Rickard, M. D. 1975. The development cycle of a species of *Sarcocystis* occurring in dog and sheep, with observations on pathogenicity in the intermediate host. Z. Parasitenkd. 46:111–123.

Munday, B. L., Humphrey, J. D., and Kila, V. 1977. Pathology produced by, prevalence of, and probable life cycle of a species of *Sarcocystis* in the domestic fowl. Avian Dis. 21:697–703.

Murray, M., Jarrett, W. F. H., Jennings, F. W., and Miller, H. R. P. 1971. Structural changes associated with increased permeability of parasitized mucous membranes to macromolecules. In: S. M. Gaafer (ed.), Pathology of Parasitic Diseases, pp. 197–205. Purdue University Studies, Lafayette, IN.

Nagey, M. S., and Mathey, W. J. 1972. Interaction of *Escherichia coli* and *Eimeria brunetti* in chickens. Avian Dis. 16:864–873.

Nation, P. N., and Wobeser, G. 1977. Renal coccidiosis in wild geese in Saskatchewan. J. Wildl. Dis. 13:370–375.

Nicolle, C., and Manceaux, L. 1908. Sur une infection a corps de Leishman (au organismes voisins) du gondi. C. R. Acad. Sci. (D) (Paris) 147:763–766.

Niilo, L. 1970a. Experimental winter coccidiosis in sheltered and unsheltered calves. Can. J. Comp. Med. 34:20–25.

Niilo, L. 1970b. The effect of dexamethasone on bovine coccidiosis. Can. J. Comp. Med. 34:325–328.

Norton, C. C., and Hein, H. E. 1976. *Eimeria maxima*: A comparison of two laboratory strains with a fresh isolate. Parasitology 72:345–354.

Oikawa, H., and Kawaguchi, H. 1974. Effect of coccidial infection on acetylcholine-induced contraction of the digestive tract in chickens: I. *Eimeria tenella* and *E. acervulina* infection. Jpn. J. Vet. Sci. 36:433–440.

Olafson, P., and Monlux, W. S. 1942. *Toxoplasma* infection in animals. Cornell Vet. 32:176–190.

Owen, D. 1975. *Eimeria falciformis* (Eimer, 1870) in specific pathogen free and gnotobiotic mice. Parasitology 71:293–303.

Pellérdy, L. P. 1974. Coccidia and Coccidiosis. 2nd Ed. Verlag Paul Parey, Berlin.

Piper, R. C., Cole, C. R., and Shadduck, J. A. 1970. Natural and experimental ocular toxoplasmosis in animals. Am. J. Ophthalmol. 69:662–668.

Platz, S. 1977. Infektiosität von *Eimeria tenella*-oocysten nach Heibverrottung kokzidienhaltigen Hubnerkotes. Dtsch. Tieraerztl. Wochenschr. 84:178–180.

Pohlenz, J., Moon, H. W., Cheville, N. F., and Bemrick, W. J. 1978. Cryptosporidiosis as a problem factor in neonatal diarrhea of calves. J. Am. Vet. Med. Assoc. 172:452–457.

Pols, J. W. 1960. Studies on bovine besnoitiosis with special reference to the aetiology. Onderstepoort J. Vet. Res. 28:265–356.

Pout, D. D. 1967a. Villous atrophy and coccidiosis. Nature 213:306–307.

Pout, D. D. 1967b. The reaction of the small intestine of the chicken to infection with *Eimeria spp.* In: The Reaction of the Host to Parasitism, pp. 28–38. Veterinary Medical Review. N. G. Elwert Universitats-und Verlagsbuchhandlung, Marsburg-Lahn.

Pout, D. D. 1973. Coccidiosis of lambs: I. Observations on the naturally acquired infection. Br. Vet. J. 129:555–567.

Pout, D. D. 1974a. Coccidiosis of lambs: III. The reaction of the small intestinal mucosa to experimental infections with *Eimeria arloingi* "B" and *E. crandallis*. Br. Vet. J. 130:45–52.

Pout, D. D. 1974b. Coccidiosis of lambs: IV. The clinical response to infections of *Eimeria arloingi* "B" and *E. crandallis* in laboratory reared lambs. Br. Vet. J. 130:54–60.

Pout, D. D. 1976. Coccidiosis of sheep: A review. Vet. Rec. 98:340–341.

Pout, D. D., and Catchpole, J. 1974. Coccidiosis of lambs: V. The clinical response to long-term infection with a mixture of different species of coccidia. Br. Vet. J. 130:388–399.

Pout, D. D., Norton, C. C., and Catchpole, J. 1973. Coccidiosis of lambs: II. The production of faecal oocyst burdens in laboratory animals. Br. Vet. J. 129:568–582.

Powell, H. S., Holscher, M. A., Heath, J. E., and Beasley, F. F. 1976. Bovine cryptosporidiosis (a case report). Vet. Med. Small Anim. Clin. 71:205–206.

Preston-Mafham, R. A., and Sykes, A. H. 1967. Changes in permeability of the mucosa during intestinal coccidiosis infections in the fowl. Experientia 23:972–973.

Preston-Mafham, R. A., and Sykes, A. H. 1970. Changes in body weight and intestinal absorption during infections with *Eimeria acervulina* in the chicken. Parasitology 61:417–424.

Quinn, P. J., and McCraw, B. M. 1972. Current status of *Toxoplasma* and toxoplasmosis: A review. Can. Vet. J. 13:247–262.

7

Randall, C. J., and Norton, C. C. 1973. Acute intestinal coccidiosis in geese. Vet. Rec. 93:46–47.

Ravenel, J. M., Suggs, J. L., and Legerton, C. W. 1976. Human coccidiosis. Recurrent diarrhea of 26 years duration due to *Isospora belli*: A case report. J. S.C. Med. Assoc. 72:217–219.

Reimann, H. P., Burridge, M. J., Behymer, D. E., and Franti, C. E. 1975. *Toxoplasma gondii* antibodies in free-living African mammals. J. Wildl. Dis. 11:529–533.

Reynolds, E. S., Walls, K. W., and Pfeiffer, R. I. 1966. Generalized toxoplasmosis following renal transplantation: Report of a case. Arch. Intern. Med. 118:401–405.

Rommel, M. 1970. Verlanf der *Eimeria scabra*- und *E. polita*-infection in vollemfäuglichen ferkeln und Läuferschweinen. Berl. Muench. Tieraerztl. Wochenschr. 83:181–186.

Rose, M. E. 1967. The influence of age of host on infection with *Eimeria tenella*. J. Parasitol. 53:924–929.

Rose, M. E., and Long, P. L. 1969. Immunity to coccidiosis: Gut permeability changes in response to sporozoite invasion. Experientia 25:183–184.

Rose, M. E., Long, P. L., and Bradley, J. W. A. 1975. Immune responses to infections with coccidia in chickens: Gut hypersensitivity. Parasitology 71:357–368.

Ruff, M. D. 1974. Reduced transport of methionine in intestines of chickens infected with *E. necatrix*. J. Parasitol. 60:838–843.

Ruff, M. D., Johnson, J. K., Dykstra, D. D., and Reid, W. M. 1974a. Effects of *Eimeria acervulina* on intestinal pH in conventional and gnotobiotic chickens. Avian Dis. 18:96–104.

Ruff, M. D., and Reid, W. M. 1975. Coccidiosis and intestinal pH in chickens. Avian Dis. 19:52–58.

Ruff, M. D., Reid, W. M., and Johnson, J. K. 1974b. Lowered blood carotenoid levels in chickens infected with coccidia. Poult. Sci. 53:1801–1809.

Ruff, M. D., Whitlock, D. R., and Smith, R. R. 1976. *Eimeria acervulina* and *E. tenella*: Effect on methionine absorption by the avian intestine. Exp. Parasitol. 39:244–251.

Ruff, M. D., and Wyatt, R. D. 1978. Influence of dietary aflatoxin on the severity of *Eimeria acervulina* infection in broiler chickens. Avian Dis. 22:471–480.

Ryley, N. G., and Ryley, J. F. 1978. Effects of saturated sodium chloride solution on coccidial oocysts. Parasitology 77:33–39.

Sanger, V. L., Chamberlain, D. M., Chamberlain, K. W., Cole, C. R., and Farrell, R. L. 1953. Toxoplasmosis: V. Isolation of *Toxoplasma* from cattle. J. Am. Vet. Med. Assoc. 123:87–91.

Sanger, V. L., and Cole, C. R. 1955. Toxoplasmosis: VI. Isolation of *Toxoplasma* from milk, placentas and newborn pigs of asymptomatic carrier sows. Am. J. Vet. Res. 16:536–539.

Sasaki, Y., Iida, T., and Oomura, K. 1974. Experimental *Toxoplasma* infection of pigs with oocysts of *Isospora bigemina* of feline origin. Jpn. J. Vet. Sci. 36:459–465.

Schmitz, J. A., and Smith, D. R. 1975. *Cryptosporidium* infection in a calf. J. Am. Vet. Med. Assoc. 167:731–732.

Sharma, V. D., and Fernando, M. A. 1975. Effect of *Eimeria acervulina* infection on nutrient retention with special reference to fat malabsorption in chickens. Can. J. Comp. Med. 39:146–154.

Sharma, V. D., Fernando, M. A., and Summers, J. D. 1973. The effect of dietary crude protein level on intestinal and cecal coccidiosis in chickens. Can. J. Comp. Med. 37:195–199.

Skene, R. C., Fernando, M. A., and Remmler, O. 1981. Coccidia of Canada geese (*Branta canadensis*) at Kortwright Waterfowl Park, Guelph, Ontario, Canada, with description of *Isospora anseris* n. sp. Can. J. Zool. 59:493–497.

Smit, J. D. 1961. Toxoplasmosis in dogs in South Africa: Seven case reports. J. S. Afr. Vet. Med. Assoc. 32:339–346.

Southern, P. M. 1972. Habitual abortion and toxoplasmosis. Obstet. Gynecol. 111:219–226.

Sprinz, H. 1962. Morphological response of intestinal mucosa to enteric bacteria and its implication for sprue and Asiatic cholera. Fed. Proc. 21:57–64.

Stein, A. S., and Marquardt, W. C. 1973. *Eimeria nieschulzi*: Glucose absorption in infected rats. Exp. Parasitol. 34:262–267.

Stephens, J. F., Borst, W. J., and Barnett, B. D. 1974. Some physiological effects of *Eimeria acervulina*, *E. brunetti*, and *E. mivati* infections in young chickens. Poult. Sci. 53: 1735–1742.

Stockdale, P. H. G. 1977. The pathogenesis of the lesions produced by *Eimeria zuernii* in calves. Can. J. Comp. Med. 41:338–344.

Stockdale, P. H. G., and Fernando, M. A. 1975. The development of the lesions caused by second generation schizonts of *Eimeria necatrix*. Res. Vet. Sci. 19:204–208.

Stockdale, P. H. G., and Niilo, L. 1976. Production of bovine coccidiosis with *Eimeria zuernii*. Can. Vet. J. 17:35–37.

Stray-Pederson, B., and Lorentzen-Styr, A. 1977. Uterine *Toxoplasma* infections and repeated abortions. Am. J. Obstet. Gynecol. 128:716–721.

Stuart, B. P., Lindsay, D. S., and Ernst, J. V. 1978. Coccidiosis as a cause of scours in baby pigs. Proceedings of the Second International Symposium on Neonatal Diarrhea, pp. 371–382. University of Saskatchewan, Saskatoon, Saskatchewan, Canada.

Syrkis, I., Fried, M., Elian, I., Pietruska, D., and Lengy, J. 1975. A case of severe human coccidiosis in Israel. Isr. J. Med. Sci. 11:373–377.

Takhar, B. S., and Farrell, D. J. 1979a. Energy and nitrogen metabolism of chickens infected with either *Eimeria acervulina* or *Eimeria tenella*. Br. Poult. Sci. 20:197–211.

Takhar, B. S., and Farrell, D. J. 1979b. Energy and nitrogen metabolism of chickens subjected to infection and reinfected with *Eimeria acervulina*. Br. Poult. Sci. 20:213–224.

Teutsch, S. M., Juranek, D. D., Sulzer, E., Dubey, J. P., and Sykes, R. K. 1979. Epidemic toxoplasmosis associated with infected cats. N. Engl. J. Med. 300:695–699.

Theologides, A., and Kennedy, B. J. 1969. Toxoplasmic myocarditis and pericarditis. Am. J. Med. 47:169–174.

Trier, J. S., Moxey, P. C., Schimmel, E. M., and Robles, E. 1974. Chronic intestinal coccidiosis in man: Intestinal morphology and response to treatment. Gastroenterology 66:923–932.

Turk, D. E. 1972. Protozoan parasitic infections of the chick intestine and protein digestion and absorption. J. Nutr. 102:1217–1222.

Turk, D. E. 1973. Calcium absorption during coccidial infections in chicks. Poult. Sci. 52:854–857.

Turk, D. E., and Stevens, J. F. 1970. Effects of serial inoculations with *Eimeria acervulina* or *Eimeria necatrix* upon zinc and oleic acid absorption in chicks. Poult. Sci. 49:523–526.

Turner, G. V. S. 1978. Some aspects of the pathogenesis and comparative pathology of toxoplasmosis. J. S. Afr. Vet. Assoc. 49:3–8.

Van der Wagen, L. C., Behymer, D. E., Rieman, H. P., and Franti, C. E. 1974. A survey for *Toxoplasma* antibodies in northern California livestock and dogs. J. Am. Vet. Med. Assoc. 164:1034–1037.

Van Pelt, R. W., and Dieterich, R. A. 1973. Staphylococcal infection complicated by toxoplasmosis in a young harbour seal. J. Wildl. Dis. 9:258–261.

Visco, R. J., and Burns, W. C. 1972a. *Eimeria tenella* in bacteria free and conventionalized chicks. J. Parasitol. 58:323–331.

Visco, R. J., and Burns, W. C. 1972b. *Eimeria tenella* in monoflora and diflora chicks. J. Parasitol. 58:576–585.

Visco, R. J., and Burns, W. C. 1972c. *Eimeria tenella* in bacteria free chicks of relatively susceptible strains. J. Parasitol. 58:586–588.

Visco, R. J., Corwin, R. R., and Selby, L. A. 1978. Effect of age and sex on the prevalence of intestinal parasites in cats. J. Am. Vet. Med. Assoc. 172:797–800.

Waldeland, H. 1976a. Toxoplasmosis in sheep: The relative importance of the infection as a

7

cause of reproductive loss in sheep in Norway. Acta. Vet. Scand. 17:412–425.

Waldeland, H. 1976b. Toxoplasmosis in sheep: The prevalence of *Toxoplasma* antibodies in lambs and mature sheep from different parts of Norway. Acta. Vet. Scand. 17:432–440.

Watson, W. A. 1972. Toxoplasmosis in human and veterinary medicine. Vet. Rec. 91: 254–258.

Wickham, N., and Carne, H. R. 1950. Toxoplasmosis in domestic animals in Australia. Aust. Vet. J. 26:1–3.

Williams, R. B. 1973. Effects of different infection rates on oocyst production of *Eimeria tenella* in the chicken. Parasitology 67: 279–288.

Wobeser, G. 1974. Renal coccidiosis in mallard and pintail ducks. J. Wildl. Dis. 10:249–255.

Work, K., Ericksen, L., Fennestad, K. L., Moller, T., and Siim, J. C. 1970. Experimental toxoplasmosis in pregnant sows. Acta. Pathol. Microbiol. Scand. (B) 78:129–139.

Wyatt, R. D., Ruff, M. D., and Page, R. K. 1975. Interaction of aflatoxin with *Eimeria tenella* infection and monensin in young broiler chickens. Avian Dis. 19:730–740.

Yvore, P., and Mainguy, P. 1972. Influence de la coccidiose duodénale sur la tenseur en carotenoides du serum chez le poulet. Ann. Rech. Vet. 3:381–387.

8

Host Immune Responses

M. Elaine Rose

In the past, host immune responses to coccidia of different genera have been separately reviewed, with most emphasis on *Eimeria* and *Toxoplasma*. Now that the coccidian nature of *Toxoplasma*, *Besnoitia*, *Hammondia*, *Sarcocystis*, and *Frenkelia* is assured, an attempt is made here to present an integrated and, where possible, a comparative approach to the immunological aspects of the host-parasite relationships obtaining in coccidia of different genera. Of necessity, the review deals mainly with *Toxoplasma* and *Eimeria*, because these have been the genera most studied and there is little information as yet available for some of the other genera, whose medical and veterinary significance has only recently been recognized. The development of immunity to the *Eimeria* species of domesticated gallinaceous birds and, to a lesser extent, of the domesticated ruminants is well chronicled because of the economic importance of coccidiosis in these animals. In the case of *Toxoplasma*, until the last decade, attention was concentrated on the intermediate hosts because of the medical significance of toxoplasmosis in man. These aspects of immunity to the coccidia have been extensively reviewed (Frenkel, 1973; Rose, 1973, 1978; Remington and Krahenbuhl, 1976, in press) and are not covered here in such detail, although reference will be made to them when appropriate. This review concentrates on work published since 1972 and includes studies on immunity in both final and intermediate hosts of the heteroxenous coccidia. The discussion is limited to host immunity of an acquired nature; resistance to infection due to other factors and the possibilities of controlling disease by prophylactic immunization are considered elsewhere.

ACQUISITION OF IMMUNITY

Eimeria

In hosts of veterinary or experimental significance, the subject has been well researched and several reviews are available (Cuckler, 1970; Euzéby, 1973; Rose, 1973, 1978). Consequently, only a summary is presented here.

Expression of Immunity The great majority of *Eimeria* species seem to be fairly immunogenic, and a single infection of a normally immunocompetent host will induce some degree of immunity to reinfection, although exceptions have been noted (Ernst et al., 1968; Todd and Hammond, 1968; Versényi and Pellérdy, 1970). Immunity to the *Eimeria* is usually considered with respect to its effect on secondary and subsequent inoculations of oocysts, and evidence on the question of whether the primary infection is qualitatively affected is conflicting. It is generally accepted that in the *Eimeria*, asexual reproduction after any given inoculation of oocysts is normally confined to a fixed and predetermined number of cycles, little influenced by host response. Such a view is substantiated by the close similarities in developmental cycles in vivo and in vitro (when such comparisons can be made), and also by recent work carried out in immunologically compromised animals. Apart from increased oocyst production, primary infections with *Eimeria falciformis* or *Eimeria nieschulzi* in T lymphocyte-deficient mice or rats, and with *Eimeria acervulina* or *Eimeria maxima* in B cell-deficient chickens did not differ from those in the respective controls (Mesfin and Bellamy, 1979a; Rose and Hesketh, 1979; Rose et al., 1979b). Patency was not appreciably prolonged, but the animals were completely susceptible to reinfection. Therefore, in these animals one must assume that, although the extent of the primary infection is limited by immune responses, the termination of asexual cycling which leads to gamete production is unaffected. However, it is possible that covert asexual cycling, not leading to gamete production, may have persisted undetected. Contrary findings were obtained with *Eimeria vermiformis* infections in one strain of nude mice; here, oocyst production was greatly prolonged (Rose, unpublished) and the results resembled those found with *Eimeria mivati* in chickens treated with corticosteroids (Rose, 1970; Long and Rose, 1970) and with *Eimeria zuernii* and *Eimeria bovis* in calves (Niilo, 1970). Prolongation of patency has also been noted in sheep given multispecific infections (Catchpole et al., 1976). Species differences in the parasites and/or strain differences in the hosts may account for the conflicting results obtained. The possible persistence of asexual stages of some species of *Eimeria* in the intestine, or elsewhere, after apparent cessation of infection (Long and Millard, 1976) requires further investigation.

Immunity to a challenge inoculum is manifest as a reduction in clinical signs and in the multiplication of the parasite, the extent depending upon a number of factors. These include the species of *Eimeria* (even within the same host), the magnitude and mode of administration of the immunizing infection (Mesfin and Bellamy, 1979b), interval between immunization and challenge, and age of hosts; these have been fully discussed in previous reviews (see Rose, 1973, 1978; Rose and Long, 1980). A reduction in pathogenic effects is sometimes more readily demonstrable than a decreased production of oocysts. This may be due to the suppression of the development of a proportion of the parasites which, by reducing the "crowding factor" (Tyzzer, 1929), results in a greater yield of oocysts (Williams, 1973) than would be obtained from a susceptible host. However, in some instances, a reduction in pathogenic effects has been evident in the absence of a true decrease in oocyst production (Joyner and Norton, 1973, 1976) or lesions (Long et al., 1980).

In partially immune animals, a proportion of parasites completes the life cycle and produces normal viable oocysts, although the period of patency is usually curtailed and the prepatent period may be extended (reviewed by Rose, 1973).

Stages of the Parasite which Induce Immunity The results of work based almost entirely on *Eimeria* species of the chicken suggest that different stages of the life cycle differ in their abilities to induce protective immune responses. The sexual stages seem to have little, if any, immunizing potential, even against homologous stages (Horton-Smith et al., 1963b; Rose, 1967; Rose and Hesketh, 1976); this seems to reside, instead, in the developing asexual stages. Similarly, it is thought that the sexual stages of *Sarcocystis* have little immunizing ability (see below). The sporozoite is probably not very immunogenic, because withdrawal of drugs that inhibit this stage is usually followed by relapse of the infection and complete susceptibility of the host (Long and Millard, 1968). In the case of at least one species (*Eimeria tenella*), it seems likely that the first generation schizont also is comparatively poor in inducing immunity, because infection with a strain selected for the absence of the second generation schizont (McDougald and Jeffers, 1976) leads to little or no immunity to challenge (Johnson et al., 1979). It is probable that in this species and in *E. maxima*, the second schizont stage is most concerned with the induction of immunity.

There have been no reports of successful immunization resulting from the administration of nonviable organisms. Different stages have been given by a variety of routes including intraenteric, with or without adjuvants, and have stimulated both cell-mediated and antibody responses, but no immunity to infection of the gut, although intravenously induced sporozoite infections were neutralized (see, for reviews, Rose, 1973; Rose and Long, 1980). It is possible that antigens which induce protective immunity are presented only by metabolizing parasites.

Stages Affected by Immunity Immunity affects the parasites of a challenge inoculum at a very early stage. It is generally agreed that excystation proceeds normally, and there is a consensus that at least a proportion of sporozoites invades the epithelial cells of the host. Some authors believe that there is a complete (Morehouse, 1938; Edgar, 1944, quoted by Leathem and Burns, 1967) or partial (Augustin and Ridges, 1963; Hammond et al., 1964; Leathem and Burns, 1967) inhibition of entry into the epithelial cells of the host. However, until recently, attempts to demonstrate any antisporozoite or antimerozoite activity in extracts of the mucosa or of the intestinal contents of immune animals were unsuccessful (Augustin and Ridges, 1963; Horton-Smith et al., 1963a; Herlich, 1965; Leathem and Burns, 1967). Antiparasite (Orlans and Rose, 1972) and sporozoite and merozoite neutralizing (Davis et al., 1978) activity is now known to be present in the intestinal contents of immune animals (see below). However, other authors have noted little or no difference between immune and susceptible animals in the numbers of parasites present in the first 24 hr after the inoculation of oocysts (Rose and Hesketh, 1976; Mesfin and Bellamy, 1979b).

The invasive stages which do penetrate are prevented from developing (Tyzzer et al., 1932; Horton-Smith et al., 1963a; Leathem and Burns, 1967; Niilo, 1967; Mesfin and Bellamy, 1979b), but viability is retained for at least 24–48 hr, and development may be resumed on transfer to a nonimmune host (Leathem and Burns, 1967; Rose and Hesketh,

8

1976). An elegant theory of the mechanism involved in inhibition of cell penetration and subsequent development has been proposed by Davis and Porter (1979) and is discussed below. Thus, immunity to reinfection with *Eimeria* species seems to be expressed against the invasive stage (which in the normal course of events would be the sporozoite), and in the following sequence of events: (*a*) possibly a reduction in the numbers of sporozoites which penetrate the cells; (*b*) an initially reversible inhibition of development; and (*c*) death or discarding of the parasite with the host cell in the course of normal epithelial cell turnover. In animals in which immunity is incomplete or waning, other invasive stages are affected and abnormal development of a proportion of schizonts may occur (Leathem and Burns, 1967, 1968; Blagburn et al., 1979).

Specificity of Immunity The exquisite specificity of immunity to *Eimeria* species is well known (for reviews, see Rose, 1973, 1978) and is commonly used to establish the identity of species (Levine, 1938, 1942; Moore and Brown, 1951; Long et al., 1976) and to determine the prevalence of various species of coccidia in commercial poultry houses. It is now evident that, at least in some species of coccidia of the domestic fowl, there is considerable strain variation in immunity (Joyner, 1969; Long, 1974; Norton and Hein, 1976; Jeffers, 1978; Long and Millard, 1979); this is relevant in any consideration of prophylactic immunization against coccidiosis (Rose and Long, 1980). Strain variation has been shown to occur in *E. acervulina* (Joyner, 1969) and is particularly noteworthy in the very immunogenic species *E. maxima* (Long, 1974; Norton and Hein, 1976; Jeffers, 1978; Long and Millard, 1979), whereas it has not yet been demonstrated in *E. tenella* (Jeffers, 1978). It is possible that the immunological diversity of *E. acervulina* and *E. maxima* may account for their known prevalence in the litter of broiler houses (Jeffers, 1974; Long et al., 1975). In addition to species and strain specificity, there is also evidence for the specificity of immunity against different stages of the same species (see Rose, 1973). However, both common and specific antigens are demonstrable between and within species (Černá, 1970; Černá and Zalmanová, 1972; Sokolić et al., 1974; Kouwenhoven and Kuil, 1976; Tanielian et al., 1976).

Stability of Immunity The duration of immunity to the *Eimeria* in the absence of reinfection is difficult to ascertain because of the ubiquity of the organism and the difficulty in eliminating extraneous infection. A consideration of the available literature (surveyed in Rose, 1973, 1978) suggests that immunity wanes with time (Mesfin and Bellamy, 1979b) and that its duration is dependent upon the mode of immunization and the age of host when given the immunizing inocula. It is not known with certainty whether the immunity is sterile, or of the premunition type (as in the case of *Toxoplasma*), active only while parasite stages remain within the host. According to Mesfin and Bellamy (1979b), the persistence of apparently trapped degenerate oocysts (Rose, 1961; Pierce et al., 1962; Hammond et al., 1963) could not be correlated with immunity. Completely immune animals do not shed oocysts for prolonged periods of time, but this does not preclude the possibility of the parasite remaining in an asexual form within the host, either in the intestine or elsewhere. The resumption of shedding of a very small number of oocysts, which occurred in a proportion of immune chickens treated with corticosteroids within a period too short to allow of reinfection by the ingestion of oocysts (Long and Millard, 1976), suggests that this could be a possibility. Observations in the field relating outbreaks of clinical coccidiosis to some form of stress, such as transport, parturition, or intercurrent disease, would support this theory, but only in the demonstrable absence of reinfection. Clearly, further work is necessary to determine the sterility of immunity to *Eimeria*.

Isospora

Until the recent interest in the coccidia of dogs and cats, immunity to *Isospora* infections has rarely been investigated. This was due to their economic insignificance and comparative lack of pathogenicity, although *Isospora belli* of man (see Dubey, 1977) and *Isospora ohioensis* (together with a similar coccidium) of the dog have been reported to cause disease (Dubey, 1978a; Dubey et al., 1978). The scant information available on immunity to the *Isospora* is summarized here, but it is highly likely that this aspect of the host-parasite relationship will receive more attention in the future. The data presented relate only to infections of the final host, although it is now known that stages of at least some species of the dog and cat can invade and persist in the extraintestinal tissues of other hosts (Frenkel and Dubey, 1972; Dubey, 1975a,b, 1978b; Mehlhorn and Markus, 1976; Dubey and Mehlhorn, 1978).

Expression of Immunity Resistance to reinfection of cats and dogs by *Isospora* species was first described by Andrews (1926) and seems to be very similar to that noted for the *Eimeria* species of various hosts.

The possibility of immunological regulation of primary infections should be considered and may be indicated by the longer duration of oocyst production after a primary inoculum of *I. ohioensis* in pups compared with older animals (Dubey, 1978a).

Immunity to reinfection seems to be readily acquired (Chessum, 1972; Dubey and Frenkel, 1974; Dubey, 1978a) but is influenced by age at primary inoculation (Dubey and Frenkel, 1974; Dubey, 1978a).

Careful comparisons of the immunizing abilities of different species have not yet been made; thus, it is not known whether they are as diverse as those of *Eimeria* species. *I. ohioensis* of dogs seems to be a highly immunogenic species, because complete immunity to challenge was apparent within 1 week of the immunizing inoculum (Dubey, 1978a). Somewhat conflicting reports are available for *Isospora felis* of cats (Chessum, 1972; Dubey and Frenkel, 1974), but this species, and to a lesser extent, *Isospora rivolta* of the cat, also induce a good immune response.

Stages Inducing and Affected by Immunity There are no publications relating to the immunogenicity of different stages of the life cycle of *Isospora* species or to the identity of the stages affected by the immune response. However, extrapolating from information available for other coccidia, it would seem likely that asexual multiplicative stages are involved.

Specificity of Immunity Exhaustive tests for the specificity of immunity have not been carried out. However, the indications are that, as with *Eimeria*, there is little or no cross-protection between species. Thus, Chessum (1972) found that cats immunized with *I. felis* were susceptible to infection with oocysts of *I. rivolta*, and Dubey et al. (1978) considered that cross-protection experiments would have been helpful in establishing the separate identities of *I. ohioensis* and a similar organism.

Stability of Immunity The limited information available suggests that immunity to *Isospora* species is, as in the case of *Eimeria* species, of limited duration and of the order of some months. Thus, adult cats did not pass oocysts of *I. felis* or *I. rivolta* when given challenge inocula at intervals of 8–26 weeks (Chessum, 1972; Dubey and Frenkel, 1974), and pups were substantially immune to challenge with *I. ohioensis* 1–2 months after an initial inoculum (Dubey, 1978a).

It is not known whether the immunity is sterile or dependent upon the persistence of cryptic stages. Because extraintestinal stages of some of the *Isospora* species are thought to occur in the final host (Dubey and Frenkel, 1972a; Dubey, 1978b, 1979), the possibility of chronic inapparent infections cannot be ruled out. In testing for this, Dubey and Frenkel (1974) found that a proportion of cats that recovered from infection with *I. felis* or *I. rivolta* shed oocysts of these species when treated with corticosteroids. However, they could not be certain that this was due to reactivation of a latent infection, because the time interval between treatment and oocyst discharge allowed for the possibility of extraneous infection, which is highly likely in the case of this species (see Dubey, 1978c). Brief recrudescence of the production of oocysts of *I. felis* and *I. rivolta* has been reported subsequent to the inoculation of *Toxoplasma* oocysts (Campana-Rouget et al., 1974).

Toxoplasma

There is a wealth of information on immunity to *Toxoplasma*, the bulk of which relates to the response of the intermediate host. Because this has been well and frequently reviewed, it is dealt with only summarily here. However, an attempt is made to survey the literature on immunity in the final host, the cat. As pointed out by Dubey et al. (1977), its effect on oocyst shedding is of considerable significance in public health.

Expression of Immunity In the absence of reinfection or of events which may activate latent (probably extraintestinal) infection (discussed below), intestinal infection with *Toxoplasma*, as with *Eimeria* and *Isospora*, seems to be self-limiting (Sheffield and Melton, 1974). However as with the related genera, primary infection, judged by oocyst shedding, is subject to host influence. Thus, cats treated with corticosteroids during a primary infection continued to shed oocysts until they died of toxoplasmosis (Dubey and Frenkel, 1974). In addition, older cats tended to excrete fewer oocysts than those infected when under 2 months of age (Dubey et al., 1977).

A very effective immunity is found to challenge inocula. Cats initially infected as adolescents or adults do not shed oocysts when reinoculated, and partial immunity, evidenced by lowered oocyst production and a shorter patent period, is seen in cats initially infected when under 2 months old (Kühn and Weiland, 1969; Dubey et al., 1970, 1977; Piekarski and Witte, 1971; Chessum, 1972; Dubey and Frenkel, 1972b, 1974; Frenkel and Dubey, 1972; Sheffield and Melton, 1974; Overdulve, 1974, 1978; Dubey, 1976a). This immunity is very rapidly developed, since the period of patency is not significantly extended nor are the numbers of oocysts greatly increased, when cats are infected on several successive days (Dubey and Frenkel, 1974; Overdulve, 1978).

Extraintestinal stages of *Toxoplasma* in cats also seem to be affected by the host immune response. Toward the end of the primary infection in the intestine, when oocyst production has almost ceased, trophozoites and proliferative forms tend to disappear from the other organs also (Dubey and Frenkel, 1972b). This effect is greater in older cats than in those infected when less than 2 months old; Dubey et al. (1977) have suggested that older cats are able to curtail the rate of multiplication of *Toxoplasma* in the visceral organs. The immunological control of extraintestinal development is also suggested by the development of acute toxoplasmic lesions in a majority of cats treated with corticosteroids, although obligatory enteroepithelial stages were not found in other tissues, except for the bile duct and gall bladder, in which they may be present even in untreated cats (Dubey and Frenkel, 1974).

Immunity to *Toxoplasma* in the intermediate host has been the subject of many reviews. This brief account is based on the most recent, which include those by Frenkel (1973), Remington and Krahenbuhl (1976, in press), and Dubey (1977).

Animals naturally infected with *Toxoplasma* often show no clinical signs, and the majority of those that do usually recover due to the acquisition of immunity. For laboratory investigations, immune animals may be obtained by the use of strains of *Toxoplasma* of reduced virulence, or by drug treatment to control the infection (see Remington and Krahenbuhl, 1976). In some instances, cessation of treatment is followed by death of the animal, indicating that the immunity developed is insufficient to control the proliferation of the parasite and this may occur particularly in neural tissues (see Frenkel, 1973; Dubey, 1977). The parasite, however, persists as cysts in many tissues of the body, probably for many years, and may be reactivated (see below), or congenitally transmitted (Remington and Krahenbuhl, 1976). Aged mice are more susceptible to infection (Gardner and Remington, 1977, 1978a,b).

Immunity to reinfection with a virulent strain arises early after the immunizing inoculum, from as soon as 1–2 weeks in mice and guinea pigs, although this will depend upon the immunizing strain and the route of challenge (see Frenkel, 1973; Remington and Krahenbuhl, 1976). The effect of immunity on the organisms of a challenge inoculum, in terms of reduction in the numbers recoverable, is evident from as early as 1 hr after challenge in hamsters; thereafter, this greatly increases (Chinchilla and Frenkel, 1978). However, in immune mice, the presence of parasites in the circulation for several days after challenge was demonstrable by the transfer of infection to susceptible mice, (Werner, 1977).

Stages Inducing and Affected by Immunity There has been little investigation into the stages of the life cycle involved in the immune response to *Toxoplasma*, and the results relating to the final host have been inconclusive (Dubey and Frenkel, 1974).

In the intermediate host, the identity of the immunizing antigens has not been established; however, the injection of killed tachyzoites induces the formation of antibodies (Krahenbuhl et al., 1972b; Masihi and Werner, 1976) and has been reported to afford some protection against challenge by some (Cutchin and Warren, 1956; Jacobs, 1956; Foster and McCulloch, 1968) but not others (Nakayama, 1965; Huldt, 1966a; Stadtsbaeder et al., 1975). Studies in vitro have indicated that, as with *Eimeria* species, tachyzoites are found in similar numbers in peritoneal cells whether these are derived from normal or immune animals, and the parasites in the "immune" cells seem normal in all respects, except that they do not divide (Jones et al., 1975). Thus, the effect of immunity here, as with *Eimeria* species, seems to be an inhibition of development rather than a toxoplasmacidal one.

Parasites of the immunizing inoculum persist (as mentioned above), and those of the virulent challenge inoculum may also be recovered from tissues for a prolonged time (Werner and Egger, 1973). It is interesting that the tissue distribution of the immunizing and challenge strains may differ (de Roever-Bonnet, 1963, 1964; Nakayama, 1964). Thus, the immune state in the case of *Toxoplasma* infections differs from that occurring with *Eimeria* species in that in the latter, parasites from the immunizing inoculum are not known to persist and it seems likely that those of the challenge inocula are very quickly eliminated.

Specificity of Immunity There seems to be a high degree of specificity in immunity to *Toxoplasma* in the cat. There are no indications of cross-protection against related genera such as *Hammondia* (Frenkel and Dubey, 1975) or *Isospora* (Piekarski and Witte, 1971). Indeed, latent infection with *Toxoplasma* tends to relapse if oocysts of *I. felis* are administered (Chessum, 1972; Dubey, 1976b, 1978c). As with *Eimeria* species, there is evidence for strain variation in immunity: cats infected with one strain of *Toxoplasma* tend to be susceptible to challenge with a heterologous strain (Dubey et al., 1970; Piekarski and Witte, 1971; Dubey and Frenkel, 1974). There is also no evidence to suggest that the immunization of cats with the BCG (bacille Calmette-Guérin) strain of *Mycobacterium tuberculosis,* widely used to induce nonspecific resistance to a variety of intracellular microbes (see below), in any way affects subsequent infection with *Toxoplasma* (Dubey, 1978c).

In the intermediate hosts, immunity does not seem to be so specific, since in mice and hamsters protection against fatal *Toxoplasma* infection is afforded by *Hammondia hammondi* (Frenkel and Dubey, 1975; Dubey, 1978d); Christie and Dubey (1977) and Dubey (1976a, 1978d) have suggested that immunization with the nonpathogenic *H. hammondi* should be considered as prophylaxis against toxoplasmosis. This immunity is apparently due to the possession of common antigens by the two genera and not to a nonspecific effect, since there is only very marginal cross-protection between *H. Hammondi* and *Besnoitia jellisoni* (Dubey, 1978d) and between *Toxoplasma* and *B. jellisoni* (Frenkel and Caldwell, 1975). The antigenic relationships between these organisms is confirmed by the results of serological tests (Lunde and Jacobs, 1965; see Dubey, 1978d). This cross-protection is therefore different from that nonspecifically induced by BCG or *Corynebacterium parvum* and active against a variety of intracellular parasites including, according to some reports, *Toxoplasma gondii* (Tabbara et al., 1975).

Cross-protection between strains of *Toxoplasma* of low and high virulence is made use of in the study of immunity to reinfection, but is not always complete. Differences in host response to individual strains is an additional factor which has to be taken into account (see Frenkel, 1973).

Stability of Immunity In the cat, immunity to *Toxoplasma* persists for some time, certainly for up to 23 months, although it may be waning after a period of about 20 months (Dubey and Frenkel, 1974; Sheffield and Melton, 1974; Overdulve, 1978). In the intermediate host, immunity persists for a very long time, perhaps for life. It is clear that immunity to *Toxoplasma* is not a sterile immunity but an infection (premunition) immunity. Whether immunity would persist in the absence of chronic infection is not known, but it is likely that it would diminish (Frenkel, 1973), as in the case of *Eimeria* infections. Immunity to *Toxoplasma* is not stable, and recrudescence of infection occurs in carrier animals. Thus, in cats, treatment with corticosteroids leads to renewed oocyst shedding and the activation of extraintestinal lesions, resulting in death (Dubey and Frenkel, 1974). Superinfection with other organisms (e.g., *I. felis*, Chessum, 1972; Dubey, 1976b, 1978c), or illness (Overdulve, 1978), may also cause reactivation of latent infection with shedding of oocysts. Not all cats treated with corticoste-

8

roids shed oocysts, and the sequence of stages leading to oocyst production has not been determined but could have arisen from hitherto dormant (non-gametocyte) stages in the intestine, or from stages reaching the intestine from ruptured cysts in the tissues (Dubey and Frenkel, 1974).

The cause of the activation of *Toxoplasma* infection with renewed production of oocysts by the superimposition of infection with *I. felis* is not known. In cats that had been immunized with *I. felis* either before (Dubey, 1978c) or after (Chessum, 1972) infection with *T. gondii*, there was no reexcretion of *Toxoplasma* oocysts after the second inoculum of *I. felis*. Thus, reactivation of *Toxoplasma* occurred only when infection with *I. felis* could be established, and not when this was prevented by specific immunization. Superimposed infection with *I. rivolta* had no effect (Chessum, 1972; Dubey, 1976b).

In the intermediate host, reactivation of latent infection with *Toxoplasma* (by immunosuppression) is well known and is a potential hazard to patients with organ transplants or carcinomata (see Remington and Krahenbuhl, 1976; Ruskin and Remington, 1976). Treatment of experimental animals with a variety of immunosuppressive agents including cortisone (Frenkel, 1957; Stahl et al., 1966; Huldt, 1966b), 6-mercaptopurine (Stahl et al., 1965, 1966), antilymphocyte serum (Nakayama and Aoki, 1970), antithymocyte serum (Strannegard and Lycke, 1972), as well as splenectomy (Stahl et al., 1966) has led to exacerbation of infection, altered pathogenesis, reduced clearance of organisms, reactivation of chronic infection, and a failure to develop immunity to reinfection.

Besnoitia

Because definitive hosts of *Besnoitia* species have only recently been recognized (Peteshev et al., 1974, quoted by Dubey, 1977; Wallace and Frenkel, 1975; Frenkel, 1977), the literature on immunity is almost entirely confined to work carried out in the intermediate hosts, principally laboratory rodents infected with *B. jellisoni*. This has been reviewed by Frenkel (1973), who has extensively investigated the immune responses of hamsters and mice to this infection, in parallel with work on *Toxoplasma* (see below).

In its definitive host, the cat (Wallace and Frenkel, 1975), *Besnoitia wallacei* does not seem to be very immunogenic, since oocysts were shed two or three times after repeated feeding of *Besnoitia* cysts, and in numbers that were similar or only slightly reduced (Frenkel, 1977). No extraintestinal development in the cat has been reported.

The response of the intermediate host has been studied in animals infected with avirulent strains, or which have recovered from drug-controlled infections; in these animals, immunity is very similar to that found with *Toxoplasma* (see Frenkel, 1973). Thus, resistance to reinfection is fairly rapidly acquired; however, the organisms persist in the host as cysts, and immunity is of the premunition type. As with most of the coccidia, immunity has a high degree of specificity.

Interspecific immunity has been compared in mice and rats infected with *B. jellisoni* or *B. wallacei* and, although some results were inconclusive, cross-immunity seems to be very slight (Frenkel, 1977). Likewise, cross-protection between related genera is minimal (Hoff and Frenkel, 1974; Frenkel and Caldwell, 1975; Dubey, 1978d), and so are cross-reactions in serological tests (see Frenkel, 1977). As with other coccidia, heterospecific cellular resistance plays little part in immunity to *Besnoitia* infections (see below).

As might be expected from its premunition nature, established immunity to *Besnoitia* in the intermediate host is unstable and can be disrupted, for example, with corticosteroid treatment (Frenkel and Lunde, 1966).

Hammondia

H. hammondi is a newly described coccidium (Wallace, 1973, 1975; Frenkel and Dubey, 1975). Consequently, there is little information available on host immunity. In the cat (final host), immunity to challenge, judged by oocyst shedding, was apparent after a single infection, complete after one or two infections, and developed quickly (Dubey, 1975c; Frenkel and Dubey, 1975). There seems to be no cross-protection against infection with *T. gondii* (Frenkel and Dubey, 1975), although the patent period of the latter in cats immune to *H. hammondi* was shortened slightly (Wallace, 1975). Also, there was no cross-reactivity in serological tests (Frenkel and Dubey, 1975; Wallace, 1975).

Immunity to the stages present in intermediate hosts is not so specific. Thus, infection with *H. hammondi* provides mice and hamsters with considerable protection against challenge with *T. gondii* (Frenkel and Dubey, 1975; Christie and Dubey, 1977; Dubey, 1978d), although there are differences between strains of *H. hammondi* (Christie and Dubey, 1977). *T. gondii* was recoverable from the brains of mice immunized with *H. hammondi* (Christie and Dubey, 1977) and vice versa (Wallace, 1975), indicating that the challenge organisms are not destroyed, but that their multiplication is retarded. It seems likely that in the intermediate host, the stages of the two genera are antigenically similar; this is confirmed by cross-reactivity in serological tests (Wallace, 1973, 1975; Frenkel and Dubey, 1975; Christie and Dubey, 1977; Weiland et al., 1979).

In the intermediate host, immunity is probably of the premunition type, since organisms are recoverable from animals capable of resisting challenge infection. In the cat, the type of immunity is not known with certainty; extraintestinal stages have not been detected (Frenkel and Dubey, 1975; Christie and Dubey, 1977). However, it seems that organisms persist in the small intestine of the cat, although they have not yet been found (Frenkel and Dubey, 1975; Dubey, 1975c). The persistence of chronic infection and its relapse after corticosteroid treatment, or spontaneously (Dubey, 1975c), suggests that immunity in the cat also may be a premunition type, which seems to be more labile than immunity to *T. gondii* (Dubey, 1975c; Frenkel and Dubey, 1975).

8

Sarcocystis

Recent reviews on *Sarcocystis* (Tadros and Laarman, 1976; Markus, 1978) have included discussions of the immunological aspects of the host-parasite relationships and have provided the foundation for the account given here.

Almost all the information available relates to immunity in the final host and is confined to observations on the fate of challenge with sarcocysts. There is a general consensus that the intestinal infection of the final host induces very little, if any, immunity to reinfection with sarcocysts. This has been shown for cats and dogs (Rommel et al., 1972; Heydorn and Rommel, 1972; Markus et al., 1974; Fayer, 1974, 1977; Tadros and Laarman, 1976; Ruiz and Frenkel, 1976) and humans (Aryeetey and Piekarski, 1976). Tadros and Laarman (1976), however, have noted that reinfection of weasels with a vole *Sarcocystis* within 3 months of an initial infection resulted in a shortened period of patency in which aberrant and nonviable sporocysts were passed.

Little is known about protective immunity in the intermediate host; however, superinfection of sheep and mice has been reported (Munday and Rickard, 1974; Ruiz and Frenkel, 1976).

The poor immunogenicity of *Sarcocystis* in the final host contrasts with the heteroxenous coccidia discussed above. However, it is interesting that, so far as is known, only gametogony stages occur in the definitive host, and in *Eimeria* species, gametocyte stages are known to be very poorly immunogenic. Markus (1978) has pointed out that there is very little cellular reaction in the intestine of carnivores infected with *Sarcocystis* species (Fayer, 1974; Munday et al., 1975) and has suggested that this might be responsible for the poor immune response.

There seems to be no cross-immunity between the different species of *Sarcocystis* in the same host and between *Sarcocystis* species and related genera. Since infection of the final host with *Sarcocystis* species induces little, if any, protection against homologous reinfection, it is not surprising that superinfection with *Toxoplasma* can occur (Piekarski and Witte, 1971; see Frenkel, 1973; Markus, 1978). Infections with related genera which are capable of inducing homologous immunity (*Toxoplasma*, *Hammondia*, *Besnoitia*, and *Isospora*) do not protect against subsequent challenge with *Sarcocystis* (Rommel et al., 1972; Heydorn and Rommel, 1972; Ruiz and Frenkel, 1976).

The results of serological tests have indicated that different species of *Sarcocystis* have antigens in common and that cross-reactions occur with *Frenkelia* and, to a lesser extent, *Besnoitia* (Tadros and Laarman, 1976; Černá and Kolářová, 1978), although this depends upon the test used (see Frenkel, 1977). However, no antigenic relationship is apparent between *Sarcocystis* and *Toxoplasma* when tested by sera from both final and intermediate hosts (Munday et al., 1975; Aryeetey and Piekarski, 1976; Ruiz and Frenkel, 1976; Tadros and Laarman, 1976; Frelier et al., 1977; Lunde and Fayer, 1977; Černá and Kolářová, 1978; see Markus, 1978, for earlier references).

Frenkelia

Studies on the immunological aspects of infections with *Frenkelia*, to date, have been restricted to serological testing designed to establish its identity and relationship to other coccidian genera (see above).

IMMUNOLOGICAL RESPONSES OF THE HOST

Infection with coccidia induces a variety of immune responses in the host. Those developed by the intermediate host to *Toxoplasma* infection have been very extensively investigated, both from the viewpoint of their diagnostic value in human infections and from an understanding of the mechanisms involved in protective immunity. The latter has also applied to infections of the intermediate host with *Besnoitia* species, studied because of their similarities to *Toxoplasma*. Immune responses to *Eimeria* infections have also been studied in some detail, mainly in chickens, because of their economic importance in domestic poultry and the comparative ease of maintaining these animals coccidia-free. However, more recently, there has been an encouraging increase in the use of laboratory rodents. As far as the other genera (and the final hosts of the heteroxenous organisms) are concerned, information is scant but more is accruing, particularly for the systems in which they may have a diagnostic application, for example, *Sarcocystis* infections in cattle. This review does not attempt to cover comprehensively material that has already been adequately reviewed in the last decade, but concentrates on recent work, particularly from a comparative aspect.

Antibody Responses

Humoral Antibodies Circulating antibody responses to intestinal infections with coccidia have been investigated in the *Eimeria* and, to a lesser extent, in *Toxoplasma* and *Sarcocystis*. In the intermediate host, the subject has been well researched in infections with *Toxoplasma* and *Besnoitia* and, more recently, considerable attention has been paid to the serology of *Sarcocystis*.

Eimeria Work on *Eimeria* has been reviewed (Rose, 1973) and only a summary is given here.

Circulating antibodies to *Eimeria* infections are demonstrable in a variety of hosts, including rabbits (Heist and Moore, 1959; Rose, 1959, 1961; Černá, 1966a,b, 1967, 1969, 1970, 1974b), rats (Liburd et al., 1973; Rose and Hesketh, 1979), mice (Černá and Žalmanová, 1972; Mesfin and Bellamy, 1979a), cattle (Andersen et al., 1965), chickens (Rose and Long, 1962; Itagaki and Tsubokura, 1963; Long et al., 1963; Burns and Challey, 1965; Černá, 1968;

Baldelli et al., 1971; Morita et al., 1972; Rose, 1974a; Movsesijan et al., 1975; Abu Ali et al., 1976; Kouwenhoven and Kuil, 1976; Kuil and Dankerts-Brand, 1976; Tanielian et al., 1976; Davis et al., 1978), and turkeys (Augustin and Ridges, 1963). The tests have included agglutination, lysis, fluorescent labeling and dye penetration of sporozoite or merozoite stages, and (with soluble antigens) complement fixation and precipitation. Cytophilic and/or opsonizing antibodies have been demonstrated in the circulation of birds infected with *E. tenella* (Rose, 1974a), probably accounting for the increased activity of macrophages (Huff and Clark, 1970; Patton, 1970; Rose, 1974b), which coincides with the antibody response detectable by other tests. At a similar stage of immunization with *E. tenella*, an Arthus-type response, transferable by serum, can be demonstrated (Rose, 1977). Some indication of such a response has also been found in rabbits infected with *Eimeria stiedae* (Klesius et al., 1976a). In addition, the neutralization of invasive stages, shown by reduced infectivity and inhibition of cell penetration, has demonstrated the presence of antiparasite factors in the circulation of infected animals (Long and Rose, 1972; Rose, 1974a; Davis et al., 1978; Davis and Porter, 1979).

Antibodies are detectable usually within 1 week of the inoculation of oocysts (Rose and Long, 1962; Cerná, 1967; Liburd et al., 1973), although this depends on many factors, for example, the size of the inoculum (Burns and Challey, 1965). Antibodies seem to persist longer in the circulation of mammals than in birds, for example, in rabbits infected with *E. stiedai* or *Eimeria magna* (Rose, 1961; Cerná, 1966b) or calves infected with *E. bovis* (Andersen et al., 1965), compared with chickens infected with *E. tenella* (Burns and Challey, 1965). However, the sensitivity of the tests used probably differs and this should be taken into account. Attempted reinfection, especially in completely immunized animals, has little, if any, effect on circulating antibodies (Rose and Long, 1962; Tanielian et al., 1976), possibly reflecting the early inhibition of parasite development in such animals. The meager data available on the specificity of the antibody response suggest that different stages of the same species, and different species of *Eimeria*, have both common and specific antigens (Rose, 1959; Rose and Long, 1962; Černá, 1974a; Kouwenhoven and Kuil, 1976; Tanielian et al., 1976; Davis et al., 1978). The comparatively feeble transient presence of circulating antibodies, particularly in birds, together with the ease of detecting infections by fecal examinations, do not indicate a use for serological methods in diagnosis, although this has been suggested (Tanielian et al., 1974).

Toxoplasma Circulating antibody responses of the cat to infection with *Toxoplasma* have been investigated chiefly by the use of the dye test, the indirect fluorescent antibody test (IFA), and the indirect hemagglutination test (IHA). Antibodies are detectable within 2 weeks of the initial inoculum, but titers are low compared with those found in the intermediate hosts, and in some cats are never developed (Dubey et al., 1970; Piekarski and Witte, 1971; Dubey and Frenkel, 1972b; Miller et al., 1972; Dubey, 1973). It has been suggested that circulating antibodies may not be formed if the infection is confined to the gut (Dubey and Frenkel, 1972b). Failure to respond to repeated inocula with increased antibody titers, especially in cats that were refractory to challenge, has also been noted (Dubey and Frenkel, 1972b, 1974; Sheffield and Melton, 1974; Overdulve, 1974, 1978) and parallels the findings in *Eimeria* infections. Thus, low or negative titers are no indication of susceptibility to infection. Antibodies are maternally transferred to kittens, which nevertheless seem to be susceptible to infection (Dubey, 1973; Overdulve, 1978). Antibodies may persist for up to 5 months (in multiply infected animals), but are reduced by treatment with corticosteroids (Dubey and Frenkel, 1974). However, similar treatment of hamsters infected with *Besnoitia* did not cause a diminution of the antibody titer within the 20 days of the experiment (Frenkel and Lunde, 1966). As with *Eimeria* infections, serological tests are likely to be of very limited use for the diagnosis of toxoplasmosis of cats (Frenkel and Dubey, 1972; Frenkel, 1978).

There is a vast literature on antibody responses to *Toxoplasma* in the intermediate host, and the serology of the infection has been much studied because of its importance in the diagnosis of disease in man. Recent reviews include those of Frenkel (1973), Jacobs (1976), and Remington and Krahenbuhl (1976). A wide range of serological tests has been used; their respective merits, especially for diagnosis, have been fully discussed (see Jacobs, 1976; Walls et al., 1977; Denmark and Chessum, 1978; Frenkel, 1978; Walls, 1978; Walls and Barnhart, 1978) and are not dealt with here. In experimental animals, antibodies appear within days of infection, depending upon the test used, the species of host, and other factors such as dose, strain of parasite, and age of host. As might be expected in the case of a chronic infection, antibodies persist in the circulation for a considerable time, although results vary according to the test used. The variation in response according to host species has been discussed by Frenkel (1973), who noted that birds, with the exception of the pigeon, failed to produce antibodies detectable by the dye test. This may be due to the peculiar requirements of most avian antibody-antigen reactions. The characterization of the immunoglobulins participating in the antibody response to *Toxoplasma* infection has been reviewed by Remington and Krahenbuhl (1976), who also discussed the specificity of the various diagnostic tests.

The production of antibody to *T. gondii* on a cellular level has recently been examined by Masihi and Werner (1976), who showed immunocyto adherence (ICA) of *Toxoplasma* to mouse spleen cells, which could be abolished by treatment with anti-mouse immunoglobulin serum. In a later study they showed that, although the majority of the antigen-binding cells were B lymphocytes, some were identified as T lymphocytes and adherent cells (Masihi and Werner, 1978a). Magliulo et al. (1976) have also demonstrated ICA with rat spleen cells and with circulating human lymphocytes in an indirect test in which antigen-coated erythrocytes were used. ICA is detectable early in infection and it has been suggested that the test could be useful in the diagnosis of active *Toxoplasmosis* in humans.

The active production of antibodies may be suppressed by passively administered, or maternally transferred antibodies (Araujo and Remington, 1974, 1975; Masihi and Werner, 1977) and a state of tolerance induced (Araujo and Remington, 1972, 1974). Such suppression

8

could result in erroneous interpretations of serological tests in the newborn (Araujo and Remington, 1975).

Besnoitia Antibodies to *B. wallacei* have been detected in the final host, the cat, by means of indirect fluorescent antibody tests with cysts as antigen (Wallace and Frenkel, 1975; Frenkel, 1977). In response to infection, titers increased significantly from the end of patency and tended to remain positive throughout the period of testing (70 days). In general, titers in the final host were similar to those found in the intermediate host (Frenkel, 1977).

More information is available concerning antibody responses to *Besnoitia* species in intermediate hosts, because *B. jellisoni* has been fairly extensively studied in laboratory rodents. Tests for antibodies have included the dye, the IHA, and the IFA. Antibodies to *B. jellisoni* are rapidly developed after infection of hamsters, reach peak values within 2–3 weeks, and remain high throughout chronic infection (Frenkel and Lunde, 1966). High titers of antibodies to *B. wallacei* seem to be developed more slowly in mice and rats and took 8–11 weeks to reach a peak. Interspecific cross-reactions were slight; there was very little or no (depending upon the test) cross-reaction with *Toxoplasma* and slight cross-reaction with *Sarcocystis* antigens (Wallace and Frenkel, 1975; Frenkel, 1977).

Hammondia Cats develop little or no antibody to *H. hammondi* when tested by a fluorescent antibody test with cystozoites from mice as antigen. Wallace (1975) reported that three of seven cats, bled at intervals for 2 or more months after infection, produced titers of only 1:4–1:6. None of these sera cross-reacted with *Toxoplasma* antigens (Wallace, 1975; Frenkel and Dubey, 1975). In the intermediate host, in contrast, fairly high titers were found in infected mice and these cross-reacted with *Toxoplasma* antigen (Wallace, 1973, 1975). The extent of cross-reaction with *Toxoplasma* antigen depends upon the numbers of *H. hammondi* oocysts fed and the genus of the intermediate host (Frenkel and Dubey, 1975). This cross-reaction may lead to erroneous diagnosis of toxoplasmosis and indicates the close relationship of these parasites.

Sarcocystis The following summary of antibody responses to infection with *Sarcocystis* is based on the reviews of Tadros and Laarman (1976) and Markus (1978), which should be consulted for further details.

No antibodies were detected in the sera of cats (Markus, 1974) or of dogs (Lunde and Fayer, 1977) with intestinal infections of species of bovine *Sarcocystis*, when tested by the IFA or IHA tests, respectively. Ruiz and Frenkel (1976), however, reported that infection of cats with *Sarcocystis muris* caused an increase in the titers of antibodies measured by the IFA test, from 1:16–1:32 to 1:512–1:4,000. Results with human sera (intestinal infections) have been similarly conflicting: Aryeetey and Piekarski (1976) did not observe an increase in antibody titer in volunteers successfully infected with *Sarcocystis* of cattle origin, whereas Tadros and Laarman (1976) reported that titers of known carriers of *Isospora hominis* were generally higher than those of the random population, but only slightly above those of vegetarians. Such antibodies were also detected in the sera of newborn and young babies and were considered to have been acquired passively. Markus (1973) and Thomas and Dissanaike (1978) have also reported finding positive titers to bovine *Sarcocystis* antigen with the IFA test.

In intermediate hosts, antibodies to *Sarcocystis* have been detected (complement fixation or IFA) from 15 days after inoculation in lambs and sheep (sporocysts from dogs or cats) and mice (sporocysts of *S. muris* from cats) (Munday and Rickard, 1974; Munday et al., 1975; Ruiz and Frenkel, 1976). The serological responses of cattle infected with sporocysts from canine feces were extensively investigated by Fayer and Lunde (1977) and Lunde and Fayer (1977), who used soluble antigen in agar gel diffusion and IHA tests. Positive results were obtained with both tests from 30–45 days after inoculation, and very high titers were obtained at 90 days with the IHA test, which was also used in a survey of apparently healthy dairy cows and in clinical cases of *Sarcocystis* (Frelier et al., 1977). Early in infection, antibody activity resided mainly in the IgM fraction of serum but, by 10–13 weeks, the distribution had changed and it was very much greater in the IgG fraction (Fayer and Lunde, 1977).

The specificity of the tests mentioned here has already been discussed. Because man is both a final and (rarely) an intermediate host for *Sarcocystis*, Markus (1979) has questioned the value of a serological test for the diagnosis of sarcocystis in humans in view of the high incidence of intestinal infection. However, in those animals that are solely intermediate hosts (herbivores), serological tests should be useful, particularly the precipitin tests described by Lunde and Fayer (1977).

Secretory Antibodies It is likely that infections of mucous surfaces would stimulate the secretory immunologic system. However, to date, the possible involvement of antibodies of the IgA class in immunity to coccidian infections has been investigated only with *Eimeria* and *Toxoplasma*.

Early attempts to demonstrate antiparasite effects in *Eimeria* infections of chickens by extracts of feces (coproantibodies) or of cecal contents or mucosa met with little success (Augustin and Ridges, 1963; Horton-Smith et al., 1963b; Herlich, 1965; see Rose, 1973), although the presence of precipitating antibodies of the IgG class, thought to have leaked from the serum, has been detected in cecal contents during and shortly after infection (Movsesijan et al., 1975). There is now evidence for the involvement of IgA secretory antibodies in the immune response of chickens to infection with *E. tenella*. Davis et al. (1978) showed that, in the course of a primary infection with *E. tenella*, the numbers of IgA-containing cells in the ceca increased, and the IgA concentration of the cecal contents was also temporarily increased on days 10 and 7 after a primary or secondary infection, respectively. In vitro extracts of cecal contents, in which IgA was the predominant immunoglobulin, inhibited the penetration of cultured cells by sporozoites and impaired their subsequent development. Evidence for a secretory immune response in infections with other genera of coccidia is scant. Chessum (1978) recovered anti-

8

bodies to *Toxoplasma* from a cat, challenged via a Thiry-Vella loop constructed 2 yr after the initial infection with *Toxoplasma*. Antibodies of both IgG and of a heavier molecular weight, possibly IgA and/or IgM, were found.

In intermediate hosts, a serum IgA response has been noted (Rottini and Favento, 1969; Rottini et al., 1971; Garrido and Boveda, 1972), particularly after the administration of antigen by the oral route (Strannegård, 1967a).

Cellular Immune Responses

In many intracellular infections, responses mediated by T lymphocytes (TCMI) are considered to be the major factor in the development of immunity. Direct evidence for this in the case of infections with *Eimeria* has been, until recently, fairly scanty; the belief that cell-mediated immunity was the overridingly important component in immunity was founded more on the inability to account fully for the mediation of immunity by other mechanisms and/or the transfer of immunity with suspensions of (uncharacterized) lymphocytes, than on data provided by experimentation. However, more evidence is now accumulating for this type of response. In the case of infections with *Toxoplasma* and *Besnoitia*, TCMI has been much implicated, especially that form that involves the participation of macrophages. The responses considered here will be those mediated by circulating specifically sensitized T lymphocytes which, on contact with antigen, proliferate and release soluble mediators or lymphokines. These mediators may then act alone, or together with other (nonspecific) cells, for example, macrophages, to produce their effects. Not considered here (however, see below) is the role of T cells as helper cells in the production of antibodies; studies in athymic animals have shown that T cells do have such a function in both *Eimeria* (Mesfin and Bellamy, 1979a; Rose and Hesketh, 1979) and *Toxoplasma* infections (Hof et al., 1976; Masihi and Werner, 1978a).

Eimeria

Delayed-type Hypersensitivity The induction of delayed-type hypersensitivity (DTH), the classical expression of TCMI, in *Eimeria* infections is now well established in rabbits and calves (Klesius et al., 1976a, 1977), and in chickens (Rose, 1977). Different antigens were used in these studies; however, similar results were obtained and the reactions fulfilled all the criteria for classical delayed-type responses. The intensity of the response and/or the number of reactors (Klesius et al., 1976a) increased with time after infection. In chickens (Rose, 1977), the response, although detectable within 1 week of infection, remained at a comparatively low level during the time that circulating antibodies (and an Arthus-type response) could be demonstrated. When these were no longer present, they increased to remain high throughout the period of testing. Birds were substantially immune throughout this time, but noninfected birds which had been injected with parasite antigens in Freund's complete adjuvant and had similar DTH responses were susceptible to infection. Thus, the induction of DTH to the antigen used did not, per se, provide protective immunity.

T Lymphocyte-mediated Immune Responses TCMI responses to infection with *Eimeria* species have also been demonstrated in vitro. Positive results have been obtained in macrophage migration inhibition tests with spleen fragments (Morita et al., 1973) or with peritoneal exudate cells (Rose, 1974c) from birds infected with *E. tenella* or *E. maxima*, but these did not always correlate with delayed-type hypersensitivity (Rose, 1977). Blastogenesis of peripheral blood lymphocytes from cattle infected with *E. bovis* in the presence of *E. bovis* and *E. stiedai* antigens has been described, and the results have been correlated with those of skin tests (Klesius et al., 1977; Klesius and Kristensen, 1977).

Delayed-type skin reactivity has been transferred with transfer factor prepared from the lymphocytes of rabbits infected with *E. stiedai* (Klesius et al., 1976a) and from cattle infected with *E. bovis* (Klesius et al., 1975, 1976b; Klesius and Kristensen, 1977). The effect of transfer factor on challenge infection is described below.

It has been suggested that interferon, another mediator of cellular immunity, plays a role in immunity to *Eimeria* infections. The evidence has been reviewed (Rose, 1973) and is mentioned only briefly here. Fewer oocysts are produced by chickens treated with interferon inducers (Long and Milne, 1971), and infections with *Eimeria* species both in vivo (Long and Milne, 1971) and in vitro (Fayer and Baron, 1971) have been shown to induce an antiviral substance.

Toxoplasma and Besnoitia

Delayed-type Hypersensitivity Delayed-type hypersensitivity to *Toxoplasma* antigens was first reported by Frenkel (1948) in infected humans and guinea pigs. The test has been found useful in the diagnosis of chronic toxoplasmosis in man (Remington et al., 1962; Schaegel, 1965) in whom DTH develops late in infection (Ludlam, 1960; see review by Remington and Krahenbuhl, 1976). Krahenbuhl et al. (1971) studied the onset and development of DTH responses in guinea pigs infected with *Toxoplasma* and found that positive results at a low level were present as early as 1 week after infection, thereafter increasing in intensity until 10–15 weeks. An Arthus-type response was not seen until 3 weeks and became more evident up to 15 weeks. Antigens prepared from killed organisms, when administered with Freund's complete adjuvant, but not alone, induce DTH and a certain degree of resistance to infection (Krahenbuhl et al., 1972b). DTH responses can also be demonstrated in hamsters infected with *Besnoitia* (Frenkel, 1967).

T Lymphocyte-mediated Immune Responses in Vitro TCMI responses to *Toxoplasma* have also been demonstrated in vitro. There is a report of the direct killing of *T. gondii* by

8

peritoneal exudate lymphocytes of immunized mice after challenge with a virulent strain. From electron microscopal evidence, this killing is produced by a process of lysis (Pelster, 1975).

Demonstration of the production of lymphokines by stimulated lymphocytes is summarized as follows. The inhibition of macrophage migration by cells from infected guinea pigs was shown by Tremonti and Walton (1970); Krahenbuhl et al. (1971) also have studied its onset and development in guinea pigs. Positive results were obtained with peritoneal exudate cells within 1 week of infection; these correlated well with the results of tests for DTH. Furthermore, the migration inhibition factor (MIF), which was active with homologous macrophages, was obtained from spleen cells of infected animals cultured with *Toxoplasma* antigen (Krahenbuhl et al., 1971; Krahenbuhl and Remington, 1971). The production of MIF, active with guinea pig macrophages, from the peripheral blood leukocytes of infected humans has also been demonstrated (Gaines et al., 1972). Specific lymphocyte transformation by *Toxoplasma* antigens has been described in rabbits (Huldt, 1967), guinea pigs (Krahenbuhl and Remington, 1971), humans (Tremonti and Walton, 1970; Krahenbuhl et al., 1972a), and mice (Gardner and Remington, 1978b). In humans, greater transformation has been noted in leukocytes from patients with chronic rather than active toxoplasmosis; however, in mice, lymphocyte transformation was noted within 1 week after infection, and positive results were obtained in rabbits tested 7–9 weeks after infection. Lymphocyte transformation in mice was not affected by aging (Gardner and Remington, 1978b); it is very persistent in humans, being present up to 19 yr after the initial infection (Krahenbuhl et al., 1972a). It can thus be related to the persistence of *Toxoplasma* cysts and DTH.

The involvement of macrophages and monocytes as effector cells in TCMI responses to infection with *Toxoplasma* and *Besnoitia* is well documented. Earlier work (reviewed by Remington and Krahenbuhl, 1976) indicated an increase in phagocytosis of *Toxoplasma* by macrophages from immune animals and a reduction in intracellular multiplication (Vischer and Suter, 1954; Huldt, 1966b). This effect may be partly due to the action of antibodies (Jones et al., 1975; Stadtsbaeder et al., 1975; Anderson et al., 1976a; Lindberg and Frenkel, 1977a), but is also seen in their absence and is now acknowledged to be largely due to the action of soluble lymphocyte products or mediators in specifically arming the macrophage, together with a minor component of nonspecific activity. There has been considerable disagreement in the literature concerning the latter. One school of thought has proposed that the reaction of macrophages against *Toxoplasma* and *Besnoitia* is similar to that seen with the intracellular microbes *Listeria*, *Mycobacterium*, *Salmonella*, and *Brucella*, that is, specifically induced but nonspecifically expressed by the "activated" macrophage (Mackaness, 1964; Mackaness and Blanden, 1967; Ruskin et al., 1969), and that a common mechanism of immunity mediates resistance to intracellular pathogens (Ruskin and Remington, 1968a). The proponents of this theory have shown that mice infected with *B. jellisoni* are resistant to infection with *T. gondii* and vice versa, as well as to a variety of unrelated organisms (Ruskin and Remington, 1968a,b; Remington and Merigan, 1969; Gentry and Remington, 1971) and tumors (Hibbs et al., 1971). Activity against the heterologous organisms was also demonstrable in vitro by macrophages obtained from animals infected with *B. jellisoni* or *T. gondii* (Ruskin and Remington, 1968b; Remington et al., 1972).However, the results of other workers have indicated that in the case of *Besnoitia* and *Toxoplasma* infections, the immune response to one parasite does not lead to an immune response against the other, either in vivo or in vitro (Hoff and Frenkel, 1974; Lindberg and Frenkel, 1977a), nor does immunization with BCG or *Listeria* protect against *Toxoplasma* (Frenkel and Caldwell, 1975; Jones et al., 1975; Swartzberg et al., 1975; Dubey, 1978c). Attempts have been made to reconcile the conflicting reports by differentiating between the killing action of activated macrophages and of macrophages armed by lymphokines (Lindberg and Frenkel, 1977a; McLeod and Remington, 1977a,b). It has been suggested that, although nonspecifically activated macrophages may be appreciably more inhibitory than normal macrophages, specifically armed macrophages are very much more inhibitory (Chinchilla and Frenkel, 1978). However, the role of activated macrophages in resistance to infection has been questioned as a result of the lack of correlation between protection in vivo and killing of organisms in vitro. Thus, immunization of mice with *C. parvum* conferred little protection against challenge with virulent *Toxoplasma*, despite the presence of a population of macrophages which was highly active against this organism in vitro (Swartzberg et al., 1975). The experiments of Sethi et al. (1975) have indicated that in activated macrophages, the lag phase which precedes multiplication of *Toxoplasma* is markedly prolonged, but thereafter growth proceeds unchecked.

Jones et al. (1975) showed that macrophages from immune animals could directly inhibit the intracellular multiplication of *Toxoplasma* in early infection, but that thereafter they had to be exposed to specifically sensitized lymphocytes and antigen, or to the lymphokine (*Toxoplasma*-inhibiting factor (ToxoIF)) produced by this interaction. According to these authors and others (Hoff and Frenkel, 1974), specifically induced *Toxoplasma*-immune lymphocytes did not confer anti-*Toxoplasma* activity on macrophages from nonimmune animals; however, this has now been demonstrated for human monocytes and macrophages (Borges and Johnson, 1975; Anderson et al., 1976b) and for mouse (Sethi et al., 1975; Shirahata et al., 1975, 1976) and hamster (Chinchilla and Frenkel, 1978) macrophages. The mediator (ToxoIF, ToxoGIF (*Toxoplasma* growth-inhibiting factor), or anti*Toxoplasma*-arming factor (ATAF)) has been the subject of much recent study, and a similar but distinct mediator has also been identified in *Besnoitia* infection (Chinchilla and Frenkel, 1978). The mediator has been shown to be the product of protein synthesis (Shirahata et al., 1977; Chinchilla and Frenkel, 1978; Sethi and Brandis, 1978; Shirahata and Shimizu, 1979) by T lymphocytes (Borges and Johnson, 1975; Sethi et al., 1975; Shirahata et al., 1977), with an activity specific for species, but not for strains of host cells, and for the microbe involved (Chinchilla and Frenkel, 1978; Sethi and Brandis, 1978). Some authors (Anderson and Remington, 1974; Anderson et al., 1976b) have reported that the products of lymphocytes stimulated by a heterologous antigen or a mitogen are also effective, but this has not been confirmed by others (Borges and Johnson, 1975; Shirahata et al., 1977; Sethi and Brandis, 1978). A molecular weight of 4,000–5,000 has been given (Chinchilla and Frenkel, 1978) and also one of 50,000–100,000 (Sethi and Brandis, 1978; Shirahata and Shimizu, 1979). According to Chinchilla and Frenkel (1978), the specificity of the induction and expression of the mediator exclude it from being identified as interferon. They have suggested that it has a similarity to transfer factor. Reports on its effectiveness in cells other than macrophages differ: Chinchilla and Frenkel (1978) found it to be active in cultured fibroblasts and kidney cells, whereas Sethi and Brandis (1978) reported it to be ineffective in mouse embryo fibroblasts. However, Sethi and Brandis (1978) noted the presence of soluble factor(s) in the sera of mice convalescing from toxoplasmosis, which did have such activity.

The mechanism of inhibition of multiplication of *Toxoplasma* within the cell as a result of interaction with the lymphokine has not been identified. However, changes have been described in protein synthesis and cyclic nucleotides (Jones et al., 1977). Pinocytic rates are reduced in *Toxoplasma*-immune macrophages but, according to Jones and Len (1976), this does not correlate with the ability of the cell to inhibit the multiplication of *T. gondii*.

Shirahata and Shimizu (1979) have pointed out the similarities between ToxoGIF and type II interferon (as well as migration inhibition factor), and Sethi and Brandis (1978) have also noted the interferon-like properties of ATAF. Earlier workers claimed that infections with *T. gondii* and *B. jellisoni* induce the production of interferon in vivo (Freshman et al., 1966; Rytel and Jones, 1966; Remington and Merigan, 1969) and that treatment with interferon protected chick and mouse cell monolayers against infection with *T. gondii* (Remington and Merigan, 1968). More recent reports (Schmunis et al., 1973; Ahronheim, 1979) have, however, cast doubts on the interactions of interferon (of both leukocyte and fibroblast origin) with *T. gondii*. Clearly, the exact nature, identities, and actions of the lymphokines produced by stimulated T lymphocytes in *Toxoplasma* and *Besnoitia* infections remain to be established.

Accessory Cells and Factors

There has been little investigation of the possible participation of accessory cells other than macrophages in immune responses to coccidian infections, although it is now well established that many such cells are involved in immunity, particularly in the expression of TCMI responses. The rapid changes in numbers of polymorphonuclear cells (PMN), both in the circulation and at the site of stimulation, in immune animals after specific challenge with *Eimeria* species has been noted earlier. PMN also actively phagocytose sporozoites of *Eimeria* species in vitro (Rose and Lee, 1977), but their significance, if any, in immunity has not been determined. There is also an intestinal mast cell response to infections with *Eimeria* species, but its association with immunity is doubtful (Rose et al., 1980b).

According to Jones and Hirsch (1972), and in contrast to a brief report by Holland and Sleamaker (1970), PMN do not engulf live *Toxoplasma*, but dead or antibody-coated organisms are taken up with resulting degranulation of the PMN. No information is available for other accessory cells in this system.

The ''accessory'' factor or ''activator'' necessary for the lysis of *T. gondii* by antibody, although considered by some to be allied to the properdin system, is probably complement (see Jacobs, 1976). Accessory factor has been studied mainly in relation to serodiagnosis. However, the results of a recent investigation in mice seemed to show that a deficiency of the fifth component of complement led to a reduction in mortality (Araujo et al., 1975). This was observed not only in inbred congenic mice of known complement type, but also in random-bred mice selected for complement type. The significance of this observation is unclear.

MECHANISMS OF PROTECTIVE IMMUNITY

Despite the identification and investigation of a variety of immune responses induced by infections with coccidia, the exact means by which the host is able to control primary infections and to resist reinfection has not been established. In this section, an attempt is made to determine (insofar as this is possible), or to speculate on, the relative contributions of the responses, discussed above, to protective immunity. The results of experiments on the attempted transfer of protection from immune to susceptible hosts, and those carried out on animals with various immunodeficiencies, will form the basis for these assessments. Such experiments have been performed mainly with animals infected with *Eimeria* species, and with *T. gondii* or *B. jellisoni* in the intermediate host.

Eimeria

Role of Antibodies The demonstration of a circulating antibody response to infections with *Eimeria* and the deleterious effect of such antibodies on parasites in vitro, together with the transmission of immunity from one isolated cecum to another (Burns and Challey, 1959; Horton-Smith et al., 1961), encouraged renewed attempts to transfer immunity with serum or globulin fractions, or to demonstrate maternally transferred immunity. Such experiments have met with some success: the administration of daily doses of serum or globulin fractions from immunized animals to susceptible rats (Wittchow, 1972; Rose and Hesketh, 1979) or chickens (Rose, 1971, 1974d) partially protected them from challenge infections. An examination of the

8

kinetics of such protective activity in serum showed that, in chickens, its presence was limited to the time of onset of complete immunity and could not be demonstrated in animals in which this was well established. A similar pattern was detected in the protection of embryonating eggs laid by immunized hens, or of chicks hatched from them (Rose and Long, 1971; Rose, 1972). Few experiments have been carried out in vivo on the role of secretory antibodies on immunity to infections with *Eimeria*. However, there is some evidence for the amelioration of infection by the administration of daily doses of IgA prepared from the intestinal contents of recovered chickens (Orlans and Rose, 1972; Rose, unpublished). The results of recent experiments on chickens with severely impaired functioning of B lymphocytes have confirmed that the role of antibodies is a definite, but minor, one (Joyner and Norton, 1974; Rose and Hesketh, 1979). Primary infections were greater and resistance to reinfection was slightly lowered in chemically bursectomized chickens, with severely depleted circulating IgG and IgM and biliary IgA, but with normally functioning T lymphocytes (Rose and Hesketh, 1979). It is somewhat difficult to envisage how circulating antibodies could affect an enteric intracellular parasite, except for those parts of the life cycle when the parasite is extracellular, that is, the invasive stages. Marked changes in vascular permeability occur within a very short time of challenge of the immune gut (Rose and Long, 1969; Kouwenhoven and van der Horst, 1973; Rose et al., 1975), which could allow contact between circulating antibodies and parasite and thus bring about the lytic and neutralizing effects demonstrable in vitro. Circulating antibodies may also be effective via their opsonic or cytophilic properties in enhancing the uptake of parasites by macrophages. Whereas damaged parasites (by antibodies, or other means) are destroyed within macrophages, there is, at present, no evidence to suggest that the increased uptake of normal parasites by immune macrophages is necessarily followed by death of the parasite. Their appearance by electron microscopal observation is unchanged (Rose and Lee, 1977) and they will resume development if transferred to a susceptible host (Rose and Long, 1976). Since, unlike some other coccidian genera, they do not normally multiply within macrophages (Long and Rose, 1976), any growth inhibitory property of immune macrophages, as is seen in the case of *T. gondii* (see below), cannot be examined. Antibodies, perhaps complexed with antigen and complement, may be involved in the rapid inflammatory response observed in the intestines of immune birds after challenge with *E. maxima* (Rose et al., 1979a). Recent observations (Rose, unpublished) suggest that this response, which can be correlated with changes in circulating lymphocytes (principally T cells) and heterophils, may be transferable with serum from immunized birds, and is possibly analogous to the serum-mediated emigration of neutrophils, which has been noted in the intestines of immunized pigs after enteric challenge (Bellamy and Nielsen, 1974). The attraction and retention of intravenously injected, radioisotope-labeled lymphocytes at the site of challenge of immune intestines has also been recorded (Rose et al., 1980a). Whether this has any significance in protective immunity has not been established, but would seem to be highly likely. Secretory antibodies may gain access to the cell and exert direct antiparasite effects therein. In addition to this, Davis and Porter (1979) have proposed that secretory antibodies in the apical mucin, by hindering the penetration of invasive stages, causes them to be exposed to nonspecific lytic factors normally present in the intestine. They have suggested that this causes the removal of surface membrane "differentiation antigens" so that even if sporozoites subsequently succeed in penetrating host cells, they are unable to differentiate. Renewal of the surface proteins on removal of sporozoites to susceptible hosts would account for the resumption of development previously discussed.

Role of Cell-mediated Immune Responses The transfer of large numbers of lymphocytes from immune animals greatly reduces infections with *Eimeria* species in recipient rats (Rommel and Heydorn, 1971; Liburd et al., 1973; Ogilvie and Rose, 1977) and chickens (Rose and Hesketh, in press). In rats, mesenteric lymph node and thoracic duct lymphocytes and, in chickens, similar numbers of peripheral blood lymphocytes were equally effective, but more spleen cells were required. Compared with other systems, the numbers of cells required were large and, because they were not fractionated, the identity of the effective cells has not been determined. Furthermore, both antibody (Liburd et al., 1973) and cell-mediated (Rose and Hesketh, in press) responses were transferred. An essential role for T lymphocytes in the mediation of immunity has, however, been demonstrated by experiments on athymic mice (Mesfin and Bellamy, 1979a) and rats (Rose and Hesketh, 1979; Rose et al., 1979b). Primary infections in these animals were enhanced, and they were completely susceptible to reinfection. In contrast to their euthymic litter mates, they did not produce circulating antibodies (Mesfin and Bellamy, 1979a; Rose and Hesketh, 1979), and rats were partially protected by convalescent serum (Rose and Hesketh, 1979). Thus, T lymphocytes were shown to act as helper cells in antibody production to membrane antigens of sporozoites. However, the results of serum transfers discussed above, taken together with the fairly substantial resistance to reinfection of bursectomized birds with very reduced (IgG and IgM) or absent (IgA) immunoglobulins, suggest that T lymphocytes must be additionally involved in some manner other than as helper cells for Ig production. This could be as effector cells which release some mediator(s) to act, either alone or through the agency of accessory cells. One mediator of cellular immunity which has been shown to moderate infections with *Eimeria* species in vivo is dialyzable transfer factor (Lawrence, 1969). Transfer factor prepared from the peripheral blood, mesenteric lymph node, and spleen lymphocytes of rats immunized by infection with *E. nieschulzi* was injected intraperitoneally into test rats 2 days before a challenge inoculum of homologous oocysts. Oocyst output resulting from this inoculum was significantly lower than that found in untreated controls, and immunity to a second inoculum was complete (Liburd et al., 1972). Klesius and co-workers (1975, 1976b, 1977, 1978, 1979) have similarly tested the effects of transfer factor prepared from the lymph nodes of cattle and rabbits infected with *E. bovis* and *E. stiedai*, respectively. Intraperitoneal injection of the bovine transfer factor-reduced infections with

E. bovis in calves (Klesius et al., 1977) and with *Eimeria ferrisi* in mice (Klesius et al., 1978), but infection with *E. stiedai* in rabbits, was not affected by either bovine or rabbit transfer factor, although CMI responses were transferred (Klesius et al., 1977). Bovine transfer factor was most effective in mice when given 2 days before oocyst inoculation; injections given simultaneously with, or 2 days after inoculation with oocysts did not protect (Klesius et al., 1978). Oral administration in the drinking water was as effective as intraperitoneal injection (Klesius et al., 1979). The patterns of oocyst production seen in animals partially protected with transfer factor mimic those found in animals partly immunized by infection (Liburd et al., 1972; Klesius et al., 1978). The results merit further investigation, but it is difficult to reconcile the effect of bovine transfer factor on infection with *E. ferrisi* in mice with the generally accepted species specificity of immunity to the *Eimeria* and the failure to demonstrate a nonspecific component (even when specifically elicited) in the effector mechanisms of immunity (Rose, 1975). Other mediators, such as those which act via macrophages or directly on parasitized cells, as in *T. gondii* infections, have not yet been demonstrated, but may be produced and active in protective immunity.

Toxoplasma and *Besnoitia*

Because host-parasite relationships are so similar for these two organisms, which have often been studied in parallel, they are considered together here.

Role of Antibodies The toxoplasmacidal effect of antibodies in vitro in the presence of complement is well known (Strannegård, 1967b; Klainer et al., 1973; Endo and Kobayashi, 1976; Wellensiek et al., 1976) and has long been utilized in serodiagnostic tests for toxoplasmosis. Recent studies have confirmed this effect of antibodies (Anderson and Remington, 1974; Jones et al., 1975; Anderson et al., 1976a) and also have shown that treatment with heat-inactivated serum from immune animals, although not lethal to the organisms, renders them incapable of initiating progressive infections within normal macrophages or fibroblasts in vitro (Vischer and Suter, 1954; Anderson and Remington, 1974; Jones et al., 1975; Stadtsbaeder et al., 1975; Sethi et al., 1975). However, attempts to transfer immunity passively with serum have had little or no success (Nakayama, 1965; Frenkel and Lunde, 1966; Foster and McCulloch, 1968; Gill and Prakash, 1970; Strannegard and Lycke, 1972), although Frenkel (1967) and Hoff and Frenkel (1974) noted that antiserum to *B. jellisoni* had a minor protective effect in hamsters and augmented the much greater protection afforded by transferred lymphoid cells (Frenkel, 1967). Krahenbuhl et al. (1972b) also observed slight but significant protection against *T. gondii* infections in mice with transferred antibodies. More recently, the multiple transfer of immune serum has been shown to confer significant protection against toxoplasmosis in mice (Masihi and Werner, 1978b) and to cause the rupture of cysts with the liberation of zoites (Werner et al., 1978). The protective effect of antibodies has also been confirmed by the enhancement of infection in mice treated with cyclophosphamide to inhibit the responses of B lymphocytes, an effect which was reversed by passively transferred antiserum (Hafizi and Modabber, 1978). These observations apparently contrast with the relapse of *B. jellisoni* infections in hamsters treated with corticosteroids (Frenkel and Lunde, 1966) and contrast with the increased susceptibility to *T. gondii* in guinea pigs treated with hexestrol (Kittas and Henry, 1979), both in the presence of high titers of circulating antibodies. However, because treatment with corticosteroids has been shown to inhibit the ability of macrophages to limit the growth of antibody-modified organisms in vitro (Lindberg and Frenkel, 1977a), the relapsing infections in the presence of high titers of antibodies noted by Frenkel and Lunde (1966) could be due, at least in part, to failure of macrophage function. It is possible that some of the negative results previously reported in experiments on passive transfer may have been due to the use of heterologous antisera or those produced by the injection of killed parasites (Nakayama, 1965; Gill and Prakash, 1970; Strannegard and Lycke, 1972). Thus, in the intermediate host, antibodies may exert their effects against *Toxoplasma* and *Besnoitia* infection by destroying extracellular organisms and by contributing to the inhibitory effects of macrophages.

Role of Cell-mediated Immune Responses In claiming a paramount role for CMI responses in immunity to infection with *T. gondii* or *B. jellisoni*, the adoptive transfer of immunity with lymphoid cells is often quoted. However, the literature is very sparse and the effective cells have not been defined. The overriding importance of CMI responses could not be deduced from the results of Frenkel's (1967) experiments (pertaining mainly to *Besnoitia* infections), because immunity, and both T (delayed hypersensitivity) and B (antibody production) lymphocyte activity, were transferred with 10^7 spleen cells or 7×10^7 mesenteric lymph node cells. However, the results of other experiments indicate an essential role for T lymphocytes in immunity. Thus, treatment of mice with antithymocyte serum (ATS) causes enhancement of current *Toxoplasma* infection and activation of chronic infection (Strannegård and Lycke, 1972). Also nude (athymic) mice were unable to develop immunity to infection with *Toxoplasma* and, unlike the hairy controls, died after withdrawal of chemotherapy (Hof et al., 1976; Lindberg and Frenkel, 1977b). Reconstitution with thymus cells enabled the athymic mice to become immune (Lindberg and Frenkel, 1977b). According to Nakayama and Aoki (1970), treatment with antilymphocyte serum depressed the numbers of circulating lymphocytes and enhanced infection with *Toxoplasma*; nevertheless, antibodies were produced. However, there is good evidence for the T dependence of *Toxoplasma* antigen in the lack of circulating antibodies in the serum of nude mice at a time when high titers were present in their hirsute counterparts (Hof et al., 1976; Masihi and Werner, 1978a). Nevertheless, attempts to increase survival in nude mice by passive transfer of antibody were unsuccessful (Lindberg and Frenkel,

1977b). Therefore, in the expression of immunity to *Toxoplasma* and *Besnoitia*, the activity of T lymphocytes in CMI responses seems to be of greater significance than their function as helper cells in antibody production. However, there is ample evidence in vitro, but little in vivo, for the antiparasite effects of such activity. Chinchilla and Frenkel (1978) have shown that a single injection of the mediator produced by incubating lymphocytes with freeze-thaw disrupted organisms does protect hamsters against infection with *Besnoitia*. In hamsters challenged with 10 or 100 organisms, protection was greater than that obtained with high titer antiserum; however, this was reversed when challenge consisted of 1,000 organisms. This mediator was shown in vitro to be effective with fibroblasts and kidney cells, as well as with macrophages. Concerning the role of the latter, the results of experiments in vitro involving heterologous challenge (discussed above) have indicated that the activation of macrophages has a minor effect on parasite growth; this has been borne out by experiments in vivo. There is little protection against *Toxoplasma* or *Besnoitia* in animals stimulated with *Bordetella pertussis*, *C. parvum*, *Mycobacterium bacillus* Calmette-Guérin (BCG), *Listeria monocytogenes*, or heterologous parasites (Frenkel and Caldwell, 1975; Swartzberg et al., 1975; Hof et al., 1976). More surprisingly, in view of the results obtained in vitro with specifically armed macrophages, functional blockade of macrophages did not enhance primary infections with *T. gondii* in mice, but did thus affect infections with *L. monocytogenes* (Hahn, 1974). However, the mediator described by Chinchilla and Frenkel (1978) is not dependent upon the macrophage for its expression.

CONCLUDING REMARKS

Much remains to be ascertained concerning the expression of immunity and of the immune responses of the final host, to several of the heteroxenous genera of coccidia. It seems likely that, insofar as the enteric stages are concerned, there would be great similarities to the *Eimeria*, while immunity to extraenteric stages in the final host and to the parasite in the intermediate host would resemble that found in *Toxoplasma* and *Besnoitia* infections.

The mechanisms involved in protective immunity to those coccidia (*Eimeria*, *Toxoplasma*, and *Besnoitia*) that have been investigated in some detail are, as might be expected, complex and probably consist of the entire range of immune responses, acting alone or together, depending possibly upon the stage of infection and of the immune status of the host. All these coccidian genera induce very similar immune responses; there seems to be little doubt that, in all cases, immunity depends upon the proper functioning of T lymphocytes. T lymphocytes seem to be necessary for the formation of antibodies to coccidia, and these antibodies are involved to some extent in protective immunity, perhaps directed mainly against parasites when extracellular. However, the greater part of resistance to reinfection is probably mediated by

T lymphocytes acting in some other capacity. In the case of stages of the heteroxenous coccidia in the intermediate host, lymphokines capable of inhibiting the intracellular multiplication of the parasite have been identified; it seems most likely that these would also be effective against intracellular stages in the final host and against intracellular stages of the monoxenous genera. Little is known of the functioning of accessory cells other than macrophages in resistance to coccidia, but it is probable that, as in infections with other pathogens, these too are involved to some extent.

Despite these varied and effective host responses, the coccidia are highly successful parasites maintaining, in normal circumstances, a well balanced relationship with their hosts. Most probably, the maintenance of this relationship depends not only upon the ability of the host to limit infection with the parasite, but also upon the ability of the parasite to evade the suppressive activity of the host. Thus, *T. gondii*, if undamaged by antibodies or other factors, is able to survive and multiply in macrophages by inhibiting the fusion of secondary lysosomes with the phagocytic vacuole (Jones et al., 1972; Khavkin and Freidlin, 1977) by a mechanism as yet unknown. It is likely that *B. jellisoni* also is equipped with this facility for combating host defense. Another factor which may be effective in limiting the host response to the benefit of parasite survival is the depressive effect of infection on host immune responses, shown in mice infected with *T. gondii* (Strickland et al., 1972, 1973). Such evasive tactics have not yet been demonstrated for other coccidia, but it would be surprising if they, or similar mechanisms, did not obtain, at least in the case of those genera that persist in the host. The *Eimeria* and *Isospora*, with their brief and defined life cycles, may not have such a need for dodging host surveillance; however, the prevalence of some highly immunogenic species is puzzling. Strain variation is a possible answer to this conundrum, and has been discussed above. Clearly, the host-parasite relationships in this group of organisms will continue to provide fascinating and diverse problems for solution.

LITERATURE CITED

Abu Ali, N., Movsesijan, A., Sokolić, A., and Tanielian, Z. 1976. Circulating antibody response to *Eimeria tenella* oral and subcutaneous infections in chickens. Vet. Parasitol. 1:309–316.

Ahronheim, G. A. 1979. *Toxoplasma gondii*: Human interferon studies by plaque assay. Proc. Soc. Exp. Biol. Med. 161:522–526.

Andersen, F. L., Lowder, L. J., Hammond, D. L., and Carter, P. B. 1965. Antibody production in experimental *Eimeria bovis* infections in calves. Exp. Parasitol. 16:23–35.

Anderson, S. E., Bautista, S. C. and Remington, J. S. 1976a. Specific antibody-dependent killing of *Toxoplasma gondii* by normal macrophages. Clin. Exp. Immunol. 26:375–380.

Anderson, S. E., Bautista, S., and Remington, J. S. 1976b. Induction of resistance to *Toxoplasma gondii* in human macrophages by soluble lymphocyte products. J. Immunol. 117:381–387.

Anderson, S. E., and Remington, J. S. 1974. Effect of normal and activated human macrophages on *Toxoplasma gondii*. J. Exp. Med. 139:1154–1174.

Andrews, J. M. 1926. Coccidiosis in mammals. Am. J. Hyg. 6:784–798.

Araujo, F. G., and Remington, J. S. 1972. Immune response to intracellular parasites: Suppression by antibody. Proc. Soc. Exp. Biol. Med. 39:254–258.

Araujo, F. G., and Remington, J. S. 1974. Induction of tolerance to an intracellular protozoan (*Toxoplasma gondii*) by passively administered antibody. J. Immunol. 113:1424–1428.

Araujo, F. G., and Remington, J. S. 1975. IgG antibody response to *Toxoplasma gondii* in newborn rabbits. J. Immunol. 115:335–338.

Araujo, F. G., Rosenberg, L. T., and Remington, J. S. 1975. Experimental *Toxoplasma gondii* infection in mice: The role of the fifth component of complement. Proc. Soc. Exp. Biol. Med. 149:800–804.

Aryeetey, M. F., and Piekarski, G. 1976. Serologische *Sarcocystis*-studien an Menschen und Ratten. Z. Parasitenkd. 50:109–124.

Augustin, R., and Ridges, A. P. 1963. Immunity mechanisms in *Eimeria meleagrimitis*. In: P. C. G. Garnham, A. E. Pierce, and I. Roitt (eds.), Immunity to Protozoa, pp. 296–335. Blackwell Scientific Publications, Ltd., Oxford.

Baldelli, B., Frescura, T., Ambrosi, M., Saravanos, K. A., and Polidori, G. A. 1971. Studio degli anticorpi fluorescenti nella coccidiosi sperimentale dei polli da *Eimeria tenella*. Parasitologia XIII:99–103.

Bellamy, J. E. C., and Nielsen, N. O. 1974. Immune-mediated emigration of neutrophils into the lumen of the small intestine. Infect. Immun. 9:615–619.

Blagburn, B. L., Chobotar, B., and Smith, R. T. 1979. Clinical and histologic observations of actively induced resistance to *Eimeria ferrisi*, Levine and Ivens, 1965 (Protozoa: Eimeriidae) in the mouse (*Mus musculus*). Z. Parasitenkd. 59:1–14.

Borges, J. S., and Johnson, W. D. 1975. Inhibition of multiplication of *Toxoplasma gondii* by human monocytes exposed to T-lymphocyte products. J. Exp. Med. 141:483–496.

Burns, W. C., and Challey, J. R. 1959. Resistance of birds to challenge with *Eimeria tenella*. Exp. Parasitol. 8:515–526.

Burns, W. C. and Challey, J. R. 1965. Serum lysins in chickens infected with *Eimeria tenella*. J. Parasitol. 51:660–668.

Campana-Rouget, Y., Dorchies, Ph., and Gourdon, L. 1974. Phenomènes de competition entre *Isospora* et *Toxoplasma* chez le chat. Proceedings of the 3rd International Congress on Parasitology, Munich. Vol. 1, pp. 108–109. Facta Publication, Vienna.

Catchpole, J., Norton, C. C., and Joyner, L. P. 1976. Experiments with defined multispecific coccidial infections in lambs. Parasitology 72:137–147.

Černá, Ž. 1966a. Anwendung der indirekten fluoreszenz antikörpereaktion zum Nachweis der Antikörper bei Kaninchen kokzidiose. Zentralbl. Bakteriol. Parasitenk. Abt. I Orig. 199:264–267

Černá, Ž. 1966b. Studies on the occurrence and dynamics of coccidial antibodies in *Eimeria stiedae* and *E. magna* by the indirect fluorescent antibody test. Folia Parasitol. (Prague) 13:332–342.

Černá, Ž. 1967. The dynamics of antibody against *Eimeria tenella* under the fluorescent microscope. Folia Parasitol. (Prague) 14:13–18.

Černá, Ž. 1968. Paraffin and esterwax, two suitable embedding media for indirect fluorescence antibody tests. Folia Parasitol. (Prague) 15:173–177.

Černá, Ž. 1969. Antibodies in chicks infected by *E. tenella* and *E. acervulina* detected by an indirect fluorescent reaction. Acta Vet. (Brno) 38:37–41.

Černá, Ž. 1970. The specificity of serous antibodies in coccidioses. Folia Parasitol. (Prague) 17:135–140.

Černá, Ž. 1974a. Antigenic relationships between *Eimeria acervulina*, *E. tenella* and *E. maxima*. Proceedings of the 3rd International

Congress on Parasitology, Munich. Vol. 2, p. 1135. Facta Publication, Vienna.

Černá, Ž. 1974b. The use of fluorescent antibody technique in coccidia. Proceedings of the 3rd International Congress on Parasitology, Munich. Vol. 2, pp. 1133–1134. Facta Publication, Vienna.

Černá, Ž., and Kolářová, I. 1978. Contribution to the serological diagnosis of *Sarcocystosis*. Folia Parasitol. (Prague) 25:289–292.

Černá, Ž., and Žalmanová, Ž. 1972. An attempt to analyse antigens from sexual and asexual stages of coccidia by the indirect fluorescence antibody reaction. Folia Parasitol. (Prague) 19:179–181.

Chessum, B. S. 1972. Reactivation of *Toxoplasma* oocyst production in the cat by infection with *Isospora felis*. Br. Vet. J. 128:33–36.

Chessum, B. S. 1978. Coproantibody response to oral infection with *Toxoplasma gondii*. Proceedings of the 4th International Congress on Parasitology, Warsaw, pp. 24–25. Published by the Organizing Committee.

Chinchilla, M., and Frenkel, J. K. 1978. Mediation of immunity to intracellular infection (*Toxoplasma*, *Besnoitia*) within somatic cells. Infect. Immun. 19:999–1012.

Christie, E., and Dubey, J. P. 1977. Cross immunity between *Hammondia* and *Toxoplasma* infections in mice and hamsters. Infect. Immun. 18:412–415.

Cuckler, A. C. 1970. Coccidiosis and histomoniasis in avian hosts. In: G. J. Jackson, R. Herman, and I. Singer (eds.), Immunity to Parasitic Animals, Vol. 2, pp. 371–397. Appleton-Century-Crofts Educational Division, Meredith Corp., New York.

Cutchin, E., and Warren, J. 1956. Immunity patterns in the guinea pig following *Toxoplasma* infection and vaccination with killed *Toxoplasma*. Am. J. Trop. Med. Hyg. 5:197–209.

Davis, P. J., Parry, S. H., and Porter, P. 1978. The role of secretory IgA in anticoccidial immunity in the chicken. Immunology 34: 879–888.

Davis, P. J., and Porter, P. 1979. A mechanism for secretory IgA-mediated inhibition of the cell penetration and intracellular development of *Eimeria tenella*. Immunology 36:471–477.

Denmark, J. R., and Chessum, B. S. 1978. Standardization of enzyme-linked immunosorbent assay (ELISA) and the detection of *Toxoplasma* antibody. Med. Lab. Sci. 35:227–232.

de Roever-Bonnet, H. 1963. Mice and golden hamsters infected with an avirulent and virulent *Toxoplasma* strain. Trop. Geogr. Med. 15:45–60.

de Roever-Bonnet, H. 1964. *Toxoplasma* parasites in different organs of mice and hamsters infected with avirulent and virulent strains. Trop. Geogr. Med. 16:337–345.

Dubey, J. P. 1973. Feline toxoplasmosis and coccidiosis: A survey of domiciled and stray cats. J. Am. Vet. Med. Assoc. 162:873–877.

Dubey, J. P. 1975a. *Isospora ohioensis* sp.n. proposed for *I. rivolta* of the dog. J. Parasitol. 61:462–465.

Dubey, J. P. 1975b. Experimental *Isospora canis* and *Isospora felis* infection in mice, cats and dogs. J. Protozool. 22:416–417.

Dubey, J. P. 1975c. Immunity to *Hammondia hammondi* infection in cats. J. Am. Vet. Med. Assoc. 167:373–377.

Dubey, J. P. 1976a. A review of *Sarcocystis* of domestic animals and of other coccidia of cats and dogs. J. Am. Vet. Med. Assoc. 169: 1061–1078.

Dubey, J. P. 1976b. Reshedding of *Toxoplasma gondii* oocysts by chronically infected cats. Nature 262:213–214.

Dubey, J. P. 1977. *Toxoplasma*, *Hammondia*, *Besnoitia*, *Sarcocystis* and other tissue cyst-forming coccidia of man and animals. In: J. P. Kreiber (ed.), Parasitic Protozoa, Volume III. Gregarines, Haemogregar-ines, Coccidia, Plasmodia and Haemoproteids, pp. 101–237. Academic Press, New York.

Dubey, J. P. 1978a. Pathogenicity of *Isospora ohioensis* infection in dogs. J. Am. Vet. Med. Assoc. 173:192–197.

Dubey, J. P. 1978b. Life cycle of *Isospora ohioen-*

sis in dogs. Parasitology 77:1–11.

Dubey, J. P. 1978c. Effect of immunization of cats with *Isospora felis* and BCG on immunity to re-excretion of *Toxoplasma gondii* oocysts. J. Protozool. 25:380–382.

Dubey, J. P. 1978d. A comparison of cross protection between BCG, *Hammondia hammondi*, *Besnoitia jellisoni* and *Toxoplasma gondii* in hamsters. J. Protozool. 25:382–384.

Dubey, J. P. 1979. Life cycle of *Isospora rivolta* (Grassi, 1879) in cats and mice. J. Protozool. 26:433–441.

Dubey, J. P., and Frenkel, J. K. 1972a. Extraintestinal stages of *Isospora felis* and *I. rivolta* (Protozoa: Eimeriidae) in cats. J. Protozool. 19:89–92.

Dubey, J. P., and Frenkel, J. K. 1972b. Cyst-induced toxoplasmosis in cats. J. Protozool. 19:155–177.

Dubey, J. P., and Frenkel, J. K. 1974. Immunity to feline toxoplasmosis: Modification by administration of corticosteroids. Vet. Pathol. 11:350–379.

Dubey, J. P., Hoover, E. A., and Walls, K. W. 1977. Effect of age and sex on the acquisition of immunity to toxoplasmosis in cats. J. Protozool. 24:184–186.

Dubey, J. P., and Mehlhorn, M. 1978. Extraintestinal stages of *Isospora ohioensis* from dogs in mice. J. Parasitol. 64:689–695.

Dubey, J. P., Miller, N. L., and Frenkel, J. K. 1970. Characterization of the new fecal form of *Toxoplasma gondii*. J. Parasitol. 56:447–456.

Dubey, J. P., Weisbrode, S. E., and Rogers, W. A. 1978. Canine coccidiosis attributed to an *Isospora ohioensis*-like organism: A case report. J. Am. Vet. Med. Assoc. 173:185–191.

Endo, T., and Kobayashi, A. 1976. *Toxoplasma gondii*: Electron microscopic study on the dye test reaction. Exp. Parasitol. 40:170–178.

Ernst, J. V., Hammond, D. M., and Chobotar, B. 1968. *Eimeria utahensis* sp.n. from kangaroo rats (*Dipodomys ordii* and *D. microps*) in northwestern Utah. J. Protozool. 15:430–432.

Euzéby, J. 1973. Immunologie des coccidioses de la poule. Cah. Med. Vet. 42:3–40.

Fayer, R. 1974. Development of *Sarcocystis fusiformis* in the small intestine of the dog. J. Parasitol. 60:660–665.

Fayer, R. 1977. *Sarcocystis leporum* in cottontail rabbits and its transmission to carnivores. J. Wildl. Dis. 13:170–173.

Fayer, R., and Baron, S. 1971. Activity of interferon and an inducer against development of *Eimeria tenella* in cell culture. J. Protozool. 18(suppl):12.

Fayer, R., and Lunde, M. N. 1977. Changes in serum and plasma proteins and in IgG and IgM antibodies in calves experimentally infected with *Sarcocystis* from dogs. J. Parasitol. 63:438–442.

Foster, B. G., and McCulloch, W. F. 1968. Studies on active and passive immunity in animals inoculated with *Toxoplasma gondii*. Can. J. Microbiol. 14:103–110.

Frelier, P., Mayhew, I. G., Fayer, R., and Lunde, M. N. 1977. *Sarcocystis*: A clinical outbreak in dairy calves. Science 195:1341–1342.

Frenkel, J. K. 1948. Dermal hypersensitivity to *Toxoplasma* antigens (toxoplasmosis). Proc. Soc. Exp. Biol. Med. 68:634–639.

Frenkel, J. K. 1957. Effects of cortisone, total body irradiation and nitrogen mustard on chronic latent toxoplasmosis. Am. J. Pathol. 33:618–619.

Frenkel, J. K. 1967. Adoptive immunity to intracellular infection. J. Immunol. 98:1309–1319.

Frenkel, J. K. 1973. Toxoplasmosis: Parasite life cycle, pathology, and immunology. In: D. M. Hammond and P. L. Long (eds.), The Coccidia: *Eimeria*, *Isospora*, *Toxoplasma*, and Related Genera, pp. 343–410. University Park Press, Baltimore.

Frenkel, J. K. 1977. *Besnoitia wallacei* of cats and rodents with a reclarificatin of other cyst-forming isosporoid coccidia. J. Parasitol. 63:611–628.

Frenkel, J. K. 1978. Toxoplasmosis in cats: Diagnosis, treatment and prevention. Comp. Immunol. Microbiol. Infect. Dis. 1:15–20.

Frenkel, J. K., and Caldwell, S. A. 1975. Specific

immunity and nonspecific resistance to infection: *Listeria*, Protozoa and viruses in mice and hamsters. J. Infect. Dis. 131:201–209.

Frenkel, J. K., and Dubey, J. P. 1972. Toxoplasmosis and its prevention in cats and man. J. Infect. Dis. 126:664–673.

Frenkel, J. K., and Dubey, J. P. 1975. *Hammondia hammondi* gen. nov. sp. nov., from domestic cats, a new coccidian related to *Toxoplasma* and *Sarcocystis*. Z. Parasitenkd. 46:3–12.

Frenkel, J. K., and Lunde, M. N. 1966. Effects of corticosteroids on antibody and immunity in *Besnoitia* infection of hamsters. J. Infect. Dis. 116:414–424.

Freshman, M. M., Merigan, T. C., Remington, J. S., and Brownlee, I. E. 1966. *In vitro* and *in vivo* antiviral action of an interferon-like substance induced by *Toxoplasma gondii*. Proc. Soc. Exp. Biol. Med. 123:862–866.

Gaines, J. D., Araujo, F. G., Krahenbuhl, J. L., and Remington, J. S. 1972. Simplified *in vitro* method for measuring delayed hypersensitivity to latent intra-cellular infection in man (toxoplasmosis). J. Immunol. 109:179–182.

Gardner, I. D., and Remington, J. S. 1977. Age-related decline in the resistance of mice to infection with intra-cellular pathogens. Infect. Immun. 16:593–598.

Gardner, I. D., and Remington, J. S. 1978a. Aging and the immune response: I. Antibody formation and chronic infection in *Toxoplasma gondii* infected mice. J. Immunol. 120: 939–943.

Gardner, I. D., and Remington, J. S. 1978b. Aging and the immune response: II. Lymphocyte responsiveness and macrophage activation in *Toxoplasma gondii* infected mice. J. Immunol. 120:944–949.

Garrido, J. A., and Boveda, I. C. 1972. Aplicacion de la immunofluorescencia al estudio de las fracciones immunoglobulinicas en el diagnostico de la toxoplasmosis adquirida y congenita. Rev. Clin. Esp. 125:37–42.

Gentry, L. O., and Remington, J. S. 1971. Resistance against cryptococcus conferred by intra-cellular bacteria and protozoa. J. Infect. Dis. 123:23–31.

Gill, H. S., and Prakash, O. 1970. A study on the active and passive immunity in experimental toxoplasmosis. Indian J. Med. Res. 58: 1157–1163.

Hafizi, A., and Modabber, F. Z. 1978. Effect of cyclophosphamide on *Toxoplasma gondii* infection: Reversal of the effect by passive immunization. Clin. Exp. Immunol. 33:389–396.

Hahn, H. 1974. Effects of dextran sulfate 500 on cell-mediated resistance to infection with *Listeria monocytogenes* in mice. Infect. Immun. 10:1105–1109.

Hammond, D. M., Andersen, F. L., and Miner, M. L. 1963. The site of the immune reaction against *Eimeria bovis* in calves. J. Parasitol. 49:415–424.

Hammond, D. M., Andersen, F. L., and Miner, M. L. 1964. Response of immunized and non-immunized calves to cecal inoculation of first generation merozoites of *Eimeria bovis*. J. Parasitol. 50:209–213.

Heist, C. E., and Moore, T. D. 1959. Serological and immunological studies of coccidiosis of rabbits. J. Protozool. 6(suppl):7.

Herlich, H. 1965. Effect of chicken antiserum and tissue extracts on the oocysts, sporozoites and merozoites of *Eimeria tenella* and *Eimeria acervulina*. J. Parasitol. 57:847–851.

Heydorn, A.-O., and Rommel, M. 1972 Beiträge zum Lebenszyklus der Sarkosporidien: II. Hund und Katze als Überträger der Sarkosporidien des Rindes. Berl. Muench. Tieraerztl. Wochenschr. 85:121–123.

Hibbs, J. B., Lambert, L. H., and Remington, J. S. 1971. Resistance to murine tumors conferred by chronic infection in intracellular protozoa, *Toxoplasma gondii* and *Besnoitia jellisoni*. J. Infect. Dis. 124:587–592.

Hof, H., Emmerling, P., Hohne, K., and Seiliger, H. P. R. 1976. Infection of congenitally athymic (nude) mice with *Toxoplasma gondii*. Ann. Microbiol. 127B:503–507.

Hoff, R. L., and Frenkel, J. K. 1974. Cell-mediated immunity against *Besnoitia* and *Toxoplasma* in specifically and cross-immunized hamsters and in cultures. J. Exp. Med. 139:560–580.

8

Holland, P., and Sleamaker, K. 1970. Motile phagocytic defense against protozoa and fungi. J. Reticulo-endothel. Soc. 7:635.

Horton-Smith, C., Beattie, J., and Long, P. L. 1961. Resistance to *Eimeria tenella* and its transference from one caecum to the other in individual fowls. Immunology 4:111–121.

Horton-Smith, C., Long, P. L., and Pierce, A. E. 1963a. Behaviour of invasive stages of *Eimeria tenella* in the immune fowl (*Gallus domesticus*). Exp. Parasitol. 13:66–74.

Horton-Smith, C., Long, P. L., Pierce, A. E., and Rose, M. E. 1963b. Immunity to coccidia in domestic animals. In: P. C. C. Garnham, A. E. Pierce, and I. Roitt (eds.), Immunity to Protozoa, pp. 273–295. Blackwell Scientific Publications, Ltd., Oxford.

Huff, D., and Clark, D. T. 1970. Cellular aspects of the resistance of chickens to *Eimeria tenella* infections. J. Protozool. 17:35–39.

Huldt, G. 1966a. Experimental toxoplasmosis: Effect of inoculation of *Toxoplasma* in seropositive rabbits. Acta Pathol. Microbiol. Scand. 67:592–604.

Huldt, G. 1966b. Experimental toxoplasmosis: Studies of the multiplication and spread of *Toxoplasma* in experimentally infected rabbits. Acta Pathol. Microbiol. Scand. 68: 401–423.

Huldt, G. 1967. *In vitro* studies of some immunological phenomena in experimental rabbit toxoplasmosis. Acta Pathol. Microbiol. Scand. 70:129–146.

Itagaki, K., and Tsubokura, M. 1963. Serological studies on coccidium in the fowl: II. Precipitation in infected chicken serum. Jpn. J. Vet. Sci. 25:187–192.

Jacobs, L. 1956. Propagation, morphology and biology of *Toxoplasma*. Ann. N. Y. Acad. Sci. 64:154–179.

Jacobs, L. 1976. Serodiagnosis of Toxoplasmosis. In: S. Cohen and E. H. Sadun (eds.), Immunology of Parasitic Infections, pp. 94–106. Blackwell Scientific Publications, Ltd., Oxford.

Jeffers, T. K. 1974. *Eimeria acervulina* and *E. maxima*: Incidence, distribution and anticoccidial drug resistance of isolants in major broiler producing areas. Avian Dis. 18: 331–342.

Jeffers, T. K. 1978. Genetics of coccidia and the host response. In: P. L. Long, K. N. Boorman, and B. M. Freeman (eds.), Avian Coccidiosis, pp. 57–125. Proceedings of the 13th Poultry Science Symposium, 1977. British Poultry Science Ltd., Longman Group Ltd., Edinburgh.

Johnson, J., Reid, W. M., and Jeffers, T. K. 1979. Practical immunization of chickens against coccidiosis using an attenuated strain of *Eimeria tenella*. Poult. Sci. 58:37–41.

Jones, T. C., and Hirsch, J. G. 1972. The interaction between *Toxoplasma gondii* and mammalian cells: II. The absence of lysosomal fusion with phagocytic vacuoles containing living parasites. J. Exp. Med. 136:1173–1194.

Jones, T. C., and Len, L. 1976. Pinocytic rates of macrophages from mice immunized against *Toxoplasma gondii* and macrophages stimulated to inhibit *Toxoplasma in vitro*. Infect. Immun. 14:1011–1013.

Jones, T. C., Len, L., and Hirsch, J. G. 1975. Assessment *in vitro* of immunity against *Toxoplasma gondii*. J. Exp. Med. 141:466–482.

Jones, T. C., Masur, M., Len, L., and Fu, T. L. T. 1977. Lymphocyte macrophage interaction during control of intracellular parasitism. Am. J. Trop. Med. Hyg. 26:187–193.

Jones, T. C., Yeh, S., and Hirsch, J. G. 1972. The interaction between *Toxoplasma gondii* and mammalian cells: Mechanism of entry and intracellular fate of the parasite. J. Exp. Med. 136:1157–1172.

Joyner, L. P. 1969. Immunological variation between two strains of *Eimeria acervulina*. Parasitology 59:725–732.

Joyner, L. P., and Norton, C. C. 1973. The immunity arising from continuous low level infections with *Eimeria tenella*. Parasitology 67:333–340.

Joyner, L. P., and Norton, C. C. 1974. The effect of early cyclophosphamide treatment on the subsequent immunity of chickens to coccidial infections. Parasitology 68:323–330.

Joyner, L. P., and Norton, C. C. 1976. The immunity arising from continuous low level infection with *Eimeria maxima* and *Eimeria acervulina*. Parasitology 72:115–125.

Khavkin, Th. N., and Freidlin, L. S. 1977. A fluorescence phase contrast study of the interaction between *Toxoplasma gondii* and lysosomes in living cells. Z. Parasitenkd. 52:19–21.

Kittas, C., and Henry, L. 1979. Effect of sex hormones on the immune system of guinea-pigs and on the development of toxoplasmic lesions in non lymphoid organs. Clin. Exp. Immunol. 36:16–23.

Klainer, A. S., Krahenbuhl, J. L., and Remington, J. S. 1973. Scanning electron microscopy of *Toxoplasma gondii*. J. Gen. Microbiol. 75:111–118.

Klesius, P. H., Elston, A. L., Chambers, W. H., and Fudenberg, H. H. 1979. Resistance to coccidiosis (*Eimeria ferrisi*) in C57BL/6 mice: Effects of immunization and transfer factor. Clin. Immunol. Immunopathol. 12:143–149.

Klesius, P. H., Kramer, T., Burger, D., and Malley, A. 1975. Passive transfer of coccidian oocyst antigen and diphtheria toxoid hypersensitivity in calves across species barriers. Transplant. Proc. 7:449–452.

Klesius, P. H., Kramer, T. T., and Frandsen, J. C. 1976a. *Eimeria stiedai*: Delayed hypersensitivity response in rabbit coccidiosis. Exp. Parasitol. 39:59–68.

Klesius, P. H., and Kristensen, F. 1977. Bovine transfer factor: Effect on bovine and rabbit coccidiosis. Clin. Immunol. Immunopathol. 7:240–252.

Klesius, P. H., Kristensen, F., Elston, A. L., and Williamson, O. C. 1977. *Eimeria bovis*: Evidence for a cell-mediated immune response in bovine coccidiosis. Exp. Parasitol. 41:480–490.

Klesius, P. H., Kristensen, F., Ernst, J. V., and Kramer, T. T. 1976b. Bovine transfer factor: Isolation and characteristics. In: M. S. Ascher, A. A. Gottlieb, and C. H. Kirkpatrick (eds.), Transfer Factor: Basic Properties and Clinical Applications, pp. 311–319. Academic Press, New York.

Klesius, P. H., Qualls, D. F., Elston, A. L., and Fudenberg, H. H. 1978. Effects of bovine transfer factor (TFd) in mouse coccidiosis (*Eimeria ferrisi*). Clin. Immunol. Immunopathol. 10:214–221.

Kouwenhoven, B., and Kuil, H. 1976. Demonstration of circulating antibodies to *Eimeria tenella* by the indirect immunofluorescent antibody test using sporozoites and second stage schizonts as antigen. Vet. Parasitol. 2:283–292.

Kouwenhoven, B., and van der Horst, C. J. G. 1973. Histological observations with respect to the immune mechanism in *Eimeria acervulina* infection in the domestic fowl. Z. Parasitenkd. 42:11–21.

Krahenbuhl, J. L., Blazkovec, A. A., and Lysenko, M. G. 1971. *In vivo* and *in vitro* studies of delayed-type hypersensitivity to *Toxoplasma gondii* in guinea pigs. Infect. Immun. 3: 260–267.

Krahenbuhl, J. L., Gaines, J. D., and Remington, J. S. 1972a. Lymphocyte transformation in human Toxoplasmosis. J. Infect. Dis. 125: 283–288.

Krahenbuhl, J. L., and Remington, J. S. 1971. *In vitro* induction of non-specific resistance in macrophages by specifically sensitized lymphocytes. Infect. Immun. 4:337–343.

Krahenbuhl, J. L., Ruskin, J., and Remington, J. S. 1972b. The use of killed vaccines in immunization against an intra-cellular parasite: *Toxoplasma gondii*. J. Immunol. 108: 425–431.

Kühn, D., and Weiland, G. 1969. Experimentelle *Toxoplasma*-infektionen bei der Katze: 1. Wiederhalte Überträgung von *Toxoplasma gondii* durch Kot von mit Nematoden infizierten Katzen. Berl. Muench. Tieraerztl. Wochenschr. 82:401–404.

Kuil, H., and Dankert-Brands, S. 1976. The effect of medication with metichlorpindol and methylbenzoquate on the development of antibodies as measured by the indirect fluorescent antibody (IFA) test against *Eimeria tenella* and *Eimeria maxima*. Vet. Parasitol. 2: 293–298.

Lawrence, H. S. 1969. Transfer factor. Adv. Immunol. 11:196–266.

Leathem, W. D., and Burns, W. C. 1967. Effects of the immune chicken on the endogenous stages of *Eimeria tenella*. J. Parasitol. 53:180–185.

Leathem, W. D., and Burns, W. C. 1968. Duration of acquired immunity of the chicken to *Eimeria tenella* infection. J. Parasitol. 54:227–232.

Levine, P. P. 1938. *Eimeria hagani* n.sp (Protozoa: Eimeriidae) a new coccidium of the chicken. Cornell Vet. 28:263–266.

Levine, P. P. 1942. A new coccidium pathogenic for chickens. *Eimeria brunetti* n.sp (Protozoa: Eimeriidae). Cornell Vet. 32:430–439.

Liburd, E. M., Armstrong, W. D., and Mahrt, J. L. 1973. Immunity to the protozoan parasite *Eimeria nieschulzi* in inbred CD-F rats. Cell. Immunol. 7:444–452.

Liburd, E. M., Pabst, H. F., and Armstrong, W. D. 1972. Transfer factor in rat coccidiosis. Cell. Immunol. 5:487–489.

Lindberg, R. E., and Frenkel, J. K. 1977a. Cellular immunity to *Toxoplasma* and *Besnoitia* in hamsters: Specificity and the effects of cortisol. Infect. Immun. 15:855–862.

Lindberg, R. E., and Frenkel, J. K. 1977b. Toxoplasmosis in nude mice. J. Parasitol. 63: 219–221.

Long, P. L. 1974. Experimental infection of chickens with two species of *Eimeria* isolated from the Malaysian jungle fowl. Parasitology 69:337–347.

Long, P. L., Johnson, J., and Wyatt, R. D. 1980. *Eimeria tenella*: Clinical effects in partially immune and susceptible chickens. Poult. Sci. 59:2221–2224.

Long, P. L., and Millard, B. J. 1968. *Eimeria*: Effect of metichlorpindol and methylbenzoquate on endogenous stages in the chicken. Exp. Parasitol. 23:331–338.

Long, P. L., and Millard, B. J. 1976. The detection of occult coccidial infections by inoculating chickens with corticosteroid drugs. Z. Parasitenkd. 48:287–290.

Long, P. L., and Millard, B. J. 1979. Immunological differences in *Eimeria maxima*: Effect of a mixed immunizing inoculum on heterologous challenge. Parasitology 79:451–457.

Long, P. L., Millard, B. J., Joyner, L. P., and Norton, C. C. 1976. A guide to laboratory techniques used in the study and diagnosis of avian coccidiosis. Folia Vet. Lat. 6:201–217.

Long, P. L., and Milne, B. S. 1971. The effect of an interferon inducer on *Eimeria maxima* in the chicken. Parasitology 62:295–302.

Long, P. L., and Rose, M. E. 1970. Extended schizogony of *Eimeria mivati* in betamethasone-treated chickens. Parasitology 60: 147–165.

Long, P. L., and Rose, M. E. 1972. Immunity to coccidiosis: Effect of serum antibodies on cell invasion by sporozoites of *Eimeria in vitro*. Parasitology 65:437–445.

Long, P. L., and Rose, M. E. 1976. Growth of *Eimeria tenella in vitro* in macrophages from chicken peritoneal exudates. Z. Parasitenkd. 48:291–294.

Long, P. L., Rose, M. E., and Pierce, A. E. 1963. Effects of fowl sera on some stages in the life cycle of *Eimeria tenella*. Exp. Parasitol. 14: 210–217.

Long, P. L., Tompkins, R. V., and Millard, B. J. 1975. Coccidiosis in broilers: Evaluation of infection by the examination of broiler house litter for oocysts. Avian Pathol. 4:287–294.

Ludlam, G. B. 1960. Laboratory diagnosis of toxoplasmosis. Proc. R. Soc. Med. 53:113–116.

Lunde, M. N., and Fayer, R. 1977. Serologic tests for antibody to *Sarcocystis* in cattle. J. Parasitol. 63: 222–225.

Lunde, M. N., and Jacobs, L. 1965. Antigenic relationship of *Toxoplasma gondii* and *Besnoitia jellisoni*. J. Parasitol. 51:273–276.

McDougald, L. R., and Jeffers, T. K. 1976. *Eimeria tenella* (Sporozoa, Coccidia): Gametogony following a single asexual generation. Science 192:258–259.

Mackaness, G. B. 1964. The immunological basis of acquired cellular immunity. J. Exp. Med. 120:105–120.

Mackaness, G. B., and Blanden, R. V. 1967. Cellular immunity. Prog. Allergy 11:89–140.

McLeod, R., and Remington, J. S. 1977a. Influence of infection with *Toxoplasma* on macrophage function and role of macrophages in resistance to *Toxoplasma*. Am. J. Trop. Med. Hyg. 26:170–186.

McLeod, R., and Remington, J. S. 1977b. Studies on the specificity of killing of intracellular pathogens by macrophages. Cell. Immunol. 34:156–174.

Magliulo, E., Concia, E., Azzini, M., Bonizzoni, D., and Scevola, D. 1976. *Toxoplasma gondii*: A new diagnostic approach based on the specific binding of erythrocytes to lymphoid cells. Exp. Parasitol. 39:143–149.

Markus, M. B. 1973. Serology of toxoplasmosis, isosporosis, and sarcosporidiosis. N. Engl. J. Med. 89:980–981.

Markus, M. B. 1974. Serology of human sarcosporidiosis. Trans. R. Soc. Trop. Med. Hyg. 68:415–416.

Markus, M. B. 1978. *Sarcocystis* and sarcocystosis in domestic animals and man. Adv. Vet. Sci. Comp. Med. 22:159–193.

Markus, M. B. 1979. Antibodies to *Sarcocystis* in human sera. Trans. R. Soc. Trop. Med. Hyg. 73:346–347.

Markus, M. B., Killick-Kendrick, R., and Garnham, P. C. C. 1974. The coccidial nature and life cycle of *Sarcocystis*. J. Trop. Med. Hyg. 77:248–259.

Masihi, K. N., and Werner, H. 1976. Rosette forming cells during immune response to *Toxoplasma gondii* in mice. Infect. Immun. 13:1678–1683.

Masihi, K. N., and Werner, H. 1977. Kinetics of antibody mediated suppression of humoral immune response to *Toxoplasma gondii* at a cellular level. Zentralbl. Bakteriol. Hyg. I. Abt. Orig. A. 237:405–410.

Masihi, K. N., and Werner, H. 1978a. Types of cells involved in antigen-stimulated and spontaneous rosette formation with *Toxoplasma gondii*. J. Immunol. 121:2056–2059.

Masihi, K. N., and Werner, H. 1978b. The effect of passively transferred heterologous serum on *Toxoplasma gondii* in NMRI mice: Influence of the treatment on course of infection and cyst formation. Zentralbl. Bakteriol. Hyg. I. Abt. Orig. A. 240:135–142.

Mehlhorn, H., and Markus, M. B. 1976. Electron microscopy of stages of *Isospora felis* of the cat in the mesenteric lymph node of the mouse. Z. Parasitenkd. 51:15–24.

Mesfin, G. M., and Bellamy, J. E. C. 1979a. Thymic dependence of immunity to *Eimeria falciformis* var *pragensis* in mice. Infect. Immun. 23:460–464.

Mesfin, G. M., and Bellamy, J. E. C. 1979b. Effects of acquired resistance on infection with *Eimeria falciformis* var *pragensis* in mice. Infect. Immun. 23:108–114.

Miller, N. L., Frenkel, J. W., and Dubey, J. P. 1972. Oral infections with *Toxoplasma* cysts and oocysts in felines, other mammals, and in birds. J. Parasitol. 58:928–937.

Moore, E. N., and Brown, J. A. 1951. A new coccidium pathogenic for turkeys, *Eimeria adenoeides* n.sp (Protozoa: Eimeriidae). Cornell Vet. 41:124–135.

Morehouse, N. F. 1938. The reaction of the immune intestinal epithelium of the rat to reinfection with *Eimeria nieschulzi*. J. Parasitol. 24:311–317.

Morita, C., Tsutsumi, Y., and Soekawa, M. 1972. Detection of humoral antibody for *Eimeria tenella* infection by a dye test. Zentralbl. Vet. Med. (B) 19:782–784.

Morita, C., Tsutsumi, Y., and Soekawa, M. 1973. Migration inhibition test of splenic cells of chickens infected with *Eimeria tenella*. J. Parasitol. 59:199.

Movsesijan, M., Sokolić, A., Tanielian, Z., and Abu Ali, N. 1975. Circulating and local antibodies in *Eimeria tenella* infection. Acta Vet. (Beograd) 25:59–64.

Munday, B. L., Barker, I. K., and Rickard, M. D. 1975. The developmental cycle of a species of *Sarcocystis* occurring in dogs and sheep, with observations on pathogenicity in the intermediate host. Z. Parasitenkd. 46:111–123.

Munday, B. L., and Rickard, M. D. 1974. Is *Sarcocystis tenella* two species? Aust. Vet. J. 50:558–559.

Nakayama, I. 1964. Persistence of the virulent

8

RH strain of *Toxoplasma gondii* in the brains of immune mice. Keio J. Med. 13:7–12.

Nakayama, I. 1965. Effects of immunization procedures in experimental toxoplasmosis. Keio J. Med. 14:63–72.

Nakayama, I., and Aoki, T. 1970. The influence of antilymphocyte serum on the resistance of mice to the *Toxoplasma* infection. (English summary) Jpn. J. Parasitol. 19:573–582.

Niilo, L. 1967. Acquired resistance to reinfection of rabbits with *Eimeria magna*. Can. Vet. J. 8:201–208.

Niilo, L. 1970. The effect of dexamethasone on bovine coccidiosis. Can. J. Comp. Med. Vet. Sci. 34:325–328.

Norton, C. C., and Hein, H. 1976. *Eimeria maxima*: A comparison of two laboratory strains with a fresh isolate. Parasitology 72:345–354.

Ogilvie, B. M., and Rose, M. E. 1977. The response of the host to some parasites of the small intestine: Coccidia and nematodes. Les Colloques de l'Institut National de la Santé et de la Recherche Médicale, INSERM 72:237–248.

Orlans, E., and Rose, M. E. 1972. An IgA-like immunoglobulin in the fowl. Immunochemistry 9:833–838.

Overdulve, J. P. 1974. Immunity to toxoplasmosis in cats. Proceedings of the 3rd International Congress on Parasitology, Munich. Vol. 1, pp. 302–303. Facta Publications, Vienna.

Overdulve, J. P. 1978. Excretion of *Toxoplasma gondii* by non-immunized and immunized cats: Its role in the epidemiology of toxoplasmosis. Proc. K. Ned. Akad. Wetensch. (Amsterdam) Series C 8:1–18.

Patton, W. H. 1970. *In vitro* phagocytosis of coccidia by macrophages from the blood of infected chickens. J. Parasitol. 56:260.

Pelster, B. 1975. Zelluläre Immunreaktionen bei der weissen Maus nach superinfektion mit *Toxoplasma gondii*. Z. Parasitenkd. 48:95–110.

Peteshev, V. M., Galuzo, I. G., and Polomoshnov, A. P. 1974. Cats: Definitive hosts of *Besnoitia* (*Besnoitia besnoiti*). Izv. Akad. Nauk SSSR (Biol.) 1.

Piekarski, G., and Witte, H. M. 1971. Experimentelle und histologische studien zur Toxoplasma-infektion der Hauskatze. Z. Parasitenkd. 36:95–121.

Pierce, A. E., Long, P. L., and Horton-Smith, C. 1962. Immunity to *Eimeria tenella* in young fowls (*Gallus domesticus*). Immunology 5: 129–152.

Remington, J. S., Burnett, C. G., Meikel, M., and Lunde, M. 1962. Toxoplasmosis and infectious mononucleosis. Arch. Int. Med. 110: 744–753.

Remington, J. S., and Krahenbuhl, J. L. 1976. Immunology of *Toxoplasma* infection. In: S. Cohen and E. H. Sadun (eds.), Immunology of Parasitic Infections, pp. 236–267. Blackwell Scientific Publications, Ltd., Oxford.

Remington, J. S., and Krahenbuhl, J. L. Immunology of *Toxoplasma* infection. In: A. J. Nahmiss and R. J. O'Reilly (eds.), Immunology of Human Infections, Plenum Press, New York. In press.

Remington, J. S., Krahenbuhl, J. L., and Mendenhall, J. W. 1972. A role for activated macrophages in resistance to infection with *Toxoplasma*. Infect. Immun. 6:829–834.

Remington, J. S., and Merigan, T. C. 1968. Interferon: Protection of cells infected with an intracellular protozoan (*Toxoplasma gondii*). Science 161:804–806.

Remington, J. S., and Merigan, T. C. 1969. Resistance to virus challenge in mice infected with protozoa or bacteria. Proc. Soc. Exp. Biol. Med. 131:1184–1188.

Rommel, M., and Heydorn, A.-O. 1971. Versuche zur Übertragung der Immunität gegen *Eimeria*-Infektionen durch Lymphozyten. Z. Parasitenkd. 36:242–250.

Rommel, M., Heydorn, A.-O., and Gruber, F. 1972. Beiträge zum Lebenszyklus der Sarkosporidien: I. Die Sporozyste von *Sarcocystis tenella* in den Fäzes der Katze. Berl. Muench. Tieraerztl. Wochenschr. 85:101–105.

Rose, M. E. 1959. Serological reactions in *Eimeria stiedae* infection of the rabbit. Immunology 2:112–122.

Rose, M. E. 1961. The complement-fixation test

in hepatic coccidiosis of rabbits. Immunology 4:346–353.

Rose, M. E. 1967. Immunity to *Eimeria tenella* and *Eimeria necatrix* infections in the fowl: I. Influence of the site of infection and the stage of the parasite. II. Cross-protection. Parasitology 57:567–583.

Rose, M. E. 1970. Immunity to coccidiosis: Effect of betamethasone treatment of fowls on *Eimeria mivati* infections. Parasitology 60:137–146.

Rose, M. E. 1971. Immunity to coccidiosis: Protective effect of transferred serum in *Eimeria maxima* infections. Parasitology 62:11–25.

Rose, M. E. 1972. Immunity to coccidiosis: Maternal transfer in *Eimeria maxima* infections. Parasitology 65:273–282.

Rose, M. E. 1973. Immunity. In: D. M. Hammond and P. L. Long (eds.), The Coccidia: *Eimeria, Isospora, Toxoplasma,* and Related Genera, pp. 295–341. University Park Press, Baltimore.

Rose, M. E. 1974a. Immunity to *Eimeria maxima*: Reactions of antisera *in vitro* and protection *in vivo*. J. Parasitol. 60:528–530.

Rose, M. E. 1974b. Immune responses in infections with coccidia: Macrophage activity. Infect. Immun. 10:862–871.

Rose, M. E. 1974c. Immune responses to the *Eimeria*: Recent observations. Proceedings of the Symposium on Coccidia and Related Organisms, 1973. University of Guelph, Guelph, Ontario, Canada, pp. 92–118.

Rose, M. E. 1974d. Protective antibodies in infections with *Eimeria maxima*: The reduction of pathogenic effects *in vivo* and a comparison between oral and subcutaneous administration of antiserum. Parasitology 68:285–292.

Rose, M. E. 1975. Infections with *Eimeria maxima* and *Eimeria acervulina* in the fowl: Effect of previous infection with the heterologous organism on oocyst production. Parasitology 70:263–271.

Rose, M. E. 1977. *Eimeria tenella*: Skin hypersensitivity to injected antigen in the fowl. Exp. Parasitol. 42:129–141.

Rose, M. E. 1978. Immune responses of chickens to coccidia and coccidiosis. In: P. L. Long, K. N. Boorman, and B. M. Freeman (eds.), Avian Coccidiosis, pp. 297–337. Proceedings of the Thirteenth Poultry Science Symposium, 1977. British Poultry Science Ltd., Longman Group Ltd., Edinburgh.

Rose, M. E., and Hesketh, P. 1976. Immunity to coccidiosis: Stages of the life cycle of *Eimeria maxima* which induce, and are affected by, the response of the host. Parasitology 73:25–37.

Rose, M. E., and Hesketh, P. 1979. Immunity to coccidiosis: Further investigations in T lymphocyte and B lymphocyte deficient animals. Infect. Immun. 26:630–637.

Rose, M. E., and Hesketh, P. Immunity to coccidiosis: Adoptive transfer with peripheral blood lymphocytes and spleen cells. Parasitic Immunol. In press.

Rose, M. E., Hesketh, P., and Ogilvie, B. M. 1979a. Peripheral blood leucocyte response to coccidial infection: A comparison of the response in rats and chickens and its correlation with resistance to reinfection. Immunology 36:71–79.

Rose, M. E., Hesketh, P., and Ogilvie, B. M. 1980a. Coccidiosis: Localisation of lymphoblasts in the infected small intestine. Parasite Immunol. 2:189–199.

Rose, M. E., and Lee, D. L. 1977. Interactions *in vitro* between sporozoites of *Eimeria tenella* and host peritoneal exudate cells: Electron microscopal observations. Z. Parasitenkd. 54:1–7.

Rose, M. E., and Long, P. L. 1962. Immunity to four species of *Eimeria* in fowls. Immunology 5:79–92.

Rose, M. E., and Long, P. L. 1969. Immunity to coccidiosis: Gut permeability changes in response to sporozoite invasion. Experientia 25:183–184.

Rose, M. E., and Long, P. L. 1971. Immunity to coccidiosis: Protective effects of transferred serum and cells investigated in chick embryos infected with *Eimeria tenella*. Parasitology 63:299–313.

Rose, M. E., and Long, P. L. 1976. Immunity to coccidiosis: Interactions *in vitro* between *Eimeria tenella* and chicken phagocytic cells. In: H. van den Bossche (ed.), Biochemistry of Parasites and Host-Parasite Relationships, pp. 449–455. Elsevier/North Holland Biomedical Press, Amsterdam.

Rose, M. E., and Long, P. L. 1980. Vaccination against coccidiosis in chickens. In: A. E. R. Taylor and R. Muller (eds.), Vaccination against Parasites. Symposia of the British Society for Parasitology. Vol. 18, pp. 57–74. Blackwell Scientific Publications, Ltd., Oxford.

Rose, M. E., Long, P. L., and Bradley, J. W. A. 1975. Immune responses to infections with coccidia in chickens: Gut hypersensitivity. Parasitology 71:357–368.

Rose, M. E., Ogilvie, B. M., and Bradley, J. W. A. 1980b. The intestinal mast cell response in rats and chickens in response to coccidiosis with some properties of chicken mast cells. Arch. Allergy Appl. Immunol. 63: 21–29.

Rose, M. E., Ogilvie, B. M., Hesketh, P., and Festing, M. F. W. 1979b. Failure of nude (athymic) rats to become resistant to reinfection with the intestinal coccidian parasite *Eimeria nieschulzi* or the nematode *Nippostrongylus brasiliensis*. Parasite Immunol. 1:125–132.

Rottini, G. D., and Favento, R. 1969. Antibodies IgM, IgA and IgG in human toxoplasmosis. Giorn Microbiol. 17:163–177.

Rottini, G., Favento, R., and Mauro, F. 1971. Dynamics of IgM, IgA and IgG: Antibody response in cases of human ophthalmic toxoplasmosis. Zentralbl. Bakteriol. Hyg. I. Abt. Orig. A. 218:507–516.

Ruiz, A., and Frenkel, J. L. 1976. Recognition of cyclic transmission of *Sarcocystis muris* by cats. J. Infect. Dis. 133:409–418.

Ruskin, J., McIntosh, J., and Remington, J. S. 1969. Studies on the mechanisms of resistance to phylogenetically diverse intracellular organisms. J. Immunol. 103:252–259.

Ruskin, J., and Remington, J. S. 1968a. Role for the macrophage in acquired immunity to phylogenetically unrelated intracellular organisms. Antimicrob. Agents Chemother. 474–477.

Ruskin, J., and Remington, J. S. 1968b. Immunity and intracellular infection: Demonstration of resistance to *Listeria* and *Salmonella* in mice infected with a protozoan (*Toxoplasma*). Science 160:72–74.

Ruskin, J., and Remington, J. S. 1976. Toxoplasmosis in the compromised host. Ann. Intern. Med. 84:193–199.

Rytel, M. W., and Jones, T. C. 1966. Induction of interferon in mice infected with *Toxoplasma gondii*. Proc. Soc. Exp. Biol. Med. 123: 859–862.

Schaegel, T. F. 1965. The value of routine testing in the etiologic diagnosis of uveitis. Am. J. Ophthalmol. 60:648–653.

Schmunis, G., Weissenbacher, M., Chowchuvech, E., Sawicki, L., Galin, M. A., and Baron, S. 1973. Growth of *Toxoplasma gondii* in various tissue cultures treated with In.Cn or interferon. Proc. Soc. Exp. Biol. Med. 143: 1153–1157.

Sethi, K. K., and Brandis, H. 1978. Characteristics of soluble T-cell derived factor(s) which can induce non-immune murine macrophages to exert anti-toxoplasma activity. Z. Immun. Forsch. 154:226–242.

Sethi, K. K., Pelster, B., Suzuki, N., Piekarski, G., and Brandis, H. 1975. Immunity to *Toxoplasma gondii* induced *in vitro* in non-immune mouse macrophages with specifically immune lymphocytes. J. Immunol. 115: 1151–1158.

Sheffield, H. G., and Melton, M. L. 1974. Immunity to *Toxoplasma gondii* in cats. In: Proceedings of the Third International Congress of Parasitology, Munich. Vol. 1, pp. 106–107. Facta Publications, Vienna.

Shirahata, T., and Shimizu, K. 1979. Some physicochemical characteristics of an immune lymphocyte product which inhibits the multiplication of *Toxoplasma* within mouse macrophages. Microbiol. Immunol. 23:17–30.

Shirahata, T., Shimizu, K., Noda, S., and Suzuki, N. 1977. Studies on production of biologically active substance which inhibits the intracellular multiplication of *Toxoplasma* within mouse macrophages. Z. Parasitenkd. 53:31–40.

Shirahata, T., Shimizu, K., and Suzuki, N. 1975. An *in vitro* study on lymphocyte-mediated immunity in mice experimentally infected with *Toxoplasma gondii*. Jpn. J. Vet. Sci. 37:235–243.

Shirahata, T., Shimizu, K., and Suzuki, N. 1976. Effects of immune lymphocyte products and serum antibody on the multiplication of *Toxoplasma* in murine peritoneal macrophages. Z. Parasitenkd. 49:11–23.

Sokolić, A., Movsesijan, M., Abu Ali, N., and Tanielian, Z. 1974. Immunofluorescence in studies on *Eimeria tenella*. Acta Vet. Yugoslav. 24:55–60.

Stadtsbaeder, S., Nguyen, B. T., and Calvin-Préval, M. C. 1975. Respective role of antibodies and immune macrophages during acquired immunity against toxoplasmosis in mice. Ann. Immunol. (Paris) 126:461–474.

Stahl, W., Matsubayashi, H., and Akao, S. 1965. Effects of 6-mercaptopurine on cyst development in experimental toxoplasmosis. Keio J. Med. 14:1–12.

Stahl, W., Matsubayashi, H., and Akao, S. 1966. Modification of subclinical toxoplasmosis in mice by cortisone, 6-mercaptopurine and splenectomy. Aust. J. Trop. Med. Hyg. 15: 869–874.

Strannegård, Ö. 1967a. The formation of *Toxoplasma* antibodies in rabbits. Acta Pathol. Microbiol. Scand. 71:439–449.

Strannegård, Ö. 1967b. Kinetics of the *in vitro* immunoinactivation of *Toxoplasma gondii*. Acta Pathol. Microbiol. Scand. 71:450–462.

Strannegård, Ö., and Lycke, E. 1972. Effect of anti-thymocyte serum on experimental toxoplasmosis in mice. Infect. Immun. 5:769–774.

Strickland, G. T., Pettit, L. E., and Voller, A. 1973. Immunodepression in mice infected with *Toxoplasma gondii*. Am. J. Trop. Med. Hyg. 23:452.

Strickland, G. T., Voller, A., Pettit, L. E., and Fleck, D. G. 1972. Immunodepression associated with concomitant *Toxoplasma* and malaria infections in mice. J. Infect. Dis. 126:54–60.

Swartzberg, J. E., Krahenbuhl, J. L., and Remington, J. S. 1975. Dichotomy between macrophage activation and degree of protection against *Listeria monocytogenes* and *Toxoplasma gondii* in mice stimulated with *Corynebacterium parvum*. Infect. Immun. 12: 1037–1043.

Tabbara, K. F., O'Connor, G. R., and Nozik, R. A. 1975. Effect of immunization with attenuated *Mycobacterium bovis* on experimental toxoplasmic retinochoroiditis. Am. J. Ophthalmol. 79:641–647.

Tadros, W., and Laarman, J. J. 1976. *Sarcocystis* and related coccidian parasites: A brief general review, together with a discussion on some biological aspects of their life cycles and a new proposal for their classification. Acta Leiden. 44:1–107.

Tanielian, Z., Abu Ali, N., Channoum, B. J., Sokolić, A., Borojević, D., and Movsesijan, M. 1976. Circulating antibody response in chickens to homologous and heterologous antigens of *Eimeria tenella*, *E. necatrix* and *E. brunetti*. Acta Parasitol. Yugoslav. 7:79–84.

Tanielian, Z., Abu Ali, N., Sokolić, A., and Movsesijan, M. 1974. Immunodiagnosis of *Eimeria tenella* infection using dried blood. Acta Vet. (Beograd) 24:75–78.

Thomas, V., and Dissanaike, A. S. 1978. Antibodies to *Sarcocystis* in Malaysians. Trans. R. Soc. Trop. Med. Hyg. 72:303–306.

Todd, K. S., and Hammond, D. M. 1968. Life cycle and host specificity of *Eimeria callospermophili* Henry, 1932, from the Uinta ground squirrel *Spermophilus armatus*. J. Protozool. 15:1–8.

Tremonti, L., and Walton, B. C. 1970. Blast transformation and migration inhibition in toxoplasmosis and leishmaniasis. Am. J. Trop. Med. Hyg. 19:49–56.

Tyzzer, E. E. 1929. Coccidiosis in gallinaceous

birds. Am. J. Hyg. 10:269–383.

Tyzzer, E. E., Theiler, H., and Jones, E. E. 1932. Coccidiosis in gallinaceous birds: II. A comparative study of species of *Eimeria* of the chicken. Am. J. Hyg. 15:319–393.

Versényi, L., and Pellérdy, L. 1970. Pathological and immunological investigations of the anseris-coccidiosis of the domestic goose (*Anser anser dom*). Acta Vet. Acad. Sci. Hung. 20:103–107.

Vischer, W. A., and Suter, E. 1954. Intracellular multiplication of *Toxoplasma gondii* in adult mammalian macrophages cultivated *in vitro*. Proc. Soc. Exp. Biol. Med. 86:416–419.

Wallace, G. D. 1973. *Sarcocystis* in mice inoculated with *Toxoplasma*-like oocysts from cat feces. Science 180:1375–1377.

Wallace, G. D. 1975. Observations on a feline coccidium with some characteristics of *Toxoplasma* and *Sarcocystis*. Z. Parasitenkd. 46: 167–178.

Wallace, G. D., and Frenkel, J. K. 1975. *Besnoitia* species (Protozoa, Sporozoa, Toxoplasmatidae): Recognition of cyclic transmission by cats. Science 188:369–371.

Walls, K. W. 1978. Serodiagnosis of toxoplasmosis. Lab. Management, January, 1978, 26–31.

Walls, K. W., and Barnhart, E. R. 1978. Titration of human serum antibodies to *Toxoplasma gondii* with a simple fluorometric assay. J. Clin. Microbiol. 7:234–235.

Walls, K. W., Bullock, S. L., and English, D. K. 1977. Use of enzyme linked immunosorbent assay (ELISA)and its microadaptation for the serodiagnosis of toxoplasmosis. Microbiol. Immunol. 5:273–277.

Weiland, G., Rommel, M., von Seyerl, F. 1979. Zur Serologischen Verwandtschaft zwischen *Toxoplasma gondii* und *Hammondia hammondi*. Berl. Muench. Tieraerztl. Wochenschr. 92:30–32.

Wellensiek, H. J., Krupe, H., Benke, W., and Wade, I. 1976. The Sabin-Feldman phenomenon: Immunocytolysis of *Toxoplasma gondii* caused by properdin dependent alternate pathway activation of human complement. Z. Immunitaetsforsch. Immunobiol. 152:123.

Werner, H. 1977. Über die Wirkung von *Toxoplasma*-Antikörpern auf *T. gondii* nach Reinfektion. II. Mitteilung: Untersuchungen über das Auftreten von Toxoplasmen im peripheren Blut nach Primät-und Sekundärinfektion. Zentralbl. Bakteriol. Hyg. I. Abt. Orig. A. 238:122–127.

Werner, H., and Egger, I. 1973. Über die Schutzwirkung von *Toxoplasma*-Antikörpern gegen Reinfektion. Z. Tropenmed. Parasitol. 24: 174–180.

Werner, H., Masihi, K. N., and Meingassner, J. G. 1978. Investigation on the effect of immune serum therapy on cysts of *Toxoplasma gondii* in latent infected mice. Zentralbl. Bakteriol. Hyg. I. Abt. Orig. A. 242:405–413.

Williams, R. B. 1973. The effect of *Eimeria acervulina* on the reproductive potentials of four other species of chicken coccidia during concurrent infections. Br. Vet. J. 129:xxiv–xxxi.

Wittchow, W. 1972. Vergleichende Untersuchungen zur passiven und adoptiven Übertragung der Immunität gegen *Eimeria nieschulzi*. Dissertation. Free University, Berlin.

9

Chemotherapy of Coccidiosis

Larry R. McDougald

9

The use of drugs to treat coccidiosis dates from the discovery by Levine (1939) that sulfanilamide would cure coccidiosis in chickens. Chemotherapeutic treatment was not applied in the agricultural industries until after World War II, mostly because of the high cost of early sulfonamides. An eminent poultry health specialist lamented that sulfaguanidine, although active against *Eimeria* in sheep, would never be used in agriculture because it was so expensive (Bushnell and Twiehaus, 1945). However, changes in the chemical industry after the war allowed dramatic price adjustments and led Grumbles et al. (1948) to recommend continuous medication with sulfaquinoxaline for the prevention of coccidiosis in chickens. By 1979, more than 30 drugs had been used for prevention of coccidiosis in chickens; only a few of them have been used for cattle or sheep. The principle of preventive medication has been well accepted, widely used, and highly successful in the management of coccidiosis in poultry. Infections in other farm livestock, however, are handled almost entirely by treatment of clinically ill animals. For the related disease, sarcocystosis, chemotherapy is still at the experimental stage.

New drugs for control of coccidiosis have been products of necessity, shadowing the expanding poultry industry during the emergence of ''agribusiness'' on a global scale. Coccidiosis control programs in poultry were the product of a set of circumstances not duplicated with most other domestic animals, including the direct nature of the parasite life cycle and the intensive nature of the poultry industry. Of extreme importance is the technical capability of feed mills to blend feeds with low concentrations of relatively toxic drugs. The intensive rearing of large numbers of chickens in enormous houses creates conditions favorable to rapid multiplication of parasites which have short, direct life cycles. Treatment of coccidiosis after an outbreak has reached clinical proportions is usually too late to prevent economic loss. The sophistication of this industry has also allowed the delicate quantitation in economic terms of the effects of medication programs. A few points in feed conversion or a small amount of extra weight may make the difference between profit and loss when extrapolated to millions of birds.

Lest the reader be impressed by the long list of drugs and the evidence for their effectiveness, we must caution that none of these drugs can be accepted at face value. The "faithful shadow" of drug resistance has rendered some drugs useless; others have undesirable side effects. Ehrlich (1909) compared chemotherapeutic agents to antibodies which, like "magic bullets," would find the target by themselves. He cautioned that in chemotherapy we must concentrate all our powers and abilities on making the aim accurate, so as to strike hard at the parasites and lightly at the body cells. The search will continue for the magic bullets, or the perfect drug, because all of our drugs have some faults and many compounds are discovered which never graduate to the realm of useful products (Ryley and Betts, 1973). Application of these tools for prevention of disease has today reduced coccidiosis in poultry to a nuisance value in most instances. However, we must have the continuous discovery of new and effective products to ensure future success in coccidiosis control by chemotherapy.

HOW ANTICOCCIDIAL DRUGS ARE USED

Prevention versus Treatment

Almost from the inception of commercial use of anticoccidial drugs in poultry, preventive programs have been preferable to treatment of sick birds. Coccidiosis is not clinically recognizable until the tissue damage associated with second or third generation schizogony occurs. The onset of morbidity and mortality is usually very soon after the first signs are seen. Because coccidia have a self-limiting life cycle, the acute phase of infection may pass before therapeutic treatment can be started. Even more important, many anticoccidial drugs act very early in asexual development and may have no effect on mature second generation schizonts or sexual stages (reviewed by Reid, 1973). Often, very high (toxic) doses of drugs are necessary for therapeutic treatment. Only the sulfonamides and amprolium have been used therapeutically, although other drugs act late enough in the life cycle to be of some value. Horton-Smith (1957) called attention to the fact that the large doses of drugs necessary for therapy were toxic to the chickens, a factor in itself to be reckoned with. There is general agreement that initiation of treatment after clinical signs appear is of value because not all birds in a flock become infected at the same time. If treatment is started as soon as clinical signs appear, the infection may be contained before the entire flock becomes sick. The logistics of such timely action, the availability of diagnostic personnel, the toxicity of high doses of drugs required, and the very structure of the poultry industry negate the usefulness of treatment as an economical or practical means of coccidiosis control in lieu of preventive medication.

Continuous Medication Programs

Medication to Market Age The most trouble-free drug program is one of continuous medication until broilers are slaughtered, or until laying birds are put into cages. This is not possible in many countries, because of governmental regulatory restrictions, which require that the drug is not fed for 3–5 days prior to slaughter (broilers) or past 16 weeks (layers). This is necessary to allow any residues of the drugs to be excreted before human consumption of meat or eggs. Some drugs (e.g., amprolium) are considered safe, and can be fed continuously. Unfortunately, these are not the ''drugs of choice'' in most instances. Some countries do not impose withdrawal periods, requiring instead that the safety of residues be demonstrated through extended tests.

Medication with a Short Withdrawal Period The withdrawal period required in the U. S. is usually 2 days longer than the time at which the last detectable residue is found in edible tissues. The required withdrawal is usually 3–5 days, but may be extended by the user for other reasons, for example, to minimize costs of the health program or to allow ''compensatory growth'' (McDougald and McQuistion, 1980a). Withdrawal longer than 5 days is often accompanied by increased incidence of coccidiosis problems.

Shuttle Programs The switching of drugs during a broiler growout, usually at 28 days, was devised to minimize build-up of drug resistance by reducing the exposure to each drug. Whether this was a scientifically valid concept, however, has not been demonstrated, because there was already widespread resistance to drugs recommended for this program. New drugs under development at this time may finally allow scientific testing of this concept after it has been used for 10 years. The original premise of delayed drug resistance was investigated by McLoughlin and Chute (1975), who found that resistance developed to two or more drugs in about the same number of generations whether the drugs were given alternately or in series of five generations on each drug.

Shuttle programs became popular with some poultrymen in the 1970s as a means of saving money on medication costs during the finishing periods (Eckman et al., 1974). By switching to a less expensive drug in the finishing ration, considerable savings would result, provided no serious exposure to coccidia ensued. If coccidia are resistant to the second drug, however, this practice may allow serious outbreaks of coccidiosis as the birds are being moved to slaughter, and also promote contamination of the premises with oocysts. The principle of the shuttle program may be valid, provided that the coccidia present are sensitive to both drugs. In addition to two-drug shuttles, some poultrymen use a third drug during the first 7–14 days (commonly called a prestarter program). Furazolidone has been used during this period to combat early chick mortality from various bacterial infections and to protect against early coccidiosis exposure. There is no published data in support of this program.

Rotation of Drugs In the 1960s, drugs were normally used for 6 months to 2 yr, after which other drugs were used. This was called "rotation." Often, an improvement in bird performance accompanied this change, suggesting that better control of coccidia was obtained after rotation to the new drug. Presumably, the coccidia had become slightly resistant to the first drug and were better controlled by the second. Rotation was less popular after the introduction of monensin in 1971. The lack of drug resistance to the ionophores and lack of suitable alternative drugs for several years limited this practice in the U. S. In other countries, however, rotation of products is still fairly popular, especially where there are distinct dry seasons and wet seasons. The strongest drugs are used during the period of heaviest infection risk (the wet season), and weaker drugs are used during the dry period.

Restricted Feeding and Other Related Programs Various programs have been devised with the intent of allowing an immunizing exposure in young layer pullets while protecting against a damaging outbreak. This concept has been widely used in replacement layer stock, which may be kept on the floor (instead of in cages) during lay, therefore allowing continuous risk of exposure to coccidia. There is presently a trend toward rearing layer replacement pullets in large wire-floored cages, which may limit the importance of coccidiosis in these birds. For the present, however, the problem is growing because there are not any new programs (drugs or other) to replace older drugs that have become ineffective because of drug resistance. There is a dearth of good data in the scientific literature on some of the programs in current use. Recently, however, Long et al. (1979) produced evidence that feeding suboptimal concentrations of monensin on alternate days in a restricted feeding program allowed the production of large numbers of oocysts, while preventing a clinical outbreak, thereby allowing natural immunization. More information of this type is needed, particularly with the new drugs.

Because the successful application of these concepts depends on management and minimizing exposure rather than on the drug, drug sponsors are unwilling to devote resources to develop such unsure, unwieldy programs. The FDA does not allow claims for encouraging the development of immunity in replacement pullets by feeding suboptimal drug concentrations. Because no alternative claim has been proposed, the new drugs might never be recommended for replacement pullets. Poultrymen today must use drugs and treatment programs that are 20 yr old, which only work under some conditions.

9

Combination of Drugs

Anticoccidial drugs have been used in combination for several reasons, but most commonly for synergism and to expand the species spectrum.

Synergism The best known example of synergism is the potentiation of sulfonamides with the 2,4-diaminopyrimidines (Lux, 1954; Joyner and Kendall, 1956). These two classes of drugs act at different points in folic acid synthesis and activation. By blocking two steps in the same pathway, a synergistic effect is obtained. Similarly, the combination of quinolones and clopidol is synergistic (Challey and Jeffers, 1973; Greuel et al., 1975; Ryley, 1975). In the latter case, there was widespread resistance to these drugs before combination. This combination was not effective in such instances. Widespread resistance to the sulfonamides (Hodgson et al., 1969) casts doubt on the value of the potentiated mixtures.

Drug Combinations to Extend the Species Spectrum of Activity Often, drugs are discovered that have a very narrow species spectrum. Some drugs have activity only against *Eimeria tenella.* This phenomenon may be a result of the initial testing methods: *E. tenella* is the most common parasite used for ''drug screening'' (Shumard and Callender, 1970). In the early days, activity against *E. tenella* might have led to some commercial life. However, at present, activity of drugs against all six important species is expected. Combinations of drugs have helped to fill this need. Amprolium, for instance, was originally sold as a single ingredient, but is was soon combined with ethopabate to extend the species spectrum. Amprolium had good activity against *E. tenella* and *Eimeria necatrix,* and some activity against *Eimeria maxima.* Combinations with ethopabate extended the spectrum to include the upper intestinal species, *Eimeria acervulina* and *Eimeria mivati;* this combination was used successfully for several years. Other drugs combined with amprolium outside the U.S. included sulfaquinoxaline and pyrimethamine.

Nitromide and aklomide, primarily effective against *E. tenella* and *E. necatrix,* were combined with a sulfonamide to provide some activity against *E. acervulina,* and roxarsone was combined to give some growth promotion and to strengthen the activity against *E. tenella.* Combination of sulfanitran with butynorate, dinsed, and roxarsone (now discontinued) was intended to control *E. tenella, E. necatrix,* and *E. acervulina,* tapeworm infections, and roundworm infections, and to promote growth, feed efficiency, and pigmentation. Such combined products may be offered as a convenience to the user (to save time and trouble in maintaining an inventory of three or four drugs for these uses), rather than extending the spectrum of a drug.

Drug combinations in general have played an important role in the field, mainly because the combinations allowed the imperfections of individual drugs to be balanced with another. These combinations are less important today, because broad-spectrum drugs are available, and there is no advantage to this approach. Experimental work and practical experience has indicated other possible reasons for using drug mixtures (Yvore, 1969), but these have not as yet been exploited.

The reason that some drugs are active against some species but not others is obscure. Ryley and Wilson (1972a) compared the activity of some drugs in vitro and in vivo against *E. tenella* and *Eimeria brunetti* and found that there were discrepancies between the two test systems. Presumably, the dynamics of absorption, tissue distribution, and excretion play an important role in the action of drugs in vivo.

MODE OF ACTION OF DRUGS

Action against Specific Endogenous Stages

Scientists quickly recognized that drugs did not kill parasites indiscriminately, but had specific actions at more or less specific times in the life cycle of the parasites. This aspect of drug action was termed "time of peak activity" by Reid (1973). This phenomenon has been investigated mostly by the medication of experimental infections for abbreviated periods coinciding with certain stages of development of the parasites, and by histopathology of experimental infections (Cuckler and Ott, 1947; McLoughlin and Wehr, 1960; Johnson and Van Ryzin, 1962). In addition. in vitro experiments allow direct observation of the effect of drugs on parasites (McDougald and Galloway, 1976).

The reason that certain of the endogenous stages are sensitive to certain drugs is obscure, but must be a result of the occurrence of metabolic events that are more important to one developmental stage of the coccidia than to other stages. Most of the drugs have activity against the asexual stages (sporozoite, trophozoite, schizonts 1 and 2), but some also affect sporogony (Figure 1). Effects on gametogony are difficult to demonstrate because pathology associated with late schizogony obscures any beneficial effects of a drug later in the cycle. The time of action of drugs was reviewed extensively by Reid (1972, 1973) and is mentioned below, as each drug is discussed. Most of the current knowledge on this subject comes from the chicken coccidia and may not necessarily apply to mammalian coccidia. We know, for instance, that monensin affects the early developmental stages of *E. tenella.* However, with naturally infected sheep and rabbits, activity must be manifested against later schizogonous or gametogonic stages. Thus, this information should be used with caution when dealing with species and host differing from the original description.

Coccidiostats versus Coccidiocides

Anticoccidial drugs were originally called coccidiostats, because the earliest drugs seemed to arrest the development of coccidia (in a reversible way) rather than kill them. Other drugs have

9

Figure 1. Life cycle of *E. tenella* and indication of endogenous stages of coccidia most sensitive to various anticoccidial drugs (adapted from Reid, 1973, and various other sources).

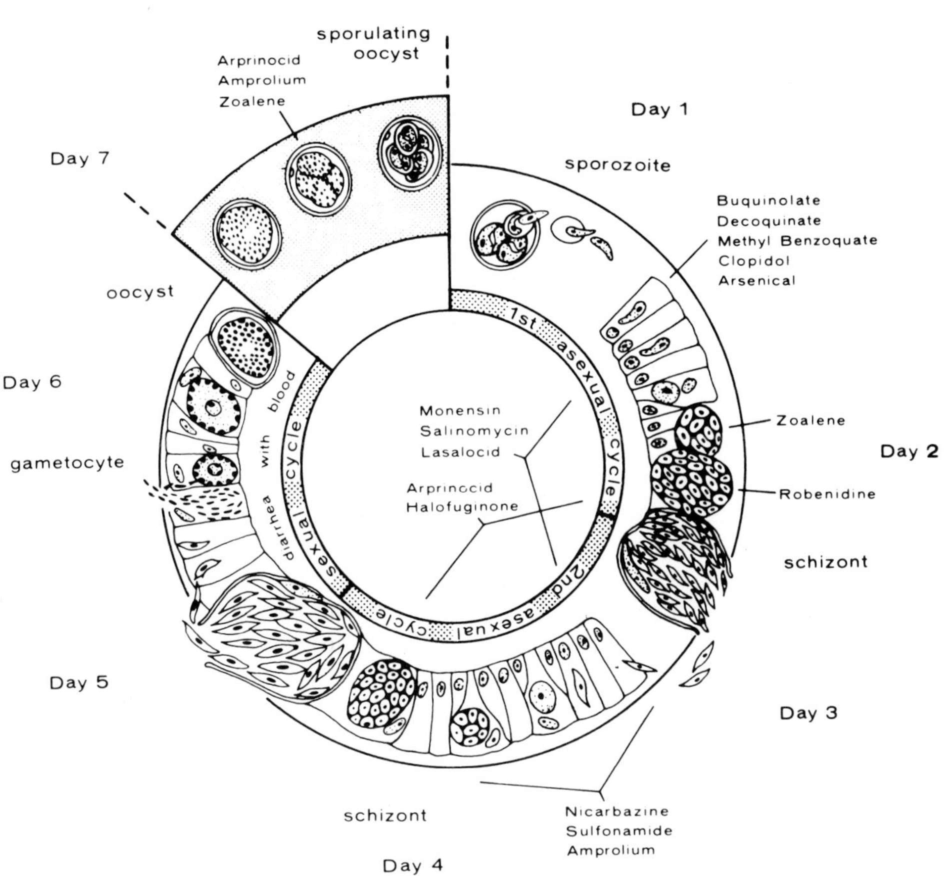

been discovered that have a coccidiocidal action (most of the parasites are killed). The coccidiostatic nature of some drugs is best demonstrated in experiments with abbreviated periods of medication, such as described by McLoughlin and Wehr (1960). The appearance of numerous oocysts in the feces of chicks several days after medication is discontinued is considered good evidence of coccidiostatic action. The coccidiocidal—coccidiostatic action of several drugs was investigated by Reid et al. (1969). The whole question of this action is complicated by the fact that parasites may merely be arrested by short periods of medication but may be killed if the drug is given long enough (McLoughlin and Wehr, 1960), and that the different species of coccidia respond differently to drugs (McQuistion and McDougald, in press,a).

In general, drugs that are predominantly coccidiocidal have enjoyed more commercial success than those predominantly coccidiostatic, but such generalizations are dangerous. Each drug must stand on its own merit; this action is only one of many characteristics that must be considered.

Biochemical Action of Anticoccidial Drugs

The way in which drugs are toxic to parasites must ultimately rest in biochemical actions. However, the actual mechanisms of anticoccidial drug action have received very little attention until recently. Fortunately, much information can be gleaned from studies on bacteria and other organisms. The action of the sulfonamides in the folate synthetic pathway is best known and is used as a classic example of antimetabolites. The sulfonamides are analogs of *p*-aminobenzoate (PABA), a basic building block of folate, and compete for incorporation into the molecule. Simultaneous administration of PABA will counter the anticoccidial value of sulfonamides (Horton-Smith and Boyland, 1946). Ethopabate, also structurally similar to PABA, acts slightly later in folate synthesis (Rogers et al., 1964). The 2,4-diaminopyrimidines and related drugs inhibit the conversion of folate or dihydrofolate to tetrahydrofolate, the active cofactor (Lux, 1954). In support of this rationale, the purified dihydrofolate reductase from *E. tenella* was about 12 times more sensitive to inhibition by pyrimethamine than was the chicken liver enzyme, thus accounting for the therapeutic index of 1.5 for the drug in the chicken (Wang et al., 1975).

The mode of action of amprolium is presumably a result of thiamine antagonism, because it is structurally related to thiamine and can be neutralized by simultaneous feeding of a small amount of thiamine. The precise mechanism of this antagonism is not clear. Because amprolium does not have a suitable group at the 5 position, it cannot compete with thiamine for phosphorylation (Rogers, 1962). McQuistion and McDougald (1979) have shown empirically that absorption of amprolium is very poor. Thus, competition for absorption at the intestinal mucosa is more likely.

9

The 4-hydroxyquinolone anticoccidials reversibly inhibit electron transport in the mitochondria (Wang, 1975, 1976). The inhibition is near cytochrome *b*. Resistant coccidia had mitochondria 100 times less sensitive to quinolone inhibition. The pyridone drug, clopidol, does not have the same mechanism of action, despite structural similarity. Synergism between quinolones and clopidol and lack of collateral cross-resistance suggest that the two drugs affect closely related pathways. Wang (1978) considered this to be evidence for two electron transport pathways; however, there is no other evidence for such systems.

The polyether, monocarboxylic acid, ionophorous antibiotics are well known for their ion-carrying action and the resulting shifts in ionic gradients across membranes (Pressman, 1976). Just how these drugs kill coccidia is not known. Wang (1978) noted the effects on ion gradients in host cells and suggested that coccidia might lack any mechanism for active transport. This model would allow selective toxicity of these drugs against the coccidia.

Analogs of the nucleosides, uridine and adenosine, have anticoccidial activity. The drug tiazuril, a uridine analog, has not been studied extensively, but presumably is active because of its effects on nucleic acid metabolism. Arprinocid, an adenosine analog, was studied extensively by Wang et al. (1979a,b; 1980; 1981) and Wang and Simashkevich (1980). The principal mode of action seems to involve inhibition of purine salvage pathways (Wang et al., 1979a) rather than direct antagonism. Action of the parent compound depends on conversion in the host to the *N*-oxide. In the host, an equilibrium is established between arprinocid and its *N*-oxide. Necessity for metabolism through the P–450 system in the liver for activity accounts for its lack of activity in vitro in chicken kidney cells.

Little is known of the mode of action of other drugs, although some of them have been used for 25 years or more. Knowledge of the precise nature of the action of drugs would be valuable, because it would not only give important information about the drugs, but would also provide tools for elucidation of the metabolism and biochemistry of coccidia.

METHODS OF DRUG DISCOVERY

Characteristics of a Useful Drug

A description of the "ideal coccidiostat" was given by Edgar (1970) at a symposium on anticoccidial drugs held at the University of Georgia. All of the present-day drugs violate some of the points that were made. Generally speaking, the anticoccidial drug is a chemical compound which has selective toxicity (Albert, 1973) to coccidia, rather than to chickens. Many compounds are discovered which in the laboratory have some demonstrable activity against one or more species of *Eimeria* in chickens (Ryley and Betts, 1973), but only a few avoid gross violations of Edgar's 15 points, and actually become useful drugs.

Discovery of numerous commercially useful anticoccidial drugs can be largely attributed to the use of the "target" animal species (the chicken) in primary evaluation of new compounds, a circumstance rarely duplicated in other fields.

Testing of Candidate Drugs for Commercial Use

Despite intensive activity by several pharmaceutical houses during the past 30 years, published descriptions of testing programs are uncommon, partly because of the need to protect proprietary information and partly because the subject does not command much general scientific interest.

Much of the early literature on anticoccidial drugs is difficult to evaluate because of differences in application (treatment versus prevention, feed versus water, etc.), lack of information on species other than *E. tenella,* and emphasis on parameters such as mortality rather than performance.

Several symposia have been held where techniques for research with drugs and coccidia were discussed. Among these were: New York Academy of Sciences, 1949; American Feed Manufacturer's Association, 1959; The University of Georgia, 1969; Guelph University, 1973; INRA, Tours, France, 1972; Nottingham University (British Poultry Science Symposium No. 13), 1977; and the Czech Academy of Sciences, Prague, Czechoslovakia, 1979. Except for the 1959 meeting, the proceedings of most of these symposia are available in major libraries. Interested persons will find them very informative on the development and "state of the arts" of drug testing.

Researchers have attempted to perfect laboratory testing models to the point that results of laboratory tests would accurately predict the probability of commercial success. Such efforts were frustrated by the commercial success of drugs that gave mediocre results in laboratory tests, but were commercial successes, and by the commercial failure of other drugs that were extremely effective in the laboratory. Disappointing experiences with promising drugs have prompted reexamination of testing models and emphasis on other phases of testing. The floor-pen test, with exposure of chickens to coccidia by indirect methods, has become essential in the drug development process, as a means of bridging the gap between laboratory and field trials.

Screening of Uncharacterized Chemical Compounds

Drug Testing in Vitro With the discovery of cell culture and embryo culture techniques for culturing coccidia (Patton, 1965; Strout et al., 1965; Long, 1965), it became possible to test drugs in vitro. One of the first reports of anticoccidial drugs in cell cultures is that of Strout and Ouellette (1973), who tested several known drugs and not only found them active but also observed the effects of the drugs on endogenous development. They suggested that this technique would be useful as a screening tool, as well as for observation of the mode of action of

9

drugs. Use of the in vitro technique for drug testing was described further by McDougald (1973), Latter (1974), Itagaki et al. (1974), and Ryley and Wilson (1976). This technique has been used for primary evaluation of industrial compounds, usually as a part of a complete testing program.

In vitro techniques have been used not only for screening of unknown compounds, but also for studying the mode of action of drugs against specific endogenous stages (McDougald and Galloway, 1973, 1976; Ryley and Wilson, 1974; Smith and Strout, 1979), the concentration of drugs in the blood of chickens (McDougald and Galloway, 1977), elucidation of the role of metabolites in the action of drugs (Latter and Wilson, 1979), antimetabolites (Ryley and Wilson, 1972b), and drug resistance (Chapman, 1974, 1976).

The endpoints of activity of known drugs in cell culture, as reported by various investigators, are in general agreement, although discrepancies do exist. The activity of certain compounds apparently depends on the exact culture conditions used, the strain of coccidia used, the endpoint, and even the diluent used to suspend the drugs (McDougald, 1978a). The type of host cell employed may be of great importance when testing drugs that are metabolized in vivo to an active metabolite (Latter and Wilson, 1979). Not surprisingly, the relative activity of drugs in vitro does not correlate particularly well with activity in vivo (Ryley and Wilson, 1972a).

Proponents of in vitro drug testing programs point to the enormous number of compounds that can be tested, the small amount of compound necessary for each test, and the ability to test substances that would be impractical for chick diet tests. Critics of the test systems argue that many compounds active in birds are inactive in vitro, compounds may be too toxic to cells to distinguish activity, and compounds can be tested only against one or two species of coccidia. The view of critics of the value of in vitro test systems is probably summed up by the words of Ryley and Wilson (1976) who declared that "it is probably better to have in vitro results than no results at all."

The use of chick embryo infections to test drugs was first investigated by Ryley (1968), who rejected the utility of the technique on the basis that not all compounds were active if embryo death from coccidiosis was used as an endpoint of activity. Long (1970) and Long and Millard (1973), however, studied the problem more extensively and proposed a system whereby other parameters, such as lesion score, could be used to judge activity. With this system, two species of *Eimeria* can be used, and the response can be judged quantitatively. Various workers have apparently overcome the early objections, because the technique is used today in some industrial research establishments.

The reliability and accuracy of in vitro techniques for screening new compounds for anticoccidial activity is probably no better or worse than reliability of in vitro techniques in microbiology. The primary pitfall seems to arise from the fact that some drugs are inactive until metabolized to an active moiety in the target animal species. An excellent example of this phenomenon is arprinocid, which is apparently metabolized to an active form in the host (Wang and Simashkevich, in press). This problem can be overcome, in the case of arprinocid, by the use of embryonic, chick liver cells, which apparently have the enzymatic machinery necessary to make the conversion, compared with kidney cells, which do not (Latter and Wilson, 1979). Another possible example is the macrolide antibiotics, tylosin and tiamulin, which have no in vitro activity against coccidia (McDougald, unpublished observations) but are active in vivo when given at high doses (Shumard, 1965; Cruthers et al., 1980).

Primary Evaluation in Vivo ("Drug Screening") Experimental compounds are mixed into chicken feed at a measured rate, usually 200–500 ppm, then fed to small groups (five to ten) of young chickens (1–3 weeks old). After 1 or 2 days of medication, or sometimes at the onset of medication, the chicks are infected with coccidia. *E. tenella* is the most common parasite used, but occasionally *E. acervulina, E. brunetti,* or even *E. necatrix* may be used. After 5–7 days, the test is terminated, and the results are determined on the basis of weight gains, lesion scores or dropping scores, or mortality. These techniques were reviewed extensively by Shumard and Callender (1970) and Morehouse and Baron (1970). Often a compound will be eliminated by results of a test with a single group of five birds. Through repetitive testing in large batteries of cages, literally thousands of compounds may be evaluated. Less than 1% of them will show any activity at all.

Secondary Testing and Drug Development Compounds that are active in primary tests must be subjected to an extensive series of tests to confirm and define the activity against different species of *Eimeria* and to confirm the optimal concentration. Raines (1978) summarized the efficacy studies necessary for registration of drugs in the U. S. Studies to support registration of a drug must be well replicated, utilize adequate numbers of birds per group, and involve infections with all six major *Eimeria* species of the chicken. Control groups include infected and uninfected, nonmedicated groups, and usually a reference drug. Testing at this stage may continue indefinitely, extending beyond the drug development program. If the new compound has survived this series of tests, studies conducted under simulated practical conditions are planned. These studies, called "floor-pen tests" (Brewer and Kowalski, 1970) were developed with the purpose of measuring performance of chickens in economic terms (weight gain and feed conversion ratio), while providing some exposure to coccidiosis. This type of testing has become very popular and is generally regarded as an important intermediate step between the laboratory and the field. In this type of test, groups of 30–200 chicks are started in pens with a floor of wood shavings or other bedding, then grown to 8 weeks of age. Weights and feed consumption of birds at 28 days and 56 days, mortality, and lesion scores of representative birds selected from each pen provide the basis for assessing the results of this type of study. Coccidiosis exposure in floor-pen trials may be accomplished by infected seeder birds

9

(Brewer and Kowalski, 1970), by the use of old (contaminated litter, or by mixing oocysts with the feed or water (Kilgore et al., 1979). Testing methods were also reviewed by Cuckler (1972).

Field Trials At some point in the development of a new drug, extended studies must be done under the practical conditions of commercial poultry farms. Because of the known interactions of anticoccidial drugs with environmental conditions, composition of feedstuffs, nutrition, breed of chickens, and other feed additives, the candidate drugs should be tested under as wide a variety of conditions as possible. Even so, side effects may not be discovered until after the drug is available commercially. Drugs may have ''fatal flaws'' that are not discovered during the developmental process, but which lead to demise of the product after only a few months of use.

Safety and Toxicity Aside from the question of effectiveness, extensive studies must be done to demonstrate the safety of potential drugs to the target animals, and also to humans who consume meat from treated animals. The amount of work necessary to prove safety for registration of a drug in the U. S. or Europe has grown considerably in recent years. While it was once adequate to define the acute and subacute toxicity of potential drugs, lack of carcinogenicity, teratogenicity, and influence on R-transfer in bacteria must now be demonstrated. Also, the ocular, nasal, and dermal absorption, the sensitizing potential, the route of absorption, the metabolic fate, and the route of excretion of the compound must be determined. Many of these tests must be repeated if the product form is changed. The requirements for these tests in the European Common Market (Ferrando, 1977) have been subject to frequent modification.

DRUGS USED FOR COCCIDIOSIS CONTROL IN THE POULTRY INDUSTRY

Accurate comparison of the drugs used for coccidiosis control is not possible because: (*a*) the drugs were never compared directly in well controlled experiments, and (*b*) some of the older drugs were tested with different methodology and with different objectives. This review is intended to put most drugs in the proper perspective, based not only on the published literature, but also on field experience, which is common knowledge or generally accepted. Only those drugs that have been used or show promise of commercial use are considered here. A comprehensive review of the voluminous literature would be beyond the scope of this chapter, and would be counterproductive to the purpose of obtaining an accurate view of the state of the science. References are cited, where appropriate, for representative documentation. The chemical structures of these drugs are found in Figure 2 and are referenced in the text by

Figure 2. Anticoccidial drugs used in the poultry industry.

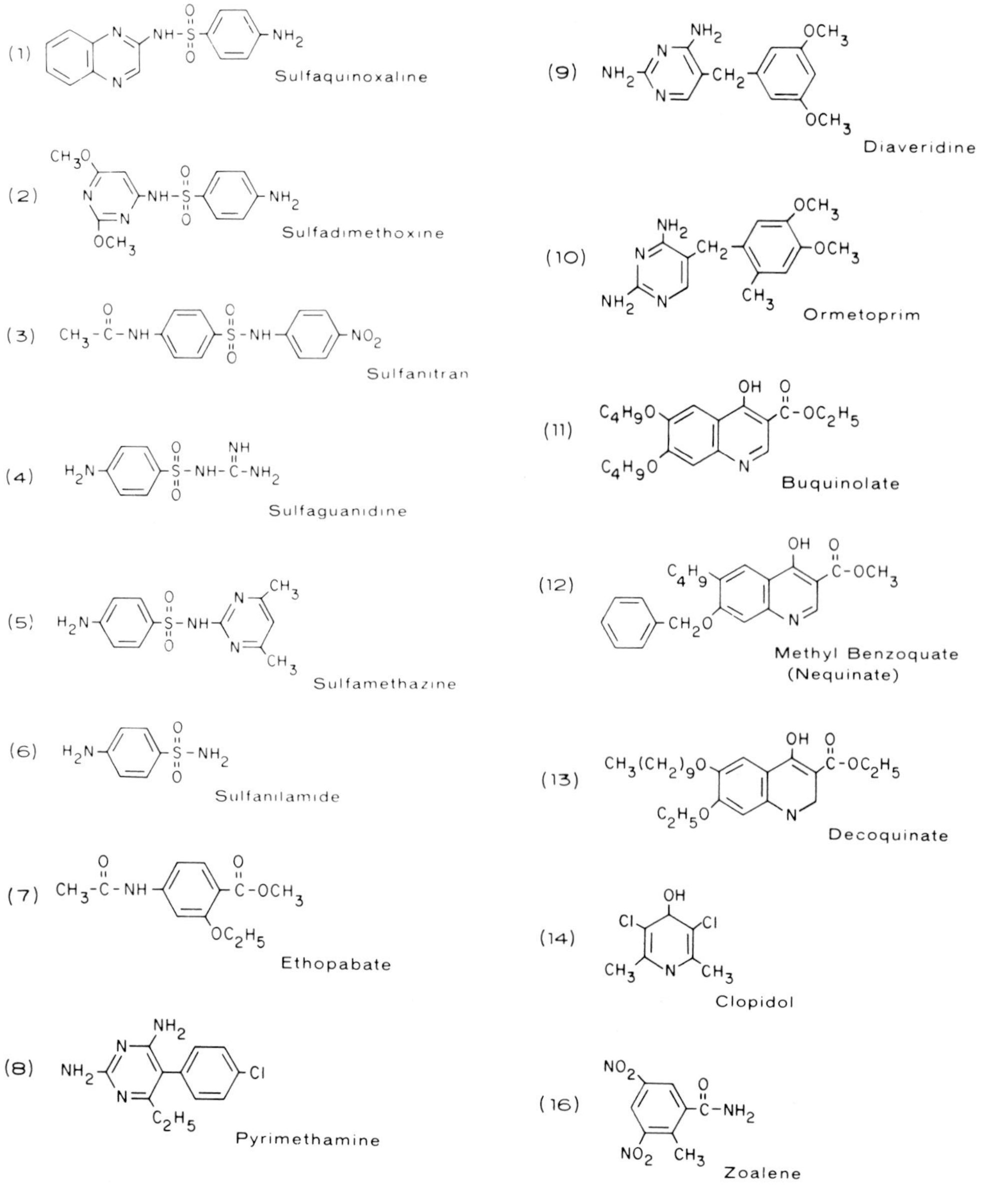

(1) Sulfaquinoxaline

(2) Sulfadimethoxine

(3) Sulfanitran

(4) Sulfaguanidine

(5) Sulfamethazine

(6) Sulfanilamide

(7) Ethopabate

(8) Pyrimethamine

(9) Diaveridine

(10) Ormetoprim

(11) Buquinolate

(12) Methyl Benzoquate (Nequinate)

(13) Decoquinate

(14) Clopidol

(16) Zoalene

9

(18) Roxarsone

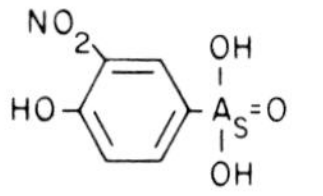

(15) Nitromide

(17) Aklomide

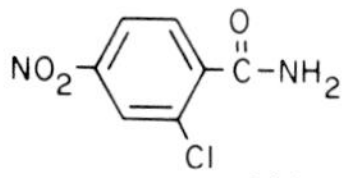

(19) Arsanilic Acid

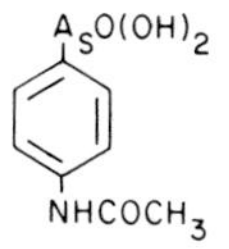

(20) Arsenosobenzine

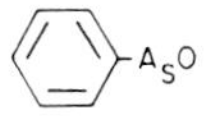

(21) Glycarbylamide

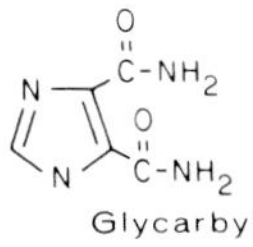

(22) Nitrophenide

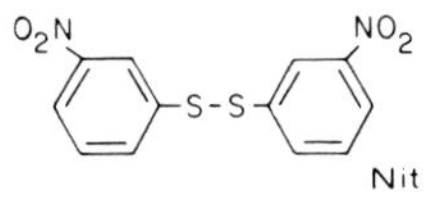

(23) (Nicarbazin). Dinitrocarbanilide component

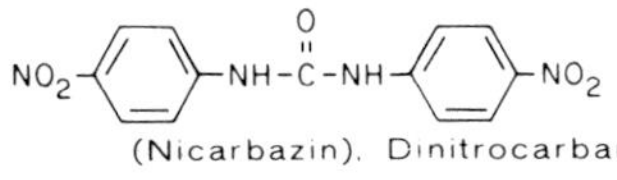

(24) Pyrimidine component

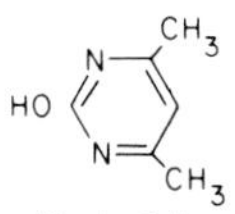

(25) Nitrofurazone

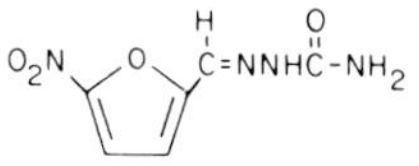

(26) Furazolidone

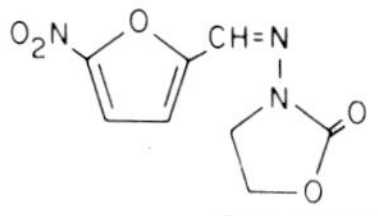

(27) Nihydrazone

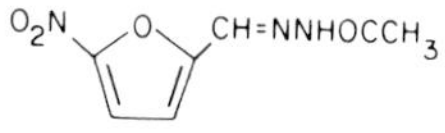

(28) Amprolium

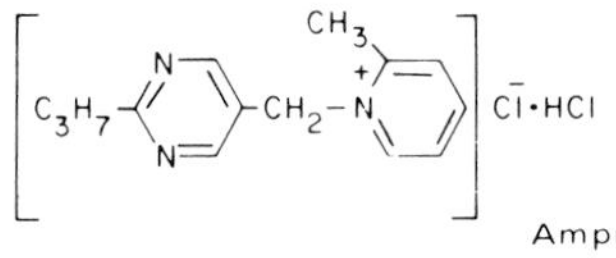

Figure 2. *continued*

(29) Monensin

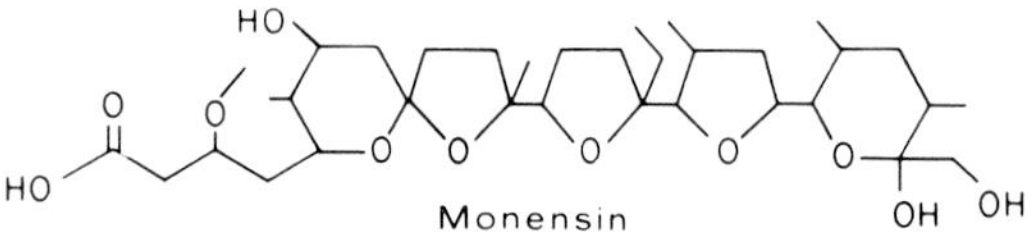

(30) Lasalocid

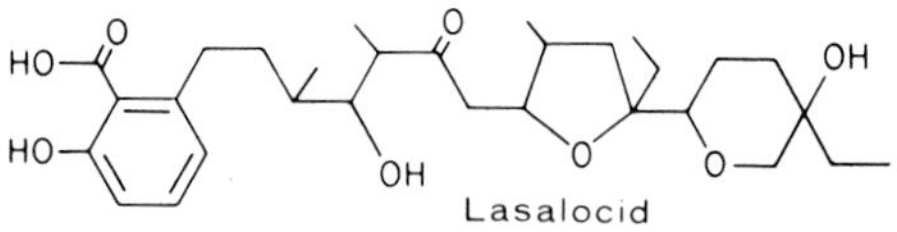

(31) Salinomycin

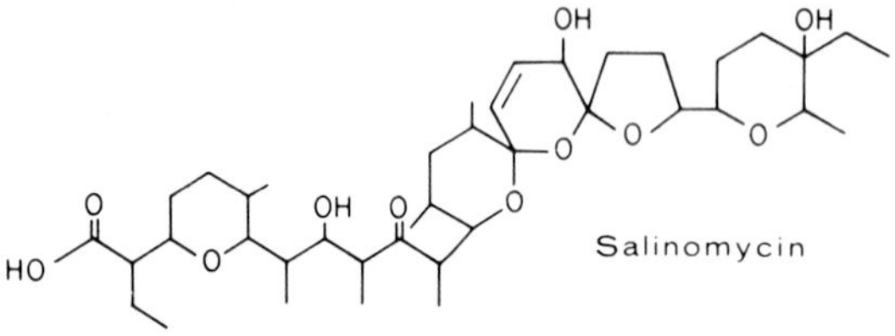

(32) Narasin

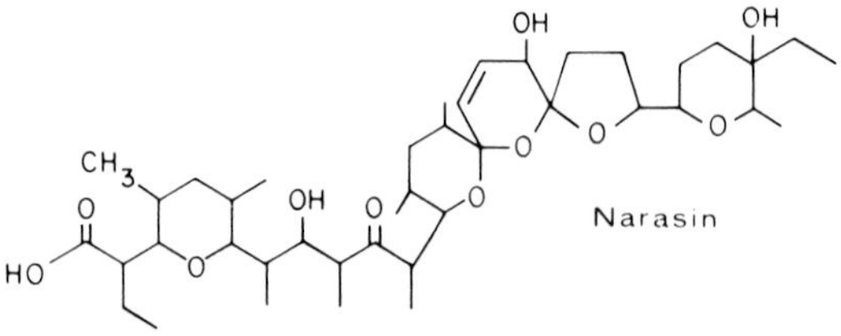

(33) Robenidine

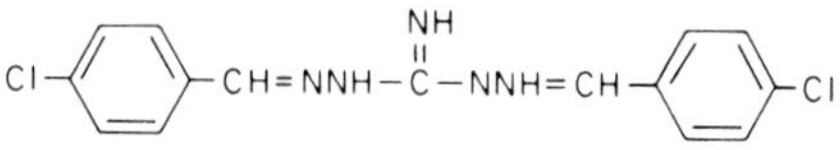

(34) Halofuginone

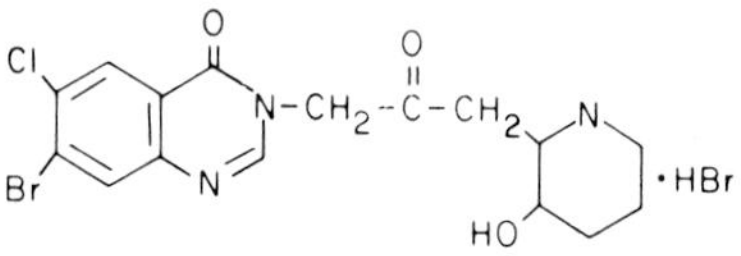

(35) Arprinocid

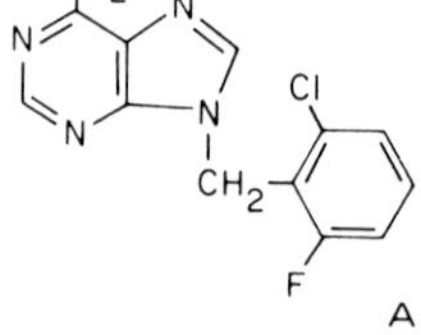

(36) Dinsed

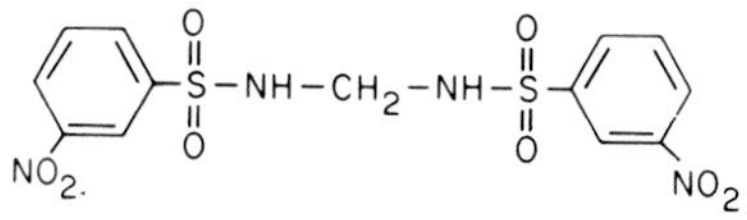

(37) Butynorate

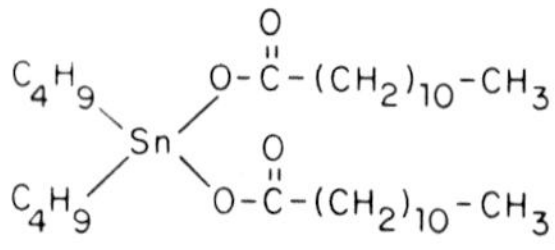

(38) Methiotriazamine

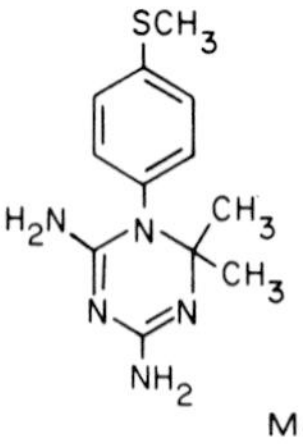

(39) Bithionol

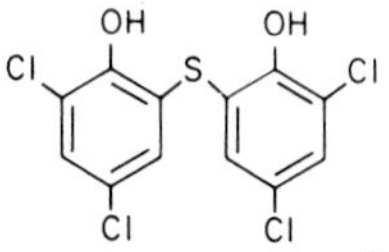

9

numbers after the name of the drug, for example, (1). Structural formulae were compiled mostly from The Merck Index (1976), which also contains additional information on the chemistry of these compounds. For information on government-approved uses of these drugs in animal feeds in the U. S., consult the Feed Additive Compendium (1980).

Folate Antagonists and Inhibitors

With the discovery by Levine (1939) that sulfanilamide (6) was effective against *E. tenella,* chemotherapy became a major field of study, although this drug was never used on a practical basis in poultry. The synthetic chemists were very active in later years, with the result that numerous sulfonamides and other folate antagonists or inhibitors were discovered and used against coccidiosis. Sulfaquinoxaline (1) was the first drug recommended for continuous (prophylactic) medication (Grumbles et al., 1948), and was used extensively through the 1950s. At least three classes of inhibitors or antagonists are active in the folate pathway: 1) the sulfonamides, which inhibit incorporation of PABA into the molecule, 2) ethopabate, which blocks a subsequent step in synthesis, and 3) the 2,4-diaminopyrimidines, which block the reduction of folate to the active cofactor, tetrahydrofolate. Combinations of two of these types of drugs, especially the sulfonamides and the diaminopyrimidines, are synergistic against the parasites (Lux, 1954).

Although new products in this group were introduced as recently as 1970 (Orton and Hambly, 1971), their importance has been limited to treatment of outbreaks of pathogenic *Eimeria* species and for minor uses, because of the availability of more effective, better tolerated drugs for prophylactic use in broilers. Drug resistance today limits the usefulness of these drugs (Hodgson et al., 1969).

Sulfonamides Numerous sulfonamides and related drugs have been used against coccidiosis, including sulfaquinoxaline (1), sulfadimethoxine (2), sulfanitran (3), sulfaguanidine (4), and sulfamethazine (5). The sulfonamides are active in vivo against the first and second generation schizonts (Horton-Smith and Taylor, 1945; Cuckler and Ott, 1947; Horton-Smith, 1948), and possibly against the sexual stages as well. This broad action, plus the water solubility, has made these drugs more useful for treatment of established infections than some other drugs. The action may be coccidiocidal or coccidiostatic, depending on the dose and duration of treatment.

The sulfonamides have a fairly broad spectrum of activity against *Eimeria* species, but activity at prophylactic concentrations is fairly low against some species (Joyner et al., 1963). Because of the large doses of sulfonamides used for therapeutic applications, toxicity can often be observed after use (Joyner et al., 1963). Hemorrhagic syndrome, kidney damage, and growth depression may occur. Even though sulfonamides were used mostly for therapy rather than for continuous prophylaxis, drug resistance and unresponsive strains of coccidia were isolated from the field as early as 1954 (Waletzky et al., 1954); by 1964, sulfonamide resistance was common in England (Warren et al., 1966; Hodgson et al., 1969). Presumably, resistance to sulfonamides would negate any synergism imparted by combination with other drugs.

The effective concentrations of the sulfonamides varies greatly. Sulfaquinoxaline has been recommended for prophylactic use at 0.0125%, but sulfamethazine and sulfaguanidine were used as high as 1% in the diet (Hawkins and Kline, 1945; Barber, 1948), for treatment. The sulfonamides also have been widely applied for antibacterial activity. Sulfanitran (3) has been used only in combination with the dinitrobenzamides, or with dinsed (36) and butynorate (37) as Polystat. This combination is discussed further under "Drugs Used for Turkey Coccidiosis" (below).

Diaminopyrimidines The 2,4-substituted diaminopyrimidines, typified by pyrimethamine (8), diaveridine (9), and ormetoprim (10), inhibit the conversion of folic acid to tetrahydrafolic acid. These drugs have been used in combination with sulfonamides, because they synergize or potentiate the anticoccidial action (Lux, 1954). Whitsyn-S, a combination of pyrimethamine and sulfaquinoxaline, was used extensively for treatment of outbreaks, and for antibacterial therapy. Diaveridine was also combined with sulfaquinoxaline as Darvisul (Ball and Warren, 1965). Rofenaid, a mixture of ormetoprim and sulfadimethoxine, was introduced as a feed additive in the late 1960s (Mitrovic et al., 1969), but has never been used extensively. Application for antibacterial therapy today accounts for most of the use of these drugs.

Ethopabate (7) Another drug which interferes with folate synthesis at yet another point is ethopabate. This drug is unique in the spectrum of activity, that is, it has good activity against *E. acervulina* and some strains of *E. maxima* and *E. brunetti,* in contrast with lack of useful activity against *E. tenella* (Rogers et al., 1964; McManus et al., 1967). Ethopabate was used only in combination with amprolium (Amprol Plus), first at 4 ppm, and later at 40 ppm.

4-Hydroxyquinolones

Three drugs, buquinolate (11), methyl benzoquate (12), and decoquinate (13), have been sold commercially, but with limited success. These drugs give spectacular results in laboratory tests against most avian coccidia (Ryley, 1967a; Edgar and Flanagan, 1968). In the field, however, drug resistance was a serious problem almost immediately, often developing during the first broiler crop with the drug. Methyl benzoquate, at 10–30 ppm, was first sold in Europe and South America in the late 1960s, and in the U. S. in 1972. Buquinolate, at 80 ppm, was sold in the U. S. in 1968 and achieved good market penetration before drug resistance appeared suddenly and dramatically; the drug was commercially "dead" within 6 months. Decoquinate, at

33 ppm, was sold in the early 1970s, with a unique market support plan. Coccidia were isolated from farms prior to use of the drug for drug sensitivity tests. Then a decision was made whether use of the drug could be advised. Additionally, the drug was used extensively in shuttle programs and withdrawal feeds. The work with field isolates was later published and remains one of the most extensive and valuable collections of data on field isolates of coccidia (Jeffers, 1974a,b,c). The toxicity of these drugs is so low that several times the recommended dose can be tolerated without ill effects.

The quinolones are almost entirely coccidiostatic against the sporozoite or early trophozoite stages (Ryley, 1967b; Long and Millard, 1968). In laboratory tests, withdrawal of medication leads to relapse of infection within 6–7 days, even if the medication had been given for several weeks. In some laboratory experiments, relapse after quinolone withdrawal was so severe as to cause mortality.

The quinolones inhibit mitochondrial energy metabolism at the cytochrome level (Wang, 1975, 1976). The cytochrome from *E. tenella* is more sensitive to inhibition with the quinolones than chick liver cytochromes. The cytochromes from drug-resistant *E. tenella* are 100 times less sensitive. Wang (1975) hypothesized that the mitochondrial basis for activity of these drugs could account for the potential for rapid drug resistance, because there are many mitochondria per cell and thus more opportunities for mutation. Despite structural similarities with clopidol, the 4-hydroxyquinolones have a different mechanism of action, because there is no cross-resistance (Jeffers and Challey, 1973), and because Clopidol has no quinolone-like effect on mitochondrial respiration (Wang, 1975). Drug resistance develops quickly under laboratory conditions (McManus et al., 1968; McLoughlin and Chute, 1971; Chapman, 1975).

Clopidol

A pyridone compound, clopidol (14) was used prophylactically at 0.125% in the feed. It affects the sporozoite or trophozoites and is almost completely coccidiostatic in action (Ryley, 1967b; Long and Millard, 1968). Renewed development of *Eimeria* parasites has been reported after medication periods as long as 60 days (Long and Millard, 1968). There is very little known about the biochemical action of clopidol. Because of structural similarity to the quinolones, lack of cross-resistance, and synergism of combinations of the two, it is presumed that the mode of action involves a different point in the same metabolic pathway. Clopidol is very effective against all species of *Eimeria* in chickens (Stock et al., 1967; Reid and Brewer, 1967), although some researchers reported mediocre results with various strains of *Eimeria* (Long and Millard, 1967; Norton and Joyner, 1968). Activity against *E. acervulina* was reported, although considerable problems were encountered with field strains of this species during the first years of its use. Introduced in 1968 in the U. S., this drug has never assumed a major market share. At first, price was a limitation. Then the problem with *E. acervulina* had to be dealt with. There was a suspected connection with necrotic enteritis, although this was never proved (McDougald et al., 1972). Structural similarity to the herbicide, picloram, caused problems when manure from birds fed clopidol was spread on cropland; tomatoes and tobacco were particularly sensitive to this compound (Minchinton et al., 1973). Clopidol is extremely safe for chickens; feeding 10 times the recommended dose is possible without problems. Also, the drug is not toxic to other animals.

As with many other drugs, clopidol was prone to resistance development. The usefulness of this drug on most broiler farms has been almost completely neutralized by drug resistance, as seen in studies of recent field isolates (Jeffers, 1974a,c; Mathis et al., 1979). Present day use of clopidol is essentially limited to shuttle programs or withdrawal feeds (e.g., clopidol may be used in the last 1–3 weeks of the broiler growout).

Nitrobenzamides

Nitromide Nitromide (15) was the first of the nitrobenzamides to be sold; it commanded a major market share from 1958 until the late 1960s. This drug has been sold in a variety of combinations with sulfanitran and roxarsone, to extend the activity spectrum or for growth promotion (Morehouse and McGuire, 1959). Nitromide used alone at 0.025% has fairly good activity against *E. tenella* and *E. necatrix.* Some activity against *E. acervulina* was obtained by the addition of 0.03% sulfanitran. The primary action is probably similar to that of zoalene, against the first generation schizont (Joyner, 1960). Nothing is known about the biochemical action of these compounds.

Zoalene (16) Introduced in 1960, this drug did not assume a major share of the market until the late 1960s, that is, from 1968 until the end of 1971, when monensin became widely used. Zoalene was usually sold alone (for use at 0.0125%) or later combined with roxarsone (0.005%). Zoalene has about the same properties as nitromide, except for a lower recommended dose (0.0125%), because it differs by only one methyl group. In critical comparisons with other drugs, Zoalene was not as effective as amprolium, nicarbazin, Unistat (nitromide + sulfanitran), or nihydrazone, against a mixture of five species of *Eimeria* (Morrison et al., 1967). In floor pens, however, Zoalene was superior to nihydrazone, Trithiadol, nicarbazin, and arsenosobenzene (Edgar et al., 1961).

Treatment for 5–6 days may be coccidiostatic, but longer treatments become coccidiocidal (Joyner, 1960). Activity is strongest against asexual stages, especially against the first generation schizont. Resistance was developed by serial passage in laboratory experiments (McLoughlin and Gardiner, 1962). Because Zoalene resistance is widespread in the field, the use of Zoalene today is limited mostly to some breeder or layer replacement pullets, where immunity to coccidiosis is important.

Aklomide (17) The last of the nitrobenzamides to be commercialized, aklomide was used at 0.02%, usually in combination with sulfanitran (Novastat). This drug differs from other nitrobenzamide drugs by chloro substitution at the 2,4 positions. The activity spectrum and basic efficacy of aklomide is about the same as the other nitrobenzamides (Ball and Parnell, 1963). It had no particular advantage over nitromide or Zoalene, and was never extensively used.

Organic Arsenicals

Of historical interest are the organic arsenicals, which include roxarsone (18), arsanilic acid (19), and arsenosobenzene (20). These compounds have limited activity against *E. tenella.* There is some evidence that combination of roxarsone with other drugs strengthens slightly the control of *E. tenella* (Kowalski and Reid, 1972) and *E. brunetti* (Kowalski and Reid, 1975). In recent years, the primary application of the arsenicals has been for growth promotion; the weak growth response to these drugs which can sometimes be measured can be justified if the drug is inexpensive. Environmental problems associated with disposal of manufacturing wastes may ultimately eliminate the arsenicals, as well as other heavy-metal drugs.

Roxarsone was recommended for use at 0.0102–0.035% as an anticoccidial; the primary target was the sporozoite (Morehouse and McCay, 1951). Arsenosobenzene, probably the most active of the arsenical products, was used at 0.002%. Gardiner and McLoughlin (1963) found that arsenosobenzene prevented mortality and did not interfere with growth. This drug was "strongly active" against *E. tenella,* but only moderately active against *E. necatrix* (Peterson, 1960).

Glycarbylamide (21)

The record for the shortest commercial life of an anticoccidial is held by glycarbylamide. Excellent activity was described from laboratory experiments; however, soon after introduction, there was widespread appearance of resistant coccidia. Glycarbylamide resistance develops rapidly under laboratory conditions (Ball, 1966; Cuckler et al., 1969) and probably more rapidly under field conditions. The rapidity of emergence of resistance in the field suggests that there existed naturally occurring, resistant coccidia, even where the drug has never been used.

At the recommended dose of 0.003% in feed, this drug was very safe, because young birds tolerated 1.0% in the feed. Cuckler et al. (1958) judged this drug to be four times as potent as nicarbazin, and active against *E. tenella, E. necatrix,* and *E. acervulina.* Glycarbylamide compared favorably to several other drugs against *E. tenella* and *E. necatrix* (Peterson, 1960).

Glycarbylamide is coccidiostatic and active against early asexual stages. Structural considerations suggested that the mode of action involved purine antagonism, but this was not proved.

Nitrophenide (22)

Only one compound of a group of bis-nitrophenols was sold in the 1950s; this drug had little efficacy against *Eimeria* species. For control of *E. tenella,* 0.02% in the feed was recommended, but with *E. necatrix,* 0.05% was necessary to prevent mortality (Waletzky et al., 1949; Peterson and Hymas, 1950).

The primary target of this drug was the second generation schizonts, which were degenerate and abnormal after treatment of birds 49–96 hr after infection (Gardiner et al., 1952). There was some effect on sporulation of oocysts of *E. tenella* (Brackett and Bliznick, 1949), even when treatment was delayed for 96–144 hr (Gardiner et al., 1952).

Nicarbazin

This drug is an "equimolar complex" of a dinitrocarbanilide (23) and a pyrimidine (24). Both compounds are necessary for anticoccidial activity (Ott et al., 1956). Nicarbazin was the first drug with truly "broad-spectrum" activity, being very active or fairly active against all important species of *Eimeria* in chickens (Cuckler et al., 1955; Horton-Smith and Long, 1959a; McManus, 1970). At 0.0125% in the diet, nicarbazin is primarily coccidiocidal, with action in vivo directed against the developing second generation schizont (McLoughlin and Wehr, 1960). Despite good anticoccidial activity, the drug was beset with problems from side effects. The first significant problem was the toxicity to laying hens, especially those with brown-shelled eggs (Polin et al., 1958; Lucas, 1958). The electrostatic properties of the drug caused contamination of layer feeds in feed mills, which was sufficient to damage production in laying flocks. In broiler chickens, the drug will sometimes depress growth slightly, with the effect measurable at any time from 1–8 weeks. The use of nicarbazin has been associated with excessive mortality from heat stress, especially in birds 4–8 weeks old (Buys and Rasmussen, 1978; McDougald and McQuistion, 1980b). These problems have limited the use of nicarbazin during the last 25 years, which is probably a reason that resistance to this drug is not particularly widespread in the U.S. today. It is probably the most effective of the older drugs, in terms of overall effectiveness and lack of drug resistance. Because of the potential for growth depression, nicarbazin is normally used in "shuttle" programs (for example, nicarbazin is fed for 3–4 weeks, then another drug is substituted for the remainder of the growout), rather than as the sole drug for 7–8 weeks.

Nitrofurans

The nitrofurans have been extensively used in animal agriculture as antibacterials and anticoccidial drugs (see Paul, 1956, and Harwood, 1960, for review). Many of the early tests of anti-

9

coccidial activity of nitrofurans were directed toward treatment of established infections rather than prophylaxis. The limited scope and availability of these tests makes comparison with other drugs difficult. Nitrofurazone (25) and furazolidone (26) were used for coccidiosis control mostly during the 1950s. Nitrofurazone was "partially protective" at 0.0108–0.0112% against cecal coccidiosis (Gardiner and Farr, 1954), but was recommended at 0.0055% (Harwood and Stunz, 1953). Johnson (1956) found good activity at 0.02%. Several workers found little efficacy with 0.01% nitrofurazone against *E. tenella* (Davies and Kendall, 1955). The action of nitrofurans against the second generation schizonts is mostly coccidiostatic, although abnormal, vacuolated schizonts were seen in stained sections (Johnson and Van Ryzin, 1962). Nihydrazone (27) was also later recommended at 0.0125% (Johnson and O'Conner, 1965). These drugs had some activity against *E. tenella* and *E. necatrix,* but very little activity against other species (Harwood, 1956). Jeffers (1974a) noted that nitrofurazone (0.011%) was not sufficiently active to prevent clinical coccidiosis and thus could not be tested in drug sensitivity tests. Furazolidone (0.0055%) is still used in broilers, but primarily as an antibacterial (see "Shuttle Programs," above).

In the early 1960s, a product called Bifuran was sold, consisting of a mixture of nitrofurazone and furazolidone (final concentration, 0.0056/0.0008/% in feed). This product was promoted as more effective than the separate component drugs, and to be better tolerated by the chickens. It was recommended for use with pullets in which immunity to coccidiosis was important. Peterson (1960), however, considered Bifuran the poorest of seven drugs that he tested. Reddy et al. (1971) found little anticoccidial effectiveness of Bifuran against *E. tenella.* Recent regulatory action against the nitrofurans as suspect carcinogens casts serious doubt on their future in animal health.

Amprolium

The thiamine antagonists, represented principally by amprolium (28), are most active against *E. tenella* and *E. necatrix,* and to a lesser extent, *E. maxima.* Amprolium was introduced in 1960, and was later combined with ethopabate (in the U.S.), sulfaquinoxaline (in England and various countries), and even pyrimethamine, to extend and strengthen its spectrum of activity (Rogers et al., 1964; Ball, 1964; Horton-Smith and Long, 1965; Yvore and Aycardi, 1968). Ethopabate is discussed under "Folate Antagonists" (above). This drug and its combinations were used widely during the 1960s and early 1970s. Probably one of the safest anticoccidial drugs to be used extensively, it could be fed at several times the recommended dose with no ill effects. By 1966, drug resistance was becoming a problem (Hodgson et al., 1969); therefore, today, the utility of this drug is limited. Replacement-pullet farms have even more resistant coccidia than broiler farms, probably a result of the continued widespread use of amprolium in replacement pullets, while the drug was essentially replaced in broilers by newer drugs after

1971. Additional documentation of the efficacy of amprolium with ethopabate is found in studies directed at other drugs (Morrison et al., 1961; Clarke, 1964; Brewer and Reid, 1967; Hodgson, 1968; Reid et al., 1972).

Polyether Ionophorous Antibiotics

The group of antibiotics characterized by the ability to facilitate movement of ions across cell membranes was discovered almost 30 years ago (Berger et al., 1951), but the anticoccidial activity of these compounds was not known until isolation and characterization of monensin (Shumard and Callender, 1967). Several other compounds of this type have been isolated from fermentation products of *Streptomyces* and other fungi. Of these, monensin (29), lasalocid (30), and salinomycin (31) are used commercially. After its introduction in the U.S. in 1971, monensin quickly became the product of choice for broiler chickens and has since set records for market penetration in most world markets. With over 80% of the market in the U.S. for several years, monensin has been fed to more chickens than any anticoccidial drug in history. The continuing success of this group of compounds is a result of 1) broad-spectrum activity against six species of coccidia in chickens, and 2) lack of serious problems with drug resistance (Jeffers, 1978).

Chemistry These compounds contain numerous ether groups and a single organic acid group. In solution, they assume a specific configuration as a result of hydrogen bonding; the center is negatively charged because of juxtaposed oxygen atoms and acts as a sort of "magnetic basket" to trap cations (Figure 3). The exterior is neutral and hydrophobic because it consists mainly of hydrocarbons. This configuration allows the molecule to interact with physiologically important cations, such as sodium and potassium, and render them fat-soluble. Rather than forming a tight bond, the ions are captured and released subject to concentration gradients. As a result, the ions move freely across cell membranes. There are several other types of antibiotics and synthetic chemicals that cause translocation of ions across membranes, but thus far useful activity has been reported only for the polyether ionophorous antibiotics. The chemical and biological properties of the ionophores were reviewed by Pressman (1976).

Some of the ionophores have different ion spectra (e.g., lasalocid will react with divalent cations). However, there is no evidence that useful activity depends on reaction with any ions other than sodium and potassium. Although these are very complex molecules, the total synthesis of monensin and lasalocid have been reported (Nakata et al., 1978; Schmid et al., 1979). The metabolism and excretion of monensin in the steer and rat have been extensively studied (Donoho et al., 1978; Herberg et al., 1978). Monensin administered orally to steers was excreted rapidly in the feces, because metabolism and excretion is via the liver and bile. Some of the identified metabolites were products of *o*-demethylation, hydroxylation, or decarboxylation.

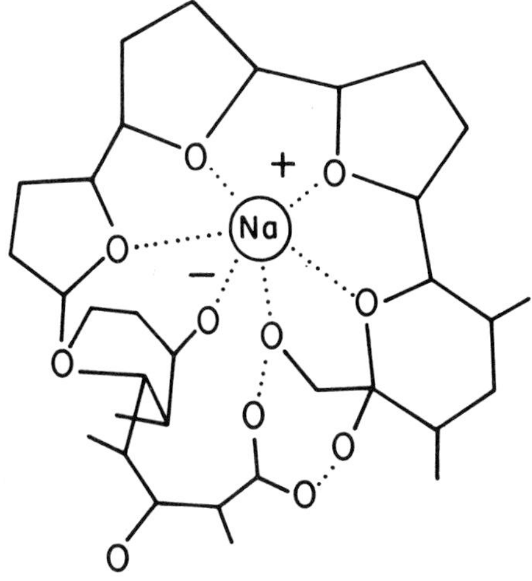

Figure 3. Stereochemical configuration of an ionophorous anticoccidial drug complexed with a monovalent cation (adapted from Pinkerton and Steinrauf, 1970).

Toxicity and Side Effects The polyether drugs are generally very toxic to animals. The LD_{50} of monensin, for instance, ranges from 2 mg/kg in the horse to 185 mg/kg in the chicken. Some of the side effects are harmful, but others are used to good advantage. For example, monensin apparently has efficacy against necrotic enteritis in chickens. Monensin has been linked to a slight growth depression under some conditions, with a resulting compensatory feed consumption and extra growth during the withdrawal period (McDougald and McQuistion, 1980a). Monensin is thought to interact with dietary ingredients, which results in feathering problems in genetically slow-feathering birds, but this relationship has never been elucidated. Kingston (1977) observed delayed feathering in birds fed 80 ppm of monensin; however, Charles and Kiker (1974) were unable to produce this effect. Lasalocid is associated with excess water excretion and a resulting problem in litter management, especially during warm weather. Considerable research has been done on this problem, but the mechanism of this action is not well understood (McDougald, 1980). Concurrent medication of chickens with monensin or saliomycin and tiamulin (a macrolide antibiotic) results in toxicity. Apparently, both compounds compete for excretion through the same metabolic process in the liver, and simultaneous administration produces higher blood levels with the ionophores (Meingassner et al., 1979). Monensin has also been associated with hyperexcitability in leghorn pullets.

In vitro studies reveal similar damage to sporozoites of *E. tenella* with monensin, lasalocid, and salinomycin, as well as with other polyether drugs. This and other evidence suggests that, while some of the side effects may be related to effects on various ions, the effects on the parasites are probably related to the same ion or group of ions (McDougald, 1978b; McQuistion and McDougald, in press,b). These drugs are most active against the earliest endogenous stages of coccidia. Little activity can be demonstrated against schizogony or gametogony in chickens, although monensin has been used successfully to treat natural, established infections in sheep. The reason for these contradictory findings is not clear. The ionophores are coccidiocidal, with little or no evidence of relapse after drug withdrawal. In vitro experiments have been used to document the way in which coccidia are killed and morphologically disrupted by these drugs.

Efficacy Several ionophorous antibiotics have now been tested in laboratory and/or floor-pen experiments. Not surprisingly, most of these compounds are active against coccidia in vitro and in vivo, although the effective dose and therapeutic index vary widely.

Monensin (29) Shumard and Callender (1967) tested monensin at 40–1,000 ppm against six species of *Eimeria* in broiler chicks and concluded that 121 ppm (110 g/ton) was optimal. Additional testing in floor pens confirmed the utility of monensin at 88–132 ppm under practical conditions. Monensin (121 ppm) was tested in a series of floor-pen experiments with broiler chickens (Reid et al., 1972) and was found superior to amprolium, clopidol, and Zoalene in control of coccidiosis (reduced mortality and reduced lesion scores) and in improved performance of the birds (live weights at 8 weeks and feed conversion). The recommended dose of monensin was adjusted in 1974 to include a range of 100–121 ppm. This coincided with a change in product form to a ''biomass'' product (containing the mycelial mass), replacing the original product which was premixed from purified crystalline antibiotic. The most extensive documentation of the efficacy of 100 ppm of monensin is provided by two reports. Callender (1978) reported extensive battery efficacy titration with monensin and obtained an excellent dose-response curve with lesions scores and mortality over the range of 80–200 ppm. Ruff et al. (1976) titrated the activity of monensin over the range of 40–140 ppm in floor-pen experiments with broilers, by using mild and heavy coccidiosis pressure. They obtained good results at 40–100 ppm when mild coccidial exposure was used, but 84–102 ppm gave the best results with heavy exposure. This was apparently the first use of floor-pen trials for dose-titration studies. Extensive evaluation of monensin under field conditions was reported by Clarke et al. (1974) from studies in Europe.

In 1974, monensin was approved for use in pullets intended as commercial layers. Reid et al. (1977) studied the anticoccidial protection and also the development of immunity to coccidiosis in birds that were receiving medication. They concluded that immunity was seriously delayed by feeding 120 ppm of monensin (because of the high efficacy of the drug against coccidia), but that lower doses of drug allowed progressively more immunity. Where immunity to coccidiosis is important, they recommended that the drug be fed at the lowest possible dose (consistent with anticoccidial protection) for the shortest practical period of time. Long et al. (1979) also recommended low doses of monensin for replacement pullets.

Additional reports on the efficacy of monensin in chickens are found in the drug resistance studies of Jeffers (1974a,b, 1978), McLoughlin and Chute (1974), and Chapman (1976), and in various efficacy studies with other drugs (Danforth et al., 1977a,b; Mitrovic et al., 1977; Frigg and Schramm, 1977; Ruff et al., 1978).

Salinomycin (31) This drug was isolated in Japan (Kinashi et al., 1973) from a culture of *Streptomyces albus* and was subsequently found active against *E. tenella* (Miyazaki et al., 1974). When titrated against single and mixed infections of six species of *Eimeria* in broiler chicks, the compound had significant anticoccidial activity at the 60–100 ppm doses (Danforth et al., 1977a). Doses of 50 ppm or less were only partly effective, and 100 ppm of salinomycin seemed to be similar to 121 ppm of monensin. Migaki and Babcock (1979) compared salinomycin directly with monensin and lasalocid at optimal and suboptimal concentration. Salinomycin at 60 ppm was equivalent to 100 ppm of monensin and 75 ppm of lasalocid. At one-half these dosages, however, the protection by salinomycin was better than that afforded by the other two drugs. In floor-pen experiments with broilers, salinomycin at 60–100 ppm was as effective as monensin in reducing lesion scores and allowing good performance of the birds at 8 weeks (Danforth et al., 1977b). In both reports, it was noted that control of upper intestinal lesions was very good. Tüller and Mödder (1979) compared salinomycin (60 ppm) with amprolium and clopidol, with results in favor of salinomycin. Chappel and Babcock (1979) conducted a series of field trials in four geographic locations with pens of 65–125 birds. Mild coccidiosis exposure was provided by "used" litter. Weight gains were usually best in lasalocid and salinomycin treatments, as compared with monensin and unmedicated treatments, although there was little evidence of coccidiosis. Morrison et al. (1979) compared salinomycin with monensin and halofuginone and found all three drugs efficacious against coccidial infections, although all three drugs seem to depress weights slightly in 56-day tests. Salinomycin was introduced in Japan in 1978, in certain other Asian countries in 1979, and in Latin America in 1980.

Lasalocid (RO 2-2985, X537A) (30) One of the first ionophores discovered (Berger et al., 1951), this drug was described by Brossi (1969) as having shown broad-spectrum anticoccidial activity. Mitrovic and Schildknecht (1974) fed 75 ppm of lasalocid to broiler chicks infected with single or mixed infections of six species of *Eimeria* and found a "high degree of anticoccidial activity." Similarly, in 8-week floor-pen trials, broiler chickens were protected against coccidiosis and had superior performance compared to birds fed other drugs. There was no drug resistance in a strain of *E. tenella* passaged 20 times in the presence of lasalocid, nor was there cross-resistance with other drugs (Mitrovic and Schildknecht, 1975). An additional series of trials with single and mixed infections of laboratory and field strains of *Eimeria* demonstrated a high degree of activity, as judged by reduced mortality, lower lesion scores, and good performance (Mitrovic and Schildknecht, 1974). Lasalocid compared favorably with other drugs and was completely compatible with growth-promoting agents (Mitrovic et al., 1975). Similar activity was demonstrated with 23 field isolates of coccidia in Europe (Frigg and Schramm, 1977). The original dosage of 75 ppm was modified in 1976 to include the range of 75–125 ppm in feed. The reasons for this change are inapparent from the literature, but were done primarily to obtain better efficacy against upper intestinal species of coccidia. Some recent publications have described tests with 75–125 ppm of lasalocid (Mitrovic and Schildknecht, 1976, Mitrovic et al., 1977). Reid et al. (1975) also found good activity of lasalocid in laboratory and floor-pen experiments with broilers. Lasalocid generally does not tend to depress weights of chicks in laboratory experiments like other ionophores and often allows better weight gains in 8-week floor-pen experiments than other drugs (Reid et al., 1975; Marusich et al., 1977; McDougald and McQuistion, 1980a).

Narasin (32) This compound differs from salinomycin by only one methyl group (Berg and Hamill, 1978); thus, it is not surprising that the ionophorous properties (Wong et al., 1977) are fairly similar. Anticoccidial activity was described by Weppelman et al. (1977) and Ruff et al. (1979) from laboratory cage experiments and in floor pens (Ruff et al., 1980). Feeding with 60, 80, or 100 ppm of narasin significantly improved weight gain and reduced lesion scores of infected birds. Narasin was effective against six species of coccidia in separate and mixed infections. Under the conditions of these tests, 80 ppm of narasin was more effective than 99 ppm of monensin. Based on the results of tests with a strain of coccidia selected for low sensitivity to monensin, Weppelman et al. (1977) suggested that there would be cross-resistance among the polyether ionophores. Ruff et al. (1980) tested narasin in three floor-pen experiments. Feeding narasin at 40–120 ppm effectively controlled coccidiosis, and protection with narasin was better than that with monensin. Treatment with 40 ppm of narasin reduced coccidial exposure enough to limit the development of immunity in the birds. These results suggested that 60–80 ppm of narasin was equivalent to 100 ppm of monensin. Plans for the commercial introduction of narasin are not known.

Lonomycin (Emericid) This compound was isolated independently by at least three drug companies (Otake et al., 1975; Ninet et al., 1976; Ohshima et al., 1976). Lonomycin was highly effective in laboratory trials against single and mixed infections of *E. acervulina, E. brunetti, E. maxima, E. necatrix,* and *E. tenella* when fed at 31–125 ppm in the feed (Cruthers et al., 1978) Cruthers et al. (1978) found the best results at 62 ppm.

Robenidine

The bis-(benzylidineamino) guanidines, represented by robenidine (33), are a small but interesting group of compounds (Kantor et al., 1970; Reid et al., 1970; Kennett et al., 1974). Robenidine was first developed for use at 66 ppm in the feed, with the intention of complete elimination of oocyst passage and the thought that drug resistance would not be possible. We now know that such reasoning was faulty—the more intense the drug pressure, the more certain the drug resistance. At 66 ppm, the drug was spectacularly effective against the six important species of *Eimeria* in chickens, but other problems precluded use at this dose. Under practical conditions, robenidine imparts a medicinal taste to edible tissues of the chicken. When the drug is given at 33 ppm, however, there is no undesirable taste, provided a suitable withdrawal period is observed before slaughter. The drug was sold in the U.S. for use at 33 ppm with a 5-day withdrawal period. However, despite excellent activity in the laboratory, the commercial life of this drug was limited because of drug resistance, which became a serious problem after about 1 yr. This problem was not documented in the literature, except for the report of Lee and Fernando (1977), regarding one strain of *E. tenella,* and the report of Joyner and Norton (1975), regarding a strain of *E. maxima* that actually grew better in the presence of the drug. Laboratory results had suggested that drug resistance would be very slow to emerge with this drug. Therefore, the quick emergence of resistant strains on poultry farms was unexpected and was a defeat for our concept of drug testing and our understanding of such phenomena. There is no doubt that robenidine has the best activity spectrum, is well tolerated by chickens, and has the best innate activity of any drugs discovered in recent years.

Robenidine is most effective against the maturing first generation schizont (Ryley and Wilson, 1971; Lee and Millard, 1972), but some activity against the sexual stages was reported in product literature. The action is predominantly coccidiocidal. There is little information on the biochemical action of this drug against parasites; in rat liver mitochondria, oxidative phosphorylation is inhibited (Wong et al., 1972).

Halofuginone (34)

This drug is derived from an extract of the *Dichroa febrifuga.* The original extract, febrifugine, was known for antimalarial and anticoccidial activity, but was never marketed because of a very narrow safety margin at the dose of 3 ppm. Halofuginone has since been offered for sale in Europe and Latin America, and development is underway in the U.S. At a feed concentration of 3 ppm, the drug is effective against all six *Eimeria* species in chickens (Fouré and Bennejean, 1974; Yvore et al., 1974; Greuel and Raether, 1979), and is well tolerated by growing birds up to 8 weeks of age (Bedrnik et al., 1979). Recent field isolates of coccidia in the U.S. are very sensitive to the drug. In studies on European coccidia (Greuel et al., 1978), halofuginone was not completely effective against all strains tested. There is little available information on the potential for drug resistance from laboratory or field experience.

Halofuginone is active against all asexual stages of *Eimeria,* but probably most effective against developing first generation schizonts. Under different conditions and with different species, it may seem predominantly coccidiocidal or coccidiostatic. There is no information on the biochemical action of halofuginone.

Nucleoside Analogs

Anticoccidial activity was found in various nucleoside analogs, including the azauracils (Chappel et al., 1974) and the benzyl purines (Miller et al., 1977). The only drug to emerge from these groups is arprinocid (35), a benzyl purine. One of the most interesting features of this drug is the effect on sporulation of oocysts passed in the presence of the drug. The drug is effective against *E. tenella* at 60 ppm, but concentrations as low as 30 ppm will significantly reduce sporulation of the oocysts (Tamas et al., 1978). The biochemical action of the drug apparently involves the purine salvage pathway (Wang et al., 1979a,b), but the exact mechanism of action is not simple. The drug is apparently metabolized in vivo to the 1-*N*-oxide, which is the active form (Wang and Simashkevich, 1980). Further work suggested that the arprinocid-1-*N*-oxide might act by dilating the rough endoplasmic reticulum in the parasite (Wang et al., 1981). Arprinocid affects any stage of Eimeria in vitro. However, in the chicken, medication during first generation schizogony is most effective. Medication for a short time produces a coccidiostatic effect, but prolonged medication is coccidiocidal.

Arprinocid has broad-spectrum activity in poultry (Olson et al., 1978). Activity against *E. tenella* is apparently not as complete as with the other species, but this is offset by the effects on oocyst sporulation (Kilgore et al., 1978). At 60 ppm, arprinocid is well tolerated in 8-week feeding studies (Ruff et al., 1978). Arprinocid was first developed for market in Latin America, and is now available in Europe; use in the U. S. is planned. Little is known about the potential for drug resistance; most field isolates in the U.S. respond well to arprinocid. However, this matter can only be determined by large scale field experience.

Butynorate (37) This drug was sold separately for use at 0.0375% in the feed (Tinostat), but mostly for use at 0.02% in various combinations with dinsed (36), roxarsone (18), and other drugs (Polystat, etc.) for coccidiosis and/or worm control. As a result of FDA pressure for additional toxicology and metabolism data, Polystat was phased out in 1980. Peterson (1960) compared Polystat to other drugs. He judged it less effective against *E. tenella* and *E. necatrix* than nicarbazin, glycarbylamide, and Zoalene, but more effective than arzene, trithiadol, and bifuran. Under field conditions, butynorate was not sufficiently active to interfere with natural immunizing infections of *E. tenella* and *E. necatrix* (White-Stevens et al., 1955). Recent surveys indicate considerable drug resistance among field isolates.

Trithiadol This drug is a mixture containing 10% methiotriazamine (38), 50% bithionol (39), and 40% calcium sulfate. Prophylactic activity against *E. tenella, E. necatrix, E. acervulina,* and *E. maxima* was reported from laboratory experiments when 0.06–0.09% of the active mixture was given continuously in the feed (Arnold and Coulston, 1959). Peterson (1960) considered Trithiadol the weakest of several drugs he tested. It was compared with glycarbylamide and nicarbazin by McLoughlin et al. (1960). The weak activity of this drug led to the recommendation that it be used in conjunction with a commercial immunization program (Edgar et al., 1958; Stuart et al., 1963). Unfortunately, this drug did not have enough activity to compete with other drugs on the market, and was soon withdrawn.

DRUGS USED FOR TURKEY COCCIDIOSIS

Compared with other poultry, the turkey industry is quite small. The bulk of the industry is found in the U.S., where the total market for an anticoccidial drug is 1.1 million dollars, based on the price of currently used drugs. It is considered by some researchers that turkeys develop an age resistance to coccidiosis (Warren et al., 1963). Therefore, they are given anticoccidial drugs for only 8–10 weeks, which coincides with the age at which turkeys may be moved to outside pastures or to larger facilities. Recent work has shown that turkeys are still susceptible to damaging infections at 12 weeks (provided prior exposure has not been sufficient to immunize) and that the birds may need more anticoccidial protection (McDougald and McQuistion, 1978).

The 1980 Feed Additive Compendium lists amprolium, butynorate, nitrofurazone, sulfadimethoxine with ormetoprim, sulfaquinoxaline, and Zoalene as approved for use in turkeys. Documentation in the scientific literature of the usefulness of these drugs in turkeys is incomplete (or totally lacking). Although there have been no surveys of drug resistance in turkey coccidia, there is evidence of widespread resistance from field experience. This is to be expected, because the drugs for turkeys have been used for many years, and have produced resistance in *Eimeria* from chickens.

Sulfonamides

Sulfaquinoxaline was recommended at 0.05% for treatment of *Eimeria adenoeides* outbreaks (Clarkson and Gentles, 1958) or for prevention of *Eimeria meleagrimitis* (Moore, 1954). Later, 0.0125% sulfaquinoxaline was reported effective (Horton-Smith and Long, 1959b, 1961; Ball and Warren, 1963a,b). Sulfaquinoxaline is approved at 0.175% in the U.S. for prevention or at 0.05% for intermittent treatment. Sulfadimethoxine (0.0125%) was as effective as sulfaquinoxline or amprolium (Mitrovic, 1968). Ball and Warren (1965) combined sulfaquinoxline with diaveridine for synergistic effect. Also, a mixture of ormetoprim and sulfadimethoxine (0.01%) was as effective as amprolium and better than Zoalene (Mitrovic et al., 1970) against *E. meleagrimitis, E. adenoeides,* and *Eimeria gallopavonis.* Joyner and Norton (1970), however, tested three field isolates of *E. meleagrimitis* and found them unresponsive to sulfaquinoxaline, probably because of drug resistance.

Amprolium

This drug is approved for use at 0.125% in the U.S. Amprolium was effective against *E. adenoeides, E. meleagrimitis,* and *E. gallopavonis* (Cuckler et al., 1961; Ball and Warren, 1963a,b; Mitrovic et al., 1970; McDougald, 1979).

Monensin

This drug was effective in laboratory and floor-pen experiments against *E. meleagrimitis, E. adenoeides,* and *E. gallopavonis* (Anderson et al., 1976; McDougald, 1976; Reid et al., 1978) at concentrations of 60–100 ppm. Mitrovic et al. (1979) considered 80 ppm to be optimum. While this drug has not yet been approved for use in turkeys in the U.S., application by veterinary prescription has been used successfully in other countries.

Lasalocid

Efficacy against the major species of *Eimeria* was optimum at 125 ppm (Mitrovic et al., 1979), although some benefit was obtained with 50–150 ppm.

Arprinocid

When this drug was fed at 120 ppm, in a floor-pen study, good efficacy and performance were obtained (McDougald and Johnson, 1979). Lower drug doses did not protect as well as amprolium (125 ppm). This drug is not yet approved in the U.S.

Halofuginone

Edgar and Flanagan (1979) obtained almost complete control of four major *Eimeria* symptoms with 3 ppm of halofuginone. The drug was well tolerated.

Butynorate

Butynorate (0.07%) has been used extensively in the U.S. for turkeys, but there is no documentation in the scientific literature of its efficacy. Polystat, a combination of butynorate (37), sulfanitran (3), dinsed (36), and roxarsone (18) was also used. It was one of the most effective drugs tested by McDougald and Johnson (1979), comparing favorably to amprolium and monensin. Polystat is no longer available in the U.S.

Other Drugs

Nitrofurazone (0.0055%) is not approved for use in the U.S. However, it is recommended for "controlling lesions due to secondary bacterial invasions concurrent with coccidiosis outbreaks..." (Feed Additives Compendium, 1980). There is no evidence of useful activity against turkey coccidiosis. Nicarbazin is not approved for use in turkeys in the U.S. Cuckler et al. (1956) showed that mortality from *E. meleagrimitis* and *E. adenoeides* could be prevented with nicarbazin in turkeys (0.0125%), but Horton-Smith and Long (1959b) did not find significant protection.

Zoalene is approved in the U.S. at 0.0125–0.01875%, although there is little evidence that it is used. The only documentation of its efficacy is casual mention that it was not as effective as other drugs (Mitrovic, 1968) or was ineffective (Joyner and Norton, 1970).

Various other drugs have been mentioned in the literature, but they had little activity or were never used commercially.

DRUGS USED FOR MAMMALIAN COCCIDIOSIS

Differences in both the techniques for rearing poultry and other livestock and in the farms and ranches make consideration of mammalian coccidiosis distinct from avian coccidiosis. Cattle

9

and sheep often spend the greater part of their lives on relatively sparse, dry pasture in widely dispersed herds. In such instances, contact with damaging infection pressure is limited to watering areas, calving or lambing pens, and other confinements. In recent years, the demand for grain-fed beef has caused the development of large feedlot complexes, where thousands of animals are grouped in relatively small pens.

In this instance, cattle may be subjected to the same conditions that produce damaging outbreaks of coccidiosis in poultry. This change in management of cattle has brought about the same type of disease problems experienced in the poultry industry (Whitlock, 1974). There are many species of coccidia in cattle and sheep, of which the most damaging ones are *Eimeria bovis* and *Eimeria zuernii* in cattle and *Eimeria ovinoidalis* (formerly *Eimeria ninakohlyakimovae)* and *Eimeria ahsata* in sheep. Because most farmers and ranchers see coccidiosis as an occasional problem, more a nuisance than a threat (often mistakenly), the disease is handled by spot treatment outbreaks where clinical signs are seen. The market for drugs for such application is too small for pharmaceutical companies to search for drugs with mammalian coccidiosis as the primary target. Thus, all the drugs used for mammalian coccidiosis were previously used for avian coccidiosis. Only a few drugs are approved for treatment of cattle and fewer for sheep.

Cattle

Until recently, sulfonamides and nitrofurans have been the only drugs available for use in cattle. Drugs have rarely been used for prophylaxis because of the general feeling that outbreaks of coccidiosis were rare, and that the disease usually responded to sulfonamide therapy. The growth of the cattle-fattening industry in the U.S. in a way that parallels the poultry industry, however, is shifting emphasis for control of this disease. The results of a recent survey suggest that coccidiosis is a far greater problem in cattle than previously recognized. Fitzgerald (1975) estimated that 3.85×10^6 cattle are treated for coccidiosis annually in the U.S. Of this number, 80,000 die from the disease. Despite the reasons given for therapeutic treatment rather than preventive programs, the scientific literature demonstrates repeatedly that administration of drugs to clinically ill animals produces less of a response and requires larger (toxic) doses of drugs.

Sulfonamides Soon after the discovery of anticoccidial activity in sulfonamides, the efficacy of sulfaguanidine in natural and experimental bovine coccidiosis was found (Boughton, 1943; Boughton and Davis, 1943). "Favorable results" were obtained when sulfaguanidine (0.1 g/kg of body wt) was given for 3 weeks, starting 2 days after inoculation of *E. bovis* oocysts, and doses of 5 g/calf were given on days 13–21. When 180 g/calf was given to naturally exposed calves over a 12-week period (30 g over 3 days, on alternate weeks) "the coccidial infections of treated calves were less severe than those of the controls." Conversely, sulfaguanidine

(10 g/calf/day) given for four 3-day periods at 2-week intervals (12 days total) failed to control natural or "superimposed" infections (Peardon et al., 1963). Peardon et al. (1965) later obtained protection against morbidity and mortality with sulfamethazine in experimentally infected calves with 1.5 grains/lb of body wt given intravenously for 4 consecutive days.

Activity of the sulfonamides against *E. bovis* seems to be concentrated on the 3rd week of infection. Thus, Hammond et al. (1959) found that a single treatment with sulfamethazine (1.5 grains/lb of body wt) on day 13 or 0.15 grains/lb on days 10–18 effectively prevented clinical signs in calves. There was no protection by sulfaquinoxaline or sulfamerazine (½–1 grain/lb for 3 days), but if treatment was given on days 13–16, good protection was obtained (Hammond et al., 1956). Sulfamethazine (21.5 mg/kg) on days 12 and 14 postinfection was effective against *E. bovis* (Hammond et al., 1960). Horton-Smith (1958) considered sulfamethazine the drug of choice for treating *E. bovis* or *E. zuernii* infections.

Nitrofurans Hammond et al. (1960) treated calves with 15 mg/kg of nitrofurazone on days 2–15 or on days 12–18 postinfection. The drug was inactive and toxic in the first treatment and only partly effective in the second treatment. They concluded that nitrofurazone was of dubious value in treatment of bovine coccidiosis. Nitrofurazone given prophylactically at 7.5–10 mg/kg daily for 6 weeks was not effective in experimentally infected (*E. zuernii* and *E. bovis*) calves (Hammond et al., 1965). Four daily doses of 30 mg/kg starting 15 days post-infection, however, was effective.

Amprolium Amprolium has been the only important drug approved for use against coccidiosis in the U.S. in recent years. Treatment with amprolium at 145 mg/kg for 21 days was very effective against experimental infections of *E. bovis* (Hammond et al., 1966) in young calves, but treatment for 5 days was not as effective. A 19-day series of doses of 25 mg/lb twice daily was highly effective, but treatment for 14 days was not (Hammond et al., 1967). A series of 21 daily, prophylactic treatments with 5 or 25 mg/kg was effective against *E. bovis,* but daily treatment with 1 mg/kg was not effective (Slater et al., 1970). Treatment of naturally infected animals with 50 or 100 mg/kg (single dose) gave clinical improvements; doses up to 750 mg/kg were well tolerated (Horak et al., 1969). Jolley et al. (1971) treated calves with 25 mg/kg/day for 30 days and obtained good protection against *E. bovis.* When older calves were treated with 5 or 10 mg/kg/day for 5 days, some protection was seen; those treated for 21 days were well protected.

The beneficial effects of amprolium in clinically ill calves are especially interesting, because they are almost immediately visible. Practitioners using this drug have reported "miraculous" cures of extremely ill animals. This drug has become the product of choice for treatment of clinically ill cattle, although a wider dosage range would be helpful.

Monensin Although monensin is not approved for treatment of coccidiosis in cattle in the U.S., the drug is widely used for improvement in feed efficiency in feedlot and pasture cattle. Conveniently, the dose approved for feed efficiency coincides with the reported dose for coccidiosis prevention (McDougald, 1978c). Therefore, the drug can be legally used for continuous feeding. When given prophylactically, monensin allowed good weight gains and completely eliminated oocyst passage at 30 g/ton (less than 1 mg/kg/day), and only a few oocysts were passed at 15 g/ton against experimental *E. bovis* infections in young calves. The drug was well tolerated at 30 g/ton (33 ppm) in the feed. Fitzgerald and Mansfield (1973) dosed calves with monensin at 0.25, 1.0, or 2.0 mg/kg of body wt in medicated feed and found efficacy and toleration of the drug best at 1.0 mg/kg. The use of this drug for improvement of feed efficiency in feedlot cattle should significantly reduce problems with coccidiosis as a direct "side benefit."

Other Ionophorous Antibiotics Lasalocid, a related compound used for coccidiosis in chickens, is also active in cattle. Fitzgerald and Mansfield (1979) gave calves lasalocid in the feed at 19, 37, and 112 ppm and infected them with *E. bovis.* Oocysts passage was eliminated only by the 112-ppm treatment.

Salinomycin, given in daily doses of 0.33, 0.66, or 1.0 mg/kg of body wt, from before infection until day 21, was effective in experimentally infected *(E. bovis)* calves (Benz and Ernst, 1979). When salinomycin was used therapeutically at 2 mg/kg/day, oocyst passage was significantly reduced if treatment was given on days 8–12 postinfection, but not on days 3–7 or 13–17 postinfection. Other related antibiotics are probably also effective, but tests have not been reported.

Decoquinate The Feed Additive Compendium (1980) lists decoquinate as approved for use at 0.5 mg/kg of body wt/day for prevention of coccidiosis in cattle. Miner and Jensen (1976) obtained complete control of *E. bovis* in young calves when decoquinate was given at the rate of 0.5–0.8 mg/kg for 21–28 days. There was no relapse for 23 days after treatment was discontinued, in contrast to experience with this drug in chickens.

Miscellaneous Drugs Peardon et al. (1963) tested glycarbylamide, framycetin, and Zoalene, and found them ineffective. Little value was found for chloroquine sulfate or dichlorophen (Peardon et al., 1965). Ethopabate at 143 mg/kg was ineffective against *E. bovis* (Hammond et al., 1966), as was 1 mg/lb of nitrophenide (Swanson et al., 1954). Nicarbazin (5–25 g/calf) controlled coccidiosis, but was very toxic, while 1 g/calf for 30 days was ineffective (Hammond et al., 1958). Treatment for only 3 days was also ineffective and toxic.

Lincomycin (1 g/calf/day) for 21 days prevented clinical infection with *E. bovis* (Arakawa et al., 1967). Treatment only on days 17–20 postinfection reduced oocyst passage, as did injections of 2 g of lincomycin. Daily doses of 0.5 g for 4 days reduced clinical effects and prevented mortality when given at various times postinfection (Peardon et al., 1965). Clopidol has been used for treatment of young calves during periods of dangerous exposure; 10 mg/kg of body wt (total dose per calf was 1 g) was the recommended dose (Sevcik et al., 1980).

Sheep

Sheep coccidiosis, like that of cattle, has been treated after diagnosis of clinically ill animals, rather than by prophylaxis. The disease may reach epidemic proportions with high mortality in large lambing operations where large numbers of ewes with young lambs are concentrated.

Sulfonamides Sulfaguanidine given at 2 g/day for 8 weeks prevented coccidiosis from natural exposure. The lambs grew well (Foster et al., 1941) when fed "low doses" during a period of "expected clinical infection." Some, but not complete, protection was obtained. Sulfaguanidine at 0.1 mg/kg of body wt gave a "marked improvement" in clinically ill, naturally infected lambs, if given early in the infection (Hawkins et al., 1943). Tarlatzis et al. (1955) gave sulfaguanidine (250 mg/kg) to sheep for 7 days, which reduced, but did not eliminate, mortality. Several reports have since appeared detailing use of sulfonamides in sheep.

Amprolium Ross (1968) gave amprolium (67.5 mg/kg) for 14 days to naturally infected sheep. The infection was reduced and weight gain improved, but considerable oocyst passage continued after treatment. When Horak et al. (1969) gave 44 or 58 mg/kg for 22 days to naturally infected sheep, weight gains were better than those obtained with sulfamethazine. Baker et al. (1972) treated feedlot lambs with amprolium in the feed at 50 mg/kg/day for 21 days. Coccidiosis was suppressed during treatment, but oocyst production resumed after treatment. They concluded that amprolium was very useful for prevention of problems in the feedlot, because the immune response prevented any damage from the infection relapse after treatment. Studies with feedlot lambs in Canada gave good results when 10 ppm of amprolium was mixed with the rations, which the lambs were fed continuously for 117 days (Horton and Stockdale, 1979). Badiola and Schindler (1980) gave amprolium to lambs for 45–60 days at 100 or 200 ppm. The 100-ppm treatment had a positive effect on feed conversion; 200 ppm was completely effective and eliminated oocyst production and significantly improved weight gain.

Nitrofurazone Administration of nitrofurazone (10 mg/kg) for 7 days prevented mortality in naturally infected sheep, whereas sulfaguanidine did not completely prevent mortality (Tarlatzis et al., 1955). Nitrofurazone at 0.0165% in the feed for 21 days controlled mortality

(Shumard, 1959a); 0.008% or 0.0133% in the drinking water after onset of clinical symptoms prevented mortality and reduced morbidity (Shumard, 1959b).

Monensin Although the primary activity in avian hosts is against the trophozoite, administration of monensin at 20 g/ton in a complete feed to naturally infected sheep terminated oocyst passage in about 7 days (Bergstrom and Maki, 1974; Fitzgerald and Mansfield, 1978). Prophylactic treatment with monensin in the feed at 10–30 g/ton effectively eliminated most oocyst passage and other clinical signs and allowed good weight gains and feed conversion (Bergstrom and Maki, 1976; Fitzgerald and Mansfield, 1978; McDougald and Dunn, 1978). Treatments over 40 g/ton tended to depress weight gain. The effect of monensin was mostly coccidiocidal; there was no relapse after medication was stopped (McDougald and Dunn, 1978).

When feedlot lambs were given a ration containing 11 ppm of monensin for 117 days, the lambs were protected against *E. ninakohlyakimovae (E. ovinoidalis)* and *E. ahsata.* Body weight gains and feed efficiencies were better in lambs receiving monensin, as compared with unmedicated or amprolium-treated lambs (Horton and Stockdale, 1979).

Monensin was not well tolerated when given in daily doses or given in one capsule, rather than in the feed (Leek et al., 1976). Leek et al. (1976) found some control of coccidiosis when monensin was given by gelatin capsule at doses of 1 mg/kg prophylactically or 2 mg/kg therapeutically (after onset of symptoms); however, weight gains were not optimal. Because of the insolubility of this compound and the toxicity when administered in large doses, it is unlikely that monensin will be recommended for treatment of clinically ill animals. The drug seems most valuable for prophylaxis, where is can be mixed into a complete feed or feed supplement. It is not presently approved for use in sheep in the U.S.

Lasalocid When feedlot lambs were fed rations containing 100 ppm of lasalocid for 100 days and infected with *Eimeria* species, the lambs were protected against coccidia infections, gained 6 kg more weight than controls, and had significantly better feed conversion (Foreyt et al., 1979).

Goats

There are few reports of controlled studies with chemotherapeutics in goats, and conditions of tests vary so widely as to make comparison of results impossible.

Nitrofurazone (10 mg/kg) prevented mortality when fed for 7 days, but sulfaguanidine (250 mg/kg) did not (Tarlatzis et al., 1955). Amprolium is effective in various doses and treatment regimens. Thus, Fitzsimmons (1967) gave goats doses of 25 or 50 mg/kg and found reduced oocyst output of naturally infected goats. Marlow (1968) treated naturally infected goats with 33 or 66 mg/kg of amprolium, with good results. Treated goats had good weight gains while untreated goats did poorly. Horak et al. (1969) found that amprolium at 50 mg/kg reduced the infection but was coccidiostatic, while 100 mg/kg seemed to be curative and allowed good weight gain.

Canines and Felines

Most of the reports on drugs for canine and feline coccidiosis represent clinical experience of veterinary practitioners who have experimentally treated a few animals with sulfonamides or other drugs that were available. In most instances, some "response" was recorded. Controlled studies with experimental infections are rare or nonexistent. Sulfadimethoxine at 12.5–25 mg/kg of body wt for 14 days cured dogs of *Eimeria bigenina* and *Isospora felis* (species diagnosis probably erroneous) (Fish et al., 1965). Similarly, sulfadimethoxine at 25 mg/lb of body wt two to three times daily for 5–7 days (Knight, 1962) or daily for 14 days (Fish, 1964) gave good results. Framycetin (25 mg/kg), administered for 5 days, cured a dog of *Isospora* species infection. Rachman and Pallock (1961) treated naturally infected dogs with sulfaguanidine (1 grain/lb) and nitrofurazone (2 mg/lb) for 1 week. The nitrofurazone-treated dogs eventually recovered without additional treatment; however, the sulfaguanidine-treated dogs required further treatment with nitrofurazone. Untreated dogs had diarrhea throughout the study. There are reports in the literature relating similar experiences.

Swine

Coccidia of swine are common and widespread and can cause disease in laboratory infections. However, there is no good evidence of practical problems associated with *Eimeria* infections. Alicata and Willett (1946) produced coccidiosis in pigs with *Eimeria scabra* and *Eimeria debliecki*. They treated the infections successfully with sulfaguanidine (100 mg/lb of body wt) given prophylactically for 7–10 days, or as therapeutic treatment for 3 days, from the start of patency. Recently, various workers have discovered that *Isospora suis* infections in piglets may be severe, causing diarrhea and even mortality. Such infections have been treated successfully with amprolium, administered to 3-day-old piglets or to sows before farrowing (Sangster et al., 1978).

Rabbits

Sulfonamides and Related Drugs Sulfaquinoxaline was effective against *Eimeria stiedai* when given continuously at 0.03% in the feed (Lund, 1954) or at 0.3% in the water for 14 days (Hagen, 1961; DeVos, 1970). Chapman (1948) obtained good results by treating experimentally infected rabbits *(E. stiedai)* with sulfaquinoxaline at 0.05% in the drinking water. Sulfamerazine fed at 0.02% reportedly prevented liver lesions from *E. stiedai* (Peterson, 1950). A mixture of sulfadimethoxine and diaveridine (3:1) was effective at 100 mg/kg of food (6–10 mg/kg of body wt), considerably better than sulfaquinoxaline, formasulfathiazole, or sulfachloropyrazine given alone (Dürr and Lammler, 1970). The mixture given at 125 mg/100 g of food for 3 days or 100 mg/kg of feed or water for 8 days prevented liver coccidiosis (Dürr and Lammler, 1970). Lower doses given for 35 days were also considered acceptable. Sulfadimethoxine alone given at 0.2 g/kg of body wt for 1 day, then at 0.1 g/kg for 3 days, or sulfathiazole given at 0.4 g/kg for 5 days, reduced oocyst passage and produced better weight gain than infected controls (Dondukov, 1969). Sulfadimethoxine (75 mg/kg) was better than sulfamonomethoxine when both were given for 7 days to treat a natural outbreak of *E. stiedai.* Seven-day treatment was better than treatment for 3 days. Pellerdy (1969) recommended the use of sulfonamides to prevent acute coccidiosis while natural immunity developed from the limited exposure to natural infections.

Sulfaquinoxaline is presently the most widely used drug for rabbit coccidiosis. Continuous medication in the feed for prevention has been the most successful program. The elimination of hepatic coccidiosis is desirable, because of loss of the livers for food purposes. Programs for the use of drugs for prevention and treatment of rabbit coccidiosis have not been discussed in the literature in the broader sense; most literature deals with laboratory efficacy tests and field reports.

Amprolium DeVos (1970) reported that amprolium (0.025%) was partly effective in reducing mortality when given for 5 days in water, but not as effective as sulfamethazine (0.2%), sulfachloropyrazine (0.3%), or sulfadimethoxine (0.2%). Similarly, Cvetovic and Tomanovic (1967) prevented acute coccidiosis with 0.0125% amprolium fed for 20 days. Fitzgerald (1973) fed pelleted feed containing 0.02% amprolium as the sole ration and obtained no protection. The rabbits receiving amprolium developed severe infections.

Nitrofurans Hagen (1961) found nitrofurazone and furazolidone "partly effective" for treatment of coccidiosis. Déom and Mortelmans (1954) treated naturally infected rabbits with nitrofurazone for 5 days and obtained good control of *Eimeria irresidua* and *Eimeria perforans.* There have also been other reports of clinical experience with the use of nitrofurans to treat coccidiosis outbreaks in rabbits, but often with no controls. Usually these reports claim some success in reducing mortality or morbidity. However, it is clear that nitrofurans are relatively weak, compared to other drugs.

Monensin In all reported work with monensin, the drug was given continuously by medicated feed in a prophylactic design. Fitzgerald (1973) gave rabbits feed containing monensin at 0.005–0.02%. Coccidiosis was controlled, including *E. stiedae* and intestinal species, although there was some weight depression from the effects of the drug. Gwyther (1976) fed rabbits monensin at 0.004% and 0.002%. Complete prevention of coccidia was obtained at the high dose, and weights after 21 days were not different from uninfected controls. The lower dose was well tolerated and gave fairly good control of *Eimeria,* but allowed some hepatic lesions. It seems that additional dose titration is needed to determine the best dose.

Robenidine With natural, mixed infections, 33 ppm of robenidine was effective in controlling coccidiosis. However, experimental infections required higher doses of the drug for complete control (Peeters et al., 1979). Weight gains and feed conversions were improved by feeding the drug for 4 weeks; liver lesions were greatly reduced. There was no effect on oocyst viability. Peeters et al. (1979) cited poor results with clopidol (200 ppm) and with a sulfaquinoxaline/pyrimethamine (82.5 ppm) combination, concluding that new drugs are needed for rabbits.

CHEMOTHERAPY OF SARCOCYSTOSIS AND TOXOPLASMOSIS

The use of drugs to treat infections of the cyst-forming coccidia in man and animals has been limited mostly to drugs that were available for other uses. The sulfonamides have been used extensively in humans. There are no drugs approved for use in livestock, although some are used experimentally. In recent years, considerable attention has been focused on treatment of the cat, because of the importance of this species in propagating infective oocysts. Unfortunately, the drugs generally reduce, but do not eliminate, passage of *Toxoplasma* oocysts by infected cats. An extensive review of the voluminous literature on experiments with drugs in humans and animals naturally or experimentally infected with *Toxoplasma* or *Sarcocystis* is beyond the scope of this book. The chemotherapy of infections and the problems of epidemiology in humans were reviewed by Frenkel (1971) and in animals, by Dubey (1977). There are several other recent reviews which also emphasize the epidemiological problems associated with treatment and control of the cyst-forming coccidia. Other information on this subject will be found in Chapter 11 (Fayer and Reid, this volume).

LITERATURE CITED

Albert, A. 1973. Selective Toxicity. 5th Ed. Chapman & Hall, Publishers, London.

Alicata, J. E., and Willett, E. L. 1946. Observations on the prophylactic and curative value of sulfaguanidine in swine coccidiosis. Am. J. Vet. Res. 7:94–100.

Anderson, W. A., Reid, W. M., and McDougald, L. R. 1976. Efficacy of monensin against

9

turkey coccidiosis in laboratory and floor pen experiments. Avian Dis. 20:387–394.

Arakawa, A., Kohls, R. E., and Todd, A. C. 1967. Effect of lincomycin hydrochloride upon experimental bovine coccidiosis. Am. J. Vet. Res. 28:653–657.

Arnold, A., and Coulston, F. 1959. A laboratory evaluation of the effect of the coccidiostat Trithiadol on the growth of chickens. Toxicol. Appl. Pharmacol. 1:475–486.

Badiola, C., and Schindler, P. 1980. The control of coccidiosis in lambs by amprolium. Proceedings of the Symposium on Coccidia and Further Prospects of Their Control, November 28–30, 1979. Prague.

Baker, N. F., Walters, G. T., and Fisk, R. A. 1972. Amprolium for control of coccidiosis in feedlot lambs. Am. J. Vet. Res. 33:83–86.

Ball, S. J. 1964. Synergistic action of sulfaquinoxaline and 2-amino-4-dimethylamino-5-(4-chlorophenyl)-6-ethylpyrimidine in caecal coccidiosis in chickens. J. Comp. Pathol. Therap. 74:487–499.

Ball, S. J. 1966. The development of resistance to glycarbylamide and 2-chloro-4-nitrobenzamide in *Eimeria tenella* in chicks. Parasitology 56:25–37.

Ball, S. J., and Parnell, E. W. 1963. Anticoccidial activity of halogenonitrobenzamides. Nature 199:612.

Ball, S. J., and Warren, E. W. 1963a. The effect of long-term medication with sulfaquinoxaline and amprolium upon serial infection of *Eimeria adenoeides* and *E. meleagrimitis* in turkeys. Br. Vet. J. 119:549–558.

Ball, S. J., and Warren, E. W. 1963b. The effect of sulfaquinoxaline and amprolium against *Eimeria adenoeides* and *E. meleagrimitis* in turkeys. Res. Vet. Sci. 4:39–47.

Ball, S. J., and Warren, E. W. 1965. Treatment of experimental avian coccidiosis with a soluble combination of sulfaquinoxaline and diaveridine. Vet. Rec. 77:1252–1256.

Barber, C. W. 1948. Sulfaguanidine and sulfamethazine in the control of experimental avian coccidiosis caused by *Eimeria tenella.* Poult. Sci. 27:60–66.

Bedrnik, P., Brož, J., Sevčik, B., and Jurkovič, P. 1979. Comparison of the efficiency of anticoccidials lasalocid, halofuginone and monensin, in condition of floor-pen experiment in combination with the growth promotant nitrovin. Arch. Geflügelk. 43:7–10.

Benz, G. W., and Ernst, J. V. 1979. Efficacy of salinomycin in treatment of experimental *Eimeria bovis* infections in calves. Am. J. Vet. Res. 40:1180–1186.

Berg, D. H., and Hamill, R. L. 1978. The isolation and characterization of narasin, a new polyether antibiotic. J. Antibiot. 31:1–6.

Berger, J., Rachlin, A. I., Scott, W. E., Sternbach, L. H., and Goldberg, M. W. 1951. The isolation of three new crystalline antibiotics from *Streptomyces.* J. Am. Chem. Soc. 73:5295–5298.

Bergstrom, R. C., and Maki, L. R. 1974. Effect of monensin in young crossbred lambs with naturally occurring coccidiosis. J. Am. Vet. Med. Assoc. 165:288–289.

Bergstrom, R. C., and Maki, L. R. 1976. Coccidiostatic action of monensin fed to lambs: Body weight gains and feed conversion efficacy. Am. J. Vet. Res. 37:79–81.

Boughton, D. C. 1943. Sulfaguanidine therapy in experimental bovine coccidiosis. Am. J. Vet. Res. 4:66–72.

Boughton, D. C., and Davis, L. R. 1943. An experiment with sulfaguanidine in the treatment of naturally acquired bovine coccidiosis. Am. J. Vet. Res. 4:150–154.

Bracket, S., and Bliznick, A. 1949. The effect of small doses of drugs on oocyst production of infections with *Eimeria tenella.* Ann. N. Y. Acad. Sci. 52:595–610.

Brewer, R. N., and Kowalski, L. M. 1970. Coccidiosis: Evaluation of anticoccidial drugs in floor-pen trials. Exp. Parasitol. 28:64–71.

Brewer, R. N., and Reid, W. M. 1967. Efficacy of buquinolate against six species of coccidia. Poult. Sci. 46:642–646.

Brossi, A. 1969. Some recent results on the chemotherapy of amebiasis, coccidiosis and malaria. Pure Appl. Chem. 19:171–185.

Bushnell, L. D., and Twiehaus, M. J. 1945. Poultry diseases and their control: Coccidiosis. Bulletin No. 326 of the Agricultural Experiment Station, Kansas State University, Manhattan, KS, p. 78.

Buys, S. B., and Rasmussen, R. W. 1978. Heat stress mortality in nicarbazin fed chickens. J. S. Afr. Vet. Assoc. 49:127–128.

Callender, M. E. 1978. The testing of anticoccidial drugs in the laboratory. In: P. L. Long, K. N. Boorman, and B. M. Freeman (eds.), Avian Coccidiosis, pp. 413–422. British Poultry Science, Ltd., Edinburgh.

Challey, J. R., and Jeffers, T. K. 1973. Synergism between 4-hydroxyquinoline and pyridone coccidiostats. J. Parasitol. 59:502–504.

Chapman, H. D. 1974. Use of chick embryo infections for the study of drug resistance in *Eimeria tenella*. Parasitology 69:283–290.

Chapman, H. D. 1975. *Eimeria tenella* in chickens: Development of resistance to quinolone anticoccidial drugs. Parasitology 71:41–49.

Chapman, H. D. 1976. Further studies on the use of chicken embryo infections for the study of drug resistance in *Eimeria tenella*. Parasitology 73:275–282.

Chapman, M. P. 1948. The use of sulfaquinoxaline in the control of liver coccidiosis in domestic rabbits. Vet. Med. 43:375–379.

Chappel, L. R., and Babcock, W. E. 1979. Field trials comparing salinomycin (Coxistac, monensin and lasalocid in the control of coccidiosis in broilers. Poult. Sci. 58:304–307.

Chappel, L. R., and Babcock, W. E. 1979. Field trials comparing salinomycin (Coxistac), monensin and lasalocid in the control of coccidiosis in broilers. Poult. Sci. 58:304–307.

Charles, O. W., and Kiker, J. 1974. Nutritional aspects of the feathering dermatitis syndrome in broilers. Poult. Sci. 53:(abstr.)1634.

Clarke, M. L. 1964. A mixture of diaveridine and sulfaquinoxaline against *Eimeria acervulina, E. brunetti* and *E. maxima* infections together with a field survey of coccidiosis in Southeast England. Vet. Rec. 76:818–821.

Clarke, M. L., Diaz, M., Guilloteau, B., Hudd, D. L., and Stoker, J. W. 1974. European field evaluation of monensin, a new anticoccidial agent. Avian Pathol. 3:25–35.

Clarkson, M. J., and Gentles, M. A. 1958. Coccidiosis in turkeys. Vet. Rec. 70:211–214.

Cruthers, L. R., Hatchkin, H. D., Sarra, L. J., Perry, D. D., and Linkenheimer, W. H. 1980. Efficacy of tiamulin against an experimental infection of broilers with Eimeria acervulina and Eimeria tenella. Avian Dis. 24:241–246.

Cruthers, L. R., Szanto, J., Linkenheimer, W. H., Maplesden, D. C., and Brown, W. E. 1978. Anticoccidial activity of lonomycin (SQ 12,525) in chicks. Poult. Sci. 57:1227–1233.

Cuckler, A. C. 1972. Methods for evaluating coccidiostats in battery tests, floor pen and field trials. Folia Vet. Lat. 2:668–685.

Cuckler, A. C., Chapin, L. R., Malanga, C. M., Rogers, E. F., Becker, H. J., Clark, R. L., Leanza, W. J., Pessolano, A. A., Shen, T. Y., and Sarett, L. H. 1958. Antiparasitic drugs: II. Anticoccidial activity of 4,5-imidazoledicarboxamide and related compounds. Proc. Soc. Exp. Biol. Med. 98:167–170.

Cuckler, A. C., Coble, W. R., McManus, E. C., and Ott, W. H. 1961. Amprolium: 6. Efficacy for turkey coccidiosis. Poult. Sci. 40:(abstr.) 1392.

Cuckler, A. C., McManus, E. C., and Campbell, W. C. 1969. Development of resistance in coccidia. Acta Vet. (Brno.) 38:87–99.

Cuckler, A. C., Malanga, C. M., Basso, A. J., and O'Neill, R. C. 1955. Antiparasitic activity of substituted carbanilide complexes. Science 122:244–245.

Cuckler, A. C., Malanga, C. M., and Ott, W. H. 1956. The antiparasitic activity of nicarbazin. Poult. Sci. 35:98–109.

Cuckler, A. C., and Ott, W. H. 1947. The effect of sulfaquinoxaline on the developmental stages of *Eimeria tenella*. J. Parasitol. 33(suppl.):10.

9

Cvetovic, Lj., and Tomanovic, B. 1967. Investigation of amprolium in the prophylaxis of intestinal coccidiosis in rabbits. (In Serbo-Croatian) Vet. Glasn. 7:607–612.

Danforth, H. D., Ruff, M. D., Reid, W. M., and Johnson, J. 1977a. Anticoccidial activity of salinomycin in floor pen experiments with broilers. Poult. Sci. 56:933–938.

Danforth, H. D., Ruff, M. D., Reid, W. M., and Miller, R. L. 1977b. Anticoccidial activity of salinomycin in battery raised broiler chicks. Poult. Sci. 56:926–932.

Davies, S. F. M., and Kendall, S. B. 1955. An experimental assessment of the value of nitrofurazone used continuously as a coccidiostatic drug. Vet. Rec. 67:867–870.

Déom, J., and Mortelmans, J. 1954. Observations sur la coccidiose du mouton et de la chévre an Congo Belge. Essais therapeutiques. Ann. Soc. Belg. Med. Trop. 36:47–52.

DeVos, A. J. 1970. Coccidiosis of rabbits at Onderstepoort. J. S. Afr. Vet. Med. Assoc. 41:189–194.

Dondukov, I. J. 1969. Efficacy of sulfadimethoxine and norsulphazole plus phthalazole in rabbit coccidiosis. (In Russian) Veterinariia (Moscow) 1:51–52.

Donoho, A., Manthey, S., Occolowitz, J., and Zornes, L. 1978. Metabolism of monensin in the steer and rat. Agric. Food Chem. 26:1090–1095.

Dubey, J. P. 1977. *Toxoplasma, Hammondia, Besnoitia, Sarcocystis* and other tissue cyst-forming coccidia of man and animals. In: J. P. Kreier (ed.), Parasitic Protozoa, pp. 102–238. Academic Press, New York.

Dürr, U., and Lammler, G. 1970. Prophylactic studies with sulfonamides in intestinal coccidiosis in rabbits. (In German) Zentralbl. Vet. Med. B. 17:554–563.

Eckman, M. K., Smith, P. E., and Clarke, W. E. 1974. Current practices in the utilization of anticoccidials in the broiler industry: Shuttle programs. Pract. Nutr. 8:1–6.

Edgar, S. A. 1970. Coccidiosis: Evaluations of coccidiostats under field conditions: Statement of problems. Exp. Parasitol. 28:90–94.

Edgar, S. A., Bond, D. S., and Seibold, C. T. 1961. Efficacy of several coccidiostatic drugs for the control of coccidiosis in chickens and turkeys. Poult. Sci. 40:(abstr.)1397.

Edgar, S. A., Coulston, F., and Waller, E. F. 1958. Trithiadol, a new medication for the chemotherapeutic and immunologic control of chicken coccidiosis. Proceedings of the 11th World's Poultry Congress, Mexico City. pp. 422–429.

Edgar, S. A., and Flanagan, C. 1968. Coccidiostatic effects of buquinolate in poultry. Poult. Sci. 47:95–104.

Edgar, S. A., and Flanagan, C. 1979. Efficacy of Stenerol (halofuginone): III. For the control of coccidiosis in turkeys. Poult. Sci. 58:1483–1489.

Ehrlich, P. 1909. Ueber den jetzigen Stand der Chemotherapie. Ber. Dtsch. Chem. Ges. 42:17–47.

Feed Additive Compendium. 1980. Miller Publishing Co., Minneapolis, MN.

Ferrando, R. 1977. Animal feed additive regulation in the common market: Rules, principles and ideas. Folia Vet. Lat. 7:183–197.

Fish, J. G., Jr. 1964. Sulfadimethoxine for enteric infections. Mod. Vet. Pract. 45:37–38.

Fish, J. G., Jr., Morgan, D. W., and Horton, C. R. 1965. Clinical experiences with sulfadimethoxine in small animal practice. Vet. Med. 60:1201–1206.

Fitzgerald, P. R. 1973. Efficacy of monensin or amprolium in the prevention of hepatic coccidiosis in rabbits. J. Protozool. 19:332–334.

Fitzgerald, P. R. 1975. The significance of bovine coccidiosis as a disease in the United States. Bovine Pract., November, 1975. pp. 28–33.

Fitzgerald, P. R., and Mansfield, M. E. 1973. Efficacy of monensin against bovine coccidiosis in young Holstein-Friesian calves. J. Protozool. 20:121–126.

Fitzgerald, P. R., and Mansfield, M. E. 1978. Ovine coccidiosis: Effect of the antibiotic monensin against *Eimeria ninakohlyakimovae* and other naturally occurring coccidia of sheep. Am. J. Vet. Res. 39:7–10.

Fitzgerald, P. R., and Mansfield, M. E. 1979. Efficacy of lasalocid against coccidia in cattle. J. Parasitol. 65:824–825.

Fitzsimmons, W. M. 1967. Amprolium as a coccidiostat for goats. Vet. Rec. 80:24–26.

Foreyt, W. J., Gates, N. L., and Wescott, R. B. 1979. Effects of lasalocid and monensin against experimentally induced coccidiosis in confinement-reared lambs from weaning to market weight. Am. J. Vet. Res. 40:97–100.

Foster, A. O., Christensen, J. F., and Haberman, R. T. 1941. Treatment of coccidial infections in lambs with sulfaguanidine. Proc. Helminth. Soc. Wash. 8:33–38.

Fouré, N., and Bennejean, G. 1974. A new prophylactic coccidiostat: Stenorol. Proceedings and Abstract of the 15th World's Poultry Congress and Exposition, New Orleans, August 11–16, 1974. pp. 92–94. (In French)

Frenkel, J. K. 1971. Toxoplasmosis: Mechanism of infection, laboratory diagnosis and management. Curr. Top. Pathol. 54:28–75.

Frigg, M., and Schramm, H. 1977. Comparative anticoccidial activity of lasalocid sodium (Avatec) in chicks: Efficacy against European strains of coccidia. Arch. Geflügelk. 41:31–34.

Gardiner, J. L., and Farr, M. M. 1954. Nitrofurazone for the prevention of experimentally induced *Eimeria tenella* infections in chickens. J. Parasitol. 40:42–49.

Gardiner, J. L., Farr, M. M., and Wehr, E. E. 1952. The coccidiostatic action of nitrophenide on *Eimeria tenella*. J. Parasitol. 38:517–524.

Gardiner, J. L., and McLoughlin, D. K. 1963. The comparative activity of certain coccidiostats in experimental *Eimeria tenella* infections. Poult. Sci. 42:932–935.

Greuel, Van E., Braunius, W. W., and Kühnhold, W. 1978. Zur Wirksamkeit verschiedener Coccidiostatica gegen Eimeria-Feldisolate. Arch. Geflügelk. 42:16–22.

Greuel, E., Kuil, H., and Robl, R. 1975. Synergism between metaclorpindol and methyl benzoquate against *E. acervulina*. Z. Parasitenkd. 46:163–165.

Greuel, Van E., and Raether, W. 1979. Kokzidiostatischer Effekt von Stenorol (Halofuginon) gegen die wichligsten *Eimeria*-Arten des Huhnes in Batterieversuch. Arch. Geflügelk 43:220–227.

Grumbles, L. C., Delaplane, J. P., and Higgins, T. C. 1948. Continuous feeding of low concentrations of sulfaquinoxaline for the control of coccidiosis in poultry. Poult. Sci. 27:605–608.

Gwyther, M. J. 1976. The efficacy of monensin and sulfaquinoxaline against the rabbit liver coccidium, *Eimeria stiedae*. Master's thesis, Clemson University, Clemson, SC.

Hagen, K. W., Jr. 1961. Hepatic coccidiosis in domestic rabbits treated with 2 nitrofuran compounds and sulfaquinoxaline. J. Am. Vet. Med. Assoc. 138:99–100.

Hammond, D. M., Clark, G. W., Miner, M. L., Trost, W. A., and Johnson, A. E. 1959. Treatment of experimental bovine coccidiosis with multiple small doses and single large doses of sulfamethazine and sulfabromomethazine. Am. J. Vet. Res. 20:708–713.

Hammond, D. M., Fayer, R., and Miner, M. L. 1966. Amprolium for control of experimental coccidiosis in cattle. Am. J. Vet. Res. 27:199–206.

Hammond, D. M., Ferguson, D. L., and Miner, M. L. 1960. Results of experiments with nitrofurazone and sulfamethazine for controlling coccidiosis in calves. Cornell Vet. 50:351–362.

Hammond, D. M., Kuta, J. E., and Miner, M. L. 1967. Amprolium for control of experimental coccidiosis in lambs. Cornell Vet. 57:611–623.

Hammond, D. M., Sayin, F., and Miner, M. L. 1965. Nitrofurazone as a prophylactic agent against experimental bovine coccidiosis. Am. J. Vet. Res. 26:83–89.

Hammond, D. M., Senger, C. M., Thorne, J. L., Shupe, J. L., Fitzgerald, P. R., and Johnson, A. E. 1958. Experience with nicarbazin in coc-

9

cidiosis *(Eimeria bovis)* in cattle. Cornell Vet. 48:260–268.

Hammond, D. M., Shupe, J. L., Johnson, A. E., Fitzgerald, P. R., and Thorne, J. L. 1956. Sulfaquinoxaline and sulfamerazine in the treatment of experimental infections with *Eimeria bovis* in calves. Am. J. Vet. Res. 17:463–470.

Harwood, P. D. 1956. Clinical applications of nitrofurans - past and present. Proceedings of the 1st National Symposium on Nitrofurans in Agriculture. September 28–29, 1956. pp. 12–23. Michigan State University, East Lansing.

Harwood, P. 1960. Recent scientific and regulatory developments on nitrofurans. Proceedings of the 3rd National Symposium on the Use of Nitrofurans in Agriculture, September 8–9. pp. 8–34. University of Kentucky, Lexington.

Harwood, P. D., and Stunz, D. I. 1953. A search for drug-fast strains of *Eimeria tenella.* J. Parasitol. 39:268–271.

Hawkins, P. A., Cole, C. L., and Thorp, F., Jr. 1943. The effects of sulfaguanidine and sulfasuxidine in a natural outbreak of ovine coccidiosis. Vet. Med. 38:337–339.

Hawkins, P. A., and Kline, E. E. 1945. The treatment of cecal coccidiosis with sulfamethazine. Poult. Sci. 24:277–281.

Herberg, R., Manthey, J., Richardson, L., Cooley, C., and Donoho, A. 1978. Excretion and tissue distribution of [14C]monensin in cattle. Agric. Food Chem. 26:1087–1090.

Hodgson, J. N. 1968. A new anticoccidial drug (M & B 15,497): Activity studies against laboratory and field strains of *Eimeria.* Br. Vet. J. 124:209–218.

Hodgson, J. N., Ball, S. J., Ryan, K. C., and Warren, E. W. 1969. The incidence of drug resistant strains of *Eimeria* in chickens in Great Britain, 1966. Br. Vet. J. 125:31–35.

Horak, I. G., Raymond, S. M., and Louw, J. F. 1969. The use of amprolium in the treatment of coccidiosis in domestic ruminants. J. S. Afr. Vet. Med. Assoc. 40:293–299.

Horton, G. M. J., and Stockdale, P. H. G. 1979. Effects of amprolium and monensin on oocyst discharge, feed utilization and rumen metabolism of lambs with coccidiosis. Am. J. Vet. Res. 40:966–970.

Horton-Smith, C. 1948. The effect of sulfamezathine on the second generation schizonts of *Eimeria tenella.* Trans. R. Soc. Trop. Med. Hyg. 42:11–12.

Horton-Smith, C. 1957. Additives for disease control in poultry and turkeys. Vet. Rec. 69:164–177.

Horton-Smith, C. 1958. Coccidiosis in domestic mammals. Vet. Rec. 70:256–262.

Horton-Smith, C., and Boyland, E. 1946. The treatment of cecal coccidiosis with sulfapyrazine. Poult. Sci. 25:390–391.

Horton-Smith, C., and Long, P. L. 1959a. The effects of different anticoccidial agents on the intestinal coccidioses of the fowl. J. Comp. Pathol. Therap. 69:192–207.

Horton-Smith, C., and Long, P. L. 1959b. The anticoccidial activity of glycarbylamide. Br. Vet. J. 115:55–62.

Horton-Smith, C., and Long, P. L. 1961. Effect of sulfonamide medication on the life cycle of *Eimeria meleagrimitis* in turkeys. Exp. Parasitol. 11:93–101.

Horton-Smith, C., and Long, P. L. 1965. The treatment of coccidial infections in fowls by a mixture of amprolium and sulfaquinoxaline in the drinking water. Vet. Rec. 77:586–591.

Horton-Smith, C., and Taylor, E. L., 1945. Sulfamethazine in the drinking water as a treatment for cecal coccidiosis in chickens. Vet. Rec. 57:35–36.

Itagaki, K., Tsubokura, M., and Otsuki, K. 1974. Studies on methods for evaluation of anticoccidial drugs *in vitro.* Jpn. J. Vet. Sci. 36:195–202.

Jeffers, T. K. 1974a. *Eimeria tenella:* Incidence, distribution and anticoccidial drug resistance of isolants in major broiler-producing areas. Avian Dis. 18:74–84.

Jeffers, T. K. 1974b. *Eimeria acervulina* and

E. maxima: Incidence and anticoccidial drug resistance of isolants in major broiler-producing areas. Avian Dis. 18:331–342.

Jeffers, T. K. 1974c. Anticoccidial drug resistance: Differences between *Eimeria acervulina* and *Eimeria tenella* strains within broiler houses. Poult. Sci. 53:1009–1013.

Jeffers, T. K. 1978. Sensitivity of recent field isolates of coccidia to monensin. Poult. Sci. 56:(abstr.)1725.

Jeffers, T. K., and Challey, J. R. 1973. Collateral sensitivity to 4-hydroxyquinolines in *Eimeria acervulina* strains resistant to metaclorpindol. J. Parasitol. 59:624–630.

Johnson, C. A. 1956. Studies on the efficacy of soluble Furacin against cecal coccidiosis. Poult. Sci. 35:1149–1150.

Johnson, C. A., and O'Conner, J. R. 1965. The anticoccidial activity of nihydrazone. Poult. Sci. 41:1654.

Johnson, C., and Van Ryzin, R. J. 1962. The mode of action of nitrofurazone and nihydrazone against cecal coccidiosis in chickens. Poult. Sci. 39:1263.

Jolley, W. R., Hammond, D. M., and Miner, M. L. 1971. Amprolium treatment of six- to twelve-month-old calves experimentally infected with coccidia. Proc. Helmint. Soc. Wash. 38:117–122.

Joyner, L. P. 1960. The coccidiostatic activity of 3,5-dinitro-orthotoluamide against *Eimeria tenella.* Res. Vet. Sci. 1:363–370.

Joyner, L. P., Davies, S. F. M., and Kendall, S. B. 1963. Chemotherapy of coccidiosis. In: R. J. Schnitzer and F. Hawking (eds.), Experimental Chemotherapy, Vol. 1, pp. 445–486. Academic Press, New York.

Joyner, L. P., and Kendall, S. B. 1956. Synergism in the chemotherapy of *Eimeria tenella.* Nature (Lond.) 176:975.

Joyner, L. P., and Norton, C. C. 1970. The response of recently isolated strains of *Eimeria meleagrimitis* to chemotherapy. Res. Vet. Sci. 11:349–353.

Joyner, L. P., and Norton, C. C. 1975. Robenidine dependence in a strain of *Eimeria maxima.* Parasitology 70:47–51.

Kantor, S., Kennett, R. L., Jr., Waletzky, E., and Tomcufcik, A. S. 1970. 1,3-Bis(*p*-chlorobenzylidine-amino) guanidine hydrochloride (robenzidine): New poultry anticoccidial agent. Science 168:373–374.

Kennett, R. L., Kantor, S., and Gallo, A. 1974. Efficacy studies with robenidine, a new type of anticoccidial, in the diet. Poult. Sci. 53:978–986.

Kilgore, R. L., Bramel, R. G., Brokken, E. S., and Miller, R. A. 1979. A comparison of methods for exposing chickens to coccidiosis in floor pen trials. Poult. Sci. 58:67–71.

Kilgore, R. L., Bramel, R. G., Brokken, E. S., Olson, G., Cox, J. L., and Leaning, W. H. D. 1978. Efficacy of arprinocid (MK-302) against *Eimeria* species in broilers. Poult. Sci. 57:907–911.

Kinashi, H., Otake, N., Yonehara, H., Sato, S., and Saito, T. 1973. The structure of salinomycin, a new member of the polyether antibiotics. Tetrahedron Lett. 49:4955–4958.

Kingston, D. J. 1977. The influence of monensin sodium upon feather development of male chickens. Aust. Vet. J. 53:251–252.

Knight, R. G. 1962. Chemotherapeutic and antibody treatment of canine coccidiosis. Vet. Med. 57:52–53.

Kowalski, L. M., and Reid, W. M. 1972. Roxarsone: Efficacy against *Eimeria brunetti* infections in chickens. Poult. Sci. 51:1586–1589.

Kowalski, L. M., and Reid, W. M. 1975. Effects of roxarsone on pigmentation and coccidiosis in broilers. Poult. Sci. 54:1544–1549.

Latter, V. S. 1974. An improved *in vitro* chemotherapeutic screen for intracellular protozoa *(Eimeria tenella* and *Trypanosoma cruzi).* (Br. Soc. Parasitol. Proc.) Parasitology 69:xxii–xxiii.

Latter, V. S., and Wilson, R. G. 1979. Factors influencing the assessment of anticoccidial activity in cell culture. Parasitology 79:169–175.

Lee, E. H., and Fernando, M. A. 1977. Drug resistance in coccidia: A robenidine-resistant strain of *Eimeria tenella.* Can. J. Comp. Med. 41:466–470.

Lee, D. L., and Millard, B. J. 1972. Fine structural changes in *Eimeria tenella* from infections in chick embryos and chickens, after exposure to the anticoccidial drug robenidine. Parasitology 65:309–316.

Leek, R. G., Fayer, R., and McLoughlin, D. K. 1976. Effect of monensin on experimental infections of *Eimeria ninakohlyakimovae* in lambs. Am. J. Vet. Res. 37:339–341.

Levine, P. P. 1939. The effect of sulfanilamide on the course of experimental avian coccidiosis. Cornell Vet. 29:309–320.

Long, P. L. 1965. Development of *Eimeria tenella* in avian embryos. Nature 208:509–510.

Long, P. L. 1970. *Eimeria tenella:* Chemotherapeutic studies in chick embryos with a description of a new method (chorioallantoic membrane foci counts) for evaluating infection. Z. Parasitenkd. 33:329–338.

Long, P. L., and Millard, B. J. 1967. The effect of meticlorpindol on *Eimeria* infections of the fowl. Vet. Rec. 81:11–15.

Long, P. L., and Millard, B. J. 1968. *Eimeria:* Effect of metaclorpindol and methyl benzoquate on endogenous stages in the chicken. Exp. Parasitol. 23:331–338.

Long, P. L., and Millard, B. J. 1973. Eimeria infection of chicken embryos: The effect of known anticoccidial drugs against *E. tenella* and *E. mivati.* Avian Pathol. 2:111–125.

Long, P. L., Millard, B. J., and Smith, K. M. 1979. The effect of some anticoccidial drugs on the development of immunity to coccidiosis in field and laboratory conditions. Avian Pathol. 8:453–467.

Lucas, J. M. S. 1958. The effect of nicarbazin on growth rate, sexual maturity, egg production, fertility and hatchability. J. Comp. Pathol. Therap. 68:300–307.

Lund, E. E. 1954. The effect of sulfaquinoxaline on the course of *Eimeria stiedae* infections in the domestic rabbit. Exp. Parasitol. 3:497–503.

Lux, R. E. 1954. The chemotherapy of *Eimeria tenella:* 1. Diaminopyrimidines and dihydrotriazines. Antibiot. Chemother. 4:971–977.

McDougald, L. R. 1973. Practical applications of the coccidiosis *in vitro* technique. Proc. Symposium International Sur Les Coccidioses. Tours, France, Sept. 11–12.

McDougald, L. R. 1976. Anticoccidial action of monensin in turkey poults. Poult. Sci. 55:2442–2447.

McDougald, L. R. 1978a. The growth of avian *Eimeria* in vitro. In: P. L. Long, K. N. Boorman, and B. M. Freeman (eds.), Avian Coccidiosis, pp. 135–184. British Poultry Science, Ltd., Edinburgh.

McDougald, L. R. 1978b. Coccidiosis control with ionophorous antibiotics. Proceedings of the XVIth World's Poultry Congress, Rio de Janeiro, September 17–21. pp. 875–885.

McDougald, L. R. 1978c. Monensin for the prevention of coccidiosis in calves. Am. J. Vet. Res. 39:1748–1749.

McDougald, L. R. 1979. Efficacy and compatibility of amprolium and carbarsone against coccidiosis and blackhead in turkeys. Poult. Sci. 58:76–80.

McDougald, L. R. 1980. Compensatory growth in broiler chickens as related to the use of ionophorous antibiotic anticoccidial drugs. Proceedings of the 1980 Georgia Nutrition Conference, University of Georgia, Athens, GA, pp. 11–16.

McDougald, L. R., and Dunn, W. C. 1978. Efficacy of monensin against coccidiosis in lambs. Am. J. Vet. Res. 39:1459–1462.

McDougald, L. R., and Galloway, R. B. 1973. *Eimeria tenella:* Anticoccidial drug activity in cell cultures. Exp. Parasitol. 34:189–196.

McDougald, L. R., and Galloway, R. B. 1976. Anticoccidial drugs: Effects on infectivity and survival intracellularly of *Eimeria tenella* sporozoites. Exp. Parasitol. 40:314–319.

McDougald, L. R., and Galloway, R. B. 1977. *Eimeria tenella:* Inhibition of development in cell culture by serum from chickens fed anticoc-

cidial drugs. Z. Parasitenkd. 54:95–100.
McDougald, L. R., and Johnson, J. K. 1979. Floor pen studies on the anticoccidial efficacy of arprinocid in turkeys. Poult. Sci. 58:72–75.
McDougald, L. R., and McQuistion, T. E. 1978. Innate and acquired immunity *vs.* anticoccidial medication in managing coccidiosis in turkeys. Avian Dis. 22:765–770.
McDougald, L. R., and McQuistion, T. E. 1980a. Compensatory growth in broilers after withdrawal of anticoccidial drugs. Poult. Sci. 59:1001–1005.
McDougald, L. R., and McQuistion, T. E. 1980b. Mortality from heat stress in broiler chickens influenced by anticoccidial drugs. Poult. Sci. 59:2421–2423.
McDougald, L. R., Reid, W. M., Taylor, E. M., and Mabon, J. L. 1972. Effects of anticoccidial and growth promoting agents on intestinal motility in broilers. Poult. Sci. 51:416–418.
McLoughlin, D. K., and Chute, B. 1971. Efficacy of decoquinate against eleven strains of *Eimeria tenella* and development of a decoquinate resistant strain. Avian Dis. 15:425–429.
McLoughlin, D. K., and Chute, M. B. 1974. The efficacy of monensin against one sensitive and thirteen drug resistant strains of *Eimeria tenella.* Poult. Sci. 53:770–772.
McLoughlin, D. K., and Chute, M. B. 1975. Sequential use of coccidiostats effect on development by *Eimeria tenella* of resistance to amprolium, nicarbazin, Unistat and zoalene. Avian Dis. 19:424–433.
McLoughlin, D. K., and Gardiner, J. L. 1962. Drug resistance in *Eimeria tenella:* II. The experimental development of a zoalene-resistant strain. J. Parasitol. 48:341–346.
McLoughlin, D. K., Gardiner, J. L., and Chester, D. K. 1960. The activity of glycarbylamide, trithiadol and nicarbazin against *Eimeria tenella* in chickens. Poult. Sci. 39:1328–1332.
McLoughlin, D. K., and Wehr, E. E. 1960. Stages in the life cycle of *Eimeria tenella* affected by nicarbazin. Poult. Sci. 39:534–538.
McManus, E. C. 1970. The efficacy of nicarbazin against *Eimeria mivati, E. hagani* and *E. praecox.* Res. Vet. Sci. 11:101–102.
McManus, E. C., Campbell, W. C., and Cuckler, A. C. 1968. Development of resistance to quinolone coccidiostats under field and laboratory conditions. J. Parasitol. 54:1190–1193.
McManus, E. C., Oberdick, M. T., and Cuckler, A. C. 1967. Response of six strains of *Eimeria brunetti* to two antagonists of para-aminobenzoic acid. J. Protozool. 14:379–381.
McQuistion, T. E., and McDougald, L. R. 1979. *Eimeria tenella:* Anticoccidial action of drugs in birds with surgically closed ceca. Z. Parasitenkd. 59:107–113.
McQuistion, T. E., and McDougald, L. R. Studies on the anticoccidial action of arprinocid and halofuginone against *Eimeria* spp. (Sporozoa: Coccidia) of the chicken. Vet. Parasitol. In press,a.
McQuistion, T. E., and McDougald, L. R. The effect of combining subtherapeutic concentrations of different ionophorous antibiotics on anticoccidial action in chickens. J. Comp. Pathol. In press,b.
Marlow, C. H. B. 1968. Amprolium as a coccidiostat for Angora goats. J. S. Afr. Vet. Med. Assoc. 39:93.
Marusich, W. L., Ogrinz, E. F., Camerlengo, N., and Mitrovic, M. 1977. Effect of diet on the performance of broiler chickens fed lasalocid in combination with growth promotants. Poult. Sci. 56:1297–1304.
Mathis, G., Johnson, J., and McDougald, L. R. 1979. Drug resistance of field isolates of coccidia. Poult. Sci. 58:(abstr.)1083.
Meingassner, J. G., Schmook, F. P., Czok, R., and Mieth, H. 1979. Enhancement of the anticoccidial activity of polyether antibiotics in chickens by tiamulin. Poult. Sci. 58:308–313.
The Merck Index. 1976. M. Windholz (ed.), 9th Ed. pp. 1–1313 and Appendix. Merck & Co., Rahway, NJ.
Migaki, T. T., and Babcock, W. E. 1979. Safety

evaluation of salinomycin in broiler chickens reared in floor pens. Poult. Sci. 58:481–482.

Miller, B. E., McManus, E. C., Solson, G., Schlein, K. D., Van Iderstine, A. A., Graham, D. W., Brown, J. E., and Rogers, E. F. 1977. Anticoccidial and tolerance studies in the chicken with two 6-amino-9-(substituted benzyl) purines. Poult. Sci. 56:2039–2044.

Minchinton, I. R., Jones, D. L., and Sang, J. P. L. 1973. Poultry manure phytotoxicity. J. Sci. Food Agric. 24:1437–1448.

Miner, M. L., and Jensen, J. B. 1976. Decoquinate in the control of experimentally induced coccidiosis of calves. Am. J. Vet. Res. 37:1043–1045.

Mitrovic, M. 1968. Sulfadimethoxine in prevention of turkey coccidiosis. Poult. Sci. 47:314–319.

Mitrovic, M., and Schildknecht, E. G. 1974. Anticoccidial activity of lasalocid (X-537A) in chicks. Poult. Sci. 53:1448–1455.

Mitrovic, M., and Schildknecht, E. G. 1975. Lasalocid: Resistance and cross-resistance studies in *Eimeria tenella*-infected chicks. Poult. Sci. 54:750–756.

Mitrovic, M., and Schildknecht, E. 1976. Anticoccidial activity of lasalocid and other anticoccidials against recent field isolates in chicks. Poult. Sci. 55:(abstr.)2068.

Mitrovic, M., Schildknecht, E. G., and Fusiek, G. 1969. Anticoccidial activity of sulfadimethoxine potentiated mixture (RO-5-0013) in chickens. Poult. Sci. 48:210–216.

Mitrovic, M., Schildknecht, E. G., and Fusiek, G. 1970. Anticoccidial activity of sulfamethoxine potentiated mixture (Rofenaid) in turkeys. Poult. Sci. 49:(abstr.)56.

Mitrovic, M., Schildknecht, E. G., and Marusich, W. L. 1975. Comparative anticoccidial activity and compatibility of lasalocid in broiler chickens. Poult. Sci. 54:757–761.

Mitrovic, M., Schildknecht, E. G., and Marusich, W. L. 1979. Lasalocid in prevention of turkey coccidiosis. Poult. Sci. 58:1154–1159.

Mitrovic, M., Schildknecht, E., and Trainor, C. 1977. Effects of lasalocid and monensin in combination with roxarsone on lesion reduction and oocyst suppression in chicks infected with *Eimeria tenella* field isolates. Poult. Sci. 56:979–984.

Miyazaki, Y., Shibuya, M., Sugawara, H., Kawaguchi, O., Hirose, C., Nagatsu, J., and Esumi, S. 1974. Salinomycin, a new polyether antibiotic. J. Antibiot. 27:814–821.

Moore, E. N. 1954. Species of coccidia affecting turkeys. Proceedings of the 91st Annual Meeting of the American Veterinary Medical Association, August, 23–26, pp. 300–304.

Morehouse, N. F., and Baron, R. R. 1970. Evaluation of coccidiostats by mortality, weight gains and fecal scores. Exp. Parasitol. 28:25–29.

Morehouse, N. F., and McCay, F. 1951. On the chemotherapeutic action of 3-nitro-4-hydroxyphenyl-arsonic acid against the coccidium *Eimeria tenella* in chickens. Iowa Acad. Sci. 58:507–516.

Morehouse, N. F., and McGuire, W. C. 1959. The use of 3,5-dinitrobenzamide and its N-substituted derivatives against coccidiosis in chickens. Poult. Sci. 38:410–423.

Morrison, W. D., Ferguson, A. E., Connell, M. C., and McGregor, J. K. 1961. The efficacy of certain coccidiostats against mixed avian coccidial infections. Avian Dis. 5:222–228.

Morrison, W. D., Ferguson, A. E., Connell, M. C., and McGregor, J. K. 1967. Efficacy of various drugs for the prevention of experimentally induced coccidiosis in chickens. Poult. Sci. 46:391–396.

Morrison, W. D., Ferguson, A. E., and Leeson, S. 1979. Efficacy of salinomycin and stenorol against various species of *Eimeria* and effect on chick performance. Poult. Sci. 58:1160–1166.

Nakata, T., Schmid, G., Vranesic, B., Okigawa, M., Smith-Palmer, T., and Kishi, Y. 1978. A total synthesis of lasalocid A. J. Am. Chem. Soc. 100:2933–2935.

Ninet, L., Benazet, F., Depaire, H., Florent, J., Lunel, J., Mancy, D., Abraham, A., Cartier, J. R., DeChezelles, N., Godard, C., Moreau,

M., Tissier, R., and Lallemand, J. Y. 1976. Emericid, a new polyether antibiotic from *Streptomyces hygroscopicus* (DS 24367). Experientia 32:219–321.

Norton, C. C., and Joyner, L. P. 1968. Coccidiostatic activity of meticlorpindol: Effectiveness against single infections of five species of *Eimeria* in the domestic fowl. Vet. Rec. 83:317–323.

Ohshima, M., Ishizaki, N., Abe, K., Ukawa, M., Marumoto, Y., Navatsuka, K., Horiuchi, T., Tonooka, Y., Yoshino, S., and Kanda, N. 1976. Antibiotic DE-3936, a polyether antibiotic identical with lonomycin: Toxonomy, fermentation, isolation and characterization. J. Antibiot. 29:354–365.

Olson, G., Tamas, T., Smith, D. A., Weppelman, R. M., Schleim, K., and McManus, E. C. 1978. Battery efficacy studies with arprinocid against field strains of coccidia. Poult. Sci. 57:1245–1250.

Orton, C. T., and Hambly, L. R. 1971. Efficacy studies on potentiated sulfadimethoxine as a chicken coccidiostat. Poult. Sci. 50:1341–1346.

Otake, N., Koenuma, M., Kinashi, H., Sato, S., and Saito, Y. 1975. The crystal and molecular structure of the silver salt of lysocellin, a new polyether antibiotic. J. Chem. Soc. Chem. Commun. pp. 92–93.

Ott, W. H., Kuna, S., Porter, C. C., Cuckler, A. C., and Fogg, D. E. 1956. Biological studies on nicarbazin, a new anticoccidial agent. Poult. Sci. 35:1355–1367.

Patton, W. H. 1965. *Eimeria tenella:* Cultivation of the asexual stages in cultured animal cells. Science 150:767–769.

Paul, H. E. 1956. Research background on the nitrofurans. Proceedings of the 1st National Symposium on Nitrofurans in Agriculture. September 28–29, pp. 6–11. Michigan State University, East Lansing.

Peardon, D. L., Bilkovich, F. R., and Todd, A. C. 1963. Trials of candidate bovine coccidiostats. Am. J. Vet. Res. 24:743–748.

Peardon, D. L., Bilkovich, F. R., Todd, A. C., and Hoyt, H. H. 1965. Trials of candidate bovine coccidiostats. Efficacy of amprolium, lincomycin, sulfamethazine, chloroquine sulfate and di-phenthane-70. Am. J. Vet. Res. 26:683–687.

Peeters, J. E., Halen, P., and Meulemans, G. 1979. Efficacy of robenidine in the prevention of rabbit coccidiosis. Br. Vet. J. 135:349–354.

Pellerdy, L. P. 1969. Problems of rabbits: Coccidiosis. Parasitol. Hung. 2:175–186.

Peterson, E. H. 1950. The prophylaxis and therapy of hepatic coccidiosis in the rabbit by the administration of sulfonamides. Vet. Med. 45: 170–172.

Peterson, E. H. 1960. A study of anticoccidial drugs against experimental infections with *Eimeria tenella* and *necatrix.* Poult. Sci. 39:739–745.

Peterson, E. H., and Hymas, T. A. 1950. Sulfaquinoxaline, nitrofurazone and nitrophenide in the prophylaxis of experimental *Eimeria necatrix* infection. Am. J. Vet. Res. 11:278–283.

Pinkerton, M., and Steinrauf, L. K. 1970. Molecular structure of monovalent metal cation complexes of monensin. J. Mol. Biol. 49:533–546.

Polin, D., Ott, W. H., and Zeissig, A. 1958. Field studies on the effect of nicarbazin on egg quality. Poult. Sci. 37:898–909.

Pressman, B. C. 1976. Biological application of ionophores. Annu. Rev. Biochem. 45:501–530.

Rachman, M., and Pallock, S. 1961. Treatment of canine coccidiosis. Vet. Med. 56:75–76.

Raines, T. V. 1978. Guidelines for the evaluation of anticoccidial drugs. In: P. L. Long, K. N. Boorman, and B. M. Freeman (eds.), Avian Coccidiosis pp. 339–346. 13th Poultry Science Symposium British Poultry Science, Ltd., Edinburgh.

Reddy, M. S., Reddy, C. V., and Reddy, K. R. 1971. Effect of coccidiostats and antibiotics upon artificially induced coccidiosis. Indian J. Poult. Sci. 6:17–22.

Reid, W. M. 1972. Anticoccidials used in the poultry industry: Time of action against the coccidial life cycle. Folia Vet. Lat. 2:641–667.

Reid, W. M. 1973. Anticoccidials: Differences in day of peak activity against *Eimeria tenella.* Proceedings of the Symposium on Coccidia and Related Organisms, pp. 119–134. University of Guelph, Guelph, Ontario.

Reid, W. M., Anderson, W. A., and McDougald, L. R. 1978. Anticoccidial protection and development of immunity to turkey coccidiosis while using monensin. Avian Pathol. 7:569–576.

Reid, W. M., and Brewer, R. N. 1967. Efficacy studies on metaclorpindol as a coccidiostat. Poult. Sci. 46:638–642.

Reid, W. M., Dick, J., Rice, J., and Stino, F. 1977. Effects of monensin-feeding regimens on flock immunity to coccidiosis. Poult. Sci. 56: 66–71.

Reid, W. M., Johnson, J., and Dick, J. W. 1975. Anticoccidial activity of lasalocid in control of moderate and severe coccidiosis. Avian Dis. 19: 12–18.

Reid, W. M., Kowalski, L. M., and Rice, J. 1972. Anticoccidial activity of monensin in floor pen experiments. Poult. Sci. 51:139–146.

Reid, W. M., Kowalski, L. M., Taylor, E. M., and Johnson, J. 1970. Efficacy evaluations of robenzidine for control of coccidiosis in chickens. Avian Dis. 14:788–796.

Reid, W. M., Taylor, E. M., and Johnson, J. K. 1969. A technique for demonstration of coccidiostatic activity of anticoccidial agents. Trans. Am. Microsc. Soc. 88:148–159.

Rogers, E. F. 1962. Thiamine antagonist. Ann. N. Y. Acad. Sci. 98:412–429.

Rogers, E. F., Clark, R. L., Becker, H. J., Pessolano, A. A., Leanza, W. J., McManus, E. C., Andriuli, F. J., and Cuckler, A. C. 1964. Antiparasitic activity of 4-amino-2-theoxybenzoic acid and related compounds. Proc. Soc. Exp. Biol. Med. 117:488–492.

Ross, D. B. 1968. Successful treatment of coccidiosis in lambs. Vet. Rec. 83:189–190.

Ruff, M. D., Reid, W. M., Dykstra, D. D., and Johnson, J. K. 1978. Efficacy of arprinocid against coccidiosis of broilers in battery and floor pen trials. Avian Dis. 22:32–41.

Ruff, M. D., Reid, W. M., Johnson, J. K., and Anderson, W. A. 1979. Anticoccidial activity of narasin in battery raised broiler chickens. Poult. Sci. 58:298–303.

Ruff, M. D., Reid, W. M., and Rahn, A. P. 1976. Efficacy of different feeding levels of monensin in the control of coccidiosis in broilers. Am. J. Vet. Res. 37:963–967.

Ruff, M. D., Reid, W. M., Rahn, A. P., and McDougald, L. R. Anticoccidial activity of narasin in broiler chickens reared in floor pens. Poult. Sci. 59:2008–2013.

Ryley, J. F. 1967a. Methyl benzoquate, a new wide-spectrum coccidiostat for chickens. Br. Vet. J. 123:513–520.

Ryley, J. F. 1967b. Studies on the mode of action of quinolone and pyridone coccidiostats. J. Parasitol. 53:1151–1160.

Ryley, J. F. 1968. Chick embryo infections for the evaluation of anticoccidial drugs. Parasitology 58:215–220.

Ryley, J. F. 1975. Lerbek, a synergistic mixture of methyl benzoquate and clopidol for the prevention of chicken coccidiosis. Parasitology 70:377–384.

Ryley, J. F., and Betts, M. J. 1973. Chemotherapy of chicken coccidiosis. Adv. Pharmacol. Chemother. 2:221–293.

Ryley, J. F., and Wilson, R. G. 1971. Studies on the mode of action of the coccidiostat robenidine. Z. Parasitenkd. 37:85–93.

Ryley, J. F., and Wilson, R. G. 1972a. Comparative studies with anticoccidials and three species of chicken coccidia *in vivo* and *in vitro.* J. Parasitol. 58:664–668.

Ryley, J. F., and Wilson, R. G. 1972b. Growth factor antagonism studies with coccidia in tissue culture. Z. Parasitenkd. 40:31–34.

Ryley, J. F., and Wilson, R. G. 1974. Anticoccidial activity of an azauracil derivative. Parasitology 68:68–69.

Ryley, J. F., and Wilson, R. G. 1976. Drug screening in cell culture for the detection of anticoccidial activity. Parasitology 73:137–148.

Sangster, L. T., Stuart, B. B., Williams, D. J., and Bedell, D. M. 1978. Coccidiosis associated with scours in baby pigs. Vet. Med./Small Amin. Clin. pp. 1317–1319. (Oct.).

Sevcik, B., Jurkovic, P., Bedrnik, P., and Firmanova, A. 1980. Use of clopidol in prevention of calf coccidiosis. Proceedings of the Symposium on Coccidia and Further Prospects of Their Control. November 28–30, 1979. Prague.

Schmid, G., Fukuyama, T., Akasaka, K., and Kishi, Y. 1979. Total synthesis of monensin: 1. Stereocontrolled synthesis of the left half of monensin. J. Am. Chem. Soc. 101:259–260.

Shumard, R. F. 1959a. Experimentally induced ovine coccidiosis: I. Use of nitrofurazone in the feed. Vet. Med. 54:421–425.

Shumard, R. F. 1959b. Experimentally induced ovine coccidiosis: II. Use of water soluble nitrofurazone as a therapeutic. Vet. Med. 54:477–479.

Shumard, R. F. 1965. Therapeutic effectiveness of tylosin in experimental coccidia infections in chickens. J. Parasitol. 51:(abstr.)54.

Shumard, R. F., and Callender, M. E. 1967. Monensin, a new biologically active compound: VI. Anticoccidial activity. Antimicrob. Agents Chemother. 1968. 369–377.

Shumard, R. F., and Callender, M. E. 1970. Anticoccidial drugs: Screening methods. Exp. Parasitol. 28:13–24.

Slater, R. L., Hammond, D. M., and Miner, M. L. 1970. *Eimeria bovis:* Development in calves treated with thiamine metabolic antagonist (amprolium) in feed. Trans. Am. Microsc. Soc. 89:55–65.

Smith, C. K., III, and Strout, R. G. 1979. *Eimeria tenella:* Accumulation and retention of anticoccidial ionophores by extracellular sporozoites. Exp. Parasitol. 48:325–330.

Stock, B. L., Stevenson, G. T., and Hymas, T. A. 1967. Coyden coccidiostat for control of coccidiosis in chickens. Poult. Sci. 46:485–492.

Strout, R. G., and Ouellette, C. A. 1973. *Eimeria tenella:* Screening of chemotherapeutic compounds in cell cultures. Exp. Parasitol. 33:477–485.

Strout, R. G., Solis, J., Smith, S. C., and Dunlop, W. R. 1965. *In vitro* cultivation of *Eimeria acervulina* (Coccidia). Exp. Parasitol. 17:241–246.

Stuart, E. E., Bruins, H. W., and Keenum, R. D. 1963. The immunogenicity of a commercial coccidiosis vaccine in conjunction with trithiadol and zoalene. Avian Dis. 12:12–18.

Swanson, L. E., Stone, W. M., Jr., and Dennis, W. R. 1954. Control of internal parasites of cattle. Annual Report of the Agricultural Experiment Station, University of Florida, June 30, pp. 140–141.

Tamas, T., Olson, G., Smith, D. A., and Miller, B. M. 1978. Effect of 6-amino-9-(substituted benzyl) purines on oocyst sporulation. Poult. Sci. 57:386–391.

Tarlatzis, C., Panetsos, A., and Dragonas, P. 1955. Furacin in the treatment of ovine and caprine coccidiosis. J. Am. Vet. Med. Assoc. 126:391–392.

Tüller, Van R., and Mödder, R. 1979. Salinomycin in Futter für Jungmasthukner. Arch. Geflugelk. 43:192–194.

Waletzky, E., Hughes, C. O., and Brandt, M. C. 1949. The anticoccidial activity of nitrophenide. Ann. N. Y. Acad. Sci. 52:543–557.

Waletzky, E., Neal, R., and Hable, I. 1954. A field strain of *Eimeria tenella* resistant to sulfonamides. J. Parasitol. 40(suppl.):24. (abstr.).

Wang, C. C. 1975. Studies of the mitochondria from *Eimeria tenella* and inhibition of electron transport by quinolone coccidiostats. Biochim. Biophys. Acta 396:210–219.

Wang, C. C. 1976. Inhibition of the respiration of *Eimeria tenella* by quinolone coccidiostats. Biochem. Pharmacol. 25:343–349.

Wang, C. C. 1978. Biochemical and nutritional aspects of coccidia. In: P. L. Long, K. N. Boor-

9

man, and B. M. Freeman (eds.), Avian Coccidiosis, pp. 135–184. British Poultry Science, Ltd., Edinburgh.

Wang, C. C., and Simashkevich, P. M. 1980. A comparative study of the biological activities of arprinocid and arprinocid-1-*N*-oxide. Biochem. Parasitol. 1:335–345.

Wang, C. C., Simashkevich, P. M., and Fan, S. S. 1981. The mechanism of anticoccidial action of arprinocid-1-*N*-oxide. J. Parasitol. 67:137–149.

Wang, C. C., Simashkevich, P. M., and Stotish, R. L. 1979a. Mode of anticoccidial action of arprinocid. Biochem. Pharmacol. 28:2241–2248.

Wang, C. C., Stotish, R. L., and Poe, M. 1975. Dihydrofolate reductase from *Eimeria tenella:* Rationalization of chemotherapeutic efficacy of pyrimethamine. J. Protozool. 22:564–568.

Wang, C. C., Tolman, R. L., Simashkevich, P. M., and Stotish, R. L. 1979b. Arprinocid, an inhibitor of hypoxanthine-guanine transport. Biochem. Pharmacol. 28:2249–2260.

Warren, E. W., Ball, S. J., and Fagg, J. R. 1963. Age resistance by turkeys to *Eimeria meleagrimitis* Tyzzer, 1929. Nature 200:238–240.

Warren, E. W., Ball, S. J., and MacKenzie, D. R. 1966. The incidence of drug-resistant strains of *Eimeria* species in chickens in Great Britain 1964/65. Br. Vet. J. 122:534–543.

Weppelman, R. M., Olson, G., Smith, D. A., Tamas, T., and Van Iderstine, A. 1977. Comparison of anticoccidial efficacy, resistance, and tolerance of narasin, monensin and lasalocid in chicken battery trials. Poult. Sci. 56:1550–1559.

White-Stevens, R., Zeibel, H. G., and Smith, F. 1955. The effects of antibiotic feeding in relation to coccidiostatic drugs in broiler production. Poult. Sci. 34:1227–1228.

Whitlock, J. H. 1974. An experimental basis for environmental medicine. Perspect. Biol. Med. 17:455–481.

Wong, D. T., Berg, D. H., Hamill, R. L., and Wilkinson, J. R. 1977. The ionophorous properties of narasin, a new polyether monocarboxylic acid antibiotic, in rat liver mitochondria. Biochem. Pharmacol. 26:1373–1376.

Wong, D. T., Horng, J. S., and Wilkinson, J. R. 1972. Robenzidine, an inhibitor of oxidative phosphorylation. Biochem. Biophys. Res. Commun. 46:621–627.

Yvore, P. 1969. Possibilities and limits of anticoccidial prevention with pyrimidine and quinolone derivatives. Acta Vet. (Brno.) 38:119–127.

Yvore, P., and Aycardi, J. 1968. Comparative efficacy of new coccidiostats against *Eimeria tenella* and *Eimeria acervulina* (French strains). (In French) Rech. Vet. 1:167–191.

Yvore, P., Foure, N., Aycardi, J., and Bennejean, G. 1974. The efficacy of stenorol (RU19110) in chemoprophylaxis of avian coccidiosis. Recl. Med. Vet. Ec. Alfort 150:495–503.

10

Anticoccidial Drug Resistance

H. David Chapman

Wherever drugs have been used widely by man, the problem of drug resistance has been encountered. The quotation from Schnitzer and Grunberg (1957)—"drug resistance has accompanied the development of chemotherapy like a faithful shadow and the history of chemotherapy is also a history of drug resistance"—is nowhere more true than in the poultry industry which, in the post-war period, has been dependent upon drugs for the control of coccidiosis. The practice of prophylactic medication in which drugs are included in the ration of the broiler chicken from 1-day-old until slaughter has inevitably resulted in widespread drug resistance. In such an environment, the parasite is exposed throughout its life cycle to agents designed to promote its demise.

In this chapter, drug resistance is reviewed. Previous reviews include those of Cuckler et al. (1969), Joyner, (1970), McLoughlin (1970a), Ryley and Betts (1973), Chapman (1978a), and Ryley (1980).

DEFINITION OF DRUG RESISTANCE

Bryson and Szybalski (1955) define resistance as "the temporary or permanent capacity of a cell and its progeny to remain viable or multiply under environmental conditions that would destroy or inhibit other cells." In the light of our knowledge of the mode of action of anticoccidial drugs, this definition would have to be modified. Thus, certain anticoccidials have a coccidiostatic rather than a coccidiocidal effect. The parasite remains viable in the presence of a coccidiostatic drug and upon drug withdrawal is capable of continuing its life cycle. Such a parasite would not be regarded as drug-resistant. Completion of the life cycle and production of infective oocysts in the presence of a drug would be a requirement for a definition of resistance in *Eimeria.*

In 1963, a special committee of the World Health Organization defined resistance as follows: "Ability of a parasite strain to multiply or to survive in the presence of concentrations of a drug that normally destroy parasites of the same species or prevent their multiplication. Such resistance may be relative (yielding to increased doses of the drug tolerated by the host) or complete (withstanding maximum doses tolerated by the host)." (Quoted by Peters, 1970.) Drug failure was defined as: "Absence or insufficiency of drug action after administration of a normally effective dose. It is important to discriminate between such causes of drug failure as deficient absorption, unusual rate of degradation or excretion of the drug, and resistance of the parasite."

This definition draws attention to the fact that drug resistance is merely one cause of drug failure. Ryley and Betts (1973) consider that a distinction should be made between limited efficacy and a change in sensitivity resulting from exposure to a particular drug. Jeffers (1974a,b),

10

for example, has attributed the failure of Amprol Plus and Zoamix to control *Eimeria acervulina* and *Eimeria maxima,* and Nidrafur to control *Eimeria tenella,* to limited drug efficacy rather than drug resistance. Difficulty is encountered, however, in deciding what constitutes "limited efficacy." With most older anticoccidials and even with recent drugs such as the ionophores, the commercially recommended level is a compromise between activity against the parasite and toxicity toward the host. Such drugs do not completely suppress the development of strains of *Eimeria* that have had no prior exposure to the drug in question. The sensitivity of coccidia to all anticoccidials depends upon the dose of the drug; complete suppression of parasite development may or may not be found at the concentration used in current practice. To demonstrate the development of drug resistance, it is necessary, therefore, to show a change in the parasite by comparing sensitivity before and after exposure to the anticoccidial.

STRAIN VARIATION IN SENSITIVITY TO ANTICOCCIDIAL DRUGS

The effect of drugs upon strains within a single species may vary, irrespective of whether the strains have been exposed to the drug (Ryley and Betts, 1973). McManus et al. (1967) found differences in the effect of ethopabate upon six strains of *Eimeria brunetti.* Only three of these strains were from countries where ethopabate has not been used, but these isolates were the most insensitive to ethopabate. In order to demonstrate intrinsic variation in sensitivity, it is essential that the history of the strain be known. Therefore, variation in the effect of older drugs may be difficult to prove. Long (1963), for instance, showed a difference in the sensitivity of two strains of *Eimeria necatrix* to the sulfonamides. However, it was possible that the variation was attributable to resistance acquired before the strains were isolated.

Even with knowledge of the history of the strain, proof of the absence of exposure to the drug may be difficult to obtain. Chapman (1975) found that, whereas the Weybridge and Elberfeld strains of *E. tenella* developed resistance to methyl benzoquate after six passages, and the Beltsville after five, the Houghton strain of this species developed resistance after a single passage. A culture of the Houghton strain that had been frozen in liquid nitrogen since 1969 took eight passages to develop resistance. Chapman (1975) suggested that drug tolerance had been acquired by the Houghton strain since 1969, probably a result of accidental contamination with resistant parasites. Shirley et al. (1977) showed that *E. acervulina* var. *diminuta* from the jungle fowl was more sensitive than *E. acervulina* var. *mivati* to amprolium and sulfaquinoxaline. This difference in sensitivity may represent intrinsic variation. Alternatively, the reduced sensitivity of *E. acervulina* var. *mivati* may have been due to exposure of the parasite to drugs, or to contamination with resistant organisms.

According to Jeffers (1978a), naturally occurring intraspecific variation to anticoccidial drugs among strains allegedly free of exposure is much less than that resulting from laboratory selection for drug resistance. In order to detect such differences, the response of strains to varying doses of drug should be compared. Few such studies have been carried out.

EXPERIMENTAL DEVELOPMENT OF DRUG RESISTANCE

Laboratory Methods

Isolation of a resistant strain often requires numerous serial passages of the parasite in chickens. Ensuring that the parasite does not become contaminated in the process is, therefore, a particular hazard. Norton and Joyner (1975) have proposed several precautions to be taken in such experiments. These include: 1) housing birds in isolators; 2) passage of coccidia in birds treated with different drugs not to be undertaken concurrently; and 3) preparation of different oocyst suspensions to be carried out separately in time and space with strict hygienic precautions.

In our laboratory, isolators have been used to propagate laboratory strains of coccidia; however, their use on a regular basis for the development of drug-resistant lines has not been possible.

Preparation of Oocysts

Most experimental work on drug resistance has been carried out with *E. tenella,* although Norton and Joyner (1975) used *E. maxima.* An advantage of *E. tenella* is that oocysts may be collected from the cecum. Oocysts may be recovered from this organ more readily with less risk of contamination than from the feces. The technique employed in our laboratory is to remove the cecum 8 days after inoculation of chickens, homogenize in phosphate buffer at pH 8.0, and incubate with 2% crude Trypsin (1:250, Difco, Ltd.) at 39°C for 30 min. The suspension is then filtered through gauze and centrifuged, and the oocysts are recovered by three centrifugal flotations in saturated salt solution. Oocysts are then sporulated in 2% potassium dichromate at 28°C for 48 hr.

Ration

Workers in the United Kingdom have found it convenient to use a standard ration; details of its composition have been given by Ryley and Betts (1973). In our laboratory, the ration is supplemented with vitamin K. There are several advantages of using a supplemented diet when *E. tenella* is the species being studied. One advantage is that mortality due to infection is reduced. In drug resistance experiments, it is a requirement that chickens should survive to produce sufficient oocysts for subsequent passages.

A further advantage of using a vitamin K-supplemented diet is that it promotes the retention of oocysts of *E. tenella* in the necrotic core of the cecum. According to Ryley et al. (1976), isolation of oocysts of *E. tenella* from intestinal tissue is not as satisfactory as isolation of oocysts from fecal preparations, because lower numbers of oocysts are recovered. While this may be true on a vitamin K-deficient diet, our experience suggests that approximately 40% of the total oocysts produced by chickens fed a supplemented diet may be retained in the cecum. Approximately 80% of these oocysts can be recovered by the methods described above, whereas recovery of oocysts from fecal preparations may be as low as 40% (Ryley et al., 1976).

Finally, drug resistance experiments are more meaningful if carried out with diets similar to those in normal use and such rations do not exclude essential vitamin supplements.

Experimental Design

Resistance is developed by passaging the parasite repeatedly in chickens fed the drug. This technique has been used to develop resistance to most anticoccidial agents and has also been used to obtain pyrimethamine-resistant *Toxoplasma gondii* in monkey kidney cultured cells (Cook, 1958).

New resistance phenotypes may also result from genetic recombination experiments, but such phenotypes require that strains be already resistant to the drugs studied (Chapter 3).

The procedure used by McLoughlin (1970a), in which the concentration of drug is progressively increased as resistance develops, has been used by most workers. McLoughlin divides the parent strain of *Eimeria* into two lines, one passaged in the presence of drug, the other passaged for the same number of times in the absence of drug. This latter line serves as a standard against which the response of medicated chickens to infection may be measured and serves to check that the pathogenicity of the parasite has not changed as a result of passage through chickens.

Choice of Drug Concentration

The drug concentration employed to initiate the selection of resistance is determined by the effect upon the parasite of the dose of the compound. Where oocysts are produced in adequate numbers in the presence of the recommended level of drug, then selection should commence at that concentration. Chapman (1976b), for example, was able to passage *E. tenella* in chickens fed monensin at 100 ppm.

Most workers have found it necessary to begin selection using concentrations of drug lower than the recommended level in order to recover sufficient oocysts for subsequent passage (McLoughlin, 1970a; Norton and Joyner, 1975; Chapman, 1975, 1976a,b). The sequence of passages found necessary by Chapman (1976a) in the development of resistance to robenidine is illustrated in Table 1. Three experiments were carried out in which different numbers of birds were given either 0.2×10^6 or 2.0×10^6 oocysts for each passage. The initial passage was carried out at 8 ppm. The concentration of drug was doubled as soon as sufficient oocysts could be recovered to permit passage at the higher concentration.

According to Weppelman et al. (1977), however, it is possible to select resistance without resorting to lower levels of drug (see below). They were able to develop resistance to the optimal concentration of glycarbylamide and amquinolate, but were unsuccessful with amprolium, nicarbazin, robenidine, and monensin. Fouré and Bennejean (1973) obtained strains of *E. acervulina* resistant to buquinolate, decoquinate, and clopidol after a single passage at the recommended concentration. Chapman (1978a) has attempted to develop resistance to the recommended levels of amprolium, robenidine, and clopidol by giving birds large doses of *E. tenella;* however, he was unsuccessful. Although occasionally oocysts were recovered from medicated birds, the parasite did not prove to be resistant.

In experiments of this type, it is essential that the strains used be sensitive to the drug in question. Contamination of the Houghton strain of *E. tenella* with resistant organisms has already been mentioned (above). Another possible example of accidental acquisition of resistance is suggested in a comparison of two reports. Mortality in chickens infected with the Azabu strain of *E. tenella* was prevented by 62 ppm of amprolium (Ryley and Wilson, 1972); however, in a later study, despite propagation of the parasite in the absence of drug by "conventional" methods, 250 ppm was required to prevent mortality, suggesting that tolerance had accidentally been acquired (Ryley and Wilson, 1976).

Table 1. Development of resistance by *E. tenella* (Houghton) to robenidine

	Experiment 1 10 birds/group Dose 0.2×10^6 oocysts		Experiment 2 5 birds/group Dose 2.0×10^6 oocysts		Experiment 3 45 birds/group Dose 2.0×10^6 oocysts	
Drug level (ppm)	Oocysts first recovered	Line-resistant	Oocysts first recovered	Line-resistant	Oocysts first recovered	Line-resistant
			Number of passages			
8	5	6	1	3	1	3
16	5	6	3	3–4	2	3
33	10	11–12	4	4	2	3
66	14	16	5	6	3	5
132			6	12	4	6
264			12	13–14	5	6

Reprinted by permission from: Chapman, H. D. *Eimeria tenella:* Experimental studies on the development of resistance to robenidine. *Parasitology* 73:265–273 (1976a).

10

Standardization of Techniques

Norton and Joyner (1975) consider that experimental conditions should be standardized if comparisons are to be made between different drugs. However, they were unsuccessful in achieving this in experiments with *E. maxima* and attributed it to the difficulty in obtaining a repeatable schedule of inoculations and drug administration. Variable yields of oocysts, particularly in earlier passages, necessitated the use of different doses at each passage. Norton and Joyner (1975) also found that poor sporulation was often associated with poor oocyst production. Occasionally it proved necessary to reduce the drug concentration; with dinitolmide, the strain was lost completely. They succeeded in giving a standard infective dose with robenidine, but the experiment could not be repeated.

Chapman (1976a, 1978b) was able to standardize the experimental inoculum in experiments in which *E. tenella* was made resistant to robenidine, clopidol, and amprolium. This was achieved by simultaneous passage of the parasite at several drug levels to ensure that the resistant line of coccidia was not lost. Passages were continued using doubling concentrations, until the desired degree of resistance was achieved. Lines were considered resistant when the weight gains of medicated infected birds were similar to nonmedicated, infected birds. Passage at each level was continued until this degree of resistance was attained, irrespective of whether passages were at the same time being carried out at higher levels. By this method, it was possible to compare the development of resistance to different drugs using the same infective dose at each passage.

A disadvantage of the method is the duplication of experimental groups. A direct comparison of experiments was not possible because of the risk of cross-contamination; therefore, each experiment was carried out separately. A line of coccidia not exposed to drugs was included in each experiment; this line retained its sensitivity even where considerable numbers of passages were carried out. Noninoculated control birds remained uninfected during the course of experiments.

Observations from Different Laboratories

Comparison of results obtained in different laboratories is difficult because of the varying experimental methods used. Despite this, results for a number of drugs correlate reasonably well, as can be seen in the summary presented in Table 2. Factors that might be expected to influence the rate of development of resistance, such as the numbers of birds used and the size of the infective dose, are included. All workers have found that resistance to the quinolones could be readily induced. Resistance to drugs such as sulfaquinoxaline, nicarbazin, amprolium, clopidol, and robenidine was induced with greater difficulty; for monensin and lasalocid,

Table 2. A summary of experiments on the development of resistance to various anticoccidial drugs

Drug	Species	No. of passages[a]	Drug concentration	No. of birds	Infective dose	Reference
Sulfaquinoxaline	*E. tenella*	5–10	0.025	10	5×10^3	Cuckler and Malanga, 1955
Sulfaquinoxaline	*E. tenella*	6	0.0125			Horton-Smith, 1958
Sulfaquinoxaline	*E. acervulina*	10	0.003	10	2×10^5	Cuckler and Malanga, 1955
Sulfaquinoxaline	*E. maxima*	21	0.064	5		Norton and Joyner, 1975
Nitrofurazone	*E. tenella*	15	No resistance	10	5×10^3	Cuckler and Malanga, 1955
Nitrofurazone	*E. tenella*	9	0.048	10–15	$10–5 \times 10^4$	Joyner, 1957
Nitrofurazone	*E. tenella*	22	0.02			Horton-Smith, 1958
Nitrofurazone	*E. tenella*	12	0.0055	20	10^5	Gardiner and McLoughlin, 1963a
Nitrophenide	*E. tenella*	15	No resistance	10	5×10^3	Cuckler and Malanga, 1955
Nicarbazin	*E. tenella*	15	No resistance	10	5×10^3	Cuckler and Malanga, 1955
Nicarbazin	*E. tenella*	22	0.0125 (partial resistance)			Horton-Smith, 1958
Nicarbazin	*E. tenella*	17	0.0125	20	10^5	McLoughlin and Gardiner, 1967
Nicarbazin	*E. tenella*	10	No resistance	10	10^5	McLoughlin and Chute, 1975
Nicarbazin	*E. tenella*	2	No resistance	195	5×10^6	Weppelman et al., 1977
Glycarbylamide	*E. tenella*	9	0.003	19–21	$5 \times 10^4–10^5$	McLoughlin and Gardiner, 1961
Glycarbylamide	*E. tenella*	24	No resistance			Siegmann, 1966
Glycarbylamide	*E. tenella*	26	0.002	4–5	$10^3–4 \times 10^3$	Ball, 1966
Glycarbylamide	*E. tenella* (11/12 isolates)	2–6	0.006	10	5×10^3	Cuckler et al., 1969
Glycarbylamide	*E. tenella* (10/15 isolates)	2	0.006	75	5×10^6	Weppelman et al., 1977
Zoalene	*E. tenella*	7	0.0125	20	10^5	McLoughlin and Gardiner, 1962
Zoalene	*E. tenella*	24	No resistance			Siegmann, 1966
Zoalene	*E. tenella*	25	No resistance	10	$2.5–3.5 \times 10^4$	Burow and Hartwigk, 1969
Zoalene	*E. maxima*		No resistance	5		Norton and Joyner, 1975
Amprolium	*E. tenella*	18	No resistance			Siegman, 1966
Amprolium	*E. tenella*	65	0.0125	10–20	10^5	McLoughlin and Gardiner, 1968
Amprolium	*E. tenella*	17	0.0125	15	5×10^4	Klimes, 1969
Amprolium	*E. tenella*	12	No resistance			Cuckler et al., 1969
Amprolium	*E. tenella*	2	No resistance	190	5×10^6	Weppelman et al., 1977
Amprolium	*E. tenella*	10	0.00625	10	2×10^5	Chapman, 1978b
Amprolium	*E. tenella*	13	0.0125	5	2×10^6	Chapman, 1978b
Amprolium	*E. tenella*	6	0.0125	45	2×10^6	Chapman, 1978b
Amprolium	*E. acervulina*	8	No resistance			Cuckler et al., 1969

Amprolium and ethopabate	*E. maxima*	8	No resistance			Cuckler et al., 1969
Novastat	*E. tenella*	2	0.1	10	10^5	McLoughlin and Chute, 1973a
Unistat	*E. tenella*	2	0.1	10	10^5	McLoughlin and Chute, 1975
Amquinolate	*E. tenella* (1/2 isolates)	4	0.0015			McManus et al., 1968
Amquinolate	*E. tenella* (5/12 isolates)	2	0.003	60	5×10^6	Weppelman et al.,1977
Amquinolate	*E. acervulina*	8	0.0015			McManus et al., 1968
Amquinolate	*E. maxima*	4	0.0015			McManus et al., 1968
Amquinolate	*E. brunetti*	4	0.0015			McManus et al., 1968
Buquinolate	*E. tenella*	<6	0.0082	10	10^5	McLoughlin, 1970b
Buquinolate	*E. maxima*	7	0.1		$0–10^3$	Norton and Joyner, 1975
Decoquinate	*E. tenella*	<6	0.003	10	10^5	McLoughlin and Chute, 1971
Methyl benzoquate	*E. tenella*	5	0.002	1	10^5	McLoughlin and Chute, 1973b
Methyl benzoquate	*E. tenella*	8	0.004	10	10^5	Chapman, 1975
Methyl benzoquate	*E. tenella*	2	0.0125	5	2×10^6	Chapman, 1978b
Methyl benzoquate	*E. maxima*	5	0.01	5		Norton and Joyner, 1975
Clopidol	*E. tenella*	17	0.0125	10	10^5	McLoughlin and Chute, 1973c
Clopidol	*E. tenella*	13	0.0125	10	2×10^5	Chapman, 1978b
Clopidol	*E. tenella*	7	0.0125	5	2×10^6	Chapman, 1978b
Clopidol	*E. tenella*	7	0.0125	45	2×10^6	Chapman, 1978b
Clopidol	*E. maxima*	14–19	0.1	5	0.5×10^3–3.2×10^5	Norton and Joyner, 1975
Lerbek	*E. maxima*		No resistance			Joyner and Norton, 1978
Quinolones and clopidol	*E. acervulina*	1	Various	48	2×10^6	Fouré and Bennejean, 1973
Monensin	*E. tenella*	12	No resistance	10	10^5	Chapman, 1976b
Monensin	*E. tenella*	2	No resistance	135	5×10^6	Weppelman et al., 1977
Lasalocid	*E. tenella*	12	No resistance	10	10^5	Chapman, 1976b
Lasalocid	*E. tenella*	15	No resistance	10	2.5×10^4	Mitrovic and Schildknecht, 1975
Robenidine	*E. tenella*	5	0.0015		10^6–5×10^6	Ryley and Betts, 1973
Robenidine	*E. tenella*	16	0.0066	10	2×10^5	Chapman 1976a
Robenidine	*E. tenella*	13–14	0.0264	5	2×10^6	Chapman, 1976a
Robenidine	*E. tenella*	6	0.0264	45	2×10^6	Chapman 1976a
Robenidine	*E. tenella*	2	No resistance	135	5×10^6	Weppelman et al., 1977
Robenidine	*E. tenella*					McLoughlin and Chute, 1978
Robenidine	*E. maxima*	8–16	0.0264	5		Norton and Joyner, 1975

[a]Number of passages for resistance to develop to the concentration of drug indicated. This includes the passage in which resistance is evaluated.

resistance was not induced. Observations on other drugs such as nitrofurazone, glycarbylamide, and Zoalene are contradictory and difficult to interpret.

FACTORS AFFECTING THE RATE OF DEVELOPMENT OF RESISTANCE

Group Size and Experimental Inoculum

In propagating a resistant line and selecting drug-resistant mutants, chances of success are proportional to the size of the parasite population, which will be increased by an increase in the number of birds used and in the number of oocysts given. An investigation of the factors affecting the emergence in *E. tenella* of resistance to methyl benzoquate, clopidol, robenidine, and amprolium was carried out by Chapman (1975, 1976a, 1978b). Results are summarized in Table 3. The effect of three levels of selection with different numbers of animals and with inocula of different sizes was investigated. Resistance developed more readily in experiments in which a larger number of coccidia was exposed to the drug, either by increasing the number of oocysts in the inoculum or by increasing the number of birds in a group. Norton and Joyner (1975) also obtained evidence that resistance to clopidol in *E. maxima* developed more rapidly with a larger infective dose. These observations confirm the need for standarization where different drugs are compared.

Drug Selection Pressure

According to Jeffers, (1978a), "the rate of development of primary drug resistance is dependent upon several factors, the most important being the type of drug and the selection pressure which the drug exerts on the coccidial population." Few studies to illustrate this have been carried out.

Chapman (1976a) found that the emergence of resistance to robenidine in *E. tenella* was dependent upon the dose of the drug. Results are illustrated in Table 4. Groups of 10 birds were inoculated with 0.2×10^6 oocysts of *E. tenella* and passaged in chickens treated with either 2 or 4 ppm of robenidine. Lines were passaged separately at higher concentrations, when oocysts could be recovered. The response to 33 ppm of robenidine was measured after the 16th passage. Weight gains of medicated birds given the lines passaged at 2, 4, and 8 ppm were not significantly different from the noninoculated, nonmedicated controls ($P < 0.05$). The lower weight gain of medicated birds given the lines passaged at 16 and 33 ppm, and of nonmedicated, inoculated birds, were highly significantly different ($P < 0.001$). These results show that whereas lines passaged at 2, 4, and 8 ppm of robenidine were not resistant to 33 ppm

Table 3. Experimental development of resistance by *E. tenella* (Houghton) to methyl benzoquate, clopidol, robenidine, and amprolium

Drug	10 birds/group Dose 0.2×10^6 oocysts[a]	5 birds/group Dose 2.0×10^6 oocysts	45 birds/group Dose 2.0×10^6 oocysts
		Number of passages	
Methyl benzoquate	8 (40)[b]	2 (125)	
Clopidol	13 (125)	7 (125)	7 (125)
Robenidine	16 (66)	12 (132)	6 (132)
Amprolium	11–19[c] (125)	13 (125)	6 (125)

[a]Dose to develop resistance to methyl benzoquate is 0.1×10^6 oocysts.
[b]Drug concentrations to which parasites are made resistant are in parentheses.
[c]Partial resistance only.
Reprinted by permission from: Chapman, H. D. Drug resistance in coccidia. In: P. L. Long, K. N. Boorman and B. M. Freeman (eds.), *Thirteenth Poultry Science Symposium*, pp. 387–412. British Poultry Science, Ltd., Edinburgh (1978a).

of the drug, lines passaged at 16 and 33 ppm were resistant. The degree of resistance, therefore, depends upon the drug selection pressure.

Similar results were found by Chapman (1974) in studies on the development of resistance by an embryo-adapted strain of *E. tenella* to methyl benzoquate and buquinolate. Cuckler et al. (1969) also found that resistance to glycarbylamide developed in those lines of *E. tenella* exposed to the highest concentration of the compound.

Jeffers (1978a) has discussed the theoretical basis for the genetic selection of drug resistance. He has used two simple models of gene action to demonstrate that the rate of

Table 4. The weight gain (g) of chickens medicated with 33 ppm of robenidine and inoculated with different lines of *E. tenella,* which had been passaged at different concentrations of drug

Drug level (ppm)	Sensitive line[a]	Weight gain (g) Passaged lines[b]				
		2	4	8	16	33
33	83.4	85.6	81.0	75.2	48.6	36.6
0	21.6	28.2	27.2	26.8	25.1	26.0

[a]Passaged in the absence of drug.
[b]Lines passaged separately until the 16th passage at the concentrations of drug indicated.
Weight gain of the noninoculated, nonmedicated controls = 81.4 g.
Table modified from Chapman (1976a).

genetic change in anticoccidial drug resistance is dependent on the selection pressure. Jeffers (1978a) assumes that "drug resistance results from alterations in the frequency of genes determining that trait in the coccidial population" and, therefore, "the magnitude of changes in resistance is proportional to the magnitude of changes in gene frequency."

While it is true that the lower the proportion of oocysts surviving, the greater the selection pressure on any variant, it is also true that a rare variant will stand a greater chance of being lost completely from the population when the selection is very intense.

ORIGIN OF RESISTANCE

Resistance usually arises by the selection of preexisting mutants, although induced mutation as a result of exposure to drugs or phenotypic adaptation by parasites to the drug remain possibilities (Bryson and Szybalski, 1955). It is assumed that resistance results from the expression of genes present in the parasite (Jeffers, 1978a), but this remains unproven. Spontaneous mutants resistant to the drug in question may be present in the parent population, and the drug provides selective pressure in favor of these resistant organisms. Some drugs have mutagenic properties and thus increase the probability of various mutants. The only compound reported to have this effect in *Eimeria* is acriflavine which, according to McLoughlin and Chute (1968), caused a restoration of drug sensitivity in a line of *E. tenella* made resistant to amprolium. The evidence for this, however, is contradictory (Ryley and Betts, 1973)

The distinction between resistance arising as a result of adaptation of organisms in the presence of a drug, or as a result of spontaneous mutation occurring prior to exposure to the drug, is made by application of the fluctuation test of Luria and Delbruck (1943). A procedure based upon this test was used by Weppelman et al. (1977), which enabled them to calculate the frequency with which mutants of *E. tenella* resistant to amquinolate and glycarbylamide occur within the parasite population. Their procedure for the isolation of resistant mutants involved three serial passages in the presence of a drug. A selection passage in which the drug-resistant organisms were selected from a largely sensitive population of coccidia was followed by an enrichment passage in which these resistant organisms were increased in numbers, and a test passage in which the degree of resistance was evaluated. According to Weppelman et al. (1977), resistance resulted from the selection of preexisting mutants. By restricting the dose of oocysts, the possibility that random mutants might arise during the enrichment passage was minimized and the selection of resistant phenotypes could be attributed to the selection passage alone.

Resistance may appear in a single step, as has been shown with *E. tenella* to methyl benzoquate (Chapman, 1976a, 1978b) and with *Plasmodium berghei* to pyrimethamine (Schoenfeld et al., 1974), and may be due to a single mutation.

More frequently, resistance arises as a result of a series of small discrete steps involving successive mutations at multiple loci (Franklin and Snow, 1975). This is the mechanism suggested for the development of resistance to clopidol, robenidine, and amprolium by *E. tenella* (Chapman, 1978b) and to arsenical drugs by trypanosomes (Hawking and Walker, 1966).

According to Weppelman et al. (1977), selection of resistance in a single step is a possibility for all drugs, providing that sufficient oocysts can be included in the first infective dose. These authors were unable to select resistance to amprolium, nicarbazin, robenidine, or monensin in this way, but argued that the number of parasites exposed to these drugs was insufficient to select resistant mutants. They estimated the frequency of such mutants as less than 5×10^{-9}. Chapman (unpublished observations) has unsuccessfully attempted to select resistance by a single passage of *E. tenella* at the recommended levels of amprolium, clopidol, or robenidine, despite giving large numbers of birds large doses of oocysts. Failure to develop resistance in birds medicated at the outset with the recommended concentration of drug may be due to the low frequency of resistant mutants in the population, as proposed by Weppelman et al. (1977). Alternatively, selection of mutants resistant to suboptimal concentrations of drug may be essential for the subsequent selection of mutants resistant to higher concentrations.

CORRELATION OF LABORATORY AND FIELD EXPERIENCE

Because the development of resistance depends upon the number of birds used and the size of the infective dose, there is likely to be difficulty in attempting to extrapolate from laboratory experiments to conditions in the field. Whereas in the former, small numbers of birds are given high doses of coccidia, in the latter, large numbers of birds are exposed to low doses of coccidia. According to Jeffers (1978b), an additional complication is an overlap of generations in the field, whereas in the laboratory, generations are discrete. According to Ryley (1980), the development of resistance described in the work of Chapman (1976a, 1978b) was so dependent upon experimental conditions as to eliminate any predictive aspect from the work. Ryley would like to be able to predict from laboratory studies how long a new compound is likely to retain its efficacy in the field. The results obtained by Chapman (1976a, 1978b), rather than lacking any predictive value, indicate that no laboratory test will precisely reflect the circumstances under which resistance develops in the field.

The correlation between laboratory observations and field experience is reasonable for a number of drugs. For example, resistance to glycarbylamide and the quinolones developed more rapidly than to amprolium in laboratory experiments (Table 2). Within 6 months of the introduction of glycarbylamide and quinolones in the field, strains were reported resistant, but partial resistance to amprolium was not described until 3 yr after introduction of the drug (Probo Prostowo and Edgar, 1965). Several unsuccessful attempts have been made to develop resistance to monensin in the laboratory (Chapman, 1976b; Elancoban Technical Manual, Elanco Products, Ltd.). After used in the field for 10 yr, there has not been a full description of resistance to this drug.

Ryley and Betts (1973) have noted a lack of correlation between laboratory and field experience with Zoalene. According to these authors, isolation of field strains sensitive to Zoalene is a rarity. In a survey of field isolates received in 1966, Hodgson et al. (1969) found that the incidence of resistance to Zoalene in *E. acervulina, E. brunetti, E. maxima,* and *E. tenella* was 87%, 45%, 41%, and 26%, respectively. Subsequent samples received in the same year showed an increase in the frequency of resistance.

Despite these observations, Zoalene has been used commercially for many years. McLoughlin and Gardiner (1961, 1962) observed resistance to both glycarbylamide and Zoalene after nine passages, and Betts (M. J. Betts, personal communication) found that resistance to Zoalene developed as readily as to glycarbylamide. Whereas the demise of glycarbylamide was precipitated by the appearance of resistant strains, Zoalene is still commercially viable. It should be noted, however, that not all workers have succeeded in developing resistance to Zoalene (Table 2). A source of confusion here is that the lack of correlation noted by Ryley and Betts (1973) assumes that the commercial life of a drug is related to its propensity to develop drug resistance. Whereas this may have been the case for glycarbylamide and the quinolones, it is likely that Zoalene continued to be used for some time, despite the presence of resistant strains.

In a classification of anticoccidial drugs according to their propensity for inducing resistance, Reid (1975) has listed them in descending order of magnitude: glycarbylamide, quinolones, clopidol, robenidine, amprolium and Zoalene, nicarbazin and monensin (absent). These are subjective impressions based upon field experience by drug users (Reid, personal communication). Clearly, caution should be exercised before attempting to extrapolate from laboratory experience to the field.

EFFECT OF DRUG CONCENTRATION UPON THE DEVELOPMENT OF RESISTANCE UNDER FIELD CONDITIONS

It is commonly believed that the use of subtherapeutic drug levels has resulted in the more rapid emergence of resistant strains (Yvoré, 1976). Thus, Ryley and Betts (1973) regarded programs of medication involving progressive reduction of the concentration of drugs both "shortsighted and irresponsible" because this might lead to drug-resistant lines.

Because it was essential to use subtherapeutic levels of certain drugs to select resistant strains of *Eimeria,* (see below), Chapman (1976a) suggested that in the field resistance may initially arise in the presence of a low concentration of drug, and that maintenance of adequate drug concentration should reduce this possibility.

10

Jeffers (1978a), however, believes that the notion that the use of reduced drug concentrations will enhance the development of resistance is erroneous, and stems from a lack of understanding of the relationship between the rate of development of resistance and the reduction in coccidial reproduction caused by the drug. Accordingly, increasing the intensity of selection will result in more rapid change in gene frequency and, therefore, resistance will arise more rapidly in the presence of effective concentrations of the drug. If Jeffers' theory is correct, then reduced concentrations of drugs may not result in more rapid resistance. Reductions in the concentrations of drugs would have the advantage of reducing costs. However, in the broiler industry the scope for this is limited, because with most drugs, lower concentrations result in reduced efficacy.

Jeffers (1978a) also observes, however, that "no resistance will develop against a drug which exercises complete control over the organism, thereby preventing reproduction" and that "elimination of oocyst production coincident with anticoccidial control can be demonstrated under laboratory conditions using high concentrations of certain drugs." Clearly, the resolution of these two contrary theories depends upon the reproduction of the parasite under practical conditions where birds are fed the optimal concentration of drug. Jeffers (1978a) is of the opinion that under practical conditions, elimination of oocyst production is impossible: "There are no proprietary drugs which completely prevent oocyst production when used at approved concentrations to medicate poultry flocks."

In laboratory experiments, it is possible to demonstrate complete suppression of oocyst production of sensitive strains of coccidia by using the recommended level of certain drugs. Because there is no reason to suppose that the mode of action of anticoccidial agents differs in the laboratory, the failure of drugs to suppress completely oocyst production in the field may be due to the ingestion of suboptimal quantities of the drug. Assuming a normal distribution curve of food and, hence, drug intake, in a given population of birds, the majority will ingest the appropriate amount of drug, but some will receive less; there will be a gradient of drug intake up to the optimum with increasing numbers of birds at each point on the gradient. Thus, the inclusion of a drug at the optimal concentration will allow some birds to receive differing suboptimal amounts of the drug and, hence, allow the selection of mutants resistant to increasing concentrations of the drug.

From a practical point of view, it is important that drug levels should be such as to minimize the development of drug resistance commensurate with toxicity and cost.

PRACTICAL APPROACH TO THE PROBLEM OF DRUG RESISTANCE

With the exception of the ionophores and a few drugs that have only recently been introduced, resistance has emerged to all compounds used against coccidiosis (however, see later). Possible ways of circumventing this problem are discussed below.

Increase in the Concentration of Drug

In the case of some drugs (clopidol, robenidine, and amprolium), it has been shown that coccidia made resistant to the recommended levels of drug may be suppressed by higher concentrations. However, with other drugs (methyl benzoquate), strains once made resistant have been shown not to be controlled by any level of drug tested (Chapman 1976a, 1978b). In practice, increasing the concentration of drug may not be feasible, because as Ryley (1980) has pointed out, even recently introduced drugs such as halofuginone, arprinocid, and the ionophores are being used at levels close to those that are toxic to the bird.

Alternations of Different Drugs

Alternation of anticoccidials or "shuttle" programs have been proposed as a means of reducing the problem of drug resistance. They are discussed elsewhere in this volume (McDougald, Chapter 9; Fayer and Reid, Chapter 11). Ideally, any resistant forms appearing during the use of the first drug would be eliminated by the second, and so on. Few experimental studies to investigate this have been carried out. McLoughlin and Chute (1975) passaged a sensitive strain of *E. tenella* in chickens medicated serially with nicarbazin, Zoalene, amprolium, or Unistat. After 40 passages (10 for each drug), the strain recovered was resistant to Zoalene, amprolium, and Unistat, but sensitive to nicarbazin, indicating that changing drugs does not prevent the development of resistance. The result of alternations of drugs would seem to be the selection of strains of coccidia resistant to several drugs.

In the field, drugs may be alternated from one crop of birds to the next or within a single crop; in the case of the latter, it is convenient to change the drugs when the rations change, that is, the starter, grower, and finisher foods. Usually an "efficient" drug, such as monensin, is employed as the second drug at the time when exposure to coccidia is highest. The legally required withdrawal period for such drugs plus their relatively high cost make it convenient for the producer to use a less "efficient" agent as the third drug in the shuttle.

Long and Millard (1978) compared a "two-drug shuttle" in which monensin (0–6 weeks) was followed by Zoalene (6–8 weeks), with a "three-drug shuttle" in which clopidol (0–3 weeks) was followed by monensin (3–6 weeks) and Zoalene (6–8 weeks). Trials were carried out in four broiler farms each comprising five to six broiler houses containing approximately $5–15 \times 10^3$ birds. Feed conversion was significantly lower ($P < 0.05$) on two farms with the three-drug shuttle, but no significant difference was found in the other two farms.

Strains of *E. maxima* from farms included in the study of Long and Millard (1978) were isolated; they all proved resistant to clopidol (Chapman, 1980). It may be concluded, therefore, that under the conditions of this experiment, no advantage resulted from the use of an "effective" anticoccidial in the first 3 weeks. In the United Kingdom, broilers started on clean litter in cleaned houses do not develop meaningful coccidial infections until after 3 weeks. However, where it is standard practice to start broilers in unclean houses on used litter, the presence of an "effective" anticoccidial in the first 3 weeks may be of greater importance.

Use of Combinations of Drugs

Drug combinations are discussed in this volume by McDougald (Chapter 9) and Fayer and Reid (Chapter 11). An advantage might be to reduce the problem of drug resistance because the chance of selecting mutants resistant to a combination of drugs is less than if those drugs are used alone (Bryson and Szybalski, 1955). The only experimental evidence to support this contention was found by Norton and Joyner (1975), who obtained resistance to 10 ppm of methyl benzoquate or 125 ppm of clopidol after three passages of *E. maxima,* but who were unable to develop resistance to a mixture of 8.35 ppm of methyl benzoquate and 100 ppm of clopidol (Joyner and Norton, 1978).

The usefulness of combinations of drugs will depend upon whether resistance has already developed to the individual ingredients, because many compounds will already have been employed independently.

Mixtures based upon amprolium and ethopabate (Amprol Plus, Amprolmix, UK), amprolium, sulfaquinoxaline, and ethopabate (Pancoxin), and amprolium, sulfaquinoxaline, ethopabate, and pyrimethamine (Supacox) are in use. A survey carried out by Hodgson et al. (1969) over a 3-year period (1964–1967) revealed a significant increase in the incidence of strains of *E. maxima* resistant to a mixture of 80 ppm of amprolium, 60 ppm of sulfaquinoxaline, and 5 ppm of ethopabate. The concentration of amprolium currently included in Pancoxin has been increased from 80–100 ppm but, according to the manufacturer, this was necessary due to a change in the incidence of pathogenic species of coccidia, increased exposure to coccidia resulting from greater stocking densities, the use of restricted feeding programs in breeders, and the increasing incidence of other diseases which may affect drug intake and lower the resistance of poultry to coccidiosis (MSD Technical Booklet, Merck Sharp & Dohme, Ltd.).

Chapman (1980) has recently investigated the incidence of resistance to Amprolmix, Pancoxin, and Supacox in isolates of *E. maxima* obtained from sites where broiler or breeder birds were reared. All isolates were resistant to Amprolmix and Pancoxin. Isolates from breeder farms caused significantly lower weight gains than isolates from broiler farms when given to chickens medicated with these drugs. This could be correlated with the more recent use of Pancoxin on the breeder sites. Most isolates were sensitive to Supacox, although one resistant and two partially resistant isolates were found from broiler sites on all of which Supacox had been used for 10 months before the isolates had been obtained. This finding suggests that with prolonged usage, resistance will ultimately become widespread to this drug also.

The sensitivity of field isolates of *E. maxima* to another combination of drugs, that is, 8.35 ppm of methyl benzoquate and 100 ppm of clopidol (Lerbek) was also examined and compared with the individual drugs (Chapman, 1980). Resistance to methyl benzoquate was widespread in isolates obtained from broiler sites; however, isolates obtained from breeder sites were sensitive and this correlated with the use of a quinolone drug. In contrast, all isolates proved resistant to clopidol, despite the fact that the drug had never been used at breeder sites. Chapman (1980) suggested that the resistance shown by these isolates may have been partly due to lack of efficacy of clopidol at 100 ppm.

Isolates resistant to methyl benzoquate and clopidol were also resistant to Lerbek. However, where sensitivity to methyl benzoquate was present, Lerbek was effective. Joyner and Norton (1978) found that Lerbek was ineffective against a line made resistant to both drugs. They also showed that, while it was difficult to induce resistance to a mixture of methyl benzoquate and clopidol, a strain resistant to both drugs could be produced by commencing selection in a strain resistant to one component. In the field, such selection has already occurred, because Jeffers (1974a) has reported a high incidence of strains resistant to both quinolones and clopidol.

Ideally, a drug combination should be introduced before resistance has developed to its individual ingredients. However, as Ryley and Betts (1973) have pointed out, this would require the simultaneous development of two compounds with separate modes of action. At present, it is not possible to speculate whether this would result in a reduced rate of emergence of resistance. It would be interesting, however, to investigate the speed with which resistance appears to drug mixtures in countries where the ingredients of such mixtures have never been employed.

STABILITY OF RESISTANCE

Restoration of Sensitivity

The utility of changing anticoccidial agents as a practical approach to the problem of drug resistance would be enhanced if it could be demonstrated that resistance, once developed, was unstable. Unfortunately, published studies on the restoration of drug sensitivity in drug-resistant strains of *Eimeria* are contradictory (Ryley and Betts, 1973; Chapman, 1978a).

Whereas McLoughlin (1971) found that strains of *E. tenella* made resistant to amprolium, nicarbazin, and Zoalene had their sensitivity restored when they were passaged in the presence of other drugs, Jeffers and Challey (1973) were unable to restore sensitivity in a strain of *E. tenella* made resistant to amprolium, buquinolate, decoquinate, or sulfaquinoxaline by passaging the strain in the presence of monensin. They found that passage of a strain of *E. tenella* in the presence of clopidol resulted in loss of resistance to decoquinate, but that passage of the strain in the presence of decoquinate did not result in loss of resistance to clopidol. The high incidence of resistance to both decoquinate and clopidol among isolates of

E. acervulina obtained from flocks in which these drugs were in use suggests that sensitivity had not been restored in the field.

A distinction needs to be drawn between loss of resistance as a result of passage in the presence of another drug and loss of resistance in the absence of such drug pressure. The stability of resistance following passage in the absence of drug has been demonstrated (Gardiner and McLoughlin, 1963b; Ball, 1968; Cuckler et al., 1969; Ryley, 1980), but, to date, there are no reports of the loss of resistance in the absence of drug selection pressure. The instability of resistance to chloroquine following passage in the absence of drug has, however, been demonstrated in *P. berghei* (Peters, 1965).

Competition between Resistant and Sensitive Parasites

Rosario et al. (1976) found that chloroquine-resistant mutants of *Plasmodium chabaudi* may supplant sensitive forms of the parasite even in the absence of drug. In contrast, studies using mixtures of pyrimethamine-resistant and -sensitive forms indicate an apparent disadvantage to the resistant parasite. Augustine et al. (1977) found that the reproductive capacities of lines of *E. tenella* made resistant to amprolium or buquinolate were not different from a sensitive line; this was also shown to be the case for strains of *E. tenella* resistant to decoquinate, clopidol, and robenidine (Chapman, 1978a). Under these circumstances, competition between resistant and sensitive strains of *Eimeria* in the absence of drug pressure would not result in an advantage to either strain.

It seems obvious that progressive dilution of a resistant population with sensitive organisms will result in a reduction of proportions of resistant parasites present. Two studies have been carried out which demonstrate this (Jeffers, 1976; McLoughlin and Chute, 1979). Jeffers (1976) showed that the introduction of massive numbers of a drug-sensitive attenuated strain of *E. tenella* into pens containing decoquinate-resistant parasites resulted in a marked reduction in the proportion of drug-resistant oocysts in the litter. The efficacy of decoquinate was restored, but the resistant population was not eliminated. Presumably, the resumption of medication would have resulted in a return to a high incidence of resistant parasites. An advantage was that birds developed immunity to *E. tenella* without the appearance of clinical coccidiosis.

DRUG RESISTANCE IN THE FIELD

Surveys of the incidence of coccidia have revealed that the parasites are widely distributed (Jeffers, 1974a,b,c; Long et al., 1975); it is likely that this is largely determined by resistance to anticoccidial drugs.

Studies of the incidence of drug resistance in field strains of *Eimeria* have been reviewed by Ryley and Betts (1973), Chapman (1978a), and Ryley (1980). The results of extensive surveys by Jeffers (1974a,b,c) of the incidence of resistance among strains of coccidia isolated from the field are summarized in Table 5. The incidence of resistance to the quinolones, buquinolate and decoquinate, and to clopidol was high, as was that of strains of *E. acervulina* resistant to nicarbazin, the latter a somewhat surprising finding in view of the difficulty in inducing resistance to this drug in the laboratory (Table 2). The correlation between the incidence of resistance in the field and laboratory observations has already been discussed.

Resistance to monensin was found only in a few isolates of *E. maxima*. Some indication of a low degree of resistance of this species to monensin has also been obtained in a recent study by Chapman (1979). It would be prudent to continue surveillance of the sensitivity of field isolates, because the ionophores are, at present, the main chemotherapeutic agents used in the treatment of coccidiosis in the broiler chicken.

SIGNIFICANCE OF DRUG RESISTANCE

Dependence upon anticoccidial drugs is likely to continue into the foreseeable future. Other chapters in this book consider alternatives to chemotherapy for the control of coccidiosis. The industry has been fortunate that a steady stream of new products has been available, thus per-

Table 5. Incidence of resistance to anticoccidial drugs in three species of *Eimeria*

			Incidence of resistance[a]		
Trade name	Compound	Concentration %	*E. acervulina*[b]	*E. maxima*[b]	*E. tenella*[c]
Amprol plus	Amprolium +	0.0125	NT[d]	NT[d]	27.3% (198)
Amprol plus	ethopabate	0.0004			
Bonaid	Buquinolate	0.011	76.0% (196)	62.5% (8)	62.5% (72)
Coban	Monensin	0.0121	0% (165)	6.7% (30)	0% (61)
Coyden	Clopidol	0.0125	40.7% (484)	28.9% (38)	27.4% (201)
Deccox	Decoquinate	0.003	59.7% (484)	73.7% (38)	41.3% (201)
Nicarb	Nicarbazin	0.0125	57.6% (66)	NT	NT
Robenz	Robenidine	0.0033	0% (40)	25.0% (12)	NT

[a]Number of isolates tested is in parentheses.

[b]Data reprinted by permission from: Jeffers, T. K. *Eimeria acervulina* and *E. maxima:* Incidence and anticoccidial drug resistance of isolants in major broiler-producing areas. *Avian Diseases* 18:331–342 (1974).

[c]Data reprinted by permission from: Jeffers, T. K. *Eimeria tenella:* Incidence, distribution and anticoccidial drug resistance of isolants in major broiler-producing areas. *Avian Diseases* 18:74–84 (1974).

[d]NT = not tested due to limited efficacy of the drug against the species concerned.

10

mitting a degree of choice in the selection of compounds for medication. Whether this will continue is doubtful. However, at present, the introduction of ionophorous antibiotics and the possibility that resistance may not develop to these compounds permits an optimistic outlook.

The standards of husbandry and hygiene in the United Kingdom are such that flocks are not necessarily exposed to a coccidial challenge sufficient to result in clinical coccidiosis. This may explain the commercial success of products which, in the laboratory, have only limited activity.

The economics of the poultry industry are such that a slight drop in the food conversion of chickens is of commercial significance. Low levels of infection might therefore be economically important. Is it possible to estimate the potential economic losses due to drug resistance when coccidia are present at subclinical levels?

The approximate saving to the broiler industry in the United Kingdom by the adoption of effective anticoccidial drugs may be calculated. Field trials carried out in 1974 and involving 330,000 broilers resulted in improvements in food conversion of 0.135 when monensin was compared with clopidol (Elanco Technical Manual). There was a significant reduction in food conversion involving birds at five commercial broiler farms when the anticoccidial medication was changed from Supacox or clopidol to monensin (Long et al., 1975). Each farm comprised 4 to 12 houses containing between $7–18 \times 10^3$ chickens. Supacox had been used at one farm and clopidol at four. Average food conversion immediately prior to the change of drug was 2.47, but following the change was 2.29. This represents a reduction in food intake of 7.3%. If this improvement is due to better control of coccidiosis, as indicated by Chapman's (1979, 1980) isolation from three of the farms of strains of *E. maxima* resistant to clopidol but sensitive to monensin, then a change to a more effective drug could result in a total saving of £20 million ($40 million) per annum in the United Kingdom (assuming 380×10^6 broiler birds weighing approximately 2kg/bird and food at £150($300)/ton). An annual saving of 7.3% of food costs indicates the losses attributable to drug resistance that might be expected in the absence of effective anticoccidials.

LITERATURE CITED

Augustine, P. C., Vetterling, J. M., and Doran, D. J. 1977. *Eimeria tenella:* Growth characteristics of drug-resistant strains in chicks and cell culture. Proc. Helminth. Soc. Wash. 44:147–149.

Ball, S. J. 1966. The development of resistance to glycarbylamide and 2-chloro-4-nitrobenzamide in *Eimeria tenella* in chicks. Parasitology 56:25–37.

Ball, S. J. 1968. The stability of resistance to glycarbylamide and 2-chloro-4-nitrobenzamide in *Eimeria tenella.* Res. Vet. Sci. 9:149–151.

Bryson, V., and Szybalski, W. 1955. Microbial drug resistance. Adv. Genet. 7:1–46.

Burow, H., and Hartwigk, H. 1969. Versuche zur

Induktion einer Zoalenresistenz mit *Eimeria (E.) tenella.* Berl. Muench. Tieraerztl. Wochenschr. 85:288–292.

Challey, J. R., and Jeffers, T. K. 1973. Synergism between 4-hydroxyquinoline and pyridone coccidiostats. J. Parasitol. 59:502–504.

Chapman, H. D. 1974. Use of chick embryo infections for the study of drug resistance in *Eimeria tenella.* Parasitology 69:283–290.

Chapman, H. D. 1975. *Eimeria tenella* in chickens: Development of resistance to quinolone anticoccidial drugs. Parasitology 71:41–49.

Chapman, H. D. 1976a. *Eimeria tenella:* Experimental studies on the development of resistance to robenidine. Parasitology 73:265–273.

Chapman, H. D. 1976b. *Eimeria tenella* in chickens: Studies on resistance to the anticoccidial drugs monensin and lasalocid. Vet. Parasitol. 2:187–196.

Chapman, H. D. 1978a. Drug resistance in coccidia. In: P. L. Long, K. N. Boorman, and B. M. Freeman (eds.), Avian Coccidiosis, pp. 387–412. Thirteenth Poultry Science Symposium. British Poultry Science, Ltd., Edinburgh.

Chapman, H. D. 1978b. *Eimeria tenella:* Experimental studies on the development of resistance to amprolium, clopidol and methyl benzoquate. Parasitology 76:177–183.

Chapman, H. D. 1979. Studies on the sensitivity of recent field isolates of *E. maxima* to monensin. Avian Pathol. 8:181–186.

Chapman, H. D. 1980. Studies on the sensitivity of field isolates of *Eimeria maxima* to combinations of anticoccidial drugs. Avian Pathol. 9:67–76.

Cook, M. K. 1958. The development of a pyrimethamine-resistant line of Toxoplasma under *in vitro* conditions. Am. J. Trop. Med. Hyg. 7:400–402.

Cuckler, A. C., McManus, E. C., and Campbell, W. C. 1969. Development of resistance in coccidia. Acta Vet. (Brno) 38:87–99.

Cuckler, A. C., and Malanga, C. M. 1955. Studies on drug resistance in coccidia. J. Parasitol. 41:302–311.

Elanco. Elancoban Technical Manual. 1974. Resistance studies. 33–37.

Fouré, N., and Bennejean, G. 1973. Les chimiorésistances acquises vis à vis des derives des quinoleines et de la pyridine. In: Symposium International sur les Coccidioses, Tours.

Franklin, T. J., and Snow, G. A. 1975. Biochemistry of Antimicrobial Action. 2nd Ed. Chapman and Hall London.

Gardiner, J. L., and McLoughlin, D. K. 1963a. Drug resistance in *Eimeria tenella:* IV. The experimental development of a nitrofurazone-resistant strain. J. Parasitol. 49:947–950.

Gardiner, J. L., and McLoughlin, D. K. 1963b. Drug resistance in *Eimeria tenella:* III. Stability of resistance to glycarbylamide. J. Parasitol. 49:657–659.

Hawking, F., and Walker, P. J. 1966. Analysis of the development of arsenical resistance in trypanosomes *in vitro.* Exp. Parasitol. 18:63–86.

Hodgson, J. N., Ball, S. J., Ryan, K. C., and Warren, E. W. 1969. The incidence of drug-resistant strains of *Eimeria* in chickens in Great Britain, 1966. Br. Vet. J. 125:31–35.

Horton-Smith, C. 1958. Resistance to anticoccidial drugs experimentally induced in a laboratory strain of *Eimeria tenella.* Eleventh World's Poultry Congress, Mexico City. Sect. III, pp. 483–489. Mexicana: Prensa

Jeffers, T. K. 1974a. *Eimeria acervulina* and *E. maxima:* Incidence and anticoccidial drug resistance of isolants in major broiler-producing areas. Avian Dis. 18:331–342.

Jeffers, T. K. 1974b. *Eimeria tenella:* Incidence, distribution and anticoccidial drug resistance of isolants in major broiler-producing areas. Avian Dis. 18:74–84.

Jeffers, T. K. 1974c. Anticoccidial drug resistance: Differences between *Eimeria acervulina* and *E. tenella* strains within broiler houses. Poult. Sci. 53:1009–1013.

Jeffers, T. K. 1976. Reduction of anticoccidial drug resistance by massive introduction of drug-sensitive coccidia. Avian Dis. 20:649–

653.

Jeffers, T. K. 1978a. Genetics of coccidia and the host response. In: P. L. Long, K. N. Boorman, and B. M. Freeman (eds.), Avian Coccidiosis, pp. 50–125. Thirteenth Poultry Science Symposium. British Poultry Science, Ltd., Edinburgh.

Jeffers, T. K. , 1978b. *Eimeria tenella:* Sensitivity of recent isolates to monensin. Avian Dis. 22: 157–161.

Jeffers, T. K., and Challey, J. R. 1973. Collateral sensitivity to 4-hydroxy-quinolines in *Eimeria acervulina* strains resistant to meticlorpindol. J. Parasitol. 59:624–630.

Joyner, L. P. 1957. Induced drug-fastness to nitrofurazone in a laboratory strain of *Eimeria tenella.* Vet. Rec. 69:1415–1418.

Joyner, L. P. 1970. Coccidiosis: Problems arising from the development of anticoccidial drug resistance. Exp. Parasitol. 28:122–128.

Joyner, L. P., and Norton, C. C. 1978. The activity of methyl benzoquate and clopidol against *Eimeria maxima:* Synergy and drug resistance. Parasitology 76:369–377.

Klimes, B. 1969. Resistance development of *Eimeria tenella* to some coccidiostats (6-azauracil, nitrofurazone, zoalene and amprolium). Acta. Vet. (Brno) 38:101–108.

Long, P. L., and Millard, B. J. 1978. Coccidiosis in broilers: The effect of monensin and other anticoccidial drug treatments on oocyst output. Avian Pathol. 7:373–381.

Long, P. L., Tompkins, R. V., and Millard, B. J. 1975. Coccidiosis in broilers: Evaluation of infection by the examination of broiler house litter for oocysts. Avian Pathol. 4:287–294.

Luria, S. E., and Delbruck, M. 1943. Mutations of bacteria from virus sensitivity to virus resistance. Genetics 28:491–511.

McLoughlin, D. K. 1970a. Coccidiosis: Experimental analysis of drug resistance. Exp. Parasitol. 28:129–136.

McLoughlin, D. K. 1970b. Efficacy of buquinolate against ten strains of *Eimeria tenella* and the development of a resistant strain. Avian Dis. 14:126–130.

McLoughlin, D. K. 1971. Drug resistance in *Eimeria tenella:* X. Restoration of drug sensitivity following exposure of resistant strains to other coccidiostats. J. Parasitol. 57:383–385.

McLoughlin, D. K., and Chute, M. B. 1968. Drug resistance in *Eimeria tenella.* VII. Acriflavin-mediated loss of resistance to amprolium. J. Parasitol. 54:696–698.

McLoughlin, D. K., and Chute, M. B. 1971. Efficacy of decoquinate against eleven strains of *Eimeria tenella* and development of a decoquinate resistant strain. Avian Dis. 15:342–345.

McLoughlin, D. K., and Chute, M. B. 1973a. Efficacy of Novastat against twelve strains of *Eimeria tenella* and the development of a Novastat resistant strain. Avian Dis. 17: 582–585.

McLoughlin, D. K., and Chute, M. B. 1973b. Efficacy of nequinate against thirteen strains of *Eimeria tenella* and the development of a nequinate resistant strain. Avian Dis. 17:717–721

McLoughlin, D. K., and Chute, M. B. 1973c. Efficacy of clopidol against twelve strains of *Eimeria tenella* and the development of a clopidol resistant strain. Avian Dis. 17:425–429.

McLoughlin, D. K., and Chute, M. B. 1975. Sequential use of coccidiostats: Effect on development by *Eimeria tenella* of resistance to amprolium, nicarbazin, Unistat and zoalene. Avian Dis. 19:424–428.

McLoughlin, D. K., and Chute, M. B. 1978. Robenidine resistance in *Eimeria tenella.* J. Parasitol. 64:874–877.

McLoughlin, D. K., and Chute, M. B. 1979. Loss of amprolium resistance in *Eimeria tenella* by admixture of sensitive and resistant strains. Proc. Helminth. Soc. Wash. 46:138–141.

McLoughlin, D. K., and Gardiner, J. L. 1961. Drug resistance in *Eimeria tenella:* I. The experimental development of a glycarbylamide-resistant strain. J. Parasitol. 47:1001–1006.

McLoughlin, D. K., and Gardiner, J. L. 1962. Drug resistance in *Eimeria tenella:* II. The experimental development of a zoalene resistant strain. J. Parasitol. 48:341–346.

McLoughlin, D. K., and Gardiner, J. L. 1967. Drug resistance in *Eimeria tenella:* V. The experimental development of a nicarbazin-resistant strain. J. Parasitol. 53:930–932.

McLoughlin, D. K., and Gardiner, J. L. 1968. Drug resistance in *Eimeria tenella:* VI. The experimental development of an amprolium-resistant strain. J. Parasitol. 54:582–584.

McManus, E. C., Campbell, W. C., and Cuckler, A. C. 1968. Development of resistance to quinoline coccidiostats under field and laboratory conditions. J. Parasitol. 54:1190–1193.

Merck Sharp & Dohme, Ltd. 1969. Amprotection, a new concept in the control of coccidiosis. MSD Technical Booklet pp. 2–4.

Mitrovic, M., and Schildknecht, E. G. 1975. Lasalocid: Resistance and cross-resistance studies in *Eimeria tenella* infected chicks. Poult. Sci. 54:750–756.

Norton, C. C., and Joyner, L. P. 1975. The development of drug-resistant strains of *Eimeria maxima* in the laboratory. Parasitology 71: 153–165.

Peters, W. 1965. Drug resistance in *Plasmodium berghei* Vincke and Lips, 1948: 1. Chloroquine-resistance. Exp. Parasitol. 17:80–89.

Peters, W. 1970. Chemotherapy and Drug Resistance in Malaria. Academic Press, New York.

Probo Prostowo, B., and Edgar, S. A. 1965. Drug resistance of chicken coccidia and effect of amprolium and ethopabate in controlling infection caused by field isolates. Poult. Sci. 44:1408.

Reid, W. M. 1975. Progress in the control of coccidiosis with anticoccidials and planned immunisation. Am. J. Vet. Res. 36:593–596.

Rosario, V. E., Hall, R., Walliker, D., and Beale, G. H. 1976. Competition between drug-resistant and drug-sensitive malaria parasites. Trans. R. Soc. Trop. Med. Hyg. 70:287.

Ryley, J. F. 1980. Drug resistance in coccidia. Adv. Vet. Sci. Comp. Med. 24:99–120.

Ryley, J. F., and Betts, M. J. 1973. Chemotherapy of chicken coccidiosis. Adv. Pharmacol. Chemother. 11:221–293.

Ryley, J. F., Meade, R., Hazelhurst, J., and Robinson, T. E. 1976. Methods in coccidiosis research: Separation of oocysts from faeces. Parasitology 73:311–326.

Ryley, J. F., and Wilson, R. G. 1972. Comparative studies with anticoccidials and three species of chicken coccidia *in vivo* and *in vitro.* J. Parasitol. 58:664–668.

Ryley, J. F., and Wilson, R. G. 1976. Laboratory studies with some older anticoccidials. Parasitology 73:287–309.

Schnitzer, R. J., and Grunberg, E. 1957. Drug Resistance of Microorganisms. Academic Press, New York.

Schoenfeld, C., Most, H., and Entner, N. 1974. Chemotherapy of rodent malaria: Transfer of resistance vs. mutation. Exp. Parasitol. 36: 265–277.

Shirley, M. W., Millard, B. J., and Long, P. L. 1977. Studies on the growth, chemotherapy and enzyme variation of *Eimeria acervulina* var. *diminuta* and *E. acervulina* var. *mivati.* Parasitology 75:165–182.

Siegmann, O. 1966. Serien passage eines *Eimeria tenella* stommes durch Küken unter der Einwirkung von Amprolium, Glycamid, Nitrofurazon und DOT (zoalen). Berl. Meunch. Tieraerztl. Wochenschr. 79:268–271.

Weppelman, R. M., Battaglia, J. A., and Wang, C. C. 1977. *Eimeria tenella:* The selection and frequency of drug resistant mutants. Exp. Parasitol. 42:56–66.

Yvoré, P. 1976. Revue sur la prevention des coccidioses en aviculture. Avian Pathol. 5:237–252.

11

Control of Coccidiosis

Ronald Fayer and W. Malcolm Reid

Efforts to control or prevent economic losses due to coccidiosis have long been the objective of research on behalf of poultry, cattle, and other livestock producers; similar efforts to control toxoplasmosis have been made with public health objectives. Although progress has been made in both fields, further research to improve present control methods is needed. ''Control'' in this chapter means reduction of infectious units with the objective of preventing clinical illness. Complete elimination of coccidia from either hosts or the environment is considered impractical or uneconomical, except in specialized laboratory procedures.

After consideration of some of the distinctive life cycles and biological characteristics of the coccidia, possible control measures and suggestions for future research are reviewed in this chapter.

LIFE CYCLES AND TRANSMISSION OF COCCIDIA

All species of coccidia belonging to the genera *Eimeria*, *Isospora*, *Cystoisospora*, *Hammondia*, *Toxoplasma*, *Besnoitia*, *Sarcocystis*, and *Frenkelia* are intracellular protozoan parasites transmitted from one host to another by the exogenous oocyst stage. All genera except *Eimeria* and *Isospora* also have a tissue cyst stage in a second host (intermediate or paratenic host). The encysted organisms live at a reduced metabolic rate and thus may prolong the life cycle and ensure its completion at a later stage. An understanding of the various coccidian life cycles and the role of the oocyst and tissue cyst within these cycles is required before considering possible control measures. In the following sections, the one-host parasites (*Eimeria*) are considered first, followed by two-host parasites of other genera. Figures 1–4 illustrate the cycles described in the text and show where various control methods can interrupt them.

Life Cycles

Eimeria Numerous *Eimeria* (Figure 1) species are found in fishes, amphibians, reptiles, and birds and a few species have been found in arthropods and prochordates. Many species and high rates of infection are found in domesticated birds and livestock. *Eimeria* have historically been considered host-specific, each species of *Eimeria* infecting only one host species. The life cycle is initiated when a host swallows a sporulated oocyst and sporozoites are released (excyst) in the gut. Nearly all species develop in the intestinal epithelium, where each species has a marked preference for development in specific sites. Intracellular sporozoites undergo nuclear and cytoplasmic changes, which result in asexual multiplication by schizogony. Merozoites that develop within the schizonts may initiate one or more subsequent generations of schizonts. Merozoites from the final generation (usually the second or third)

11

develop into male or female gametes (microgametes = male, macrogametes = female). Fertilized macrogametes become oocysts, which enter the intestinal lumen and are shed in the feces of the host. Oocysts are unsporulated when shed and are not infective for another host. They sporulate outside the body and eventually contain four sporocysts, each with two sporozoites. Only sporulated oocysts can initiate infection when swallowed by an appropriate host. The life cycle is considered self-limiting; if new oocysts are not swallowed, the original organisms will have developed through all the endogenous stages and the host will be freed of infection.

Isospora and Cystoisospora *Isospora* (Figure 1) species are found primarily in amphibians, reptiles, passeriform birds, carnivores, and primates, but have also been reported in slugs, some galliform birds, marsupials, insectivores, rodents, and ungulates. The life cycle of *Isospora* species is essentially the same as that of *Eimeria* species. After the host has swallowed sporulated oocysts, excystation, schizogony, gametogony, fertilization, and oocyst formation take place in the gastrointestinal tract and unsporulated oocysts are shed in the feces. Sporulation of oocysts takes place outside the body. Sporulated oocysts are infective for the definitive host alone.

Cystoisospora canis and *Cystoisospora ohioensis* of dogs and *Cystoisospora felis* and *Cystoisospora rivolta* of cats have historically been identified as *Isospora* species. A typical eimerian-isosporan life cycle is completed in the gastrointestinal tract after a carnivore swallows sporulated oocysts. However, if *Cystoisospora* oocysts are swallowed by vertebrates other than the definitive host, sporozoites released from these oocysts invade and remain within extraintestinal tissues (Dubey and Mehlhorn, 1978). When such infected tissues are eaten by the appropriate carnivore host, the sporozoite is freed and initiates an intestinal infection as if it had just been released from an oocyst.

Sarcocystis and Frenkelia *Sarcocystis* and *Frenkelia* (Figure 2) were initially identified by their distinctive cysts in the intermediate hosts. The cysts of each genus are identifiable by their general size, shape, and location within the host. Cysts of *Sarcocystis* species, which range in length from less than 100 μm to several millimeters, may be threadlike, spindle, or egg-shaped. They are usually found in cardiac and skeletal muscles, but are sometimes located in nervous tissue. Sarcocysts are found in fish (trout), a few reptiles, a considerable number of birds and seven orders of mammals. Especially high rates of infection are found in domestic animals, including cattle, sheep, swine, camels, llamas, and carabao. Cysts of *Frenkelia*, which vary in diameter from approximately 100–400 μm, are round or multilobulated and are found in the central nervous system of rodents. When cysts of either genus are eaten by the definitive host (a predator), bradyzoites (terminology defined by Frenkel, 1973) freed from the cysts enter intestinal cells, where they develop directly into male or female gametocytes; fertilized female gametes develop into oocysts which sporulate in situ and are shed in the feces as sporulated oocysts or individual sporocysts. When an appropriate intermediate host (a prey animal)

Figures 1-4. Diagrammatic drawings of life cycle stages of coccidia within (endogenous) and outside (exogenous) the host's body. Activities necessary for the parasite to complete its life cycle are labeled, as applicable (e.g., digestion of cyst, excystation, merogony, gametogony, fertilization, excretion, and sporulation). Methods of transmission from host to host are labeled, as applicable (e.g., ingestion, carnivorism, milk, vertical transmission, blood transfusion, and organ transplant). Known effective methods for killing or inhibiting the parasite stages are listed outside the life cycle in *blocks* surrounded by *solid lines* with *markers* pointing to affected stages. Presumed methods for control are similarly

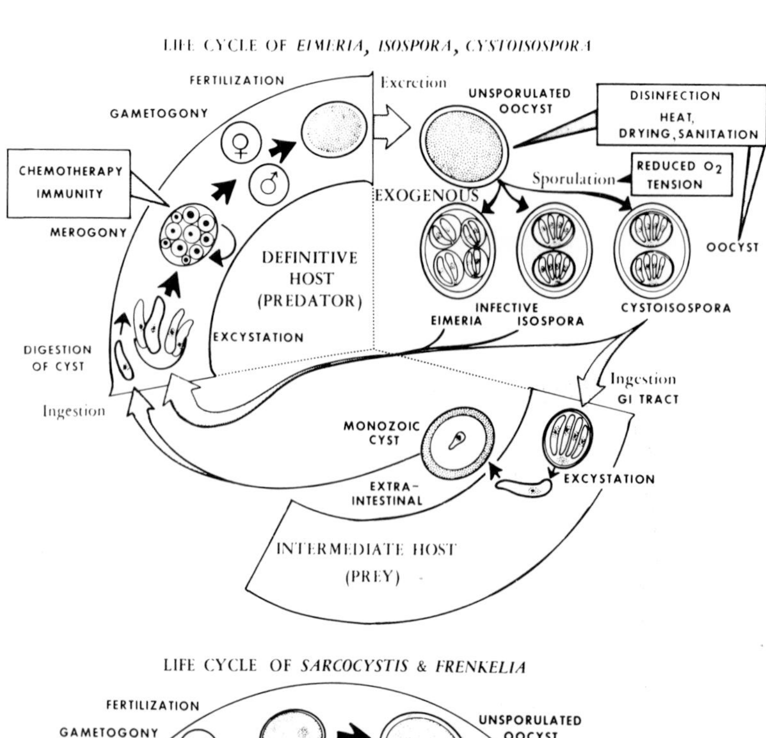

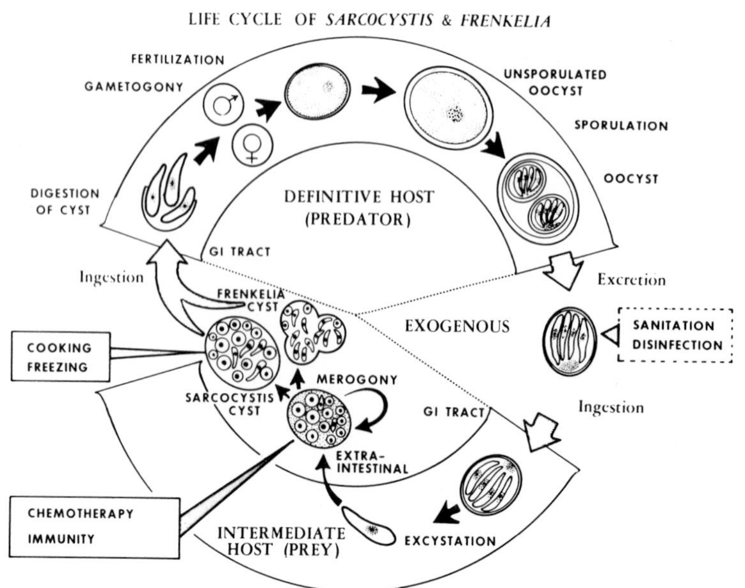

11

located and surrounded by *broken lines*. 1. Life cycles of three genera of coccidia. *Eimeria* and *Isospora* have only direct life cycles, whereas *Cystoisospora* may use an intermediate host. 2. Life cycles of *Sarcocystis* and *Frenkelia*. Coccidia in these genera require a predator as a definitive host and a prey animal as an intermediate host. 3. Life cycles of *Besnoitia* and *Hammondia*. Coccidia of these genera require a predator as a definitive host and a prey animal as an intermediate host. 4. Life cycle of *Toxoplasma*. This genus requires a felid as a definitive host, but nearly all other vertebrates may serve as intermediate hosts and may acquire infections from the felid or from other intermediate hosts.

LIFE CYCLE OF *BESNOITIA* & *HAMMONDIA*

FERTILIZATION
Excretion
GAMETOGONY
UNSPORULATED OOCYST
DISINFECTION
HEAT
DRYING
SANITATION
MEROGONY
OOCYST
DEFINITIVE HOST (PREDATOR)
EXOGENOUS
Digestion of Cyst
Ingestion
GI TRACT
GI TRACT
HAMMONDIA CYST
Ingestion
EXCYSTATION
CHEMOTHERAPY (BESNOITIA)
EXTRA-INTESTINAL
BESNOITIA CYST
INTERMEDIATE HOST (PREY)

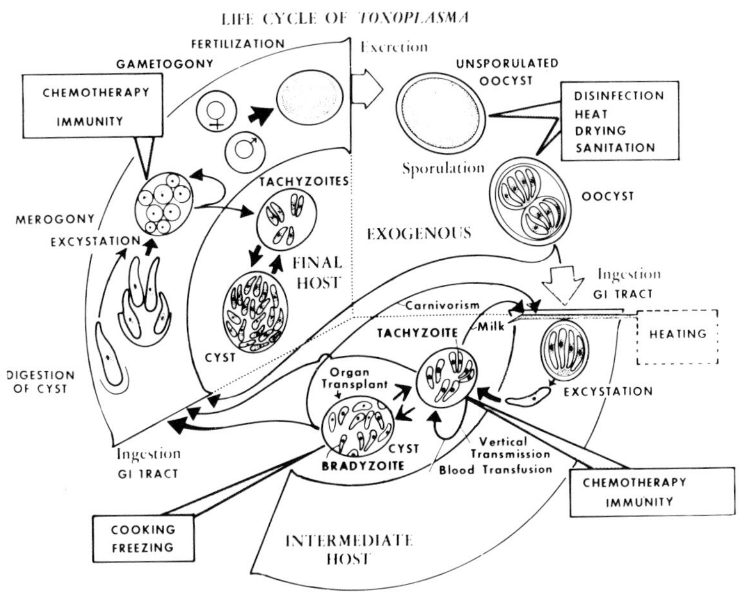

swallows oocysts or sporocysts, sporozoites excyst and initiate schizogony in vascular endothelial cells throughout the body (*Sarcocystis*) or in liver parenchymal cells (*Frenkelia*). Merozoites may initiate subsequent generations of schizonts, may multiply in the bloodstream (*Sarcocystis*), or may initiate cyst development. Cysts begin with the formation of rounded noninfectious metrocytes, which multiply and eventually give rise to infectious bradyzoites, which are present in mature cysts (Frenkel, 1973).

Besnoitia *Besnoitia* (Figure 3) species were originally identified by their distinctive thick-walled cysts in reptiles, marsupials, rodents, and ungulates. When cysts of *Besnoitia wallacei* from mice, *Besnoitia besnoiti* from cattle, or *Besnoitia darlingi* from opossums are eaten by the definitive host (cats), bradyzoites from the cysts enter intestinal cells and develop into schizonts. Depending on the species, a second generation of schizonts may develop. Merozoites from the schizonts then develop into male or female gametes; fertilized female gametes develop into oocysts that are shed in the feces and sporulate outside the body. When an appropriate intermediate host swallows the sporulated oocysts, sporozoites excyst in the gut. Sporozoites are thought to give rise to tachyzoites (see Frenkel, 1973) in extraintestinal tissues and then form cysts containing numerous bradyzoites in fibroblastic cells in various organs of the body.

Hammondia *Hammondia* (Figure 3) cysts, like those of *Sarcocystis*, are found principally in skeletal and cardiac muscles and in the brain of the intermediate hosts. In addition, *Hammondia* are found in spleen, lymph nodes, liver, and lungs. Cysts of *Hammondia hammondi* have very slender bradyzoites, but no metrocytes (see Frenkel, 1973). When cysts of *H. hammondi* (in mice, hamsters, guinea pigs, monkeys, dogs, rabbits, and pigs) are eaten by cats, when cysts of *Hammondia pardalis* (in mice) are eaten by various felids, or when cysts of *Hammondia heydorni* (in cattle and dogs) are eaten by dogs, schizonts and then gametes develop in the intestine and unsporulated oocysts are shed in the feces and sporulate outside the body. When an appropriate intermediate host swallows the sporulated oocysts, sporozoites excyst in the gut and cysts develop in several organs of the body.

Toxoplasma *Toxoplasma gondii* (Figure 4) is probably the most ubiquitous protozoan parasite with approximately 200 vertebrate species acting as intermediate hosts (Levine, 1977). The final host, a wild or domestic felid, becomes infected by swallowing sporulated oocysts or by eating animal tissues or products infected with either tachyzoites or tissue cysts containing bradyzoites. Such stages initiate schizogony in the feline intestine. This is followed by gametogony, fertilization, oocyst development, and shedding of unsporulated oocysts in the feces; oocysts sporulate outside the body. Most domestic or feral cats probably become infected by eating rodents or passeriform birds or by being fed uncooked lamb, pork, beef, or offal from these animals. Intermediate hosts, including humans, acquire infections by swallowing

11

sporulated oocysts from cat feces that have contaminated pastures, farm feed storage facilities, sand boxes, gardens, litter pans, implements, or food products; by eating uncooked meat, milk, or eggs containing infective stages; by receiving blood transfusions or organ transplants containing the organism; and by transmission from mother to fetus. In intermediate hosts, sporozoites, tachyzoites, and bradyzoites initially proliferate by endodyogeny as clusters of tachyzoites in various organs of the body (acute phase of disease). Tachyzoites may eventually give rise to cysts containing bradyzoites that remain viable for the life of the host. Bradyzoites may reinitiate tachyzoite development, if the host becomes immunosuppressed.

Transmission of Coccidia by Biological and Mechanical Means

The oocyst stage shed in the feces of an infected host is the primary source of infection in all coccidia. Except for coccidia belonging to *Sarcocystis* and *Frenkelia* and a few unusual species of *Eimeria* in domestic animals, newly shed oocysts are not infective until they sporulate outside the body. Depending upon the environment, oocysts can become infective and remain so for a year or more (Farr and Wehr, 1949; Koutz, 1950; Kogan, 1960; Yilmaz and Hopkins, 1972; Frenkel et al., 1975). After physical or biological dispersal, these oocysts eventually must be swallowed by susceptible hosts to initiate new infections.

Infected animals that are clinically ill or subclinically infected can contaminate the immediate environment with oocysts. When such animals are transported to new environments, the infective agent is also carried. Feces containing oocysts can be dropped directly, kicked, splashed, or blown into feed or water. Animals can soil their body, coat, or feathers with feces containing oocysts, which are then swallowed when they lick, preen, or peck. Humans may transport oocysts on soiled shoes, clothes, or hands, and on tools, various mechanical equipment, or vehicles. Pets or animals such as rats and mice may also transport oocysts after becoming soiled with feces from the infected hosts. Because oocysts are smaller than dust particles, rain droplets and wind can easily disperse them.

Wild birds that eat seeds from cattle feces may ingest cattle oocysts, which pass through their digestive tract unchanged and are shed in their feces at other locations. Invertebrates such as earthworms, flies, and cockroaches may also transport oocysts on their body or within their intestinal tract.

The concept that *Eimeria* of domestic animals are host-specific is widely accepted and probably holds true for most species. However, findings that some *Eimeria* infect several host species and that others develop in extraintestinal sites indicate that exceptions exist and must be considered when determining the epizootiological aspects of coccidiosis. Several species of cattle (*Bos taurus*) coccidia have been reported in other ruminants. *Eimeria auburnensis*, *Eimeria brasiliensis*, *Eimeria bovis*, and *Eimeria bukidnonensis* were found in feces from American bison (*Bison bison*) (Ryff and Bergstrom, 1975). *Eimeria zuernii* were reported from the wisent (*Bison bonasus*), white-tailed deer (*Odocoileus virginianus*), roe deer (*Capreolus*

capreolus), and elk (*Cervus canadensis*) without morphological description (Levine, 1973). Levine (1973) believed that these undescribed parasites were actually other species. However, Levine (1970) indicated that in 1969 Sayin transmitted four species of *Eimeria*, including *E. zuernii*, from the water buffalo (*Bubalus bubalis*) to the ox (*B. taurus*). If wild ruminants share coccidia of domestic ruminants, transmission could occur at common feeding and watering areas. This is most probable during periods of environmental stress, such as drought or blizzard conditions. Wild and domestic sheep are known to share several species of *Eimeria*, but the epizootiology is not known.

Characteristics of Oocysts and Tissue Cysts

Effects of Physical Factors Infectivity and viability decrease with age and with concurrent decrease in the sporozoite storage polysaccharide. Climatic factors such as temperature, moisture, and oxygen tension can greatly affect the viability and longevity of unsporulated and sporulated oocysts, as well as affect the sporulation process. Unsporulated oocysts seem less able to survive extreme changes in climatic factors than sporulated oocysts (Horton-Smith and Long, 1954).

Oocyst viability can also be reduced under laboratory conditions by a wide variety of physical and chemical agents. Again, unsporulated oocysts usually are more susceptible to the effects of these agents than are sporulated oocysts. Control measures aimed at preventing sporulation may be more effective than attempts to kill sporulated oocysts.

Sporulation for most *Eimeria* species is maximum between 28 and 31°C. Low temperatures (0–5°C) may only retard or temporarily inhibit sporulation of oocysts of some *Eimeria*; sporulation continues upon warming (Christensen, 1939; Edgar, 1954). Viability and sporulation of oocysts of *Toxoplasma*, *Hammondia*, and other coccidia were greatly reduced when feces containing oocysts were kept from 1 to several weeks at refrigerator temperatures before sporulation was begun (Fayer, unpublished; A. K. Prestwood, personal communication; Levine, 1973). Freezing temperatures are more detrimental to unsporulated than to sporulated oocysts. For example, unsporulated oocysts of *T. gondii* were killed after 1 day at −21°C, whereas sporulated oocysts survived for 28 days at −21°C (Frenkel and Dubey, 1972). During the winter in Byelorussia, unsporulated chicken oocysts were killed when temperatures reached −38.5°C, whereas some sporulated oocysts survived (Kogan, 1959).

Oocysts of some species survive freezing temperatures better than those of other species. Although freezing temperatures (−18°C for 12 hr) killed 100% of *Eimeria tenella* oocysts (Coudert and Yvoré, 1973), *Eimeria arloingi*, *Eimeria ninakohlyakimovae*, and *Eimeria parva* withstood freezing temperatures (−25°C for 7 days) during the winter in Wyoming without appreciable loss of viability (Landers, 1953). Fernando (Chapter 7, this volume) has made similar observations and discusses the effects of overwintering on the epidemiology of coccidiosis.

11

Oocysts are generally more sensitive to high than to low temperatures. Temperatures of 35°C or greater reduce or permanently inhibit sporulation of several *Eimeria* species (Long, 1959) and *Toxoplasma* (reviewed by Marquardt, 1960; Dubey et al., 1970). Investigations of oocyst survival suggest that, as temperature increases from 37–80°C, the killing time decreases (Horton-Smith and Long, 1954; Duncan, 1959; Prassad, 1959; Horchner and Dalchow, 1972; Yvoré, 1976). However, some species seem better able to withstand elevated temperatures than others. Chang (1937) found *E. arloingi* (*ovina*) (from sheep) four times more resistant at 46°C than *Eimeria separata* (from rats).

Steam heat, applied under pressure, is the best method of sterilization of oocysts in the laboratory. Applying a blowtorch to the floors of poultry houses did not kill all the oocysts, unless it was applied long enough to char the wood (Horton-Smith and Taylor, 1939). Poultry producers heap built-up litter into piles, producing temperatures of 63°C to destroy oocysts (Horton-Smith and Long, 1954).

Oocysts require moisture for survival; drying kills them (Brotherston, 1948). Although *E. tenella* oocysts survived for 49–52 days at 90% relative humidity, they survived only 32 days at 61% relative humidity (Farr and Wehr, 1949). Oocysts of *T. gondii* in feces survived progressively longer from 19, 37, or 58%, to 80% relative humidity (Frenkel and Dubey, 1972). Relative humidities of 75% or less greatly reduced sporulation of *E. zuernii* (Marquardt et al., 1960).

Oxygen is required for sporulation of oocysts. Oocysts of *T. gondii* and *E. zuernii* do not sporulate in the absence of oxygen, and sporulation is impaired when oxygen tension is reduced below 10% of normal (Marquardt et al., 1960; Dubey et al., 1970).

Sunlight is detrimental to oocyst survival. Oocysts of *Eimeria acervulina* in chicken feces on outdoor soil plots survived better in shade than in direct sunlight (Farr and Wehr, 1949). Similar findings have been reported for *T. gondii* oocysts (Yilmaz and Hopkins, 1972; Frenkel et al., 1975). Unsporulated oocysts seem more susceptible than sporulated oocysts to destruction by sunlight. Unsporulated *E. zuernii* oocysts survived only 4 hr and sporulated oocysts survived 8 hr after exposure to sunlight (Marquardt et al., 1960).

Numerous studies have been conducted on unsporulated and sporulated oocysts to determine the effects of a variety of sources of radiation other than sunlight; these include gamma radiation, x-rays, ultraviolet light, ultrasonic waves, and low accelerating voltage electrons. These energies, thought to attenuate oocysts, do not destroy oocysts in animal-rearing facilities.

Although tissue cysts of *Toxoplasma*, *Sarcocystis*, and *Besnoitia* mature, age, and deteriorate with time, some are thought to remain infectious for the life of the host. Organisms within cysts remain infectious for a period after the death of the host. Refrigeration of infected tissues may prolong the survival time to several weeks, as indicated by the survival of cyst organisms for 68 days at 4°C (Jacobs et al., 1960), the infectivity of *Toxoplasma* in meat in a hospital in France (Desmonts et al., 1965), and the infectivity of *Sarcocystis* in meat purchased in retail food stores (Fayer, 1975; Leek and Fayer, 1978; Box and McGuinness, 1978). Both

Toxoplasma and *Sarcocystis* cyst organisms are killed by cooking and by freezing (Jacobs et al., 1960; Dubey, 1974; Gestrich, 1974; Fayer, 1975).

Effects of Chemical Factors Coccidial oocysts are highly resistant to most harsh chemicals. Tests on the resistance of *Eimeria*, *Isospora*, and *Toxoplasma* oocysts to chemical disinfectants have been reviewed by Schneider et al. (1973), Yvoré (1976), and Dubey (1977). The most effective disinfectants are either toxic or so caustic as to require specialized conditions to be effective. They are generally unsuitable for use in animal-rearing facilities. A variety of commercial disinfectants have been found effective against *Eimeria* or *Toxoplasma* oocysts under specialized test conditions (Ayeni et al., 1972; Ito et al., 1975). Active ingredients include: carbon disulfide alone, or mixed with chlorinated hydrocarbons; mixtures containing cresols, chlorinated cresols, and chlorinated hydrocarbons; a mixture of phenols, monocyclic and polycyclic aromatic hydroxy compounds; and halogenated, condensated products of alkylated and arylated hydroxytoluenes and cresol sulfonic acids. The most effective chemicals against oocysts seem to be small molecular weight compounds such as formalin, gaseous or aqueous ammonia (Horton-Smith et al., 1940), and methylbromide (Long et al., 1972). To be effective, most gaseous disinfectants must be in direct contact with the oocysts, preferably in a hermetically sealed environment. Oocysts in large quantities of feces, or in litter in buildings or outdoors, are difficult to treat effectively.

The exogenous stage of *Sarcocystis* is usually shed in feces as sporulated sporocysts rather than as the unsporulated oocysts shed by other coccidia. The viability of *Sarcocystis bovicanis* sporocysts was reduced when sporocysts were stored in chemicals commonly used for storage of *Eimeria* and *Toxoplasma* oocysts, including 2% sulfuric acid in water, 2.5% potassium dichromate in water, and 1% sodium hypochlorite in water (Leek and Fayer, 1979).

Tissue cysts of *Toxoplasma* in mouse brain suspensions were killed by ethyl alcohol, propyl alcohol, peracetic acid, 2.5% potassium dichromate, and 6% sodium hypochlorite (Exner et al., 1972; Dubey, 1977). Tachyzoites and bradyzoites were killed by 0.7% hydrochloric acid solution (Pettersen, 1979). Water killed bradyzoites (Jacobs et al., 1960). Nothing is known about the effects of disinfectants on oocysts or tissue cysts of *Besnoitia*, *Hammondia*, or *Frenkelia*.

CHEMOTHERAPEUTIC CONTROL

In Poultry

The use of anticoccidial drugs as preventives is a practice that has been almost universally adopted and is one of the great successes in disease control for the ever expanding poultry industry. Programs of preventive medication against coccidiosis will rank as one of the most im-

11

portant advances in parasite control. Rearing of thousands of birds under one roof results in a tremendous build-up of oocysts. Without continuous medication, disastrous coccidiosis outbreaks frequently occur. Chemoprophylaxis is the only method of coccidiosis control presently used for floor-reared broilers. Anticoccidial drugs presently available for use in poultry are shown in Table 1. Neither sanitation nor immunization are employed as a substitute for chemoprophylaxis.

With the exception of monensin, which has dominated the anticoccidial market for broilers during the past 9 yr, none of the drugs introduced since 1961 have had much continuing market success (Figure 5). Reasons include drug resistance and high manufacturing and clearance costs. Marketing returns for robenidine (Robenz) were limited to about 2 yr; for buquinolate (Bonaid), 3 yr; clopidol (Coyden), 5 yr; and decoquinate, 2 yr. However, the latter two drugs are still being used on a limited scale (about 1–2% market penetration). Nequinate (Statyl), although introduced to the U.S. market in 1972, was completely unsuccessful. Two other promising 4-hydroxyquinolines were withdrawn by the developers (Merck and Squibb) shortly before marketing because of probable drug resistance problems. Glycamide introduced by Merck about 1949 was quickly withdrawn after drug resistance appeared within the 1st yr of marketing. The combination of sulfadimethoxine plus ormetoprim (Rofenaid) was priced too high to be accepted by the poultry industry. However, it has achieved limited sales because of its combined antibacterial and anticoccidial activity. The ionophore lasalocid (Avatec), introduced in 1974, has had 4–10% of the broiler market in different periods.

In contrast, several older drugs are still being marketed in limited quantities for broilers. These include nicarbazin (with 24 yr in the market), amprolium in combinations with ethopabate (16 yr), and Zoalene (14 yr). Unistat and Polystat, recently withdrawn from the market, were each successful for about 12 yr.

To be successful, any new drug must 1) show competitive advantages over currently available drugs, 2) be slow in permitting drug-resistant strains to emerge, and 3) be backed by large capital expenditures to carry out required governmental clearance procedures.

Similar programs involving continuous feeding of anticoccidials have also been adapted for use with breeder and layer stock, if brooding is started on the floor. Complete coccidiosis control has not always been as successful with breeders and layers as with broilers. Management programs must consider effects of drug withdrawal and the level of flock immunity which has become established. Preventive drugs used at too high a level or for too long a time may inhibit development of immunity.

Some producers of replacement stock have tried continuous medication throughout the life of the bird. Cost factors and resistance to amprolium have limited this approach. Amprolium is the only marketed product currently approved by the U.S. Food and Drug Administration for feeding to egg-producing stock. Advantages and disadvantages of combination programs are discussed under "Immunological Control" and "Control by Sanitation and Management Practices."

Table 1. Anticoccidial drugs available (or planned) for use as preventives for chicken and turkey

Generic, chemical, or government-approved name	Trade name	Manufacturer's or government-approved feed level(s) (ppm)	Manufacturer(s)	Remarks
Aklomide[a]	Aklomix, Novastat	250	Salsbury	Manufacture to be discontinued
Amprolium[a]	Amprol, Amprolmix	125–250[b]	Merck Sharp & Dohme	Approved for turkeys; often marketed in combinations with ethopabate (4 or 40 ppm), sulfaquinoxaline, and pyrimethamine
Arprinocid	Arpocox	70	Merck Sharp & Dohme	A new experimental anticoccidial; effective against oocysts
Arsanilic acid or sodium arsalilate[a]	Pro-Gen	400	Abbott	Currently used as a growth promotant; limited anticoccidial activity
Beclotiamine	Cocciden	40–125	Sankyo	Similar in activity to amprolium
Buquinolate	Bonaid	82.5	Norwich	Marketing has been discontinued (drug resistance)
Butynorate (dibutyltindilaurate)	Tinostat, Polystat	375	Salsbury	Approved only for turkeys; the combination with Polystat (sulfanitran, dinsed, and roxarsone) being discontinued
Chlortetracycline[a]	Aureomycin	440[b]	American Cyanamid	An antibiotic sometimes substituted as an anticoccidial in spite of weak, anticoccidial activity
Clopidol, meticlorpindol, clopindol	Coyden	125, 250	Dow, AL Laboratories	Synergistic activity with nequinate
Decoquinate[a]	Deccox	30	May & Baker, Hess & Clark, Rhône-Poulenc	Selectively marketed where drug-resistant strains are absent
Dimethalium	Actonate		Takeda	Similar in activity to amprolium
Dinitolmide	DOT, Zoalene, Whitsyn T	40–250	Dow, Salsbury	Antibacterial approved for turkeys; still useful if resistant strains are absent
Furazolidone[a]	NF-180	55	Smith & Kline, Hess & Clark, Rhône-Poulenc	Sometimes used as an antibiotic substituting for an anticoccidial; provides limited anticoccidial activity

Halofuginone	Stenerol	3	Roussell	Has been introduced in Europe and Latin America
Lasalocid[a]	Avatec	75–125	Hoffmann-La Roche	An ionophore introduced in 1976
Monensin[a]	Coban, Elancoban	100, 120	Eli Lilly	The most extensively used broiler anticoccidial 1972–1980; limited use in turkeys
Narasin	Monteban	Similar range to salinomycin	Eli Lilly	An experimental ionophore which has had extensive field testing
Nequinate, methyl benzoquate	Statyl, Lerbek combined with clopidol)	10, 30	ICI, Dow	Withdrawn from the market (drug resistance)
Nicarbazin[a]	Nicarb, Nicrazin	100–200	Merck Sharp & Dohme	An older anticoccidial (introduced 1956) with fewer drug resistance problems
Nihydrazone	Zonifur	110	Norwich	Manufacturing discontinued
Nitrofurazone[a]	nfz, Amifur	55[b]	Hess & Clark, Rhône-Poulenc, Smith & Kline	Approved for use with turkeys
Nitromide[a]	Unistat	250	Salsbury	Similar in activity to dinitolmide; used in combinations
Oxytetracycline[a]	Terramycin	200[b]	Pfizer	Used for antibiotic activity; substituted for but not approved for anticoccidial activity
Robenidine[a]	Robenz, Cycostat	33	American Cyanamid	Discontinued marketing (drug resistance)
Roxarsone[a]	3-Nitro	50[b]	Salsbury, Rhône-Poulenc	Approved only for "growth promotion," synergistic with some anticoccidials
Sulfadimethoxine[a]	Rofenaid	125[b]	Hoffmann-La Roche	Combined with the synergistic ormetoprim; approved for turkeys
Sulfaquinoxaline	SQ, Sulquin, Sulfacox	150–250	Merck Sharp & Dohme	No longer used for prevention of coccidiosis in chickens; danger of inducing hemorrhagic syndrome; approved for turkeys
Salinomycin		50, 60, 75(?)	Kaken, Pfizer, A. H. Robins, Hoechst	A new ionophore

[a]Currently listed in the 1980 Feed Additive Compendium (Gates, 1979).
[b]Also approved for treatment using higher levels or different combinations.

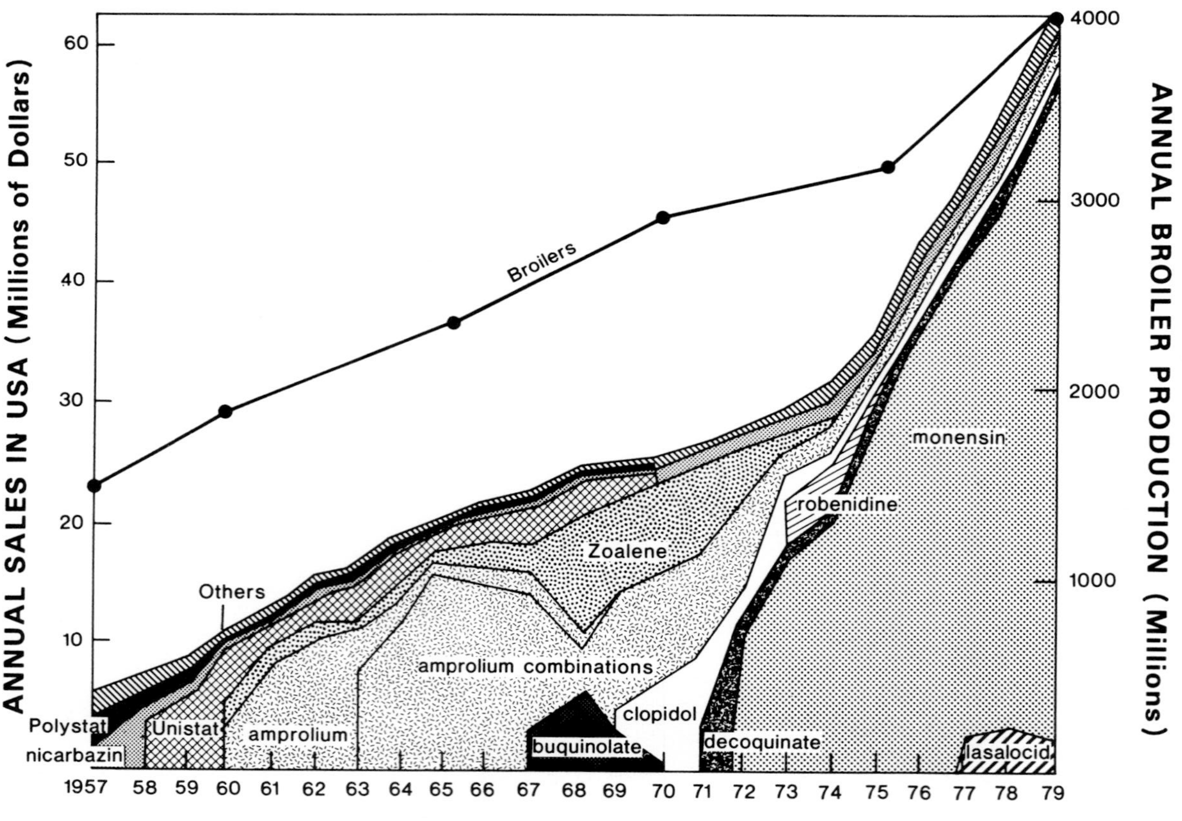

Figure 5. Comparative market life of broiler anticoccidials after introduction to the U.S. market. Amounts were obtained by rough calculations with data derived from numerous sources.

11

Coccidiosis is rare in caged layers or caged broilers. Sanitation is the method of control emphasized. Treatment with anticoccidials has often been used in the event of outbreaks.

With turkeys, chemotherapy is widely employed, because coccidiosis from at least three species frequently causes weight loss and, occasionally, mortality. Preventive anticoccidial programs are frequently, but not universally, used with turkeys up to 8 weeks of age. Medication is rarely continued beyond 8 weeks; birds are not severely infected beyond this age, possibly because of some age resistance. As a result of the higher costs of continuous medication of turkeys compared to chickens, a treatment approach is more often relied upon for control. However, treatment is no more reliable with turkeys than with chickens (see McDougald, Chapter 9, this volume). Because of the more limited market potential, pharmaceutical companies have put less effort into obtaining governmental clearance for turkey anticoccidials than they have for chicken anticoccidials. Several drugs cleared by governmental agencies for use in chickens are being used (illegally) in turkeys.

In Mammals

Cattle arriving in feedlots usually carry coccidia without any clinical signs of infection. Most clinical signs of coccidiosis begin within 30–60 days after cattle arrive. The signs probably result from the stresses of shipping, new diet, new environment, handling, and crowding. The two species most often identified as pathogens are *E. bovis* and *E. zuernii*. Preventive medication for these species should begin as soon as cattle enter the feedlot and should continue for 21–28 days. Anticoccidial drugs available for use in mammals in the U.S. are listed in Table 2.

Range cattle experience epizootics called "summer coccidiosis" and "winter coccidiosis" under especially harsh conditions of either heat and drought or frozen, snow-covered range. During these times, cattle huddle close to a sparse water supply and to limited feed. Feces accumulate within this restricted area. As soon as such conditions are recognized, producers should begin preventive medication.

Feedlot sheep are subjected to very much the same conditions as feedlot cattle. The time of occurrence of clinical coccidiosis is so well known that prophylactic feeding of drugs can be effectively confined to a well defined period. Several anticoccidials can be effective when given in feed for 21 days beginning 5 days after sheep enter the feedlot.

Lambing is another stressful period for sheep; it is a time when preventive medication should be used. Ewes brought into confined lambing sheds saturate the environment with feces containing oocysts, which accumulate in large numbers by the time lambs are born. Ewes should be given preventive medication beginning a week before or as soon as they enter the lambing area to reduce the number of oocysts they shed. Lambing areas should be cleaned of feces shortly before lambing begins.

Table 2. Anticoccidial drugs approved by the FDA for mammals

Disease	Host	Drug	Use level
Coccidiosis	Cattle	SQ[a] Na in water	13 mg/kg
		SQ solution in water	13 mg/kg
		Amprolium in feed	
		prophylaxis	5 mg/kg for 21 days
		treatment	10 mg/kg for 5 days
		Decoquinate in feed	
		prophylaxis	0.5 mg/kg for at least 28 days
	Sheep	SQ Na in water	13 mg/kg
		SQ powder	25%
	Pig	SQ powder	25%
	Rabbit	SQ in feed	
		prophylaxis	0.25% for 30 days
		treatment	0.1% for 2 weeks
	Cat	None	
	Chinchilla	None	
	Dog	None	
	Goat	None	
	Horse and mule	None	
	Mink	None	
Sarcocystosis		None	
Toxoplasmosis		None	

[a]SQ = sulfaquinoxaline.

Peracute coccidiosis of neonatal pigs resulting in piglet death before oocysts are shed has recently been recognized in farrowing barns in Georgia and the corn belt states of Iowa, Illinois, Indiana, Nebraska, and South Dakota (J. E. Fox, personal communication). Diagnosis is based on histological examination, and *Isospora suis* has been identified as the pathogen. Amprolium (Corid) has been used experimentally to treat pigs with coccidiosis; scours in baby pigs are treated either by giving the drug to 3-day-old pigs for 5 consecutive days, or by giving the drug to sows before they farrow to reduce the number of oocysts shed (Sangster et al., 1978).

Most commercial dairy goat producers immediately separate kids from does and thus prevent coccidiosis. In contrast, small breeders and family goatkeepers usually separate kids from does at 1–3 days and later return the kid to the same pen or area. Under these management conditions, kids often experience coccidia problems. Angora goat raisers usually bring does in from pasture to nursery pens, which become highly contaminated with oocysts. Kids are born and does are then turned out, while kids remain in the pens. The small breeders, family goatkeepers, and Angora goat raisers should use preventive medication for kids when high exposure to oocysts is likely. Treatment with sulfonamides and amprolium has been successful.

11

In horses, only one species of coccidia (*Eimeria leukarti*) has been found in the Western Hemisphere. Its life cycle, pathogenicity, and response to medication are not well documented.

Commercial rabbitries experience losses from both hepatic and intestinal coccidiosis caused by *Eimeria stiedai* and *Eimeria perforans*, respectively. Fitzgerald (1972) found monensin, but not amprolium, to be effective for hepatic coccidiosis. Monensin at 20 or 40 ppm and sulfaquinoxaline at 250 ppm prevented mortality (Gwyther, 1976). For prophylaxis of intestinal coccidiosis, Vetkovic and Tomanovic (1967) found amprolium to be effective. For both intestinal and hepatic coccidiosis on three broiler rabbit farms in Italy, Gallazzi (1976) found a therapeutic program of sulfaquinoxaline and pyrimethamine in drinking water effective; however, furazolidone, clopidol, amprolium plus sulfaquinoxaline, ethopabate plus pyrimethamine, and sulfaquinoxaline plus clopidol were ineffective at the levels tested. In rabbits experimentally infected with *Eimeria intestinalis*, *Eimeria magna*, *E. perforans*, and *E. stiedae*, robenidine significantly improved feed conversion and weight gain, but clopidol, amprolium plus ethopabate, methyl benzoquate, and amprolium plus ethopabate, sulfaquinoxaline, and pyrimethamine were not nearly as efficacious (Peeters and Halen, 1979a,b; Peeters et al., 1979).

Most canine and feline coccidia are usually nonpathogenic or only mildly pathogenic, although severe outbreaks have occurred in young animals in kennels. For coccidiosis of dogs, Brunnthaler (1977) found amprolium as well as spiramycin effective; however, sulfadimethoxine, sulfaguanidine, and sulfadiazine-trimethoprim were less effective. Sulfaquinoxaline treatment has led to prolonged prothrombin times and death (Patterson and Gremn, 1975).

For prophylaxis of *Isospora rivolta* in cats, Matsui et al. (1977) found sulfamonomethoxine effective; for treatment of *Isospora felis*, Wilkinson (1977) found sulfadimethoxine effective.

Although coccidiosis is not now as serious a problem in mink raised in wire pens as it once was in mink raised on the ground, three species were found to be quite prevalent on 29 ranches in Wisconsin (Foreyt and Todd, 1976). Therefore, eight anticoccidials were tested for efficacy in mink (Foreyt et al., 1977); in juveniles, amprolium and sulfaquinoxaline reduced almost all oocyst production, and in adults, monensin, a commercial antibiotic-sulfonamide mixture, and lasalocid almost completely inhibited oocyst production.

Treatment has been attempted for *Isospora belli* in humans. Quinacrine hypochloride, nitrofurantoin, tetracycline, and metronidazole were ineffective for controlling clinical illness. A combination of either pyrimethamine and sulfadiazine or trimethoprim and sulfamethoxazole (co-trimoxazole) resulted in complete cures (Trier et al., 1974; Syrkis et al., 1975; Westerman and Christensen, 1979).

Neither chemoprophylaxis nor chemotherapy is administered to poultry and livestock for *T. gondii*, nor is chemoprophylaxis used for humans. Acute human toxoplasmosis is most often treated in the U.S. with combinations of pyrimethamine and sulfas (usually sulfadiazine), which act synergistically. However, because pyrimethamine inhibits folinic acid biosynthesis, it is potentially toxic to bone marrow as well as potentially teratogenic; alternative compounds should therefore be tried (Araujo and Remington, 1974). Norrby et al. (1975) found that a combination of trimethoprim and a sulfa was more effective than pyrimethamine and a sulfa against toxoplasmosis; however, Seah (1975) found the opposite.

Clindamycin, a semisynthetic antibiotic produced by chlorination of lincomycin, protects mice from lethal acute toxoplasmosis, prevents congenital transmission in mice, removes *Toxoplasma* from peripheral tissues of mice with acute and chronic infection (except for cysts in the central nervous system), and effectively treats acute experimental retinochoroiditis in rabbits (McMaster et al., 1973; Araujo and Remington, 1974; Tabbara et al., 1974; Tate and Martin, 1977). The drug has been used in humans with ocular toxoplasmosis. However, the efficacy of clindamycin for treatment of acute and chronic toxoplasmosis in humans has not yet been determined by controlled human studies. Because other treatments are so toxic, spiramycin has been recommended for treatment during pregnancy (Couvreur, 1976).

Of the many anticoccidial drugs that have been used so effectively to treat coccidiosis in poultry and livestock, few have even been tested for *Toxoplasma* in cats. Sulfadiazine alone or combined with pyrimethamine reduces the number of toxoplasma oocysts shed when given orally, by injection, or in feed (Frenkel, 1975; Sheffield and Melton, 1976). Sulfamerazine, sulfamethazine, sulfalene, and sulfadoxone may be substituted for sulfadiazine in combination with pyrimethamine (Frenkel, 1978). Although Oshima and Kumada (1974) reported that SDDS (2-sulfamoyl-4,4′-diaminodiphenylsulfone) prevented oocyst shedding in cats given continuous medication in their feed, Dubey and Yeary (1977) found that neither toxic levels of pyrimethamine-sulfadiazine nor high levels of SDDS or clindamycin completely suppressed oocyst shedding by cats.

Theoretically, the incorporation of anticoccidial drugs into cat feed could prevent or greatly reduce oocyst shedding. However, the paucity of data on parasite strain differences, drug resistance, and possible interference with immunity makes chemoprophylaxis an impractical approach. For example, 11 strains of *Toxoplasma* were tested with six antitoxoplasmic drugs; highly variable strain responses were found (Oshima and Hoshino, 1977). Drug resistance of *Toxoplasma* to sulfamethoxazole and sulfadiazine has been reported (Lai et al., 1974). Under these circumstances, reliance on medicated food could result in a mistaken belief that *Toxoplasma* infection is unquestionably being prevented.

Furthermore, some anticoccidials prevent development of coccidial stages that immunize the host. The intermittent feeding of medicated food may result in the accumulation of partially developed stages that complete the life cycle and yield numerous oocysts when medication is withdrawn. In contrast, one or more low level natural infections result in an immunized cat that sheds few if any oocysts.

11

Despite the extremely high prevalence of *Sarcocystis* species in cattle and sheep, acute infections are infrequently diagnosed. Treatment of a naturally acquired infection has yet to be reported. Amprolium used prophylactically is effective in reducing the severity of acute experimental sarcocystosis associated with endothelial schizonts and blood-borne merozoites in cattle and sheep (Fayer and Johnson, 1975; Leek and Fayer, 1980a). Salinomycin also is effective in preventing acute experimental sarcocystosis in sheep while allowing for the development of immunity (Leek and Fayer, 1980b). Although quinine sulfate prevented entry of *Sarcocystis* zoites into cultured cells (Fayer et al., 1972), this compound has not been tested in vivo. Anticoccidials have not been tested for efficacy against *Sarcocystis* stages in carnivores.

Despite the economic importance of *B. besnoiti* in cattle in parts of Europe, Asia, and Africa, no reports of attempted chemical treatment of cattle could be found in the literature. Goats artificially infected with bovine strains of *B. besnoiti* grown in rabbits recovered from clinical signs after treatment with 1% antimony solution (Lee et al., 1979). Laboratory infections of hamsters with *Besnoitia jellisoni* and mice with *B. darlingi* were controlled with sodium sulfadiazine in drinking water (Smith and Frenkel, 1977). Although quinine sulfate prevented entry of *B. jellisoni* zoites into cultured cells (Fayer et al., 1972), this compound has not been tested in vivo. Anticoccidials have not been tested for efficacy against *Besnoitia* stages in carnivores.

Future Prospects

Only for broiler production has the potential anticoccidial drug market been large enough to activate major screening and developmental programs by pharmaceutical manufacturers. The potential U.S. market, estimated at 10 million dollars in 1960 and 20 million dollars in 1970, had increased to 70 million dollars in 1980, reflecting not only expansion of the industry but also inflationary trends. The cost of obtaining government clearance continues to rise, with ever increasing requests for more sophisticated experimentation. Unfortunately, agencies in different countries may require repetition of expensive experimentation already executed elsewhere. Personnel in regulatory agencies, who are frequently hesitant about giving approval for marketing new drugs, often postpone approvals by requiring additional experimental results after all data originally requested have been submitted. Delays and postponements, often requiring years of waiting to go to market, have discouraged many formerly successful manufacturers of anticoccidials. Several have closed down their anticoccidial study units, while others plan to discontinue work in the near future.

If new proposals for "Sensitivity of Method" advocated by the Food and Drug Administration (Birkhead, 1979) are promulgated, the additional research cost would average 4.3 million dollars for every new drug. Furthermore, similar studies for each metabolite, of which

60 or more may be possible, might also be required. Reid and McDougald (1981) conclude that costs of governmental clearances have been a major factor in reducing by two-thirds the 25 pharmaceutical companies now screening for new anticoccidials.

The rapid emergence of drug resistance (Chapman, Chapter 10, this volume) may result in a relatively short market life (Figure 5). Several anticoccidials have had a useful market life of less than 3 yr. Glycamide and nequinate do not appear on the chart, because they had a market life of less than 6 months. Several successful pharmaceutical companies have abandoned further testing programs for new drugs (Joyner, 1964; Reid, 1978b).

The poultry industry would like to find a new approach to coccidiosis control. With the cost of anticoccidials doubling or tripling in the past 10 yr, users have often expressed disfavor with present-day programs, which seem to offer only increasing costs for their coccidiosis control programs.

Future prospects for new anticoccidials useful in cattle are similarly discouraging. Fitzgerald (1975), in an extensive survey of the significance of cattle coccidiosis in the United States, concluded that about 3.85 million cattle are treated and about 80,000 die annually of coccidiosis mostly caused by *E. zuernii*. Treatment, usually with sulfa drugs rather than prophylactic medication, is the approach most widely used except after large herd outbreaks, when prophylactic medication may be used. The market for preventive anticoccidial drugs is not large enough to interest pharmaceutical companies in initiating primary screening to control cattle coccidiosis. Amprolium and decoquinate are the only two drugs currently approved in the U.S. with anticoccidial claims (Gates, 1979) for use with cattle. Such approvals were not sought until some commercial success was established using the same drugs on chickens. Although monensin and other ionophores have shown anticoccidial activity in calves, commercial success of monensin resulted from increases in feed efficiency with the use of this drug.

Although economic losses due to coccidiosis occur with turkeys, ducks, geese, rabbits, game birds, guinea fowl, sheep, hogs, goats, and other livestock, the potential market for anticoccidial drugs is too small to interest many pharmaceutical companies. Risk capital is required to screen, manufacture, and obtain governmental clearances for such a product. The last hurdle is now recognized as requiring much more time, effort, and capital than all other requirements combined. Similarly, there is insufficient market potential to motivate public health interest in anticoccidial drugs useful in control of sarcocystosis, toxoplasmosis, and coccidiosis as it affects humans, cats, dogs, and other pets. In many cases, drugs that would prevent or control these coccidia are known or could readily be found, but the costs of clearance are too high for the manufacturer to proceed with clearance procedures.

Efforts of the U.S. Food and Drug Administration (Raines, 1979) to simplify drug clearance procedures for "minor uses," including use in species of game birds, ducks, trout, salmon, goats, and sheep (major species are cattle, horses, swine, chickens, turkeys, dogs, and cats), have so far produced no new applications from the pharmaceutical industry.

11

Prospects are not bright for the increasing use of chemotherapy as an approach to the control of coccidiosis within the animal industry. The industry should encourage research that may lead to alternative methods of coccidiosis control if chemotherapy should no longer remain a viable method.

IMMUNOLOGICAL CONTROL

In Poultry

Various methods have been used to induce sufficient immunity to provide flock protection against mortality and morbidity in poultry (Reid, 1972, 1978a; Rose, 1978; Rose and Long, 1980; reviewed by Rose, Chapter 8, this volume). Acquired immunity is built up in chickens and turkeys by repeated successive exposures to small numbers of oocysts. Oocysts that are picked up while the bird is feeding continually reinforce this immunity. Infections that are so mild that they may be subclinical are sometimes called *coccidiasis* to differentiate them from clinical coccidiosis. Low level exposure, subsequent cycling, and immunity occur so naturally and frequently in many flocks that the poultryman is usually unaware of the presence of oocysts until an outbreak follows a sudden build-up in oocyst numbers. After feeding of preventive anticoccidial drugs has ceased, the poultry producer may be fully dependent upon this immunity for protection against coccidiosis, because previously uninfected older chickens are equally as susceptible as young chicks to heavy oocyst infections (Hegde and Reid, 1969). Since immunity is developed independently against each species, exposure to all species is required to provide complete flock protection.

The most extensive program of planned immunization in the U.S. has been that of Edgar (1956, 1958, 1964; Libby et al., 1959; Stuart, 1960). Although adoption outside the U.S. has been limited, the program has been recommended in Rhodesia (Huchzermeyer, 1968, 1972). This program has been adopted by a number of primary breeders for use in genetically valuable breeding stock. A suspension of live oocysts (sold under the trade name of Coccivac), consisting of a mixture of eight species, is given in the feed or drinking water to chicks between 4 and 14 days of age. Although the primary exposure starts the process of immunization, development of substantial immunity depends upon oocyst production arising from the initial exposure, as a source of infective oocysts that reinfect the birds. The number of infections required to produce complete immunity varies with the species. The initial infection requires only a few oocysts (100–700 oocysts/species; Edgar, 1958; Edgar, personal communication). The desirability of using small numbers of oocysts has been confirmed by Joyner and Norton (1973, 1976), who found that immunity to *Eimeria maxima*, *E. acervulina*, and *E. tenella* was stronger after repeated daily doses of one or five oocysts over a period of 20 days than after a single dose of a larger number of oocysts. Immunity produced by "trickle infection" becomes reinforced by reinfection from oocysts in the litter.

Adoption of this planned immunization program has been limited to use with breeders and floor layers. Some failures have been reported when used for broilers where supervision was inadequate. Precautions require that: 1) all birds obtain some of the oocysts, overnight water or feed starvation be recommended, and a suspending agent be used in water to provide equal oocyst distribution; 2) the litter be moist enough (20–30% humidity recommended) to permit freshly shed oocysts to sporulate and provide subsequent exposures; and 3) use of anticoccidial drugs that may reduce coccidial cycling be limited or entirely avoided. Even anticoccidial drugs that permit passage of some oocysts (e.g., Zoalene or amprolium) may interrupt the life cycle sufficiently to prevent development of good immunity. Limited adoption of the program is a result of: 1) failure to develop immunity by a few birds within a flock, 2) failure of some species to become established and produce oocysts, and 3) improper storage of oocysts in vaccine bottles, that is, overheating in parcel post shipments or in automobiles and freezing during refrigerated storage, which destroy oocysts.

The loss of a single bird out of a large flock, with a diagnosis of cecal coccidiosis, may represent an inconsequential economic loss, but confidence of the poultryman is usually lost. Concomitant infections with Marek's disease (Biggs et al., 1968; Rice and Reid, 1973) or infectious bursal disease (Anderson et al., 1977) may interfere with immunity to coccidiosis.

Planned immunization programs would prove more acceptable to poultry producers if immunogenic, attenuated coccidia were available for mass vaccination programs (Long, 1978). Efforts of various investigators to use irradiation in developing nonpathogenic strains have been unsuccessful. More recently, nonpathogenic strains of *E. tenella* have been established by egg embryo cultivation (Long, 1974) and by selecting strains that mature early (Jeffers, 1974). Although Long and Millard (1977) have also established nonpathogenic strains of other species, nonpathogenic strains of *Eimeria necatrix* have not been developed. Genetically stable attenuated strains have not been isolated from the field.

A planned immunization program for turkeys would be useful, because they are usually raised on litter or on range. Experimental immunization has been demonstrated for *Eimeria* species affecting turkeys (Edgar, 1958).

If chemotherapeutic control needs to be abandoned because of drug resistance, cost of drug clearance, or public pressures against use of medication, planned immunization seems to be the most promising alternative for coccidiosis control.

In Mammals

Much progress has recently been made toward understanding the immunological response of the host to infection with species of *Eimeria*, *Toxoplasma*, *Besnoitia*, *Sarcocystis*, and *Hammondia* (see Rose, Chapter 8, this volume). However, practical applications are few. In general, immunity is acquired as a result of active infections. These infections need not result in severe disease but can be mild and still produce some resistance to reinfection. Species of *Eimeria* differ in their immunizing ability, some requiring only one infection and others requiring several to produce immunity. Immunity is species-specific with little or no cross-protection among species. The duration of immunity is not easy to determine. It is generally concluded that immunity decreases with time after an infection and must be "boosted" by repeated infections, unless organisms persist in tissues for a prolonged time with little proliferation (premunition).

There are neither vaccines nor commercially available immunization programs similar to Coccivac for any of the mammalian species of *Eimeria*, *Isospora*, *Toxoplasma*, *Sarcocystis*, *Frenkelia*, or *Hammondia*.

A live vaccine has been prepared and used effectively against bovine besnoitiosis (Bigalke et al., 1974). *Besnoitia* zoites from cysts obtained from the blue wildebeest were serially passaged 78 times in rabbits by subinoculation of blood collected during the acute phase of the disease. Zoites were then grown in lamb kidney cell cultures and stored as a frozen stabilate for subsequent vaccine production in either lamb or monkey kidney cell cultures. A single inoculation of cattle with this vaccine provided complete protection against clinical infection for 1–4 yr.

GENETIC RESISTANCE OF THE HOST AS A CONTROL METHOD

The prospects for control of coccidiosis through selective breeding are best applied to the domestic fowl, in which genetic variation in host response to coccidia is abundant and the effectiveness of selection in altering host resistance has been clearly demonstrated (Chobotar and Scholtyseck, Chapter 4, this volume; Jeffers, 1978). The extensive work of 21 investigators in 25 papers illustrates these theoretical possibilities. Genetic improvement in the overall performance of poultry requires the breeder to practice selection on several traits in tandem, or simultaneously, using a selection index in which each trait is weighted according to its relative economic importance. The number of traits selected for reduces the intensity of selection practiced on each individual trait. Thus, breeders must concentrate on a limited number of the traits most directly related to the profitability of poultry production. One major breeder of egg-producing stock (Hy-Line Poultry Farms, 1963) was unsuccessful in marketing a selected coccidiosis-resistant strain. It is unlikely that poultry breeders will soon make further efforts to control coccidiosis by genetic selection. The current complacency of poultry breeders stems from confidence in the effectiveness of current chemotherapeutic, immunological, and management control measures (Jeffers, 1978).

Virtually no attempts have been made to control coccidiosis, toxoplasmosis, besnoitiosis, or sarcocystosis in mammals via genetic resistance of the host. Presently, investigators are at the stage of documenting differences in susceptibility to infection. Results of experimental infections suggest there are strain-dependent differences in susceptibility of mice to *Eimeria* (Klesius and Hinds, 1979). Similar strain-dependent differences in susceptibility to *Toxoplasma* were observed in mice (Araujo et al., 1976; Kamei et al., 1976), and a specific gene locus affecting susceptibility to *T. gondii* was suggested (Williams et al., 1978). Susceptibility of sheep to *T. gondii* seemed to be related to the presence of hemoglobin type B (Waldeland, 1976). However, after examination of more sheep, no association was found between susceptibility to *T. gondii* and hemoglobin type (Waldeland, 1977).

CONTROL BY NUTRITIONAL SUPPLEMENTATION OR ANTAGONISM

The nutritional state of animals can influence various infectious diseases (reviewed by Scrimshaw et al., 1968). Specific information relating to coccidiosis is not abundant. There are no nutrients known that act prophylactically against coccidia when added to the diet; however, supplementation with some vitamins may be beneficial to the host. When chicken rations containing normal levels of vitamin A are supplemented with additional vitamin A, neither morbidity nor mortality from coccidiosis is reduced, but recovery following coccidiosis may be hastened (Reid, 1972). Mortality from *E. tenella* and *E. necatrix* is reduced when vitamin K is added to deficient diets, but this has no effect on the nonhemorrhagic species *E. acervulina*, *Eimeria brunetti*, and *E. maxima* (Baldwin et al., 1941; Ryley and Hardman, 1977). The minimum requirement for this vitamin (0.53 mg/kg) reduces clotting time and prevents hemorrhage. The highest requirement for vitamin K is in young chicks, in breeder hens to prevent hatching of vitamin-deficient chicks, and perhaps during periods of medication with sulfonamides (Reid, 1972). Shortly after discovery of the dramatic effects of added vitamin K on *E. tenella* infection, one manufacturer considered requesting a drug claim for coccidiosis control. However, reduction in the price of the product, the effects being limited to species of coccidia causing bleeding, and the rapid activity of nutritionists in universally adding the vitamin (often in excess amounts) to diet formulae made this approach unfeasible.

The severity of coccidiosis from *E. tenella* in chickens seems to be related to the protein content of feed. In groups of chickens fed 0.5, 10, 15, 20, and 30% protein diets before inoculation with oocysts, fatalities increased progressively up to 15% (Britton et al., 1964). It was hypothesized that as more protein became available for trypsin digestion, excystation (a trypsin-dependent process) increased, producing greater numbers of sporozoites causing more severe coccidiosis. High mortality was observed by Sharma et al. (1973) in *E. tenella*-infected chickens fed high protein diets, and greater production of oocysts was observed in *E. acervulina*-infected chickens fed high protein diets. After a natural outbreak of intestinal coccidiosis,

11

a high protein diet seemed more beneficial; hens fed 13% protein rather than 17% protein had a greater drop in egg production (Harms et al., 1967).

Vitamin-deficient diets were fed to chickens to determine the requirements of coccidia for 19 vitamins (Warren, 1968). For normal development, *E. acervulina* and *E. tenella* required thiamine, riboflavin, biotin, nicotinic acid, and folic acid in the diet. Studies of growth of *E. tenella* in cell cultures have provided information concerning nutrients necessary for development. In one or more of these studies, glutamine, *p*-aminobenzoic acid, nicotinamide, pyridoxine, thiamine, vitamin A, biotin, folic acid, calcium pantothenate, pyridoxal, riboflavin, ascorbic acid, calciferol, α-tocopherol, menadione, choline chloride, inositol, and vitamin B_{12} were found necessary for development through second generation schizogony (Ryley and Wilson, 1972; Sofield and Strout, 1974; Doran and Augustine, 1978; Latter and Holmes, in press). Many of these nutrients are also necessary for gamete development.

It follows that if coccidia are dependent on specific vitamins, either the lack of these vitamins or the presence of vitamin antagonists will interfere with their normal development. Several coccidiostats function in just such a way—as vitamin antagonists. Sulfonamides antagonize *p*-aminobenzoic acid, pyrimethamine antagonizes folinic acid, and amprolium antagonizes thiamine (for a more thorough review, see McDougald, Chapter 9, this volume).

As new anticoccidials are introduced, their chemical mode of action needs to be elucidated concerning any other effects on nutritional interaction with other feed ingredients. Research in this area should be a challenge to physiologists, biochemists, and nutritionists.

These authors do not anticipate any early practical developments in coccidiosis control from nutritional studies; however, any possibilities of new discoveries should be encouraged. Experience with vitamin K suggests that new discoveries would be rapidly incorporated into nutritional practices.

CONTROL BY SANITATION AND MANAGEMENT PRACTICES

For Poultry

Before prophylactic medication became the major method of coccidiosis control in broiler production, most recommendations coming from state and federal experiment stations emphasized control by sanitation, disinfection, or quarantine. Sanitation in this context refers chiefly to management practices that prevent chickens from coming into contact with feces contaminated with oocysts. Although these practices alone were entirely inadequate in avoiding disastrous outbreaks, some sanitation practices are still useful in modern operations.

Coccidiosis outbreaks seldom occur in laying hens maintained in cages, but have occasionally occurred when feces accumulated on the cage floors or in watering devices. Oocysts may also build up when feces contaminate watering and feeding systems. Insufficiently heated, reprocessed feces fed back to chickens as a dietary supplement have also produced outbreaks.

Anticoccidial drugs are not required to prevent coccidiosis if cages are kept clean. To prevent coccidiosis and for other management reasons, many poultrymen would like to convert their broiler and breeder flocks entirely to cage-type operations. Discouraging results have been obtained from experiments in several large-scale broiler operations. Because of equipment costs, downgrading due to leg problems and breast blisters, problems of recovering dead birds from cages, labor costs of maintenance, and the necessary change from rearing to laying cages, most companies rely on floor-rearing methods. For breeder flocks, similar difficulties plus fertility problems have discouraged most producers from converting to cage operations. A few poultrymen have combined artificial insemination with cage layer management.

Poultry producers trained to use good sanitary programs show much concern for oocyst contamination in floor or litter-reared broiler operations, especially when effective anticoccidial drugs are used. However, coccidiosis problems occur more frequently in new houses with new litter than in old houses with reused litter. Extreme sanitary measures may delay exposure to coccidia and as a result, development of immunity; this delay may result in coccidiosis outbreaks when the birds are mature and most valuable. Use of anticoccidial drugs on day-old chicks may also prevent the development of immunity leading to outbreaks of coccidiosis after drug withdrawal. Unexposed older chickens are just as susceptible to coccidiosis as younger birds (Hegde and Reid, 1969).

Litter should be inspected for wet spots, because many outbreaks occur after leaks in roofs or waterers. Wet spots provide ideal conditions for rapid oocyst sporulation. Wet litter should be cleaned out and replaced with dry litter. Older sanitation practices such as a daily removal of feces have become totally impractical with expansion to large flocks in a single house. Although footbaths of disinfectant installed at the entrance of a poultry house aid in control of bacterial and viral diseases, they are completely ineffective against oocysts; exposure time is too short, most disinfectants used are too weak to kill oocysts, and attendants usually avoid their use except during official inspections.

Although many other poultry diseases are best controlled by quarantine and slaughter, or by disinfection and sanitation, the widespread occurrence of oocysts in large poultry buildings makes such an approach for coccidiosis control impractical.

For Mammals

Sanitation is important in preventing coccidiosis in mammals. Overstocking or crowding of animals not only creates stress, which may reduce resistance to coccidiosis, but also results in a rapid buildup of feces; it should therefore be avoided. Pastures, yards, pens, and feedlots should be cleaned as often as possible and should have good drainage. Concrete or gravel floors are easier to clean and provide better drainage than dirt floors. Buildings should be self-cleaning or designed to facilitate frequent cleaning, and should be kept dry. Feed and water containers should be high enough off the ground to prevent or reduce kicking or splashing of feces into feed or water. Feed should never be placed directly on the ground or floor. Some feeds, such as corn silage or chopped alfalfa, are moist enough to support sporulation and survival of oocysts and, when used, feeders should be emptied and cleaned regularly. Bedding such as hay and wood chips retain moisture and ensure survival of oocysts; thus, regular cleaning is necessary. All animals known to have coccidiosis should be promptly isolated from the others.

Age segregation minimizes the number of oocysts that young susceptible animals might acquire from adult carriers that show no clinical signs. Dairy calves are usually isolated from their dams within 1–3 days after birth. Adult female sheep, goats, and swine are often brought into confined housing immediately before, during, or shortly after giving birth. The offspring are housed with the adults for a few days before the age groups are separated. This period before separation is very important in the spread of coccidiosis. The housing facilities often are heavily contaminated with oocysts shed by the numerous confined adults; acute or peracute coccidiosis develops in the young during or shortly after this time. To reduce the number of oocysts in this environment, adults should receive anticoccidial medication from 1 week before giving birth through the period they are housed with the young. All feces should be removed at 2-day intervals. Under less intensive rearing conditions such as ranging of animals, age segregation may not be necessary. Usually the vastness of the range serves to dilute the oocysts. However, animals do congregate around water and feed bunkers. Therefore, feces should not be allowed to accumulate in these areas.

Hay lofts and other feed storage facilities, feeders and waterers, should be kept free of cats, dogs, birds, and rodents. Careful inspection of feed storage facilities on farms revealed relatively large quantities of cat feces in feed rooms and grain bins and in ground feed (Penkert, 1973; Plant et al., 1974). On a farm where toxoplasmosis has occurred, feed contaminated with cat feces containing oocysts of *Toxoplasma* could serve as a source of infection for rats, mice, birds, livestock, other cats, and humans (Frenkel et al., 1975). Two outbreaks of bovine sarcocystis in the U.S. were traced to canine feces in hay fed to calves (Frelier et al., 1977; Giles et al., 1980). The dogs had been housed in barns or had free access to barns where hay was stored.

FUTURE RESEARCH IDEAS THAT MAY LEAD TO IMPROVED CONTROL

Chemotherapy

1. Improve tissue culture techniques to aid in mode-of-action studies and to simplify drug screening techniques (Long, 1970).

2. Establish a preservation and distribution center where various species and strains could be made available to pharmaceutical companies and research centers. Cryostatic methods now available for long-term preservation make this possibility more practical than when first suggested (Hunter, 1959).
3. Conduct physiological and biochemical studies on mode-of-action of available anticoccidial drugs with the goal of a rational synthesis of new products rather than continue to select them from blind mass screening attempts.
4. Continue the search for combinations of presently available anticoccidials, which may supplement the list of synergistically active mixtures (McDougald, 1977, 1978).
5. Continue to search for less toxic drugs effective against tachyzoites of *T. gondii* in humans.
6. Test existing anticoccidials for efficacy against acute toxoplasmosis, besnoitiosis, and sarcocystis.

Immunity

1. Develop a rapid, practical method of determining flock immunity, which would entail handling the birds only once. This could replace the labor-consuming procedure of present-day challenge methods. A preliminary study using a delayed hypersensitivity skin test (Giambrone et al., 1980) needs further confirmation and verification.
2. Determine the immune mechanism responsible for development of the distinctive tissue-oriented immunity that gives protection after "trickle exposure" to coccidia.
3. Find immunogenic, nonpathogenic strains of pathogenic species of *Eimeria*, which can be used in a vaccination program (Rose, 1976).
4. Develop and improve practical inexpensive methods of introducing flock immunity for protection against coccidiosis (Edgar, 1958; Davis and Reynolds, 1979; Rose and Long, 1980).
5. Reduce or eliminate shedding of *Toxoplasma* oocysts by cats by finding avirulent, highly immunogenic strains of *Toxoplasma*: determine the number of immunotypes within the species; determine the suitability of planned infection with oocysts combined with drug treatment for clinical and domestic use; and determine suitability of live versus killed injectable vaccines.
6. For protection of humans against *Toxoplasma* at vulnerable times such as pregnancy, develop vaccine from avirulent highly immunogenic strains. Explore the possibilities of producing protective antibodies to systemic coccidia (*Toxoplasma*, initially) through hybridoma techniques (lymphocytes from an animal sensitized to a specific antigen).
7. For protection of cattle, sheep, and pigs against sarcocystis, determine whether protective immunity is developed after infection with various species. Find avirulent immunogenic strains.

8. Explore possibilities of producing immunogenic antigens through recombinant DNA techniques.

Genetics

1. Through breeding methods, select host strains that are less susceptible to coccidiosis and other diseases (Jeffers and Shirley, Chapter 3, this volume).

Nutritional Supplementation

1. Continue basic nutritional studies on differences among breeds in response to anticoccidial drug treatments.
2. Increase knowledge of dietary requirements of coccidia in vivo and in vitro (Latter and Holmes, in press).

Sanitation and Combinations of Methods

1. Continue the search for practical, efficient and inexpensive methods of animal production by which livestock can be produced without contact with feces.

LITERATURE CITED

Anderson, W. I., Reid, W. M., Lukert, P. D., and Fletcher, O. J., Jr. 1977. Influence of infectious bursal disease on the development of immunity to *Eimeria tenella*. Avian Dis. 21:637–641.

Araujo, F. G., and Remington, J. S., 1974. Effect of clindamycin on acute and chronic toxoplasmosis in mice. Antimicrob. Agents Chemother. 5:647–651.

Araujo, F. G., Williams, D. M., Grumet, F. C., and Remington, J. S. 1976. Strain-dependent differences in murine susceptibility to *Toxoplasma*. Infect. Immun. 13:1528–1530.

Ayeni, A. O., Dingeldein, E., and Durr, U. 1972. Studies on the inactivation of coccidian oocysts. Acta Vet. Acad. Sci. Hung. 22: 111–122.

Baldwin, F. M., Wiswell, O. B., and Jankiewicz, J. A. 1941. Hemorrhage control in *Eimeria tenella* infected chicks when protected by antihemorrhagic factor, vitamin K. Proc. Soc. Exp. Biol. Med. 48:278–280.

Bigalke, R. D., Basson, P. A., McCully, R. M., Bosman, P. P., and Schoeman, J. H. 1974. Studies in cattle on the development of a live vaccine against bovine besnoitiosis. J. S. Afr. Vet. Assoc. 45:207–209.

Biggs, P. M., Long, P. L., Kenzy, S. G., and Rootes, D. G. 1968. Relationship between Marek's disease and coccidiosis. Vet. Rec. 83:284–289.

Birkhead, H. 1979. What's wrong with FDA's SOM? Anim. Nutr. Health 34(Oct.):5, 26, 28.

Box, E. D., and McGuinness, T. B. 1978. *Sarcocystis* in beef from retail outlets demonstrated by digestion technique. J. Parasitol. 64:

161–162.
Britton, W. M., Hill, C. H., and Barber, C. W. 1964. A mechanism of interaction between dietary protein levels and coccidiosis in chickens. J. Nutr. 82:306–310.
Brotherston, J. G. 1948. The effect of relative dryness on the oocysts of *E. tenella* and *E. bovis*. Trans. R. Soc. Trop. Med. Hyg. 42:10–11.
Brunnthaler, F. 1977. Coccidiosis in dogs. Prakt. Tieraerztl. 58:849–851.
Chang, K. 1937. Effects of temperature on the oocysts of various species of Eimeria (Coccidia, Protozoa). Am. J. Hyg. 26:337–351.
Christensen, J. F. 1939. Sporulation and viability of oocysts of *Eimeria arloingi* from the domestic sheep. J. Agric. Res. 59:527–534.
Coudert, P., and Yvoré, P. 1973. Sensibilité des oocysts d'*Eimeria* a la température. J. Rech. Avic. Cunic. (Dec.):269–272.
Couvreur, J. 1976. Comment je raite une toxoplasmose chez la femme enceinte. Gaz. Med. Fr. 83:3659–3663.
Davis, P. J., and Reynolds, J. F. (Unilever). Immunisation of poultry to coccidiosis—by mixing spore-carrying oocysts with the feedstuff. U.K. patent DT 2849-226. 1979.
Desmonts, G. S., Couvreur, J., Alison, F., Baudelot, J., Gerbeaux, J., and Lelong, M. 1965. Étude epidemiologique sur la toxoplasmose: de l'influence de la cuisson des viandes de boucherie sur la frequence de l'influence humaine. Rev. Fr. Etud. Clin. Biol. 10: 952–958.
Doran, D. J., and Augustine, P. 1978. *Eimeria tenella*: Vitamin requirements for development in primary cultures of chicken kidney cells. J. Protozool. 25:544–546.
Dubey, J. P. 1974. Effect of freezing on the infectivity of *Toxoplasma* cysts to cats. J. Am. Vet. Med. Assoc. 165:534–536.
Dubey, J. P. 1977. *Toxoplasma*, *Hammondia*, *Besnoitia*, *Sarcocystis* and other tissue cyst-forming coccidia of man and animals. In: J. P. Kreier (ed.), Parasitic Protozoa, pp. 101–237. Vol. 3. Academic Press, New York.
Dubey, J. P., and Mehlhorn, H. 1978. Extraintestinal stages of *Isospora ohioensis* from dogs in mice. J. Parasitol. 64:689–695.
Dubey, J. P., Miller, N. L., and Frenkel, J. K. 1970. The *Toxoplasma gondii* oocyst from cat feces. J. Exp. Med. 132:636–662.
Dubey, J. P., and Yeary, R. A. 1977. Anticoccidial activity of 2-sulfamoyl-4,4-diaminodiphenylsulfone, sulfadiazine, pyrimethamine and clindamycin in cats infected with *Toxoplasma gondii*. Can. Vet. J. 18:51–57.
Duncan, S. 1959. The effects of some chemical and physical agents on the oocysts of the pigeon coccidium *Eimeria labbeana* (Pinto, 1928). J. Parasitol. 45:193–197.
Edgar, S. A. 1954. Effect of temperature on the sporulation of oocysts of the protozoan *Eimeria tenella*. Trans. Am. Microsc. Soc. 73:237–242.
Edgar, S. A. 1956. Coccidiosis immunization. Iowa State Coll. Vet. 17:9–11, 17.
Edgar, S. A. 1958. Coccidiosis of chickens and turkeys and control by immunization. Avicultura Moderna del XI Congreso Mundial. pp. 415–421, 769. La Prensa Medica Mexicana. Mexico City.
Edgar, S. A. Stable coccidiosis immunization. U.S. Patent 3,147,186. 1964.
Exner, G., Wachtel, D., and Schweitzer, H. 1972. Untersuchungen uber die Wirksamkeit von Desinfektion-smitteln auf Toxoplasmazysten. Off. Gesundh.-Wesen 27:2198–2200.
Farr, M. M., and Wehr, E. E. 1949. Survival of *Eimeria acervulina*, *E. tenella* and *E. maxima* oocysts on soil under various conditions. Ann. N.Y. Acad. Sci. 52:468–472.
Fayer, R. 1975. Effects of refrigeration, cooking and freezing on *Sarcocystis* in beef from retail food stores. Proc. Helminth. Soc. Wash. 42:138–140.
Fayer, R., and Johnson, A. J. 1975. Effect of amprolium on acute sarcocystosis in experimentally infected calves. J. Parasitol. 61: 932–936.
Fayer, R., Melton, M. L., and Sheffield, H. G. 1972. Quinine inhibition of host cell penetra-

tion by *Toxoplasma gondii*, *Besnoitia jellisoni* and *Sarcocystis sp. in vitro*. J. Parasitol. 58: 595–599.

Fitzgerald, P. R. 1972. Efficacy of monensin or amprolium in the prevention of hepatic coccidiosis in rabbits. J. Protozool. 19:332–334.

Fitzgerald, P. R. 1975. The significance of bovine coccidiosis as a disease in the United States. Bovine Practitioner (Nov.):28–33.

Foreyt, W. J., and Todd, A. C. 1976. Prevalence of coccidia in domestic mink in Wisconsin. J. Parasitol. 62:496.

Foreyt, W. J., Todd, A. C., and Hartsough, G. R. 1977. Anticoccidial activity of 8 compounds in domestic mink. Am. J. Vet. Res. 38:391–394.

Frelier, P., Mayhew, I. G., Fayer, R., and Lunde, M. N. 1977. Sarcocystosis: A clinical outbreak in dairy calves. Science 195:1341–1342.

Frenkel, J. K. 1955. Effects of hormones on the adrenal necrosis produced by *Besnoitia jellisoni*.

Frenkel, J. K. 1973. Toxoplasmosis: Parasite life cycle, pathology, and immunology. In: D. M. Hammond and P. L. Long (eds.), The Coccidia: *Eimeria*, *Isospora*, *Toxoplasma*, and Related Genera, pp. 343–410. University Park Press, Baltimore.

Frenkel, J. K. 1975. Toxoplasmosis in cats and man. Feline Practice (Jan.–Feb.): 28, 29, 33, 36, 37, 40, 41.

Frenkel, J. K. 1978. Toxoplasmosis in cats: Diagnosis, treatment and prevention. Comp. Immunol., Microbiol., Infect. Dis. 1:15–20.

Frenkel, J. K., and Dubey, J. P. 1972. Rodents as vectors for feline coccidia, *Isospora felis* and *Isospora rivolta*. J. Infect. Dis. 125:69–72.

Frenkel, J. K., Ruiz, A., and Chinchilla, M. 1975. Soil survival of *Toxoplasma* oocysts in Kansas and Costa Rica. J. Trop. Med. Hyg. 24: 439–443.

Gallazzi, D. 1976. Use of anticoccidials in table rabbit rearing. Coniglicoltura 13:31–36.

Feed Additive Compendium. 1980. Miller Publishing Co., Minneapolis, MN.

Gestrich, R. 1974. Investigations on survival and resistance of *Sarcocystis fusiformis* cysts in beef. Proceedings of the 3rd International Congress on Parasitology, Munich. p. 117.

Giambrone, J. J., Klesius, P. H., and Edgar, S. A. 1980. Avian coccidiosis: Evidence of a cell-mediated immune response. Poult. Sci. 59: 38–40.

Giles, R. C., Tramontin, R., Kadel, W. L., Whitaker, K., Miksch, D., Bryant, D. W., and Fayer, R. 1980. Sarcocystosis in cattle in Kentucky. J. Am. Vet. Med. Assoc. 176:543–548.

Gwyther, M. J. 1976. The efficacy of monensin and sulfaquinoxaline against the rabbit liver coccidian, *Eimeria stiedae*. Master's thesis, Department of Poultry Science, Clemson University, Clemson, S.C.

Harms, R. H., Simpson, C. F., Bradley, R. E., and Damron, B. L. 1967. Influence of coccidiosis on protein requirement of laying hens. Proceedings of the 64th Annual Convention of the Association of Southern Agricultural Workers (New Orleans), p. 308.

Hegde, K. S., and Reid, W. M. 1969. Effects of six single species of coccidia on egg production and culling rate of susceptible layers. Poult. Sci. 48:928–932.

Horchner, F., and Dalchow, W. 1972. Zur Hitzesterilisation parasitarer Dauerstadien. Berl. Muench. Tieraerztl. Wochenschr. 85:32–34.

Horton-Smith, C., and Long, P. L. 1954. Preliminary observations on the physical conditions of built-up litter and their possible effects on the parasite populations. Tenth World's Poultry Congress, Edinburgh, pp. 266–272.

Horton-Smith, C., and Taylor, E. L. 1939. The efficacy of the blowlamp for the destruction of coccidial oocysts in poultry houses. Vet. Rec. 51:839–842.

Horton-Smith, C., Taylor, E. L., and Turtle, E. E. 1940. Ammonia fumigation for coccidial disinfection. Vet. Rec. 52:829–832.

Huchzermeyer, F. W. 1968. Coccidiosis vaccine in chickens. Rhod. Agric. J. 65:19–21.

Huchzermeyer, F. W. 1972. Problems in the control of coccidiosis with Rhodesian climatic conditions. Rhod. Agric. J. 69(2):49–50.

Hunter, J. E. (Chairman). 1959. Considerations for the evaluation of coccidiostats. Proceedings of the 19th Annual Meeting of the American Feed Manufacturers' Council.

Hy-Line Poultry Farms. 1963. Control coccidiosis—and protect egg profits. Profit Pointers 2:1–4.

Ito, S., Tsunoda, K., Shimada, K., Taki, T., and Matsui, T. 1975. Disinfectant effects of several chemicals against *Toxoplasma* oocysts. Jpn. J. Vet. Sci. 37:229–234.

Jacobs, L., Remington, J. S., and Melton, M. L. 1960. The resistance of the encysted form of *Toxoplasma gondii*. J. Parasitol. 46:11–21.

Jeffers, T. K. 1974. Immunization against *Eimeria tenella* using an attenuated strain. 15th World's Poultry Congress, New Orleans, pp. 105–107.

Jeffers, T. K. 1978. Genetics of coccidia and the host response. In: P. L. Long, K. N. Boorman, and B. M. Freeman (eds.), Avian Coccidiosis, pp. 51–125. Thirteenth British Poultry Science Symposium, British Poultry Science, Ltd., Edinburgh.

Joyner, L. P. 1964. Coccidiosis in the domestic fowl. A review of the disease in Britain and its chemotherapeutic control during the last decade. Vet. Bull. 34(6):311–315.

Joyner, L. P., and Norton, C. C. 1973. The immunity arising from continuous low-level infection with *Eimeria tenella*. Parasitology 67:333–340.

Joyner, L. P., and Norton, C. C. 1976. The immunity arising from continuous low-level infections with *Eimeria maxima* and *Eimeria acervulina*. Parasitology 72:115–125.

Kamei, K., Sato, K., and Tsunematsu, Y. 1976. A strain of mouse highly susceptible to *Toxoplasma*. J. Parasitol. 62:714.

Klesius, P. H., and Hinds, S. E. 1979. Strain differences in murine susceptibility to coccidia. Infect. Immun. 26:1111–1115.

Kogan, Z. M. 1959. Survival of the sporulated and unsporulated oocysts of chick coccidiae and the hibernation in different conditions. (In Russian) Zool. Zh. Ukr. 38:684–693.

Kogan, Z. M. 1960. Survival of chick oocysts after repeated hibernation under natural conditions in the Byelorussia. (In Russian) Zool. Zh. Ukr. 39:617–618.

Koutz, F. R. 1950. The survival of oocysts of avian coccidia in the soil. The Speculum 3(3):1–5.

Lai, C. H., Tizard, I. R., and Ingram, D. G. 1974. Development of a sulphonamide-resistant strain of *Toxoplasma gondii*. Trans. R. Soc. Trop. Med. Hyg. 68:257–258.

Landers, E. J. 1953. The effect of low temperatures upon the viability of unsporulated oocysts of ovine coccidia. J. Parasitol. 39:547–552.

Latter, V. S., and Holmes, L. S. Identification of some nutrient requirements for the *in vitro* cultivation of *Eimeria tenella*. Coccidiosis Symposium, Prague, November, 1979. In press.

Lee, H. S., Lee, H. B., and Moon, M. H. 1979. Studies on control and therapeutics of *Besnoitia besnoiti* (Marotel, 1912): Infection in Korean native cattle. Kor. J. Anim. Sci. 21:281–288.

Leek, R. G., and Fayer, R. 1978. Infectivity of *Sarcocystis* in beef and beef products from a retail food store. Proc. Helminth. Soc. Wash. 45:135–136.

Leek, R. G., and Fayer, R. 1979. Survival of sporocysts of *Sarcocystis* in various media. Proc. Helminth. Soc. Wash. 46:151–154.

Leek, R. G., and Fayer, R. 1980a. Amprolium for prophylaxis of ovine *Sarcocystis*. J. Parasitol. 66:100–106.

Leek, R. G., and Fayer, R. 1980b. Experimental *Sarcocystis ovicanis* infection in lambs: Salinomycin chemoprophylaxis and protective immunity. J. Parasitol. (suppl.) 66:43–44.

Levine, N. D. 1970. The Coccidian Parasites (Protozoa, Sporozoa) of Ruminants. Illinois Biological Monographs, No. 44. University of Illinois Press, Urbana.

Levine, N. D. 1973. Protozoan Parasites of Domestic Animals and of Man. 2nd Ed. Burgess Publishing Co., Minneapolis, MN.

Levine, N. D. 1977. Taxonomy of *Toxoplasma*.

J. Protozool. 24:36–41.

Libby, D. A., Bickford, R. L., and Glista, W. A. 1959. Vaccinating chickens for coccidiosis. Feedstuffs 31(26):18–24, 73–74.

Long, P. L. 1959. A study of *Eimeria maxima* Tyzzer, 1929, a coccidium of the fowl (*Gallus gallus*). Ann. Trop. Med. Parasitol. 53: 325–333.

Long, P. L. 1970. Coccidiosis: Development of new techniques in coccidiostat evaluation. Exp. Parasitol. 28:151–155.

Long, P. L. 1974. Further studies on the pathogenicity and immunogenicity of an embryo-adapted strain of *Eimeria tenella*. Avian Pathol. 3:255–268.

Long, P. L. 1978. The problem of coccidiosis: General considerations. In: P. L. Long, K. N. Boorman, and B. M. Freeman (eds.), Avian Coccidiosis, pp. 4–28. Symposium of the 13th British Poultry Science, Ltd., Edinburgh.

Long, P. L., Brown, W. B., and Goodship, G. 1972. The effect of methyl bromide on coccidial oocysts determined under controlled conditions. Vet. Rec. 90:562–567.

Long, P. L., and Millard, B. J. 1977. *Eimeria*: Immunisation of young chickens kept in litter pens. Avian Pathol. 6:77–92.

McDougald, L. R. Coccidiocidal combinations. U.S. Patent 4,061,755. 1977.

McDougald, L. R. Coccidiocidal combinations. U.S. Patent 4,083,962. 1978.

McMaster, P. R. B., Powers, K. G., Finerty, J. F., and Lunde, M. N. 1973. The effect of two chlorinated lincomycin analogues against acute toxoplasmosis in mice. Am. J. Trop. Med. Hyg. 22:14–17.

Marquardt, W. C. 1960. Effect of high temperature on sporulation of *Eimeria zuernii*. Exp. Parasitol. 10:58–65.

Marquardt, W. C., Senger, C. M., and Seghetti, L. 1960. The effect of physical and chemical agents on the oocysts of *Eimeria zuernii* (Protozoa, Coccidia). J. Protozool. 7:186–189.

Matsui, T., Morii, T., Iijima, T., Ito, S., and Tsunoda, K. 1977. Effect of sulfamonomethoxine against *Isospora rivolta* in cats. Jpn. Vet. J. 4:235–239.

Norrby, R., Eilard, T., Svedhem, A., and Lycke, E. 1975. Treatment of toxoplasmosis with trimethoprim-sulphamethoxazole. Scand. J. Infect. Dis. 7:72.

Oshima, S., and Hoshino, M. 1977. Different susceptibilities of Toxoplasma strains to anti-toxoplasmic drugs. Jpn. J. Parasitol. 26: 127–131.

Oshima, S., and Kumada, M. 1974. Prevention of *Toxoplasma* oocyst excretion by cat with 2-sulfamoyl-4,4′-diaminodiphenylsulfone (SDDS). Jpn. J. Parasitol. 23:20–24.

Patterson, J. M., and Gremn, H. H. 1975. Hemorrhage and death in dogs following the administration of sulfaquinoxaline. Can. Vet. J. 16:265–268.

Peeters, J. E., and Halen, P. 1979a. Effekt van enkele coccidiostatika op darm coccidiose bij het konijn: 1. Amprolium/ethopabate en metichlorpindol. Vlaams Diergeneesk. Tijdschr. 48:299–306.

Peeters, J. E., and Halen, P. 1979b. Effekt van enkele coccidiostatika op darm coccidiose bij het konijn: 2. Pancoxin-plus en Statyl. Vlamms Diergeneesk. Tijdschr. 48:387–395.

Peeters, J. E., Halen, P., and Meulemans, G. 1979. Efficacy of robenidine in the prevention of rabbit coccidiosis. Br. Vet. J. 135:349–354.

Penkert, R. A. 1973. Possible spread of toxoplasmosis by feed contaminated by cats. J. Am. Vet. Med. Assoc. 162:924.

Pettersen, E. K. 1979. Destruction of *Toxoplasma gondii* by HCl solution. Acta Pathol. Microbiol. Scand. (B). 87:217–220.

Plant, J. W., Richardson, N., and Moyle, G. C. 1974. *Toxoplasma* infection and absorption in sheep associated with cat feces. Aust. Vet. J. 50:19–21.

Prassad, H. 1959. Studies on the effect of temperature on sporulation and viability of avian coccidian *Eimeria acervulina* Tyzzer, 1929. Ceylon Vet. J. 7:38–40.

Raines, T. V. 1979. Safety and effectiveness data

supporting the approval of minor use of new animal drugs. Federal Register 44(141): 42714–42717. July 20, 1979.

Reid, W. M. 1972. Coccidiosis. In: M. S. Hofstad, B. W. Calnek, C. F. Helmboldt, W. M. Reid, and H. W. Yoder, Jr. (eds.), Diseases of Poultry, pp. 944–989. Iowa State University Press, Ames.

Reid, W. M. 1978a. Protozoa-coccidia. In: M. S. Hofstad, B. W. Calnek, C. F. Helmboldt, W. M. Reid, and H. W. Yoder, Jr. (eds.), Diseases of Poultry, pp. 783–815. Iowa State University Press, Ames.

Reid, W. M. 1978b. Prospects for the control of coccidiosis. In: P. L. Long, K. N. Boorman, and B. M. Freeman (eds.), Avian Coccidiosis, pp. 501–526. British Poultry Science, Ltd., Edinburgh.

Reid, W. M., and McDougald, L. R. 1981. New drugs to control coccidiosis: To be or not to be. Feedstuffs 53(2):27–30.

Rice, J. T., and Reid, W. M. 1973. Coccidiosis immunity following early and late exposure to Marek's disease. Avian Dis. 17:66–71.

Rose, M. E. 1976. Coccidiosis: Immunity and the prospects for prophylactic immunisation. Vet. Rec. 98:481–484.

Rose, M. E. 1978. Immune responses of chickens to coccidia and coccidiosis. In: P. L. Long, K. N. Boorman, and B. M. Freeman (eds.), Avian Coccidiosis, pp. 297–336. British Poultry Science, Ltd., Edinburgh.

Rose, M. E., and Long, P. L. 1980. Vaccination against coccidiosis in chickens. In: A. E. R. Taylor and R. Muller (eds.), Vaccines Against Parasites, Vol. 18. British Society of Parasitology, Blackwell Scientific Publications, Ltd., Oxford.

Ryff, K. L., and Bergstrom, R. C. 1975. Bovine coccidia in American bison. J. Wildl. Dis. 11:412–414.

Ryley, J. F., and Hardman, L. 1977. The use of vitamin K-deficient diets in the screening and evaluation of anticoccidial drugs. Parasitology 75:xv.

Ryley, J. F., and Wilson, R. G. 1972. Growth factor antagonism studies with coccidia in tissue culture. Z. Parasitenkd. 40:31–34.

Sangster, L. T., Stuart, B. P., Williams, D. J., and Bedell, D. M. 1978. Coccidiosis associated with scours in baby pigs. Vet. Med./Small Anim. Cin. 1317–1319. (Oct.).

Schneider, D., Ayeni, A. O., and Durr, U. 1973. Zur Resistenz von Kokzidienoocysten gegen Chemikalien. Dtsch. Tieraerztl. Wochenschr. 80:541–564.

Scrimshaw, N. S., Taylor, C. E., and Gordon, J. E. 1968. Interactions of nutrition and infection. WHO Chron. Geneva.

Seah, S. K. K. 1975. Chemotherapy in experimental toxoplasmosis: Comparison of the efficacy of trimethoprim-sulfur and pyramethamine-sulfur combination. J. Trop. Med. Hyg. 25:379–383.

Sharma, V. D., Fernando, M. A., and Summers, J. D. 1973. The effect of dietary crude protein level on intestinal and cecal coccidiosis in chickens. Can. J. Comp. Med. 37:195–199.

Sheffield, H. G., and Melton, M. L. 1976. Effect of pyrimethamine and sulfadiazine on the intestinal development of *Toxoplasma gondii* in cats. Am. J. Trop. Med. Hyg. 25:379–383.

Smith, D. D., and Frenkel, J. K. 1977. *Besnoitia darlingi* (Protozoa: Toxoplasmatinae): Cyclic transmission by cats. J. Parasitol. 63: 1066–1071.

Sofield, W. L., and Strout, R. G. 1974. Amino acids essential for *in vitro* cultivation of *Eimeria tenella*. J. Protozool. 21(suppl.):434.

Stuart, E. E. 1960. Cocccidiosis vaccination. Avian Dis. 4:305.

Syrkis, I., Fried, M., Elian, I., Pietrushka, D., and Lengy, J. 1975. A case of severe human coccidiosis in Israel. Israel J. Med. Sci. 11: 373–377.

Tabbara, K. F., Nozik, R. A., and O'Conner, G. R. 1974. Clindamycin effects on experimental ocular toxoplasmosis in the rabbit. Arch. Ophthalmol. 92:244–247.

Tate, G. W., and Martin, R. G. 1977. Clindamycin in the treatment of human ocular toxoplasmosis. Can. J. Ophthalmol. 12:188–195.

Trier, J. S., Moxey, P. C., Schimmel, E. M., and Robles, E. 1974. Chronic intestinal coccidiosis in man: Intestinal morphology and response to treatment. Gastroenterology 66:923–935.

Vetkovic, L., and Tomanovic, B. 1967. Investigation of amprolium prophylaxis for intestinal coccidiosis in rabbits. Vet. Glasn. 7:607–612.

Waldeland, H. 1976. Toxoplasmosis in sheep: *Toxoplasma gondii* in muscle tissue, with particular reference to dye test titers and hemoglobin type. Acta Vet. Scand. 17: 403–411.

Waldeland, H. 1977. Toxoplasmosis in sheep: Influence of various factors on the antibody contents. Acta Vet. Scand. 18:237–247.

Warren, E. W. 1968. Vitamin requirements of the coccidia of the chicken. Parasitology 58:137–148.

Westerman, E. L., and Christensen, R. P. 1979. Chronic *Isospora belli* infection treated with co-trimoxazole. Ann. Intern. Med. 91: 413–414.

Wilkinson, G. T. 1977. Coccidial infection in a cat colony. Vet. Rec. 100:156–157.

Williams, D. M., Grumet, F. C., and Remington, J. S. 1978. Genetic control of murine resistance to *Toxoplasma gondii*. Infect. Immun. 19: 416–420.

Yilmaz, S. M., and Hopkins, S. H. 1972. Effects of different conditions on duration of infectivity of *Toxoplasma gondii* oocysts. J. Parasitol. 58:938–939.

Yvoré, P. 1976. Revue sur la prevention des coccidioses en aviculture. Avian Pathol. 5: 237–252.

Glossary

Acephaline Gregarine A superseded term for an aseptate gregarine.
Anisogamy Occurrence of structurally dissimilar gametes.
Apical Complex The anterior structures of certain cells in the phylum Apicomplexa, including the polar ring(s), conoid, rhoptries, micronemes, and subpellicular microtubules.
Apical Ring A superseded term for polar ring.
Aseptate Gregarine A gregarine that is not divided into segments by septa.
Bradyzoite A slowly developing merozoite. This term is used especially for the merozoites in the last generation meronts (sarcocysts) of *Sarcocystis* and for the merozoites in pseudocysts of *Toxoplasma*, *Besnoitia*, and *Frenkelia*.
Cephalin A superseded term for a gregarine gamont attached to or within its host cell, i.e., still with an epimerite or mucron.
Cephaline Gregarine A superseded term for a septate gregarine. A young gregarine with an epimerite.
Clear Globule See Globule, Clear.
Conoid An electron-dense, hollow structure in the form of a truncated cone inside the polar ring(s) at the anterior end of certain stages of most Apicomplexa; composed of spirally coiled microtubules.
Convoluted Tubule A superseded term for microneme.
Copula The structure formed by two gametes coming together before they fuse to form a zygote.
Cryptozoite The primary exoerythrocytic meront of certain members of the suborder Haemospororina.
Cyst A resistant stage of an organism formed by the organism's laying down a wall around itself.
Cystozoite A bradyzoite.
Cytostome An opening in the body through which particulate food is taken; visible with the light microscope. The term is sometimes used erroneously for micropore.
Cytozygote A superseded term for oocyst.
Dense Body A superseded term for rhoptry.
Deutomerite The posteriormost segment of a septate gregarine. It generally contains the nucleus.
Dicystid Gregarine A gregarine that has two segments—either an epimerite or mucron and a second nonsegmented section (a protodeutomerite), or only a protomerite and deutomerite.

Dormozoite A superseded term for hypnozoite.

Eimerianeme A superseded term for rhoptry.

Endodyocyte A trophozoite formed by endodyogeny.

Endodyogeny Formation of two daughter individuals (merozoites or metrocytes) by internal budding.

Endogenous Cycle That part of the life cycle that takes place within a host.

Endogeny Formation of two or more daughter individuals by internal budding.

Endopolygeny Formation of more than two daughter cells (merozoites or metrocytes) by internal budding.

Endozoite (1) An endodyozoite or an endopolygenite (i.e., a merozoite formed by endogeny). (2) A tachyzoite.

Epicyte The external layer of the cytoplasm in gregarines.

Epimerite The anteriormost organelle of a gregarine. It contains no nucleus and attaches the gregarine to its host cell. It generally breaks off when the gregarine becomes detached from the host cell. It appears under the light microscope to be divided from the rest of the gregarine by a septum.

Erythrocytic Phase That part of the life cycle that takes place within the host's erythrocytes.

Euryxenous Having a broad host range (considered as occurring in more than one host family). Most avian malaria parasites are euryxenous. The asexual part of the life cycle of *Toxoplasma* is euryxenous.

Exflagellation Formation of microgametes from the microgamont in the Haemospororina.

Exoerythrocytic (EE) Stage A stage in the life cycle of Haemospororina occurring outside the host's erythrocytes.

Exogenous Cycle That part of the life cycle that takes place outside the host.

Fibril, Subpellicular A superseded term for a subpellicular microtubule.

Gamete A mature sex cell that unites with a cell of the opposite sex to form a zygote.

Gametocyst A cyst formed by the union of two gregarine gamonts and the secretion of a wall around them. Gametes are formed within it; these then fuse to form oocysts in which sporozoites develop. Gametocytes are commonly formed by gregarines.

Gametocyte A gamont (i.e., a cell that will form gametes).

Gametogony Formation of gametes, often by schizogony.

Gamont An individual that will produce one or more gametes. A gametocyte. In the Apicomplexa, the gamonts are thought to be haploid.

Gamontocyst A gametocyst.

Globule, Clear A body in some coccidian sporozoites, apparently composed of protein plus amylopectin, that stains with eosin and has a homogeneous, clear structure. Often called a refractile body.

Globule, Eosinophilic A clear globule (stainable with eosin).

Granule, Refractile   A superseded term for polar granule.

Heterogenetic Parasite   A parasite in whose life cycle there is alternation of generations. The great majority of Apicomplexa are thought to be heterogenetic, with both sexual and asexual generations.

Heteroxenous   Having two or more types of host in its life cycle.

Homoxenous   Having one type of host in its life cycle.

Hypnozoite   A nonmultiplying sporozoite in a transport host.

Isogamy   Occurrence of structurally similar gametes.

Isolate   Viable organisms present in experimental animal hosts, avian embryos, or cultured cells, after the introduction of a sample. An isolate is not necessarily monospecific and not all the organisms in the sample will necessarily appear.

Lankesterelloneme   A superseded term for rhoptry.

Line   Laboratory derivative of a strain maintained in a different location or under different conditions from the original.

Macrogamete   A relatively large gamete, considered female.

Macrogametocyte   A macrogamont.

Macrogamont   A gamont that will turn into a macrogamete. Macrogamonts are young macrogametes. All coccidian macrogamonts are thought to be already haploid.

Meiocyte   A superseded term for zygote.

Merogony   Formation of merozoites.

Meront   An asexual stage in the life cycle that forms merozoites.

Merozoite   A cell produced by merogony in the asexual part of an apicomplexan life cycle. It forms either a new meront (schizont) or a gamont.

Mesoxenous   Having a moderate host range (considered as occurring in more than one host family within a host order).

Metacryptozoite   A merozoite formed by merogony from a meront produced by a cryptozoite. It may develop either into another cryptozoite or an erythrocytic stage.

Metrocyte   A mother cell. A pre-merozoite found in the last generation meronts of *Sarcocystis* and *Frenkelia*. It is rather stout and ellipsoidal, with a deeply folded cell surface and few if any elements of an apical complex. It divides repeatedly by endodyogeny to form fresh metrocytes which become progressively more elongate and eventually become typical merozoites with a complete apical complex.

Microgamete   A relatively small gamete, considered male.

Microgametocyte   A microgamont.

Microgamont   That which produces microgametes, the number depending on the parasite group.

Microneme   An elongate, electron-dense organelle extending longitudinally in the anterior part of the body of certain stages in certain Apicomplexa, possibly attached to and giving rise to the rhoptries.

Micropore An opening in the body visible only with the electron microscope through which particulate food or other material can be taken into the body. Sometimes called a cytostome, micropyle, or cuticular pore.

Micropyle An opening (or the position of an opening) in the wall of an oocyst.

Microtubule, Subpellicular A slender, electron-dense, hollow structure extending back from the polar ring region just beneath the pellicle, of the sporozoite, merozoite, and sometimes other stages of the Apicomplexa. Visible only with the electron microscope.

Monocystid Gregarine A superseded term for aseptate gregarine.

Monogenetic Parasite A parasite in whose life cycle there is no alternation of generations.

Monoxenous Parasite (1) A parasite with a single species of host in its life cycle. (2) A parasite with one type of host in its life cycle. (This definition is often used, but "homoxenous" is preferable.)

Mucron An attachment organelle of aseptate gregarines. It is similar to an epimerite, but is not set off from the rest of the gregarine body by a septum.

Oocyst A cyst formed around a zygote.

Ookinete A motile zygote.

Paired Organelle A superseded term for rhoptry.

Paranuclear Body Clear globule. Probably composed of amylopectin plus protein.

Phanerozoite A merozoite that has been formed in an erythrocyte and has then entered another kind of cell to develop exoerythrocytically. Found in bird malaria.

Plastic Granule A granule in coccidian macrogametes that will later participate in forming the oocyst wall. A wall-forming body.

Polar Granule A structure found in the oocysts of some coccidia, formed by the first (reduction) division of the zygote in the oocyst.

Polar Ring An electron-dense ring at the anterior end of certain stages of the Apicomplexa. Visible only with the electron microscope.

Polycystid Gregarine A superseded term for septate gregarine.

Primite The anterior member of a pair of gregarine gamonts in syzygy.

Protomerite The segment of a septate gregarine between the epimerite and deutomerite. It ordinarily contains no nucleus.

Pseudocyst (1) A cyst-like stage the wall of which has been formed by the host cell and not the parasite. (2) A superseded term for a residuum occurring in the gametocysts of some gregarines.

Refractile Body A clear globule.

Residual Body A superseded term for residuum.

Residuum, Gametocyst The material remaining in a gametocyst after formation of gametes and zygotes. (In gregarines sometimes erroneously called a pseudocyst.)

Residuum, Oocyst The material (exclusive of the polar granule(s) remaining in an oocyst after formation of sporocysts or sporozoites.

Residuum, Sporocyst The material remaining in a sporocyst after formation of sporozoites.

Rhoptry The electron-dense, tubular or saccular organelle, often enlarged at the posterior end, extending back from the anterior region in the sporozoite, merozoite, and sometimes other stages of the Apicomplexa. Superseded synonyms: toxoneme, paired organelle, lankesterelloneme, eimerianeme, dense body.

Ring Stage A young trophozoite produced from a merozoite of the Haemospororina, which has just invaded an erythrocyte; so-called because when stained with a Romanowsky stain, it has a red nucleus and blue cytoplasm surrounding a vacuole, so that it looks like a signet ring.

Rod-shaped Granule A superseded term for microneme.

Rostrum A mucron (in the gregarines).

Sample Part of a coccidial (or protozoan) population collected on a unique occasion at a specified location.

Sarcocyst The last generation meront of *Sarcocystis* in the muscles of the intermediate host.

Sarconeme A superseded term for microneme.

Satellite The posterior member of a pair of gregarine gamonts in syzygy. In a few species there may be more than one satellite.

Schizogony Formation of daughter cells by multiple fission. If the daughter cells are merozoites, the schizogony may be called merogony. If they are sporozoites, the schizogony may be called sporogony. If they are gametes, the schizogony may be called gametogony. (Some people limit the term "schizogony" to merogony.)

Schizont A stage in the life cycle that divides by schizogony (multiple fission). This name is often used only for the asexual stage that produces merozoites, i.e., for a meront.

Schizozoite A cell produced from a schizont by schizogony. This term is usually given only to merozoites.

Segmenter A late meront in the Haemospororina, which is producing or has produced merozoites.

Septate Gregarine A gregarine that is divided into segments by one or more septa (which can be seen with the light microscope but which are not necessarily complete when viewed with the electron microscope).

Species A group of actually or presumably potentially interbreeding populations, which is reproductively isolated from other groups. Naming of a species should conform with the International Code of Zoological Nomenclature.

Sporadin A superseded term for the gamont of a gregarine after it has lost its epimerite or mucron. Often called a trophozoite, and sometimes a sporont.

Spore A resistant stage formed within a cell. In the Apicomplexa, this term is often applied to the sporocysts of coccidia or oocysts of gregarines, and in this case apparently signifies a walled body containing or producing one or more uninucleate bodies (sporozoites) capable of developing into other stages in the life cycle. This usage is erroneous.

Sporoblast A cell formed by division of the sporont of coccidia; an immature sporocyst.

Sporocyst A cyst formed with the oocysts of most coccidia; it contains the sporozoites. (The term sporocyst is often applied to the oocyst of gregarines, but these protozoa actually have no sporocyst.)

Sporoduct A tube in a gregarine gametocyst through which the oocysts ("spores") pass to the outside. Sometimes called a spore duct.

Sporogony Formation of sporocysts and sporozoites by division of a zygote. The first division is by meiosis and subsequent divisions are by mitosis.

Sporont The stage in the life cycle of coccidia that will form sporocysts, i.e., the zygote within the oocyst wall.

Sporozoite A motile infective stage produced by sporogony.

Sporulation The process of sporozoite formation, i.e., sporogony.

Stabilate Coccidia viably preserved (e.g., by freezing in liquid nitrogen) on a unique occasion.

Stenoxenous Having a narrow host range (considered as occurring in a single host family). Most coccidia and mammalian malaria parasites are stenoxenous.

Stieda Body A knob-like thickening at one end of the sporocyst wall in coccidia.

Stock Coccidia originally propagated from an isolate by serial passage in vivo or in vitro without any implication of homogeneity or characterization. Several stocks may be derived from the isolate at different times under different conditions.

Strain A population of a single species from a stock or isolate and maintained by serial passage, having regularly reproducible characters, which are stable within the limits of the tests used. In the Coccidia, strains are normally established from single oocysts or sporocysts.

Subspecies An allopatric group of organisms with clearly definable differences from other groups within the same species. Naming of a subspecies should conform with the International Code of Zoological Nomenclature.

Substiedal Body A dependent body attached to the Stieda body and extending into the sporocyst.

Syngamy Union of the cytoplasm and nucleus of two gametes.

Syzygy Association of gamonts prior to gametocyst and gamete formation, primarily in the gregarines.

Tachyzoite A fast-developing merozoite. Term used especially in referring to the merozoites found in the group stages (aggregates) of *Toxoplasma* in the acute stage of the infection.

Toxoneme A superseded term for rhoptry.

Tricystid Gregarine A gregarine that has three segments—epimerite, protomerite, and deutomerite.

Trophozoite The vegetative stage of a protozoon. This term is also used for the gamont ("cephalin") of a gregarine while it is still within or attached to a host cell.

Vacuole, Parasitophorous The vacuole in a host cell that contains a stage of an apicomplexan parasite.

Vacuole, Periparasitic A parasitophorous vacuole.

Vestigial Body A superseded term for residuum.

Wall-forming Body A body in a coccidian macrogamete that will later participate in formation of the oocyst wall. A plastic granule.

Zoite A merozoite, sporozoite, or endodyocyte.

Zoitoblast A sporoblast.

Zoitocyst A coccidian sporocyst.

Zoospore A term used primarily by botanists to refer to motile propagative cells, often formed within a cyst.

Zygocyst An oocyst.

Zygote The cell formed by the union of the cytoplasm and nuclei of two gametes.

The glossary has been compiled mainly by Norman D. Levine.

Index